Vol. 1

Frontiers in Nanobiomedical Research

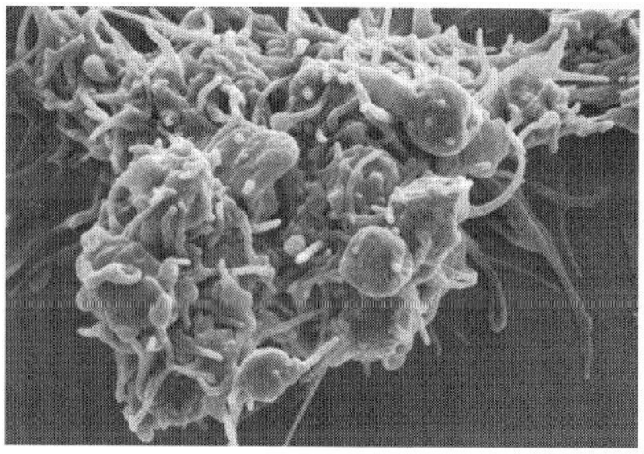

HANDBOOK OF IMMUNOLOGICAL PROPERTIES OF ENGINEERED NANOMATERIALS

Frontiers in Nanobiomedical Research

ISSN: 2251-3965

Published

Vol. 1: Handbook of Immunological Properties of Engineered Nanomaterials
edited by Marina A. Dobrovolskaia and Scott E. McNeil
(SAIC-Frederick, Inc., USA)

Forthcoming titles

Nanobiomedical Research: Fundamentals, Main Applications and Future Developments (In 4 Volumes)
edited by Vladimir P. Torchilin (Northeastern University, USA)

Tissue Regeneration: Where Nanostructure Meets Biology
edited by Qing Liu (3D Biotek LLC, USA)

Cancer Therapeutics and Imaging: Molecular and Cellular Engineering and Nanobiomedicine
edited by Kaushal Rege (Arizona State University, USA)

Nano Vaccines
edited by Balaji Narasimhan (Iowa State University, USA)

Nano Pharmaceuticals
edited by Rajesh N. Dave (New Jersey Institute of Technology, USA)

Thermal Aspects in Nanobiomedical Systems and Devices
by Dong Cai (Boston College, USA)

Nano Mechanochemistry in Biology
edited by Jeffrey Ruberti (Northeastern University, USA)

Nanomaterial Probes of Biological Processes and Systems
edited by David Mast (University of Cincinnati, USA)

Handbook of Biomaterials
edited by Donglu Shi (University of Cincinnati, USA) and
Xuejun Wen (Clemson University, USA)

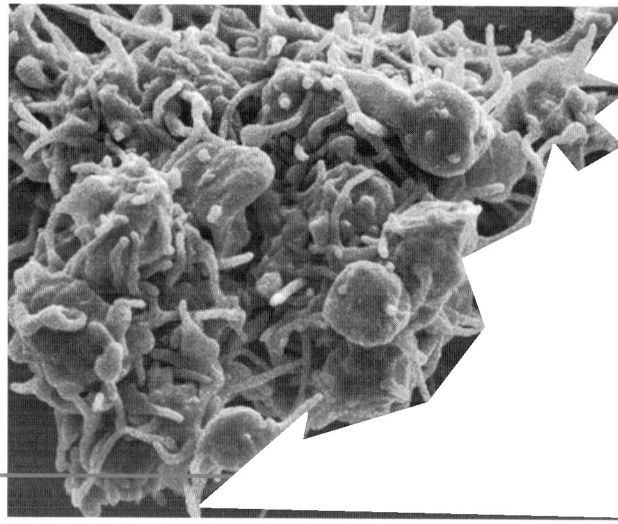

Vol. 1

Frontiers in Nanobiomedical Research

HANDBOOK OF IMMUNOLOGICAL PROPERTIES OF ENGINEERED NANOMATERIALS

editors

Marina A Dobrovolskaia
Scott E McNeil

SAIC-Frederick, Inc., USA

World Scientific

NEW JERSEY • LONDON • SINGAPORE • BEIJING • SHANGHAI • HONG KONG • TAIPEI • CHENNAI

Published by

World Scientific Publishing Co. Pte. Ltd.

5 Toh Tuck Link, Singapore 596224

USA office: 27 Warren Street, Suite 401-402, Hackensack, NJ 07601

UK office: 57 Shelton Street, Covent Garden, London WC2H 9HE

British Library Cataloguing-in-Publication Data
A catalogue record for this book is available from the British Library.

Frontiers in Nanobiomedical Research — Vol. 1
HANDBOOK OF IMMUNOLOGICAL PROPERTIES OF ENGINEERED NANOMATERIALS

ISBN 978-981-4390-25-5

Typeset by Stallion Press
Email: enquiries@stallionpress.com

Printed by FuIsland Offset Printing (S) Pte Ltd Singapore

Contents

Preface xix

List of Contributors xxiii

Chapter 1 Immunological Properties of Engineered Nanomaterials: An Introduction 1
Marina A. Dobrovolskaia and Scott E. McNeil

1. Introduction 1
2. Interaction with Blood Components 4
   2.1. Hemolysis 5
   2.2. Coagulation 6
   2.3. Complement activation 7
3. Nanoparticle Uptake by Phagocytic Cells 8
4. Immunogenicity 10
   4.1. Inflammation 11
   4.2. Antigenicity 12
   4.3. Adjuvant properties 12
5. Immunosuppression 13
6. Designing Immunotoxicity Studies 16
7. Future Directions 17

References 17

Chapter 2 Importance of Physicochemical Characterization Prior to Immunological Studies 25
Jeffrey D. Clogston and Anil K. Patri

1. Introduction 25
2. Key Parameters that Influence Nanomaterial Immunological Properties 27
3. Instrumentation for Physicochemical Characterization 31

4. Methodologies for Nanomaterial Separation 33
5. A Closer Look 36
5.1. Dynamic light scattering 36
5.2. Zeta potential 46
6. Summary 50
References 51

Chapter 3 Impact of Nanoparticle Sterilization on Analytical Characterization 53
Nanda Subbarao

1. Introduction 53
2. Why Is Sterilization a Challenge for Nanotherapeutics? 54
3. Previous Studies Reviewing the Effects of Various Sterilization Procedures on Nanomaterials 54
4. Available Methods for Sterilization 55
4.1. Autoclaving 55
4.2. Filtration 57
4.3. Gamma irradiation 59
4.4. Gaseous ethylene oxide 68
4.5. High hydrostatic pressure sterilization 69
4.6. Formaldehyde 69
5. Comparison of Different Methods for Sterilization 69
6. Impact on Nanotherapeutic Analytics 70
7. Conclusions 70
References 71

Chapter 4 Endotoxin and Engineered Nanomaterials 77
Marina A. Dobrovolskaia and Scott E. McNeil

1. Introduction 77
2. What Endotoxin Is and Why It Is Important to Study in Nanoformulations 80
3. Methods Used for Endotoxin Detection and Quantification 83
3.1. Traditional methods 83
3.2. Alternative methods 87
4. Estimating Endotoxin in Nanoformulations 93
4.1. Nanoparticle interference with traditional methods 93
4.2. Evaluation of interference 93
4.3. How to choose an appropriate LAL format 96
4.4. Application of non-traditional methods 99

5. Use of Traditional Endotoxin-Neutralizing Reagents in Nanoparticle Research 102
5.1. Case study 1 102
5.2. Case study 2 104
6. Minimizing Contamination and Interference 105
7. Purification and Depyrogenation of Nanoformulations 107
8. Conclusions and Future Directions 109
References 110

Chapter 5 Surface Adsorbates on Nanomaterials and Their Possible Roles in Host Inflammatory and Toxicological Processing 117
Clinton F. Jones, David G. Castner, and David W. Grainger

1. Introduction 118
2. Adsorption as a Natural Consequence of Surfaces 120
3. Host Processing of Nanomaterials 125
4. Polydimethylsiloxane Adsorbates 128
5. Polyaromatic Hydrocarbon Adsorbates 130
6. Catalyst Residues from Carbon Nanotube Synthesis 132
7. Adsorbed or Conjugated Polymer Coatings 133
8. Biomolecule Adsorbates 135
9. Lipopolysaccharide (Bacterial Endotoxin) 138
10. Contaminant Detection and Assessment 140
11. Conclusions 142
Acknowledgments 143
References 143

Chapter 6 Nanoparticle Interaction with Plasma Proteins as It Relates to Biodistribution 151
Lennart Treuel and G. Ulrich Nienhaus

1. Introduction 152
2. Nanoparticle Separation from Plasma 154
3. Protein Identification 155
4. Mechanistic Aspects of Protein Corona Formation 159
5. Effect of Protein Binding on Cellular Uptake and Biodistribution 161
6. The Protein Corona in Medical Applications 163

7. Final Remarks 164
Acknowledgments 164
References 164

Chapter 7 The Effects of Engineered Nanomaterials on Erythrocytes 173
Bridget Wildt, Richard A. Malinauskas, and Ronald P. Brown

1. Introduction 174
2. Erythrocyte Structure and Function 174
3. *In Vitro* Techniques Used to Investigate NM Interactions with RBCs 175
   3.1. Hemolysis 175
   3.2. Oxidative stress 181
   3.3. RBC aggregation and agglutination 182
   3.4. NM membrane interaction and cellular uptake 183
   3.5. NM-induced changes in RBC morphology and mechanical properties 185
4. *In Vivo* Studies of RBC and NM Interactions 186
5. *In Vitro–In Vivo* Correlation 188
6. Case Study: Interaction of AgNPs with RBCs 189
7. Points to Consider Regarding NM–RBC Interaction 191
8. Summary 192
9. Suggestions for Future Directions 194

Acknowledgements 195
References 195

Chapter 8 The Effects of Engineered Nanomaterials on Cultured Endothelial Cells 207
Jan Simak

1. Introduction 208
2. The Pathophysiology of Possible Adverse Effects of Nanomaterials on Endothelial Cells 208
3. Methods for Evaluating the *In Vitro* Effects of Nanomaterials on Endothelial Cells 215
   3.1. Endothelial cell cultures 222
   3.2. Nanomaterial uptake 223
   3.3. Morphological and structural changes 223
   3.4. Viability/cytotoxicity and proliferation 223
   3.5. Apoptosis, necrosis, and autophagy 224

3.6. NO and ROS production and redox status ... 225
3.7. EC activation markers ... 227
3.8. EC mobility and morphogenesis, monolayer permeability, and vasoactivity assays ... 227
4. Studies on the *In Vitro* Interactions of Nanomaterials with Endothelial Cells ... 228
4.1. Metal nanomaterials ... 229
4.2. Carbon nanomaterials ... 237
4.3. Silica-based nanomaterials ... 241
4.4. Dendrimer nanomaterials ... 243
4.5. Polymer nanomaterials ... 244
4.6. Liposomes ... 246
4.7. Other nanomaterials ... 247
5. Selected Protocols for *In Vitro* Endothelial Cell Studies Used in Our Laboratory ... 248
5.1. Culture of human umbilical vein endothelial cells ... 248
5.2. Cell Counting Kit-8 viability assay ... 248
5.3. Lactate dehydrogenase assay ... 249
5.4. Analysis of ROS production ... 249
5.5. Flow cytometry annexin V/PI and TUNEL apoptosis assays ... 250
5.6. Flow cytometric assay of EC activation markers ... 250
5.7. Free intracellular Ca^{2+} assay ... 251
5.8. Autophagy assay: Western blot analysis of LC3-II ... 251
5.9. Autophagy assay: Immunofluorescence microscopy of LC3-II ... 252
5.10. Premo Autophagy Sensor ... 253
5.11. LysoTracker Red dye uptake assay ... 253
6. Conclusion ... 253
References ... 254

Chapter 9 The Effect of Engineered Nanomaterials on the Plasma Coagulation System ... 263

Jan Simak

1. Introduction ... 264
2. The Pathophysiology of the Possible Adverse Effects of Nanomaterials on the Kallikrein–Kinin System, the Plasma Coagulation System, and Fibrinolysis ... 265
2.1. The kallikrein–kinin system ... 265
2.2. The plasma coagulation system and fibrinolysis ... 266

3. Methods for Evaluating the *In Vitro* Effects of Nanomaterials on the Kallikrein–Kinin System, the Plasma Coagulation System and Fibrinolysis 270
3.1. The preparation of a nanomaterial sample for testing 272
3.2. The plasma and blood samples for testing 273
3.3. Clot-based plasma coagulation tests 273
3.4. Thromboelastography 275
3.5. Synthetic substrate-based assays 276
3.6. ELISA and other immunoassays 277
3.7. Fibrinolytic and thrombolytic activity assays 277
4. Studies of the *In Vitro* Interactions of Nanomaterials with the Plasma Coagulation System, the Kallikrein–Kinin System, and Fibrinolysis 277
5. Conclusion 284
References 285

Chapter 10 The Effects of Engineered Nanomaterials on Platelets 293
Jan Simak

1. Introduction 294
2. The Pathophysiology of the Possible Adverse Effects of Nanomaterials on Platelets 294
2.1. Blood platelets 294
2.2. Possible adverse effects of nanomaterial interactions with platelets 298
3. Methods for Evaluating the *In Vitro* Effects of Nanomaterials on Platelets 300
3.1. The preparation of a nanomaterial sample for testing 305
3.2. Platelets for testing 305
3.3. Cell counter-based assays 306
3.4. Microscopy 307
3.5. Aggregometry 307
3.6. Flow cytometry 309
3.7. Solid phase assays (ELISA) 311
3.8. Photometry, luminometry, and scintillation assays 311
3.9. Biochemical status: pH, pO_2, pCO_2, HCO_3^-, lactate, glucose, and ATP 312
3.10. Dynamic light scattering 312
3.11. Platelet function analyzers and thromboelastography 312

3.12. Perfusion chambers 313
3.13. Quartz crystal microbalance with dissipation 313
3.14. Platelet proteomic, genomic, and transcriptomic analyses 314
3.15. *In vivo* animal models and clinical trials 315
4. Studies on the *In Vitro* Interactions of Nanomaterials with Platelets . 316
4.1. Carbon nanomaterials 316
4.2. Silica nanomaterials 322
4.3. Metal nanomaterials 323
4.4. Polymer nanomaterials 327
4.5. Dendrimer nanomaterials 331
4.6. Liposomes 332
4.7. Platelet coating with nanoparticles 339
4.8. Tissue scaffolds 339
4.9. *In vitro* interactions of nanomaterials with platelets: Conclusions 340
5. Selected Protocols for *In Vitro* Platelet Studies Used in Our Laboratory 341
5.1. Basic assays 341
5.2. Special investigational assays 343
6. Materials 346
7. Conclusion 347
References 348

Chapter 11 Complement Activation 357
Carolina Salvador-Morales and Robert B. Sim

1. The Complement System 358
1.1. Pathways of complement activation 358
1.2. Physicochemical characteristics of key complement proteins 360
2. Physicochemical Characteristics of Nanoparticles 364
2.1. PEG chain density 364
2.2. PEG chain length 365
2.3. Morphology of the nanocarrier 366
3. Recent Engineering Approaches to Avoid or Reduce Complement Activation 367
3.1. Influence of PEG mushroom- and brush-like configurations on complement activation 367
3.2. Modulation of complement system activation via functional groups 370

4. Methods for the *In Vitro* Study of Complement Activation by Different Nanomaterials 372
4.1. Hemolytic assay (CH_{50}) 372
4.2. ELISA kits 373
4.3. Wieslab diagnostics kits 376
4.4. 2D immunoelectrophoresis method 376
5. *In Vivo* Studies of Complement Activation by Different Nanomaterials 378
6. Concluding Remarks 379
References 380

Chapter 12 Bidirectional Interaction Between Nanoparticles and Cells of the Mononuclear Phagocytic System 385

Whitney P. Caron, Sumit Rawal, Gina Song, Parag Kumar, John C. Lay, and William C. Zamboni

1. Introduction 386
2. Description of Nanoparticles: Terminology and Basic Concepts as They Relate to MPS Uptake 386
2.1. Nanoparticle terminology 386
2.2. Interaction between nanoparticles and the MPS 388
2.3. Capture versus hijacking 389
3. Mechanisms and Factors Involved in Nanoparticle Uptake 390
3.1. Opsonization 390
3.2. Endocytosis 390
3.3. Surface receptors 393
3.4. The role of the complement system in nanoparticle uptake by the MPS 394
3.5. Complement affects nanoparticle stability, biodistribution, and PK 394
3.6. Accelerated blood clearance phenomenon 395
4. The Effects of Physicochemical Properties on Nanoparticle Uptake by the MPS 396
4.1. Modifiable factors 396
4.2. Nanoparticle composition 396
4.3. Particle size and dimension 399
4.4. Surface charge 400
5. Nanoparticle Cytotoxicity to Cells of the MPS — *In Vitro* 400
6. Nanoparticle Cytotoxicity to Cells of the MPS — *In Vivo* 402

7. Translational and Clinical Relevance 405
  7.1. The MPS in tumors 405
  7.2. Clinical effects 405
8. Final Remarks 409
References 410

Chapter 13 The Effects of Nanoparticles on Dendritic Cells 417
Valentyna Fesenkova

1. Introduction 417
2. The Role of Dendritic Cells in Linking the Innate and Adaptive Immunity and Their Effector Functions in Antitumor Immune Response 419
3. The Role of Nanoparticle Physicochemical Properties in Their Uptake by Dendritic Cells 424
4. Challenges and Examples of Nanoparticle Use for Antigen and Drug Delivery to Dendritic Cells 425
5. Future Directions 428
Acknowledgements 429
References 430

Chapter 14 The Effects of Nanoparticles on Bone Marrow Cells 433
Ekaterina Dadachova

1. Introduction 434
2. Biodistribution 434
3. Imaging 435
4. Toxicity 439
5. Immunomodulation 442
6. Radioprotection 443
7. Conclusions 445
References 445

Chapter 15 Nanoparticles, Immunomodulation and Vaccine Delivery 449
Sue D. Xiang, Martina Fuchsberger, Tanya De L. Karlson, Charles L. Hardy, Cordelia Selomulya and Magdalena Plebanski

1. Introduction 450

2. Vaccine Design and Immune Responses 454
3. Nanoparticle Modulation of Immune Cells 455
3.1. Size dependence of nanoparticle uptake 455
3.2. Nanoparticle surface charge and chemistry 457
3.3. Nanoparticle shape 458
4. Nanoparticles and Cell Signalling Pathways 459
5. Nanoparticle-based Vaccines 460
5.1. Polymeric particles 460
5.2. Chitosan 461
5.3. Non-degradable but biocompatible nanoparticles 462
5.4. Magnetic nanoparticles 462
6. Challenges of Nanoparticulate Vaccines 464
References 465

Chapter 16 Nanoparticles as Drug Delivery Vehicles for the Therapy of Inflammatory Disorders 477
Deepthy Menon, J. Gopikrishna, Dhanya Narayanan, and Shantikumar V. Nair

1. Introduction 478
2. The Inflammatory Process 478
3. Active *vs.* Passive Targeting in Inflammatory Disorders 479
4. Nanoconstructs for Inflammatory Diseases 481
5. Drug Loading and Release from Nanoconstructs 483
6. Types of Nanocarriers 483
6.1. Polymeric nanoparticles 484
6.2. Liposomes 488
6.3. Dendrimers 492
6.4. Micelles 493
6.5. Solid lipid nanoparticles 497
7. Conclusions 502
References 502

Chapter 17 Nanostructures and Allergy 517
Silvia Lorenzo-Abalde and África González-Fernández

1. Introduction 517
2. Allergy 518
3. Types of Hypersensitivity 520
3.1. Type I hypersensitivity 520

3.2. Pseudoallergy 522
3.3. Type II hypersensitivity 524
3.4. Type III hypersensitivity 525
3.5. Type IV hypersensitivity 526
4. Allergic Diseases Induced by Nanomaterials 528
4.1. Pseudoallergy 528
4.2. Asthma 529
4.3. Acute dermatitis 530
4.4. Contact dermatitis 530
5. Beneficial Anti-Inflammatory Effects 531
6. Hypersensitivity Tests and Models 532
6.1. Diagnosis of type I hypersensitivity 532
6.2. Diagnosis of type II hypersensitivity 534
6.3. Diagnosis of type III hypersensitivity 535
6.4. Diagnosis of type IV hypersensitivity 535
7. Conclusion 536
Acknowledgements 537
Conflict of Interests 537
References 537

Chapter 18 Nanoparticles and Antigenicity 547
Marina A. Dobrovolskaia

1. Introduction 547
2. Antigenic Response to Engineered Nanomaterials 549
3. Antigenic Response to Nanoparticle Surface Coatings 556
3.1. Dose 557
3.2. Interval between first and second doses 564
3.3. Size 564
3.4. Charge 564
3.5. Liposomal composition 565
3.6. PEG length and density 565
3.7. Anti-PEG IgM 565
3.8. Species specificity 566
4. Antigenic Response to Therapeutic Proteins 567
4.1. Antigenicity of therapeutic proteins in the presence of accidental particulate contaminants 569
4.2. Antigenicity of therapeutic proteins attached to nanoparticle surfaces 570

5. Summary and the Road Ahead 571
References 573

Chapter 19 *In Vitro* Assays for Monitoring Nanoparticle Interaction with Components of the Immune System.....581
Marina A. Dobrovolskaia and Scott E. McNeil

1. Introduction 582
2. *In Vitro–In Vivo* Correlation as a Measure of Predictability of *In Vitro* Assays 587
3. *In Vitro* Assays to Monitor Immunomodulatory Properties of Nanoparticles 594
 3.1. Effects of nanoparticles on the formation of CFU-GM *in vitro* 594
 3.2. Effects of nanoparticles on the cytotoxicity NK cells *in vitro* 601
 3.3. Effects of nanoparticles on the maturation of monocyte-derived dendritic cells *in vitro* 609
 3.4. Effects of nanoparticles on the production of nitric oxide by macrophages *in vitro* 615
 3.5. Effects of nanoparticles on leukocyte PCA *in vitro* 623
 3.6. Effects of nanoparticles on leukocyte proliferation *in vitro* 629
References 634

Chapter 20 Evaluating the Adverse Effects of Nanomaterials on the Immune System with Animal Models 639
Matthew J. Smith, Colleen E. McLoughlin, Kimber L. White, Jr., and Dori R. Germolec

1. Introduction 640
2. Immunotoxicological Assessments 641
 2.1. Nanomaterials and the immune system 641
 2.2. Model and study designs 643
 2.3. Methods: Inflammation, immunomodulation, and hypersensitivity 644
3. *In Vivo* Immunotoxicological Evaluations of Nanomaterials: Three Case Studies 647
 3.1. TiO_2 nanoparticles 649
 3.2. Biodegradable electrospun materials for tissue engineering and drug delivery 652
 3.3. Carbon nanomaterials 657

4. Conclusion and Future Outlook 662
Acknowledgements 662
References 663

Chapter 21 Immunotoxicity Testing for Drug–Nanoparticle Conjugates: Regulatory Considerations 671
Simona Bancos, Katherine M. Tyner, and James L. Weaver

1. Introduction 672
2. Regulation of Drug Products 672
3. Regulation of Nanomaterial-Containing Drug Products 673
4. Immunotoxicological Regulation of Nanomaterial-Containing Drug Products 674
5. Recommended Studies by ICH S8 674
   5.1. Hematological and clinical chemistry tests 676
   5.2. Histological tests and organ weights 677
   5.3. Other signs of immunosuppression 678
   5.4. Factors to consider in the WoE review 678
6. Additional Recommended Studies by ICH S8 679
7. Recommended Studies by ICH S6 680
8. Summary 681
9. Nanoparticle Research Conducted at the FDA/CDER 681
10. Conclusions 683
References 683

Index 687

Preface

The story of this book began in 2007 when we first presented our review "Immunological properties of engineered nanomaterials" in *Nature Nanotechnology*. In 2010, World Scientific Publishing released a book entitled *Nanoscience and Nanotechnology: A Collection of Reviews from Nature Journals* edited by Dr. Peter Rogers, with our review included in that book. Soon after that, Dr. Zvi Ruder, Senior Executive Editor at World Scientific, came to us with the attractive proposal of expanding our review into a book.

Instead of simply re-writing an extended version of the review on our own, we contacted top experts in numerous areas of nanoimmunotoxicology and invited them to contribute a chapter related to each of their unique area of expertise such as nanoparticle interaction with erythrocytes, endothelial cells, various types of antigen-presenting cells, complement activation, platelets and the plasma coagulation system, bone marrow cells, nanoparticle use for drug delivery and vaccine adjuvants, sterility and sterilization, the presence of surface contaminants, regulatory considerations for the screening of nanoparticle toxicity to the immune system, as well as the use of animal models to study nanoparticle immunotoxicity. Since we strongly believe that the success of immunotoxicological studies of nanoparticles vitally depends on the understanding of their physicochemical properties and composition, we have also invited renowned chemists to contribute a chapter to guide immunologists in appropriate material characterization. A generous consensus among all of these scientists to devote a chapter within their individual areas of expertise allowed us to establish an international team of experts, whose excellent works are presented in this book.

We are grateful to Dr. Nanda Subbarao of Biological Consulting Group, Inc. for sharing her expertise on nanoparticle sterility and sterilization; Drs. Clinton Jones and David Grainger of the University of Utah for their work on surface contaminants relating to nanoparticle immunotoxicity; Drs. G. Ulrich Nienhaus and Lennart Treuel for their contribution to the chapter on nanoparticle interaction with plasma proteins; Drs. Bridget Wildt, Ronald Brown, and Richard Malinauskas of the U.S. FDA CDRH for their knowledge on nanoparticle effects on erythrocytes; Dr. Jan Simak of the U.S. FDA CBER for his heroic effort in writing three extensive chapters focusing on nanoparticle interactions with platelets, endothelial cells, and the plasma coagulation system; Drs. Carolina Salvador-Morales of George Mason University and Robert Sim of the University of Oxford for their expertise on nanoparticle interaction with the complement system; Dr. William Zamboni and his team of graduate students — Whitney Caron, Sumit Rawal, Gina Song, Parag Kumar, and John Lay of the University of North Carolina for their expert opinion on nanoparticle interaction with phagocytes; Dr. Valentyna Fesenkova of the Georgia Institute of Technology for her contribution on dendritic cells; Dr. Ekaterina Dadachova for her knowledge on nanoparticle interaction with the bone marrow; Drs. Magdalena Plebanski and Sue Xiang and their colleagues — Drs. Martina Fuchsberger, Tanya De Karlson, Charles Hardy, and Cordelia Selomulya at Monash University in Australia for contributing a comprehensive chapter about nanoparticle use in vaccine delivery; Dr. Deepthy Menon and her colleagues — Drs. J. Gopikrishna, Dhanya Narayanan, and Shantikumar Nair at the Amrita Center for Nanoscience and Molecular Medicine in India for sharing their work on nanoparticle use for the delivery of anti-inflammatory drugs; Dr. África González-Fernández of the University of Vigo in Spain and her graduate student Silvia Lorenzo-Abalde for their chapter on nanoparticle allergenicity; Dr. Dori Germolec of the NIEHS and her colleagues — Drs. Matthew Smith, Colleen McLoughlin, and Kimber White, Jr. for their contribution on investigating nanoparticle immunotoxicity *in vivo*; and Drs. James Weaver, Katherine Tyner, and Simona Bancos of the U.S. FDA CDER for their expertise on regulatory considerations during nanoparticle testing for immunotoxicity.

The Nanotechnology Characterization Laboratory (NCL) was established in 2004 to promote the transition of basic research nanotechnology concepts into clinic. Following its main mission of conducting preclinical characterization of engineered nanomaterials and sharing its knowledge on nanoparticles with the research community, the NCL has characterized more than 250 different types of engineered nanomaterials. Our experiences with nanoparticle physicochemical characterization, dealing with endotoxin contamination in various types of nanoparticles, using *in vitro* methods to screen for nanoparticle immunotoxicity, as well as weighing potential concerns regarding nanoparticle antigenicity are shared in this book. We are grateful to Drs. Jeffrey Clogston and Anil Patri for their contributions and performing the formidable task of explaining the nuances of nanoparticle physicochemical characterization in a language easily comprehensible to immunologists. Furthermore, it is almost impossible to overstate the contribution of the NCL technical staff — Barry Neun, Timothy Potter, Christopher McLeland, Jamie Rodriguez, Sarah Skoczen, and former

NCL post-doc Parag Aggarwal in generating the data presented in the chapters discussing *in vitro* assays and endotoxin detection and quantification, and its role in nanoparticle immunotoxicity. We are also indebted to NCL's toxicologist Dr. Stephan Stern for being most helpful in establishing the *in vitro–in vivo* correlation for immunoassays, and to Drs. Jennifer Grossman and Rachael Crist for their assistance with the preparation of NCL chapters.

While working on this book, we have also appealed to two industry experts renowned worldwide for their contribution to the field of nanomedicine for their opinions on the contents of the book. We are grateful to Dr. Larry Tamarkin, the President and CEO of CytImmune Sciences developing PEGylated colloidal gold nanoparticles as carriers of the therapeutic protein Tumor Necrosis Factor-alpha, and Dr. Neil Desai of Celgene, the inventor of the nanoalbumin particle paclitaxel formulation Abraxane®, for their support and encouragement.

Nanoimmunotoxicology is a new and rapidly developing field. In spite of many obstacles, significant progress in our understanding of nanoparticle interaction with the components of the immune system has been achieved. Much is yet to be studied and understood. We are certain that this book will be of interest to the broad audience, to those who are just starting their work in this new and exciting field, as well as to experts representing academia, industry, and regulatory agencies. This book not only provides a comprehensive overview of the current literature and regulatory guidelines, but also includes protocols and practical considerations for studying nanoparticle immunoreactivity in both *in vitro* and *in vivo* conditions.

Once more, we sincerely thank everyone who has contributed to this book in one way or another. We hope that the reader will find it useful and enjoy it as much as we did while working on it.

References

1. Dobrovolskaia MA, McNeil SE. Immunological properties of engineered nanomaterials. *Nat Nanotechnol* 2007;2:469–478.
2. Peters R. *Nanoscience and Nanotechnology: A Collection of Reviews from Nature Journals.* World Scientific Publishing, Singapore, 2010.

Marina A. Dobrovolskaia
and Scott E. McNeil

Nanotechnology Characterization Laboratory
of SAIC-Frederick, Inc., contractor to
Frederick National Laboratory for Cancer Research
(formerly known as National Cancer Institute at Frederick)

List of Contributors

Simona Bancos
Division of Drug Safety Research
Center of Drug Evaluation and Research
Food and Drug Administration
Silver Spring, MD 20993, USA
Chapter 21

Ronald P. Brown
Office of Science and Engineering Laboratories
Center for Devices and Radiological Health
Food and Drug Administration
Silver Spring, MD 20993, USA
Chapter 7

Whitney P. Caron
Division of Pharmacotherapy and Experimental Therapeutics
UNC Eshelman School of Pharmacy
University of North Carolina at Chapel Hill
Chapel Hill, NC 27599, USA
Chapter 12

David G. Castner
National ESCA and Surface Analysis Center for Biomedical Problems
Departments of Bioengineering and Chemical Engineering
Box 351653
University of Washington
Seattle, WA 98195-1653, USA
Chapter 5

Jeffrey D. Clogston
Nanotechnology Characterization Laboratory
Advanced Technology Program
SAIC-Frederick, Inc.
National Cancer Institute at Frederick
Frederick, MD 21702, USA
Chapter 2

Ekaterina Dadachova
Departments of Radiology and Microbiology and Immunology
Albert Einstein College of Medicine
1695A Eastchester Road
Bronx, NY 10461, USA
Chapter 14

Marina A. Dobrovolskaia
Nanotechnology Characterization Laboratory
Advanced Technology Program
SAIC-Frederick, Inc.
National Cancer Institute at Frederick
Frederick, MD 21702, USA
Chapters 1, 4, 18, and 19

Valentyna Fesenkova
The Wallace H. Coulter Department of Medical Engineering
Georgia Institute of Technology
315 Ferst Drive
Atlanta, GA 30332, USA
Chapter 13

Martina Fushsberger
Department of Immunology
Central Clinical School
Faculty of Medicine
Nursing and Health Sciences
Monash University
89 Commercial Road
Melbourne, VIC 3004, Australia
Chapter 15

Dori R. Germolec
Division of the National Toxicology Program
The National Institute for Environmental Health Sciences
530 Davis Drive
Morrisville, NC 27560, USA
Chapter 20

África González-Fernández
Immunology, Biomedical Research Centre (CINBIO)
University of Vigo, Campus Lagoas Marcosende
Vigo 36310, Spain
Chapter 17

J. Gopikrishna
Amrita Centre for Nanosciences and Molecular Medicine
Amrita Institute of Medical Sciences and Research Centre
Amrita Vishwa Vidyapeetham University
Kochi-682041, Kerala, India
Chapter 16

David W. Grainger
Department of Pharmaceutics, Pharmaceutical Chemistry, and Bioengineering
University of Utah, Salt Lake City
UT 84112-5820, USA
Chapter 5

Charles L. Hardy
Department of Immunology
Central Clinical School
Faculty of Medicine
Nursing and Health Sciences
Monash University
89 Commercial Road
Melbourne, VIC 3004, Australia
Chapter 15

Clinton F. Jones
Center for Nanotechnology in Drug Delivery
Division of Molecular Pharmaceutics
UNC Eshelman School of Pharmacy
University of North Carolina at Chapel Hill
Chapel Hill, NC 27599, USA
Chapter 5

Tanya De L. Karlson
Department of Immunology
Central Clinical School
Faculty of Medicine
Nursing and Health Sciences
Monash University
89 Commercial Road
Melbourne, VIC 3004, Australia
Chapter 15

Parag Kumar
Division of Pharmacotherapy and Experimental Therapeutics
UNC Eshelman School of Pharmacy
University of North Carolina at Chapel Hill
Chapel Hill, NC 27599, USA
Chapter 12

John C. Lay
UNC School of Medicine
University of North Carolina at Chapel Hill
Chapel Hill, NC 27599, USA
Chapter 12

Silvia Lorenzo-Abalde
Immunology, Biomedical Research Centre (CINBIO)
University of Vigo, Campus Lagoas Marcosende
Vigo 36310, Spain
Chapter 17

Richard A. Malinauskas
Office of Science and Engineering Laboratories
Center for Devices and Radiological Health
Food and Drug Administration
Silver Spring, MD 20993, USA
Chapter 7

Colleen E. McLoughlin
Immunotoxicology Group
Virginia Commonwealth University
PO Box 980613
Richmond, VA 23298, USA
Chapter 20

Scott E. McNeil
Nanotechnology Characterization Laboratory
Advanced Technology Program
SAIC-Frederick, Inc.
National Cancer Institute at Frederick
Frederick, MD 21702, USA
Chapters 1, 4, and 19

Deepthy Menon
Amrita Centre for Nanosciences and Molecular Medicine
Amrita Institute of Medical Sciences and Research Centre
Amrita Vishwa Vidyapeetham University
Kochi-682041, Kerala, India
Chapter 16

Shantikumar V. Nair
Amrita Centre for Nanosciences and Molecular Medicine
Amrita Institute of Medical Sciences and Research Centre
Amrita Vishwa Vidyapeetham University
Kochi-682041, Kerala, India
Chapter 16

Dhanya Narayanan
Amrita Centre for Nanosciences and Molecular Medicine
Amrita Institute of Medical Sciences and Research Centre
Amrita Vishwa Vidyapeetham University
Kochi-682041, Kerala, India
Chapter 16

G. Ulrich Nienhaus
Institute of Applied Physics and Center for Functional Nanostructures (CFN)
Karlsruhe Institute of Technology (KIT)
76128 Karlsruhe, Germany *and*
Department of Physics
University of Illinois at Urbana-Champaign
Urbana, IL 61801, USA
Chapter 6

Anil K. Patri
Nanotechnology Characterization Laboratory
Advanced Technology Program
SAIC-Frederick, Inc.
National Cancer Institute at Frederick
Frederick, MD 21702, USA
Chapter 2

Magdalena Plebanski
Department of Immunology
Central Clinical School
Faculty of Medicine
Nursing and Health Sciences
Monash University
89 Commercial Road
Melbourne, VIC 3004, Australia
Chapter 15

Suwit Rawal
Division of Pharmacotherapy and Experimental Therapeutics
UNC Eshelman School of Pharmacy
University of North Carolina at Chapel Hill
Chapel Hill, NC 27599, USA
Chapter 12

Carolina Salvador-Morales
Bioengineering Department *and*
Krasnow Institute for Advanced Study
George Mason University
4400 University Drive, MS 1G5
Fairfax, VA 22030, USA
Chapter 11

Cordelia Selomulya
Department of Chemical Engineering
Monash University
Clayton, VIC 3800, Australia
Chapter 15

Robert B. Sim
Department of Pharmacology
University of Oxford
Mansfield Road, OX1 3QT, UK
Chapter 11

Jan Simak
Center for Biologics Evaluation and Research
Food and Drug Administration
1401 Rockville Pike, HFM-335
Rockville, MD 20852-1448, USA
Chapters 8, 9, and 10

Matthew J. Smith
Immunotoxicology Group
Virginia Commonwealth University
PO Box 980613
Richmond, VA 23298, USA
Chapter 20

Gina Song
Division of Pharmacotherapy and Experimental Therapeutics
UNC Eshelman School of Pharmacy
University of North Carolina at Chapel Hill
Chapel Hill, NC 27599, USA
Chapter 12

Nanda Subbarao
Biologics Consulting Group
400 N Washington Street Suite 100
Alexandria, VA 22314, USA
Chapter 3

Lennart Treuel
Institute of Applied Physics and Center for Functional Nanostructures (CFN)
Karlsruhe Institute of Technology (KIT)
76128 Karlsruhe, Germany *and*
Institute of Physical Chemistry
University of Duisburg-Essen
45141 Essen, Germany
Chapter 6

Katherine M. Tyner
Division of Drug Safety Research
Center of Drug Evaluation and Research
Food and Drug Administration
Silver Spring, MD 20993, USA
Chapter 21

James L. Weaver
Division of Drug Safety Research
Center of Drug Evaluation and Research
Food and Drug Administration
Silver Spring, MD 20993, USA
Chapter 21

Kimber L. White, Jr.
Immunotoxicology Group
Virginia Commonwealth University
PO Box 980613
Richmond, VA 23298, USA
Chapter 20

Bridget Wildt
Office of Science and Engineering Laboratories
Center for Devices and Radiological Health
Food and Drug Administration
Silver Spring, MD 20993, USA
Chapter 7

Sue D. Xiang
Department of Immunology
Central Clinic School
Faculty of Medicine
Nursing and Health Sciences
Monash University
89 Commercial Road
Melbourne, VIC 3004, Australia
Chapter 15

William C. Zamboni
Division of Pharmacotherapy and Experimental Therapeutics
UNC Eshelman School of Pharmacy,
UNC Institute for Pharmacogenomics and Individualized Therapy,
UNC Lineberger Comprehensive Cancer Center,
Carolina Center of Cancer Nanotechnology Excellence
University of North Carolina at Chapel Hill
Chapel Hill, NC 27599, USA, *and*
North Carolina Biomedical Innovation Network
The Research Triangle Park
Durham, NC 27709, USA
Chapter 12

Chapter 1

Immunological Properties of Engineered Nanomaterials: An Introduction

Marina A. Dobrovolskaia and Scott E. McNeil*

Nanotechnology Characterization Laboratory
Advanced Technology Program, SAIC-Frederick, Inc.
NCI Frederick, Frederick, MD 21702, USA

Clear benefits of using engineered nanoparticles for biomedical applications are often challenged by concerns about the safety of these materials. Since the main job of the immune system is to efficiently detect and eliminate foreign materials from the body, nanoparticle effects on and interaction with various components of the immune system are active areas of research in current bionanotechnology and nanomedicine. Nanoparticles can be engineered to either avoid immune recognition or to specifically interact with the immune system. Below, we will provide a top level overview of the current state of science in the area of nanoimmunotoxicology, highlight common challenges associated with research of immunological reactivity of engineered nanomaterials, identify current gaps in our understanding of nanoparticle interaction with components of the immune system, and introduce other chapters in this book.

1. Introduction

Nanoparticle interaction with components of the immune system may produce different outcomes: 1) it may inhibit or suppress the immune system (immunosuppression) or 2) it may stimulate or enhance immune function (immunostimulation). Both immunostimulation and immunosuppression can be beneficial or detrimental depending on the intended use of an engineered

*Corresponding author. E-mail: marina@mail.nih.gov

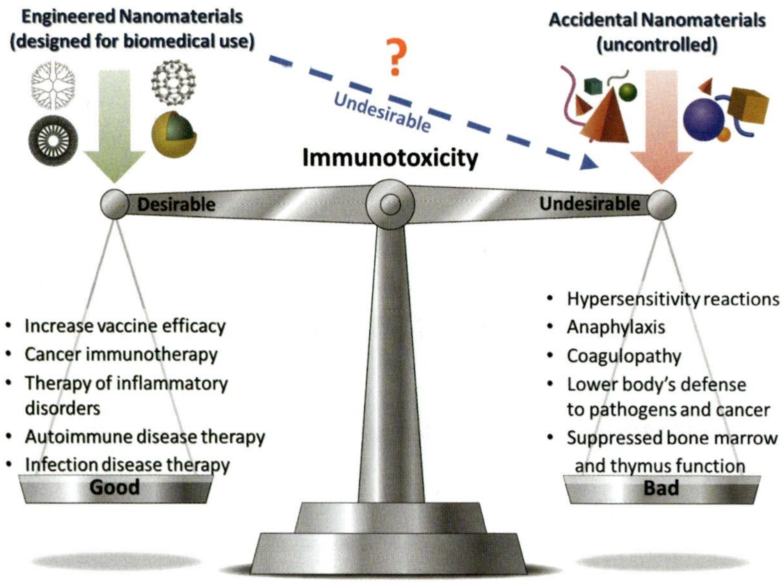

Figure 1. Nanoparticle immunotoxicity. Immunotoxicity of nanomaterials can be beneficial or detrimental depending on its desirability. Engineered nanoparticles can be designed to specifically interact or avoid interaction with the immune system. In contrast, accidental nanomaterials are more reactive and their interactions with the immune system are uncontrolled. Undesirable interactions between engineered nanomaterials and the immune system are the subject of basic research and preclinical studies.

nanomaterial (Figure 1). When immunostimulation is desired, it may aid in vaccine efficacy and therapy of cancer, while unwanted immunostimulation may lead to complications including hypersensitivity reactions, anaphylaxis, and thrombosis.[1] Likewise, desirable immunosuppression is beneficial for the treatment of inflammatory disorders and autoimmune diseases, improvement of transplant acceptance, and prevention of allergic reactions, while undesirable immunosuppression may lower the body's response to infected and malignant cells, and lower bone marrow and thymus functions.[1] It is now generally accepted that nanoparticle physicochemical properties such as size, charge, hydrophobicity, and presence of targeting moieties determine their interaction with the components of the immune system and immunotoxicity.[2]

Nanoparticles can be immunotoxic, however, so far no new, i.e., specific to nano, immunotoxicity has been reported. This is why it is currently accepted that immunotoxicity of engineered nanomaterials can be effectively assessed using the current portfolio of methods established for small and macromolecules.[3] Reviews of available data suggest that engineered nanomaterials are intrinsically no more immunotoxic than traditional drugs currently in use.

Moreover, the incorporation of conventional pharmaceutics into nanotechnology-based platforms helps to decrease immunotoxicity of these traditional formulations. For example, benefits of using recombinant tumor necrosis factor-alpha (TNF-α) for cancer therapy were halted by systemic toxicity of this protein — fever and hypotension were dose-limiting adverse reactions associated with overt immunostimulation caused by this cytokine. The maximum tolerated dose (MTD) of this protein established in phase I clinical trials was 261 mg/m^2 after slow (over 24 h) continuous intravenous (i.v.) infusion.[4] Recently, the same protein was covalently attached to the surface of polyethylene glycol (PEG)ylated colloidal gold nanoparticles (Cyt6091 or Aurimmune™) and assessed in clinical trials. It has been reported that PEG–gold–TNF formulation was well tolerated by cancer patients when administered by bolus i.v. injection at 600 mg/m^2 without reaching the MTD.[5] Another two examples are cytotoxic oncology drugs paclitaxel and doxorubicin. The clinical application of the traditional formulation of the anticancer drug paclitaxel (Taxol®) is associated with severe hypersensitivity reactions even when this formulation is administered slowly and patients are premedicated with immunosuppressive agents, while nanoparticle formulation of the same drug (Abraxane®) could be administered as a fast infusion and did not require premedication.[6] Likewise, the clinical use of the traditional formulation of another anticancer agent doxorubicin (Adriamycin®) is limited, among other side effects by disseminated intravascular coagulation (DIC), while incorporation of this small molecule drug into PEGylated nanoliposomes eliminates this toxicity.[7]

Although these examples are very encouraging, we have to keep in mind that each nanoparticle is unique and a positive experience with one formulation does not guarantee similar success with another, i.e., each nanoformulation needs a separate set of studies to understand its safety. The "shifting" of a drug's toxicity from one target organ to another due to the change in biodistribution caused by a nanocarrier is an important point to think about when evaluating nanoformulations. For example, nephrotoxicity is a common dose-limiting toxicity of the anticancer drug cisplatin.[8] When cisplatin is incorporated onto the nanotechnology platform, it may follow the nanocarrier; if the nanocarrier is distributed to another organ, e.g., spleen or liver, the toxicity associated with cisplatin may "relocate" to these organs. This is why thorough pharmacokinetic (PK) studies assessing the distribution of both the drug and the nanocarrier are needed to identify potential changes in the drug's biodistribution and new site(s) of undesirable toxicity caused by this change.[9] Avoiding the "double-hit" scenario is another important point to consider when designing nanoformulations with the aim of reducing toxicity of the legacy drug. For instance, if a nanoformulation is considered for the delivery of oligonucleotides (ODNs) for which complement activation-mediated pseudoallergy (CARPA) is a common dose-limiting toxicity,[10] it is better to consider a nanocarrier which is known not to cause CARPA. In this example, a nanocarrier in a form of PEGylated liposomes does not appear to be an optimal carrier choice for ODNs, because CARPA is a very common side effect of PEGylated liposomes.[11] Of course, all of these examples are generalized and should not be used to discourage the development of an individual nanoformulation.

During recent years of research investigating the immunocompatibility of engineered nanomaterials, several common types of immunotoxicity have been identified. They are cytokine-mediated reactions, complement activation, coagulopathy (specifically DIC-like reactions), and the exaggeration of endotoxin responses. Among the common problems experienced in early preclinical testing, endotoxin detection and removal, nanoparticle interference with traditional *in vitro* assays, the lack of standard toxic reference materials, and the need for modified versions of traditional *in vivo* tests are on top of the list. Seven years ago, the entry of key words "nanoparticles" and "immune system" or "immunotoxicity" into the PubMed database generated no results. Now, entering these words brings over 2,000 hits. A common problem limiting the understanding of the reported test results is the lack of physicochemical characterization of the studied particles. In recent years, the understanding of sample preparation issues as well as the significance of nanoparticle physicochemical properties and composition have improved. Chapter 2 of this book contributed by Drs. Jeffrey Clogston and Anil Patri provides a comprehensive overview of critical points relating to nanomaterial characterization prior to their use in immunological tests. Other common issues which may confound the results of immunotoxicological studies are nanoparticle sterility, sterilization, and contamination with endotoxin and synthesis by-products. All these challenges and practical solutions to overcome them are reviewed in detail in Chapter 3 by Dr. Nanda Subbarao, Chapter 4 by us, and Chapter 5 by Dr. David Grainger *et al.*

2. Interaction with Blood Components

If a nanoformulation is administered via i.v. or distributed into the systemic circulation after administration through other routes, then understanding its interaction with blood components becomes an essential initial step in preclinical development. Blood is composed of plasma and cells, where the plasma constitutes about 50% of blood fluid and is composed of water, mineral ions, dissolved gases (oxygen and carbon dioxide), glucose, and proteins. There are about 3,000 proteins in the plasma.[12] When nanoparticles enter the bloodstream, they bind proteins almost instantaneously.[13] The protein layer on the particle surface creates a so-called "corona," which determines particle biodistribution and clearance.[13] Charged nanoparticles bind more proteins than particles with neutral surfaces. Surface neutralization by means of coating with hydrophilic polymers such as PEG is one of the best studied approaches to protect the nanoparticle surface from protein binding and extend their circulation in the bloodstream. The role of nanoparticle physicochemical properties such as size, charge, hydrophobicity, composition, dynamics, and the significance of opsonization in nanoparticle biodistribution and clearance are discussed by Drs. Lennart Treuel and G. Nienhaus in Chapter 6.

The cells in blood include erythrocytes, monocytes, granulocytes, lymphocytes, and thrombocytes (also known as platelets). Hemolysis (damage of red blood cells), coagulation (blood clotting through effects on coagulation cascade, platelets, endothelial cells, and leukocytes), and activation of the complement system are commonly used parameters to test nanoparticle hematotoxicity.

2.1. *Hemolysis*

This term refers to the potential of a nanoparticle to damage erythrocytes. The loss of red blood cells following hemolysis may lead to anemia, and the release of iron-containing protein hemoglobin into the circulation may lead to nephrotoxicity.[14] This is why optimizing the nanoparticle formulation to prevent hemolysis is essential to avoid such complications. Nanoparticle interactions with red blood cells are largely determined by their physicochemical properties, which may differ among different classes of engineered nanomaterials. For example, the hemolytic activity of silver colloids with an identical surface charge was shown to depend on particle surface area, in that particles with a smaller size and larger surface-to-volume ratio were more toxic in the hemolysis assay than their larger counterparts.[15] The mechanism of toxicity of anionic silver nanoparticles was attributed to the greater release of silver ions from particles with a greater surface area.[15] Hemolysis caused by silica nanoparticles was recently attributed to surface charge, with cationic species being more damaging, as well as to porosity and shape in that mesoporous particles with a large aspect ratio were less hemolytic than spherical particles and mesoporous particles with a low aspect ratio.[16] Shape was also named a primary parameter in determining hematocompatibility of gold colloids, with the spherical shape being more hematocompatible.[17] It is important to note that studies investigating the role of nanoparticle shape are complicated by the presence of surfactants used during particle synthesis. For example, gold chloride and sodium citrate are two components required to synthesize gold spheres; however, the synthesis of gold nanorods requires one more component — a surfactant — to prevent particle aggregation. The purification of this surfactant is challenging, while the presence of even trace amounts in nanoparticle formulations is cell-damaging.[18] Nanotechnology has been shown to benefit the formulation of hydrophobic macromolecules with reduced hemolytic activity by improving solubility and reducing zeta potential through the preparation of polymer-based nanosuspensions.[19,20] There are a lot of experimental nuances in each study, which may make comparisons among them inaccurate. The use of surfactants during particle synthesis, variations in material purity, stability under physiological conditions, interaction with plasma proteins, and use of different methods to study hemolysis are a limited set of variables, which may influence the interpretation of test results. Chapter 7 of this book written by Dr. Bridget Wildt *et al.* provides a detailed overview of these and other questions, as well as discusses experimental approaches and suggests practical solutions to common challenges experienced with studying nanoparticle interactions with erythrocytes.

2.2. *Coagulation*

The mammalian blood coagulation system is very complex and plays a key role in maintaining hemostasis. It involves various cell types including platelets (also known as thrombocytes) and endothelial cells, and plasma proteins. Nanoparticle effects on endothelial cells and experimental approaches in studying them are reviewed in length by Dr. Jan Simak in Chapter 8. Some normal blood cells (e.g., monocytes) and cancer cells (both hematological and solid tumors) can also contribute to coagulation by activating the plasma coagulation cascade. Particle effects on plasma coagulation are discussed in Chapter 9, also by Dr. Simak. The activation of leukocytes and induction of the procoagulant activity (PCA) complex composed of phosphatidylserine and tissue factor on the cell surface recruit coagulation factors and initiates the plasma coagulation cascade. The formation of the PCA complex on the leukocyte surface was shown to depend on particle size and surface charge, in that larger particles and particles with a cationic, but not anionic and neutral, surface were capable of inducing PCA.[21] Futhermore, PCA is recognized as a marker for DIC, which is a coagulation disorder characterized by hemorrhages and abnormal blood clotting. It is a very common complication in cancer and sepsis, and is also increasingly reported for some nanomaterials. For example, it has been demonstrated that the i.v. injection of cationic PAMAM (polyamidoamine) dendrimers results in vascular congestion and abnormal bleeding consistent with the symptoms of DIC.[22] The induction of DIC by dendrimers in mice was dependent on particle size and surface charge in that larger dendrimers and dendrimers with cationic, but not anionic, surface induced DIC-like toxicity.[22] It has also been reported that the porosity of silica nanoparticles is another important contributor to their thrombogenic potential as non-porous materials were less thrombogenic than porous materials.[23] Nanoparticle interactions with platelets depend on nanoparticle composition, size, and charge.[22,24–30] Of interest is the data suggesting that nanoparticles activate platelets through uncanonical pathways, e.g., those involving matrix metalloproteinases.[24,29] The alteration of the cell membrane by some cationic nanoparticles[25,26,28] and membrane perforation[27] were described as common mechanisms of nanoparticle mediated platelet aggregation. Nanoparticle effects on platelets and protocols to study platelet activation and aggregation are described in Chapter 10 by Dr. Simak.

2.3. *Complement activation*

The complement system is made up of a group of about 30 proteins. The name of this biochemical cascade is derived from its function in the body which is to "complement" cells of the innate and adaptive immunity in protecting the host from invading pathogens. The complement system plays many protective functions including opsonization (necessary for promoting phagocytosis of antigens), chemotaxis (needed for the recruitment of neutrophils and macrophages to the site of inflammation), and cell lysis (the rupture of the membranes of foreign and infected host cells). Complement proteins are also important for supporting the adaptive immunity through the activation of dendritic cells, T cells, and B cells. The undesirable activation of complement is responsible for hypersensitivity reactions and anaphylaxis. Upon activation, complement proteins acquire protealytic activity; each activated protein in the cascade cleaves the next protein which triggers the activation of the first protein and the cleavage of the next protein, etc. There are three main pathways of activation — classical, alternative (also known as properdin), and lectin. The activation of the classical pathway is triggered by the antigen–antibody complex, that of the alternative pathway by the spontaneous hydrolysis of the C3 component and factor P (properdin), and the activation of the lectin pathway is triggered by mannose-binding lectin. It has been demonstrated that nanoparticles can activate complement through one of these three main pathways or a combination thereof,[31–44] which is primarily determined by nanoparticle surface properties and the conformation of polymer coating.[31,33,37] Another interesting fact is that the activation of the classical pathway by nanoparticles does not require a specific antibody–antigen complex, but is triggered by the direct interaction with the C1q component of complement or the non-specific binding of immunoglobulins.[32] It is generally accepted that nanoparticles intended for systemic administration should be engineered with the purpose of avoiding complement activation, in order to prevent hypersensitivity reactions and anaphylaxis. In contrast to undesirable complement activation caused by systemic injection of certain nanomaterials, the activation of complement by other nanoparticles administered via different routes (subcutaneous or intradermal) can benefit vaccine efficacy.[45,46] Although nanoparticle coating with hydrophilic polymers such as PEG is the generally recognized way to increase their circulation time and protect them from immune recognition, PEG coating of some nanomaterials (e.g., liposomes) does not protect them from complement activation. Mechanisms of complement activation by nanoparticles and the role of nanoparticle physicochemical properties in the activation are well understood through the studies with PEGylated liposomes. The details of nanoparticle interactions with the complement system as well as practical approaches to study these interactions are described in Chapter 11 by Drs. Carolina Salvador-Morales and Robert Sim. Here, we would like to emphasize the importance of paying attention to the synthesis procedures and linkers used to attach polymers and ligands onto nanoparticle surfaces. For example, a recent study demonstrated that linkers with hydroxyl and thiol groups do not activate the complement system, while linkers with methoxy groups are potent activators.[47] Another study has shown that the methylation of the phosphate oxygen moiety of methoxyPEG reduces complement activation by PEGylated liposomes.[40] From a practical standpoint, it means that the use of "complement-reactive" linkers may be acceptable for nanoparticles intended for vaccine delivery when the activation of complement is desirable, while the use of "complement-unreactive" linkers is crucial for formulations intended for systemic administration when complement activation is undesirable.

3. Nanoparticle Uptake by Phagocytic Cells

Nanoparticle recognition and uptake by phagocytic cells are determined by their physicochemical properties such as size, zeta potential, solubility, surface functionality, density, and conformation of the polymer coating (Figure 2). It

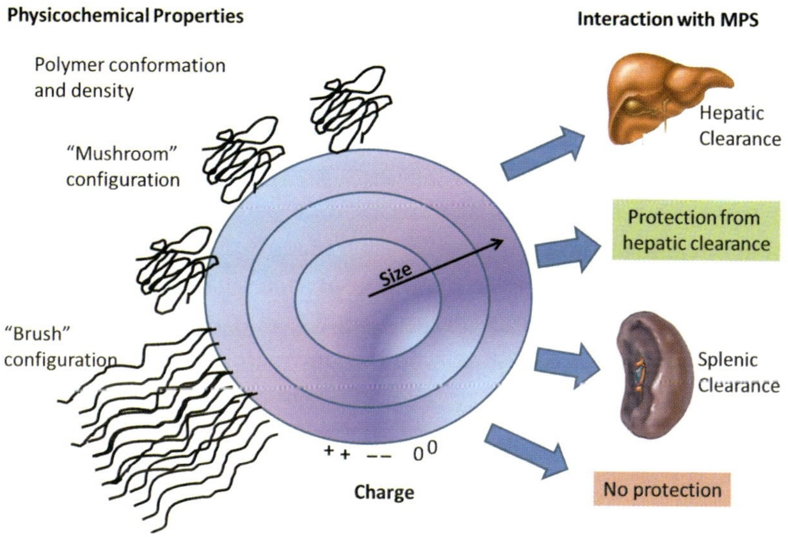

Figure 2. The role of nanoparticle physicochemical properties in their clearance by the mononuclear phagocytic system (MPS). Charged particles are cleared more readily than their neutral counterparts. Larger particles are more attractive to phagocytes than smaller particles. Polymer conformation plays a key role in determining the clearance route. Particles with small to medium size (50–200 nm) coated with a polymer in the "mushroom" configuration are less protected from MPS uptake than their counterparts coated with a polymer in the "brush" configuration, and are cleared by Kupffer cells of the liver (hepatic clearance). Increasing the particle core size to 300 nm determines their clearance by red pulp macrophages in the spleen (splenic clearance). Further increases in size regardless of polymer coating do not protect nanoparticles from MPS uptake. The figure is prepared based on Refs. 43, 50–55.

is generally agreed that larger particles are taken up more efficiently by macrophages than smaller particles of the same composition and surface functionality.[48] Charged nanoparticles (cationic or anionic) are more attractive to phagocytes then neutral particles of the same size.[49] Manipulation of nanoparticle surface properties was shown to play a key role in their clearance by specific organs of the mononuclear phagocytic system (MPS). For example, nanoparticles with sizes between 50–200 nm coated with poloxamer in a "mushroom" configuration (less dense coating) resulted in a predominant clearance by liver resident macrophages (i.e., Kupffer cells); changing the polymer conformation to "brush" (dense coating) on the surface of nanoparticles of the same composition and size resulted in its protection from hepatic clearance. Increasing the particle size to 200–300 nm regardless of polymer configuration shifted the clearance route to red pulp macrophages of the spleen, while further increases in size regardless of polymer conformation and density resulted in no protection from MPS uptake.[43,50–55]

Each nanoparticle is unique and its clearance by the cells of the MPS may occur through specific mechanisms. This is why in the field of nanomedicine, in order to design immunocompatible and efficient nanoparticle-based drug delivery systems, it is important to understand the mechanisms by which nanoparticles are cleared from circulation. In the past, it was believed that phagocytosis was restricted to large particulate matter (>0.5 μm) and therefore not applicable to nanoparticles. Several recent studies have debunked this presumption by demonstrating that nanoparticles can indeed be phagocytosed.[56–59] Of special interest is the finding that macrophages utilize multiple routes for the uptake of the same type of nanomaterials, and the blocking of one pathway allows particle entry through another route.[56]

The engineering of nanoparticle surfaces has proven useful for desirable particle delivery to immune cells and for avoiding protein binding and undesirable uptake by the MPS. For example, surface functionalization with chitosan and mannose was shown to deliver nanoparticles to macrophages and dendritic cells via specific phagocytic pathways and improve immune response against nanoparticle-bound antigens.[60–62] In contrast to vaccine delivery, where uptake by immune cells is desired, nanoparticle internalization by phagocytes is avoided in other applications by the addition of hydrophilic polymers such as PEG.[63] Nanoparticle interaction with cells of the MPS is discussed in detail by Dr. William Zamboni *et al.* in Chapter 12.

4. Immunogenicity

Immunogenicity is a broad term referring to the ability of a nanoparticle to activate the immune system. The mechanisms of activation and the outcome vary depending on nanoparticle composition, size, charge, hydrophobicity, and the presence of targeting moieties or drugs. Each nanoparticle is unique in that a minor change in its physicochemical properties or in payload may create new entities in terms of compatibility with the immune system. While the evaluation of cytotoxic effects of nanomaterials on immune cells is very useful to get an idea of acute toxicity reactions, the understanding of nanoparticle immunogenicity is another important measure of their immunocompatibility. Nanoparticles may be immunogenic and some of these effects (e.g., cytokine storm) may have acute manifestation, while other effects (e.g., adjuvanticity) require time before manifestation. Hence, it is becoming a common practice to study nanoparticle effects on immune cell function. Blood is composed of multiple cell types, each of which has a certain function. An understanding of how nanoparticles can change the function of an individual subset of immune cells is informative in preventing undesirable immunogenicity and manipulating desirable immunostimulation. Nanoparticle effects on dendritic cells are reviewed in Chapter 13 by Dr. Valentyna Fesenkova, their interaction with the bone marrow is reviewed in Chapter 14 by Dr. Ekaterina Dadachova, the manipulation of nanoparticle physicochemical properties to achieve desirable immunostimulation and benefit vaccine delivery is covered in detail in Chapter 15 by Dr. Sue Xiang *et al.*, and nanoparticle allergenicity and antigenic properties are reviewed in Chapters 17 (by Dr. África González-Fernández and Silvia Lorenzo-Albalde) and 18 (by us), respectively.

4.1. *Inflammation*

The activation of immune cells results in the secretion of a variety of mediators of inflammation — antibodies, cytokines, chemokines, prostaglandins, histamine, various kinins, reactive oxygen species, and neuropeptides. In health, balanced inflammation is required to destroy pathogens. However, overwhelming inflammation and alterations in the negative regulation of healthy inflammatory response result in undesirable immunostimulation. One form of undesirable immunostimulation are hypersensitivity reactions (allergy and anaphylaxis). Studies on nanoparticle-specific hypersensitivity reactions are limited; however, this is a very important and fast-expanding area of research due to the increasing use of engineered nanomaterials in cosmetics, biology, and medicine. Nanoparticle-mediated reactions and the role of nanoparticle physicochemical properties in mediating allergic reaction are reviewed in Chapter 17.

Immune cells have a versatile set of receptors to recognize foreign materials. One group of well-studied receptors of the innate immunity are Toll-like receptors (TLRs). Each TLR is known for its specific recognition of certain endogenous and exogenous ligands.[64] In recent years, there is an increasing number of studies describing either nanoparticle recognition by TLRs or nanoparticle use to increase/decrease biomaterial recognition by TLRs.[65–70] For example, the stimulation of cytokine expression and other inflammatory mediators in macrophages was reported to occur through TLR4 recognition of polyglutamic acid biodegradable nanoparticles,[70] through TLR7 and TLR8 stimulation by lipidoid nanoparticles carrying siRNA payload,[66] activation of TLR2 and TLR3 by polysaccharide–siRNA nanoparticles,[68] and activation of TLR9 by polystyrene nanoparticle–CpG DNA conjugates.[65] In contrast to these studies demonstrating the activation of TLR-dependent signaling, Tsai *et al.* reported that citrate-stabilized gold nanoparticles of small size (4 nm), but not their larger counterparts, attenuate TLR9 signaling by interfering with TLR9 translocation in endosomes and binding to high-mobility group 1 protein critical for the regulation of

TLR9 signaling.[69] Nanoparticles are also used as tools to study TLR signaling events.[67] However, a common limitation in these studies is endotoxin contamination. The majority of studies neither provide adequate physicochemical characterization of the nanoparticles nor do they assess endotoxin contamination. The latter is very important as endotoxin removal from commercial or in-lab synthesized nanomaterials was shown to reduce cytokine secretion and macrophage activation through TLR4.[71]

4.2. *Antigenicity*

The antigenicity of nanoparticles is drawing interest for several reasons. It has been demonstrated that some nanoparticles can serve as adjuvants due to their ability to increase the antigenicity of weak antigens.[72] Many engineered nanoparticles are used as drug delivery vehicles and thus, are intentionally introduced into human body; many of them carry biological ligands (proteins, peptides, lipids, and nucleic acids) as targeting moieties or as active therapeutic ingredients. It is generally recognized that nanoparticles with surfaces unprotected by hydrophilic polymers shielding them from immune recognition, interact with and bind to plasma proteins. There is also data suggesting that nanoparticles are not antigenic on their own,[73–76] but some of them may behave as haptens and become antigenic after binding to a protein carrier.[65,77–79] Finally, some accidental nanoparticles have been shown to enhance antigenicity of therapeutic proteins.[80–82] It is very important to understand the differences between accidental and engineered nanomaterials. Nanoparticle properties can be tuned to avoid undesirable antigenicity or enhance antigenicity when it is needed for certain applications such as vaccine adjuvants.[72] There are several examples demonstrating that nanoparticle-based formulations of therapeutic proteins, such as TNF-α, streptokinase, and factor VIII, avoid immune recognition of these proteins and the formation of protein-specific antibodies.[5,83] Another important factor to consider is that the linkers, commonly used during nanoparticle synthesis to attach polymers and ligands to particle surfaces, have various degrees of antigenicity on their own.[84] Therefore, the careful selection of the synthetic procedure and the linker is crucial in order to achieve the desirable or avoid the undesirable antigenicity. These points are reviewed in more detail in Chapter 18 of this book.

4.3. *Adjuvant properties*

Adjuvants are immunogenic materials which stimulate immune cells (both antigen-presenting cells and lymphocytes) and increase their response to antigens.[85] While strong antigens (e.g., ovalbumin) do not need adjuvants, in order to generate specific immune recognition of weak antigens (e.g., viral epitops), good adjuvants are required. Different adjuvants have the ability to induce certain types of immune responses. For example, alum is an adjuvant used in clinical practice, which stimulates Th2 cytokines and antibody response.[85] Other emulsion-type adjuvants are being investigated for their ability to stimulate Th1-type immune response.[85] Nanoparticles can serve as adjuvants and improve vaccine efficacy by stimulating antigen-presenting cells and providing the controlled release of antigens. Furthermore, nanoparticle adjuvanticity is determined by the particle's physicochemical properties such as size and surface charge.[45,72] This area is being actively explored for the development of vaccines. Nanoparticle adjuvant properties, the benefits of using them for vaccine delivery, as well as challenges and future directions in this field are discussed in Chapter 15.

5. Immunosuppression

Nanoparticles can be engineered to be immunosuppressive or to carry immunosuppressive payloads.[86–89] There are more studies describing immunomodulation by nanoparticles loaded with immunosuppressive drugs than those describing immunosuppressive nanoparticles. For example, cholesteryl butyrate-conjugated solid lipid nanoparticles were shown to inhibit neutrophil infiltration and adhesion to endothelial cells.[90] This approach is currently considered for the therapy of inflammatory ulcerative colitis. Another study reported that generation (G) 3.5 PAMAM dendrimers conjugated to glucosamine inhibits the induction of the transcription of cytokines TNF-α, IL-1β, IL-6, and IL-12, and chemokines MIP-1α/1β in human monocyte-derived macrophages and dendritic cells exposed to bacterial endotoxin.[91,92] The same dendrimers conjugated to glucosamine-6-sulfate were even more potent immunosuppressants in that the LD_{50} values of dendrimer glucosamine were 134 μM and 209 μM for T cells and macrophages, respectively, while those of dendrimer glucosamine-6-sulfate were 22 μM and 19 μM, respectively.[91,92] Unlike other sulfated molecules, these dendrimers had no anti-coagulant, heparin-like, or complement-activating activities when added to fresh whole blood, and their effects were specific to leukocytes.[92] Shaunak *et al.* also suggested that the immunosuppressive effects of dendrimer–glucosamine conjugates may be beneficial for the treatment and prevention of scar tissue formation.[91,92] Interestingly, although dendrimer–glucosamine conjugates suppressed the nuclear factor-κB (NF-κB) pathway (MyD88-dependent arm of TLR4 signaling), they did not inhibit the production of

IFN-β and maturation of dendritic cells, suggesting that the IRF3 pathway (TRIF-dependent arm of the TLR4 signalling) leading to type I interferon production is not affected by dendrimer conjugates. However, the exact molecular mechanism of the observed selectivity is still not understood. Recently, the same research group conducted a computational study demonstrating the formation of dendrimer–glucosamine–MD2 complex.[93] The study suggested that dendrimer conjugates interact electrostatically with amino acids lining the entrance into the hydrophobic pocket of the MD2 protein, interfere with the electrostatic binding of endotoxin to MD2, block the MD2 hydrophobic pocket, and prevent the formation of the endotoxin–receptor complex (CD14–MD2–TLR4).[93] Conjugates of generation 4 amine- and hydroxyl–terminated dendrimers with celastrol were successfully used for the inhibition of microglial inflammation.[94]

Data about immunosuppressive effects of nanoparticles *per se* have been reported by several researchers. John *et al.* prepared polymerized lipid nanoparticles which bind preferentially to selectins on activated endothelial cells. The particles were capable of attenuating both peribronchial inflammation and airway hyper-reactivity induced by allergen challenge, presumably because they prevent eosinophils binding to targeted selectins.[95] Chen *et al.* reported immunosuppressive properties of fullerene derivatives, which could quench inflammation by binding nitric oxide and other reactive oxygen species produced by macrophages.[96] This approach was also suggested for medical applications to benefit the therapy of inflammatory disorders. PAMAM dendrimers are one of the best studied groups of engineered nanomaterials in which architecture and surface properties can be modified to create anti-inflammatory nanoparticles. However, in this case, anti-inflammatory properties were attributed to an anti-bacterial mode of action rather than a direct effect on the immune cells. For example, Wang *et al.* have demonstrated the use of PAMAM dendrimers with amine-, carboxy-, and hydroxy-terminated surfaces in the treatment and prevention of intramniotic infections in guinea pigs.[97] Of interest are the data suggesting that various surface groups of PAMAM dendrimers have different mechanisms of action. Both amine-terminated G4 and carboxy-terminated G3.5 dendrimers interfered with the bacterial membrane; the former did so through electrostatic binding to the anionic part of the endotoxin core (lipid A moiety), while the latter chelated divalent cations in the outer membrane of bacterial cells. Hydroxy-terminated dendrimers were reported to act through interaction with the polysaccharide chains of endotoxin.[97] Another group is actively investigating the role of dendrimer surface functionalities to scavenge and inactivate the lipid A portion of endotoxin. This group has demonstrated that cationic dendrimers can be modified to space surface amines in such a way that the distance between the nearest amino groups on the dendrimer surface matches that between negatively charged phosphate groups on lipid A of lipopolysaccharide (LPS).[98] These dendrimer constructs act as "endotoxin sponges" and not only electrostatically bind LPS, but also inactivate it, making it unavailable for recognition by the immune cells.[98]

While desirable immunosuppression by nanoparticles *per se* and nanoparticle–anti-inflammatory drug conjugates are being actively promoted to create new therapeutic approaches for the treatment of inflammation-associated disorders, undesirable immunosuppression is of concern as it may decrease the body's response to infections and cancer, as well as cause thymic and bone marrow suppression. The studies focusing on the understanding of nanoparticle physicochemical properties determining their undesirable immunosuppression are limited. Partially, it can be explained by the lack of *in vitro* approaches one could use for the rapid screening of undesirable immunosuppression. Standard acute toxicity studies do not always detect immunosuppression because only strong effects are detected, while mild and moderate immunosuppression manifest only at a functional level. This is why it became a common practice in the drug development industry to supplement standard acute toxicity studies with specialized immune function tests. These specialized tests are focused on a certain function of a subset of immune cells and are proven to be very useful in identifying functional immunosuppression. For example, one recent study used a modified version of the traditional T cell-dependent antibody response (TDAR) assay to investigate immunosuppressive properties of superparamagnetic iron oxide nanoparticles. This study utilized iron oxide-based imaging agent Resovist®, approved for clinical use in Europe, but not marketed in the United States due to idiosyncratic reactions associated with the dextran coating on the particle surface.[99] In this study, the animals were dosed with Resovist® at 30 mg iron/kg dose level equivalent to that used for imaging and 1 h later, the animals were challenged with ovalbumin (OVA). The production of OVA-specific antibodies and T cell reactivity was assessed seven days after the challenge and suppressed response was observed.[99] Another recent study reported that quantum dots impair the phagocytic activity of macrophages.[100] The suppression of TDAR and phagocytic activity may lead to the decrease of the body's defense against pathogens and therefore, there should be studies to thoroughly elucidate the underlying mechanisms.

More research is indeed needed to understand undesirable immunosuppression by engineered nanomaterials and determine nanoparticle properties crucial for this type of immunomodulatory effect. Various approaches in designing nanoparticles for the treatment of inflammation-associated disorders are reviewed in Chapter 16 by Dr. Deepthy Menon *et al.*

6. Designing Immunotoxicity Studies

There is a growing interest in recent years in *in vitro* models useful for the assessment of nanoparticle immunotoxicity. The increasing body of evidence suggests that validated *in vitro* tests are very helpful in identifying toxic particles early in preclinical development and in reducing the number of animals used for the identification of promising immunocompatible nanoplatforms. The correlation between *in vitro* methods and *in vivo* toxicity data are discussed in all chapters of this book. Each individual chapter also provides protocols and methodological considerations to specific hematological and immune function assays. There are many challenges complicating both *in vitro* and *in vivo* immunotoxicity studies of nanoparticles. Nanoparticle interference with traditional *in vitro* assays is the major challenge experienced by many labs. The sources of interference include nanoparticle optical and catalytic properties.[3] To identify false-positive or false-negative interferences, the running of additional controls such as particle-free controls, reagents, nanoparticle controls, as well as inhibition/enhancement controls (IECs) is very important. The purpose of these controls and how to prepare them are described in detail elsewhere[3] and also covered in Chapter 19. Chapter 19 also provides comprehensive protocols for the assessment of immune cell function *in vitro*. Endotoxin contamination and presence of synthesis by-products are other common complications in *in vitro* immunotoxicity studies.

In vivo immunotoxicity studies are conducted using traditional approaches; however, some of these methods require modifications. For example, the standard test for contact hypersensitivity called the local lymph node assay (LLNA) is not applicable to the majority of engineered nanomaterials because this test includes the topical application of test materials, and unlike small molecules, most engineered nanomaterials do not cross the intact skin barrier. Details about modifications and applicability of standard *in vivo* immunotoxicological studies to engineered nanomaterials have been discussed earlier.[3] Considerations and common challenges with *in vivo* immunotoxicity studies are reviewed in depth by Dr. Matthew Smith *et al.* in Chapter 20.

Since many nanoparticles are being developed for drug delivery, understanding the regulatory considerations in nanoparticle immunotoxicity is essential. The applicability of ICH8 and ICH6 immunotoxicity documents to the assessment of nanoparticle immunotoxicity and other regulatory considerations are highlighted by Dr. James Weaver and his colleagues in Chapter 21.

7. Future Directions

The future of nanoimmunotoxicology includes the expansion of our current understanding of nanoparticle interactions with the components of the immune system. This should include structure–activity relationship studies aimed at understanding key physicochemical properties determining nanoparticle interactions with the immune system, and mechanistic studies to understand the mechanisms behind these interactions. What makes one formulation immunosuppressive and another formulation immunostimulatory, and what neutralizes the effects of nanoparticles on the immune system? Delving into these questions will help us better define approaches to screen for immunotoxicity. While hematocompatibility is more or less understood, nanoparticle effects on immune cell function still require more attention. We foresee that nanoparticle-mediated immunosuppression, their antigenicity, and the exaggeration of the immune response to classical inflammatory stimulants by nanomaterials are going to be active areas of research in the next five to 10 years.

References

1. Zolnik BS, Gonzalez-Fernandez A, Sadrieh N, Dobrovolskaia MA. Nanoparticles and the immune system. *Endocrinology* 2010;151:458–465.
2. Di Gioacchino M, Petrarca C, Lazzarin F, Di Giampaolo L, Sabbioni E, Boscolo P, Mariani-Costantini R, Bernardini G. Immunotoxicity of nanoparticles. *Int J Immunopathol Pharmacol* 2011;24:65S–71S.
3. Dobrovolskaia MA, Germolec DR, Weaver JL. Evaluation of nanoparticle immunotoxicity. *Nat Nanotechnol* 2009;4:411–414.
4. Steinmetz T, Schaadt M, Gahl R, Schenk V, Diehl V, Pfreundschuh M. Phase I study of 24-hour continuous intravenous infusion of recombinant human tumor necrosis factor. *J Biol Response Mod* 1988;7:417–423.
5. Libutti SK, Paciotti GF, Byrnes AA, Alexander HR, Jr., Gannon WE, Walker M, Seidel GD, Yuldasheva N, Tamarkin L. Phase I and pharmacokinetic studies of CYT-6091, a novel PEGylated colloidal gold-rhTNF nanomedicine. *Clin Cancer Res* 2010;16: 6139–6149.
6. Gradishar WJ, Tjulandin S, Davidson N, Shaw H, Desai N, Bhar P, Hawkins M, O'Shaughnessy J. Phase III trial of nanoparticle albumin-bound paclitaxel compared with polyethylated castor oil-based paclitaxel in women with breast cancer. *J Clin Oncol* 2005;23:7794–7803.
7. Schneider P, Van Dreden P, Rousseau A, Kassim Y, Legrand E, Vannier JP, Vasse M. Increased levels of tissue factor activity and procoagulant phospholipids during treatment of children with acute lymphoblastic leukaemia. *Br J Haematol* 2010;148:582–592.

8. Sanchez-Gonzalez PD, Lopez-Hernandez FJ, Lopez-Novoa JM, Morales AI. An integrative view of the pathophysiological events leading to cisplatin nephrotoxicity. *Crit Rev Toxicol* 2011;41:803–821.
9. Zolnik BS, Stern ST, Kaiser JM, Heakal Y, Clogston JD, Kester M, McNeil SE. Rapid distribution of liposomal short-chain ceramide *in vitro* and *in vivo*. *Drug Metab Dispos* 2008;36:1709–1715.
10. Ferdinandi ES, Vassilakos A, Lee Y, Lightfoot J, Fitsialos D, Wright JA, Young AH. Preclinical toxicity and toxicokinetics of GTI-2040, a phosphorothioate oligonucleotide targeting ribonucleotide reductase R2. *Cancer Chemother Pharmacol* 2011;68:193–205.
11. Szebeni J, Muggia F, Gabizon A, Barenholz Y. Activation of complement by therapeutic liposomes and other lipid excipient-based therapeutic products: Prediction and prevention. *Adv Drug Delivery Rev* 2011;63:1020–1030.
12. Anderson NL, Anderson NG, Pearson TW, Borchers CH, Paulovich AG, Patterson SD, Gillette M, Aebersold R, Carr SA. A human proteome detection and quantitation project. *Mol Cell Proteomics* 2009;8:883–886.
13. Aggarwal P, Hall JB, McLeland CB, Dobrovolskaia MA, McNeil SE. Nanoparticle interaction with plasma proteins as it relates to particle biodistribution, biocompatibility and therapeutic efficacy. *Adv Drug Delivery Rev* 2009;61:428–437.
14. Cohn SM. Alternatives to blood in the 21st century. *Crit Care* 2004;8(Suppl 2): S15–S17.
15. Choi J, Reipa V, Hitchins VM, Goering PL, Malinauskas RA. Physicochemical characterization and *in vitro* hemolysis evaluation of silver nanoparticles. *Toxicol Sci* 2011;123:133–143.
16. Yu T, Malugin A, Ghandehari H. Impact of silica nanoparticle design on cellular toxicity and hemolytic activity. *ACS Nano* 2011;5:5717–5728.
17. Sun YN, Wang CD, Zhang XM, Ren L, Tian XH. Shape dependence of gold nanoparticles on *in vivo* acute toxicological effects and biodistribution. *J Nanosci Nanotechnol* 2011;11:1210–1216.
18. Leonov AP, Zheng J, Clogston JD, Stern ST, Patri AK, Wei A. Detoxification of gold nanorods by treatment with polystyrenesulfonate. *ACS Nano* 2008;2:2481–2488.
19. de Ven HV, Van Dyck L, Weyenberg W, Maes L, Ludwig A. Nanosuspensions of chemically modified saponins: Reduction of hemolytic side effects and potential tool in drug targeting strategy. *J Controlled Release* 2010;148:e122–e123.
20. Shelma R, Sharma CP. Development of lauroyl sulfated chitosan for enhancing hemocompatibility of chitosan. *Colloids Surf B Biointerfaces* 2011;84:561–570.
21. Dobrovolskaia MA, Patri AK, Potter TM, Rodriguez JC, Hall JB, McNeil SE. Dendrimer-induced leukocyte procoagulant activity depends on particle size and surface charge. *Nanomedicine (Lond)* 2012;7:245–256.
22. Greish K, Thiagarajan G, Herd H, Price R, Bauer H, Hubbard D, Burckle A, Sadekar S, Yu T, Anwar A, *et al.* Size and surface charge significantly influence the toxicity of silica and dendritic nanoparticles. *Nanotoxicology* 2011 [Epub ahead of print].
23. Yu T, Greish K, McGill LD, Ray A, Ghandehari H. Influence of geometry, porosity, and surface characteristics of silica nanoparticles on acute toxicity: Their vasculature effect and tolerance threshold. *ACS Nano* 2012;6:2289–2301.
24. Corbalan JJ, Medina C, Jacoby A, Malinski T, Radomski MW. Amorphous silica nanoparticles aggregate human platelets: Potential implications for vascular homeostasis. *Int J Nanomedicine* 2012;7:631–639.

25. Dobrovolskaia MA, Patri AK, Simak J, Hall JB, Semberova J, De Paoli Lacerda SH, McNeil SE. Nanoparticle size and surface charge determine effects of PAMAM dendrimers on human platelets *in vitro*. *Mol Pharm* 2012;9:382–393.
26. Jones CF, Campbell RA, Franks Z, Gibson CC, Thiagarajan G, Vieira-de-Abreu A, Sukavaneshvar S, Mohammad SF, Li DY, Ghandehari H, *et al.* Cationic PAMAM dendrimers disrupt key platelet functions. *Mol Pharm* 2012, [Epub ahead of print].
27. Lacerda SH, Semberova J, Holada K, Simakova O, Hudson SD, Simak J. Carbon nanotubes activate store-operated calcium entry in human blood platelets. *ACS Nano* 2011;5:5808–5813.
28. Li X, Radomski A, Corrigan OI, Tajber L, De Sousa Menezes F, Endter S, Medina C, Radomski MW. Platelet compatibility of PLGA, chitosan and PLGA-chitosan nanoparticles. *Nanomedicine (Lond)* 2009;4:735–746.
29. Radomski A, Jurasz P, Alonso-Escolano D, Drews M, Morandi M, Malinski T, Radomski MW. Nanoparticle-induced platelet aggregation and vascular thrombosis. *Br J Pharmacol* 2005;146:882–893.
30. Semberova J, DePaoli Lacerda SH, Simakova O, Holada K, Gelderman MP, Simak J. Carbon nanotubes activate blood platelets by inducing extracellular Ca^{2+} influx sensitive to calcium entry inhibitors. *Nano Lett* 2009;9:3312–3317.
31. Al-Hanbali O, Rutt KJ, Sarker DK, Hunter AC, Moghimi SM. Concentration dependent structural ordering of poloxamine 908 on polystyrene nanoparticles and their modulatory role on complement consumption. *J Nanosci Nanotechnol* 2006;6:3126–3133.
32. Andersen AJ, Hashemi SH, Andresen TL, Hunter AC, Moghimi SM. Complement: Alive and kicking nanomedicines. *J Biomed Nanotechnol* 2009;5:364–372.
33. Hamad I, Al-Hanbali O, Hunter AC, Rutt KJ, Andresen TL, Moghimi SM. Distinct polymer architecture mediates switching of complement activation pathways at the nanosphere-serum interface: Implications for stealth nanoparticle engineering. *ACS Nano* 2010;4:6629–6638.
34. Lettiero B, Andersen AJ, Hunter AC, Moghimi SM. Complement system and the brain: Selected pathologies and avenues toward engineering of neurological nanomedicines. *J Controlled Release* 2011 [Epub ahead of print].
35. Mizrahy S, Raz SR, Hasgaard M, Liu H, Soffer-Tsur N, Cohen K, Dvash R, Landsman-Milo D, Bremer MG, Moghimi SM, *et al.* Hyaluronan-coated nanoparticles: The influence of the molecular weight on CD44-hyaluronan interactions and on the immune response. *J Controlled Release* 2011;156:231–238.
36. Moein Moghimi S, Hamad I, Bunger R, Andresen TL, Jorgensen K, Hunter AC, Baranji L, Rosivall L, Szebeni J. Activation of the human complement system by cholesterol-rich and PEGylated liposomes-modulation of cholesterol-rich liposome-mediated complement activation by elevated serum LDL and HDL levels. *J Liposome Res* 2006;16:167–174.
37. Moghimi SM, Andersen AJ, Ahmadvand D, Wibroe PP, Andresen TL, Hunter AC. Material properties in complement activation. *Adv Drug Delivery Rev* 2011;63:1000–1007.
38. Moghimi SM, Andersen AJ, Hashemi SH, Lettiero B, Ahmadvand D, Hunter AC, Andresen TL, Hamad I, Szebeni J. Complement activation cascade triggered by PEG-PL engineered nanomedicines and carbon nanotubes: The challenges ahead. *J Controlled Release* 2010;146:175–181.

39. Moghimi SM, Hamad I. Liposome-mediated triggering of complement cascade. *J Liposome Res* 2008;18:195–209.
40. Moghimi SM, Hamad I, Andresen TL, Jorgensen K, Szebeni J. Methylation of the phosphate oxygen moiety of phospholipid-methoxy(polyethylene glycol) conjugate prevents PEGylated liposome-mediated complement activation and anaphylatoxin production. *FASEB J* 2006;20:2591–2593.
41. Moghimi SM, Hunter AC. Complement monitoring of carbon nanotubes. *Nat Nanotechnol* 2010;5:382;382–383.
42. Moghimi SM, Hunter AC. Complement monitoring of nanomedicines and implants. *Adv Drug Delivery Rev* 2011;63:963–964.
43. Moghimi SM, Szebeni J. Stealth liposomes and long circulating nanoparticles: Critical issues in pharmacokinetics opsonization and protein-binding properties. *Prog Lipid Res* 2003;42:463–478.
44. Szebeni J, Moghimi SM. Liposome triggering of innate immune responses: A perspective on benefits and adverse reactions. *J Liposome Res* 2009;19:85–90.
45. Reddy ST, van der Vlies AJ, Simeoni E, Angeli V, Randolph GJ, O'Neil CP, Lee LK, Swartz MA, Hubbell JA. Exploiting lymphatic transport and complement activation in nanoparticle vaccines. *Nat Biotechnol* 2007;25:1159–1164.
46. Thomas SN, van der Vlies AJ, O'Neil CP, Reddy ST, Yu SS, Giorgio TD, Swartz MA, Hubbell JA. Engineering complement activation on polypropylene sulfide vaccine nanoparticles. *Biomaterials* 2011;32:2194–2203.
47. Arima Y, Toda M, Iwata H. Complement activation on surfaces modified with ethylene glycol units. *Biomaterials* 2008;29:551–560.
48. Fang C, Shi B, Pei YY, Hong MH, Wu J, Chen HZ. *In vivo* tumor targeting of tumor necrosis factor-alpha-loaded stealth nanoparticles: Effect of MePEG molecular weight and particle size. *Eur J Pharm Sci* 2006;27:27–36.
49. Zahr AS, Davis CA, Pishko MV. Macrophage uptake of core-shell nanoparticles surface modified with poly(ethylene glycol). *Langmuir* 2006;22:8178–8185.
50. Gbadamosi JK, Hunter AC, Moghimi SM. PEGylation of microspheres generates a heterogeneous population of particles with differential surface characteristics and biological performance. *FEBS Lett* 2002;532:338–344.
51. Laverman P, Carstens MG, Storm G, Moghimi SM. Recognition and clearance of methoxypoly(ethyleneglycol)2000-grafted liposomes by macrophages with enhanced phagocytic capacity. Implications in experimental and clinical oncology. *Biochim Biophys Acta* 2001;1526:227–229.
52. Moghimi SM, Hunter AC. Capture of stealth nanoparticles by the body's defences. *Crit Rev Ther Drug Carrier Syst* 2001;18:527–550.
53. Moghimi SM, Hunter AC, Murray JC. Long-circulating and target-specific nanoparticles: Theory to practice. *Pharmacol Rev* 2001;53:283–318.
54. Moghimi SM, Hunter AC, Murray JC. Nanomedicine: Current status and future prospects. *FASEB J* 2005;19:311–330.
55. Moghimi SM, Pavey KD, Hunter AC. Real-time evidence of surface modification at polystyrene lattices by poloxamine 908 in the presence of serum: *In vivo* conversion of macrophage-prone nanoparticles to stealth entities by poloxamine 908. *FEBS Lett* 2003;547:177–182.

56. Franca A, Aggarwal P, Barsov EV, Kozlov SV, Dobrovolskaia MA, Gonzalez-Fernandez A. Macrophage scavenger receptor A mediates the uptake of gold colloids by macrophages *in vitro*. *Nanomedicine (Lond)* 2011;6:1175–1188.
57. Kanno S, Furuyama A, Hirano S. A murine scavenger receptor MARCO recognizes polystyrene nanoparticles. *Toxicol Sci* 2007;97:398–406.
58. Nagayama S, Ogawara K, Fukuoka Y, Higaki K, Kimura T. Time-dependent changes in opsonin amount associated on nanoparticles alter their hepatic uptake characteristics. *Int J Pharm* 2007;342:215–221.
59. von Zur Muhlen C, von Elverfeldt D, Bassler N, Neudorfer I, Steitz B, Petri-Fink A, Hofmann H, Bode C, Peter K. Superparamagnetic iron oxide binding and uptake as imaged by magnetic resonance is mediated by the integrin receptor Mac-1 (CD11b/CD18): Implications on imaging of atherosclerotic plaques. *Atherosclerosis* 2007;193:102–111.
60. Cui Z, Mumper RJ. Coating of cationized protein on engineered nanoparticles results in enhanced immune responses. *Int J Pharm* 2002;238:229–239.
61. Kim S, Choi IH. Phagocytosis and endocytosis of silver nanoparticles induce interleukin-8 production in human macrophages. *Yonsei Med J* 2012;53:654–657.
62. Murphy FA, Schinwald A, Poland CA, Donaldson K. The mechanism of pleural inflammation by long carbon nanotubes: Interaction of long fibres with macrophages stimulates them to amplify pro-inflammatory responses in mesothelial cells. *Part Fibre Toxicol* 2012;9:8.
63. Jokerst JV, Lobovkina T, Zare RN, Gambhir SS. Nanoparticle PEGylation for imaging and therapy. *Nanomedicine (Lond)* 2011;6:715–728.
64. Defranco AL, Rookhuizen DC, Hou B. Contribution of Toll like receptor signaling to germinal center antibody responses. *Immunol Rev* 2012;247:64–72.
65. Chen HC, Sun B, Tran KK, Shen H. Effects of particle size on Toll-like receptor 9-mediated cytokine profiles. *Biomaterials* 2011;32:1731–1737.
66. Nguyen DN, Mahon KP, Chikh G, Kim P, Chung H, Vicari AP, Love KT, Goldberg M, Chen S, Krieg AM, *et al.* Lipid-derived nanoparticles for immunostimulatory RNA adjuvant delivery. *Proc Natl Acad Sci U S A* 2012;109:e797–e803.
67. Piazza M, Colombo M, Zanoni I, Granucci F, Tortora P, Weiss J, Gioannini T, Prosperi D, Peri F. Uniform lipopolysaccharide (LPS)-loaded magnetic nanoparticles for the investigation of LPS-TLR4 signaling. *Angew Chem Int Ed Engl* 2011;50:622–626.
68. Tincer G, Yerlikaya S, Yagci FC, Kahraman T, Atanur OM, Erbatur O, Gursel I. Immunostimulatory activity of polysaccharide-poly(I:C) nanoparticles. *Biomaterials* 2011;32:4275–4282.
69. Tsai CY, Lu SL, Hu CW, Yeh CS, Lee GB, Lei HY. Size-dependent attenuation of TLR9 signaling by gold nanoparticles in macrophages. *J Immunol* 2012;188:68–76.
70. Uto T, Akagi T, Yoshinaga K, Toyama M, Akashi M, Baba M. The induction of innate and adaptive immunity by biodegradable poly(gamma-glutamic acid) nanoparticles via a TLR4 and MyD88 signaling pathway. *Biomaterials* 2011;32:5206–5212.
71. Vallhov H, Qin J, Johansson SM, Ahlborg N, Muhammed MA, Scheynius A, Gabrielsson S. Scheynius A, Gabrielsson S. The importance of an endotoxin-free environment during the production of nanoparticles used in medical applications. *Nano Lett* 2006;6:1682–1686.

72. Fifis T, Gamvrellis A, Crimeen-Irwin B, Pietersz GA, Li J, Mottram PL, McKenzie IF, Plebanski M. Size-dependent immunogenicity: Therapeutic and protective properties of nano-vaccines against tumors. *J Immunol* 2004;173:3148–3154.
73. Andreev SM, Babakhin AA, Petrukhina AO, Romanova VS, Parnes ZN, Petrov RV. Immunogenic and allergenic properties of fulleren conjugates with aminoacids and proteins. *Dokl Biochem* 2000;370:4–7.
74. Dykman LA, Sumaroka MV, Staroverov SA, Zaitseva IS, Bogatyrev VA. Immunogenic properties of the colloidal gold. *Izv Akad Nauk Ser Biol* 2004:86–91.
75. Masalova OV, Shepelev AV, Atanadze SN, Parnes ZN, Romanova VS, Vol'pina OM, Semiletov Iu A, Kushch AA. Immunostimulating effect of water-soluble fullerene derivatives — perspective adjuvants for a new generation of vaccine. *Dokl Akad Nauk* 1999;369:411–413.
76. Roberts JC, Bhalgat MK, Zera RT. Preliminary biological evaluation of polyamidoamine (PAMAM) Starburst dendrimers. *J Biomed Mater Res* 1996;30:53–65.
77. Braden BC, Goldbaum FA, Chen BX, Kirschner AN, Wilson SR, Erlanger BF. X-ray crystal structure of an anti-Buckminsterfullerene antibody fab fragment: Biomolecular recognition of C(60). *Proc Natl Acad Sci U S A* 2000;97:12193–12197.
78. Chen BX, Wilson SR, Das M, Coughlin DJ, Erlanger BF. Antigenicity of fullerenes: Antibodies specific for fullerenes and their characteristics. *Proc Natl Acad Sci U S A* 1998;95:10809–10813.
79. Erlanger BF, Chen BX, Zhu M, Brus L. Binding of an anti-fullerene IgG monoclonal antibody to single wall carbon nanotubes. *Nano Lett* 2001;1:465–467.
80. Cao S, Jiao N, Jiang Y, Mire-Sluis A, Narhi LO. Sub-visible particle quantitation in protein therapeutics. *Pharmeur Bio Sci Notes* 2009;2009:73–79.
81. Carpenter J, Cherney B, Lubinecki A, Ma S, Marszal E, Mire-Sluis A, Nikolai T, Novak J, Ragheb J, Simak J. Meeting report on protein particles and immunogenicity of therapeutic proteins: Filling in the gaps in risk evaluation and mitigation. *Biologicals* 2010;38:602–611.
82. Liu W, Swift R, Torraca G, Nashed-Samuel Y, Wen ZQ, Jiang Y, Vance A, Mire-Sluis A, Freund E, Davis J, *et al.* Root cause analysis of tungsten-induced protein aggregation in pre-filled syringes. *PDA J Pharm Sci Technol* 2010;64:11–19.
83. Ramani K, Purohit V, Miclea R, Gaitonde P, Straubinger RM, Balu-Iyer SV. Passive transfer of polyethylene glycol to liposomal-recombinant human FVIII enhances its efficacy in a murine model for hemophilia A. *J Pharm Sci* 2008;97:3753–3764.
84. Francis GE, Fisher D, Delgado C, Malik F, Gardiner A, Neale D. PEGylation of cytokines and other therapeutic proteins and peptides: The importance of biological optimisation of coupling techniques. *Int J Hematol* 1998;68:1–18.
85. Petrovsky N, Aguilar JC. Vaccine adjuvants: Current state and future trends. *Immunol Cell Biol* 2004;82:488–496.
86. Azzi J, Tang L, Moore R, Tong R, El Haddad N, Akiyoshi T, Mfarrej B, Yang S, Jurewicz M, Ichimura T, *et al.* Polylactide-cyclosporin A nanoparticles for targeted immunosuppression. *FASEB J* 2010;24:3927–3938.
87. Ghosh K, Kanapathipillai M, Korin N, McCarthy JR, Ingber DE. Polymeric nanomaterials for islet targeting and immunotherapeutic delivery. *Nano Lett* 2012;12:203–208.
88. Keijzer C, Slutter B, van der Zee R, Jiskoot W, van Eden W, Broere F. PLGA, PLGA-TMC and TMC-TPP nanoparticles differentially modulate the outcome of nasal vaccination by inducing tolerance or enhancing humoral immunity. *PLoS One* 2011;6:e26684.

89. Pandey RS, Sahu S, Sudheesh MS, Madan J, Kumar M, Dixit VK. Carbohydrate modified ultrafine ceramic nanoparticles for allergen immunotherapy. *Int Immunopharmacol* 2011;11:925–931.
90. Dianzani C, Cavalli R, Zara GP, Gallicchio M, Lombardi G, Gasco MR, Panzanelli P, Fantozzi R. Cholesteryl butyrate solid lipid nanoparticles inhibit adhesion of human neutrophils to endothelial cells. *Br J Pharmacol* 2006;148:648–656.
91. Shaunak S. Dendrimer drugs prevent scar tissue formation. *Discovery Med* 2004;4: 464–469.
92. Shaunak S, Thomas S, Gianasi E, Godwin A, Jones E, Teo I, Mireskandari K, Luthert P, Duncan R, Patterson S, *et al.* Polyvalent dendrimer glucosamine conjugates prevent scar tissue formation. *Nat Biotechnol* 2004;22:977–984.
93. Barata TS, Teo I, Brocchini S, Zloh M, Shaunak S. Partially glycosylated dendrimers block MD-2 and prevent TLR4-MD-2-LPS complex mediated cytokine responses. *PLoS Comput Biol* 2011;7:e1002095.
94. Boridy S, Soliman GM, Maysinger D. Modulation of inflammatory signaling and cytokine release from microglia by celastrol incorporated into dendrimer nanocarriers. *Nanomedicine (Lond)* 2012 [Epub ahead of print].
95. John AE, Lukacs NW, Berlin AA, Palecanda A, Bargatze RF, Stoolman LM, Nagy JO. Discovery of a potent nanoparticle P-selectin antagonist with anti-inflammatory effects in allergic airway disease. *FASEB J* 2003;17:2296–2298.
96. Chen YW, Hwang KC, Yen CC, Lai YL. Fullerene derivatives protect against oxidative stress in RAW 264.7 cells and ischemia-reperfused lungs. *Am J Physiol Regul Integr Comp Physiol* 2004;287;R21–R26.
97. Wang B, Navath RS, Menjoge AR, Balakrishnan B, Bellair R, Dai H, Romero R, Kannan S, Kannan RM. Inhibition of bacterial growth and intramniotic infection in a guinea pig model of chorioamnionitis using PAMAM dendrimers. *Int J Pharm* 2010;395: 298–308.
98. Cromer JR, Wood SJ, Miller KA, Nguyen T, David SA. Functionalized dendrimers as endotoxin sponges. *Bioorg Med Chem Lett* 2005;15:1295–1298.
99. Shen CC, Wang CC, Liao MH, Jan TR. A single exposure to iron oxide nanoparticles attenuates antigen-specific antibody production and T-cell reactivity in ovalbumin-sensitized BALB/c mice. *Int J Nanomed* 2011;6:1229–1235.
100. Qu G, Zhang C, Yuan L, He J, Wang Z, Wang L, Liu S, Jiang G. Quantum dots impair macrophagic morphology and the ability of phagocytosis by inhibiting the Rho-associated kinase signaling. *Nanoscale* 2012;4:2239–2244.

Chapter 2

Importance of Physicochemical Characterization Prior to Immunological Studies

*Jeffrey D. Clogston and Anil K. Patri**

Nanotechnology Characterization Laboratory
Advanced Technology Program, SAIC-Frederick, Inc.
NCI-Frederick, Frederick, MD 21702, USA

Physicochemical characterization seeks to define the physical and chemical properties, composition, identification, quality, purity, and stability of the material. Since so many parameters influence nanoparticle immunological properties, thorough characterization is essential to draw meaningful conclusions. This chapter will discuss the parameters that need to be monitored and why, and the means of doing so with the appropriate instrumentation, with emphasis on dynamic light scattering and zeta potential measurements.

1. Introduction

Nanotechnology, as defined by the National Nanotechnology Initiative, is "the understanding and control of matter at dimensions between approximately 1 and 100 nanometers, where unique phenomena enable novel applications" (http://www.nano.gov/nanotech-101/nanotechnology-facts). While many novel nanotechnologies fall within this definition and demonstrate unique properties compared to their bulk-sized counterparts, other agencies and organizations have different definitions that encompass submicron-sized

*Corresponding author. E-mail: patria@mail.nih.gov

particles. Apart from the ambient nanoparticles present in nature, engineered nanomaterials are specifically designed for therapeutic applications. Many nanomaterials are currently in use for non-medical and industrial applications, with potential exposure issues and health implications. These nanomaterials may have significantly different compositions and physicochemical attributes compared to intentionally designed nanomaterials. The routes of exposure and the ensuing biological effects may be significantly different. In either case, an accurate estimation of what is being tested in biological or immunological studies has paramount importance since the conclusions drawn can substantially impact one's perception of nanomaterials. It makes a good research practice to understand biological effects that are reproducible. The concern over the effects of the particles on the immune system is real since the body recognizes these materials as "foreign" in most cases, compared to small molecule analogs. A lot is known about nanomaterials in the fields of colloidal sciences, polymers, and liposomes since they are in existence and in human use. However, in the wake of new applications, compositions, and coatings, a fresh look at the different technologies, characterization methodologies, and relationships of their properties to their immunological behavior is appropriate.

The engineered nanomaterial has found applications in cancer diagnostics, imaging, and therapy due to several reasons: therapeutic benefits, improved solubility, reduced toxicity, increased efficacy, and active and passive targeting capability.[1,2] In the latter, the EPR (enhanced permeability and retention) phenomenon first described[3] particles and macromolecules selectively accumulating in tumor tissues, presumably due to the leaky neovasculature[2] and absence of lymphatic drainage. To achieve the benefits of nanotechnology, nanomaterials can be engineered to have a specific size dimension for desired biodistribution, localization, elimination, and modified surface chemistry to escape or be recognized by the immune system depending on the desired outcome. These parameters will ultimately define the nanomaterial's biocompatibility, biodistribution, and biodegradability and hence, toxicity and efficacy.[4] To exploit nanotechnology, the relationship between physicochemical parameters and biological outcomes needs to be elucidated and requires thorough characterization prior to biological studies. This chapter will cover the parameters that need to be monitored and why, and the means of doing so with the appropriate instrumentation. It is intended for an immunologist and provides a basic primer on characterization covering only engineered nanomaterials. Ambient particles or particles that are not intended for medical applications are outside the scope of this chapter.

2. Key Parameters that Influence Nanomaterial Immunological Properties

There are a vast number of instrumentation and techniques that can be used to characterize nanomaterials, leading to many different characterization parameters. The techniques and instrumentation can be a daunting challenge even for a seasoned scientist with expertise in analytical chemistry. However, at least those that are relevant should be measured to help draw conclusions from biological data. A summary of the key physicochemical characterization (PCC) parameters regarding nanomaterials is given in Table 1. Not all of

Table 1. PCC analytical parameters.

- Size, Size distribution
- Topology, Shape
- Molecular weight
- Surface characteristics
  - Net charge
  - Zeta potential
- Functionality (Identification, Quantitation, Distribution)
  - Targeting agents
  - Imaging agents
  - Therapeutics
- Composition
  - Elemental
  - Core-shell
  - Coating
- Purity
  - Homogeneity/Inhomogeneity
  - Residual solvents
  - Free components
  - Free *vs.* bound drug
- Stability
  - Thermal
  - pH
  - Photo
  - Freeze/thaw
  - Lyophilization
  - Centrifugation
  - Short-term storage
  - Long-term storage
  - Drug release kinetics
  - Stability of the coating

these parameters may be applicable for all nanoparticle measurements; for example, the molecular weight of liposomes or metal colloids is not relevant. Based on the literature and experience, at a minimum, characterization should address the following: size (and size distribution), zeta potential, targeting/drug/imaging quantitation, coating integrity and quantitation, purity (such as residual products from reactions, surfactants, free metals, and ions), stability, and batch-to-batch consistency. Assumptions based on purification procedures should be avoided; measured values would provide more conclusive information. For example, the dialysis of a nanomaterial sample may not get rid of all impurities; in many cases, these impurities may stay encapsulated within the nanoparticle matrix and elicit responses in immunological assays. As a result, the biological effects may be attributed to the nanomaterial, where it may be due to the presence of impurities. This is a significant difference between the biological assessment of a small molecule that can be purchased from vendors with reasonable confidence regarding their purity and identity, compared to a nanomaterial.

Size is one of the most critical parameters as it influences the nanoparticle's accumulation (via EPR), biodistribution, and route of clearance.[4–9] Size can be defined in many ways and as a result, there are many techniques to measure size. Techniques such as dynamic light scattering (DLS), transmission electron microscopy (TEM), and atomic force microscopy (AFM) all yield size, but each defines it in a different way. Take polyethylene glycol (PEG)ylated colloidal gold for example. DLS measures the hydrated size, which takes into account the PEG layer as well as the water associated with the PEG coating. TEM measures only the electron-dense material, namely gold, while AFM gives height information of the hard core. The size resulting from each of these techniques will be different and it is important to know what each instrument is measuring. Taken together, characterization by multiple/orthogonal techniques is extremely useful in understanding the nanomaterial sample under consideration. For immunological assessments, a small amount of agglomerates might cause significant effects compared to the intended monomeric nanomaterial. Since microscopy (TEM, AFM) only provides a glimpse of a small portion of nanomaterial sample, one may have to rely on solution phase measurements to look for these agglomerates in appropriate buffers/media used for the biological assay. More details will be covered in a later section.

In addition to size, surface charge plays a role in a nanoparticle's biocompatibility and biodistribution.[6, 10–12] Depending on the surface attribute that is being measured, one has to choose the appropriate technique. Zeta potential is a common and easier measurement that indicates charge on a nanoparticle's surface. In addition to providing surface charge information, it can be used to track surface modifications in multistep nanoparticle synthesis. For instance, zeta potential can be used to track the PEGylation of citrate-stabilized colloidal gold, where the initial negative zeta potential should become neutral as the particle becomes PEGylated. As a general indicator of the net charge based on electrophoretic mobility, zeta potential is useful but it does not distinguish differences or polydispersity in charge distribution within a given sample. Zeta potential values also depend on the pH and ionic strength of the medium. (See the section below on zeta potential for a more in-depth discussion.) In cases such as polymers and dendrimers, an average charge distribution among amines and carboxylic acid functional groups can be measured using traditional acid–base titrations, which are more reliable compared to zeta potential measurements in these specific cases.

The quantitation of active and inactive components, whether be it a therapeutic, a targeting ligand, an imaging agent, stabilizers, or surfactants, is very important as this will determine the dosing and toxicity in an *in vivo* setting. Simply using theoretical loading or anecdotal evidence is not enough as this value will most likely be inaccurate depending on the synthetic and purification processes used. Inactive components such as surfactants, chelators, stabilizers, dopants, and oxidizing and reducing agents are most often used in nanomaterial synthesis and may cause undesirable and unintended effects in biological assays. They are usually not specified and are often ignored during characterization. Testing for the presence and quantitation of these components will help narrow down the component that might be causing the biological effect in the nanomaterial sample matrix. In addition to quantitation, stability under relevant conditions should be examined. Total drug loading as well as free drug (not associated with the nanoparticle) should be determined. Drug (either encapsulated or conjugated) release profiles should be assessed under physiological conditions (37°C in phosphate buffered saline (PBS) and plasma). The finding of premature or no release of drug under such conditions will save time and resources, which can be spent on *in vivo* studies. The same holds true for targeting ligands; the premature release or cleavage of targeting ligands results in no targeting benefits. MRI (magnetic resonance imaging) contrast agents typically use gadolinium complexed to the nanoparticle via a ligand such as bifunctional DTPA or DOTA. The stability, in terms of metal complexation (binding) and release, should be determined as the presence of free metal results in *in vivo* toxicity. Many nanoparticles are coated, for example with polymers, and the stability of the coating under physiological conditions should be assessed. For example, does the coating remain intact from synthesis to biological assessment under the storage conditions? What happens to the coating in the presence of buffer or plasma? A word on multifunctional nanoparticles is needed here. The characterization of multi-functional nanoparticles, that is particles containing a therapeutic, targeting ligand, and/or imaging component, can be made easier when precursors are available. The characterization of the final product containing all components may be quite challenging due to the nature of the nanoparticle. However, characterization at each sequential step in nanoparticle synthesis may be possible. An example would be a dendrimer containing both a conjugated drug and targeting ligand that absorb at the same wavelength. In this case, the drug is conjugated first followed by the targeting ligand. Initial drug loading can be determined on the precursor using ultraviolet–visible (UV/Vis) spectroscopy. If the drug loading does not change with each sequential synthetic step, the characterization of the next step (i.e., drug and targeting ligand containing the nanoparticle) on the precursor is made possible by correcting for the drug's contribution to the absorbance. Such sequential multistep synthesis may not be possible in many other cases where a one-step self-assembly process is used. In such a scenario, the development of appropriate characterization methodologies for the presence of the ligands, their inhomogeneity/heterogeneity, activity, purity, and reproducibility is crucial.

Stability under non-physiological or storage conditions should also be examined. Do processes such as lyophilization, freeze/thaw cycles, ultrafiltration, and centrifugation affect the integrity of the nanoparticles? This will determine the handling procedures of the nanoparticles. Thermal and pH stabilities should be addressed as both will determine the storage conditions of the nanoparticles. Should the nanoparticles be stored at room temperature or at –20°C, and in which state — dry or in solution — should they be stored? If they are in solution, is water suitable or is a buffer at a certain pH needed? To correctly answer these questions, the relevant key parameters along with instrumentation are needed. These parameters need to be chosen such that they reflect a certain *in vivo* outcome. These may be size and zeta for some nanoparticles, but may mean drug loading and release criteria for another. Whatever the case for your nanoparticle, this will define its short- and long-term storage or shelf-life. How long can the nanoparticle be stored at room temperature before its efficacy is lost? Or, how many freeze/thaw cycles can the sample undergo before it no longer remains at the same size?

The same criteria used for shelf-life determination can also be used for batch-to-batch consistency. This is one of the U.S. Food and Drug Administration (FDA)'s requirements for any material entering clinical trials. To achieve this, the synthesis of nanoparticles needs to be reproducible and sound. For example, this could mean that the concentration of a drug conjugated to a dendrimer, or that the size of liposomes needs to be consistent from batch to batch. This especially becomes a challenge for more mature concepts during scale-up and GMP manufacturing. If fewer steps are used for efficiency, one has to repeat the biological assessment with crucial assays to confirm the integrity of the material. Otherwise, it will be a costly mistake and the large-scale batch may not work the way as initially intended. Again, these criteria/PCC parameters must relate to a desired *in vivo* response, and can be used as specifications for each new synthesis batch. This in turn can be used to judge the outcome of each new batch (if it passes or fails), and ultimately determine its usability.

Many nanomaterial platforms use fluorescent tags for *in vitro* cell culture and flow cytometry experiments toward proof of principle. Amine functionalization on nanoparticles is a common way of labeling with tags (such as FITC or other markers). In such cases where targeting ligands are used, a careful assessment of residual unreacted amines on the nanomaterial should be considered as their presence may lead to non-specific interactions or binding with cell surfaces. Careful purification to remove these free fluorophores would minimize errors in conclusions. However, dialysis, in many cases, does not completely remove the residual/free unconjugated fluorophores. Since many fluorophores are insoluble in aqueous buffers, they tend to stay within the nanomaterial matrix and are released during *in vitro* cell culture experiments. Since the conclusions are attributed to fluorescence in such assays, precautions must be taken to control for such possibilities. If column purification can be utilized, gel filtration with appropriate resin (G-25/PD-10 column) would minimize the presence of free fluorophores. If column purification is not feasible due to nanomaterial trapping in the stationary phase of the column, then one has to rely on using organic phase extractions. The same holds true for bifunctional ligands such as maleimide/sulfhydryl-containing ligands most commonly employed for bioconjugations. Appropriate characterization at each stage would minimize assumptions in functionalization and conclusions drawn from the experiments.

3. Instrumentation for Physicochemical Characterization

PCC seeks to define the physical and chemical properties, composition, identification, quality, purity, and stability of the material. For small molecules (of which most traditional drugs are), these physicochemical parameters can be answered by traditional analytical techniques such as elemental analysis, mass spectrometry (MS), nuclear magnetic resonance (NMR), UV/Vis spectroscopy, infrared (IR) spectroscopy, high pressure liquid chromatography (HPLC), gas chromatography (GC), and polarimetry, to name a few.[13] With small molecules, these analytical techniques are well established and validated. However, with nanoparticles, this is not the case. The same physicochemical questions still need to be answered, in addition to nanomaterial-specific properties such as size and size distribution (polydispersity), zeta potential, and surface chemistry. These additional parameters provide a better and more complete understanding of the nanomaterial. In certain cases, some of the traditional analytical techniques can still be employed to answer such questions. For example, matrix-assisted laser desorption/ionization-time of flight (MALDI-TOF) MS is used for molecular weight determination of dendrimers, fullerenes, and polymers. NMR is used to characterize these same nanomaterials to answer for structure, purity, and functionality (i.e., drug and/or targeting ligands). Drug loading and targeting ligand concentration can be assessed by UV/Vis spectroscopy or fluorimetry if components are either chromophores or fluorophores, respectively. New techniques that are not normally employed for small molecule analysis, such as microscopy, namely AFM, scanning electron microscopy (SEM), and TEM, are used to determine crystal structure, size, size distribution, topography, and shape. TEM coupled to energy dispersive X-ray spectrometry (EDS) can be used to determine the elemental composition of the nanoparticles; this is qualitative in nature since the field of view is rather limited. Inductively coupled plasma (ICP)-MS can be used to determine the metal composition of the nanoparticle; this is a quantitative tool where the concentration of a given metallic species can be accurately determined down to parts per trillion. Light scattering (both static and dynamic) techniques can be used for size, size distribution, and molecular weight determination. DLS measures hydrodynamic size, i.e., the solvated size of a nanoparticle in solution. It can also give information on size distribution (polydispersity) of the nanomaterial. Static light scattering (SLS) measures the radius of gyration (or root mean square radius), which is the size weighted by its mass distribution about its center of mass; this only applies if the nanoparticle's radius of gyration is greater than 10 nm. The molar mass of the nanoparticle can also be measured by SLS. Laser Doppler velocimetry (LDV) measures particle velocity from the frequency shift of scattered light. If this velocity is measured under an applied electric field, it is referred to as the electrophoretic mobility and can be used with an appropriate model to calculate the zeta potential. Zeta potential does not measure surface charge but rather, provides an indication of it. For colloidal stability, zeta potential measurements can be used in making materials with the desired charge repulsion in solution.

4. Methodologies for Nanomaterial Separation

The techniques mentioned thus far are all batch-mode measurements. That is, measurements are made on samples as they are (such as in a cuvette), without any separation/fractionation. Characterization through separation is equally important and can be illustrated by the following example. Consider a drug-loaded nanoparticle. The total drug loaded can be determined through simple UV/Vis analysis. However, this would represent the total drug present; it would not be able to distinguish between the free and conjugated drug. To determine the amount of free drug, the separation of the nanoparticle and free drug is necessary before characterization. This is because free drug (i.e., drug that is not nanomaterial-bound, unencapsulated, or unconjugated) will have a different immunotoxicity profile compared to a nanomaterial-bound drug, which may be masked. Such an analysis will help with additional understanding based on drug-related effects and nanomaterial-bound drug-related effects in a biological assay.

To achieve separation, chromatographic methods such as reverse phase (RP)-HPLC and size exclusion chromatography (SEC) are usually employed. RP-HPLC is the most commonly used mode of liquid chromatography for nanoparticle component (such as drugs and small molecule ligands) characterization. The separation is based on the analyte's partitioning between the polar mobile phase and the nonpolar hydrophobic (C_{18} being the most common) stationary phase. Typically, acetonitrile or methanol in water is used as the mobile phase, with the amount of water decreasing throughout the run (and hence, polarity); thus, the polar species will first be eluted, followed by the nonpolar species.[14] The detection of the analyte after separation is equally as important as the separation. Common detectors include UV/Vis and fluorescence for absorbing and fluorescent analytes. For the case of non-UV–absorbing or non-fluorescent analytes, ELSD (evaporative light scattering detector), CAD (charged aerosol detector), and MS detectors have been used. RP-HPLC with UV/Vis detection has been employed for purity assessment of generations (G) 1 through 9 of amine-terminated PAMAM (polyamidoamine) dendrimers.[15] Liposomes have been extensively characterized by RP-HPLC in terms of lipid components and drug loading. Methods for cholesterol and DOPC in liposome-based drug formulations,[16] encapsulated doxorubicin quantification in liposomal and micellar formulations,[17] and stability of liposomes[18] have been established. One of the limitations of HPLC in nanomaterial analysis is that the particles themselves (such as metallic particles) might be adsorbed on the stationary phase and cannot be eluted without changing the sample integrity. In such instances, results would be unusuable or data from such an analysis may be less meaningful.

In SEC, particles are separated by size (molecular weight). It is used for the removal of species with molecular weights higher or lower than the nanoparticle's molecular weight. As with RP-HPLC, SEC can be coupled to different detectors, thus allowing for further characterization like size, size distribution, molecular weight, and purity. Typical techniques for nanoparticle characterization include UV/Vis, fluorescence, multi-angle laser light scattering (MALLS), measurements of refractive index (RI), DLS, ICP-MS, and viscometry. Several detectors in such techniques can be connected in a series, therefore allowing for multiple characterization analysis on a fractionated sample. In the case of MALLS coupled with a concentration detector (i.e., UV/Vis, RI, and viscometer), the absolute molecular weight can be determined without the use of calibration standards. SEC coupled with UV/Vis, MALLS, and RI detection was used to determine the molar mass distributions of generations 1 through 9 of amine-terminated PAMAM dendrimers as well as four different surface chemistries (amino, hydroxyl, carboxylate, and acetamide) for a given dendrimer generation[15] (generations 1 through 4). Similarly, the molecular weight of each precursor to a G5-acetylated FITC–folic acid–paclitaxel dendrimer was determined and the number of functional groups attached calculated.[19]

Separation by either RP-HPLC or SEC coupled with multiple detectors paints a more complete characterization of the nanomaterial. Characterization depends on the efficiency/resolution of the separation process. However, there may be cases where the nanomaterial interacts with the stationary phase (column packing), thus making RP-HPLC and/or SEC not possible. An alternative analytical separation process called asymmetric flow field flow fractionation (AFFFF) may find use in these cases. In AFFFF, samples flow through a channel (referred to as channel flow) separated by two plates. The upper plate is non-permeable while the lower plate is a porous frit with (typically) a 10-kDa molecular weight cutoff (MWCO) membrane. A cross flow is applied perpendicular to the channel flow which effects separation. Smaller particles, because of their higher diffusion rates, travel with the channel flow, while larger particles are "pinned down" on the membrane. As a result, smaller particles will be eluted first. AFFFF provides a gentler separation process as no stationary phase is present. Concerns regarding the alteration or degradation of the particle as it passes through a column under high pressure are alleviated. For example, single particles may either agglomerate or agglomerates may break up as a result of passing through a column. The same detection methods employed in SEC can be used with AFFFF, thus providing information on size, size distribution, molecular weight, and purity. For instance, AFFFF coupled with MALLS and RI have been used to determine the size distribution and loading efficiency of hemoglobin-loaded liposomes.[20] Size and size distribution characterization of citrate-stabilized and PEGylated gold nanoparticles have been accomplished by AFFFF coupled with MALLS, RI, and UV/Vis.[21] AFFFF can also be used to separate and analyze the agglomeration of biologicals such as proteins or antibodies in solution. Care must be taken such that the flow forces themselves do not cause such artifacts. This can happen when the sample is concentrated in the "focusing step" of the analysis.

Another separation technique is capillary electrophoresis (CE), which is based on differences in the charge-to-mass ratio. Also known as capillary zone electrophoresis (CZE), the separation of charged species takes place in a small capillary filled with an electrolyte solution under an electric field. CE serves as an alternative to HPLC and has higher separation efficiency. It has been used for nanoparticle characterization in terms of size and charge distribution (electrophoretic mobilities) and purity. Also, it has been used to examine the purity of C_{60} derivatives, carboxyfullerene (C_3), and dendrofullerene (DF-1), as well as the quantitation of each derivative in serum.[22] Separation of generations 1 through 5 of amine-terminated PAMAM dendrimers as well as separation of half-generation (0.5, 1.5, and 2.5) carboxylate-terminated PAMAM dendrimers have been achieved by CE.[23]

A summary of the routinely used instrumentation for nanomaterial PCC is given in Table 2. There are many more instrumentation available, but they tend to more specialized and/or nanoparticle-specific. The goal here is to provide a list of instrumentation that is commonly found and readily available in analytical labs.

Table 2. Instrumentation routinely utilized for PCC.

UV/Vis	DLS
Infrared	SEC-MALLS
Fluorescence	FFF-MALLS
NMR	HPLC
ESI & MALDI-TOF MS	LC-MS
ICP-MS	FPLC
TEM/STEM	Zeta potential
SEM	CE
AFM	PAGE

5. A Closer Look

The previous section discussed the many PCC parameters that influence nanomaterial biological effects and the different instrumentation used to address the characterization of nanoparticles. Since so many parameters have an impact on nanoparticle immunological properties, thorough characterization is essential to draw meaningful conclusions. As resources are limited for such analyses and data resulting from many studies are contradictory, a certain minimum level of characterization of particles is necessary. Perusal of the literature shows that size and zeta are the most common PCC parameters cited in journals regarding nanoparticle characterization. Size, as measured by DLS, tends to be the most routinely used technique. Hand in hand with size is zeta, which is not surprising as the same instrument is typically capable of measuring both parameters. For this reason, the remainder of the chapter will deal with a more detailed look at both DLS and zeta. The basics of each technique will be given, along with the procedures of making reliable measurements. Guidelines and tips, as well as issues associated with each technique and their remedies will be illustrated with examples.

5.1. *Dynamic light scattering*

DLS is a technique used to measure the hydrodynamic size, size distribution, and polydispersity of a sample in an aqueous solution. It is also known as photon correlation spectroscopy (PCS) and quasi-elastic light scattering (QELS). The hydrodynamic diameter (size) is defined as the diameter of an equivalent hard sphere that diffuses at the same rate as the particle. DLS is a quick and noninvasive technique that requires a low sample volume (as low as ~12 μL, depending on the cuvette used) and can be used to measure particles from 1 to 1,000 nm in size.

In DLS, the nanoparticle is illuminated by a monochromatic laser and its time-dependent intensity scattering is measured at a certain scattering angle, as shown in Figure 1(a). The measured intensity scattering will fluctuate with time (microseconds) due to Brownian motion. Using the method of correlation, the degree of non-randomness in the intensity fluctuations is converted to a correlation coefficient. This, when plotted against time, gives what is called a correlation curve as shown in Figure 1(b). The correlation curve is then fitted by an appropriate model to yield the translational diffusion coefficient, which in turn is used in the Stokes–Einstein equation to back-calculate the hydrodynamic diameter Figure 1(b). It should be noted that in DLS, the diffusion coefficient is measured and that the hydrodynamic diameter is indirectly measured.[24] The mathematics and theory of DLS are beyond the scope

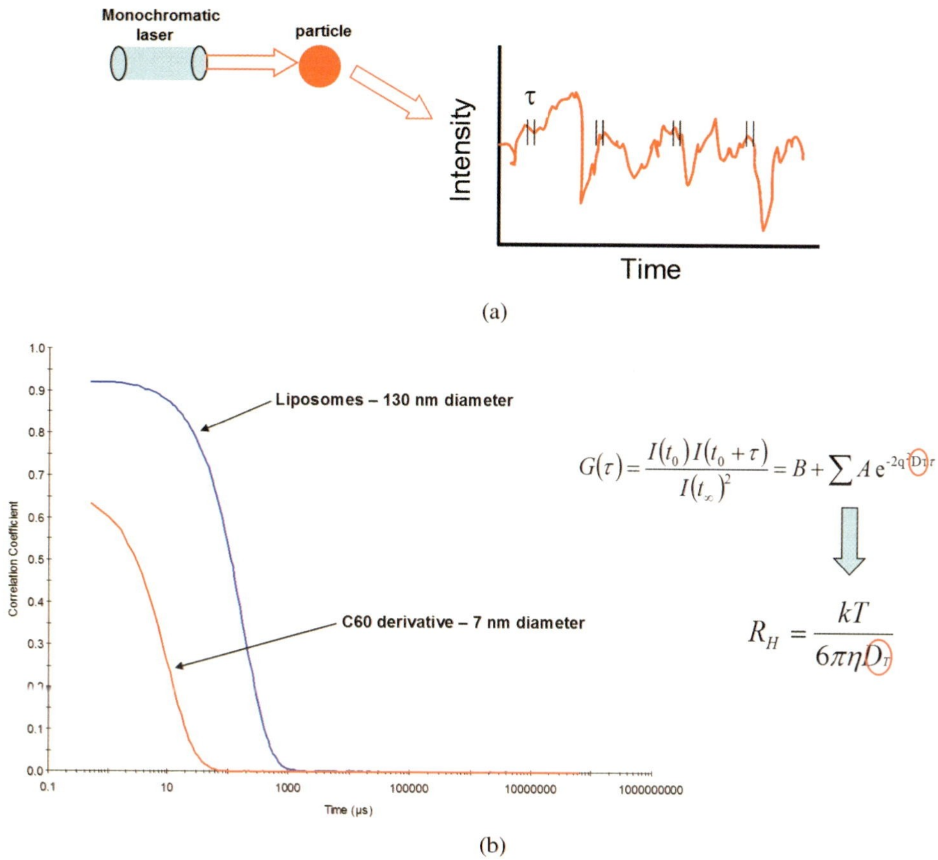

Figure 1. Schematic of how DLS works and typical correlation curves comparing two different-sized samples. (a) In DLS, the particle is illuminated by a monochromatic laser and its intensity fluctuations as a function of time are measured. Through the method of correlation, these intensity fluctuations are converted to a correlation curve. (b) Two correlation curves are plotted for two different-sized particles. The correlation curves are fitted by an appropriate model to yield the translational diffusion coefficient D_T, which in turn is used in the Stokes–Einstein equation to calculate the hydrodynamic size R_H.

of this text, but we recommend Berne and Pecora[25] for readers who are interested in further details.

Sample preparation and measurement procedures for making successful size measurements in the nanoscale range are described elsewhere in detail[24,26] and would not be discussed here. Included in those references are guidelines and precautions, as well as discussions of relevant standards and data analyses. In addition, the webpage contains a video demonstration of how to carry out a DLS measurement. As the references go into the specifics of DLS measurements, it is our intention to provide some of the practical issues associated with DLS here as it applies to nanoparticle size characterization. These are separated into two categories — general and nanoparticle-specific issues. The general issues include the use of intensity-, volume-, and number-weighted size distributions and *Z*-average in sample preparation. Nanoparticle-specific issues address multiple scattering, viscosity effects, absorbance, and rotational diffusion.

General issues in fact, deal with the correct choice of size. This has been discussed,[24] but warrants further elaboration and clarification as there tends to be confusion as to the right size to report. Cumulants analyses (ISO-recommended) assume a single monomodal and Gaussian distribution and yield only the *Z*-average and polydispersity index (PDI). However, more rigorous algorithms are available, especially when dealing with multimodal size distributions (NNLS, CONTIN); these yield intensity-, volume-, and number-weighted size distributions. Regardless of the size reported, several other parameters should be included — temperature, viscosity, solvent and its RI, concentration and any sample handling, instrument laser wavelength and scattering angle, cuvette, and path length.

With the many sizes available, which is the most appropriate choice? Figure 2 illustrates the intensity-, volume-, and number-weighted distributions for a nanoemulsion in water. All three distributions appear different and hence, give different sizes; polydispersed multimodal peaks are observed in the intensity- and volume-weighted distributions, while a single peak is observed in the number-weighted distribution. Of these, the intensity-weighted distribution should be used for reporting size. The volume-weighted distribution should be used for relative amounts, while the number-weighted distribution should be avoided because several assumptions are involved in the transformation. Intensity- and volume- weighted distributions are useful when dealing with samples with multiple peak sizes, in which the relative amounts are important. This is in contrast to a cumulants analysis which yields a mean intensity-weighted size and a measure of the broadness of the distribution (PDI). This type of analysis is ISO-recommended and results in only a single size value and no size distribution.

Figure 3(a) shows an example of *Z*-average and PDI for 30-nm colloidal gold. *Z*-average can also be used for comparative purposes; Figure 3(b) shows an example of two batches of liposomes where *Z*-average was used as a quality control parameter. As *Z*-average is an intensity-weighted average of the distribution, sample handling can affect its value.

Figure 4 illustrates the effects of filtration on a dendrimer sample on *Z*-average. The intensity distribution for the unfiltered sample [Figure 4(a)] shows multiple peaks with the major peak being the smaller ~9 nm peak (see volume-weighted distribution). The *Z*-average value is 43.5 nm, which is the

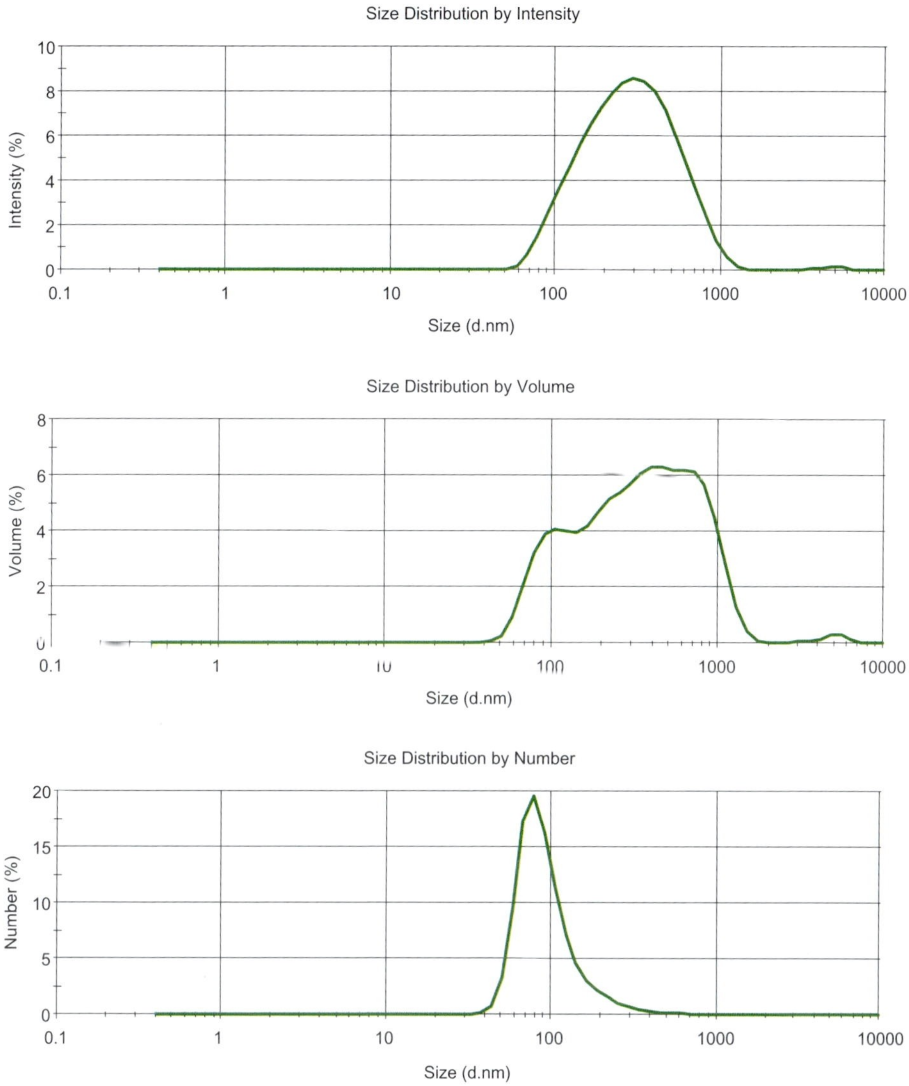

Figure 2. Intensity-, volume-, and number-weighted size distributions for a nanoemulsion diluted 1000-fold in water (RI = 1.332, 0.890 cP) at 25°C. Measurements were made in a low volume quartz cuvette (path length 10 mm) utilizing a DLS instrument with a laser wavelength of 633 nm and 173° scattering angle. Traces are an average of 10 measurements.

intensity-weighted average of the three peaks, and the PDI value is high, reflecting a very polydispersed sample. Upon filtering with a 0.02-μm filter, the larger peaks (due to aggregates) are removed and a single peak is observed in both the intensity- and volume-weighted distributions [Figure 4(b)]. The *Z*-average value is now 8.4 nm, which corresponds to the actual dendrimer

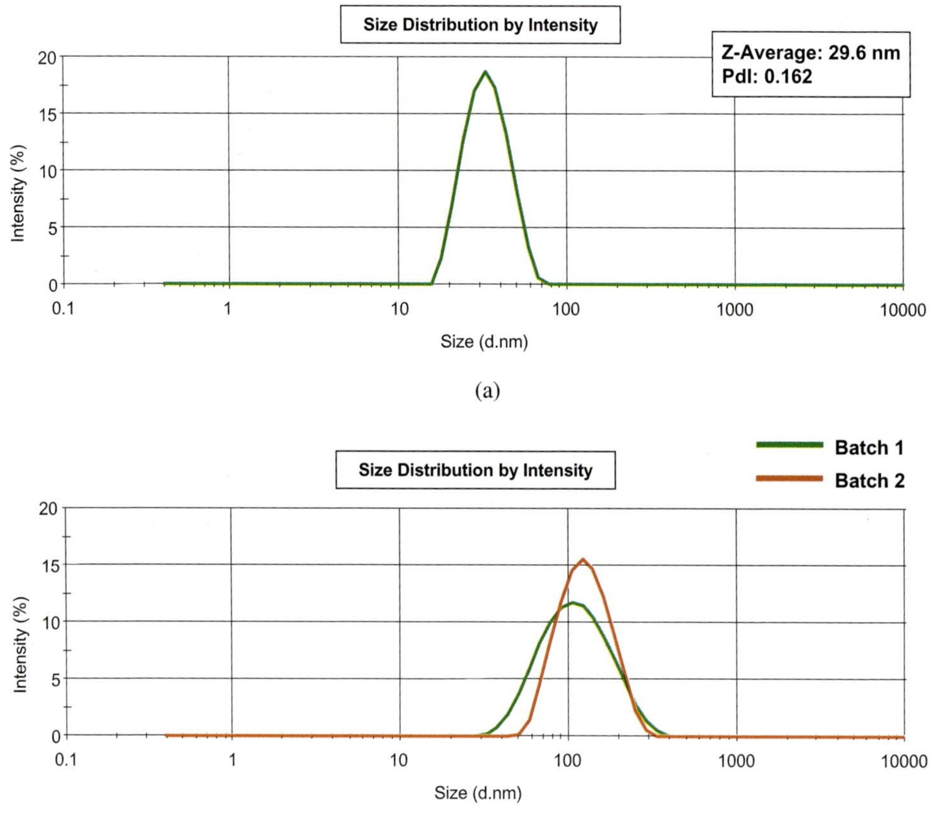

Figure 3. Intensity-weighted size distributions. (a) 30-nm colloidal gold at 100 ug/ml in 2 mM NaCl (RI = 1.332, 0.890 cP) and (b) liposomes at 1 mg/mL in saline (RI = 1.334, 0.908 cP) at 25°C. Measurements were made in a low volume quartz cuvette (path length 10 mm) utilizing a DLS instrument with a laser wavelength of 633 nm and 173° scattering angle. Traces are an average of 10 measurements.

size. The PDI value (0.058 in this case) is representative of a monodispersed sample; PDI <0.2 in practice implies a monodispersed sample. In summary, *Z*-average should be used for comparative purposes (i.e., batch-to-batch consistency), while the intensity- and volume-weighted distributions should be used for size and relative amounts, respectively. Measuring samples as they are (i.e., no filtration) and after filtration through an appropriate filter size are useful for examining samples in their native state, answering the question "what do we actually have in the vial?" If larger aggregates/agglomerates are present in the sample, they will interfere with and possibly swamp out the light scattering signal from the smaller and more important size population. This can be easily remedied by filtering the sample.

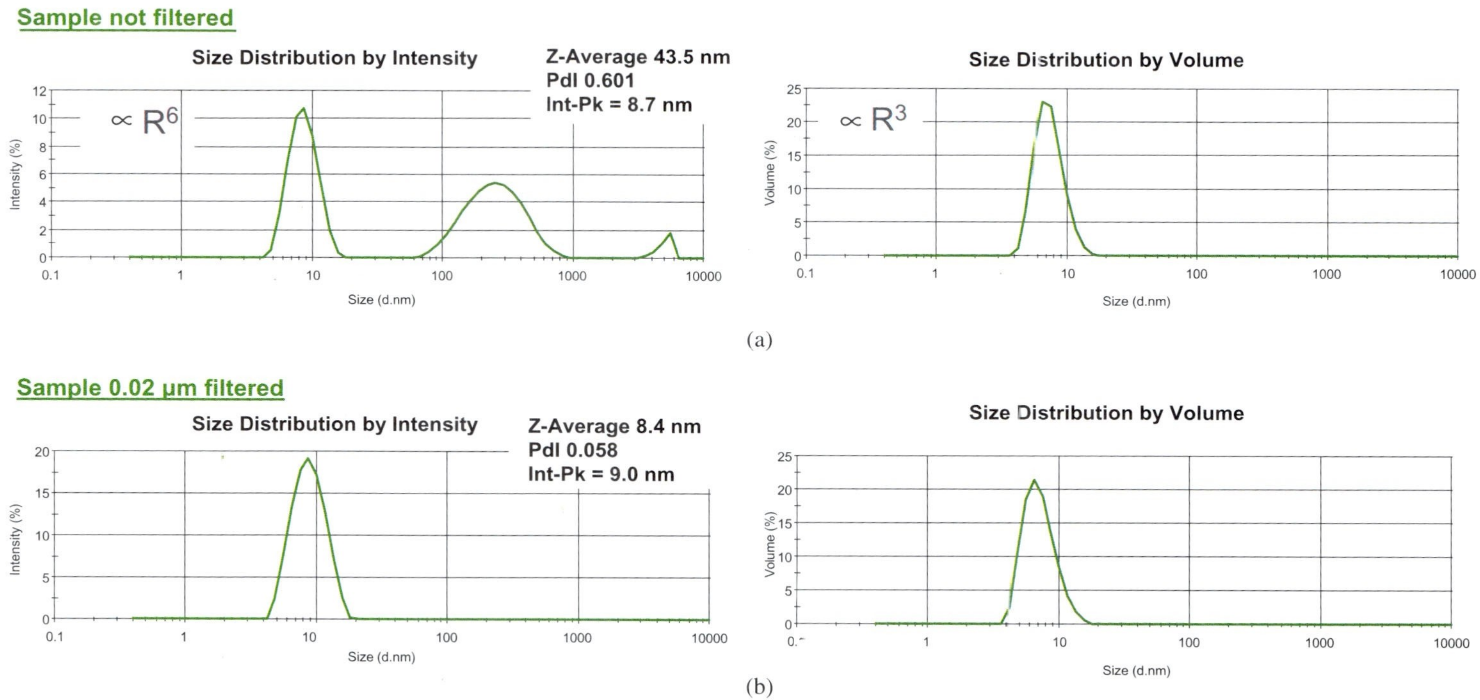

Figure 4. Intensity (left panels)- and volume (right panels)-weighted size distributions for G6–NH_2 dendrimer at 2 mg/ml in PBS (RI = 1.334, 0.911 cP). (a) Without filtration and (b) 0.02-μm filtered. Measurements were made in a low volume quartz cuvette (path length 10 mm) utilizing a DLS instrument with a laser wavelength of 633 nm and 173° scattering angle. Traces are an average of 10 measurements.

5.1.1. *Additional considerations for DLS measurements*

Issues like multiple scattering, viscosity effects, absorbance, and rotational diffusion are commonly encountered when performing DLS measurements on certain nanoparticles. Multiple scattering occurs when the concentration of the sample is too high. This is ubiquitous among nanoparticles such as nanoemulsions or liposomes, which tend to have very high stock concentrations. In multiple scattering, the photons are re-scattered from neighboring particles. This leads to an apparent increase in randomness and as a result, the particles appear to move faster. This culminates in a smaller observed size. Figure 5 depicts this phenomenon for a nanoemulsion. Three dilutions are shown and only when the sample is diluted 10,000-fold or more does the size remain unchanged. At 1,000-fold dilution of the stock, the size is smaller relative to the higher diluted samples. Therefore, it is important to run a concentration series when measuring samples to avoid multiple scattering effects.

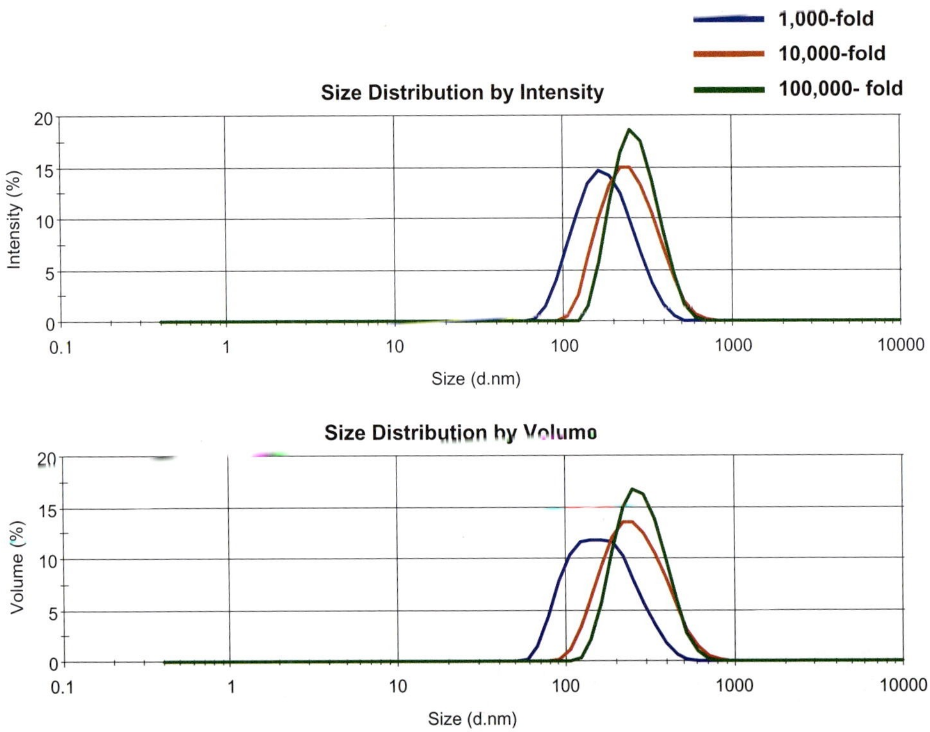

Figure 5. Intensity- and volume-weighted size distributions for a nanoemulsion at three different dilutions in water (RI = 1.332, 0.890 cP) at 25°C. Measurements were made in a low volume quartz cuvette (path length 10 mm) utilizing a DLS instrument with a laser wavelength of 633 nm and 173° scattering angle. Traces are an average of 10 measurements.

Similarly, highly concentrated samples can have an effect on bulk viscosity. Since viscosity is in the Stokes–Einstein equation [Figure 1(b)], any error will lead to an inaccurately calculated hydrodynamic size. To resolve this, it is recommended again to run a concentration series. As the sample is diluted, the viscosity approaches a value at infinite dilution. It is advisable to dilute by orders of magnitude if possible. Figure 6 shows an example of three concentrations for liposomes; note that the higher concentration sample has a slightly larger size.

DLS examines the light scattered by the particle upon illumination by a laser at a certain wavelength. However, some nanoparticles such as quantum dots or metal colloids absorb at the laser wavelength. This will result in the absorption of light and minimal light scattering. Therefore, it is important to measure the nanoparticle's absorbance and fluorescence spectra before carrying out DLS measurements. Figure 7(a) shows an example of a PEGylated quantum dot where absorption occurred at the instrument's laser

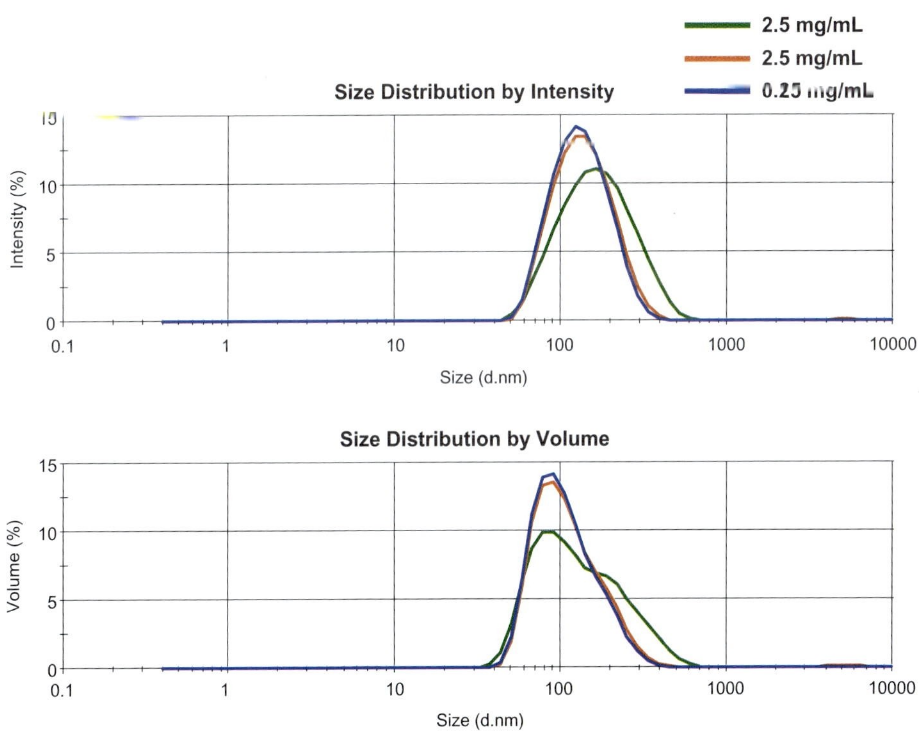

Figure 6. Intensity- and volume-weighted size distributions for liposomes at three different concentrations in saline (RI = 1.334, 0.908 cP) at 25°C. Measurements were made in a low volume quartz cuvette (path length 10 mm) utilizing a DLS instrument with a laser wavelength of 633 nm and 173° scattering angle. Traces are an average of 10 measurements.

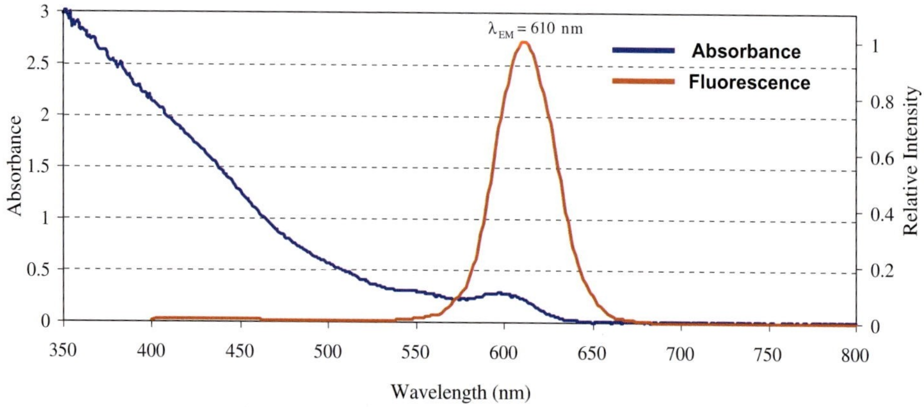

(a)

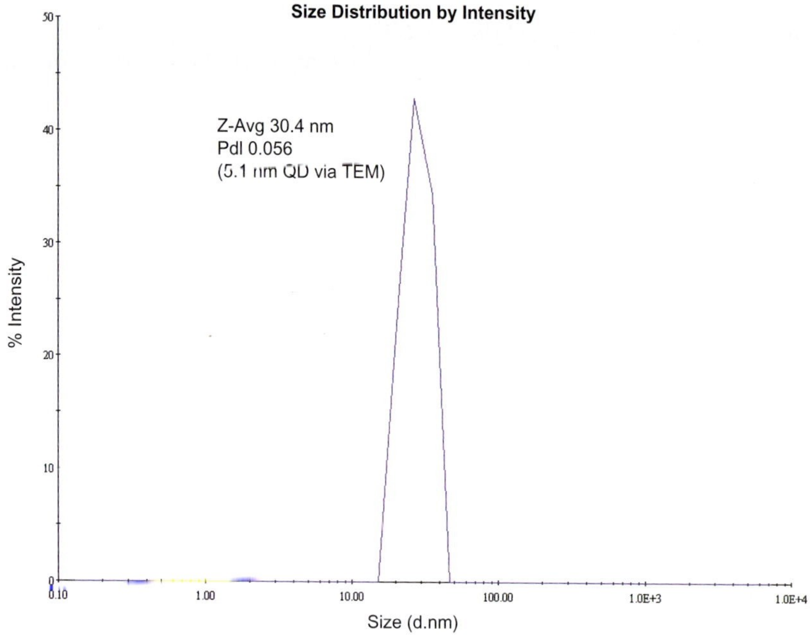

(b)

Figure 7. Nanoparticles that absorb at the laser wavelength. (a) Absorbance and fluorescence spectra for a PEGylated quantum dot. (b) Intensity-weighted size distribution for a PEGylated quantum dot at 8 μg/mL in water (RI = 1.332, 0.890 cP) at 25°C. Measurements were made in a low volume quartz cuvette (path length 10 mm) utilizing a DLS instrument with a laser wavelength of 830 nm and 90° scattering angle. Traces are an average of 10 measurements.

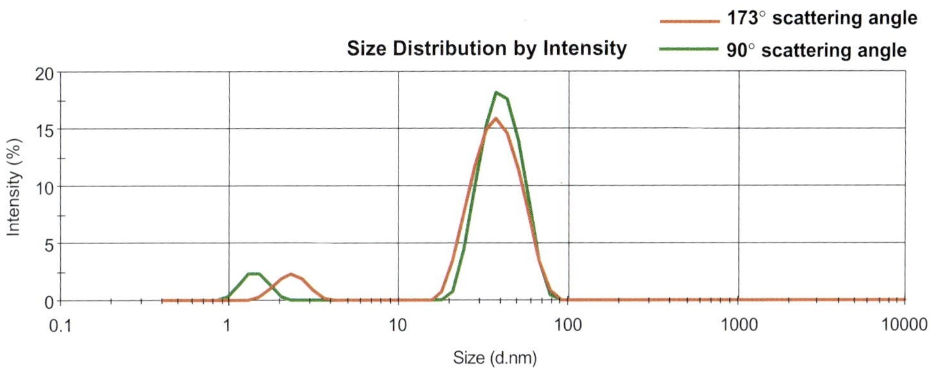

Figure 8. Intensity-weighted size distribution for 30-nm colloidal gold at 133 μg/mL in water (RI = 1.332, 0.890 cP) at 25°C. Measurements were made at two scattering angles, 173° and 90°, and at the same laser wavelength, 633 nm. Measurements were made in a low volume quartz cuvette (path length 10 mm). Traces are an average of 10 measurements.

wavelength (633 nm). Measurements were made on a separate DLS instrument where the laser wavelength was 830 nm [Figure 7(b)]. In the event that a second DLS instrument is not available, one can dilute the sample further to reduce absorption. However, there is a cutoff in which the concentration will be too low; this will be dependent on the scattering properties of the sample.

DLS measures the translational diffusion coefficient, which in turn is used to calculate the hydrodynamic size. However, with some nanoparticles such as metal colloids or nanorods, an additional peak resulting from rotational diffusion is observed. Figure 8 gives an example of 30-nm colloidal gold measured at two different scattering angles. The smaller peak, ~1–3 nm, was the result of rotational diffusion caused by the optical properties of the particle. Similar results occured for nanorods, where the rods had an additional axis of rotation. To determine if a small (and often unexplained) peak is due to rotational diffusion, the sample should be measured at two or more scattering angles. Regardless of the scattering angle, the translational diffusion peak will remain constant, while the rotational diffusion peak will shift due to its angular dependence. Therefore, from Figure 8, the larger peak (~30 nm) was due to translational diffusion, while the smaller peak was due to rotational diffusion. Additional experiments, namely polarization experiments, could be performed to further prove this, but it would require more advanced DLS setups.

DLS is a fast and relatively inexpensive technique to measure size in solution. Nevertheless, it does have its limitations. As a general rule of thumb, DLS can baseline resolve particles that are three- to five-fold different in size; Figure 9 illustrates this point. Dendrimers of the same generation but with two different

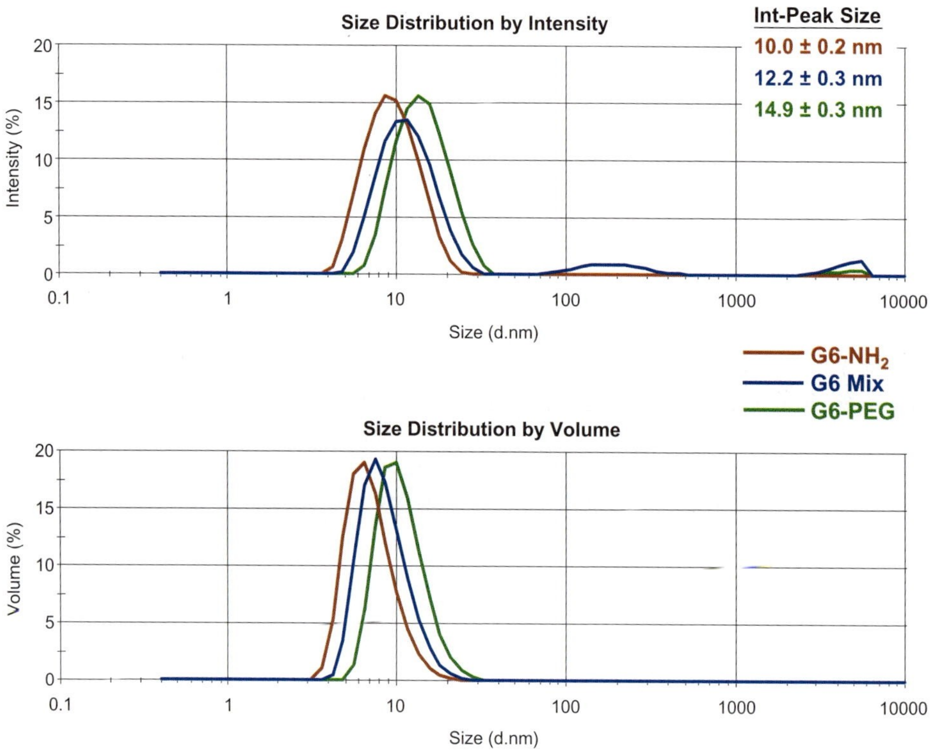

Figure 9. Intensity- and volume-weighted size distributions for G6–NH_2, G6–PEG, and a mixture of both dendrimers at 2 mg/mL in PBS (RI = 1.334, 0.911 cP) at 25°C. Measurements were made in a low volume quartz cuvette (path length 10 mm) utilizing a DLS instrument with a laser wavelength of 633 nm and 173° scattering angle. Traces are an average of 10 measurements.

surfaces were run individually and as a mixture. As can be seen, DLS was unable to resolve the individual components and an average size was observed.

5.2. *Zeta potential*

Zeta potential is defined as the potential difference between the bulk solution (dispersing medium) and the surface of hydrodynamic shear (slipping plane). A schematic of how zeta potential is defined is given in Figure 10(a). Tightly associated ions of opposite charges (relative to the particle surface charge) make up the Stern layer. Loosely associated ions make up the diffuse layer. Within this diffuse layer, there exists a boundary where the particle and ions form a single entity. When the particle moves, the ions within this boundary move together with the particle, while those outside of it stay with the bulk

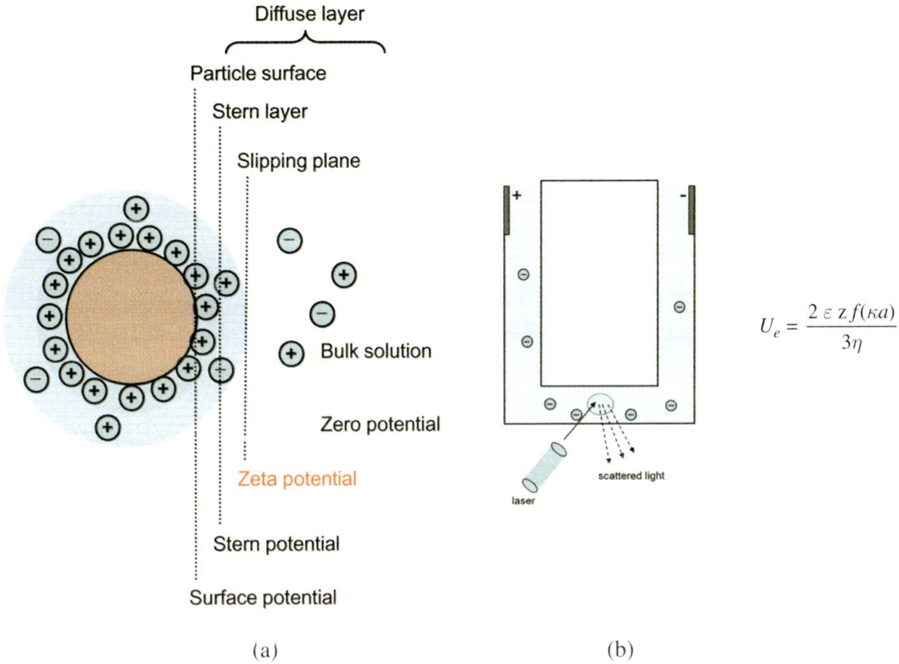

Figure 10. Measurement of the zeta potential. (a) Definition of zeta potential and (b) how it is measured. Given to the right in (b) is the Henry equation. U_e is the electrophoretic mobility, ε is the dielectric constant, η is the absolute zero-shear viscosity of the medium, $f(\kappa a)$ is the Henry function, and κa is a measure of the ratio of the particle radius to the Debye length. (Redrawn from Ref. 27).

solution. The Stern layer and the slipping plane are collectively referred to as the electrical double layer. In addition, zeta potential measurements are based on light scattering. A schematic of how zeta potential is measured is given in Figure 10(b). The sample solution is filled into a chamber containing two electrodes. An electric field is applied and causes the charged particles to migrate to the corresponding electrode with a velocity proportional to its zeta potential. Using LDV, the particle velocity is measured from the frequency shift of the scattered light. The velocity, or electrophoretic mobility, is converted to zeta potential using the Henry equation and Smoluchowski or Huckel models.[24,27]

Sample preparation and measurement procedures for performing successful zeta potential measurements are described elsewhere in detail[24,26] and would not discussed here. Included in those references are guidelines and precautions, as well as a discussion of relevant standards and data analysis. As the references go into the specifics of carrying out zeta potential measurements, it is our intention to give here some of the practical issues associated with zeta as it

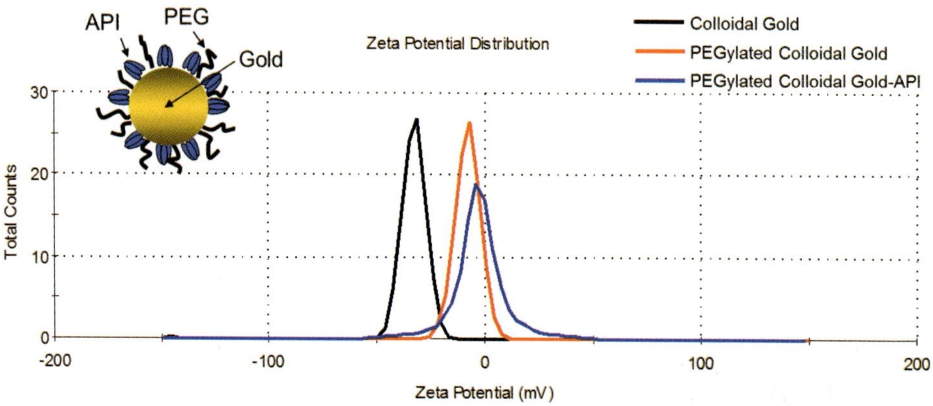

Figure 11. The averaged zeta potential distributions for colloidal gold, PEGylated colloidal gold, and PEGylated colloidal gold–active pharmaceutical ingredient (API). Samples were diluted 100-fold in 10 mM NaCl and the pH adjusted to 7.0. Measurements were made at 25°C and at an applied voltage of 150 V. A cartoon of the PEGylated colloidal gold–API is shown for illustration purposes.

applies to nanoparticle size characterization. Similar to DLS, these are broken down into two categories — general and nanoparticle-specific issues.

General issues associated with zeta potential are primarily due to the lack of reporting or researchers not being mindful of what measurement conditions are used. The temperature, pH, conductivity (ionic strength), and solvent used all affect zeta potential and must be reported along with it. For instance, Figure 11 highlights an example of the utility of zeta potential on tracking the reaction progress of the PEGylation of colloidal gold. Citrate-stabilized colloidal gold is negative, but as it is coated with PEG and the active pharmaceutical ingredient (API), it becomes neutral.

For such a comparison to be made, exact sample conditions must be used; in this case, all samples were diluted in 10 mM NaCl, with pH adjusted to 7.0, and measured at 25°C. To further illustrate this point, the effect of pH on an amine-terminated dendrimer is shown in Figure 12. The native pH of the sample was ~10 and at this pH, the dendrimer was neutral. However, as the pH dropped when HCl was added, the dendrimer became protonated and as a result, the zeta potential became positive.

A final example is given in Figure 13 where the effect of ionic strength on zeta potential is demonstrated on Doxil® (doxorubicin HCl liposomes). Measurements were made either in water or 10 mM NaCl at the same concentration and adjusted to the same pH. As can be seen from the zeta potential distribution, slight changes in ionic strength can lead to drastic changes in zeta potential. The examples given highlight the importance of measuring

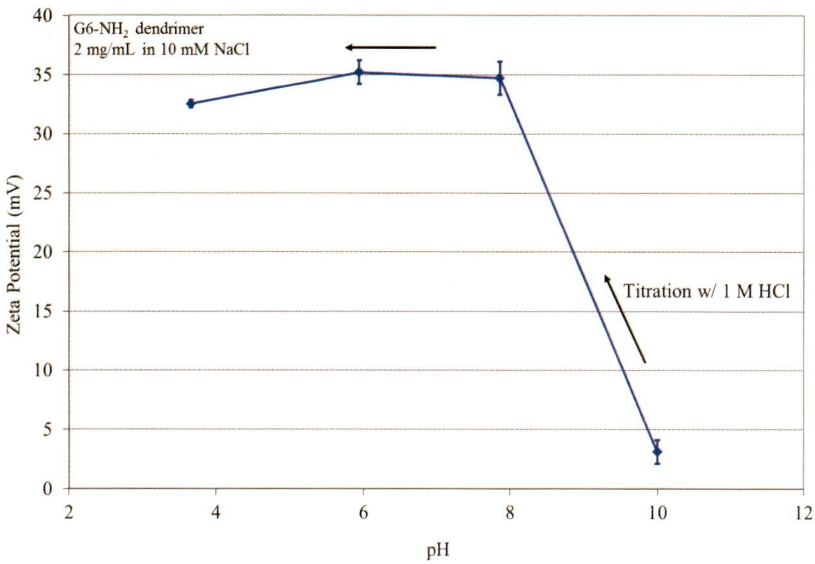

Figure 12. The effect of pH on zeta potential for a G5–NH_2 dendrimer. Samples were prepared at 2 mg/mL in 10 mM NaCl and the pH adjusted with standardized (1 M) HCl. Measurements were made at 25°C and at an applied voltage of 150 V.

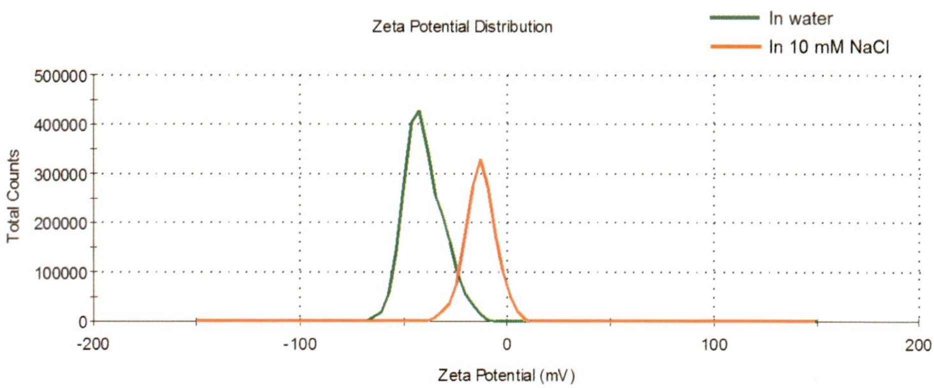

Figure 13. The average zeta potential distributions for Doxil® in water and 10 mM NaCl demonstrating the effect of ionic strength on zeta potential. Stock Doxil® was diluted 100-fold in 10 mM NaCl or water and the pH adjusted to 6.9. Measurements were made at 25°C and at an applied voltage of 150 V.

and reporting the pH and ionic strengths of the solvent with the zeta potential. Finally, regarding terminology, zeta potential is often used interchangeably with surface charge, which is incorrect [Figure 10(a)]. While surface charge cannot be directly measured, zeta potential can and gives an indication of the surface of the nanoparticle.

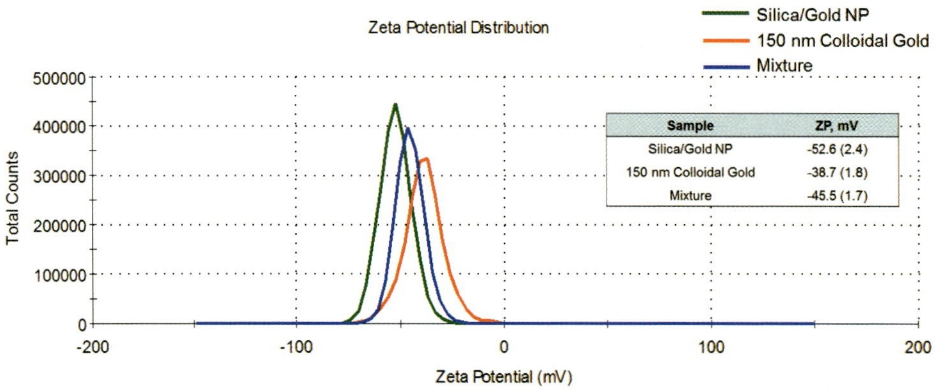

Figure 14. The average zeta potential distributions for silica/gold nanoparticle (NP), 150-nm colloidal gold, and a mixture of both. Silica/gold NP and 150-nm colloidal gold were diluted 100-fold and 10-fold, respectively, in 10 mM NaCl and the pH adjusted to 6.9. Measurements were made at 25°C and at an applied voltage of 150 V. A table summarizing the zeta potentials is given in the figure.

Since zeta potential measurements are based on light scattering, they have the same nanoparticle-specific issues associated with DLS. These include multiple scattering, viscosity effects, and absorbance, which have been discussed earlier. As with DLS, zeta potential does have its limitations in resolving power. Figure 14 shows the zeta potential distributions for silica/gold nanoparticle, 150-nm colloidal gold, and a mixture of both. From the zeta distributions and the table summarizing the zeta potentials, the zeta potentials of the individual components of the mixture of similarly charged populations cannot be resolved. While discussed elsewhere,[24,26] it is important to reiterate the importance of possible sample degradation during zeta measurements. Since a voltage is applied to the sample, it is crucial to look for visual signs of sample degradation, i.e., color change, precipitation, and blackening of the electrodes. DLS size measurements can be made after zeta potential measurements to assess size integrity, or the applied voltage can be changed accordingly to reduce sample degradation. Sample concentration is nanoparticle-specific. Typically, instruments measuring zeta can also measure size by DLS. Therefore, a good starting concentration for zeta would be the concentration which provides good and reliable DLS results.

6. Summary

Adequate PCC of nanomaterials prior to immunological studies should be considered as a requirement. At a minimum, characterization should address the following: size (and size distribution), zeta potential, targeting/drug/

imaging quantitation, purity, stability, and batch-to-batch consistency. These parameters and their relevance to characterization have been discussed along with the instrumentation and techniques needed. Characterization in batch mode (i.e., no separation) and through separation have also been discussed. In addition, guidelines and issues associated with size by DLS and zeta potential measurements have been provided. Both general and nanoparticle-specific issues associated with carrying out DLS and zeta potential measurements were addressed and illustrated with examples.

References

1. McNeil SE. Nanotechnology for the biologist. *J Leukoc Biol* 2005;78:585–594.
2. McNeil SE. Nanoparticle therapeutics: A personal perspective. *Wiley Interdiscip Rev Nanomed Nanobiotechnol* 2009;1:264–271.
3. Maeda H. Tumor-selective delivery of macromolecular drugs via the EPR effect: Background and future prospects. *Bioconjug Chem* 2010;21:797–802.
4. Adiseshaiah P, Hall JB, McNeil SE. Nanomaterial standards for efficacy and toxicity assessment. *Wiley Interdiscip Rev Nanomed Nanobiotechnol* 2009;2:99–112.
5. Dobrovolskaia MA, Patri AK, Zheng J, Clogston JD, Ayub N, Aggarwal P, Neun BW, Hall JB, McNeil SE. Interaction of colloidal gold nanoparticles with human blood: Effects on particle size and analysis of plasma protein binding profiles. *Nanomedicine* 2009;5:106–117.
6. Malik N, Wiwattanapatapee R, Klopsch R, Lorenz K, Frey H, Weener JW, Meijer EW, Paulus W, Duncan R. Dendrimers: Relationship between structure and biocompatibility *in vitro*, and preliminary studies on the biodistribution of I-125-labelled polyamidoamine dendrimers *in vivo*. *J Controlled Release* 2000;65:133–148.
7. Scholer N, Hahn H, Muller RH, Liesenfeld O. Effect of lipid matrix and size of solid lipid nanoparticles (SLN) on the viability and cytokine production of macrophages. *Int J Pharm* 2002;231:167–176.
8. Kobayashi H, Brechbiel MW. Dendrimer-based macromolecular MRI contrast agents: Characteristics and application. *Mol Imaging* 2003;2:1–10.
9. Kobayashi H, Kawamoto S, Jo SK, Bryant HL Jr, Brechbiel MW, Star RA. Macromolecular MRI contrast agents with small dendrimers: Pharmacokinetic differences between sizes and cores. *Bioconjug Chem* 2003:14;388–394.
10. Dobrovolskaia MA, McNeil SE. Immunological properties of engineered nanomaterials. *Nat Nanotechnol* 2007;2:469–478.
11. Salvador-Morales C, Zhang L, Langer R, Farokhzad OC. Immunocompatibility properties of lipid–polymer hybrid nanoparticles with heterogeneous surface functional groups. *Biomaterials* 2009;30:2231–2240.
12. Aggarwal P, Hall JB, McLeland CB, Dobrovolskaia MA, McNeil SE. Nanoparticle interaction with plasma proteins as it relates to particle biodistribution, biocompatibility and therapeutic efficacy. *Adv Drug Delivery Rev* 2009;61:428–437.
13. Patri AK, Dobrovolskaia MA, Stern ST, McNeil SE. Preclinical characterization of engineered nanoparticles intended for cancer therapeutics. In: Amiji M (ed.), *Nanotechnology for Cancer Therapy*. CRC Press, 2007, pp. 104–113.

14. Dong MW. *Modern HPLC for Practicing Scientists.* John Wiley & Sons, 2006, pp. 1–14.
15. Islam MT, Shi X, Balogh L, Baker JR, Jr. HPLC Separation of different generations of poly(amidoamine) dendrimers modified with various terminal groups. *Anal Chem* 2005; 77:2063–2070.
16. Singh R, Ajagbe M, Bhamidipati S, Ahmad Z, Ahmad I. A rapid isocratic high-performance liquid chromatography method for determination of cholesterol and 1,2-dioleoyl-sn-glycero-3-phosphocholine in liposome-based drug formulations. *J Chromatogr A* 2005; 1073:347–353.
17. Wei G, Xiao S, Si D, Liu C. Improved HPLC method for doxorubicin quantification in rat plasma to study the pharmacokinetics of micelle-encapsulated and liposome-encapsulated doxorubicin formulations. *Biomed Chromatogr* 2008;22:1252–1258.
18. Vernooij EA, Kettenes-van den Bosch JJ, Underberg WJ, Crommelin DJ. Chemical hydrolysis of DOTAP and DOPE in a liposomal environment. *J Controlled Release* 2002; 79:299–303.
19. Majoros IJ, Myc A, Thomas T, Mehta CB, Baker JR, Jr. PAMAM dendrimer-based multifunctional conjugate for cancer therapy: Synthesis, characterization, and functionality. *Biomacromolecules* 2006;7:72–579.
20. Arifin DR, Palmer AF. Determination of size distribution and encapsulation efficiency of liposome-encapsulated hemoglobin blood substitutes using asymmetric flow field-flow fractionation coupled with multi-angle static light scattering. *Biotechnol Prog* 2003; 19:1798–1811.
21. Cho TJ, Hackley VA. Fractionation and characterization of gold nanoparticles in aqueous solution: Asymmetric-flow field flow fractionation with MALS, DLS, and UV-Vis detection. *Anal Bioanal Chem* 2010;398:2003–2018.
22. Chan KC, Patri AK, Veenstra TD, McNeil SE, Issaq HJ. Analysis of fullerene-based nanomaterial in serum matrix by CE. *Electrophoresis* 2007;28:1518–1524.
23. Brothers HM, II, Piehler LT, Tomalia DA. Slab-gel and capillary electrophoretic characterization of polyamidoamine dendrimers. *J Chromatogr A* 1998;814:233–246.
24. Hackley VA, Clogston JD. Measuring the hydrodynamic size of nanoparticles in aqueous media using batch-mode dynamic light scattering. In: McNeil SE (ed.), *Methods in Molecular Biology, Vol. 697. Characterization of Nanoparticles Intended for Drug Delivery*, Humana Press, 2011, pp. 35–52.
25. Berne BJ, Pecora, R. *Dynamic Light Scattering with Applications to Chemistry, Biology, and Physics.* Dover Publications, 2000, pp. 10–37.
26. Hackley VA, Clogston JD. NIST–NCL Joint Assay Protocol PCC-1, Measuring the size of nanoparticles in aqueous media using batch-mode dynamic light scattering. Retrieved June 1, 2012, from http://ncl.cancer.gov/working_assay-cascade.asp.
27. Malvern Instruments, Zeta potential measurement using laser Doppler electrophoresis (LDE), Retrieved June 1, 2012, from http://www.malvern.com/LabEng/technology/zeta_potential/zeta_potential_LDE.htm.

Chapter 3

Impact of Nanoparticle Sterilization on Analytical Characterization

Nanda Subbarao

Biologics Consulting Group, 400 N Washington Street Suite 100, Alexandria, VA 22314, USA
nsubbarao@bcg-usa.com

Nanotherapeutics (NTs) are complex systems with multiple components, each of which could be susceptible to the damaging effects of the sterilization procedure in a different way. Sterilization by autoclaving could result in heat-induced chemical and physical changes in NTs and sterilization by gamma irradiation could induce free radical formation, which could result in immediate chemical changes as well as impact the stability of the product. A suitable battery of tests needs to be utilized during developmental studies to understand the impact of sterilization. The panel of analytical methods must be appropriate for monitoring the physical and chemical changes to NTs. Due to the impact of sterilization, it is necessary to perform all characterization studies after sterilization procedures and to establish a stability program suitable for the detection of chemical degradants during the shelf-life of NTs.

1. Introduction

Nanotherapeutics (NTs) intended for parenteral applications are required to be sterile. The manufacturing process must therefore either be a sterile process or must include terminal sterilization. Sterilization procedures used for this purpose are numerous and include autoclaving, filtration, gamma irradiation, and treatment with ethylene oxide. This review is not focused on these various sterilization procedures but rather, on the degradation induced by these sterilization procedures and the impact on analytical testing and the characterization of the NTs. The discussion in this review leads to the conclusion that it is necessary to perform product characterization and impurity analysis on the NT only after sterilization by the optimized procedure. The review also addresses the analytical methods required to test the NT to evaluate the deleterious effects of sterilization procedures. The chapter will cover solid lipid nanoparticles (SLNs), liposomes, and microspheres.

2. Why Is Sterilization a Challenge for Nanotherapeutics?

NTs are usually made of multiple components, each of which is best sterilized using a different sterilization procedure. Although more than one option for sterilization is available for most nanoparticle (NP) products, each of them has the potential for negatively impacting the chemical and physical characteristics of different components of the NT. Therefore, sterilization of an NT will often require a method which may be less than ideal, but sufficient to support the desired shelf-life of the product. The chosen sterilization procedure parameters and the formulation will need to be optimized to mitigate the sterilization-induced degradation to an acceptable level. It is crucial to test several sterilization methods for each NP in order to select one which causes the least degradation.

NPs present another unique challenge for terminal sterilization that has little to do with the sensitivity of the NP to sterilization-induced degradation. This challenge is due to the requirement that along with the sterilization step, the product has to be depyrogenated and the success of the depyrogenation has to be monitored by performing an endotoxin assay on the drug product. Unfortunately, most NPs interfere with the Litmus Amebocyte Lysate (LAL) assay, which is the most convenient assay for endotoxin.[1,2] The LAL assay interference, in effect, decreases the confidence in the assay results and in some cases, may necessitate the use of the rabbit pyrogen test which has high assay variability. For some NTs, there will also be the uncertainty as to whether endotoxin can be entirely removed from the NP structure once they are formed. Therefore, for some NTs, it could be advantageous to opt not to sterilize the NT, but to instead start with sterile, pyrogen-free raw materials and manufacture the NP drug under sterile conditions.

3. Previous Studies Reviewing the Effects of Various Sterilization Procedures on Nanomaterials

Sterilization during the NT manufacturing process has been the subject of several excellent reviews. Sterilization of liposomal preparations of vaccines has been reviewed by Mohammad *et al.*,[3] while that of various polymer NPs including nanospheres and nanocapsules has been addressed by Vauthier and Bauchemal[4] and Manjunath *et al.*[5] The sterilization of SLNs has been covered as part of a more general review of the manufacturing of SLNs.[6] As mentioned in the introduction, the focus of this review is not on the sterilization step itself, but on degradation induced by sterilization procedures and its impact on the necessary analytical testing and characterization of the NT.

4. Available Methods for Sterilization

4.1. *Autoclaving*

Sterilization by autoclaving, also referred to as moist heat sterilization, is an effective method accepted by most pharmacopeias. This method has been used in the sterilization of NPs.[7–10] However, the high temperature, usually 120°C, in the presence of steam can induce chemical changes in the NT, particularly in those containing biologics and other heat-sensitive drugs or excipients. An example is the autoclaving of polymer-based gene transfer complexes which results in the total loss of transfection capacity[11] and the yellowing of and drop in pH of formulations with glucose as an excipient. Autoclaving can also induce physical modifications such as changes in NP size and morphology, which in turn can alter the amount of drug loaded in the NP and the rate of release of the drug from the NP.

It is to be expected that NPs with a glass transition and/or a melting point below 120°C are most impacted by autoclaving. The increase in temperature during autoclaving followed by cooling will result in a rearrangement of the NP molecular structure. Such a rearrangement may be acceptable only for NPs where the internal structure of the particle does not impact the safety, efficacy, and quality of the NT, which can be a rare feature among NTs. Examples of NPs with structures negatively impacted by autoclaving include a poly(isobutylcyanoacrylate) nanocapsule where its size increased from 200 to 500 nm after sterilization. This effect was attributed to either the swelling of the polymeric membrane composed of poly(isobutylcyanoacrylate) or the expansion of the oily phase.[12] A significant increase of poly(butylcyanoacrylate) nanosphere size due to sterilization by autoclaving has also been reported by Sommerfeld *et al.*[13]

Autoclaving is not a suitable sterilization method for NPs such as liposomes where the physical structure is not stabilized by covalent bonds and the bilayer structure is easily disrupted by high temperature. Autoclaving will cause a massive loss of hydrophilic, non-bilayer–interacting compounds entrapped in the liposomes. Autoclaving can also result in the hydrolysis of the liposomal phospolipids at a rate which depends on the formulation's pH. One option available for the sterilization of liposomes by autoclaving is to perform the removal of the free drug or the encapsulation after the autoclaving step.[14]

In NP formulations which are coated with surfactants to mitigate aggregation, heating results in phase separation at a temperature above the surfactant's cloud point, where the molecule is likely to dissociate from the particle. The NP core is then free to aggregate. On cooling to room temperature, the surfactant will re-dissolve in the solution and coat the aggregated particles, preventing their dissociation. The use of surfactants with higher cloud points, such as phospholipids, can overcome this problem.[14]

Yet, where the loss in activity or changes to the NP induced by autoclaving is/are minor, autoclaving may be an acceptable method for sterilization and many examples have been described in the literature. Sterilization by autoclaving is the best option for SLNs with heat-resistant drugs and a transition temperature above the autoclaving temperature. For example, autoclaving was employed as the method for sterilization of SLNs with the prodrug 3′-azido-3′-deoxythymidine, otherwise known as AZT. The SLNs consisted of trilaurine as the core and neutral and negatively charged phospholipids as the coating. There were no significant changes in the mean diameter and the zeta potential of the NPs after autoclaving at 121°C for 20 min. The amount of incorporated pro-drug was however, slightly reduced due to hydrolysis. Autoclaved SLNs were stable for a period of 10 weeks at 20°C, but an increase in particle size and loss of the pro-drug were observed at 4°C and 37°C.[15]

Cavalli *et al.*[16] have described SLN preparations composed of oil-in-water microemulsions with the internal phase constituting different lipid matrices. These SLNs sterilized by autoclaving were stable during the process and maintained a spherical shape and narrow size distribution as confirmed by transmission electron microscopy (TEM) analysis. Collagen is thermally stable and therefore, NPs composed of collagen can be readily sterilized by autoclaving. Collagen microparticles (3 to 40 μm) were sterilized by autoclaving in a study by Rössler *et al.*[17]

NPs composed of chitosan can also be sterilized by autoclaving. For example, scanning electron microscopy (SEM), TEM, Fourier transform infrared spectroscopy (FT-IR), and X-ray diffraction (XRD) analysis all revealed that the structural integrity of chitosan–carboxymethyldextran NPs was not destroyed after high steam pressure sterilization (121°C, 30 min).[18]

More examples of SLNs with relatively stable structures at temperatures reached during autoclaving were described by Seyfoddin *et al.*[19] and Bondi and Craparo.[20]

Muller *et al.*[4] proposed to optimize the autoclaving temperature to decrease the heat-induced degradation of the NP matrix. They reported that autoclaving at 121°C cannot be performed when using sterically stabilizing polymers, e.g., the poloxamer series. Instead, the autoclaving temperature can be reduced, while prolonging the autoclaving time to retain sterilization efficacy.

4.1.1. *Impact on NT analytics*

In cases where the NT has been sterilized by terminal autoclaving, the product must be tested for heat-induced chemical degradation. The degradant formed will depend on the excipients, the active drug, and the formulation. For SLNs with a melting temperature below or in the vicinity of the autoclaving temperature, there is a need to test for changes to the NP structure and its impact on the long-term stability of the product. In particular, the particle size and distribution must be evaluated. The change in the drug load and rate of release of the drug may be impacted by the autoclaving and must also be tested.

4.2. *Filtration*

Sterile filtration is one of the most common methods for the sterilization of heat-labile therapeutics such as biologics. This method however, is limited to NTs by the viscosity of the solution. The additional limitation is that NPs need to pass through 0.2-μm filters, either by being small enough or their physical state is such that they can be extruded through the filters and re-form after passage through the filter.[21] The filtration method has also been proven to be unsuccessful, where the drug was adsorbed on the particle surface and stripped off from the particles. Nonetheless, filtration has been used as the sterilization method for some NTs in spite of these limitations, and some examples are described here. In one study, savoxapine-loaded polymer-layered silicate (PLS) NPs were successfully sterilized by microfiltration.[22]

One approach to sterilize NPs by filtration is to optimize the NT's design. Tsukada *et al.*[23] described one such study where they controlled the size of NPs to under 0.2 μm to ensure that the formulation can be effectively subjected to sterile filtration. The conditions for preparation of poly-(D,L-lactide-co-glycolide) (PLG) nanospheres by the emulsion solvent diffusion method were optimized to produce NPs which could pass through a membrane filter with 0.2 μm pore size. Such NPs were successfully produced, sterile-filtered, and passed bacterial challenge tests.[23]

In a second study, NPs were manufactured from several PLG polymers by the salting-out procedure with the process parameters optimized to obtain sub-200–nm particles. The NP suspensions with different lyoprotectants were sterilized by filtration. Freeze-drying was performed using vials covered with 0.22-μm membrane filters in order to preserve the suspensions from bacterial contamination. Sterility was assessed on the final product according to pharmacopeial requirements using the membrane filtration method and analyses showed no microbial contamination.[24]

In another similar study, sub-150–nm NPs incorporated with a photosensitizer, meso-tetra(4-hydroxyphenyl)porphyrin, were found to be suitable for sterilization by filtration using 0.22-μm membranes. All final freeze-dried products were refiltrable on a 0.22-μm membrane and had stable mean particle size and drug loading for six months, and were found to be sterile by sterility testing.[25]

Liposomes illustrate the case where NPs may be larger than 0.2 μm, but can be sterilized by filtration. This is because liposomes composed of phosholipids above the phase transition temperature can be extruded through the filters and the liposomal structures re-form after extrusion. Such an approach has been described by Darole *et al.*[6] In this study, amphotericin B was co-evaporated with an oil–surfactant mixture to produce a microemulsion, diluted in water, sterilized by filtration, and lyophilized to obtain the final product. The stability of the lyophilized product and the reconstituted product were studied, and the reconstituted product was subjected to physicochemical characterization.[26] This liposomal formulation has the advantage of the encapsulated drug being lipid-associated and will continue to be lipid-associated after filtration.

In contrast, when liposomes contain an aqueous encapsulated drug, leakage of the drug will occur when the liposomes are extruded through filters. Therefore, the filtration/sterilization step needs to be performed prior to the drug encapsulation step. Such a procedure has been described by Li and Deng.[27] In this study, the liposome-forming lipids and water-soluble carrier materials were dissolved in tert-butyl alcohol/water cosolvent systems to form an isotropic monophase solution. The resulting solution was sterilized by filtration through a 0.2-μm filter followed by lyophilization. On the addition of water, the lyophilized product spontaneously formed a homogeneous liposomal preparation which was further processed and characterized.[27]

Even when sterilization by filtration is effective, the process can be unsatisfactory due to the substantial loss of the NP matrix or the drug on the filter. Bos *et al.*[11] described such an observation during the sterilization of poly(2-(dimethylamino) ethyl methacrylate)-based gene transfer complexes. Autoclaving was found to be unsuitable for this product because it dramatically altered the physical characteristics, and also resulted in the complete loss of transfection potential. Sterile filtration on the other hand, did not change the topology and integrity of the DNA assessed by agarose gel electrophoresis, and CD (circular dichroism) spectra and the transfection potential were fully retained. However, the concentration of DNA in the complex was slightly decreased. Pre-coating of the filter with polyplexes reduced material loss. Separate sterilization of the polymer by autoclaving or filtration and DNA by filtration was found to be a viable option.

4.2.1. *Impact on NT analytics*

When sterile filtration is adopted for the sterilization of NPs, there will be some retention of the particles on the filter and size range, and the distribution of the NT will be altered. In all cases, even where the size of the particles is smaller than the filter pore size, there is a possibility of selective retention of the active or the formulation components on the filter, thereby altering the composition of the formulation. Liposomes larger than 0.2 μm may be filtered above the phospholipid phase transition temperature; however, encapsulated material in the aqueous compartment will be lost and the structure of the liposome will be altered. Massive loss of encapsulated material in the aqueous phase is to be expected.

NTs sterilized by filtration must therefore be evaluated for changes in particle size and distribution. The drug load in the NP must be re-evaluated. The composition of the formulation must be analyzed to control for the possible selective retention of components on the filters. Any changes to the viscosity after filtration could be an indication of material retention on the filter.

4.3. *Gamma irradiation*

Gamma irradiation as a procedure for sterilization has been used with success for NTs which are not heat-tolerant. The NP sample viscosity is not a concern with this method and it can even be applied to sealed packaged products, avoiding further risk of microbial contamination. However, the presence of water during gamma irradiation results in the formation of reactive free radicals such as ·OH, ·H, and hydrated electrons, which in turn can set off a cascade of free radical reactions. Gamma irradiation therefore leads to the formation of peroxides which can interact with the NP matrix, the drug, and sometimes the excipient. The impact can be on the drug itself or on the NP's chemical and physical states. Gamma irradiation could also alter the interaction of the NP with the incorporated drug. The sensitivity of the individual NP matrix polymer and encapsulated drug to gamma ray-induced degradation does not necessarily predict the sensitivity of the drug-loaded NP, and the formulation buffer components are known to influence peroxide formation. It is therefore necessary to confirm the gamma irradiation sensitivity of each individual NT formulation.

Gamma irradiation-induced reactions can degrade the potency and impurity profiles of the NT. Irradiation can also increase cross-linking or chain scissions in the NP matrix, thereby altering the amount of drug released or the drug release rate. For example, the viscosity of hydroxyethyl cellulose-containing formulations is known to decrease upon irradiation due to depolymerization. However, perhaps the most serious concern is that gamma irradiation can increase the toxicity of the formulation.

Yet, with all these drawbacks, gamma irradiation has continued to be a sterilization method of interest for NTs. Numerous studies to understand the impact of gamma irradiation on liposomes and SLNs have been published and are discussed below.

4.3.1. *Gamma irradiation of SLNs*

Chain scissions of NP materials is a major point of concern and is illustrated by Allémann *et al.*[22] This study reports that gamma irradiation did not alter the size of polylactic acid (PLA) NPs. However, polymeric chain scissions occurred in PLG NPs, causing an increase in the rate of the drug release from the NPs as well as in the rate of NP degradation.[28]

Gamma irradiation-induced degradation depends not only on the chemical nature of the polymer, but also on the physical form in which the polymer exists. Therefore, it is important to test the effect of gamma irradiation on the NPs themselves rather than on the constituent polymers. Such an observation has been made by Montanari *et al.*,[29] who studied the influence of ionizing radiation on modified release microparticulate drug delivery systems made of two types of PLG NPs. The effects of gamma irradiation on polymer stability of the raw polymers and related microspheres were evaluated. It was found that microspheres made from irradiated polymer materials showed a decreasing trend in molecular weight as a function of irradiation dose.[29] After 150 days, molecular weight decay was 25% in microspheres and 20% in raw polymers.

Even if the polymeric material is gamma irradiation-resistant, degradation can be induced by the drug, which may be sensitive to irradiation. Furthermore, oxygen in the formulation could sensitize the sample to degradation reactions. The peroxide-generating effect of gamma irradiation depends on the oxygen content of the sample, and vacuum during irradiation was shown to protect the product. The effects of gamma irradiation on the stability of microspheres made of a PLG copolymer loaded with 15% w/w of clonazepam was studied. The influence of clonazepam on PLG radiolysis mechanisms and the identification of possible irradiation markers were also investigated. Microspheres were prepared by means of a spray-drying method. Gamma irradiation was carried out either under vacuum or in air, at a dose of 25 kGy by using a ^{60}Co source. The stability of clonazepam-loaded microspheres was evaluated over a six-month period on the basis of the drug content and dissolution profile. Radiolysis mechanisms were investigated by using electronic paramagnetic resonance (EPR) analysis. The microspheres irradiated under vacuum were stable over the considered period of time. In the same study, we saw an illustration of the fact that the incorporated drug could impact the degradation of the polymeric NP material. After irradiation in air, the clonazepam release rate increased by approximately 10% and did not change further in the following period of storage. The EPR analysis showed some radicals arising from both the polymeric matrix and the active ingredient. Polymer/clonazepam spin transfer reactions suggested that clonazepam has a radio-stabilizing effect on the polymeric matrix.[30]

Other studies illustrated cases where gamma radiation induced free radicals in the encapsulated drug, which in turn induces the degradation of the NP matrix. Bittner *et al.*[31] investigated the effects of gamma irradiation on the formation of free radicals in the polymer and drug and the mechanism of chain scission with an extensive set of analytical methods. Tetracycline HCl-loaded microspheres were prepared from PLG by spray-drying. The spin probe 4-hydroxy-2,2,6,6-tetramethyl-piperidine-1-oxyl and the spin trap tert-butyl-phenyl-nitrone were co-encapsulated into the tetracycline-loaded and placebo particles. Gamma irradiation was performed at 26.9 and 54.9 kGy using a ^{60}Co source. The microspheres were characterized with respect to the formation of radicals and *in vitro* polymer degradation. EPR spectroscopy, gel permeation chromatography (GPC), differential scanning calorimetry (DSC), high-performance liquid chromatography (HPLC), gas chromatography-mass spectroscopy (GC-MS), and SEM were used for the characterization of the microspheres. Using EPR spectroscopy, gamma irradiation-induced free radicals were found within the tetracycline-loaded microspheres, while unloaded PLG did not contain radicals under the same conditions. An irradiation dose-dependent decrease in the molecular weight of PLG was observed, but the tetracycline content of the microspheres was not affected,[31] proving that the tetracycline transferred the free radicals to the polymer, which in turn was degraded. This study demonstrated that while the sensitivity of the NP and the drug to gamma irradiation can be predicted by observations on the NP and drug, respectively, by themselves, it is important to confirm the impact of the gamma irradiation on the final NT formulation.

Gamma irradiation could impact the encapsulated drug through the gamma irradiation-sensitive polymer material. 17-beta estradiol-loaded microparticles with PLG polymers were prepared by a modified spray-drying method and the effects of gamma irradiation on drug substance, polymer, and microparticles were investigated. Irradiation doses ranging from 5.1 to 26.6 kGy were applied using a ^{60}Co source. The 17-beta estradiol drug substance showed excellent stability against gamma irradiation in the investigated dose range, whereas microencapsulated estradiol seemed to be converted to conjugation products with PLG and to a lesser extent, to the degradation product 9,11-dehydroestradiol. The weight-average molecular weight of the PLG polymers decreased with increasing irradiation dose, while polydispersity indices remained nearly unchanged, compatible with a random chain scission mechanism in PLG degradation. The study concluded that the sterilization conditions need to be carefully adjusted for the final dosage form.[32]

Some cases exist where degradation in the drug substance and the NP polymer will occur independently as illustrated in a study by Montanari *et al.*[33] Here, the effects of gamma radiation (25 kGy, ^{60}Co source) on the stability of microspheres made of a PLG 50:50 copolymer and loaded with 40% bupivacaine were studied. The radiolytic mechanisms of bupivacaine and bupivacaine-loaded microspheres were investigated by EPR analysis. The relative abundance of bupivacaine radicals with PLG radicals was proportional to the electronic fractions of the components, implying that no spin transfers between bupivacaine and PLG had occurred during gamma irradiation.

It appears that even though the NP size and range are not impacted by the effects of gamma irradiation, chain scissions or increased cross-linking could alter the drug release rate from SLNs. The rate of release of the drug from gamma irradiation (25 kGy, ^{60}Co source)-sterilized ciprofloxacin HCl-loaded PLG NPs was evaluated.[34] The ratio of the particle size before and after freeze-drying, the ease of reconstitution of the NP suspensions, the drug release profiles of irradiated and non-irradiated NPs, and the antibacterial activity against *Pseudomonas aeruginosa* were tested. No significant difference in mean particle size was observed after gamma sterilization, but reconstitution proved more difficult and drug release was negatively influenced.

4.3.2. *Gamma irradiation and liposomes*

Gamma irradiation can induce the degradation of liposomes in liquid medium as well as in the lyophilized product. In the liquid phase, radiation can act directly on the liposomal bilayer or generate peroxides, which may set off a cascade degradation reaction of the phospholipids with the formation of lysophospholipids and free fatty acids among other degradants. Residual moisture associated with the bilayer headgroup in the lyophilized form can also induce the free radical degradation cascade.[35]

The most significant drawback of using gamma irradiation on liposomal NTs is the increase in toxicity. Stensrud *et al.*[36] have reported that gamma irradiation of liposomes composed of unsaturated phospholipids increased their toxicity as indicated by the cytotoxicity test and the growth inhibition test. However, the toxicity of liposomes composed of saturated phospholipids was not affected. It was therefore concluded that gamma irradiation was suitable only for the sterilization of liposomes composed of saturated lipids.[36] The relative stability of liposomes composed of saturated lipids to gamma irradiation has also been demonstrated by an investigation into the toxic effects of irradiated liposomes such as hemolysis of erythrocytes, platelet aggregation, and coagulation. The least amount of hemolysis was observed with irradiated liposomes composed of saturated or charged phospholipids.[37]

The extent of liposomal degradation due to gamma irradiation depends on its formulation. For example, the presence of free radical scavengers such as α-tocopherol can limit the amount of gamma irradiation-induced damage to lipids in liposomes,[38] while trehalose used as a cryoprotectant in solution is known to promote irradiation-induced damage.[39] The effects of gamma irradiation on liposomal preparation can also be mitigated by the appropriate choice of lipids as demonstrated in studies by Zuidam *et al.*[40]

Mohammed *et al.*[41] have reported a detailed investigation in which a freeze-dried liposomal formulation of a vaccine against *Mycobacterium tuberculosis* was successfully sterilized by means of gamma irradiation. The sterilization dose was 25 kGy and the dose rate was 2.8 Gy/min. The product was characterized by a series of tests before and after irradiation such as liposomal size, zeta potential, pH, viscosity, and moisture content. Nuclear magnetic resonance (NMR) and mass spectroscopy were utilized for the analysis of chemical degradants. The liposomes were stable for 12 months at a long-term storage temperature of 25°C and 60% relative humidity, and for six months at 40°C and 75% relative humidity. The low levels of degradation were attributed to the low levels of moisture, cationic components in the bilayer headgroup dimethyldioctadecylammonium, and the right cryoprotectants in the formulation. The study also showed that the inclusion of lysine in the formulation facilitated a superior level of antigen-specific spleen cell proliferation following radiation sterilization.[3]

4.3.3. *Mitigating the effects of gamma irradiation*

As the discussions in the above sections suggest, there are several ways of mitigating degradation due to gamma irradiation on NT formulations. Methods include the removal of water by freeze-drying, using gamma radiation-resistant excipients, and including free radical scavengers in the formulation. Furthermore, the removal of oxygen or application of vacuum during irraditation can reduce gamma irradiation-induced degradation by decreasing the formation of free radicals. The examples below illustrate efforts to control the effect of gamma irradiation on NTs.

The impact of vacuum during gamma irradiation on the stability of microspheres made of a PLG copolymer has been described under gamma irradiation of SLNs.[42] The observations point to the possibility that the control of oxygen during irradiation may improve the stability of the product.

Gamma irradiation at low temperatures can decrease degradation, indicating that temperature control can be a means of mitigating gamma ray-induced degradation of NTs. Fernández-Carballido *et al.*[43] reported on the effects of gamma irradiation on the characteristics of non-loaded PLG/PEG-derivative and indomethacin-loaded PLG/PEG-derivative microspheres with and without dry ice protection. Non-protected microspheres showed changes in their morphological surface, polymer glass transition temperature, molecular weight, and release rate of indomethacin after sterilization. However, microspheres sterilized with dry ice protection did not show any significant differences after gamma irradiation exposure.

The degradation effects of gamma rays can be reduced by optimizing the irradiation dose. Such an optimization study has been reported by Maksimenko *et al.*[44] The sterilization of doxorubicin-loaded poly(butyl cyanoacrylate) NPs prepared by anionic polymerization using either gamma irradiation or electron beam irradiation was investigated. The irradiation doses ranged from 10 to 35 kGy and *Bacillus pumilus* was used as the official test microorganism. The bioburden of the untreated formulation was found to be 100 CFU/g. Microbiological monitoring revealed that at this level of bioburden, an irradiation dose of 15 kGy was sufficient for the sterilization of the NPs. The formulation showed excellent stability with both types of irradiation in the investigated dose range. The irradiation did not influence the physicochemical parameters of the loaded and empty NPs, including the mean particle size, polydispersity, and aggregation stability. The molecular weights of the polymer as well as the polydispersity indices remained nearly unchanged. The drug substance was also stable to radiolysis.[44]

4.3.4. *Gamma irradiation remains a method of choice for NTs*

In spite of its possible negative impact on NTs, gamma irradiation remains one of the more successful methods of sterilization. In some cases, gamma irradiation is an acceptable choice because the ill effects are within a permissible window. For example, in the case described above, gamma irradiation and electron beam irradiation at a dose of 15 kGy were sufficient to sterilize doxorubicin-loaded poly(butyl cyanoacrylate) NPs. The formulation showed excellent stability to irradiation. The drug was stable to radiolysis, while the molecular weight of the NP polymer remained nearly unchanged.[44]

Hernandez *et al.*[45] reported an evaluation of the effect of gamma irradiation on the biopharmaceutical properties of PLG microspheres containing SPf66 malarial antigen. The *in vitro* results obtained showed that the gamma irradiation exposure had no apparent effect on SPf66 integrity and formulation properties such as morphology, size, and peptide loading. However, the release rate of SPf66 was slightly faster after gamma irradiation. Subcutaneous administration of irradiated and non-irradiated microspheres into mice induced a similar immune response (IgG, IgG1, and IgG2a levels).

Martins *et al.*[46] described a study in which the change caused by gamma irradiation was not significant or was at an acceptable level to support the shelf-life of the product. This study described the sterilization of NPs composed of a *Brucella ovis* antigenic extract in ovine mucosal surfaces entrapped in conventional and mannosylated poly(anhydride) NPs by the solvent displacement method. Sterilization at either 10 or 25 kGy did not modify the size, morphology, and antigen content of the NPs. Similarly, the integrity and antigenicity of the entrapped antigen were not affected by gamma irradiation. Yet, the 25-kGy gamma irradiation dose seemed to negatively influence antigen release from the carriers. However, all release patterns followed a similar tendency. Furthermore, the stability of the vaccine systems on lachrymal and nasal ovine fluids showed that gamma irradiation had no significant effects on them.[46]

In other cases, gamma irradiation is the method of choice because the alternative methods present more severe problems. For instance, different sterilization techniques such as gamma irradiation and autoclaving were evaluated in one study by Memisoglu-Bilensoy and Hincal.[21] Injectable nanospheres and nanocapsules prepared from amphiphilic β-cyclodextrin were assessed for their *in vitro* properties such as particle size distribution, zeta potential, drug entrapment efficiency, and drug release profiles after sterilization. It was found that among these techniques, gamma irradiation had no effect on particle size, drug loading, and drug release properties. Gamma irradiation altered the zeta potential on β-cyclodextrin, but this did not lead to any significant changes of NP properties. Autoclaving, on the other hand, caused massive aggregation of the NPs followed by precipitation. Sterile filtration was not feasible since NP sizes were larger than the filter pore size and the yield after sterilization was very low.

Although gamma irradiation causes physical and chemical changes to the NP structure, these changes are not detrimental in all cases as demonstrated in some studies summarized below. More significantly, irradiation has in some cases been shown not to change the toxicity profile of NPs. Some of the studies described here do not include the drug in NPs, and it must be emphasized that toxicity studies would need to be repeated or confirmed in NPs bound to the drug because the presence of the drug can alter the free radical-induced reactions and therefore, the toxicity of the irradiated NPs. Cottam *et al.*[47] reported that gamma irradiation altered the physical and chemical structures of drug-free NPs, but not in a detrimental fashion. The effect of sterilization by 2.5-Mrad gamma irradiation on the properties of polycaprolactone was also studied. Gamma irradiation significantly decreased the rate of degradation and also affected the appearance of the degraded specimens when they were examined by SEM. However, chondrocyte attachment and growth on polycaprolactone were not significantly affected.

In another study, hydrophobic and hydrophilic poly(ether–ester) copolymer microspheres were evaluated for their *in vitro* and *in vivo* biocompatibilities and degradation before and after sterilization. The *in vivo* biocompatibility of PEG terephthalate and poly(butylene terephthalate) microspheres was evaluated subcutaneously and intramuscularly for 24 weeks in rabbits. *In vivo* degradation of the microspheres was studied microscopically and compared to *in vitro* degradation. In both studies, biocompatibility of the microspheres of both polymers was shown. Extracts of these microspheres showed no cytotoxic reactivity in the *in vitro* cytotoxicity test. The sterilization of the microspheres by gamma irradiation did not affect their cytotoxicity.[48]

NPs with a mean size of 200–300 nm were prepared from alendronate covalently bound to PLG containing a free-end carboxylic group and sterilized by gamma irradiation. The PLG–alendronate conjugate did not induce either hemolysis or alterations of the plasmatic phase of coagulation, or cytotoxic effects on endothelial cells and trabecular osteoblasts.[49]

Puthli and Vavia[50] described a study to evaluate the effects of gamma sterilization on *in vitro* and *in vivo* drug release characteristics. Levonorgestrel microspheres were fabricated using PLG by a solvent evaporation technique. The microparticles passed the test for sterility and the absence of pyrogens. *In vitro* dissolution conducted on the product before and after gamma irradiation at 2.5 Mrad indicated no significant difference in drug release patterns.[50] A similar study was reported for levonorgestrel in NPs composed of biodegradable polymers of a natural origin.[51,52]

Gamma irradiation effects on aciclovir–PLG microspheres, with gelatin as the additive, were studied.[53] The microspheres were sterilized by gamma irradiation at a dose of 25 kGy. The irradiation caused no surface changes which could be observed by SEM. The microparticle mean diameter and drug loading efficiency were not affected. IR spectroscopy, DSC, and XRD showed no changes. The controlled release profiles of the drug-loaded microspheres for 73 days were not altered. However, a decrease in the molecular weight of the polymer was observed by GPC. The study concluded that the gamma sterilization method is adequate for this purpose.

Gamma ray sterilization of hydroxyethylcellulose/trehalose spherical microspheres containing the antibiotic vancomycin[54] showed that the irradiation did not alter the UV spectrum and drug release rate from the microspheres. There were no significant changes to the shape and size distribution of the microspheres.

Gamma irradiation has also been successfully used for the sterilization of microspheres with encapsulated ibuprofen.[55] The effect of the irradiation dose (25 kGy) exposure at low temperature on PLG (13,137 Da) microspheres was evaluated. Size exclusion chromatography revealed no significant changes in the polymer molecular weight. The results obtained after storing the sterilized microspheres at 4°C in a vacuum desiccator for one year showed no significant alterations in the ibuprofen release rate or in the molecular weight of the PLG.

4.3.5. *Impact on NT analytics*

NTs which have been sterilized by gamma irradiation will require extensive testing after the sample has been treated by the optimized sterilization procedure. The irradiation procedure is very likely to alter the NP matrix as well as the interactions between them. The action on the NP matrix could also alter the amount of bound drug and its rate of release. Irradiation can also dramatically modify the impurity profile of the NT. Furthermore, the stability of the NT could be significantly altered due to the formation of peroxides and methods suitable for monitoring the level of free radicals may provide useful predictive information. Most importantly, the toxicity profile of the NT might change and must be fully tested. NT characterization studies must be performed after the sample has been subjected to the optimized sterilization procedure.

4.4. *Gaseous ethylene oxide*

Ethylene oxide is often used as a sterilization agent for devices which are not heat-tolerant. However, its mechanism of action, i.e., alkylation of the protein and nucleic acid, can be a drawback because it chemically alters the drug or the NP matrix. Other disadvantages of ethylene oxide gas include its mutagenic properties and the possibility of toxic residues in treated materials, particularly those containing chloride ions. NTs sterilized with gaseous ethylene oxide must therefore be tested for residues of the gas and other toxic degradants. As in the case of gamma irradiation, treatment with ethylene oxide can set off a cascade of free radical reactions in the NP matrix material, in the drug, or both, and could result in the formation of toxic degradants. NTs which do not react with ethylene oxide (e.g., gold NPs) are best suited for sterilization by ethylene oxide.

Franca *et al.*[56] published a study of the physicochemical properties of two types of gold NPs upon sterilization by means of five different techniques. Cell viability and production of reactive oxygen species were also investigated. The results indicated that sterilization by ethylene oxide seems to be the most appropriate technique for both types of NPs. Ethylene oxide has also been used for the sterilization of PLG microspheres with all-*trans* retinoic acid,[57] and did not alter the drug release characteristics or the particle morphology.

4.4.1. *Impact on NT analytics*

NTs which have been sterilized using ethylene oxide will need to be tested with the same rigor required for gamma-irradiated NTs. The procedure could alter the NP matrix, the bound drug, as well as the interactions between them. The action on the NP matrix could also change the rate of release of the bound drug. Moreover, the impurity and toxicity profiles and stability of the NT could be impacted dramatically, and will all need to be fully tested.

4.5. *High hydrostatic pressure sterilization*

High hydrostatic pressure has been proposed as a sterilization procedure.[58] However, the method is not effective in eliminating spores and needs to be significantly improved before it can be used.

4.6. *Formaldehyde*

Although formaldehyde can, in principle, be used for sterilization, it is not a technique of choice for NTs. Sommerfeld *et al.*[13] studied the effects of sterilization of poly(butylcyanoacrylate) NP suspensions and lyophilized NP powders, and concluded that aseptic manufacturing may be preferable.

5. Comparison of Different Methods for Sterilization

There are several publications in which different NT sterilization procedures are compared when applied to the same product. These studies illustrated how different sterilization methods can affect the chemical and physical characteristics of the NPs, and the fact that all sterilization procedures will impact most NTs at least to a certain extent.

Sterilization by different methods has been compared for PLG microspheres with encapsulated transforming growth factor-β3[59] by factors such as ethylene oxide gas, radio-frequency glow discharge, or UV light. It was found that the release of the growth factor was significantly decreased by UV light, but not by either ethylene oxide gas or radio-frequency glow discharge.

Masson *et al.*[60] reported a comparative study of autoclaving, gamma irradiation, and filtration as means of sterilizing poly(ε-caprolactone) nanospheres formulated with Synperonic® PE/F68, Synperonic® PE/F127, or Cremophor® RH40. Autoclaving did not induce any modification of the polymer molecular weight or the diameter of Synperonic® nanospheres, but catalyzed some reactions with surfactants and tonicity agents. The authors concluded that this method could be used if the excipients are chosen with care. Gamma irradiation induced preservative degradation and viscosifiant depolymerization. The cross-linking of poly(ε-caprolactone) chains was observed, as reflected by a sharp increase of its molecular weight. Finally, sterile filtration was the only process which ensured the conservation of the physicochemical integrity of nanospheres. However, the smallest filter that could be employed was 0.45 μm since the NPs were of about 0.20 μm in size.

The impact of three different types of sterilization on carbon nanotubes has also been studied by micro-Raman spectroscopy. The effects of ethylene oxide, autoclaving, and UV irradiation were analyzed and it was found that NPs sterilized by UV irradiation samples were least impacted by the procedure.[61] The study concluded that it is crucial to test several methods in order to establish the specific type of sterilization to be performed for each particular NP.

6. Impact on Nanotherapeutic Analytics

The sterilization of NTs presents a challenge because common and relatively innocuous methods such as autoclaving or sterile filtration could profoundly degrade the NT. The observations thus far suggest that several different tests may need to be employed to identify the most appropriate sterilization method for a specific NT formulation, and considerable optimization of the chosen method may be required for each formulation. A minimum amount of chemical and/or physical degradation can be expected from the use of each sterilization method. Sterile filtration appears to have the least impact on formulations in cases where it can be used, and gamma irradiation and ethyl ene oxide cause the most extensive degradation, which includes the formation of cytotoxic degradants.

In addition, the sterilization procedures can modify the size distribution and morphology of the particles. The amount of bound drug, drug release rate, and formulation composition could be altered. The amount and type of impurities could also be changed dramatically. Therefore, analytical methods developed prior to the sterilization of the NT may not be suitable for the product after sterilization. The stability of the NT can be expected to be strongly impacted by the chosen sterilization method. Finally, the toxicity profile of the NT could also be subjected to changes.

Hence, NT samples which have been sterilized by the optimized procedure must be used even at the early analytical method development, testing, and characterization stages. Any extrapolation of analytical characterization and testing data obtained using pre-sterilization samples to the structure and impurity profiles of post-sterilized material must be performed with care.

7. Conclusions

NPs loaded with drugs are a complex system with multiple components, each of which could be susceptible to the damaging effects of sterilization procedures in different ways. High temperatures encountered in autoclaving are damaging to NPs consisting of matrices with low melting points, in addition to being detrimental to temperature-liable drugs, including biologics.

Filtration is useful for low viscosity formulations with small NPs which can pass through 0.2-μm filters, but the process may distort larger NPs and/or cause the loss of loaded drugs. Also, gamma irradiation and ethylene oxide are able to induce free radical cascade reactions and generate cytotoxic impurities.

The studies presented in this review demonstrated the strengths and weaknesses of each sterilization method. The implication is that during manufacture, it is crucial to choose the most suitable sterilization technology and optimize the conditions for the specific NT manufactured by procedures similar to the clinical manufacturing process. The optimized parameters need to be selected based on an acceptable level of product degradation, drug load loss, and the shelf-life of the product.

Finally, a suitable battery of tests needs to be utilized during developmental studies to understand the impact of the sterilization procedure on the short-term effects and the stability of the product during its shelf-life. The degradants formed immediately after some of these procedures are free radicals which are transient, but may initiate a cascade of reactions leading to low levels of multiple degradants that may be difficult to detect. Due to the negative consequences of sterilization on NTs, it is necessary to perform all characterizations after the sterilization procedure, and establish a stability program suitable for the detection of chemical degradants during the shelf-life of the NT.

References

1. Dobrovolskaia MA, Neun BW, Clogston JD, Ding H, Ljubimova J, McNeil SE. Ambiguities in applying traditional Limulus Amebocyte Lysate tests to quantity endotoxin in nanoparticle formulations. *Nanomedicine* 2010;5(4):555–562.
2. McNeil SE. Challenges for nanoparticle characterization. *Methods Mol Biol* 2011; 697:9–15.
3. Mohammed AR, Bramwell VW, Coombes AGA, Perrie Y. Lyophilisation and sterilisation of liposomal vaccines to produce stable and sterile products. *Methods* 2006;40:30–38.
4. Vauthier C, Bouchemal K. Methods for the preparation and manufacture of polymeric nanoparticles. *Pharm Res* 2009;26(5):1025–58.
5. Manjunath K, Reddy JS, Venkateswarlu V. Solid lipid nanoparticles as drug delivery systems. *Methods Find Exp Clin Pharmacol.* 2005;27(2):127–144.
6. Müller RH, Mäder K, Gohla S. Solid lipid nanoparticles (SLN) for controlled drug delivery — a review of the state of the art. *Eur J Pharm Biopharm* 2000;50(1):161–177.
7. Marty J, Oppenheim R, Speiser P. Nanoparticles. A new colloidal drug delivery system. *Pharm Acta Helv* 1978;53(1):17–23.
8. Al Khouri F, Roblot-Treupel L, Fessi H, Devissaguet JP, Puisieux F. Development of a new process for the manufacture of polyisobutlycyanoacrylate nanocapsules. *Int J Pharm* 1986;28:125–132.

9. Rolland A. Clinical pharmacokinetics of doxorubicin in hepatoma patients after a single intravenous injection of free and nanoparticle–bound anthracycline. *Int J Pharm* 1989;54:113–121.
10. Jin Y, Tong L, Ai P, Li M, Hou X. Self-assembled drug delivery systems. 1. Properties and *in vitro/in vivo* behavior of acyclovir self-assembled nanoparticles (SAN). *Int J Pharm* 2006;309(1–2):199–207.
11. Bos GW, Trullas-Jimeno A, Jiskoot W, Crommelin DJA, Hennink WE. Sterilization of poly(dimethylamino) ethyl methacrylate-based gene transfer complexes. *Int J Pharm* 2000;211:79–88.
12. Rollot J, Couvreur P, Roblot-Treupel L, Puisieux F. Physicochemical and morphological characterization of polyisobutylcyanoacrylate nanocapsules. *J Pharm Sci* 1986;75:361–364.
13. Sommerfeld P, Schroeder U, Sabel BA. Sterilization of unloaded polybutylcyanoacrylate nanoparticles. *Int J Pharm* 1998;164:113–118.
14. Zuidam NJ, Lee SS, Crommelin D. Sterilization of liposomes by heat treatment. *J Pharm Res* 1993;10(11):1591–1596.
15. Heiati H, Tawashi R, Phillips NC. Drug retention and stability of solid lipid nanoparticles containing azidothymidine palmitate after autoclaving, storage and lyophilization. *J Microencapsul* 1998;15(2):173–84.
16. Cavalli R, Caputo O, Carlotti ME, Trotta M, Scarnecchia C, Gasco MR. Sterilisation and freeze-drying of drug-free and drug loaded solid lipid nanoparticles. *Int J Pharm* 1997; 148:47–54.
17. Rössler B, Kreuter J, Scherer D. Collagen microparticles: Preparation and properties. *J Microencapsul* 1995;12(1):49–57.
18. Lin YS, Renbutsu E, Morimoto M, Okamura Y, Tsuka T, Saimoto H, Okamoto Y, Minami S. Preparation of stable chitosan-carboxymethyl dextran nanoparticles. *J Nanosci Nanotechnol* 2009;9(4):2558–2565.
19. Seyfoddin A, Shaw J, Al-Kassas R. Solid lipid nanoparticles for ocular drug delivery. *Drug Delivery* 2010;17(7):467–489.
20. Bondì ML, Craparo EF. Solid lipid nanoparticles for applications in gene therapy: A review of the state of the art. *Expert Opin Drug Delivery* 2010;7(1):7–18.
21. Memisoglu-Bilensoy E, Hincal AA. Sterile, injectable cyclodextrin nanoparticles: Effects of gamma irradiation and autoclaving. *Int J Pharm* 2006;311:203–208.
22. Allémann E, Leroux JC, Gurny R, Doelker E. *In vitro* extended-release properties of drug-loaded poly(DL-lactic acid) nanoparticles produced by a salting-out procedure. *Pharm Res* 1993;10:1732–1737.
23. Tsukada Y, Hara K, Bando Y, Huang CC, Kousaka Y, Kawashima Y, Morishita R, Tsujimoto H. Particle size control of poly(dl-lactide-co-glycolide) nanospheres for sterile applications. *Int J Pharm* 2009;370(1–2):196–201.
24. Konan YN, Gurny R, Allémann E. Preparation and characterization of sterile and freeze-dried sub-200 nm nanoparticles. *Int J Pharm* 2002;233(1–2):239–252.
25. Konan YN, Cerny R, Favet J, Berton M, Gurny R, Allémann E. Preparation and characterization of sterile sub-200 nm meso-tetra(4-hydroxylphenyl)porphyrin-loaded nanoparticles for photodynamic therapy. *Eur J Pharm Biopharm* 2003;55(1):115–124.
26. Darole PS, Hegde DD, Nair HA. Formulation and evaluation of microemulsion based delivery system for amphotericin B. *AAPS PharmSciTech* 2008;9(1):122–128.
27. Li C, Deng Y. A novel method for the preparation of liposomes: Freeze drying of monophase solutions. *J Pharm Sci* 2004;93(6):1403–1414.

28. Athanasiou KA, Niederauer GG, Agrawal CM. Sterilization, toxicity, biocompatibility and clinical applications of polylactic acid/polyglycolic acid copolymers. *Biomaterials* 1996;17:93–102.
29. Montanari L, Costantini M, Signoretti EC, Valvo L, Santucci M, Bartolomei M, Fattibene P, Onori S, Faucitano A, Conti B, Genta I. Gamma irradiation effects on poly(DL–lactictide-co-glycolide) microspheres. *J Controlled Release* 1998;56(1–3):219–229.
30. Montanari L, Cilurzo F, Valvo L, Faucitano A, Buttafava A, Groppo A, Genta I, Conti B. Gamma irradiation effects on stability of poly(lactide-co-glycolide) microspheres containing clonazepam. *J Controlled Release* 2001;75(3):317–330.
31. Bittner B, Mäder K, Kroll C, Borchert HH, Kissel T. Tetracycline-HCl-loaded poly(DL-lactide-co-glycolide) microspheres prepared by a spray drying technique: Influence of gamma-irradiation on radical formation and polymer degradation. *J Controlled Release* 1999;59(1):23–32.
32. Mohr D, Wolff M, Kissel T. Gamma irradiation for terminal sterilization of 17beta-estradiol loaded poly-(D,L-lactide-co-glycolide) microparticles. *J Controlled Release* 1999;61(1–2):203–217.
33. Montanari L, Cilurzo F, Conti B, Genta I, Groppo A, Valvo L, Faucitano A, Buttafava A. Gamma irradiation effects and EPR investigation on poly(lactide-co-glycolide) microspheres containing bupivacaine. *Farmaco* 2002;57(6):427–433.
34. Bozdag S, Dillen K, Vandervoort J, Ludwig A. The effect of freeze-drying with different cryoprotectants and gamma irradiation sterilization on the characteristics of ciprofloxacin HCl-loaded poly(D,L-lactide-glycolide) nanoparticles. *J Pharm Pharmacol* 2005;57(6):699–707.
35. Stensrud G, Redford K, Smistad G, Karlsen J. Effects of gamma irradiation on solid and lyophilised phospholipids. *Radiat Phys Chem* 1999;56:611–622.
36. Stensrud G, Mönkkönen J, Karlsen J. Toxicity of gamma irradiated liposomes: 2. *In vitro* effects on cells in culture, *Int J Pharm* 1999;178(1):47–53.
37. Stensrud G, Passia S, Larsena T, Sandset PM, Smistad G, Mönkkönen J, Karlsen J. Toxicity of gamma irradiated liposomes: 1. *In vitro* interactions with blood components, *Int J Pharm* 1999;178(1):33–46.
38. Sailer K, Viaggi S, Nusse M. Kinetics of radiation- and cytochrome c-induced modifications in liposomes analysed by FT-Raman spectroscopy. *Biochim Biophys Acta* 1997;1329:259–268.
39. Zuidam NJ, Lee SS, Crommelin DJ. Gamma-irradiation of non-frozen, frozen, and freeze-dried liposomes. *Pharm Res* 1995;12:1761–1768.
40. Zuidam NJ, Versluis C, Vernooy EAAM, Crommelin DJA. Gammairradiation of liposomes composed of saturated phospholipids. Effect of bilayer composition, size, concentration and absorbed dose on chemical degradation and physical destabilization of liposomes. *Biochim Biophys Acta* 1996;280:135–148.
41. Mohammed AR, Bramwell VW, Kirby DJ, McNeil SE, Perrie Y. Increased potential of a cationic liposome-based delivery system: Enhancing stability and sustained immunological activity in pre-clinical development. *Eur J Pharm Biopharm* 2010;76:404–412
42. Montanari L, Cilurzo F, Valvo L, Faucitano A, Buttafava A, Groppo A, Genta I, Conti B. Gamma irradiation effects on stability of poly(lactide-co-glycolide) microspheres containing clonazepam. *J Controlled Release* 2001;75(3):317–330.

43. Fernández-Carballido A, Puebla P, Herrero-Vanrell R, Pastoriza P. Radiosterilisation of indomethacin PLGA/PEG-derivative microspheres: protective effects of low temperature during gamma-irradiation. *Int J Pharm* 2006;313(1–2):129–135.
44. Maksimenko O, Pavlov E, Toushov E, Molin A, Stukalov Y, Prudskova T, Feldman V, Kreuter J, Gelperina S. Radiation sterilisation of doxorubicin bound to poly(butyl cyanoacrylate) nanoparticles. *Int J Pharm* 2008;356(1–2):325–332.
45. Igartua M, Hernández RM, Rosas JE, Patarroyo ME, Pedraz JL, Gamma-irradiation effects on biopharmaceutical properties of PLGA microspheres loaded with SPf66 synthetic vaccine. *Eur J Pharm Biopharm* 2008;69(2):519–526.
46. Da Costa Martins R, Gamazo C, Irache JM. Design and influence of gamma-irradiation on the biopharmaceutical properties of nanoparticles containing an antigenic complex from *Brucella ovis*. *Eur J Pharm Sci* 2009;37(5):563–572.
47. Cottam E, Hukins DWL, Lee K, Hewitt C, Jenkins MJ. Effect of sterilisation by gamma irradiation on the ability of polycaprolactone (PCL) to act as a scaffold material. *Med Eng Phys* 2009;31(2):221–226.
48. van Dijkhuizen-Radersma R, Hesseling SC, Kaim PE, de Groot K, Bezemer JM. Biocompatibility and degradation of poly(ether–ester) microspheres: *In vitro* and *in vivo* evaluation. *Biomaterials* 2002;23(24):4719–4729.
49. Pignatello R, Cenni E, Micieli D, Fotia C, Salerno M, Granchi D, Avnet S, Sarpietro MG, Castelli F, Baldini N. A novel biomaterial for osteotropic drug nanocarriers: Synthesis and biocompatibility evaluation of a PLGA-ALE conjugate. *Nanomedicine (Lond)* 2009; 4(2):161–175.
50. Puthli S, Vavia P. Formulation and performance characterization of radio-sterilized progestin-only microparticles intended for contraception. *AAPS PharmSciTech* 2009; 10(2):443–452.
51. Puthli S, Vavia P. Gamma irradiated micro system for long-term parenteral contraception: An alternative to synthetic polymers. *Eur J Pharm Sci* 2008;35(4):307–317.
52. Puthli SP, Vavia PR. Fabrication, characterization and *in vivo* studies of biodegradable gamma sterilized injectable microparticles for contraception. *Pharm Dev Technol* 2009;14(3):278–289.
53. Martínez-Sancho C, Herrero-Vanrell R, Negro S. Study of gamma-irradiation effects on aciclovir poly(D,L-lactic-co-glycolic) acid microspheres for intravitreal administration. *J Controlled Release* 2004;99(1):41–52.
54. Bartolotta A, D'Oca MC, Campisi M, De Caro V, Giandalia G, Giannola LI, Brai M, Calderaro E. Effects of gamma-irradiation on trehalose-hydroxyethylcellulose microspheres loaded with vancomycin. *Eur J Pharm Biopharm* 2005;59(1):139–146.
55. Fernández-Carballido A, Herrero-Vanrell R, Molina-Martínez IT, Pastoriza P. Sterilized ibuprofen-loaded poly(D,L-lactide-co-glycolide) microspheres for intra-articular administration: effect of gamma-irradiation and storage. *J Microencapsul* 2004;21(6): 653–665.
56. França A, Pelaz B, Moros M, Sánchez-Espinel C, Hernández A, Fernández-López C, Grazú V, de la Fuente JM, Pastoriza-Santos I, Liz-Marzán LM, González-Fernández A. Sterilization matters: Consequences of different sterilization techniques on gold nanoparticles. *Small* 2010;6(1):89–95.
57. Jeong YI, Song JG, Kang SS, Ryu HH, Lee YH, Choi C, Shin BA, Kim KK, Ahn KY, Jung S. Preparation of poly(DL-lactide-co-glycolide) microspheres encapsulating all-trans retinoic acid. *Int J Pharm* 2003;259(1–2):79–91.

58. Brigger I, Armand-Lefevre L, Chaminade P, Besnard M, Rigaldie Y, Largeteau A, Andremont A, Grislain L, Demazeau G, Couvreur P. The stenlying effect of high hydrostatic pressure on thermally and hydrolytically labile nanosized carriers. *Pharm Res* 2003;20(4):674–683.
59. Moioli EK, Hong L, Guardado J, Clark PA, Mao JJ. Sustained release of TGFbeta3 from PLGA microspheres and its effect on early osteogenic differentiation of human mesenchymal stem cells. *Tissue Eng* 2006;12(3):537–546.
60. Masson V, Maurin F, Fessi H, Devissaguet JP. Influence of sterilization processes on poly(epsilon-caprolactone) nanospheres. *Biomaterials* 1997;18(4):327–335.
61. Bellucci S, Chiaretti M, Onorato P, Rossella F, Grandi MS, Galinetto P, Sacco I, Micciulla F. Micro-Raman study of the role of sterilization on carbon nanotubes for biomedical applications. *Nanomedicine* 2010;5(2), 209–215.

Chapter 4

Endotoxin and Engineered Nanomaterials

Marina A. Dobrovolskaia and Scott E. McNeil*

Nanotechnology Characterization Laboratory
Advanced Technology Program
SAIC-Frederick Inc., NCI-Frederick
Frederick, MD 21702, USA

Endotoxin is a common contaminant in engineered nanomaterial formulations and can confound the results of efficacy and toxicity studies due to its high immunostimulatory potential. There is a growing need for methods to reliably detect, quantify, and remove endotoxin from nanoformulations prior to biological studies. This chapter provides a general overview of the endotoxin structure and its mechanism of immune cell activation, review methods traditionally used for endotoxin detection and quantification, and the applicability of these methods to engineered nanomaterial formulations. We also highlight challenges associated with endotoxin detection and quantification in nanomaterial formulations, and provide practical suggestions for overcoming these challenges.

1. Introduction

Endotoxin is an essential component of the Gram-negative bacterial cell wall.[1] It is commonly found in water, air, many laboratory reagents, on surfaces, on glassware, as well as on common laboratory tools such as balances and spatulas.[2] While appropriate for non-clinical applications, water purification systems, reverse osmosis membranes, deionized resins, glass surfaces, and chromatography columns are considered major sources of endotoxin in research materials synthesized in laboratories.[3] Endotoxin is a remarkably stable molecule; it

*Corresponding author: E-mail: marina@mail.nih.gov

tolerates high temperatures and is resistant to boiling and even autoclaving.[4] Any material, including engineered nanoparticles, may be contaminated with endotoxin during preparation, unless it is produced under aseptic conditions and with sterile, depyrogenated reagents. In addition, due to the nanoparticles' large surface-to-volume ratios, nanomaterial formulations are thought to be at particular risk of endotoxin contamination.[2]

Endotoxin is a potent immunostimulant that induces inflammatory responses at concentrations as low as the picomolar range.[1] It activates cells, which produce a broad range of secondary messenger molecules (cytokines, chemokines, eicosanoids, etc.) in response. Although almost all cell types can respond to endotoxin, the cells of the immune system (e.g., monocytes, macrophages, dendritic cells) are particularly sensitive.[5] Cytokines and other messenger molecules, which are produced by the immune cells in response to endotoxin, are responsible for many pathophysiological events, such as fever and hypotension.[1] High concentrations of endotoxin can induce immunostimulation leading to septic shock, tissue damage (primarily in the kidney, liver, and lungs), multiple organ failure, and disseminated intravascular coagulation (DIC), all of which are potentially life-threatening (Figure 1).

Endotoxin contamination is particularly problematic for pharmaceutical and medical device development. If a drug, a medical device, or a nanoparticle formulation undergoing safety investigation is contaminated with endotoxin and patient volunteers or test animals show signs of immunostimulation, it may be difficult or even impossible to determine whether the observed immunostimulation was caused by the test article (the drug, the device, or the nanoparticle formulation *per se*) or by the endotoxin. This is why it is important to reliably detect and accurately quantify endotoxin in formulations and devices, and to do so early in the developmental process. Recognizing this, regulatory bodies have established limits on endotoxin levels in pharmaceutical formulations and on the surfaces of medical devices.

Endotoxin activity is measured in endotoxin units (EU). In the United States, two documents have historically been used to guide industrial endotoxin testing for medical products — the United States Pharmacopeia (USP) General Chapter 85, "Bacterial Endotoxin Test (BET)"[6] and the Food and Drug Administration (FDA)'s "Guideline on Validation of the Limulus Amebocyte Lysate Test as an End-product Test for Human and Animal Parenteral Drugs, Biological Products, and Medical Devices" issued in 1987, and its subsequent 1991 amendment, "Interim Guidance for Human and Veterinary Drug Products and Biologicals, Kinetic LAL Techniques."[7] Effective as of July 12, 2011, the FDA has retired its guidance on endotoxin testing, and now refers drug manufacturers to the USP standard.[6] This standard directs that all drugs and biologics intended for use in medical applications via injection

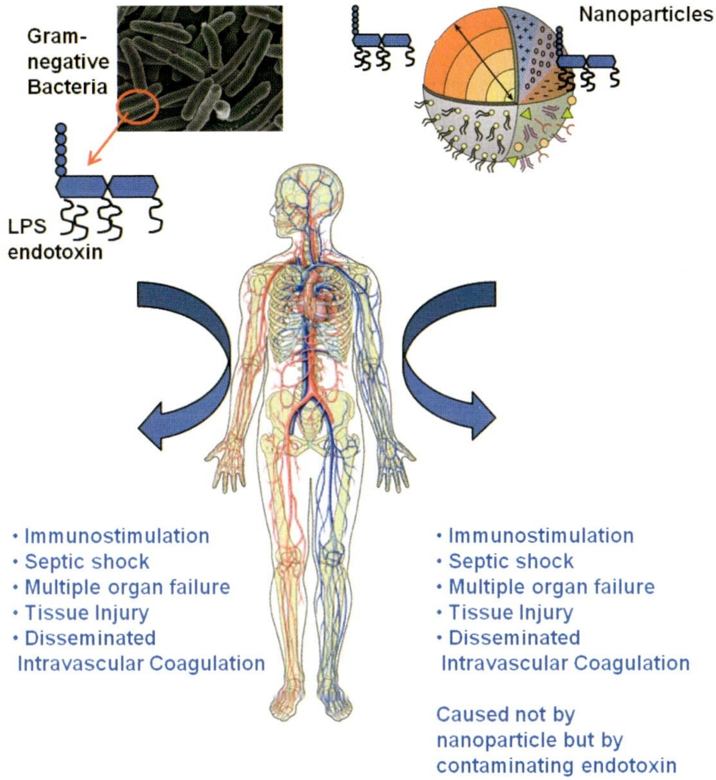

Figure 1. Endotoxin contamination of nanoparticles may cause undesirable immunotoxicity. Endotoxin is a component of the Gram-negative bacterial cell wall. When endotoxin is introduced into a body (e.g., via injection into humans or animals), it may trigger immunostimulation and adverse reactions such as septic shock, tissue injury, and multiple organ failure. Endotoxin may be introduced into nanoparticle formulations during synthesis or handling.

are required to be tested for endotoxin contamination and conform to the upper limit of 5 EU per kilogram bodyweight per hour of injection (5 EU/kg/h), except for intrathecally administrated drugs, for which the limit is 0.2 EU/kg/h.[6] The allowable upper endotoxin limit for the surfaces of medical devices is 0.5 EU/mL, except for devices intended to come into contact with cerebrospinal fluid, for which the limit is 0.02 EU/mL.[6]

This chapter describes the current understanding of endotoxin biology as it pertains to the safety of medical products. We review common approaches to endotoxin detection and quantification in pharmaceutical formulations and medical devices. We also share some of the challenges in applying traditional methods of endotoxin detection to engineered nanomaterial formulations, discuss ways to minimize endotoxin contamination of engineered nanomaterials, and suggest methods and strategies for overcoming these challenges.

2. What Endotoxin Is and Why It Is Important to Study in Nanoformulations

Endotoxin is a molecule of lipopolysaccharide (LPS), which is a part of the outer membrane of Gram-negative bacteria. The terms "endotoxin" and "lipopolysaccharide" are often used interchangeably, though endotoxin specifically refers to a crude (less pure) form of LPS containing other components of the bacterial cell wall.[3] The composition of LPS varies among various bacterial strains; however, the principal structure is the same in all bacteria, consisting of: 1) a lipid A part, composed of a disaccharide backbone and negatively charged phosphates and fatty acids, and 2) a polysaccharide part (Figure 2). The lipid A part is the biologically active, toxic part of LPS, while the polysaccharide part is responsible for LPS antigenic properties. Depending on the length of the polysaccharide part, LPS can be classified as rough (contains the inner core only), semi-rough (contains the inner core and the outer

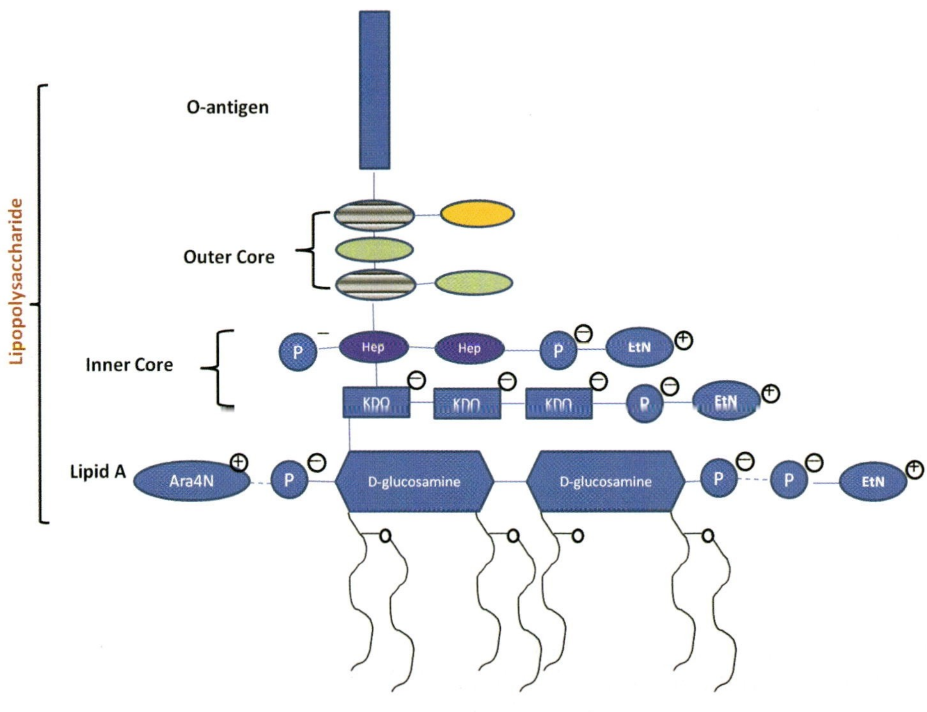

Figure 2. Schematic representation of bacterial lipopolysacharide (LPS; endotoxin) structure. This is based on the structure of LPS from *Salmonella typhimurium*. LPS from other Gram-negative bacteria may have different numbers of fatty acids and phosphates, as well as different compositions of the polysaccharide chain. The details of the chemical structure of LPS are described in Ref. 1.

Ara4N: 4-amino-4-deoxy-L-arabinose, EtN: ethanolamine, Hep: heptose, KDO: 2-keto-3-deoxyoctonic acid, P: phosphate.

core), or smooth (contains the inner core, outer core, and O-antigen). The biological activity of LPS in EU depends on the number of fatty acids and negatively charged phosphates on the lipid A part, and may vary from strain to strain.[1] An estimation of biological activity can be obtained by converting the mass of LPS into EU using the relationship that 100 pg of endotoxin is approximately 1 EU of activity. An accurate way of determining the biological activity or potency of endotoxin, however, is by the analysis of individual endotoxin in a bioassay, such as the Limulus Amebocyte Lysate (LAL) assay discussed below.[1]

The human immune system recognizes endotoxin through the extracellular endotoxin–receptor complex, which is made up of several proteins (TLR4, MD2, and CD14).[5] TLR4 is a member of the Toll-like receptor family; it does not bind endotoxin, but acts as the signal-transmitting arm of the endotoxin–receptor complex. MD2 is an extracellular adaptor protein, which binds LPS and brings it in close proximity to TLR4.[5,8] MD2 is often co-expressed and recycled onto cell membranes along with TLR4. CD14 is a GPI-anchored protein which binds endotoxin,[5,8] and also exists in a soluble form (sCD14). CD14, sCD14, and LPS binding protein (LBP) are thought to function as endotoxin recognition and binding proteins, where their function is to deliver endotoxin to MD2.[8] The binding of endotoxin by MD2 promotes the oligomerization of TLR4, which in turn triggers the recruitment of intracellular adaptor proteins and initiation of intracellular signaling (Figure 3). The ultimate result of TLR4 activation is the production of inflammatory cytokines and type I interferons. Almost all cells express TLR4, but antigen-presenting cells of the immune system (macrophages, monocytes, and dendritic cells) are the most sensitive. Secondary messenger molecules (cytokines, chemokines, nitric oxide, prostaglandins, leukotriens, etc.) produced by cells in response to endotoxin are responsible for fever, hypotension, blood coagulation, and other pathophysiological events.[1,9]

The inflammatory properties of endotoxin, particularly its ability to induce cytokine tumor necrosis factor-alpha (TNF-α), have been evaluated as cancer therapies.[10–13] Clinical trials investigating the safety and therapeutic efficacy of endotoxin as a cancer drug established the maximum tolerated dose of LPS in humans to be 50 EU/kg.[14] Although tumor regression was observed in various animal models,[10,11,13] systemic inflammation caused by LPS in clinical trials prevented its further clinical development.[14,15]

Endotoxin retains its inflammatory and anti-tumor properties when presented as a contaminant in nanomaterial formulations, and may thus confound the results of immunotoxicity and efficacy studies by causing false-positive

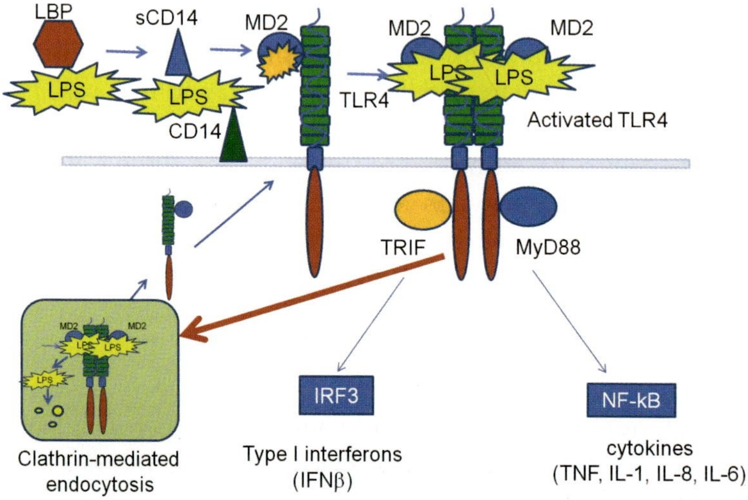

Figure 3. Simplified schematic representation of endotoxin recognition and signaling events. LPS is bound by LBP, sCD14, and CD14 and delivered to MD2. MD2 binds to LPS and brings it in close proximity to TLR4. The binding of the LPS by MD2 associated with TLR4 results in the oligomerization of TLR4, which triggers the activation of the endotoxin–receptor complex. The activation results in the recruitment of MyD88, which leads multiple signal transduction proteins to transcription factor NF-κB. The recruitment of another adaptor protein TRIF results in the activation of IRF3. Biomarkers of the MyD88-mediated arm of TLR4 signaling are inflammatory cytokines. The biomarker of TRIF (MyD88-independent) arm of TLR4 signaling is IFN-β. The activated receptor complex containing TLR4, MD2, and LPS is endocytosed through the clathrin-mediated pathway. This leads to the degradation of endotoxin and recycling of TLR4/MD2 back to the cell surface.

sCD14: soluble CD14, CD14: membrane-bound CD14, IFN: interferon, IRF: interferon regulatory factor, LPS: lipopolysaccharide, LBP: LPS binding protein, NF-κB: nuclear factor kappa B, MyD88: myeloid differentiation 88, TLR4: Toll-like receptor 4, TRIF: TIR-domain containing adapter-inducing interferon γ, TNF: tumor necrosis factor, IL: interleukin

results. Moreover, several recent studies performed in our laboratory[16] and by other investigators[17–21] have suggested that some nanoparticles may exaggerate the inflammatory properties of endotoxin. For example, cationic dendrimers can enhance the induction of leukocyte procoagulant activity (PCA) by endotoxin (Figure 4 and Ref. 16). Leukocyte PCA is an important component of the coagulation disorder DIC.[22]

To avoid ambiguous conclusions in nanoparticle preclinical safety and efficacy studies, it is important to keep endotoxin levels in nanoparticle formulations as low as possible, and below the FDA/USP mandated limits for clinical studies.

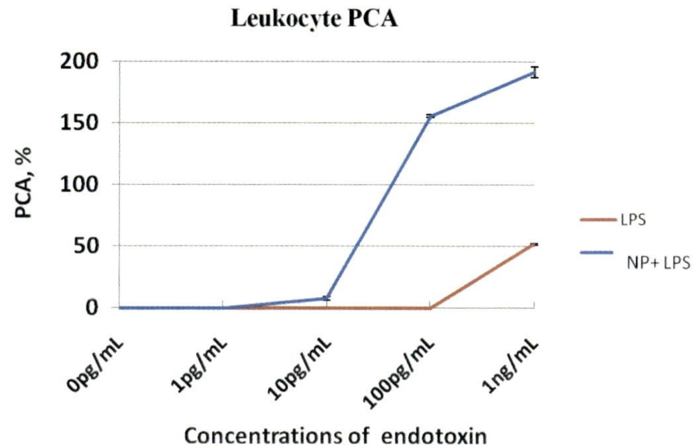

Figure 4. Enhancement of endotoxin inflammatory properties by certain nanoparticles. Leukocyte procoagulant activity (PCA) was estimated in human peripheral blood mononuclear cells (PBMCs) cultured for 24 h with various concentrations of endotoxin (LPS) alone or in combination with subcytotoxic concentrations of cationic nanoparticles (NP + LPS). Nanoparticles alone did not cause the induction of leukocyte PCA.

3. Methods Used for Endotoxin Detection and Quantification

3.1. *Traditional methods*

Since one of the common and rapid responses to systemic endotoxin exposure is elevated body temperature (fever), historically, endotoxin detection was performed using an *in vivo* test known as the rabbit pyrogen test (RPT). The RPT involves the injection of a test material into the ear vein of a rabbit, and the animal body temperature is monitored before the injection and every 30 min during 3 h after the injection (Figure 5).

Although the purpose of the test is the same, the experimental design, required number of test animals, allowed initial body temperature, determination of baseline temperature, and pyrogenicity results judgment algorithm (PRJA) are different in the United States, Japanese, Chinese, and European Pharmacopeias[23–29] (Table 1).

Not surprisingly, given the differences between the test instructions, a recent study comparing the test results for the same test substance evaluated according to protocols from different pharmacopeias revealed a great degree of variability.[23] The authors concluded that the determination of pyrogenicity based on the sum of total temperature increases is more accurate than judgment based on the animal number.[23] This study points out an important issue — even though there is an international consensus on the importance of the RPT, the

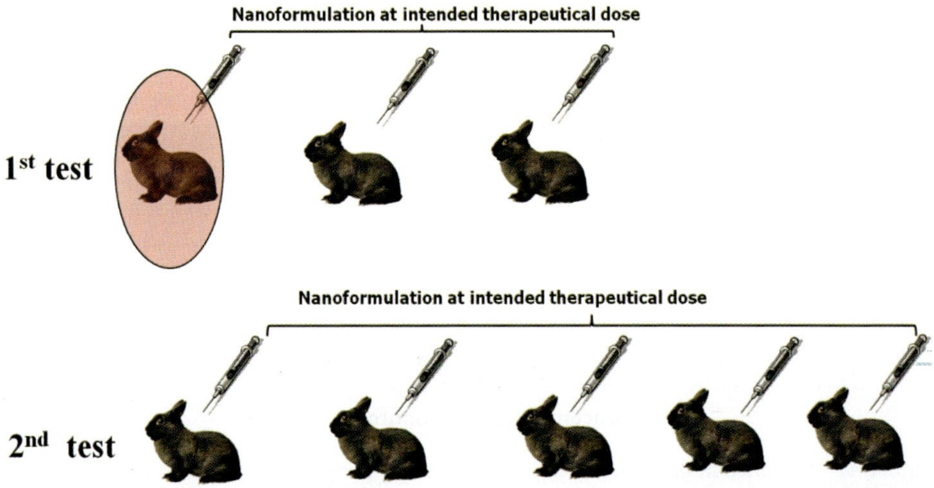

Figure 5. Schematic of the rabbit pyrogen test (RPT) conducted according to the USP General Chapter 151. The test is started with three animals. Animal body temperature is monitored before injection and every 30 min during 3 h after injection of the test formulation. If at least one animal shows a temperature rise of ≥0.5°C above the baseline temperature (for demonstration purposes, highlighted with a pink oval in the figure), the test continues with an additional five animals. At the end of the test, if three or more animals show elevated temperatures of ≥0.5°C above the baseline temperature, and the sum of all temperature rises is ≥3.3°C, the test formulation is concluded to be pyrogenic.

individual pharmacopeias describe methods for individual countries, thus the interpretation of test results may be different depending on the particular algorithm used. This observation highlights the necessity for further harmonization of this test method.[23,25]

Other concerns have been raised about the RPT. For example, a few studies have demonstrated significant differences between human and rabbit responses to certain pyrogens.[30,31] This issue called into question the relevance of the RPT for human drug product testing, and was further exaggerated by reports describing cases when some parenteral products caused fever in patients after successfully passing the RPT.[32,33] Nevertheless, the RPT is still in use and according to some reports, approximately 200,000 rabbits per year are used for pyrogen testing in the EU.[25] The RPT is still required for pyrogenicity testing of blood products in Japan and for medical devices in the United States.[34]

Since the discovery that the protein extract of amebocytes of the horseshoe crab *Limulus polyphemus*[35] clots in response to endotoxin, the LAL assay has rapidly gained popularity, and by 1980, it replaced the RPT for the most parenteral products in the United States.[6,7] In subsequent years, this shift from the RPT to the LAL assay was adapted by other countries as well.[27–29]

Table 1. Comparison of RPT designs and pyrogen result judgment algorithms from different pharmacopeias.

Pharmacopeia	Animal number	Determination of initial BT	Acceptable initial BT, °C	Positive test result
United States 32*	3 8	30 min prior to ITS	≤39.8	If at least 1 animal BT > 0.5°C If ≥ 3 animals show BT > 0.5°C and the sum temperature rise ≥ 3.3°C
European 6.0**	3 6 9 12	Mean of two measurements at 30 min interval within 40 min prior to ITS	38.0–39.8	If sum temperature rise ≥ 2.65°C If sum temperature rise ≥ 4.3°C If sum temperature rise ≥ 5.95°C If sum temperature rise ≥ 6.6°C
Japanese XIV***,¶	3 5	Mean of two measurements at 30 min interval within 40 min prior to ITS	≤39.8	If 2 or 3 animals show BT > 0.6°C If ≥ 2 animals show BT > 0.6°C
Chinese III*	3 8	Mean of two measurements at 30 min interval within 40 min prior to ITS	38.0–39.6	If at least one animal BT > 0.6°C If ≥ 4 animals show BT > 0.6°C and the sum temperature rise ≥ 3.5°C

The requirement for animal number, the number of tests, directions for determination of initial body temperature, requirements for acceptable initial body temperature, and the criteria for positive test results from the United States, European, Chinese, and Japanese Pharmacopeias are summarized. The number following the name of the individual pharmacopeia refers to the volume.

*The test is conducted in two phases; the second phase includes five animals, but judgment is based on the total number of animals from both phases (3 + 5 = 8).

**The test includes four phases; three animals in each phase, and judgment is based on total number of animals from individual phases.

***The test is conducted in two phases; the second phase includes five animals like in the USP, however unlike the USP, judgment is based on animals in the individual phase.

¶The Japanese Pharmacopeia XV requires nine animals in total (3 + 6).

BT: body temperature, ITS: injection of test sample.

The table is based on Refs. 23–29.

The LAL assay exists in three main formats known as the gel-clot, chromogenic, and turbidity tests. The gel-clot is the oldest method; the chromogenic and turbidity assays were developed later. Both the chromogenic and turbidity LAL assays are available in both endpoint and kinetic formats. The turbidity kinetic LAL assay is the most sensitive LAL assay available, with a lower limit of detection of 0.001 EU/mL. There is also the recombinant factor C (rFC) assay, which features only the endotoxin-responsive arm of the LAL cascade.[36–38] Although the rFC assay shares a similar sensitivity with the LAL[36] and is not susceptible to beta-glucan interference issues,[37,38] it is not as broadly used as the LAL assay due to the absence of regulatory guidelines.

Although the LAL assay is very sensitive to endotoxin, unfortunately, it is not a very specific assay, in that beta-glucans can also trigger the activation of the clotting cascade and result in false-positives for endotoxin. This creates an issue for products with synthesis and/or manufacture involving filtration through cellulose–acetate filters, since these are a known source of beta-glucans.[39] For example, nanoparticles filtered through such filters show clear contamination with beta-glucans. Accurate testing of such particles for endotoxin requires separating beta-glucan effects on the clotting of the LAL proteins from that due to endotoxin (Table 2). Although there are no formal thresholds for allowable amounts of beta-glucans in drugs or medical devices, recent data demonstrating immunostimulation by a combination of low, otherwise non-stimulatory doses of endotoxin and beta-glucans,[40] suggests that an awareness of beta-glucans in test formulations may prove helpful for the evaluation of immunostimulation. Many other factors can also interfere with the LAL assay, resulting in either overestimation (enhancement) or underestimation (inhibition) of endotoxin (Table 3). Another limitation of this assay

Table 2. Example of beta-glucan interference with endotoxin detection in nanoparticle formulations. Several nanoparticle formulations were tested by our laboratory using the kinetic turbidity LAL assay under regular conditions, or in the presence of the Glucashield™ buffer, which blocks LAL interaction with beta-glucans. All formulations were manufactured using cellulose–acetate filters. The formulations indicated with asterisks are cGMP formulations.

Formulation	Regular LAL testing, EU/mg	LAL testing in the presence of Glucashield™, EU/mg
1	1.170	0.431
2	0.776	0.336
3*	0.120	<0.001
4*	0.140	0.070
5*	0.160	0.080

Table 3. Sources of interference with the LAL assay and solutions to overcome them.

Inhibition		Enhancement	
Source	**Solution**	**Source**	**Solution**
EDTA (chelates Mg^{2+})	Add $MgSO_4$	Glucans	Reconstitute lysate in Glucashield™ buffer or use recombinant factor C assay; avoid cellulose filters, filtered tips and pipettes with cotton plugs
Na-citrate (chelates Mg^{2+})	Add $MgSO_4$	CMC	Same as above
Ca^{2+} (competes with Mg^{2+})	Addition of low conc of EDTA (1mM)	Trypsin	Heat inactivate 75°C for 15 min before testing
Serine protease inhibitors (inhibit cascade)	If proteinaceous in nature, heat at 75°C for 15 min to inactivate	Detergents	Case specific; avoid detergents in sample
Polypropylene (tubes and tips absorb endotoxin)	Use borosilicate glass tubes; 2nd choice: polystyrene tubes	Proteins	Inactivate with heat or digest with proteinase K
Any chelators of Mg^{2+}	Dilute samples in $MgSO_4$		
Water (leachables from storage bottles)	Test multiple batches		

CMC: carboxymethyl cellulose, EDTA: ethylene diamine tetra acetic acid, $MgSO_4$: magnesium sulfate.

is its inability to detect and estimate material-mediated pyrogenicity and other microbial pyrogens, thus the need for the RPT discussed above.

In addition, there is a concern about the preservation of growing horseshoe crabs. Although the crabs are returned to their natural habitat following the collection of hemolymph (a source of LAL reagents), this procedure results in the death of approximately 30,000 animals per year, which along with environmental pollution, use as fish bait, and habitat loss, are contributing to a decline in the horseshoe crab population.[25]

This data along with the limitations of the RPT described above have prompted many researchers to consider alternative methods for endotoxin and pyrogen detection.

3.2. *Alternative methods*

Beyond the LAL assay, there are several commercially available materials and kits for endotoxin detection. These methods include gel-staining techniques,

Western blotting, enzyme-linked immunosorbent assays (ELISAs), surface plasmon resonance (SPR), gas chromatography mass spectrometry (GC-MS), capillary electrophoresis-coupled mass spectrometry (CE-MS), and competitive electrochemical assays (CECAs). The key difference between these alternative methods and the traditional LAL assay is that the latter is based on the detection of the biological activity of endotoxin, while the alternative methods attempt to detect endotoxin itself. In the alternative methods, detection is based on the physical interaction between endotoxin and the assay reagents (e.g., dyes, antibodies, endotoxin binding proteins, peptides). In addition to these *in vitro* methods, there is an ongoing effort to replace the RPT with an *in vitro/ex vivo* test based on macrophage cytokine secretion in response to endotoxin. This test is called the macrophage activation test (MAT). The following section provides an overview of these methods and examples of their use for the detection and quantification of endotoxin in common (non-nanotechnology) samples. Examples of the use of these methods for endotoxin detection in nanoparticle formulations are discussed below after we review issues of nanoparticle interference with traditional methods.

3.2.1. *Gel-staining–based methods*

In the United States, there are three commercial sources of gel-staining reagents which are marketed for the detection of endotoxin — GlycoProfile® III (Sigma-Aldrich), Krypton® (Pierce), and Pro-Q® Emerald 300 (Molecular Probes). Such methods are based on the electrophoretic mobility of endotoxin, which is negatively charged, and separation by reducing polyacrylamide gel electrophoresis (PAGE). The gel is washed to remove excess surfactants and the endotoxin is oxidized using reagents provided as part of the kit. This step is followed by the incubation of the gel in a staining solution containing a dye which binds to the oxidized endotoxin. The main principle of the oxidation and staining steps is based on the periodic acid Shiff base method commonly used to study glycoproteins. The individual kits use proprietary modifications of this traditional method to increase the sensitivity of detection. Although good staining may be achieved, by design, these methods may also pick up glycoproteins if such proteins are present in test samples.

3.2.2. *Ligand binding-based methods*

This is a relatively large group of methods that achieve endotoxin recognition and binding by using specific proteins such as monoclonal antibodies or

LPS-binding proteins. These proteins may be used for endotoxin detection through techniques such as Western blotting, ELISA, CECA, and SPR, also known as Biacore based on the name of the instrument used for measuring SPR. The list of the proteins and peptides specific to endotoxin, their descriptions, and sources are summarized in Table 4.

Proteins and peptides with high affinity binding to endotoxin are used in many immunoassays for the capture and detection of endotoxin. For example, in ELISA (http://www.hyglos.de/en/products-services/products/elisa-based-endotoxin-detection-method.html), LBP is used to coat plates for endotoxin capture and another protein conjugated with horseradish peroxidase

Table 4. Summary of reagents which were shown to specifically bind endotoxin and can be used in protein-based endotoxin detection assays. These reagents are rather unique and not broadly available.

Reagent	Description	Source
LALF peptide	Limulus Amebocyte Lysate factor (LALF) is a protein present in LAL, which specifically binds to lipid A and neutralizes endotoxin. LALF peptide is part of this protein representing amino acids 31–52.	HC4031-01, www.cellsciences.com*
LBP	Recombinant LPS binding protein (LBP), which binds endotoxin.	CRL701 (human), CRL700 (mouse), www.cellsciences.com*
LBP peptide	A part of the LBP, representing amino acids 86–99 and binds endotoxin.	HC4030, www.cellsciences.com*
WN1 222-5	Monoclonal antibody against the LPS core, binds endotoxin.	HM6001, www.cellsciences.com*
Mab C20	Monoclonal antibody (clone 20) against the KDO part of LPS, binds KDO on the lipid A part of endotoxin.	HM2045, www.cellsciences.com*
ENP	Endotoxin-neutralizing protein, recombinant version of LALF, binds endotoxin.	www.acciusa.com*
Li-5-025	Synthetic peptide: K^{D}YSSSISSIRAC^{D}.	Not available commercially, Suzuki *et al.*[41]
Mp12	CD14 mimetic peptide: FHRWPTWPLPSP.	Not available commercially, Xu Z *et al.*[42]
Phage peptides	Peptides selected by phage display screening: ASFPPAF and SSHTISF.	Not available commercially, Guo X and Chen RR[43]

*Sources for ordering are shown for the information purposes only and not intended as an endorsement of the vendors. Reagents from other suppliers can also be used if their properties match those summarized above. ^{D}Y and C^{D} refer to the D-forms of the corresponding amino acid.

(HRP) is used for colorimetric detection of the captured endotoxin. The sensitivity of this method is 0.05 EU/mL, which is comparable to the endpoint chromogenic LAL assay (0.1 EU/mL) and gel-clot LAL assay (0.03 EU/mL) (http://www.hyglos.de/en/products-services/products/elisa-based-endotoxin-detection-method.html). In Western blotting, endotoxin is first separated electrophoretically using standard PAGE, then transferred onto a membrane and detected using an endotoxin-specific antibody followed by chemiluminescence. While the antibody for this application is available commercially, the sensitivity of the Western blot has not been reported. An endotoxin-neutralizing protein (ENP) labeled with a fluorophore was recently demonstrated to be a useful endotoxin-specific probe that measures changes in fluorescence polarization.[44] Although this method is specific to endotoxin, some researchers have raised concerns about its relatively low sensitivity, dependence on the temperature, and viscosity of the test sample.[45]

Another assay that involves the use of the ENP is CECA. In this assay, an endotoxin standard is conjugated to HRP and competes with endotoxin binding of the test sample to ENP-coated electrodes.[45,46] Although the sensitivity and design of this method have some advantages over alternatives, its utility is limited due to the absence of easily accessible instrumentation for performing the measurements. The electrodes and instrument to conduct electrochemical measurements are custom-made by the researchers who developed CECA.[45,46] Commercialization of this instrument may promote the broader use of CECA for endotoxin detection. Another interesting approach for endotoxin detection was proposed by Rangin and Basu.[47] This method is based on polymerized liposomes made of 1,3-diacetylenic lipids and endotoxin-specific ligands embedded in the lipid layer. The polydiacetylene component works as a bridge between the endotoxin-specific receptor incorporated into liposomes and the endotoxin in the test sample. The interaction between the ligand and the endotoxin results in a color change of the liposome from blue to red, and the intensity of the change is monitored spectrophotometrically.[47]

3.2.3. *Mass spectrometry-based methods*

Capillary zone electrophoresis (CE) and gas chromatography (GC) are high resolution bioanalytical techniques widely applied for the separation of complex biological mixtures. In CE, the separation of ionic species occurs based on their charge, hydrodynamic radius, and frictional forces. GC involves a liquid stationary phase and a gas mobile phase, and measures the retention time of various compounds. CE and GC have been proposed for the detection and quantification of 3-hydroxy fatty acids of the endotoxin lipid A core[48–51] and O-deacylated LPS,[52–54] respectively. To increase the sensitivity and resolution, both methods were coupled to mass spectrometry (MS). Electrospray MS,[51] and high field asymmetric waveform ion mobility MS (FAIMS)[50] have also been used in combination with CE, while tandem MS (MSMS),[54] quadrupole MS,[53] and electron impact MS[52] have been coupled with GC. While these methods are very specific to the unique charge-to-mass ratio of the endotoxin, their sensitivity is around 400 EU/mL. This is very low as compared to some traditional LAL assays, which are capable of detecting concentrations as low as 0.001 EU/mL, and some protein-based methods, which are able to detect up to 0.05 EU/mL. This is why CE-MS and GC-MS methods have been used in agricultural and environmental sciences, where the threshold for allowable endotoxin levels is much higher than it is for medical devices and drugs.

3.2.4. *Macrophage activation test*

Due to the high cost of animal studies, the large amounts of test materials required for *in vivo* tests, and the inability of the LAL assays to detect pyrogens other than Gram-negative bacterial endotoxin, an effort was undertaken to replace the RPT with an alternative method. Since the induction of cytokines, particularly IL-1 and IL-6, is recognized as a prerequisite of fever, an *in vitro* test using macrophage cell line MonoMac6 (MM 6), human whole blood, or human peripheral blood mononuclear cells (PBMCs), and the monitoring of the secretion of IL-1 or IL-6 in culture supernatants was proposed.[55,56] The test was originally developed and validated for the assessment of clinically approved drugs in Europe by the European Centre for the Validation of Alternative Test Methods (ECVAM),[24] and subsequently evaluated by the U.S. National Toxicology Program (NTP) Interagency Center for the Evaluation of Alternative Toxicological Methods (NICEATM) and the Interagency Coordinating Committee on the Validation of Alternative Methods (ICCVAM) (http://iccvam.niehs.nih.gov/docs/pyrogen/PrRevPanFinRpt.pdf; http://iccvam.niehs.nih.gov/docs/pyrogen/TMER/PyroTMER2008.pdf; http://iccvam.niehs.nih.gov/docs/pyrogen/BRD/PyroBRD2008.pdf).

The test procedure involves the incubation of the monocytic cell line MM-6, human whole blood, or purified PBMCs, with the test substance, the bacterial endotoxin (positive control), and PBS (negative control) for 24 h, followed by the collection of cell-free culture supernatants (Figure 6). The supernatants are either analyzed immediately by ELISA for the presence of cytokines IL-1β, IL-6, or TNF-α,[24,25,56] or stored at −20°C (http://iccvam.niehs.nih.gov/docs/pyrogen/PrRevPanFinRpt.pdf).

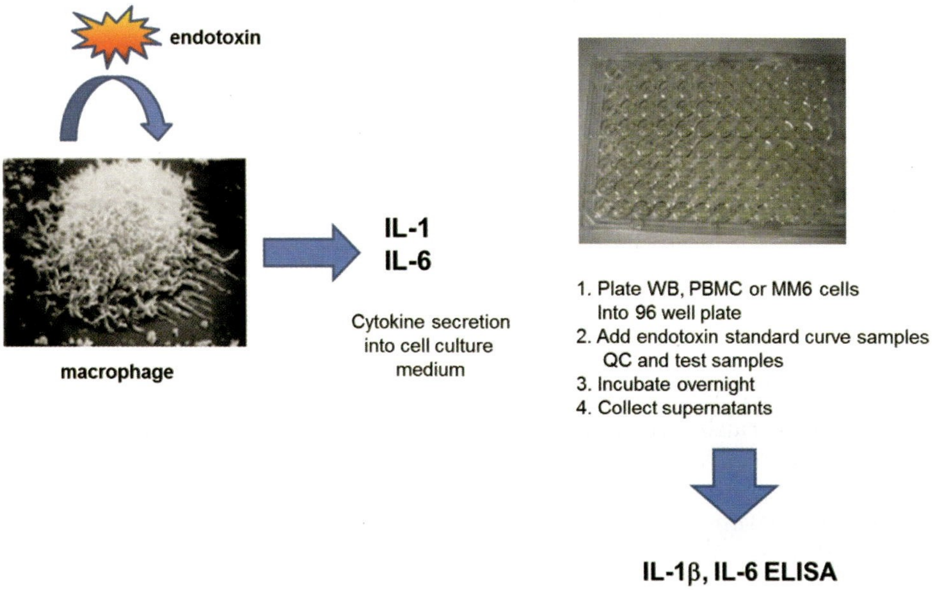

Figure 6. Schematic of the macrophage activation test (MAT). This is an *in vitro* test which involves the incubation of primary cultures (whole blood or PMBCs) or MonoMac-6 cell line with controls and test nanomaterials, followed by the collection and analysis of cell culture supernatants for the presence of cytokines IL-1β or IL-6 by ELISA.

Though the elevation of several cytokines is observed during fever response, several studies have suggested that IL-6 production by purified mononuclear cells is a better marker of pyrogencity than IL-1β and TNF-α.[24,31,40,57,58] Another study reported that measuring IL-8 secretion in whole blood is more predictive of the pyrogen response than measuring IL-6 production by purified mononuclear cells.[40] The use of human embryonic kidney (HEK) cells stably transfected with individual TLRs was considered as a test model for the MAT surrogate assay, but the idea was retired due to the variability of pyrogen-inducing substances and the diversity of signaling they initiate, making it impossible to predict the pyrogenic response for a broad panel of test substances using only one HEK–TLR transfectant.[40]

Like all the other methods described above, MAT has known limitations. For example, the sensitivity of MAT for the detection of endotoxin and its reproducibility were found appropriate in an interlaboratory study testing eight approved drugs.[24,25,58] However, taking a closer look at this method suggests that even though the sensitivity (0.5 EU/mL as described in European interlaboratory studies and 1.0 EU/mL established using MM-6 model in our laboratory) is appropriate to predict the *in vivo* pyrogen response, it is still 500 times lower than that of the most sensitive LAL assay (0.001 EU/mL).

Other limitations include wide donor variability in response to endotoxin and other pyrogens, as well as the presence of cytokine-inducing substances in fetal calf serum (FCS).[58] FCS is an essential component of *in vitro* cell culture, thus Kikkert *et al.* suggested screening multiple lots of FCS before the MAT is conducted. However, FCS is an expensive reagent, and screening multiple lots will substantially raise assay cost. In addition, a recent report pointed out that beta-glucans may influence MAT sensitivity to pyrogens.[40]

The above clearly shows that there is no one method which is ideal for the detection and quantification of endotoxin in test samples — none of the methods met all the requirements (i.e., predictive, sensitive, free of interference, and cost-effective). Below, we review the applicability of standard methods and some of the alternative methods for the detection and quantification of endotoxin in nanoparticle formulations.

4. Estimating Endotoxin in Nanoformulations

4.1. *Nanoparticle interference with traditional methods*

Engineered nanoparticles are a chemically diverse category of materials. As such, they have a very broad spectrum of properties. Some nanoparticles have catalytic properties, some scatter light or have absorbance at 405 nm (the detection wavelength of the chromogenic LAL test), and some interact with proteins and bind endotoxin. Particle size, aspect ratio, hydrophobicity, surface charge, and the presence of ligands on nanoparticle surfaces may all affect LAL testing. Many nanoparticles created in the research environment interfere with one or more LAL assays formats.[59,60] Nanoparticle interference with certain formats of the LAL assay are summarized in Figure 7.

4.2. *Evaluation of interference*

Since interference may complicate data interpretation, its detection and control are very important. The most appropriate way of evaluating interference is by including inhibition/enhancement controls (IECs) in the test method. For endotoxin detection assays, IECs are prepared by spiking known amounts of an endotoxin reference standard into test formulations. These samples are analyzed against a standard curve with the aim of understanding the ability of the assay to correctly measure the spiked standards (termed "spike recovery"). Additional samples, called quality controls (QCs), are prepared by spiking known amounts of an endotoxin reference standard into endotoxin-free water. Water does not interfere with the LAL assay, so if the assay is performing well, one should expect to see a complete recovery of the

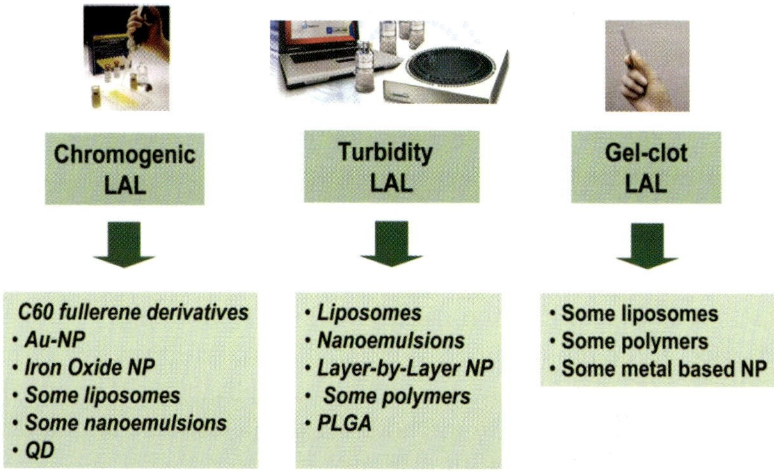

Figure 7. Nanoparticle interference with traditional LAL assays. Due to their diverse physicochemical properties, nanoparticles may interfere with chromogenic, turbidity, or gel-clot assays. Some nanomaterials interfere with more than one assay.

LAL: Limulus Amebocyte Lysate, NP: nanoparticles, PLGA: poly(lactic-co-glycolic acid), QD: quantum dots.

spiked endotoxin in the QCs. For example, a QC sample is shown as a blue bar on Figure 8(a). In this sample, 0.400 EU of the endotoxin standard was added per 1 mL of endotoxin-free water, and the total amount of endotoxin detected was 0.415 EU/mL. This corresponds to almost 100% recovery. When the same amount of endotoxin was spiked into citrate-stabilized colloidal gold nanoparticles, the total amount detected by the LAL assay (0.1 EU/mL) was less than 50% of its theoretical value (0.400 EU/mL) [Figure 8(a)]. This is called inhibition, and may be caused by a variety of effects, such as the endotoxin absorbing onto nanoparticle surfaces or nanoparticles binding proteins in the LAL extract, making it unavailable for interaction with the endotoxin. When 0.400 EU of the endotoxin reference standard was spiked into 1 mL of polymalic acid nanoparticles [Figure 8(a), PMA sample, brown bar], the amount actually detected by the LAL assay was 1.9 EU/mL. This type of interaction is called enhancement. Enhancement is often difficult to distinguish from contamination. IEC controls prepared with polyethylene glycol (PEG)ylated polymethacrylate (PMA) nanoparticles and PAMAM (polyamidoamine) dendrimers demonstrated accurate recovery [Figure 8(a)]. Now, when the IEC performances for the individual nanomaterials are known, we can conclude that the endotoxin levels measured in unspiked gold nanoparticles cannot be trusted as they may be underestimated due to inhibition. Similarly, the results of PMA nanoparticles are not reliable

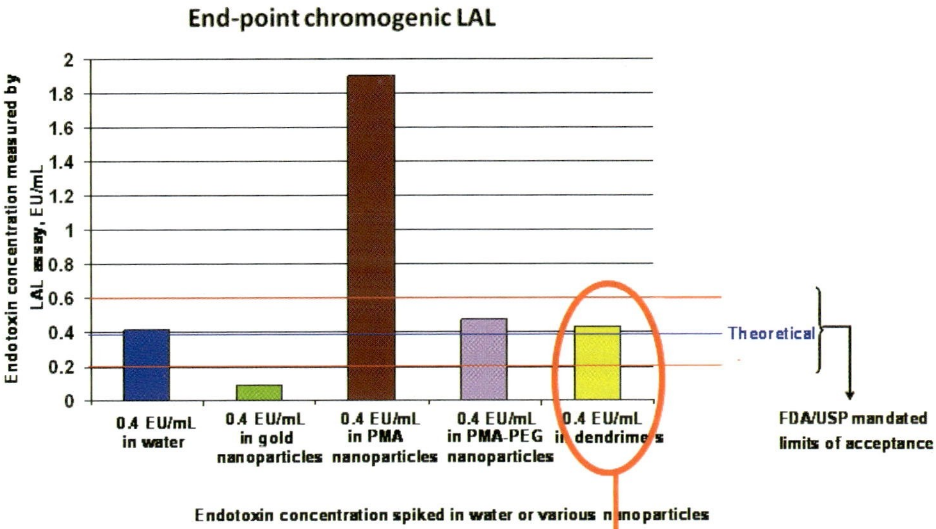

(a)

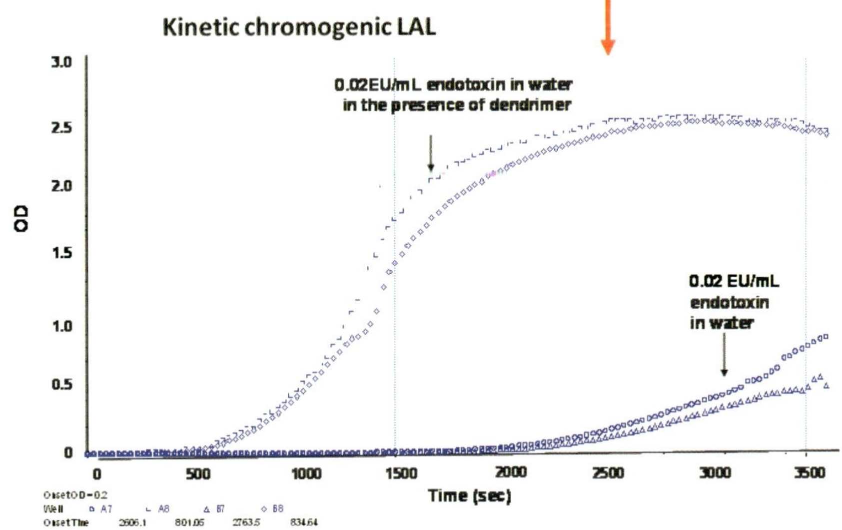

(b)

Figure 8. Application of quality and inhibition enhancement controls to monitor LAL assay performance for nanoparticles. (a) Four different nanoparticles (gold colloids, PMA, PMA–PEG, and PAMAM dendrimers) were used to prepare inhibition enhancement controls (IECs) and studied in on endpoint chromogenic LAL assay. The IECs were prepared by spiking known concentrations of an endotoxin reference standard into nanoparticle formulations. The quality control (QC) was prepared by spiking the same amount of endotoxin into LAL-grade water. All samples were evaluated against a standard curve prepared by spiking known concentrations of an endotoxin reference standard into LAL-grade water. Bars represent the

due to enhancement, and only results obtained in this test for dendrimers and PEG–PMA nanoparticles can be reported. We would also like to emphasize that separate IECs are required for each individual LAL assay, because nanoparticles that do not interfere with one LAL format may interfere with another assay. For example, in the experiment described above, PAMAM dendrimers demonstrated no interference with the endpoint chromogenic LAL assay. However, when the same particles were tested in the kinetic chromogenic LAL assay, the IEC revealed enhancement [Figure 8(b)]. The endotoxin amounts used to prepare QCs and IECs differ between endpoint chromogenic and kinetic turbidity LAL assays, because each assay has its own range and its own sensitivity. The range of the kinetic chromogenic LAL assay is narrower and the sensitivity is higher, and the concentration of endotoxin used for the QCs and IECs should fall in the middle of this range. It is therefore possible that dendrimers have potential interference with the LAL assay, but this interference is not picked up by the endpoint assay because this assay is for the detection of higher concentrations of endotoxin. The kinetic assay detects lower endotoxin concentrations, therefore the interference is detected with that assay. LAL assays with higher sensitivities are naturally also more sensitive to interference.[61] The mechanism(s) of such interference may be complex and it is often impossible to explain without additional extensive investigation. For example, some nanoparticles such as gold colloids may bind proteins and inhibit LAL by preventing interactions between endotoxin and enzymes in the lysate.[62,63] Nevertheless, for preclinical studies, being able to detect interference and thus develop a strategy to avoid it and correctly measure endotoxin concentrations, may be more important than understanding the actual mechanism of interference.

4.3. *How to choose an appropriate LAL format*

Our laboratory is frequently asked which of the available LAL assays is the best for nanoparticle samples. Unfortunately, there is no panacea as the answer is largely dependent on the properties of the individual nanoparticle formulation. Some nanoparticles interfere with one or more LAL assays (Table 5; nanoliposomes, polymer, and gold nanoparticles), while other formulations

Figure 8. (*Continued*) amount of endotoxin quantified in each sample. The blue line shows the theoretical amount of endotoxin spiked in the sample and the red lines outline the acceptable range of endotoxin spike recovery according to FDA guidelines. (b) The same dendrimers used in (a) were studied in a kinetic chromogenic LAL assay. The data is reproduced from Ref. 57 with permission from *Nanomedicine (Future Medicine)*.

Table 5. Nanoparticle performance in various LAL assays. This table shows several examples of commonly used engineered nanomaterials tested for endotoxin by three different formats of the LAL assay. For some nanoparticles, no test results can be reported due to the IEC sample revealing interference (enhancement or inhibition). Some nanoparticles interfered with one or more LAL assays, while others interfered with all three formats. Some formulations (e.g., dendrimers) demonstrated consistent results with all three formats, while others (e.g., nanocrystals) showed inconsistent results across all formats, though the IECs were acceptable.

Formulation	Gel-Clot LAL, EU/mg	Chromogenic LAL, EU/mg	Turbidity LAL, EU/mg
Colloidal Gold	Inhibition	Inhibition	0.799
Polymer A	Enhancement	Enhancement	16.1
PolymerA-PEG	Enhancement	0.53	21.4
Dendrimers	< 0.03	< 0.1	< 0.05
Nanocrystal	< 0.03	0.7	1.5
Liposome A	Enhancement	Enhancement	0.5
Liposome B	Inhibition	Inhibition	Inhibition

Gel-clot LAL has greater interference rate

Consistent between tests

Relying on gel-clot may lead to endotoxin overdose

do not interfere and produce comparable results between different LAL assays (Table 5; dendrimers and polymer–PEG nanoparticles after purification). However, even when all formal acceptance criteria are met and interference is not detected, test results for the same nanoformulation may be different depending on the format of the LAL assay used (Table 5, nanocrystal and polymer–PEG nanoparticles). Traditionally, the LAL assay can be trusted when the precision and accuracy of the calibration curve and QCs are within 25%, and the percent spike recovery for the IECs is between 50% and 200%.[6] The data presented in Table 5 is commonly observed with other engineered nanomaterials, i.e., even when formal acceptance criteria are met, the assay is still not accurate for some nanoformulations.[60,64–66] According to FDA guidelines, any of the LAL formats can be used, but in case of differences in results, the gel-clot assay would be the determining test. While this rule may make sense when testing optimized clinical grade formulations, the data from preclinical testing of nanomaterials with a broad spectrum of properties suggests that relying on the gel-clot assay may be problematic for some formulations. In the case of nanocrystals (Table 5), it does not detect endotoxin, while the turbidity and chromogenic LAL assays show an endotoxin level between 0.8 and 1.5 EU/mg. In this instance, if the nanocrystal is used at high mg/kg doses, relying on the gel-clot assay will result in an "overdose" of endotoxin.

This data raises a very important question: how one can then be assured that the results of the LAL assay for a given nanoformulation is accurate? There are no formal recommendations or guidelines available at this time. Below, we will describe the approach proposed by the NCI nanoparticle immunotoxicity workshop[59] and used in our laboratory for preclinical studies. In this approach, the choice of the LAL assay format is based on the physicochemical properties of the given test nanoformulation and analysis is performed using two different formats (Figure 9). For example, if the test nanoformulation is turbid, then analysis is performed using the chromogenic and the gel-clot LAL assays; if the formulation has an absorbance at 405 nm, then the analysis is performed using the gel-clot and the turbidity LAL assays; if there is no obvious potential interference, then any two formats can be used. At the end, the results from two different LAL assays are compared, and if they are in agreement (i.e., the test results are within 25%), endotoxin testing is concluded. However, if the test results differ by more than 25% (e.g., polymer–PEG NP in Table 5), then the results are verified using the RPT. This approach has been used in our laboratory for several years to determine if further purification/depyrogenation is required before biological tests.

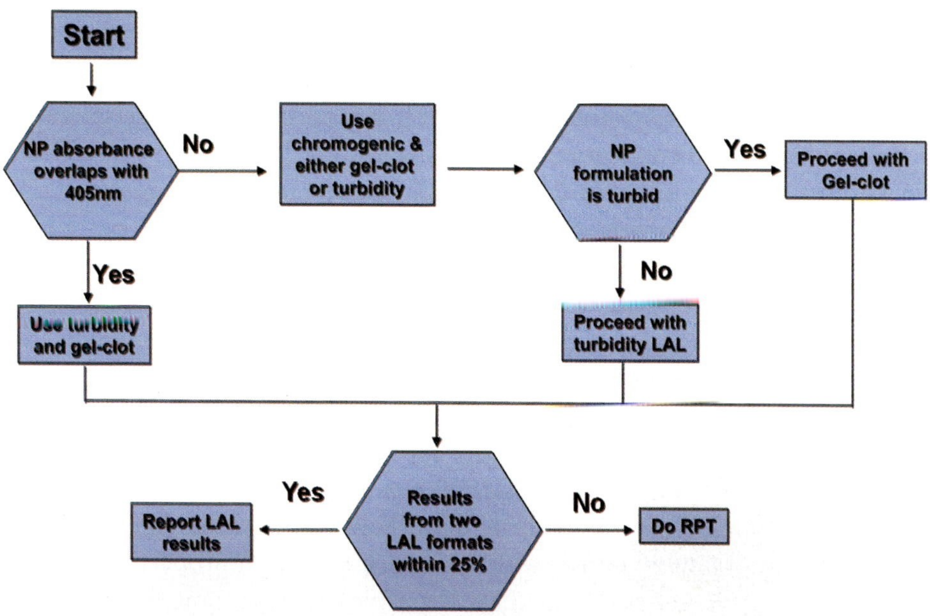

Figure 9. Flowchart for the selection of the appropriate LAL method. A thorough review of nanoparticle physicochemical properties is necessary before selecting an appropriate LAL assay format. Two different test formats are performed for each nanoparticle formulation and the results are compared. If the results are in agreement, then analysis is concluded. If the results differ by more than 25%, then the discrepancy is resolved with the RPT.

4.4. *Application of non-traditional methods*

Since many nanoparticles interfere with traditional LAL assays, we have attempted to qualify non-traditional methods for endotoxin detection in engineered nanomaterials. Our experience with using these methods and the method overviews are described further below.

4.4.1. *Gel-based assay*

In this approach, test samples are separated by traditional PAGE, followed by an oxidation step and staining using commercial dyes. The dyes are marketed as endotoxin-specific stains; however, since they operate by the interaction of the dye with the polysaccharide moiety of the endotoxin, the dyes also stain glycosylated proteins. We have identified three different commercially available stains: GlycoProfile® III from Sigma, Krypton® from Pierce, and Pro-Q® Emerald 300 from Molecular Probes. We next compared the endotoxin detection performance of these commercial stains. The standard curves of endotoxin from various commercial sources including an endotoxin reference standard as well as various nanoparticles and controls were subjected to SDS-PAGE analysis under denaturing conditions on 10–20% Tricine gels, stained, and visualized according to each manufacturer's protocol. All three stains showed similar results in terms of their ability to detect endotoxin (Figure 10). The sensitivity of these methods

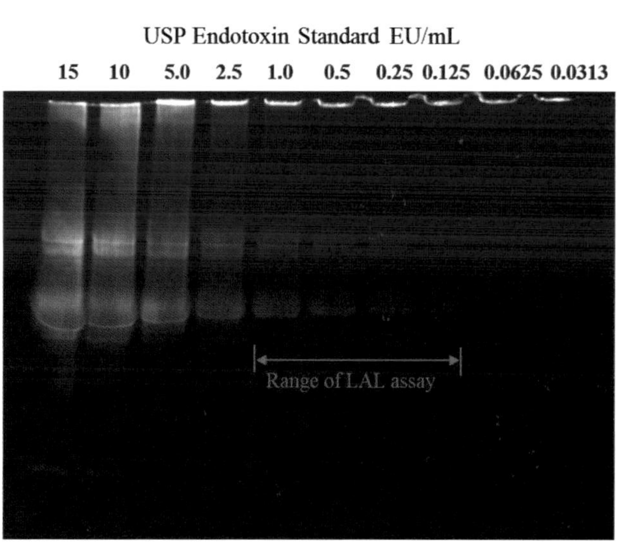

Figure 10. GlycoProfile® III stain analysis of LPS standard curve. The gel was loaded with a standard curve of LPS obtained from Cambrex and stained according to the manufacturer's protocol. Sources of reagents are named for information purposes and not to endorse any specific commercial vendor.

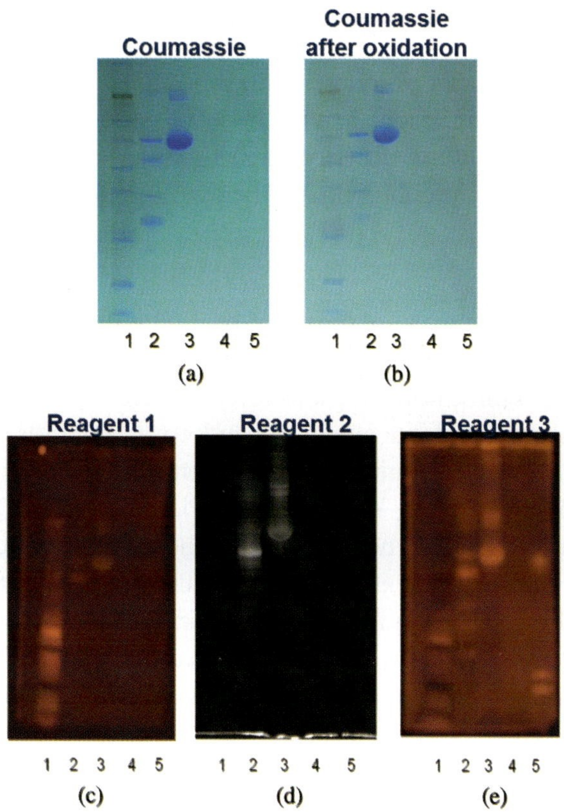

Figure 11. Analysis of LPS standard curve with various stains. (a) Colloidal blue stain, (b) colloidal blue stain after gel was fixed and oxidized with periodic acid, (c) commercial reagent 1, (d) commercial reagent 2, and (e) commercial reagent 3.
Lanes: 1) Protein ladder standard from Invitrogen, 2) Glycoprotein standard from Sigma, 3) LPS standard from Cambrex (1500 pg/mL), 4) LPS standard from Sigma (1500 pg/mL), 5) LPS ladder standard from Molecular Probes (40 μg/mL). Sources of commercial reagents are named for information purposes and not to endorse any specific commercial vendor.

was considerably below that of the LAL assays. In addition, all of the stains also showed non-specific staining of protein bands (Figure 11).

At first glance, it appeared as though the stains were very sensitive, as a nice standard curve appeared in the gel (Figure 10). However, upon closer look, the colloidal blue protein staining of duplicate gels revealed that the bands were not LPS, but in fact, protein contamination in the LPS standards that were analyzed (Figure 11). Standards from a different vendor did not contain this contamination and were used to attempt to reproduce the curves. The LPS was detectable only at higher concentrations, which was too high to be applicable for the testing of nanoparticle samples.

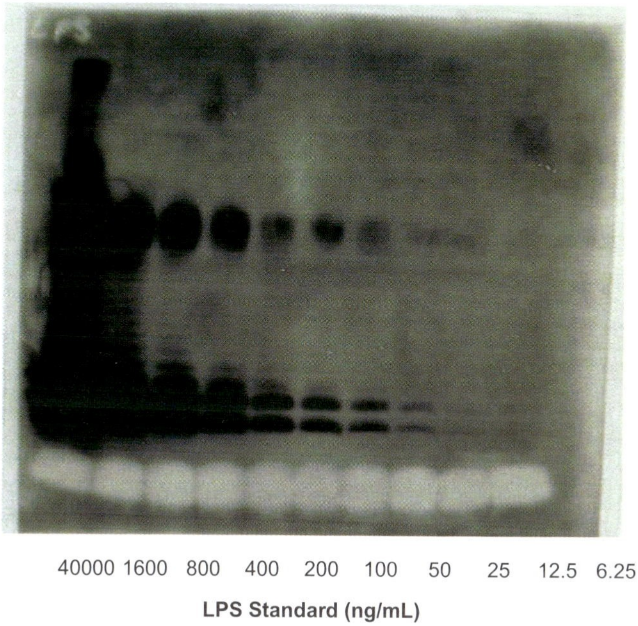

Figure 12. Western blot analysis of LPS standard curve. The blot was treated with a mouse anti-LPS primary antibody (2.5 μg/mL) from Hycult Biotechnology. The secondary antibody was an HRP-conjugated anti-mouse antibody (16 ng/mL) from Jackson ImmunoResearch. The blot was developed using an HRP substrate from Pierce. The ladder of bands represents the different forms of LPS. The LPS standard was from Molecular Probes and did not contain any additional proteins. Sources of reagents are named for information purposes and not to endorse any specific commercial vendor.

4.4.2. *Western blot-based assay*

Various primary antibodies to LPS and portions of LPS are commercially available. To evaluate these antibodies, standard curves of LPS were subjected to SDS-PAGE analysis under denaturing conditions on 10–20% Tricine gels, transferred onto PVDF (polyvinylidene fluoride) membranes, and developed according to the manufacturer's protocol. The best results were achieved using a primary anti-LPS antibody (Figure 12).

The anti-LPS antibody was the only antibody that demonstrated good sensitivity for all forms of LPS. The other antibodies which were tested — an anti-KDO antibody, an anti-lipid A antibody, and another anti-LPS antibody, had very poor recognition of all forms of LPS. The sensitivity achieved with a standard HRP substrate was considered low at concentrations of ng/mL, while that of LAL assays was in pg/mL. Therefore, while it is very specific to endotoxin, Western blot currently does not confer sufficient sensitivity to be used for endotoxin detection as compared to LAL assays.

5. Use of Traditional Endotoxin-Neutralizing Reagents in Nanoparticle Research

Endotoxin contamination of laboratory reagents and various commercial TLR ligands is a common problem in traditional immunology research. Various endotoxin-neutralizing reagents have historically been used to distinguish the inflammatory signaling of a given test reagent or a TLR ligand from a possible endotoxin contamination. Such reagents include polymyxin B (PMB), *Rhodobacter sphaeroides* LPS, and endotoxin-neutralizing antibody.[67–73] Some of them, e.g., PMB and neutralizing antibody, are species-independent, while others, e.g., *R. sphaeroides* LPS, is species-specific in that it antagonizes *E. coli* LPS in human and murine cells, but acts like an agonist in equine and hamster cells.[72,74] Due to these nuances and a broader availability, PMB is the most commonly used reagent. PMB is a polycationic, pore-forming peptide, which is used as an antibiotic to treat skin infections caused by Gram-negative bacteria. PMB is not specific to endotoxin (http://www.drugbank.ca/drugs/DB00781); however, due to its high positive charge, it electrostatically interacts with endotoxin and neutralizes its activity *in vitro*.[75] This property makes PMB a helpful tool in traditional research.[73,76] However, one has to be careful when using PMB so as to rule out endotoxin contamination of engineered nanomaterials. Below, we review a few case studies which demonstrate that PMB may interact with certain anionic nanoparticles and result in erroneous conclusions.

5.1. *Case study 1*

In this study, we used citrate-stabilized colloidal silver nanoparticles with the nominal size of 20 nm. This formulation induces the secretion of inflammatory cytokines in the MM-6 cell line *in vitro*. Analysis by gel-clot and kinetic turbidity LAL assays showed that the endotoxin level in this formulation was below the assay detection limit (< 0.03 and < 0.001 EU/mg, respectively). The induction of cytokine secretion by this formulation was not affected by the addition of 5 μg/mL PMB [Figure 13(a)]. To mimic endotoxin contamination, we spiked this formulation with several concentrations of ultrapure K12 *E. coli* LPS ranging from 1 pg/mL to 10 ng/mL. The addition of PMB to cell cultures inhibited only endotoxin-induced cytokines, but not those that were induced by the nanoparticle formulation [Figure 13(b)]. This data confirmed that if endotoxin is present in this formulation, the addition of PMB would be a useful control for detection. However, cytokine induction by silver nanoparticles may be an attribute of

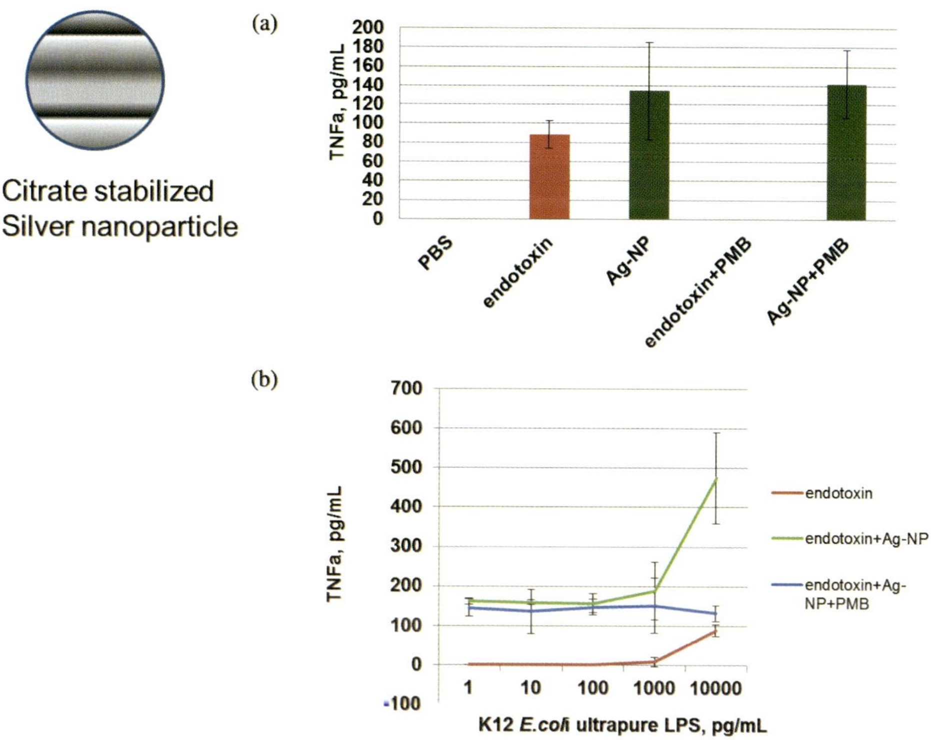

Figure 13. Use of polymyxin B (PMB) in immunological studies of anionic silver nanoparticles. (a) Cytokine secretion was studied in MM-6 cells treated for 24 h with PBS (negative control), endotoxin (10 ng/mL), and 20-nm citrate-stabilized silver nanoparticles (1 mg/mL) alone or in combination with 5 μg/mL of PMB (pH 6.8). TNF-α was measured in cell culture supernatants using Quantikine kit from R&D Systems. (b) To mimic endotoxin contamination, K12 *E. coli* LPS was added to silver nanoparticles at various concentrations ranging from 1 pg/mL to 10 ng/mL. Analysis and PMB concentrations were the same as described in (a).

Ag-NP: silver nanoparticles; endotoxin in silver nanoparticles were undetectable by gel-clot and kinetic turbidity LAL assays, PMB: polymyxin B.

silver ions released by the particles and not the property of their nano-size.[77]

If we look at the endpoint specific to the nanoscale size of silver colloids, PMB may interfere with this property. Citrate-stabilized silver nanoparticles used in our study were anionic, and since PMB interaction with endotoxin is based on charge, PMB was also bound to anionic silver nanoparticles, resulting in aggregation/agglomeration [Figures 14(a) and (b) and Table 6] and a change in zeta potential (Table 6). Therefore, one has to be careful when using PMB with anionic particles. The verification of nanoparticle physicochemical properties in the presence of PMB is needed prior to making conclusions about their biological properties.

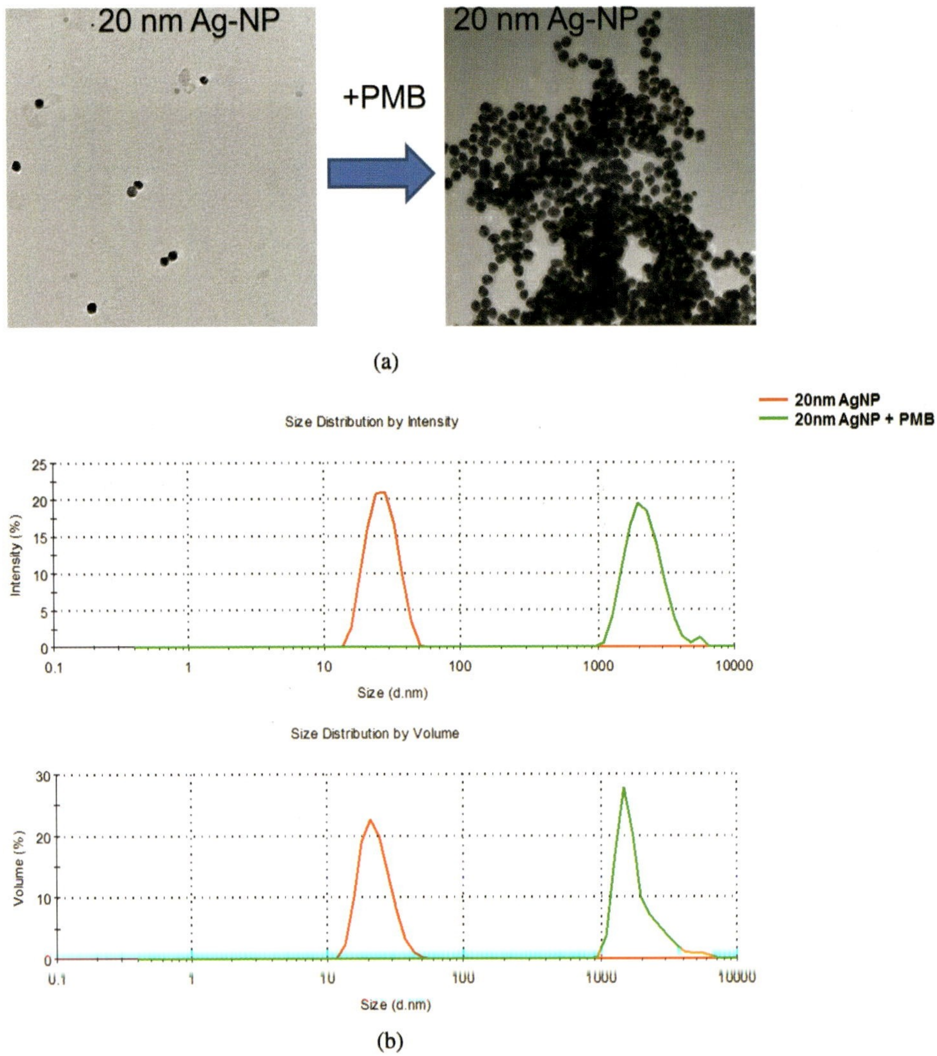

Figure 14. PMB affects the size of anionic silver colloids. (a) 20-nm citrate-stabilized silver colloids were sampled alone or in the presence of 5 μg/mL of PMB and their size was analyzed by transmission electron microscopy. (b) Particle hydrodymanic size with and without 5 μg/mL of PMB was analyzed by dynamic light scattering.

5.2. *Case study 2*

In this study, we used a formulation containing colloidal metal nanoparticles with zwitterionic coating. Endotoxin was undetectable in this formulation by the gel-clot LAL assay and approximately 0.1 EU/mg by the kinetic turbidity LAL assay. Due to the presence of both anionic and cationic ligands on the

Table 6. PMB affects the zeta potential of anionic silver colloids. Citrate-stabilized silver colloids with a nominal size of 20 nm were analyzed with and without 5 μg/mL PMB using Malvern zeta-sizer.

	Without PMB		With PMB	
Nanoparticle	**Size by DLS, nm (SD)**	**Zeta potential, mV**	**Size by DLS, nm**	**Zeta potential mV**
Ag-CIT (20 nm)	25.7 (0.1)	–47.8	—	9.4

Ag-CIT: citrate-stabilized silver nanoparticles, PMB: polymyxin B, SD: standard deviation.
* –: too large for the instrument to measure.

particle surface, the net charge of nanoparticles in this formulation was neutral. This formulation resulted in the induction of inflammatory cytokines, which was inhibited by PMB [Figure 15(a)].

At first glance, this data would be interpreted as traces of endotoxin present in this formulation are responsible for the observed cytokine induction. However, analysis of this formulation in the presence of other known endotoxin antagonists produced the opposite results. *R. sphaeroides* LPS is a known antagonist of *E. coli* LPS[67–71] and as expected, the addition of this reagent to cells treated with endotoxin resulted in an inhibition of endotoxin-mediated cytokine secretion [Figure 15(b)]. However, instead of an inhibition of cytokine secretion similar to what was observed with PMB, the addition of *R. sphaeroides* LPS to zwitterionic metal nanoparticles resulted in an enhancement of cytokine production [Figure 15(b)]. This data suggested that zwitterionic coating on the particle surface is responsible for the observed effect, because an enhancement of inflammatory response by *R. sphaeroides* LPS was also described for several other zwitterionic ligands.[78] Moreover, nanoparticles and endotoxin appeared to compete against each other in the induction of cytokines [Figures 15(c) and (d)]. Collectively, the data demonstrated that cytokine induction observed with this nanoformulation is not due to the contaminating endotoxin, and that relying only on the PMB experiment may be misleading.

6. Minimizing Contamination and Interference

The best way to minimize issues related to endotoxin contamination is to optimize the synthesis of nanomaterials in such a way that prevents contamination.[79] Following the steps below during synthesis would help to reduce endotoxin contamination in the final nanoformulation:

1) Wear gloves, spray gloves with 70% alcohol before touching tubes/reagents, etc., and change gloves often during the process.

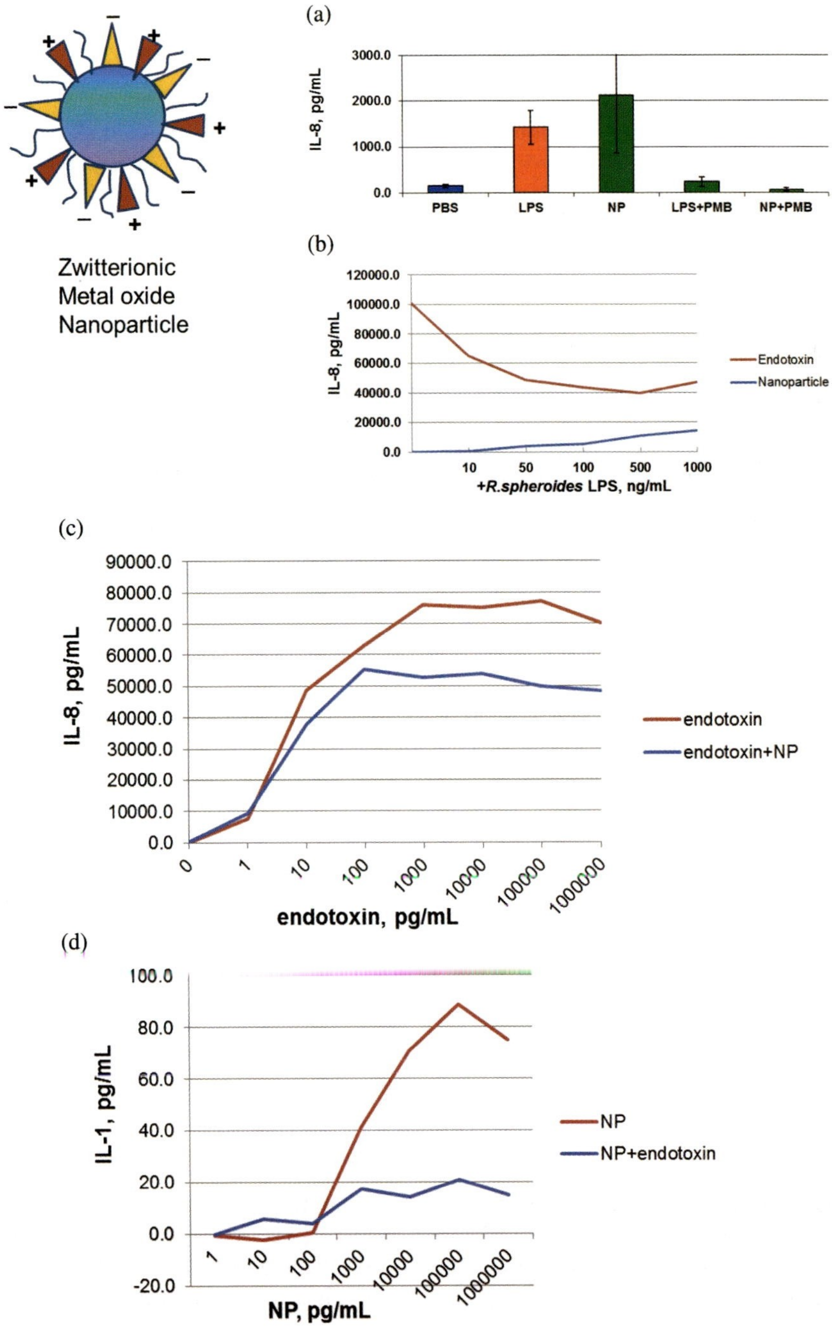

Figure 15. PMB may present misleading conclusions in certain formulations. (a) Cytokine secretion was studied in human PBMC cultures derived from healthy donor volunteers; the cells were treated for 24 h with PBS (negative control), endotoxin (10 ng/mL), or zwitterionic metal nanoparticles (1 mg/mL) alone or in combination with 5 μg/mL of PMB (pH 6.8).

2) Use pyrogen-free reagents and water (for example, Lonza LAL-grade water, ACC Pyroclear certified LAL reagent water, or any other source tested to be endotoxin-free).
3) Use sterile glassware and pipettes; to depyrogenize glassware, one can bake it at 250°C for 30 min or at 200°C overnight.
4) Ideally, the entire synthesis should be conducted in a laminar flow hood.
5) Avoid using cellulose-based filters as these may be a source of beta-glucans which interferes with the LAL assay.
6) Avoid breathing, coughing, and sneezing into tubes/reaction beakers. Avoid touching your face, wash hands and change gloves immediately if you come into contact with non-sterile items.

7. Purification and Depyrogenation of Nanoformulations

If a nanoparticle formulation contains plasmid DNA or recombinant proteins, it may have high levels of endotoxin contamination, simply because plasmid DNA and recombinant proteins are traditionally generated by fermentation using *E. coli*. For such formulations, additional endotoxin purification steps may be required prior to attaching these biological moieties to nanoparticles. If a final engineered nanomaterial formulation contains endotoxin at levels exceeding those allowable by regulatory agencies (5 EU/kg/h), purification (i.e., endotoxin removal) or depyrogenation (i.e., endotoxin inactivation) of the nanomaterial is needed prior to its use in biomedical applications or tests. Methods commonly used for endotoxin removal from recombinant proteins and nucleic acids are reviewed in detail elsewhere[80–84] and are briefly summarized in Table 7 below.

Not all of these methods are applicable to the vast number of nanomaterials. Nanoparticle stability is the main concern, and should be tested on a case-by-case basis. For example, a traditionally and commonly used

Figure 15. (*Continued*) (b) PBMCs were treated with 10 ng/mL of K12 *E. coli* ultrapure LPS or 1 mg/mL zwitterionic nanoparticles in the presence of increasing concentrations of *R. sphaeroides* LPS. (c) Cells were treated with various concentrations of K12 *E. coli* ultrapure LPS ranging from 1 pg/mL to 1 mg/mL alone or in combination with 1 mg/mL of zwitterionic nanoparticles. (d) Cells were treated with various concentrations of zwitterionic nanoparticles ranging from 1 ng/mL to 1 mg/mL alone or in combination with 1 ng/mL of K12 *E. coli* ultrapure LPS. Cytokines were measured in cell culture supernatants using MSD Th1/Th2 human inflammation kit. Cells from at least three donors were analyzed and showed similar trends. Shown is data from one donor. Each sample was analyzed in duplicate, shown is the mean response (%CV < 20).

NP: zwitterionic nanoparticles, PMB: polymyxin B, %CV: percent coefficient of variation.

Table 7. Methods for inactivation or extraction of endotoxin from biotechnology products. The methods are summarized from Refs. 64–68.

Depyrogenation	Purification
Baking at 200°C for at least 30 min	Triton X-114 extraction
Ethylene oxide	Ultrafiltration
Acid hydrolysis	Anion-exchange chromatography
Alkylation	PMB columns
Gamma irradiation	Affinity adsorption
Hydrogen peroxide gas plasma	Immunoaffinity chromatography
Soft hydrothermal process	Hydrophobic interaction chromatography

depyrogenation method is the exposure of a material to extreme temperature for prolonged time (at least 30 min at ≥200°C). This method is very efficient for inactivating endotoxin; however, this may not be tolerated by many engineered nanomaterials, especially those that contain biological components (e.g., protein cores), targeting ligands (e.g., antibodies or recombinant proteins), or drugs (e.g., therapeutic proteins). Nevertheless, this method is highly applicable for the depyrogenation of glassware and other common laboratory tools used in the synthesis of nanomaterials.

Ethylene oxide and gamma irradiation are common sterilization methods which have also been reported to inactivate endotoxin. However, many polymer-based nanoparticles do not tolerate ethylene oxide and some nanoparticles (e.g., silver colloids) do not tolerate gamma irradiation. (See Chapter 3 for more details on nanoparticle sterilization.) One may consider several scenarios when preparing nanoparticle formulations, for which excessive endotoxin levels are a concern: 1) test starting materials and depyrogenate those having high levels of endotoxin, 2) depyrogenate the final formulation, provided an appropriate (i.e., not destructive to the final nanoformulation) method has been identified, or 3) extract endotoxin from the final formulation. For example, Triton X-114 extraction described by Aida and Pubst[80] for recombinant proteins has been used successfully for endotoxin extraction from a very complex formulation consisting of a natural polymer, PEG, an endosome escape moiety, and a targeting nucleic acid moiety.[60]

Some examples of endotoxin levels in nanoformulations before and after purification or depyrogentaion are summarized in Figure 16. As one can see, the amount of endotoxin detected in the nanoformulation may vary substantially. Both purification and depyrogenation may help reduce these levels significantly (e.g., from 26.1 to 0.007 EU/mg). However, the success of purification and depyrogenation procedures is determined not by the fold of

	Before purification	After purification	Intended dose
Formulation 1:	21.4 EU/mg	0.04 EU/mg	1.0 mg/kg
Formulation 2:	26.1 EU/mg	0.007 EU/mg	0.03 mg/kg

	Before Irradiation	After Irradiation	Intended dose
Formulation 3*:	0.3 EU/mg	0.09 EU/mg*	400 mg/kg
Formulation 4:	0.08 EU/mg	<0.001 EU/mg	
Formulation 5:	5.97 EU/mg	0.17 EU/mg	1.5 mg/kg
Formulation 6:	63.3 EU/mg	0.15 EU/mg	

Figure 16. Example of endotoxin removal and inactivation in nanoformulations. Several nanoformulations were studied using a kinetic turbidity LAL assay before and after endotoxin purification or gamma irradiation. Levels of endotoxin detected in these formulations are shown before and after the depyrogenation procedures.

*Highlights the formulation in which endotoxin levels were reduced following depyrogenation procedure; however, this reduced level was still above the mandated acceptable endotoxin limit of 5 EU/kg/h.

reduction in endotoxin levels, but by the final amount of endotoxin in the nanoformulation at its intended dose level. For example, the amount of endotoxin in nanoformulation 3 (Figure 16) is not acceptable before irradiation but is acceptable after gamma irradiation, because this formulation is dosed at a low mg/kg level. In contrast, formulation 1 after gamma irradiation contains a lower amount of endotoxin per mg of nanoformulation than formulation 3, but this level is still unacceptable because nanoformulation 1 is administered at a high mg/kg level, and at that dose, the amount of endotoxin in the formulation exceeds the 5 EU/kg limit even after gamma irradiation.

8. Conclusions and Future Directions

Pyrogenicity assessment is an important initial component of preclinical characterization of engineered nanomaterials. It often coincides with the testing of the material's sterility. Pyrogens are substances that cause fever and are subjects for testing in the preclinical development of drugs, medical devices, and contrast agents intended to come into contact with blood. Two types of pyrogens exist — biological pyrogens (most commonly endotoxin or LPS) and chemical (material-mediated). While material-mediated pyrogenicity is harder to control, biological pyrogens, especially endotoxin, are ubiquitous unwanted "guests" in formulations containing recombinant biological products (e.g., peptides, antibodies, proteins). Engineered nanomaterials synthesized in academic laboratories during early development are often created using reagents and equipment that are not pyrogen-free. This is why they are more susceptible to contamination by endotoxin. Furthermore, endotoxin is very common in water, laboratory water filtration systems, reagents, and tools. Due to their large surface-to-volume ratios, nanoparticles may adsorb endotoxin to a greater extent than larger particulates. Endotoxin is a potent immunostimulant (i.e., it induces fever, hypotension, and cytokine responses at picogram levels), and when present in nanoparticle formulations, it may elicit immunostimulatory reactions and result in the unfair disqualification of a given formulation from further development. Protocols and challenges of testing nanomaterials for endotoxin have been described earlier.[16,60] Future work in the area of detection and quantifying endotoxin in nanoparticle formulations should include the development and harmonization of methods with high specificity and sensitivity to endotoxin, and are free of interfering factors. There is also a need for developing and harmonizing standard approaches and algorithms for selecting appropriate methods for a given nanoformulation.

References

1. Brade H, Opal S, Vogel SN, Morris D (eds.), *Endotoxin in Health and Disease*, Marcel Decker Inc., 1999, pp. 1–929.
2. Jones CF, Grainger DW. *In vitro* assessments of nanomaterial toxicity. *Adv Drug Delivery Rev* 2009;61:438–456.
3. Majde JA. Microbial cell-wall contaminants in peptides: A potential source of physiological artifacts. *Peptides* 1993;14:629–632.
4. Sharma SK. Endotoxin detection and elimination in biotechnology. *Biotechnol Appl Biochem* 1986;8:5–22.
5. Dobrovolskaia MA, Vogel SN. Toll receptors, CD14, and macrophage activation and deactivation by LPS. *Microbes Infect* 2002;4:903–914.
6. USP 30 NF 25. <85> Bacterial endotoxin test, Vol. 1, 2007.
7. US FDA CDER, CBER, CDRH, CVM. Guideline on validation of the Limulus Amebocyte Lysate test as an end-product endotoxin test for human and animal parenteral drugs, biological products, and medical devices, 1987.
8. Triantafilou M, Triantafilou K. The dynamics of LPS recognition: Complex orchestration of multiple receptors. *J Endotoxin Res* 2005;11:5–11.
9. Bosshart H, Heinzelmann M. Targeting bacterial endotoxin: Two sides of a coin. *Ann N Y Acad Sci* 2007;1096:1–17.
10. Berendt MJ, North RJ, Kirstein DP. The immunological basis of endotoxin-induced tumor regression. Requirement for a pre-existing state of concomitant anti-tumor immunity. *J Exp Med* 1978;148:1560–1569.
11. Berendt MJ, North RJ, Kirstein DP. The immunological basis of endotoxin-induced tumor regression. Requirement for T-cell-mediated immunity. *J Exp Med* 1978; 148:1550–1559.
12. Berendt MJ, Saluk PH. Tumor inhibition in mice by lipopolysaccharide-induced peritoneal cells and an induced soluble factor. *Infect Immun* 1976;14:965–969.

13. Freudenberg N, Joh K, Westphal O, Mittermayer C, Freudenberg MA, Galanos C. Haemorrhagic tumour necrosis following endotoxin administration. I. Communication: Morphological investigation on endotoxin-induced necrosis of the methylcholanthrene (Meth A) tumour in the mouse. *Virchows Arch A Pathol Anat Histopathol* 1984; 403:377–389.
14. Engelhardt R, Mackensen A, Galanos C. Phase I trial of intravenously administered endotoxin (*Salmonella abortus equi*) in cancer patients. *Cancer Res* 1991;51: 2524–2530.
15. Engelhardt R, Otto F, Mackensen A, Mertelsmann R, Galanos C. Endotoxin (*Salmonella abortus equi*) in cancer patients. Clinical and immunological findings. *Prog Clin Biol Res* 1995;392:253–261.
16. Dobrovolskaia MA, Patri AK, Potter T, Rodriguez J, McNeil SE. Dendrimer-induced leukocyte procoagulant activity depends on particle size and surface charge. *Nanomedicine (Lond)* 2012;7:245–256.
17. Inoue K. Promoting effects of nanoparticles/materials on sensitive lung inflammatory diseases. *Environ Health Prev Med* 2011;16:139–143.
18. Inoue K, Takano H. Aggravating impact of nanoparticles on immune-mediated pulmonary inflammation. *Sci World J* 2011;11:382–390.
19. Inoue K, Takano H, Yanagisawa R, Hirano S, Kobayashi T, Fujitani Y, Shimada A, Yoshikawa T. Effects of inhaled nanoparticles on acute lung injury induced by lipopolysaccharide in mice. *Toxicology* 2007;238:99–110.
20. Inoue K, Takano H, Yanagisawa R, Hirano S, Sakurai M, Shimada A, Yoshikawa T. Effects of airway exposure to nanoparticles on lung inflammation induced by bacterial endotoxin in mice. *Environ Health Perspect* 2006;114:1325–1330.
21. Shi Y, Yadav S, Wang F, Wang H. Endotoxin promotes adverse effects of amorphous silica nanoparticles on lung epithelial cells *in vitro*. *J Toxicol Environ Health A* 2010; 73:748–756.
22. Hiller E, Saal JG, Ostendorf P, Griffiths GW. The procoagulant activity of human granulocytes, lymphocytes and monocytes stimulated by endotoxin. Coagulation and electron microscopic studies. *Klin Wochenschr* 1977;55:751–757.
23. Du Y, Li XJ, Tan DJ. Comparison of temperature rise interpretations in the rabbit pyrogen test among Chinese, Japanese, European, and United States pharmacopeias and 2-2-2 theoretical models proposed by S. Hoffmann. *Innate Immun* 2011;17:486–495.
24. Hoffmann S, Luderitz-Puchel U, Montag T, Hartung T. Optimisation of pyrogen testing in parenterals according to different pharmacopeias by probabilistic modelling. *J Endotoxin Res* 2005;11:25–31.
25. Hoffmann S, Peterbauer A, Schindler S, Fennrich S, Poole S, Mistry Y, Montag-Lessing T, Spreitzer I, Loschner B, van Aalderen M, *et al.* International validation of novel pyrogen tests based on human monocytoid cells. *J Immunol Methods* 2005;298:161–173.
26. USP 30 NF 25. <151> Pyrogen test, Vol. 1, 2007.
27. Council of Europe Methods of analysis, Biological Tests, 2.6.8 Pyrogen. In: *European Pharmacopeia* 2007:164–165.
28. Society of Japanese Pharmacopoeia. Biological tests, 4.04 Pyrogen Test. *The Japanese Pharmacopeia* 2006;87–88.
29. Chinese Pharmacopoeia Commission, Beijing Chemical Industry Publishing House Appendix XIID Pyrogen Test *Pharmacopoeia of the People's Republic of China* 2005; 3:77–78.

30. Greisman SE, Hornick RB. Comparative pyrogenic reactivity of rabbit and man to bacterial endotoxin. *Proc Soc Exp Biol Med* 1969;131:1154–1158.
31. Nakagawa Y, Maeda H, Murai T. Evaluation of the *in vitro* pyrogen test system based on proinflammatory cytokine release from human monocytes: Comparison with a human whole blood culture test system and with the rabbit pyrogen test. *Clin Diagn Lab Immunol* 2002;9:588–597.
32. Dinarello CA, O'Connor JV, LoPreste G, Swift RL. Human leukocytic pyrogen test for detection of pyrogenic material in growth hormone produced by recombinant *Escherichia coli. J Clin Microbiol* 1984;20:323–329.
33. Poole S, Thorpe R, Meager A, Gearing AJ. Assay of pyrogenic contamination in pharmaceuticals by cytokine release from monocytes. *Dev Biol Stand* 1988;69:121–123.
34. Ochiai M, Yamamoto A, Naito S, Maeyama J, Masumi A, Hamaguchi I, Horiuchi Y, Yamaguchi K. Applicability of bacterial endotoxin test to various blood products by the use of endotoxin-specific lysates. *Biologicals* 2010;38:629–636.
35. Levin J, Bang FB. Clottable protein in Limulus: Its localization and kinetics of its coagulation by endotoxin. *Thromb Diath Haemorrh* 1968;19:186–197.
36. Alwis KU, Milton DK. Recombinant factor C assay for measuring endotoxin in house dust: Comparison with LAL, and (1–3)-beta-D-glucans. *Am J Ind Med* 2006;49:296–300.
37. Ding JL, Ho B. Endotoxin detection — from Limulus Amebocyte Lysate to Recombinant factor C. *Subcell Biochem* 2010;53:187–208.
38. Thorne PS, Perry SS, Saito R, O'Shaughnessy PT, Mehaffy J, Metwali N, Keefe T, Donham KJ, Reynolds SJ. Evaluation of the Limulus Amebocyte Lysate and recombinant factor C assays for assessment of airborne endotoxin. *Appl Environ Microbiol* 2010;76:4988–4995.
39. Henne W, Schulze H, Pelger M, Tretzel J, von Sengbusch G. Hollow-fiber dialyzers and their pyrogenicity testing by Limulus Amebocyte Lysate. *Artif Organs* 1984;8:299–305.
40. Kikkert R, Bulder I, de Groot ER, Aarden LA, Finkelman MA. Potentiation of Toll-like receptor-induced cytokine production by (1–3)-beta-D-glucans: Implications for the monocyte activation test. *J Endotoxin Res* 2007;13:140–149.
41. Suzuki MM, Matsumoto M, Yamamoto A, Ochiai M, Horiuchi Y, Niwa M, Omi H, Kobayashi T, Takagi T. Molecular design of LPS-binding peptides. *J Microbiol Methods* 2010;83:153–155.
42. Xu Z, Qian GS, Li Q, Feng QJ, Wu GM, Li KL. Screening of mimetic peptides for CD14 binding site with LBP and antiendotoxin activity of mimetic peptide *in vivo* and *in vitro. Inflamm Res* 2009;58:45–53.
43. Guo X, Chen RR. An improved phage display procedure for identification of lipopolysaccharide-binding peptides. *Biotechnol Prog* 2006;22:601–604.
44. Novitsky TJ, Ridge RJ, Sloyer JL. Detection and quantitation of endotoxin by fluorescence polarization, United States: Associates of Cape Code, Inc., 2001;6,282,807.
45. Priano G, Pallarola D, Battaglini F. Endotoxin detection in a competitive electrochemical assay: Synthesis of a suitable endotoxin conjugate. *Anal Biochem* 2007;362:108–116.
46. Priano G, Battaglini F. Use of an antimicrobial protein for endotoxin detection in a competitive electrochemical assay. *Anal Chem* 2005;77:4976–4984.
47. Rangin M, Basu A. Lipopolysaccharide identification with functionalized polydiacetylene liposome sensors. *J Am Chem Soc* 2004;126:5038–5039.

48. Li J, Cox AD, Hood DW, Schweda EK, Moxon ER, Richards JC. Electrophoretic and mass spectrometric strategies for profiling bacterial lipopolysaccharides. *Mol Biosyst* 2005;1:46–52.
49. Li J, Martin A, Cox AD, Moxon ER, Richards JC, Thibault P. Mapping bacterial glycolipid complexity using capillary electrophoresis and electrospray mass spectrometry. *Methods Enzymol* 2005;405:369–397.
50. Li J, Purves RW, Richards JC. Coupling capillary electrophoresis and high-field asymmetric waveform ion mobility spectrometry mass spectrometry for the analysis of complex lipopolysaccharides. *Anal Chem* 2004;76:4676–4683.
51. Forner-Cordero I, Navarro-Monsoliu R, Munoz-Langa J, Alcober-Fuster P, Rel-Monzo P. Use of a nanocrystalline silver dressing on lymphatic ulcers in patients with chronic lymphoedema. *J Wound Care* 2007;16:235–239.
52. Binding N, Jaschinski S, Werlich S, Bletz S, Witting U. Quantification of bacterial lipopolysaccharides (endotoxin) by GC-MS determination of 3-hydroxy fatty acids. *J Environ Monit* 2004;6:65–70.
53. Kirschner D, Que Hee SS, Clark CS. Method for detecting the 3-hydroxymyristic acid component of the endotoxin of Gram-negative bacteria in compost samples. *Am Ind Hyg Assoc J* 1985;46:741–746.
54. Pomorska D, Larsson L, Skorska C, Sitkowska J, Dutkiewicz J. Levels of bacterial endotoxin in air of animal houses determined with the use of gas chromatography-mass spectrometry and Limulus test. *Ann Agric Environ Med* 2007;14:291–298.
55. Daneshian M, von Aulock S, Hartung T. Assessment of pyrogenic contaminations with validated human whole-blood assay. *Nat Protoc* 2009;4:1709–1721.
56. Daneshian M, Wendel A, Hartung T, von Aulock S. High sensitivity pyrogen testing in water and dialysis solutions. *J Immunol Methods* 2008;336:64–70.
57. Gaines Das RE, Brugger P, Patel M, Mistry Y, Poole S. Monocyte activation test for pro-inflammatory and pyrogenic contaminants of parenteral drugs: Test design and data analysis. *J Immunol Methods* 2004;288:165–177.
58. Kikkert R, de Groot ER, Aarden LA. Cytokine induction by pyrogens: Comparison of whole blood, mononuclear cells, and TLR-transfectants. *J Immunol Methods* 2008;336:45–55.
59. Dobrovolskaia MA, Germolec DR, Weaver JL. Evaluation of nanoparticle immunotoxicity. *Nat Nanotechnol* 2009;4:411–414.
60. Dobrovolskaia MA, Neun BW, Clogston JD, Ding H, Ljubimova J, McNeil SE. Ambiguities in applying traditional Limulus Amebocyte Lysate tests to quantify endotoxin in nanoparticle formulations. *Nanomedicine (Lond)* 2010;5:555–562.
61. Roslansky PF, Novitsky TJ. Sensitivity of Limulus Amebocyte Lysate (LAL) to LAL-reactive glucans. *J Clin Microbiol* 1991;29:2477–2483.
62. Handley DA, Chien S. Colloidal gold: A pluripotent receptor probe. *Proc Soc Exp Biol Med* 1983;174:1–11.
63. Yamaguchi M, Kondo I. Immunoelectron microscopy of *Proteus vulgaris* by the plasma polymerization metal-extraction replica method: Differential staining of flagellar (H) and somatic (O) antigens by colloidal golds. *J Electron Microsc (Tokyo)* 1989;38:382–388.
64. Cliff RO, Kwasiborski V, Rudolph AS. A comparative study of the accurate measurement of endotoxin in liposome-encapsulated hemoglobin. *Artif Cells Blood Substit Immobil Biotechnol* 1995;23:331–336.
65. Harmon P, Cabral-Lilly D, Reed RA, Maurio FP, Franklin JC, Janoff A. The release and detection of endotoxin from liposomes. *Anal Biochem* 1997;250:139–146.

66. Poxon SW, Hughes JA. Characterization of endotoxin and cationic liposome interaction. *Pharm Dev Technol* 1999;4:135–143.
67. Aida Y, Kusumoto K, Nakatomi K, Takada H, Pabst MJ, Maeda K. An analogue of lipid A and LPS from *Rhodobacter sphaeroides* inhibits neutrophil responses to LPS by blocking receptor recognition of LPS and by depleting LPS-binding protein in plasma. *J Leukoc Biol* 1995;58:675–682.
68. Jagielo PJ, Quinn TJ, Qureshi N, Schwartz DA. Grain dust-induced lung inflammation is reduced by *Rhodobacter sphaeroides* diphosphoryl lipid A. *Am J Physiol* 1998;274: L26–31.
69. Jarvis BW, Lichenstein H, Qureshi N. Diphosphoryl lipid A from *Rhodobacter sphaeroides* inhibits complexes that form *in vitro* between lipopolysaccharide (LPS)-binding protein, soluble CD14, and spectrally pure LPS. *Infect Immun* 1997;65: 3011–3016.
70. Kirikae T, Schade FU, Kirikae F, Qureshi N, Takayama K, Rietschel ET. Diphosphoryl lipid A derived from the lipopolysaccharide (LPS) of *Rhodobacter sphaeroides* ATCC 17023 is a potent competitive LPS inhibitor in murine macrophage-like J774.1 cells. *FEMS Immunol Med Microbiol* 1994;9:237–243.
71. Kutuzova GD, Albrecht RM, Erickson CM, Qureshi N. Diphosphoryl lipid A from *Rhodobacter sphaeroides* blocks the binding and internalization of lipopolysaccharide in RAW 264.7 cells. *J Immunol* 2001;167:482–489.
72. Lien E, Means TK, Heine H, Yoshimura A, Kusumoto S, Fukase K, Fenton MJ, Oikawa M, Qureshi N, Monks B, *et al.* Toll-like receptor 4 imparts ligand-specific recognition of bacterial lipopolysaccharide. *J Clin Invest* 2000;105:497–504.
73. Unitt J, Hornigold D. Plant lectins are novel Toll-like receptor agonists. *Biochem Pharmacol* 2011;81:1324–1328.
74. Lohmann KL, Vandenplas M, Barton MH, Moore JN. Lipopolysaccharide from *Rhodobacter sphaeroides* is an agonist in equine cells. *J Endotoxin Res* 2003;9:33–37.
75. HsuChen CC, Feingold DS. The mechanism of polymyxin B action and selectivity toward biologic membranes. *Biochemistry* 1973;12:2105–2111.
76. Yonekawa K, Neidhart M, Altwegg LA, Wyss CA, Corti R, Vogl T, Grigorian M, Gay S, Luscher TF, Maier W. Myeloid related proteins activate Toll-like receptor 4 in human acute coronary syndromes. *Atherosclerosis* 2011;218:486–492.
77. Locht LJ, Smidt K, Rungby J, Stoltenberg M, Larsen A. Uptake of silver from metallic silver surfaces induces cell death and a pro inflammatory response in cultured J774 macrophages. *Histol Histopathol* 2011;26:689–697.
78. Yang H, Daun JM, Rose JR, Christ WJ, Gusovsky F, Chow JC. Examination of chlorpromazine and other amphipathic drugs on the activity of lipopolysaccharide antagonists, E5564 and E5531. *J Endotoxin Res* 2000;6:447–452.
79. Afonin KA, Grabow WW, Walker FM, Bindewald E, Dobrovolskaia MA, Shapiro BA, Jaeger L. Design and self-assembly of siRNA-functionalized RNA nanoparticles for use in automated nanomedicine. *Nat Protoc* 2011;6:2022–2034.
80. Aida Y, Pabst MJ. Removal of endotoxin from protein solutions by phase separation using Triton X-114. *J Immunol Methods* 1990;132:191–195.
81. Magalhães PO, Lopes AM, Mazzola PG, Rangel-Yagui C, Penna TC, Pessoa A, Jr. Methods of endotoxin removal from biological preparations: A review. *J Pharm Sci* 2007;10:388–404.

82. Petsch D, Anspach FB. Endotoxin removal from protein solutions. *J Biotechnol* 2000;76:97–119.
83. Diogo MM, Queiroz JA, Prazeres DM. Purification of plasmid DNA vectors produced in *Escherichia coli* for gene therapy and DNA vaccination applications. In: Barredo JL (ed.), *Methods in Biotechnology: Microbial Processess and Products*, Humana Press, 2005, pp. 165–178.
84. Miyamoto T, Okano S, Kasai N. Inactivation of *Escherichia coli* endotoxin by soft hydrothermal processing. *Appl Environ Microbiol* 2009;75:5058–5063.

Chapter 5

Surface Adsorbates on Nanomaterials and Their Possible Roles in Host Inflammatory and Toxicological Processing

*Clinton F. Jones**, *David G. Castner*[§,¶,||], *and David W. Grainger*[†,‡,**]

**Center for Nanotechnology in Drug Delivery*
Division of Molecular Pharmaceutics
UNC Eshelman School of Pharmacy
University of North Carolina at Chapel Hill
Chapel Hill, NC 27599, USA
[†]*Department of Pharmaceutics and Pharmaceutical Chemistry and* [‡]*Bioengineering, University of Utah*
Salt Lake City, UT 84112-5820, USA
[§]*National ESCA and Surface Analysis Center for Biomedical Problems*
[¶]*Departments of Bioengineering and* [||]*Chemical Engineering*
Box 351653, University of Washington
Seattle, WA 98195-1653, USA

Surface characteristics of nanomaterials are well recognized to contribute prominently to their unique properties and applications. It is less appreciated that the same surfaces have high intrinsic reactivities, typically distinct from identical bulk material chemistries. These size-dependent properties facilitate rapid, ubiquitous surface adsorption and contamination from many sources: storage containers; nanosynthesis components, by-products, and stabilizers; ubiquitous and adventitious environmental pollutants in clean room air and water; and diverse biomolecules that nanomaterials might encounter throughout their life cycle in biological, biomedical, and environmental exposures. Given their extremely large specific surface areas and high surface reactivities, nanoparticle surfaces are dynamic and rarely (if ever) the

**Corresponding author. E-mail: david.grainger@utah.edu

pure nanomaterial. Instead, such surfaces comprise complex mixtures of poorly characterized and controlled adsorbates, depending on the nanomaterial exposure history. Many adsorbates bring their own biological and toxicological activities. Significantly, surface adsorption may alter otherwise benign or innocuous adsorbates to render certain antigenic or inflammatory responses on a nanomaterial surface. To understand if such a scenario truly exists and the possible risks involved, rigorous nanomaterial surface assessments combined with careful immune system activation studies are necessary. However, the appropriate analytical tools to readily ascertain such interactions and mechanisms remain lacking.

1. Introduction

Many efforts to reduce material dimensionalities are motivated by the attractive material properties and functions unique to their nanometer regime. At the nanoscale, certain properties of matter become scale-dependent, including capillary forces, optical effects/color, melting points, conductivity, ionization potential, electron affinity, magnetism, and significantly, surface energy and surface reactivity. For example, while gold as a noble metal has a recognized low oxidation potential in bulk, properties of gold nanoclusters (e.g., Au_{55} clusters) are distinctly different, exhibiting highly oxidizable surface atoms.[1]

Optical, electronic, mass transfer, and thermal property changes that appear for materials in the nanoscale dimension are often exploited and studied. However, the impact of increasing surface areas with decreasing material dimensions must also be carefully considered as a truly impactful property. Sub-micron particles (e.g., nanoparticles with diameters of 1–300 nm) are the most popular and manipulated form in nanomaterials to date, present in increasing amounts in consumer and commercial products, and advocated for even further applications. Specific to biological exposures, nanomaterials in consumer products can enter the ecosystem in manufacturing, use, and disposal, producing some concerns over their life-cycle management, environmental tracking, and fate. Nanomaterials are also extremely popular in biomedical applications for *in vitro* immunolabeling, diagnostics, and imaging, as well as for *in vivo* diagnostics, advanced drug formulations and delivery systems, and imaging and tissue targeting functions.[2,3] In these applications, a primary feature is the direct contact of nanomaterial surfaces with metabolic cascades and literally tens of thousands of different biological components with redox properties, optical/chemical reactivity, and surfactancy. Basically, the introduction of many nanomaterials into biological milieus places two intrinsically reactive systems with dynamic properties into direct contact.

Specific surface areas for micron-sized materials (e.g., fumed silicas, commercial carbon blacks) are typically 60–80 m^2/g, a considerable surface-to-mass ratio providing extensive utility in their applications. Further reduction of micromaterial diameter to tens of nanometers increases the specific surface area by manyfold, yielding an impressive rise in surface area per mass. Hence, commercial CB-1 carbon blacks and single-walled carbon nanotubes (CNTs) — another major material focus for nanobiotechnology — may have specific surface areas exceeding 1300 m^2/g.[4] As material size decreases, the surface properties of the atoms increasingly dominate, producing significant changes in material reactivity.[5] Relative fractions of surface atoms to bulk atoms in any structure is called "dispersion," exhibiting a power law ranging in the nanoscale regime.[5] While less than 1% of micromaterial atoms occupy surface positions, 10% of the atoms in a 10-nm diameter metal material reside on its surface (and 60% in a 2-nm material!).[6] The excess surface energy and thermodynamic instability of materials below 20 nm not only result in the appearance of surface defects and atom/lattice rearrangements that stabilize the materials, but also produce unique nanomaterial reactivities and properties.[5] Effects of size reduction on material surface topology, porosity, texturing, and high-density fabrication are observed in sub-micron scaling features, contributing to the unique performance properties of nanomaterials. Surface structure and composition, and resulting surface reactivity, dominate nanomaterial properties. Hence, surface effects are a unique and very significant functional nano-property that require both control and careful characterization for exploitation in specific nanotechnologies.[5] Direct nanomaterial interfacial contact with biological molecules and reactive metabolic pathways produces possibilities for uncontrolled or undesired effects where specific surface interactions with biology intended for nanomaterial function might be hindered, and where host responses could induce adverse reactions.[6]

Colloidal materials have a substantial history of complex physicochemical behaviors in biological milieus.[7–11] To claim that well-studied colloidal systems (e.g., lipid materials, silicas, polystyrene latexes) can be controlled predictably in physiological milieus is a gross overstatement. Classic issues with colloidal instabilities, aggregation, and flocculation dominate these behaviors due to uncontrolled adsorption and surface reactivity. Therefore, even greater difficulties in characterizing and controlling behaviors for nano-sized colloidal materials in complex biological systems are expected and in fact, routinely encountered. High specific surface area materials exhibit substantial challenges in controlling interfacial reactivity. Thus, increasing specific surface area, as is characteristic of nanomaterials, produces new issues in such control. Two important distinguishing features of surface atoms compared to bulk atoms are: 1) lower coordination numbers and 2) increased exposure to and encounter with reactive species in the environment. These distinct surface features translate to intrinsically higher surface atom reactivities than bulk atoms, frequently manifested in certain predictable, but also other very unique, and even some exotic ways.[12]

2. Adsorption as a Natural Consequence of Surfaces

High surface energy materials of any dimension are plagued by adsorption as a spontaneous process that reduces surface free energy. This is often not a chemistry-specific phenomenon, but rather a general kinetic process involving any or all adventitious species present. Consequently, surface contamination is a thermodynamically driven process that reduces surface energy through the adsorption of adventitious air-, water-, or media-borne contaminants to yield a new adsorbate "surface" (i.e., an adlayer). Depending on the presence of other possible surface-active adsorbates, this adsorbed surface layer can be displaced, remodeled, or covered with another adsorbed layer to produce a new, even more thermodynamically stable state. Hence, the adsorption process can be dynamic and evolving, depending on the milieu, significantly changing surface composition, wettability, surface energetics and material–material two-body interactions, surface charge (ionic adsorption), and surface reactivity. Importantly, adsorption also chemically and physically alters the original material surface through adsorbate processes, including electron transfer-induced changes (redox chemistry), frequently yielding oxidized surface states in environmental systems and physical properties associated with the adsorbed layer. Few, if any, surfaces retain their original chemistry very long even under clean room conditions, and adsorption is highly ubiquitous to all surfaces even during synthesis.

Additionally, nanomaterial surface contamination may range from adventitious adsorption of adlayers to surface oxidation, corrosion, charging, or electron transfer reactions of the nanomaterial itself.[6,13] Due to dispersion, the higher intrinsic reactivity of nanophase surfaces prompts rapid and spontaneous surface adsorption of many diverse types of molecules in their milieu, reducing nanomaterial surface free energy and altering interactions with other nanomaterials (i.e., producing aggregation and flocculation) and with components in the milieu. In fact, surface adsorption is used deliberately and universally to stabilize nanomaterial solids in suspension in most commercial preparations through the intentional introduction of surface-active additives, e.g., CTAB, SDS, citrate, and charged or polar hydrophilic (co)polymers, that form stabilizing "coronas" around materials, keeping them from aggregating under shipping and storage conditions.

Due to their tremendous specific surface areas and intrinsic adsorption capacities, finely dispersed materials (e.g., diatomaceous earth, alumina, silica, activated charcoals) are exploited commercially as adsorbent beds to remove dissolved species from solutions, even in medicine. Nanomaterials satisfy this same purpose, but with substantially increased surface areas and adsorption capacities. It follows therefore that surface contamination of nanomaterials in both air and in aqueous milieus is spontaneous and substantial, by common adsorption from any and all ambient phases, and at a level proportional to 200–1000 m^2/g nanomaterial surface areas. Hence, a few hundred milligrams of CNTs contaminated with a monolayer of unintentional adsorbates (such as Fe, Co, Ni, S, or Mn atoms from common CNT manufacturing catalysts) could introduce to their surrounding solutions nearly a millimole of these adsorbed materials as surface leachates upon introduction to solutions.[12] The take-home lesson is that adsorbate layers on nanomaterials, and not the virgin nanomaterial chemistry, are increasingly thought to be the primary surface that interacts with biological components (i.e., proteinaceous, cellular, or lipid) in complex biological environments. In fact, it is likely that nanomaterials will carry significant adsorbed contamination into other systems (i.e., environmental and physiological) under most conditions. These various adsorption/desorption pathways are schematically shown in Figure 1. Nanomaterial surface chemistry should never be presumed to be either clean or unreactive when exposed to air, water, or biological environments.[12,14]

Biological component adsorption to nanomaterials is also rapid, often irreversible and non-descript since blood, for example, contains over 70 mg/ mL of thousands of soluble proteins and enzymes, small molecule solutes, dissolved gases, oxidizing and reducing agents, hundreds of lipids, fatty acids and other surfactants, and dozens of cell types, including blood platelets at a daunting concentration of 10^8/mL, orders of magnitude higher than most nanomaterial dispersion concentrations.[15] All of these components recognize and react to nanomaterials immediately upon their introduction into the bloodstream, irreversibly influencing nanomaterial physical and chemical behaviors, aggregation, and all subsequent cellular interactions. Other physiological compartments and environments are equally complex. Cell-mediated material uptake (at least at the micron-size scale) includes a diverse array of possible cell entry mechanisms, including phagocytosis, pinocytosis, endocytosis, coated pit, and caveolae-mediated entry, and is influenced by material surface properties and chemistry, surface adsorption, contamination, and aggregation. Hence, adsorption in biological systems profoundly affects cell-material processing mechanisms, altering the behaviors of materials introduced *in vitro* and *in vivo*. Therefore, the monitoring of material surface composition and its changes in the milieus of relevance for intended applications are requisite to understanding nanomaterial interactions with living

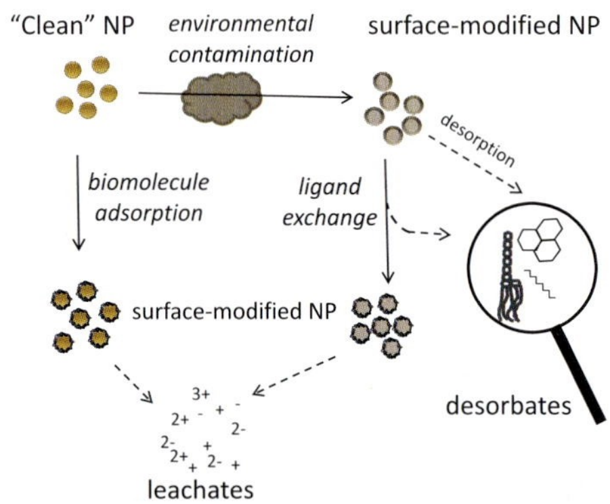

Figure 1. Environmental conditioning of nanoparticles and resulting products. Examples include contaminated nanoparticles, adsorbed stable adlayers, sloughed contaminants, and dynamically exchanging, displaced, and desorbing adlayers.

systems. Such analyses remain challenging since analytical tools to observe nanomaterial behaviors and various adsorption processes and property changes are crude and highly limited to date. Correlations to observed biological responses and behaviors in biological assays are therefore plagued by current inabilities to track nanomaterial surface and property changes within complex biological systems.[12]

Nevertheless, biological responses to nanomaterials may not be assayed with any real sense of scientific validity or confidence *before* nanomaterial properties (as supplied, with associated contamination), cleaning procedures, batch to batch nanomaterial variability, and solution properties (such as intrinsic aqueous stability, leachates, surfactant additives, aggregation, and flocculation) are thoroughly screened, considered, and reported. Essentially, this amounts to routine material science quality control, but remains critically important in nanomaterial research before meaningful results may be interpreted from subsequent *in vitro* or *in vivo* biological testing. Material composition must be ascertained with analytical specificity to understand its effects on the biology under assay. As one analytical routine, bulk material analysis involves material purity certification and trace contamination (e.g., soluble residual ions, leachates, organic reagents), aqueous leachables, electron microscopy of bulk nanophase morphology, polydispersity analysis (typically light scattering in aqueous dispersions of varying concentrations), intrinsic aggregation under shelf-life conditions, intrinsic nanomaterial solubility, and when appropriate, bulk-phase thermal analysis, conductivity, or redox behavior, and spectroscopy (i.e., fluorescence and vibrational) to provide some chemical quality control for physical and chemical states, and stability. This bulk analysis provides a rational basis to then proceed with a second, critical effort in surface analysis to understand possible surface states in the nanomaterials of interest since virgin nanomaterials do not present this bulk chemistry at their surface.[6,12] Together with nanomaterial size and bulk purity, nanophase surfaces must be continuously analyzed and tightly controlled for biomedical applications where surface contamination might elicit adverse, undesired, or unexpected biological responses. Possible sources of nanomaterial-derived toxicity in physiological systems should therefore consider both the nanomaterial and, perhaps even more, the ubiquitous intrinsic nanomaterial adsorbates presented to and processed by host immune components, in both adsorbed and desorbed states in biological milieus.

Contaminants and adventitious adsorbates from diverse sources decorate nanomaterials and exert unknown consequences on biological assays. During nanomaterial synthesis, catalysts, solvents, unreacted reagents, catalysts, or synthetic by-products may be entrained into or adsorbed onto nanomaterials.[16] During and following synthesis, materials may be exposed to contaminants in their environment (e.g., in filtration and separations, in suspending water/solvents, from ambient air, and storage/processing containers and materials). Common airborne contaminants in any laboratory, even clean rooms, include plasticizers (typically surface-active phthalates) from reagent packaging and container caps and sealants, polyaromatic hydrocarbons (PAHs), volatile polydimethylsiloxanes (PDMS), hydrocarbons and fluorinated ether oligomers from pump exhausts, and ubiquitous environmental adsorbates such as airborne carbonaceous and silicate particulates and bacterial lipopolysaccharides (LPS)[6] (Figure 2).[17,18] Reagent-grade water in contact with common labware (plastic or glass) also contains substantial surface-active contamination (rarely does purified laboratory water exhibit its theoretically pure surface tension of 72.8 mN/m at 20°C for "pure" water).

Alternatively, many "pure" adsorbates are intentionally introduced to nanomaterials as suspension stabilizers. These include block copolymers, charged polymers, small molecule surfactants (SDS, CTAB, ethoxlyated alkylesters), and highly charged polyions (e.g., citrate). Many such nanomaterial-adsorbed stabilizers are known to exchange in the presence of other competitive ligand exchange reactions. Gold nanomaterials in particular are subject to ligand exchange that can alter their aggregation states.[20,21] Other common nanomaterial additives include cell uptake-modulating and protein-repelling polymers (i.e., polyethers, celluloses, polyvinyl alcohol, poloxamer, poloxamine, polyethyleneimime, and polysorbates) or biomolecules (e.g., transferrin, albumin, DNA, other proteins or synthetic peptides). While these polymers adsorb to nanomaterials and stabilize or modify nanomaterial

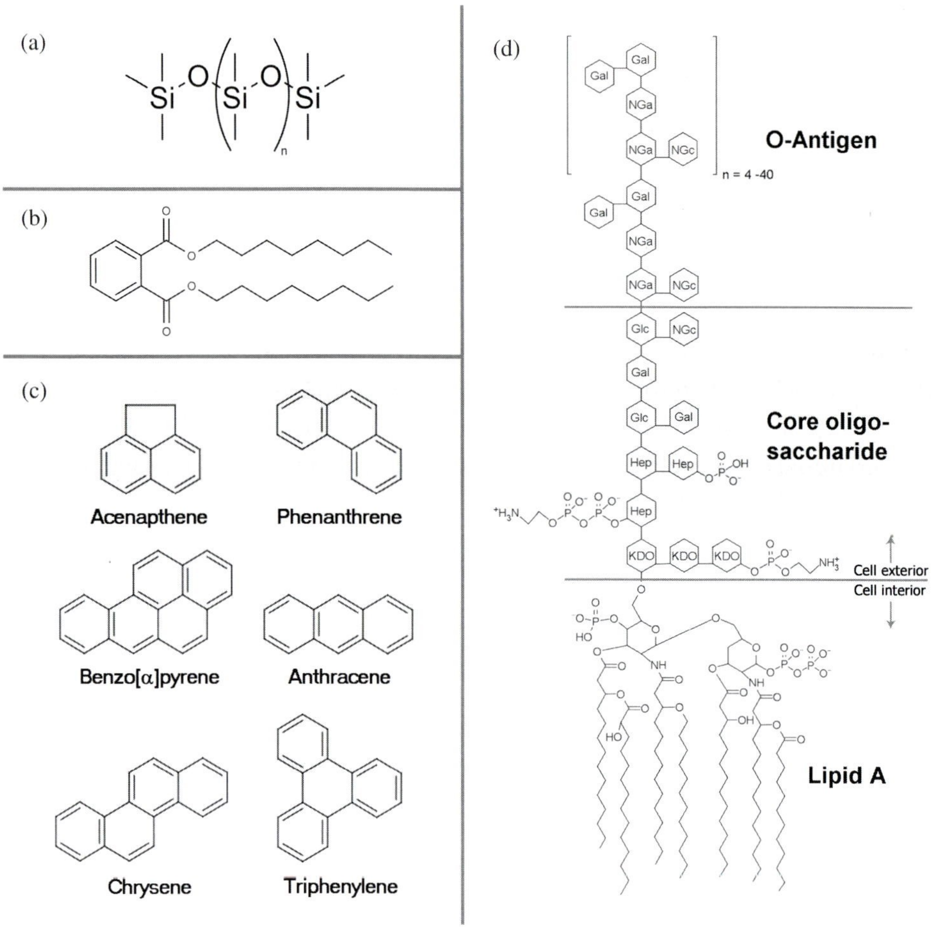

Figure 2. Selected common environmental adsorbates. Polydimethylsiloxane (PDMS) (a), dioctylpththalate (b), polyaromatic hydrocarbons (PAHs) (c), and lipopolysaccharide (d). Panels (c) and (d) were adapted with permission from Refs. 17 and 18, respectively, using ChemAxon software.

Gal: galactose; Glc: glucose; Hep: 1-glycero-d-manno-heptose; KDO: 3-Deoxy-D-manno-oct-2-ulosonic acid; NGa: N-acety1-galactosamine; NGc: N-acetyl-glucosamine.

aggregation behaviors in simple *in vitro* systems with relatively low concentrations of competitive adsorbates (i.e., low levels of plasma proteins), relatively little is known about their adsorbed stability under physiological conditions in the presence of high concentrations of proteins, lipids, and cells, alone or in any combination.

Nonetheless, understanding and controlling this nanomaterial interface in biological milieus is a key challenge to understanding nanomaterial biocompatibility, biological function and performance, risk assessment, and possible toxicity mechanisms. This rationale follows from well-known surface analytical paradigms in conventional biomaterials and medical device biocompatibility testing.[22] Biomaterial surface characterization methods commonly used for quality control and to report on surface states of implanted material chemistry and physical conditions are diverse and extensive, with many different types and sources of information reported from each technique depending on its requirements.[23,24]

This chapter lacks the breadth necessary to sufficiently describe the numerous well-known material characterization tools appropriate for material surfaces in general, or for nanomaterial purposes in particular. Several reviews on surface issues in nanomaterial characterization have been published.[6,25,26] As with any analytical technique, one must match the measurement capabilities of a selected surface interrogation method with the information sought, and apply best practices to the characterization of nanomaterials. In any case, surface analysis should routinely accompany bulk nanomaterial analysis prior to nanomaterial introduction into any *in vitro* or *in vivo* assay. Such nanomaterial quality control reporting should be expected as essential to understand nanomaterial characteristics prior to exposure to biological systems.[12]

3. Host Processing of Nanomaterials

It is highly likely that mammalian host defenses never see a bare nanomaterial because, as mentioned above, before a nanomaterial ever sees a biological system, it is already coated with a conditioning or stabilizing layer of either an intentional or adventitious origin. Upon introduction to the body, nanomaterial adlayers are then instantly subjected to a concentrated environment of surface-active biomolecules that further overcoat both bare nanomaterial surfaces and adlayers with endogenous proteins and lipids, and can displace adlayer coatings with endogenous molecules.[27] Thus, host defenses may encounter a bare nanomaterial (rare if ever), a contaminant- or stabilizer-coated nanomaterial (possible), a protein-/biomolecule-coated nanomaterial with an underlying exogenous coating (likely), or a protein-/biomolecule-coated nanomaterial with underlying exogenous species desorbed and displaced (likely). Given the unique nanoscale size of these material–biological adsorbate entities, host recognition systems could process them in multiple ways, resulting in normal clearance or alternatively, host reactivity. Significantly, adsorbate presentation on a solid nanomaterial surface of the same approximate size as many antigenic epitopes (several nanometers) could exacerbate host responses. High intrinsic nanomaterial surface areas are capable of delivering significant amounts of surface adsorbates to immune cells via both direct nanomaterial–adsorbate presentation and adlayer desorption by biological surfactants.

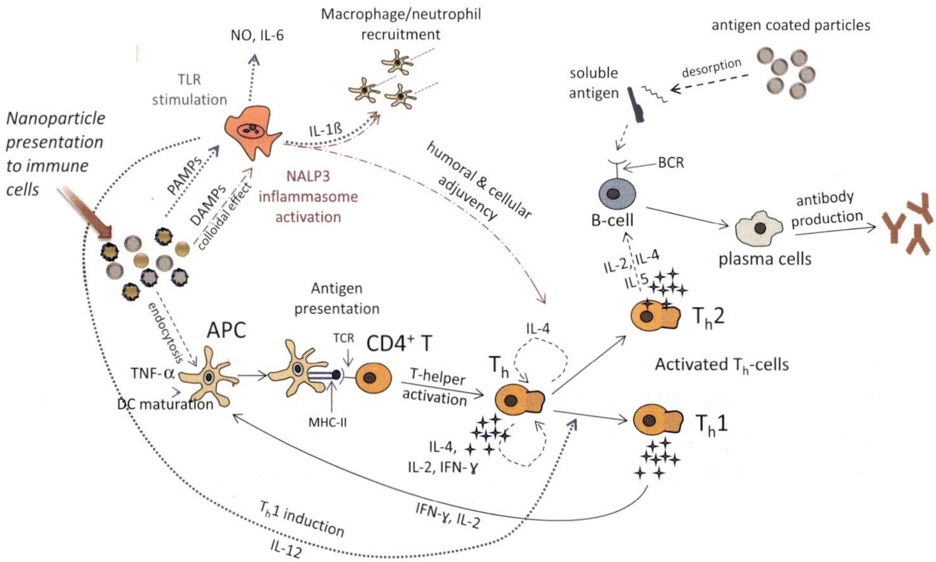

Figure 3. Possible nanoparticle processing pathways by host cellular immunological components. Nanoparticles, their desorbed adlayers, or nanoscale composites may induce a variety of immune responses. Given their size, some nanomaterials are likely to be processed as unique nanoscale composites comprising nanomaterials and adsorbed biological adlayers.

An overview of various immune processing pathways for nanomaterials and adsorbates is shown in Figure 3. In one host response scenario, the nanomaterial (coated or uncoated) may be phagocytosed by antigen-presenting cells (APCs), macrophages, dendritic cells, or neutrophils. Any peptide-based adsorbates on the material surface may be lysozomally processed and presented as digested peptide antigens by major histocompatibility complex class II (MHC-II) molecules on the APC surface. Other adlayer adsorbates carried by nanomaterials can be taken up and then desorbed within APCs and possibly presented by MHC-II molecules as antigens, although how general this processing mechanism and antigenic potential might be is very debatable. That nanomaterials can be processed intact with their adlayers as MHC-II complexes is not well described, but represents an interesting possibility. While unproven, combinations of uniquely engineered, synthetic nanomaterials with physical dimensions capable of direct immune component processing as possible antigens, high solid interfacial areas for additional adsorbate presentation and significant desorption, and unique interfacial properties and reactivities may provide unique immune stimuli, particularly when administered through previously unnatural routes (e.g., parenteral). More specifically, deliberate nanomaterial introduction into host surveillance systems by unnatural pathways in high doses (i.e., intravenous, inhaled, and subdermal) could provoke unique host immune responses.

However, this has not yet been validated. While highly variable, MHC-II antigen presentation of many molecules may activate T helper cells, leading to antibody production through the activation of B cells to plasma cells.

A second scenario involves the interaction of endogenous proteins with nanomaterials, resulting in the nucleation of protein fibrillation[28] or modification/denaturation of endogenous proteins on the nanomaterial adlayer/surface such that the protein adlayer is recognized by scavenger receptors for uptake by macrophages/endothelial cells. Moreover, protein denaturation at the particle surface could expose and present cryptic epitopes and generate haptens,[13,27,29] and the close proximity and high concentrations of proteins possible at the nanoparticle surface raise the possibility of amyloid formation that precedes disease states dependent on protein misfolding.[28]

A third scenario is the interaction of nanomaterial surface adsorbates either as presented on the nanomaterial or in their desorbed and soluble state. Adlayer molecules of pathogenic origin (pathogen-associated molecular patterns or PAMPs) may activate innate immune inflammation through the appropriate pathogen pattern recognition receptor class (TLRs, NLRs, RIG-1, MDA-5). Host cytokines released during innate inflammation may recruit phagocytes to the site of inflammation, induce cell death, generate an oxidative burst, and/or enhance an adaptive immune response. Moreover, in the case of material-presented PAMPs, the material itself acts as an adjuvant, activating the NALP-3 inflammasome to amplify the adaptive immune response to the vaccine antigen.[30] This route of action has been demonstrated for a growing list of nanomaterials that includes chitosan, poly(lactic-glycolic acid) (PLGA), polystyrene, silica, alum (aluminum oxide microparticles, the only solid-phase vaccine adjuvant approved for human use in the United States), and uric acid crystals (the only endogenous adjuvant, formed as a precipitate of accumulated uric acid released during mass necrosis).[30–32] Though the exact mechanism of NALP-3 activation and adjuvancy is not well defined, the nanomaterials themselves appear to be simply acting as an adjuvant and as such, antibodies are not likely to be generated specifically toward the nanomaterials.

Particle adjuvancy has been found to depend on antigen adsorption to particles, which is in turn affected by vaccine solution ionic strength, ionic composition (similarity to particle composition), pH, and particle and antigen isoelectric points.[33] (See Chapter 15 for more information on nanoparticle adjuvancy.) For example, adjuvant properties known for PLGA degradable polymer materials make them attractive delivery vehicles for DNA vaccines. Encapsulation, incorporation, or adsorption of antigenic DNA to PLGA–poly-L-lysine (PLL), PLGA–CTAB, or chitosan micromaterials (or some minor variation of one of these material types) have markedly improved immunogenicity, which has been attributed to a material-based adjuvant effect comparable to that of Freund's adjuvant.[34–36] As an example, the host antibody response to the hepatitis B vaccine antigen DNA adsorbed to PLGA–CTAB nanomaterials was significantly higher (i.e., double) than that of naked antigen DNA dosed alone. This humoral response was accompanied with significant increases in interferon-gamma (IFN-γ) production (five-fold over antigen alone), cytotoxic T lymphocyte activity, extended expression by APCs, and tumor protection and animal survival following tumor cell (HBsAg-expressing CT26/S) challenge.[37] Hence, similar to known alum- and particle-based host reactivity, nanomaterial–antigen presentation synergies are implicated in adjuvant effects, likely from antigen adsorption to the material surface, enhancing host immune system recognition and processing, and perhaps exacerbating sensitivity. This scenario also introduces the possibility of activated immune system components generating auto-antibodies toward nanomaterial-adsorbed endogenous — but adulterated — proteins with exposed native or altered (and possibly) cryptic sequences.[13]

Alternatively, leached metal ion species, non-pathogenic adlayers, or desorbed molecules of environmental contaminants may elicit host responses ranging from oxidative bursts, induced by intrinsic cellular toxicity, to immunosuppression and reduction of immune cell activation or their maturation by disruption of cytokine signaling cascade components.[38,39] Thus, the vast variety of possible nanomaterial adsorbates produces a broad range of resulting endogenous responses. While not widely reported or studied, immunotoxic implications of known nanomaterial adsorbates and ostensible contaminants comprise the remainder of this chapter.

4. Polydimethylsiloxane Adsorbates

PDMS [Figure 2(a)] and their smaller oligomers represent a synthetic class of silicone-based polymers and surface-active (i.e., antifoaming) additives common to labware, vacuum pump oils, lubricants and greases, lithographic print and mold release agents, sealants and water-proofing coatings, tape adhesives, and consumer products. PDMS surface contamination of synthetic nanomaterials is extremely likely because of its many forms and pervasiveness in common synthetic chemistry and fabrication laboratory areas, and the difficulty in removing it from laboratory plasticware, glassware, and air. The intrinsic surface activity and volatility of PDMS in lower molecular weight liquids combine to make it a contaminant common in laboratories — even clean rooms — ranking it as one of the most common surface contaminants observed under high-resolution analysis.[6] PDMS transfer from slightly contaminated synthetic glassware into many solvents and onto material surfaces is facile. Trace silicone detection with surface-sensitive X-ray photoelectron spectroscopy (XPS) is relatively straightforward except on siliceous substrates (i.e., silicate glasses, quartz, silica, or silicon wafers). Secondary ion mass spectrometry is even more sensitive than XPS for trace silicone detection, especially on siliceous substrates. Yet, the presence of PDMS on nanomaterials is rarely controlled or reported.[6]

Some level of PDMS and related siliceous surface contamination on nanomaterials is likely. Its removal is difficult without powerful oxidants and etchants (i.e., plasma gas treatment, ozonation, nitric acid, the RCA clean). As a contaminant adlayer on nanomaterials, its potential to alter nanomaterial interactions with biological systems is therefore considerable — PDMS adlayers mask intentionally adsorbed stabilizing species and also the ability to intentionally surface-modify materials with further adlayers (e.g., grafted polymer layers, functional groups, targeting ligands, or cell adhesion peptides). PDMS is known to adsorb plasma proteins readily, and alter cell and bacterial adhesion.[40] As it is hydrophobic and non-wetting, PDMS surface contamination changes droplet drying and deposition, fluidics, and wetting processes on surfaces (as in diagnostics and microfluidics devices), and adlayers can preclude ligand recognition of immobilized affinity partners at surfaces important to sandwich and affinity capture surfaces and targeting.

Silicon is the second most abundant element on earth, yet is only found in trace levels in the human body, and nearly always covalently attached to oxygen as silica or silicate networks. The silica content of human tissue varies from 10 mg to 200 mg/100 g dry weight. By contrast, silicone derivatives such as PDMS are oligomers and polymers of dimethylsiloxane ($-O-Si(CH_3)_2-$) units and related chemistries. PDMS are not found in nature or human tissues; there are no enymatic pathways known that produce or metabolize this pure chemistry. Therefore, silicones and PDMS are truly foreign materials in biological systems. This has led to the intense scrutiny of their biocompatibility and toxicology. In general, PDMS are known to have very little toxicity or biological effects, even at doses an order of magnitude above those observable in individuals with PDMS implants.[40] However, some interesting effects have been observed for specific variants. Cyclic PDMS (D_3, D_4, D_5, and D_6) are known to cause reversible liver enlargement that subsides upon the termination of exposure, but cause few additional effects other than slight dermal or ocular irritation.[40] Interestingly, decamethylcyclopentasiloxane penetrates skin slightly and octamethylcyclotetrasiloxane has demonstrated effects of reversible vaginal mucification and ovarian atrophy following inhalation exposure ($12\,g/m^3$) or adjuvant activity (humoral only), along with transient interferon production and a reduction in macrophage phagocytic activity following subcutaneous injection.[40] Linear decamethylsiloxane species cause more pronounced irritation with some, albeit equivocal, data for immune modulation.[40] Moreover, severe ulceration and necrosis were observed during skin exposure to PDMS precursors (non-dimethylsiloxanes), while other PDMS precursors cause minimal acute toxicity, irritation, and sensitization.[40] As is true for any non-degradable implant, PDMS implants generally trigger granulatomous reactions and fibrosis with a low incidence of solid-state sarcomas in response to PDMS gel implants.[40] Furthermore, injections of medical-grade Dow Corning 350cS PDMS fluid resulted in the observation of silicone within macrophages in regional lymph nodes and the adrenal system.[40] Thus, these compounds have the potential for bio-accumulation and immune modulation, especially at high doses. With ubiquitous environmental and industrial presence combined with high specific surface area nanomaterials, high doses are conceivable. However, more studies are needed to conclusively determine and elucidate specific effects of PDMS, either alone or with nanomaterials. Silicone and silicone gel have demonstrated adjuvant activity as an insoluble emulsifier of known antigens, but not when given separately from the antigen.[40] Silicone gel administered separately from an antigen does not exhibit adjuvant activity, but may produce a transient increase in natural killer cell activity.[40] Silicone also appears incapable of inducing classical adjuvant arthritis. Importantly, no significant links have been made between these experimental observations and any clinical immunological effect.[40]

With numerous precedents for interfacial reactivity, and also evidence for influences on biological systems, PDMS warrants careful consideration for its impact on nanomaterial properties in current biological applications. Little attention has been directed to date on significant PDMS contamination issues on material properties and behavior for high specific surface area nanomaterials.

5. Polyaromatic Hydrocarbon Adsorbates

PAHs, long known as potent carcinogens,[41] commonly adsorb to surfaces exposed to ambient air which ubiquitously contain PAHs in trace amounts, as products of combustion in transportation, manufacturing, and natural calamities. The production of carbon-based nanomaterials is now known to generate substantial toxic aromatic carbonaceous by-products including at least 15 PAHs[42,43] [Figure 2(c)]. Additionally, carbon nanomaterials are well known to actively adsorb volatile PAHs co-produced as part of the carbon nanophase manufacturing process.[44–46] Without rigorous post-synthesis processing and cleaning steps, carbon-based engineered nanomaterials are often contaminated by PAH adsorption. Unfortunately, simple water rinsing or suspension does not remove these hydrophobic adsorbate layers easily or completely. With specific surface areas approaching 1300 m^2/g,[47] CNTs represent major transport vehicles for PAHs. Given the similar chemical nature between CNTs graphitic chemistry and PAH chemistry, the detection of PAHs as an adsorbate on carbon-based nanomaterials is difficult. As an aromatic polycyclic species, PAHs share many attributes of the underlying graphitic substrates in many carbon-based nanomaterials. Standard *in situ* surface analytical methods cannot generally distinguish PAH adlayers on CNTs by chemical fingerprinting. Only thorough material solvent extraction and mass spectrometry of the rinsed phase can determine the extent of PAH surface contamination, although this is rarely reported. Oxidizing treatments of the nanomaterials (i.e., ozonolysis, plasma treatments, nitric acid washing, and RCA-type peroxide cleaning[48,49]) can oxidize PAH species to more soluble forms, facilitating their desorption from nanomaterial substrates in a "cleaning step," but this is often performed at the risk of also modifying the material surface. Hence, PAHs represent a ubiquitous and difficult adsorbate to both distinguish and remove from graphitic nanomaterials. Without standard surface treatments to both control for and remove PAHs from carbon-based nanomaterials as contaminants, it is difficult to understand the many reports for CNT behaviors in biological and cellular milieus. Few studies provide sufficient data on the identity, purity, and chemistry of the substantial surface areas of CNTs to correlate either with observed biological responses or with each other.[6]

Surface-contaminated nanomaterials may elicit substantially different biological responses *in vitro* and *in vivo* compared to pure or cleaner phases of the same nanomaterials. As produced, unpurified CNTs are reported to induce the production of pro-inflammatory cytokines and reactive oxygen species (ROS) similar to their combustion by-product material counterparts, while cleaned CNTs (with reduced PAH and heavy metal ion content) show reduced or complete absence of these effects.[50] Moreover, diesel exhaust material inhalation has been shown to increase the incidence of pathological lung infection.[51,52] Though the mechanism is unknown, this immunosuppression has been attributed to the depression of pro-inflammatory cytokine and ROS expression and antimicrobial potential of alveolar macrophages by adsorbed organic species on a given material, rather than the material itself.[52,53] In another study, benzofuran or acrolein adsorbates impaired *ex vivo* Fc-receptor–mediated carbon black particle uptake by rat alveolar macrophages.[54] It has been hypothesized that this type of effect may result from material overloading and lead to the co-expression of prostaglandins and corticosteroids, rather than being an adsorbate-mediated phenomenon.[55] However, several PAHs are known to be immunosuppressive (and still many others could have similar properties as most PAHs have not been extensively studied), but the mechanism of immunosuppression is not definitively demonstrated.[43] One popular hypothesis is that PAHs exert their immunesuppressive effects through cellular AhR receptor binding, but the binding strength for a given PAH is not an effective predictor of its degree of immunosuppression.[38] Moreover, AhR-deficient mice have demonstrated similar or greater immunosuppression in response to PAH exposure than AhR-competent mice.[38,56] Nonetheless, some evidence suggests that these planar, lipophilic molecules exert their immunosuppressive effects through the interruption of transmembrane/receptor signaling, disruption of interleukin signaling (IL-1, IL-2, and/or IL-3), and/or mobilization of and increases in intracellular calcium.[38] Additionally, many surface-active components present in biological fluids (proteins, lipids, fatty acids, other small molecule amphiphiles) are known to promote the exchange and desorption of these adlayers when they are not removed by other cleaning methods.[57,58] Hence, PAH contaminants on nanomaterials are likely off-competed and replaced by biological surface-active substances. This scenario produces new possible host exposures: 1) desorbed PAH reactions with cell membranes (adsorption, insertion, permeation, penetration), 2) partitioning into plasma proteins (e.g., albumins, apolipoproteins), and 3) direct biological reactions (cellular, proteinaceous) with the nanomaterial via its new, altered surface adlayer. Alas, little evidence is available to date to distinguish these phenomena *in vitro* or *in vivo*.

6. Catalyst Residues from Carbon Nanotube Synthesis

The chemical vapor deposition and ARC synthesis processes for generating CNTs utilize several catalytic reagents that include heavy metals.[59–61] As many of these become integrated into the carbon materials or tenaciously adsorbed as surface contamination, carbon-based engineered nanomaterials are often contaminated with a variety of non-carbon species, including transition metals.[62,63] In addition to the mainstay catalysts utilized in CNT synthesis (Fe, Co, Ni, Y, and Mo), neutron activation analysis and inductively coupled plasma mass spectroscopy (ICP-MS) have demonstrated an expansive list of CNT contaminants including metals/elements such as Al, As, Br, Ce, Cr, Cu, Gd, Ho, La, Mg, Na, Nd, Se, Sm, Sr, Tb, Th, Ti, V, W, Yb, Zn, and/or Zr, many of which may be employed in post-synthesis purification steps or during synthesis as catalyst dopants to favor specific CNT product attributes or improve catalyst lifetime.[59,62] These contaminants have been observed to constitute up to 30 wt% even in commercial "purified CNTs."[62,63] Residual post-processing catalyst removal from CNTs is routinely performed by nitric acid or mixed acid reflux for approximately two days, while amorphous carbon is oxidized and removed by 20–70-h treatment with H_2O_2 or by thermal treatment.[48,49] However, metal impurities are often left entrained within the nanotube structure or ensconced in their own graphitic shell following purification, precluding complete impurity removal without concomitant destruction (or at least significant structural alteration) of the nanotubes themselves.[64] Nevertheless, in spite of graphene encapsulation, toxicologically significant quantities of metal-based catalyst impurities may be released into incubation media, particularly under mildly acidic (pH = 5, lysosomal) conditions.[63] This metal release from purified CNTs may be due to inadequate acid treatment (cleaning) time, which would kinetically limit the dissolution of catalyst metals (or their oxides) possessing slow corrosion rates such as Ni, Ti, Zn, or Mg,[62] or inadequate rinsing during purification, allowing ionic metals to adhere to CNTs, particularly at divalent cationic sites such as carboxylic groups.[62] There is some data to suggest that the avoidance of oxidizing acids during CNT purification allows for sufficient acid treatment time, while preventing the exposure of the metal that is fully encased in carbon and completely inaccessible.[62] However, an oxidizing environment or organic solvent extraction would be needed for the removal of PAH adsorbates (noted previously) and it is not known whether the *in vivo* phagocytic oxidizing environment is sufficient for the liberation of carbon-encapsulated metals. Relatively little analysis has been performed on the consequences of catalyst release from CNTs and catalyst metal release could contribute to the observed pro-inflammatory cytokine and ROS induction and/or immunosuppressive effects of CNTs as noted above in the section on PAHs. Indeed, soluble iron (Fe^{2+}) is capable of forming ROS in physiologically relevant conditions;[62,65] Cu, Co, Al, Fe, Ni, and Cr are known to be toxic by inhalation, ingestion, or dermal exposure;[65] ultrafine combustion materials, with their characteristic transition metal or sulfuric acid core with an adlayer of semi-volatile organics, have been shown to induce oxidative stress;[13] and animal studies have demonstrated that Cd, V, Cr, Pb, and Ni decrease antibody formation, antigen processing, and lymphocyte processing.[66]

7. Adsorbed or Conjugated Polymer Coatings

Several polymer surface coatings are deliberately applied to nanomaterials with the intent to modify (i.e., improve) their specific interactions with host biology. Most of these polymers are hydrophilic and intended to help suspend

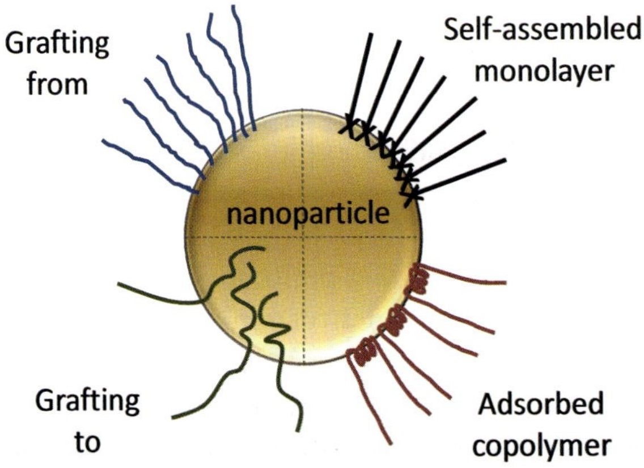

Figure 4. Common nanoparticle surface polymer functionalization schemes. These approaches include polymer growth from the surface (grafting from), polymer growth and subsequent grafting to the surface, self-assembled monolayers, and surface copolymer adsorption.

or solubilize nanomaterials in aqueous media. These polymers can be physically adsorbed to nanomaterial surfaces to form surface adlayers, or chemically conjugated to pre-existing nanomaterial surface functional groups (Figure 4). Both approaches create polymer "coronas" around the nanomaterials, dramatically altering nanomaterial surface properties, chemistry, and interactions with aqueous media. All seek to limit non-specific protein adsorption from biological milieus and improve particle suspension lifetimes in aqueous media.

Some of these polymer surface modifications aim to improve cellular uptake for imaging, drug delivery, or gene transfection. These types of particle coatings are generally simple and include polyethyleneimine, polyamines, DNA, PLL, or chitosan. Other polymer coatings on nanomaterials attempt to evade uptake and capture by the host filtration organs (i.e., the reticuloendothelial system, RES) during systemic circulation; materials such as polyethylene glycol (PEG), poloxamer block copolymers, celluloses, polyvinyl alcohol, hydrophilic polyacrylates such as polyHEMA and polyacrylic acid, and poloxamines have all been used to limit cell uptake, delay phagocytosis, and modify the host clearance of nanomaterials.[11] For more information on the specific roles of surface coatings on nanomaterial immunogenicity, please refer to Chapter 18.

PEG conjugation to nanomaterial surfaces follows countless precedents of this strategy applied to other biomaterials for the same purpose — to mask the material's surface from endogenous proteins and receptors.[11,67] However, CNTs functionalized with PEG, followed by conjugation with fluorescein isothiocyanate as an imaging agent, produced pro-inflammatory macrophage responses, though the observed lymphocyte behavior was unchanged.[68] Anti-PEG antibodies are now recognized as well,[69–73] with evidence for complement activation (see Chapter 11 for more details) and hypersensitivity reactions,[71,73–75] particularly for hydroxy-terminated PEG,[76] indicating that PEG may have limited value *in vivo*. With the exception of PEG, PLL, polyvinyl alcohol, poloxamine and celluloses (complement activators),[71,74,76,77] and some chitosan variants (known to act as adjuvants and immunostimulants),[35,36,78–80] most of these hydrophilic designer surface coatings are used with little observable immunogenic effects. However, non-immunostimulating polymer coatings are not without their problems, especially in non-covalent, surface-adsorbed states. For example, fluorescent spectroscopy has been used to demonstrate that single-walled CNTs rapidly shed their passivating Pluronic coating in serum.[81] That this surface alteration was assessed directly is admirable, supporting no known specific binding affinity between the surface chemistries of CNT and Pluronic, and Pluronic's ready removal in serum from competing surfactancy intrinsic to many serum components. PAHs and other adsorbate contaminants common to CNT surfaces (*vida supra*) beg the question as to the true chemically bound or physically adsorbed states of other added nano material and single-walled CNT polymer coatings and their stability in biological milieus. Polymer adsorbate displacement or adlayer remodeling in biological milieus may prompt other unintended property changes consequent to protein adsorption and material aggregation.[27] Poor polymer adlayer adsorption resistance to high protein surface bombardment typical of full plasma or serum (70 mg/mL protein) can bury polymer adlayers,[82] rendering any intended interaction effects of them with cells ineffective, and inducing material aggregation.

8. Biomolecule Adsorbates

Most globular proteins are highly surface-active, given their amphiphilic chemical and physical structures and abundant metastable folded states. Many proteins of diverse sizes and structures are known to adsorb rapidly to nearly all surface chemistries and known proteins are generally highly surface-active.[7,27,83] This issue remains a major unsolved challenge to the biomaterials field in general, evading solutions for over 50 years. (For more information on the impact of nanomaterial–protein adsorption/interactions on material distribution, please see Chapter 6.) Protein adsorption to nanomaterials is well documented despite various surface modifications,[29,82,84–86] occurring immediately upon or after their introduction into blood, cell culture media, food extracts, cell lysates, or other protein-containing fluids. This is expected, given that their surface treatments are similar to those applied to conventional biomaterials for decades, and the universal failure to produce a suitable solution to this problem across all materials, regardless of size or surface. Protein adsorption is a major mechanism by which all materials are recognized and removed from circulation by filtration organs of the RES — the liver, spleen, kidney, and lung.[86,87] The process of opsonization — the non-specific adsorption of plasma proteins to circulating colloids, producing recognition by the RES tissues[75,86] — is the central mechanism of this endogenous host clearance mechanism.

With the possible exception of vitronectin and similar proteins whose adsorption may attenuate complement activation,[74] protein adsorption to materials often results in both the activation of complement with subsequent cell uptake, and activation of cell-mediated immune components.[88] For example, the intraperitoneal injection of a fullerene C_{60}–albumin conjugate into mice induced a fullerene IgG antibody response.[89] Single- and double-walled CNTs introduced into human plasma were coated primarily with fibrinogen and apolipoproteins (e.g., AI, AIV, CIII), activating the classical complement cascade though direct binding of C1q. CNTs also activated the alternative complement cascade through C3b binding.[90] Nonetheless, other studies have shown minimal roles for complement in material uptake, attributed to differing protein adsorbates and nanomaterial types. Cell uptake efficiency is optimal for CNT or polystyrene nanomaterials pre-suspended in even dilute serum or dilute albumin, respectively.[2,91] Polystyrene material data obtained from an intricate *ex vivo* spleen slice model revealed a key role for adsorbed albumin, but little uptake was observed in a purified fibronectin system and no reduction in nanomaterial cell uptake was observed for serum depleted of complement, IgG, or heparin-binding proteins.[91] This may be because albumin is the most abundant major plasma protein and is highly adsorbed to almost all materials. Nanomaterial uptake is also proposed to be mediated through macrophage scavenger receptors responding to conformational alterations in adsorbed endogenous proteins.[2,91] Moreover, the activation of scavenger receptors could attenuate innate immunity by the reduction of chemokine production or by the inhibition of inflammatory mediators.[52] Additionally, some speculate that leached ions or nanomaterial interactions with endogenous proteins could result in protein unfolding that expose normally cryptic domains or sequences, deactivate protein function, decrease bioactivity or generate new haptens, and raise allergic or autoimmune reactions.[2,13,91–93]

The inhalation route of nanomaterial exposure results in nanomaterial conditioning distinct from that of other administration/exposure routes due to the unique biomolecular environment and tissue processing present in the lung. Material size scales inversely with the depth of material penetration into and deposition within the lung, resulting from aerodynamic effects upon inhalation. As such, larger materials are retained primarily in the upper respiratory system (throat and bronchi), where they are cleared by the mucociliatory escalator, while smaller materials penetrate deeper into the bronchiolar bed, and a small fraction of the smallest materials might reach farthest to the alveolar bed, where they become coated with lung surfactant.[94] From there, materials may interact with surfactant proteins (apolipoproteins A and D or surfactant glycoproteins A, B, or C) and enter alveolar epithelial (type I or II), abundant alveolar macrophage, or bronchiolar dendritic cells.[94] Alternatively, cellular uptake and the subsequent generation of inflammatory cytokines may result in altered material transport as inflammation increases permeability and opens up a paracellular route to the interstitium with eventual translocation to the blood circulation.[94,95]

One anecdote that possibly links nanomaterial–protein interactions to unusual host response is demonstrated through distinct host reactions to the model protein ovalbumin (OVA) in the presence and absence of CNTs. OVA and CNTs, co-administered intratracheally (in saline with Tween 80) to mice, resulted in significant increases in oxidative stress markers (e.g., heme oxygenase-1, HO-1), DNA modification (8-hydroxy-2′-deoxyguanosine, 8-OHdG), lipid peroxide products, and advanced glycation end-products [AGEs, pentosidine, and Nε-(carboxymethyl)lysine][96] commonly found in an oxidative *in vivo* environment. The binding of AGEs to corresponding receptors (RAGEs) found on many cell types contributes to chronic inflammatory diseases through pattern recognition receptor-mediated perpetuation of cellular activation.[97] Furthermore, expression levels of multiple allergy-related cytokines (IL-4, IL-5, IL-13, IL-17A, and IL-33) were elevated in the lung following OVA + CNT administration to these mice. These observed collective murine responses to OVA + CNT were significantly greater than the host response to OVA alone. While it is not known whether the adsorption of OVA to CNTs elicited the differences, or whether the observed responses resulted from a synergy of responses to the two entities together,[96] the findings suggested that the adsorptive presentation of OVA by CNTs increases the observed host response to OVA and that the adsorptive presentation of allergens or antigens by CNTs may exacerbate allergic or immunogenic responses. If consistent with host responses to other nanomaterials, biological adsorption to nanomaterials relative to activating adaptive immunity and interactions with pattern recognition receptors already implicated in nanomaterial immunotoxicity[16,98] should become a serious focus (Figure 3). As the RAGE receptor also regulates sepsis, this tentatively implicates endotoxin and

LPS processing by the host to such nanomaterial-based presentation of antigen and even LPS itself.

9. Lipopolysaccharide (Bacterial Endotoxin)

As a ubiquitous component of Gram-negative bacterial cell walls, LPS remains one of the most common surface-adsorbed contaminants of serious concern for all nanomaterials.[99] LPS is a peptidoglycan-decorated phosphorylated multi-chain lipid shed from bacterial cell walls [Figure 2(d)]. Given the considerable presence of these bacterial flora on skin, airborne of aerosols and particles, on all surfaces that comes in contact with the skin, in all milieus capable of supporting bacteria, and in most water sources (except expensive, designated "pyrogen-free" water), LPS is indeed ubiquitous. Therefore, LPS contaminates everything that most water sources and mammals have contacted, including laboratory glassware, utensils, clean rooms, and filtration equipment. Like proteins and silicones, its surface activity is particularly problematic; the LPS chemical structure and resulting surface activity favor its adsorption to hydrophobic surfaces by its multiple acyl lipid chains, while its phosphate, amine, and polar glycan chemistries facilitate binding to polar and charged surfaces.[99] These properties allow LPS to contaminate virtually any surface and remain tenaciously adsorbed. Its contamination of numerous biomaterials and medical devices is well documented.[26,100–103] Nonetheless, its detection is primarily done using assays that identify its soluble form.[12] *In situ* detection in adsorbed states is difficult and non-quantitative[6] (see Chapter 4 for more details).

Endotoxin strongly affects biological systems as a known pyrogen by raising the body temperature and inducing inflammatory cellular cytokine production through its presentation to many types of host cells.[95,102,104,105] Macrophages and other immune cells use Toll-like receptors (TLRs) to recognize and process LPS. Hence, reliably discriminating intrinsic LPS bioactivity, distinct from the nanomaterial's intrinsic inflammatory activity, is both important and technically arduous. Given the exaggerated areas and surface-to-volume ratios for nanophase systems, possible amounts of adsorbed endotoxin (e.g., micrograms of adsorbed endotoxin per gram of material) are significant. Titanium dioxide (TiO_2) materials have been shown to bind LPS in such a way that reduces *in vitro* cellular inflammatory response to LPS.[106] Other nanomaterials have also been demonstrated to bind endotoxin[14,107] and it is likely that all such materials will bind some of these contaminants. A select few biomaterials have been evaluated for their ability to bind and remove endotoxin from aqueous solution.[108,109] Moreover, investigations of material–endotoxin interactions related to orthopedic wear debris in hip arthroplasties reported microparticle-adsorbed endotoxin effects on implant loosening through localized inflammation.[104,110] Therefore, as endotoxin contamination is nearly ubiquitous on most nanomaterials, it is capable of producing inflammatory responses magnified over those elicited from the presence of the nanomaterial alone.[12] This also provides the potential for nanomaterial–endotoxin synergy in host responses.

These effects warrant concern over the potent inflammatory reactivity of endotoxin *in vivo* and *in vitro* with many cell lines,[102,105,110,111] particularly inflammatory cell lines (e.g., macrophages lines of various lineages and maturities[12]). The presence of adventitious endotoxin contamination in cell cultures or on nanomaterials applied to cells could activate these cells and generate inflammatory cytokine responses essentially indistinguishable from any response generated by the exposure to the nanomaterial alone. Additionally, such cellular activation could also alter other crossover cell signaling pathways in cell-based assays, generating unintended consequences or confounding results. Consequently, endotoxin contamination of nanomaterials, nanomaterial stock solutions, laboratory glassware, lab-grade and even ultrafiltered water, and cell culture materials should be routinely assayed,[12,99] controlled, and remediated before any cell response assay to nanomaterials is performed.[12,99]

Unfortunately, detecting LPS on nanomaterials is currently challenging.[6] Typical endotoxin assays only respond to solution phase LPS. The Limulus Amebocyte Lysate (LAL) assay is the current standard for detecting soluble endotoxin *in vitro*, despite inherent inconsistencies (i.e., the assay signal is relative to fixed highly controlled LPS standards that may or may not reflect diverse sources, heterogeneity, purity, or reactivity of endogenous LPS). Colorimetric variation of the LAL assay in response to LPS dosing is the most widely used detection metric in biomaterial testing, particularly in preclinical settings, and is more accurately defined as a measure of relative endotoxin activity rather than absolute endotoxin concentration.[12]

Importantly, endotoxin may only be detected by these assays if it is soluble or readily displaceable (i.e., not permanently surface-bound). Nevertheless, some methods have been developed to correlate detectable solute endotoxin to surface-bound endotoxin within the same material system.[12,101–103] Furthermore, a recent murine macrophage-based commercial tool — RAW-Blue® reporter cells — facilities LPS detection or other solutes that activate the NF-κB or AP-1 transcription factor pathways. Though this reporter cell line provides LPS detection indiscernible from other possible signaling pathway activators, it yields useful information concerning the presence of soluble agonists to the TLR family, NOD family, RIG-1 and MDA-5 receptors (RAW-Blue® cells product insert, version #11E30-MM, Invivogen, San Diego, CA).

Additional endotoxin assays are discussed in detail elsewhere[99] and in Chapter 4. However, for any of these methods to be quantitatively reliable, adsorbed endotoxin should be shown to correlate directly to known LPS exchange or desorption rates under relevant *in vitro* conditions, where LPS undetected by the assay in surface-bound form produces a reliable answer when surface-displaced in biological milieus. That is, irreversibly adsorbed endotoxin might be indirectly correlated by knowing its soluble, reversibly adsorbed LPS fraction displaced from a surface in biological milieus. This rigorous analysis is not performed on nanomaterials, despite the likelihood that LPS as a common, highly biologically reactive contaminant is present at high levels as introduced from routine material processing and manufacturing. Hence, biological tracing of adsorbed LPS and its partitioning between nanophases and biological milieus remain unknown.[12,108] Consequently, the rabbit pyrogen test has been recommended for verifying endotoxin contamination of nanomaterials to improve otherwise ambiguous results from other detection methods.[99]

10. Contaminant Detection and Assessment

Although the appreciation of the need for detailed, comprehensive surface characterization of nanomaterials exposed to environmental and biomedical applications is increasing, differences between traditional bulk material surface science and analysis of nanophases presents daunting challenges for quality control in nanomaterial applications. Surface analysis methods [e.g., XPS, low energy electron diffraction (LEED), Auger electron spectroscopy, mass spectrometries, etc.][22] have been adopted for nanomaterial use from a long history of development from surface science applied to particle-supported catalysts and microelectronics analytics for detailed characterization of nanomaterial, but typically for nanomaterials prepared and analyzed under well-defined conditions. Such analysis requires specialized instrumentation and standard, validated experimental protocols to prepare, store, and transport samples in specific, well-defined conditions (oxidized, reduced, ligand-protected, sulfided, etc.) (see Refs. 6, 111, 112 and references cited therein). Nonetheless, surface analytical tools capable of elucidating specific nanoscopic details are still limited, even under the best high vacuum analysis conditions.[114] Most of these tools do not have the spatial resolution to examine individual nanoparticles or nanomaterial features; all results thus average area-based values from ensembles of nanomaterials and from background sources in spaces beneath and around them, and from bulk nanomaterials in multilayers.

Due to the complexity intrinsic to most biological environments, surface science methods for nanomaterials in complex milieus are lacking; the direct translation of traditional ultrahigh vacuum (UHV) surface science methods to nanomaterials placed into aqueous, proteinaceous environments provides very limited useful information. Recent studies have developed quantitative XPS methods for determining the overlayer thickness on functionalized gold particles[115] and the core-shell structures of quantum dots (QDs).[116] These studies along with others have used XPS, ToF-SIMS, and NMR methods to identify the presence of surface contaminants on gold particles and QDs.[115–119] While these and other systematic surface analysis studies[120] of functionalized nanoparticles are an important starting point for developing tools for the better characterization of nanomaterial surface chemistry, they do not address ultimate challenges of characterizing the nanomaterial surface chemistry *in situ* under biologically relevant conditions. Nonlinear optical methods offer promise to contribute information about the surface chemistry of nanomaterials in aqueous solutions. For example, sum-frequency generation in a scattering geometry has been used to characterize micro- and nanomaterials in solution.[121,122] The scattering patterns provide information about the structure of the materials (core-shell, etc.), while vibrational spectra yield information about surface species.

Current reliance of the field to date on transmission electron microscopy (TEM) to describe nanomaterial properties and behaviors is insufficient and misleading since it is performed *ex situ* after dessication, and provides little surface information. The increasingly common scenario reported in the biomedical literature involves placing a very complex, poorly defined, and highly reactive nanomaterial into an *even* more complex biological system. This routine starkly contrasts the carefully controlled ways that, for example, supported materials are studied and handled for gas-phase structure–property surface catalysis experiments (i.e., for heterogeneous catalysis). The biological results for nanomaterials are often not definitive, elucidating, mechanistic, or very good science needed to inspire approaches to improving nanotechnology, or for understanding either surface adsorbate behaviors or mechanisms of how nanomaterials interact with biological or environmental systems. The situation begs the question about how to improve (nano)science to provide accurate information on nanomaterial interactions in complex biological scenarios.[6]

11. Conclusions

Nanomaterial interactions with biological, environmental, and biomedical systems are a growing concern with increasingly diverse material sources and chemistries exposed to humans in their daily routines. A large part of this concern is focused on the surfaces of nanomaterials that comprise the majority of their interactions with human response elements. Nanomaterial surface properties and their modifications over time and in various conditions are therefore critical to elucidate in nanomaterial presentation to various immune, inflammatory, and APCs. Pathways for cellular particle, particle–adsorbate, and adsorbed antigen processing are known, but the specifics of nanomaterial involvement in such host cell-based processing are severely lacking.

Nonetheless, technical challenges for characterizing nanophase materials with significantly enhanced specific surface areas and reactivities are daunting, particularly in extending analysis beyond the UHV environment and into complex real systems. Increased analytical power is essential to produce new information capable of elucidating nanomaterial surface adsorbate chemistries, adsorption/desorption phenomena including stability, dynamics, and exchange reactions in complex media, and how these nanomaterial compositions influence host responses. Tools for such sensitive nanomaterial analyses are not yet sufficiently developed to provide the quality of information necessary to discern mechanisms and distinguish material properties in biological systems. Currently detectable cell-based toxicities and immune responses can only be correlated indirectly to nanomaterials; direct, reliable, and sensitive causal information linking observed cellular reactions to specific nanomaterials can only be inferred at best. The generation of reliable nanomaterial structure–property relationships in host immune and toxicity processing requires sensitive and versatile analytical methods to detect adsorbate classes, understand cell–material interactions, and the ways in which nanomaterials present chemistry to host inflammatory mediators. Beyond isolated, anecdotal gross histological, or morphological evidence for nanomaterial toxicity and inflammatory provocation, very little definitive information exists to demonstrate nanomaterial-induced adverse host events. Hence, the link between host cell responses and nanomaterial surface properties, conditioning, and delivery of adsorbates is poorly understood. As such, no mechanistic or structure–property relationships implicated in any possible or alleged nanomaterial toxicity risks can be rationally asserted. Improving on this scientific challenge will be essential to best understand, control, and exploit the unique features of nanoscale properties and materials in their various human and mammalian applications.[6]

Acknowledgments

The authors acknowledge support from NIH grants EB-00894 (CFJ and DWG) and EB-002027 (DGC).

References

1. Turner M, Golovko VB, Vaughan OPH, *et al.* Selective oxidation with dioxygen by gold nanoparticle catalysts derived from 55-atom clusters. *Nature* 2008;454(7207):981–983.
2. Nel AE, Mädler L, Velegol D, *et al.* Understanding biophysicochemical interactions at the nano–bio interface. *Nat Mater* 2009;8(7):543–557.
3. Petros RA, DeSimone JM. Strategies in the design of nanoparticles for therapeutic applications. *Nat Rev Drug Discovery* 2010;9(8):615–627.
4. Pumera M. Electrochemistry of graphene: New horizons for sensing and energy storage. *Chem Rec* 2009;9(4):211–223.
5. Auffan M, Rose J, Bottero JY, Lowry GV, Jolivet JP, Wiesner MR. Towards a definition of inorganic nanoparticles from an environmental, health and safety perspective. *Nat Nano* 2009;4(10):634–641.
6. Grainger DW, Castner DG. Nanobiomaterials and nanoanalysis: Opportunities for improving the science to benefit biomedical technologies. *Adv Mater* 2008;20(5):867–877.
7. Gray JJ. The interaction of proteins with solid surfaces. *Curr Opin Struct Biol* 2004;14(1):110–115.
8. Norde W, Lyklema J. Protein adsorption and bacterial adhesion to solid surfaces: A colloid-chemical approach. *Colloids Surf* 1989;38(1):1–13.
9. Jacob K. Surface interactions with adsorbed macromolecules. *J Colloid Interface Sci* 1986;111(2):305–313.
10. Andrade J, Hlady V. Protein adsorption and materials biocompatibility: A tutorial review and suggested hypotheses. *Adv Polym Sci* 1986;79(1):1–63.
11. Owens DE, Peppas NA. Opsonization, biodistribution, and pharmacokinetics of polymeric nanoparticles. *Int J Pharm* 2006;307(1):93–102.
12. Jones CF, Grainger DW. *In vitro* assessments of nanomaterial toxicity. *Adv Drug Delivery Rev* 2009;61(6):438–456.
13. Nel A, Xia T, Mädler L, Li N. Toxic potential of materials at the nanolevel. *Science* 2006;311(5761):622–627.
14. Peula-García JM, Molina-Bolivar JA, Velasco J, Rojas A, Galisteo-González F. Interaction of bacterial endotoxin (lipopolysaccharide) with latex particles: Application to latex agglutination immunoassays. *J Colloid Interface Sci* 2002;245(2):230–236.
15. Barrett KE, Barman SM, Boitano S, Brooks H. *Ganong's Review of Medical Physiology*, McGraw-Hill Medical, 2009.
16. Dobrovolskaia MA, McNeil SE. Immunological properties of engineered nanomaterials. *Nat Nano* 2007;2(8):469–478.
17. Yuan T, Marshall WD. Catalytic hydrogenation of polycyclic hydrocarbons over palladium/γ-AI2O3 under mild conditions. *J Hazard Mater* 2005;126(1–3):149–157.
18. Petsch D, Anspach FB. Endotoxin removal from protein solutions. *J Biotechnol* 2000;76(2–3):97–119.

19. Woehrle GH, Brown LO, Hutchison JE. Thiol-functionalized, 1.5-nm gold nanoparticles through ligand exchange reactions: Scope and mechanism of ligand exchange. *J Am Chem Soc* 2005;127(7):2172–2183.
20. Swift J, Butts CA, Cheung-Lau J, Yerubandi V, Dmochowski IJ. Efficient self-assembly of archaeoglobus fulgidus ferritin around metallic cores. *Langmuir* 2009;25(9):5219–5225.
21. Castner DG, Ratner BD. Biomedical surface science: Foundations to frontiers. *Surf Sci* 2002;500:28–60.
22. Castner DG, Grainger DW. Surface analytical methods: *In situ* and UHV techniques. In: Healy K, Ducheyne P, Grainger D, Kirkpatrick J, Hutmacher D (eds.), *Comprehensive Biomaterials — an on-line compendium.* Vol. 3(1), Elsevier, 2011, pp. 1–22.
23. Recum AV, Jacobi JE. Surface characterization. In: LaBerge M (ed.), *Handbook of Biomaterials Evaluation: Scientific, Technical, and Clinical Testing of Implant Materials*, CRC Press, 1998, pp. 893.
24. Baer DR, Gaspar DJ, Nachimuthu P, Techane SD, Castner DG. Application of surface chemical analysis tools for characterization of nanoparticles. *Anal Bioanal Chem* 2010;396(3):983–1002.
25. Richman EK, Hutchison JE. The nanomaterial characterization bottleneck. *ACS Nano* 2009;3(9):2441–2446.
26. Lynch I, Cedervall T, Lundqvist M, Cabaleiro-Lago C, Linse S, Dawson KA. The nanoparticle-protein complex as a biological entity; a complex fluids and surface science challenge for the 21st century. *Adv Colloid Interface Sci* 2007;134–135(0):167–174.
27. Linse S, Cabaleiro-Lago C, Xue WF, *et al.* Nucleation of protein fibrillation by nanoparticles. *Proc Natl Acad Sci U S A* 2007;104(21):8691–8696.
28. Lynch I, Dawson KA, Linse S. Detecting cryptic epitopes created by nanoparticles. *Sci STKE* 2006;2006(327):pe14.
29. Sharp FA, Ruane D, Claass B, *et al.* Uptake of particulate vaccine adjuvants by dendritic cells activates the NALP3 inflammasome. *Proc Natl Acad Sci U S A* 2009;106(3):870–875.
30. Eisenbarth SC, Colegio OR, O'Connor W, Sutterwala FS, Flavell RA. Crucial role for the NALP3 inflammasome in the immunostimulatory properties of aluminium adjuvants. *Nature* 2008;453(7198):1122–1126.
31. Martinon F, Petrilli V, Mayor A, Tardivel A, Tschopp J. Gout-associated uric acid crystals activate the NALP3 inflammasome. *Nature* 2006;440(7081):237–241.
32. Gupta R, Rost B, Relyveld E, Siber G. Adjuvant properties of aluminum and calcium compounds. In: Powell M, Newman M (eds.), *Vaccine Design: The Subunit and Adjuvant Approach*, Plenum Press, 1995, pp. 229–248.
33. O'Hagan DT, Singh M, Ulmer JB. Microparticle-based technologies for vaccines. *Methods* 2006;40(1):10–19.
34. Arca HC, Günbeyaz M, Senel S. Chitosan-based systems for the delivery of vaccine antigens. *Expert Rev Vaccines* 2009;8(7):937–953.
35. Şenel S. Chitosan-based particulate systems for non-invasive vaccine delivery. In: Jayakumar R, Prabaharan M, Muzzarelli RAA (eds.), *Chitosan for Biomaterials I.* Vol. 243, Springer, 2011, pp. 111–137.
36. He X, Jiang L, Wang F, *et al.* Augmented humoral and cellular immune responses to hepatitis B DNA vaccine adsorbed onto cationic microparticles. *J Controlled Release* 2005;107(2):357–372.

37. White K, Kawabata TT, Ladics G. Mechanisms of polycyclic aromatic hydrocarbon immunotoxicity. In: Dean JH (ed.), *Immunotoxicology and Immunopharmacology.* Vol. 2, 1994, pp. 123–149.
38. Kundsin RB, Walter CW. Detection of endotoxin on sterile catheters used for cardiac catheterization. *J Clin Microbiol* 1980;11(3):209–212.
39. Bondurant S, Ernster VL, Herdman R. *Committee on the Safety of Silicone Breast Implants.* National Acadamic Press, Institute of Medicine, 2000, pp. 540.
40. Vakharia DD, Liu N, Pause R, *et al.* Polycyclic aromatic hydrocarbon/metal mixtures: Effect on PAH induction of CYP1A1 in human HepG2 cells. *Drug Metab Dispos* 2001;29(7):999–1006.
41. Plata DeL, Hart AJ, Reddy CM, Gschwend PM. Early evaluation of potential environmental impacts of carbon nanotube synthesis by chemical vapor deposition. *Environ Sci Tech* 2009;43(21):8367–8373.
42. Saboori A, Newcombe D. Environmental chemicals with immunotoxic properties. In: Newcombe DS, Rose NR, Bloom JC (ed.), *Clinical Immunotoxicology,* Raven Press, 1992, pp. 365–400.
43. Yang K, Xing B. Desorption of polycyclic aromatic hydrocarbons from carbon nanomaterials in water. *Environ Pollut* 2007;145:529–537.
44. Chen RJ, Zhang Y, Wang D, Dai H. Noncovalent sidewall functionalization of single-walled carbon nanotubes for protein immobilization. *J Am Chem Soc* 2001;123(16): 3838–3839.
45. Kah M, Zhang X, Jonker MTO, Hofmann T. Measuring and modeling adsorption of PAHs to carbon nanotubes over a six order of magnitude wide concentration range. *Environ Sci Technol* 2011;45(14):6011–6017.
46. Hiraoka T, Izadi-Najafabadi A, Yamada T, *et al.* compact and light supercapacitor electrodes from a surface-only solid by opened carbon nanotubes with 2200 m^2 g^{-1} surface area. *Adv Funct Mater* 2010;20(3):422–428.
47. Deck CP, McKee GS, Vecchio KS. Synthesis optimization and characterization of multi-walled carbon nanotubes. *J Electron Mater* 2006;35(2):211–222.
48. Bendjemil B, Borowiak-Palen E, Graff A, *et al.* Elimination of metal catalyst and carbon-like impurities from single-wall carbon nanotube raw material. *Appl Phys A* 2004; 78:311–314.
49. Pulskamp K, Diabaté S, Krug HF. Carbon nanotubes show no sign of acute toxicity but induce intracellular reactive oxygen species in dependence on contaminants. *Toxicol Lett* 2007;168(1):58–74.
50. Veranth J, Ghandehari H, Grainger DW. *Nanoparticles in the lung. Comprehensive Toxicology.* Vol. 8, Elsevier, 2010.
51. Castranova V, Ma J, Yang HM, *et al.* Effect of exposure to diesel exhaust particles on the susceptibility of the lung to infection. *Environ Health Perspect* 2001;109(Suppl 4): 609–612.
52. Yeates DB, Mauderly JL. Inhaled environmental/occupational irritants and allergens: mechanisms of cardiovascular and systemic responses. Introduction. *Environ Health Perspect* 2001;109(Suppl 4):479–481.
53. Jakab GJ, Risby TH, Sehnert SS, Hmieleski RR, Gilmour MI. Suppression of alveolar macrophage membrane-receptor-mediated phagocytosis by model particle-adsorbate complexes: Physicochemical moderators of uptake. *Environ Health Perspect* 1990;89:169–174.

54. Gilmour MI, Daniels M, McCrillis RC, Winsett D, Selgrade M. Air pollutant-enhanced respiratory disease in experimental animals. *Environ Health Perspect* 2001;109(Suppl 4): 619–622.
55. White KL. Immunosuppression by polycyclic aromatic hydrocarbons: A structure-activity relationship in B6C3F1 and DBA/2 mice. *Immunopharmacology* 1985;9(3):155–164.
56. Brewer SH, Glomm WR, Johnson MC, Knag MK, Franzen S. Probing BSA binding to citrate-coated gold nanoparticles and surfaces. *Langmuir* 2005;21(20):9303–9307.
57. Nonckreman CJ, Fleith S, Rouxhet PG, Dupont-Gillain CC. Competitive adsorption of fibrinogen and albumin and blood platelet adhesion on surfaces modified with nanoparticles and/or PEO. *Colloids Surf B Biointerfaces* 2010;77(2):139–149.
58. Öncel Ç, Yürüm Y. Carbon nanotube synthesis via the catalytic CVD method: A review on the effect of reaction parameters. *Fullerenes Nanotubes Carbon Nanostruct* 2006;14(1):17–37.
59. Jadhav N, Shah MD, Nair K, Shirote P, Bhatia NM. Methods of carbon nanotube and nanohorn synthesis: A review. *Pharm Rev.* 2007;5, from http://www.pharmainfo.net/reviews/methods-carbon-nanotube-and-nanohorn-synthesis-review.
60. Duncan R, Stolojan V, Lekakou C. Manufacture of carbon multi-walled nanotubes by the arc discharge technique. *Proc World Congr Eng* 2007;2:1391–1395.
61. Ge C, Li W, Li Y, *et al.* Significance and systematic analysis of metallic impurities of carbon nanotubes produced by different manufacturers. *J Nanosci Nanotechnol* 2011;11(3):2389–2397.
62. Liu X, Gurel V, Morris D, *et al.* Bioavailability of nickel in single wall carbon nanotubes. *Adv Mater* 2007;19(19):2790–2796.
63. Liu X, Guo L, Morris D, Kane AB, Hurt RH. Targeted removal of bioavailable metal as a detoxification strategy for carbon nanotubes. *Carbon N Y* 2008;46(3):489–500.
64. Buzea C, Pacheco, II, Robbie K. Nanomaterials and nanoparticles: Sources and toxicity. *Biointerphases* 2007;2(4):MR17–71.
65. Kowolenko M, Mc-Cabe M, Lawrence D. Metal induced alterations of immunity. In: Newcombe D, Rose N, Bloom C (eds.), *Clinical Immunotoxicology*, Ravenial, 1992: pp. 401–419.
66. Dobrovolskaia MA, Aggarwal P, Hall JB, McNeil SE. Preclinical studies to understand nanoparticle interaction with the immune system and its potential effects on nanoparticle biodistribution. *Mol Pharm* 2008;5(4):487–495.
67. Dumortier H, Lacotte S, Pastorin G, *et al.* Functionalized carbon nanotubes are non-cytotoxic and preserve the functionality of primary immune cells. *Nano Lett* 2006;6(7):1522–1528.
68. Armstrong JK, Hempel G, Koling S, *et al.* Antibody against poly(ethylene glycol) adversely affects PEG-asparaginase therapy in acute lymphoblastic leukemia patients. *Cancer* 2007;110(1):103–111.
69. Ganson NJ, Kelly SJ, Scarlett E, Sundy JS, Hershfield MS. Control of hyperuricemia in subjects with refractory gout, and induction of antibody against poly(ethylene glycol) (PEG), in a phase I trial of subcutaneous PEGylated urate oxidase. *Arthritis Res Ther* 2006;8(1):R12–R21.
70. Knop K, Hoogenboom R, Fischer D, Schubert US. Poly(ethylene glycol) in drug delivery: Pros and cons as well as potential alternatives. *Angew Chem Int Ed Engl* 2010;49(36):6288–6308.

71. Richter AW, Akerblom E. Antibodies against polyethylene glycol produced in animals by immunization with monomethoxy polyethylene glycol modified proteins. *Int Arch Allergy Appl Immunol* 1983;70(2):124–131.
72. Sroda K, Rydlewski J, Langner M, Kozubek A, Grzybek M, Sikorski AF. Repeated injections of PEG-PE liposomes generate anti-PEG antibodies. *Cell Mol Biol Lett* 2005;10(1):37–47.
73. Moghimi SM, Andersen AJ, Hashemi SH, *et al.* Complement activation cascade triggered by PEG-PL engineered nanomedicines and carbon nanotubes: The challenges ahead. *J Controlled Release* 2010;146(2):175–181.
74. Vittaz M, Bazile D, Spenlehauer G, *et al.* Effect of PEO surface density on long-circulating PLA-PEO nanoparticles which are very low complement activators. *Biomaterials* 1996;17(16):1575–1581.
75. Arima Y, Toda M, Iwata H. Complement activation on surfaces modified with ethylene glycol units. *Biomaterials* 2008;29(5):551–560.
76. Al-Hanbali O, Rutt KJ, Sarker DK, Hunter AC, Moghimi SM. Concentration dependent structural ordering of poloxamine 908 on polystyrene nanoparticles and their modulatory role on complement consumption. *J Nanosci Nanotechnol* 2006;6(9–10):3126–3133.
77. Wen ZS, Xu YL, Zou XT, Xu ZR. Chitosan nanoparticles act as an adjuvant to promote both Th1 and Th2 immune responses induced by ovalbumin in mice. *Mar Drugs* 2011;9(6):1038–1055.
78. Ghendon Y, Markushin S, Akopova I, Koptiaeva I, Krivtsov G. Chitosan as an adjuvant for poliovaccine. *J Med Virol* 2011;83(5):847–852.
79. Vauthier C, Bertholon I, Labarre D. Integrated development of glycobiologics: From discovery to applications in the design of nanoparticular drug delivery systems. In: Gad SC (ed.), *Handbook of Pharmaceutical Biotechnology*, John Wiley and Sons, 2006, pp. 125–160.
80. Cherukuri P, Gannon CJ, Leeuw TK, *et al.* Mammalian pharmacokinetics of carbon nanotubes using intrinsic near-infrared fluorescence. *Proc Natl Acad Sci U S A* 2006;103(50):18882–18886.
81. Bertholon I, Vauthier C, Labarre D. Complement activation by core-shell poly(isobutylcyanoacrylate)–polysaccharide nanoparticles: Influences of surface morphology, length, and type of polysaccharide. *Pharm Res* 2006;23(6):1313–1323.
82. Sadana A. Protein adsorption and inactivation on surfaces. Influence of heterogeneities. *Chem Rev* 1992;92(8):1799–1818.
83. Cedervall T, Lynch I, Foy M, *et al.* Detailed identification of plasma proteins adsorbed on copolymer nanoparticles. *Angew Chem Int Ed Engl* 2007;119(30):5856–5858.
84. Lynch I, Dawson KA. Protein-nanoparticle interactions. *Nanotoday* 2008;3(1–2):40–47.
85. Moghimi SM, Hunter AC, Murray JC. Long-circulating and target-specific nanoparticles: Theory to practice. *Pharmacol Rev* 2001;53(2):283–318.
86. Li S-D, Huang L. Nanoparticles evading the reticuloendothelial system: Role of the supported bilayer. *Biochim Biophys Acta* 2009;1788(10):2259–2266.
87. Ratner BD, Hoffman AS, Schoen FJ, Lemons JE. *Biomaterials Science — An Introduction to Materials in Medicine*, Elsevier, 2004.
88. Chen BX, Wilson SR, Das M, Coughlin DJ, Erlanger BF. Antigenicity of fullerenes: Antibodies specific for fullerenes and their characteristics. *Proc Natl Acad Sci* 1998;95(18):10809–10813.

89. Salvador-Morales C, Flahaut E, Sim E, Sloan J, H. Green ML, Sim RB. Complement activation and protein adsorption by carbon nanotubes. *Mol Immunol* 2006;43(3):193–201.
90. Demoy M, Andreux JP, Weingarten C, Gouritin B, Guilloux V, Couvreur P. *In vitro* evaluation of nanoparticles spleen capture. *Life Sci* 1999;64(15):1329–1337.
91. Donaldson K, Aitken R, Tran L, *et al.* Carbon nanotubes: A review of their properties in relation to pulmonary toxicology and workplace safety. *Toxicol Sci* 2006;92(1):5–22.
92. Gatti AM, Tossini D, Gambarelli A, Montanari S, Capitani F. Investigation of the presence of inorganic micro-and nanosized contaminants in bread and biscuits by environmental scanning electron microscopy. *Crit Rev Food Sci Nutr* 2008;49(3):275–282.
93. Mühlfeld C, Rothen-Rutishauser B, Blank F, Vanhecke D, Ochs M, Gehr P. Interactions of nanoparticles with pulmonary structures and cellular responses. *Am J Physiol Lung Cell Mol Physiol* 2008;294(5):L817–L829.
94. Choi HS, Ashitate Y, Lee JH, *et al.* Rapid translocation of nanoparticles from the lung airspaces to the body. *Nat Biotech* 2010;28(12):1300–1303.
95. Inoue K, Yanagisawa R, Koike E, Nishikawa M, Takano H. Repeated pulmonary exposure to single-walled carbon nanotubes exacerbates allergic inflammation of the airway: Possible role of oxidative stress. *Free Radic Biol Med* 2010;48(7):924–934.
96. Liliensiek B, Weigand MA, Bierhaus A, *et al.* Receptor for advanced glycation end products (RAGE) regulates sepsis but not the adaptive immune response. *J Clin Invest* 2004;113(11):1641–1650.
97. Klippstein R, Fernandez-Montesinos R, Castillo PM, Zaderenko AP, Pozo D. Silver nanoparticles interactions with the immune system: Implications for health and disease. In: Perez DP (ed.), *Silver Nanoparticles*, In Tech, 2010, pp. 309–324.
98. Dobrovolskaia MA, Neun BW, Clogston JD, Ding H, Ljubimova J, McNeil SE. Ambiguities in applying traditional Limulus Amebocyte Lysate tests to quantify endotoxin in nanoparticle formulations. *Nanomedicine (Lond)* 2010;5(4):555–562.
99. Giljohann DA, Seferos DS, Daniel WL, Massich MD, Patel PC, Mirkin CA. Gold nanoparticles for biology and medicine. *Angew Chem Int Ed Engl* 2010;49(19): 3280–3294.
100. Gorbet MB, Sefton MV. Leukocyte activation and leukocyte procoagulant activities after blood contact with polystyrene and polyethylene glycol-immobilized polystyrene beads. *J Lab Clin Med* 2001;137(5):345–355.
101. Gorbet MB, Sefton MV. Endotoxin: The uninvited guest. *Biomaterials* 2005; 26(34):6811–6817.
102. Ragab AA, Van De Motter R, Lavish SA, *et al.* Measurement and removal of adherent endotoxin from titanium particles and implant surfaces. *J Orthop Res* 1999;17(6): 803–809.
103. Greenfield EM, Bechtold J. What other biologic and mechanical factors might contribute to osteolysis? *J Am Acad Orthop Surg* 2008;16(Suppl 1):S56–62.
104. Huttunen K, Jussila J, Hirvonen MR, Iivanainen E, Katila ML. Comparison of mycobacteria-induced cytotoxicity and inflammatory responses in human and mouse cell lines. *Inhal Toxicol* 2001;13(11):977–991.
105. Schwab LP, Xing Z, Hasty KA, Smith RA. Titanium particles and surface-bound LPS activate different pathways in IC-21 macrophages. *J Biomed Mater Res B Appl Biomater* 2006;79B(1):66–73.

106. Gretzer C, Werthén M, Thomsen P. Apoptosis and cytokine release in human monocytes cultured on polystyrene and fibrinogen-coated polystyrene surfaces. *Biomaterials* 2002;23(7):1639–1648.
107. Darkow R, Groth T, Albrecht W, Lützow K, Paul D. Functionalized nanoparticles for endotoxin binding in aqueous solutions. *Biomaterials* 1999;20(14):1277–1283.
108. Karl DW, Magnusson JC, Carr PW, Flickinger MC. Preliminary assessment of removal of pyrogenic lipopolysaccharides with colloidal zirconia adsorbents. *Enzyme Microb Technol* 1991;13(9):708–715.
109. Cho DR, Shanbhag AS, Hong C-Y, Baran GR, Goldring SR. The role of adsorbed endotoxin in particle-induced stimulation of cytokine release. *J Orthop Res* 2002; 20(4):704–713.
110. Greenfield EM, Bi Y, Ragab AA, Goldberg VM, Nalepka JL, Seabold JM. Does endotoxin contribute to aseptic loosening of orthopedic implants? *J Biomed Mater Res A* 2005;72B(1):179–185.
111. Castner D. Chemical modification of surfaces. In: Czanderna A, Powell C, Made T (eds.), *Specimen Handling, Beam Effects and Depth Profiling*, Plenum Press, 1998, pp. 209–238.
112. Castner DG, Watson PR, Chan IY. X-ray absorption spectroscopy, X-ray photoelectron spectroscopy, and analytical electron microscopy studies of cobalt catalysts. 2. Hydrogen reduction properties. *J Phys Chem* 1990;94(2):819–828.
113. Baer D, Engelhard M, Gaspar D, *et al.* Challenges in applying surface analysis methods to nanoparticles and nanostructured materials. *J Surf Anal* 2005;12(2):101–108.
114. Techane S, Baer DR, Castner DG. Simulation and modeling of self-assembled monolayers of carboxylic acid thiols on flat and nanoparticle gold surfaces. *Anal Chem* 2011;83(17):6704–6712.
115. Zorn G, Dave SR, Gao X, Castner DG. Method for determining the elemental composition and distribution in semiconductor core-shell quantum dots. *Anal Chem* 2011;83(3):866–873.
116. Techane SD, Gamble LJ, Castner DG. X-ray photoelectron spectroscopy characterization of gold nanoparticles functionalized with amine-terminated alkanethiols. *Biointerphases* 2011;6(3):98–104.
117. Min H, Kim Y, Yu H, *et al.* Probing the surface of organic and bioconjugated nanocrystals by using mass spectrometric imaging. *Chemistry* 2008;14(28):8461–8464.
118. Morris-Cohen A, Donakowski M, Knowles K, Weiss E. The effect of a common purification procedure on the chemical composition of the surfaces of CdSe quantum dots synthesized with trioctylphosphine oxide. *J Phys Chem C* 2010;114(2):897–906.
119. Techane SD, Gamble LJ, Castner DG. Multi-technique characterization of self-assembled carboxylic acid terminated alkanethiol monolayers on nanoparticle and flat gold surfaces. *J Phys Chem C Nanomater Interfaces* 2011;115(19):9432–9441.
120. de Beer AGF, Roke S. Nonlinear Mie theory for second-harmonic and sum-frequency scattering. *Phys Rev B* 2009;79(15):155420–155428.
121. de Beer AG, Roke S. Obtaining molecular orientation from second harmonic and sum frequency scattering experiments in water: Angular distribution and polarization dependence. *J Chem Phys* 2010;132(23):234702–234708.

Chapter 6

Nanoparticle Interaction with Plasma Proteins as It Relates to Biodistribution

Lennart Treuel[*,†] *and G. Ulrich Nienhaus*[*,‡,**]

**Institute of Applied Physics*
and Center for Functional Nanostructures (CFN)
Karlsruhe Institute of Technology (KIT)
76128 Karlsruhe, Germany
†Institute of Physical Chemistry, University of Duisburg-Essen
45141 Essen, Germany
‡Department of Physics, University of Illinois at Urbana-Champaign
Urbana, IL 61801, USA

Whenever nanoparticles (NPs) come into contact with biological fluids, a layer of proteins ("protein corona") may adsorb onto their surfaces. This corona enshrouds the NPs and largely defines their biological identity. The nature of the corona and the efficiency of its formation can be a decisive factor for the biological response of an organism to NP exposure. Here, we describe methods presently available for determining the composition of the protein corona under physiological conditions and recent findings in this developing field. Mechanistic aspects of protein adsorption to NP surfaces are reviewed in the context of the current literature. Additionally, the effects of protein corona formation on cellular uptake and biodistribution of NPs are discussed. Finally, we comment on the significant role of the protein corona for contemporary nanomedical applications.

**Corresponding author. E-mail: uli@uiuc.edu

1. Introduction

While the use of nanoparticles (NPs) in scientific[1] and commercial applications is rapidly increasing,[2–4] a comprehensive risk assessment of nanotechnology is still incomplete,[5–8] largely due to a lack of mechanistic knowledge regarding the interactions of nanoscale objects with biological systems. Even for intended therapeutic applications of NPs, knowledge about their mechanisms of action and potential risks is still lacking.[9–11] Nevertheless, more than 20 NP-based therapeutics have currently been approved for clinical use by the U.S. Food and Drug Administration (FDA).[12–14]

Since NPs are of similar size as typical biomacromolecular assemblies, they may evade natural defenses of biological organisms, utilize the endocytosis machinery for intruding cells, and thereby, generate permanent cell damage.[11,15] Whenever NPs come into contact with biological systems, physical and chemical interactions take place between NP surfaces and biological components (e.g., proteins, membranes, phospholipids, endocytic vesicles, organelles, DNA, etc.) which is known as the bio-nano interface. By now, it has been well established that proteins from body fluids bind to NP surfaces[16–21] upon their exposure to an organism. Consequently, living systems interact with protein-coated rather than naked NPs.[18,22] The so-called "protein corona" forming around NPs[18,21–23] largely defines their biological identity, and the nature and stability of this corona can be decisive factors for the biological response of an organism to NP exposure.[24] Nel and co-authors[25] presented an in-depth discussion of the basic physical interactions taking place at the bio-nano interface. Table 1 summarizes the relevant forces controlling the interfacial interactions between nanomaterials and biosystems.

Understanding the formation and persistence of the protein corona is a formidable task which is, however, of paramount importance not only for the elucidation, interpretation, and assessment of biological effects of unintended exposure to NPs, but also for their intended use in nanomedicine. In the latter context, targeted delivery of NPs is often needed, which requires exquisite control over the formation of the protein layer. Essentially, this formation is a competition of proteins (and other biomolecules) for binding to the NP surface. The notions "hard" and "soft" protein corona have been introduced, referring to strongly and weakly bound proteins on the surface.[18] It has been suggested that the "soft" corona forms on short time scales (seconds to minutes) and subsequently, evolves into a "hard" corona over times in the order of hours.[26]

Protein adsorption onto NP surfaces may involve conformational changes of the proteins,[27–29] leading to a loss of biological function.[30,31] Structural

Table 1. Key forces controlling the interfacial interactions between nanomaterials and biological systems. (Adapted from Nel *et al.*[25])

Force	Origin and nature	Range (nm)	Possible impact on the interface
Hydrodynamic interactions	Convective drag, shear, lift, and Brownian diffusion are often influenced at nanoscale separations between interacting interfaces	10^2 to 10^6	Increases the collision frequencies between NPs and other surfaces responsible for transport
Electrostatic interactions	Charged interfaces attract counter-ions and repel co-ions (Coulomb forces) giving rise to the formation of an electrostatic double layer	1 to 100	Overlapping double layers are generally repulsive for equally charged objects but can be attractive for oppositely charged objects
Electrodynamic interactions	van der Waals (vdW) interactions	1 to 100	Attractive in aqueous media, substantially smaller in biological media
Solvent interactions	Lyophilic materials interact favorably with solvent molecules	1 to 10	Lyophilic materials are thermodynamically stable in the solvent and do not aggregate
	Lyophobic materials interact unfavorably with solvent molecules		Lyophobic materials are spontaneously expelled from the bulk of the solvent and forced to minimize the contact surface.
Steric interactions	Polymeric species adsorbed onto NPs give rise to repulsive interactions with other interfaces	1 to 100	Generally increase stability of individual NPs but can interfere with cellular uptake, especially when surface polymers are highly water-soluble
Polymer bridging interactions	Polymeric species adsorbed onto inorganic NPs containing charged functional groups can be attracted by oppositely charged moieties on a substrate surface	1 to 100	Generally promote aggregation, particularly when charge functionality is carboxylic acid and dispersed in aqueous media containing calcium ions

changes in the protein may lead to the exposure of "cryptic" peptide epitopes.[22,32] These surfaces may function as new antigenic sites and may elicit an immune response,[33,34] which, if launched against a self-protein, could promote autoimmune diseases.[25] At present, the driving forces and mechanistic details of protein structural changes at NP surfaces still remain elusive.[25,35–37]

2. Nanoparticle Separation from Plasma

Under physiological conditions, for example, in nanomedical applications, NP surfaces will always be exposed to a large excess of proteins.[22] The overall protein concentration in typical body fluids (e.g., in blood, lung, gut) and intracellular environments can be as high as 0.35 g/mL.[22] These fluids contain more than 3,000 different proteins[38] with widely differing individual concentrations. While highly abundant proteins will likely dominate the protein corona initially after exposure, proteins with a lower abundance but higher affinity might prevail on longer time scales.[18,25]

Studying plasma protein adsorption to NPs under physiological conditions therefore requires experimental solutions for separating corona-containing NPs from excess plasma. Common separation techniques are based on centrifugation, dialysis, gel filtration, and where applicable, magnetic separation.[19,26,39–41] Among these, centrifugation is the most prevalent method.[39] In innumerous studies, NPs were incubated with serum until NP–protein complexes had formed, and then centrifuged once or repeatedly. However, solution volumes and washing steps can severely impact on the outcome by affecting the association and dissociation of proteins. Moreover, protein concentration gradients can arise during centrifugation due to large proteins or protein aggregates sedimenting alongside the NPs. This can introduce a potential bias toward an increasing amount of large proteins found in the corona.

Dialysis-based methods are also utilized, but frequently need more samples and also suffer from the aforementioned possible dilution influence on the corona composition. Moreover, size-exclusion chromatography (SEC, gel-filtration) is used and chromatographic elution profiles of protein mixed with NPs are compared between profiles from pure protein and pure NPs.[18]

Magnetic separation of magnetic iron oxide NPs was compared with other separation techniques including centrifugation, gel-filtration, and membrane-based filtration.[41] The studies showed similar adsorption patterns, but in contrast to the other approaches, albumin was found only in negligible amounts on magnetically separated NPs. In a recent study of plasma protein adsorption onto ultrasmall superparamagnetic iron oxide (USPIO) using magnetic separation,[40] the absence of significant amounts of albumin on the NP surfaces was confirmed. This observation illustrated as to how the results of protein corona analysis can depend on the separation method and underlined that utmost care must be taken in any separation procedure.

In summary, all current methods that are based on the separation of NPs from dissolved biomacromolecules rely on the assumption that a "hard" protein corona forms around the NPs and is largely independent of concentration changes in their surroundings. The validity of this assumption strongly depends on the various NP properties discussed above and critically determines the results.

3. Protein Identification

A variety of experimental techniques is available for studying the structure of proteins in solution and in NP–protein aggregates, including circular dichroism,[23,42] Fourier transform infrared spectroscopy,[43] Raman spectroscopy, surface-enhanced Raman spectroscopy (SERS),[44] fluorescence spectroscopy,[45,46] SEC,[47–52] isothermal titration calorimetry (ITC),[53,54] and surface plasmon resonance (SPR).[18,55] Moreover, one- and two-dimensional polyacrylamide gel electrophoresis (1D-/2D-PAGE) are commonly utilized[39] to identify corona proteins by comparison of 2D protein gels to a 2D master map of the plasma.[56–63] To improve the read-out of this technique, N-terminal sequencing on individual spots was employed for the identification of individual proteins.[62–64]

Mass spectrometry (MS) can reveal the composition of individual protein spots of the gel.[17,56,65–67] In addition, specific proteins may be identified by immunoblotting and Western blotting.[68–71] Tenzer *et al.*[72] recently employed label-free liquid chromatography (LC)-MS, alongside with 1D-/2D-gel electrophoresis and immunoblotting for a quantitative composition analysis of the protein corona on amorphous silica NPs (nominal diameters of 8, 20, and 125 nm, carrying a negative surface charge) exposed to human blood serum. In this comprehensive study, they demonstrated the potential of LC-MS for studies of the protein corona composition and identified 125 proteins. The abundance of proteins in the corona did not simply reflect their relative abundance in the plasma, but revealed an enrichment of specific proteins. Figure 1 shows the classification of the proteins identified in this study according to their biological roles in the blood system. These results revealed that proteins associated with complement activation, lipoproteins, coagulation proteins, and proteins grouped as "tissue leakage" are enriched in the corona relative

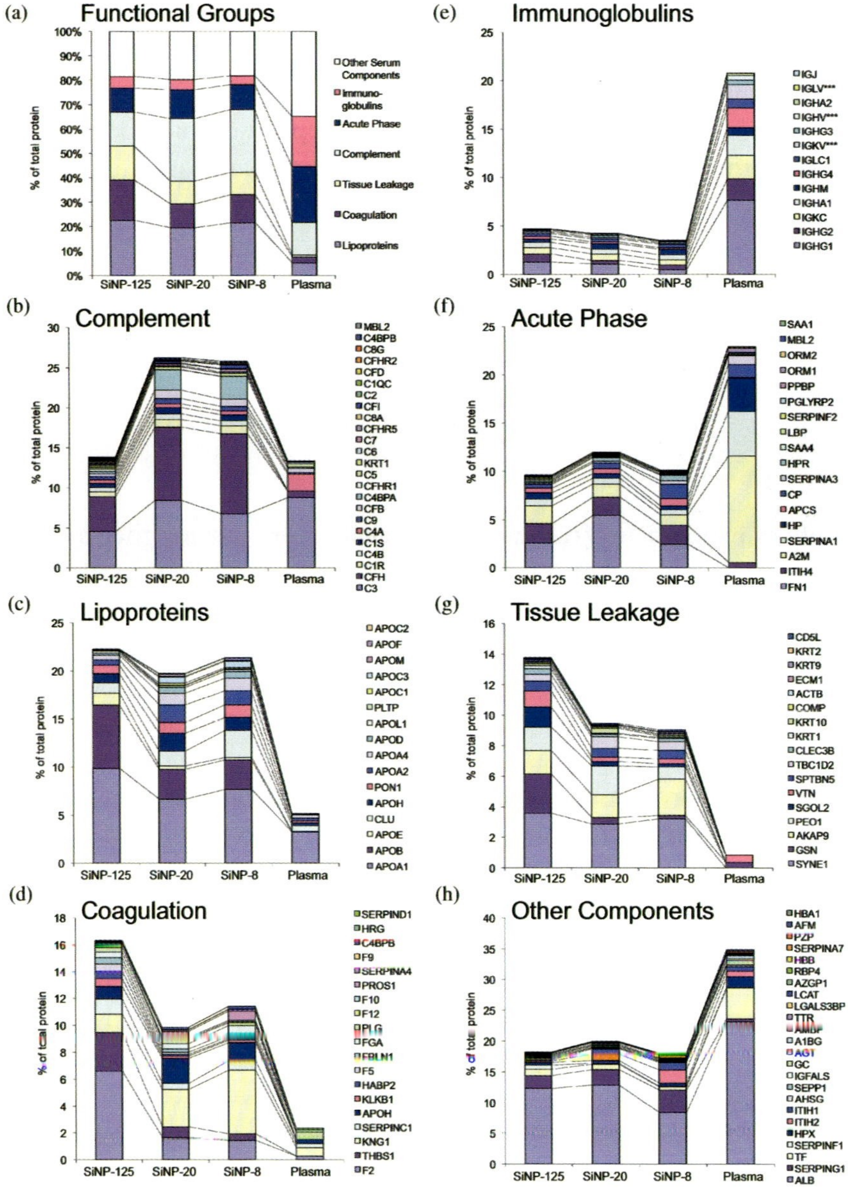

Figure 1. Classification of proteins identified in the corona around silica NPs (SiNPs) (nominal diameters of 8, 20, and 125 nm) according to their functions. (a) The identified proteins were classified according to biological processes of the blood system; their relative fractions on NPs with three different sizes were compared to crude plasma. Significant enrichments of proteins involved in (b) complement activation, (c) lipoproteins, (d) coagulation as well as (g) proteins grouped as "tissue leakage" are apparent. In contrast with this behavior, (e) immunoglobulins, (f) acute phase response proteins, and (h) serum albumin showed lower affinities for the SiNPs. (Reproduced with kind permission from Tenzer *et al.*[72])

to their abundance in plasma. On the other hand, the relative abundances of immunoglobulins, acute phase response proteins, and serum albumin were decreased in the corona. However, even though the abundance of albumin in the corona was decreased as compared to free serum, it is still a major constituent of the corona. The implications of these intriguing findings for cellular uptake and biological transport remain to be uncovered. Numerous other studies exist on the topic and Table 2 presents a summary of the most abundant proteins identified in the coronae around a wide variety of different NPs.

Table 2. Outline of various NPs and the proteins they bind, the methods used to isolate the proteins, and the methods used to separate and identify individual proteins. (Adapted from Aggarval *et al.*[39])

Nanoparticles	Identified proteins	Method of protein isolation	Method of protein separation and identification [[1]comparison to known 2D gel of plasma]	Reference
Liposomes	Albumin, fibrinogen, apolipoproteins, IgG, α1-antitrypsin, α2-macroglobulin, IgM	Gel filtration	2D-PAGE[1]	61
Single-walled carbon nanotubes	Albumin	Centrifugation	Western blot	65
Solid lipid NPs with Tween 80	Fibrinogen, IgG, IgM, apolipoproteins (including ApoE), transthyretin	Centrifugation, gel filtration	2D-PAGE[1]	73
Poly(lactic acid) NPs with PEG	Albumin, fibrinogen, apolipoproteins, IgG	Centrifugation	2D-PAGE[1] and N-terminal sequencing	62
Polyhexadecylc-yanoacrylate NPs	Albumin, apolipoproteins, IgG, transferrin	Centrifugation	2D-PAGE[1] and MS	56,68
Poly(ε-caprolactone) NPs	IgG, apolipoproteins	Centrifugation	2D-PAGE[1] and N-terminal sequencing	63
Polycyanoacrylate NPs	Albumin, IgG, IgM, fibrinogen, apolipoproteins	Centrifugation	2D-PAGE[1]	74

(*Continued*)

Table 2. (*Continued*)

Nanoparticles	Identified proteins	Method of protein isolation	Method of protein separation and identification	Reference
Polycyanoacrylate NPs	Albumin, IgG, IgM, fibrinogen, apolipoproteins	Centrifugation	2D-PAGE[1]	74
Iron oxide NPs	Albumin, IgG, IgM, fibrinogen Albumin, IgM, fibrinogen, C3b, apolipoprotein A-1 Albumin, IgG, IgM, fibrinogen, C3b IgG, IgM, fibrinogen	Centrifugation Gel filtration Static filtration Magnetic separation	2D-PAGE[1]	41
Various polymer/ copolymer composition NPs	Albumin, IgG, fibrinogen, apolipoproteins	Centrifugation	2D-PAGE[1]	75
Poly(isobutylcy-anoacrylate) with Dextran	Albumin, IgG, fibrinogen, apolipoproteins, serotransferrine, transthyretine	Centrifugation	2D-PAGE[1]	76
Single- and double-walled carbon nanotubes	Albumin, fibrinogen, apolipoproteins, C1q	"Affinity chromatography," centrifugation of beads	SDS-PAGE with N-terminal sequencing, MS, or Western blot	77
PBCA NPs with polysorbate 80	Albumin, IgG, fibrinogen, IgM, apolipoproteins	Centrifugation	2D-PAGE[1]	58
Solid lipid NPs with poloxamer or poloxamine coating	Albumin, fibrinogen, apolipoproteins	Centrifugation	2D-PAGE[1]	57
Sulfonated polystyrene NPs SiO_2 NPs	Fibrinogen, IgκC, apolipoprotein A-I, serum albumin Apolipoprotein A-I, glycoprotein, serum albumin	Centrifugation	Differential centrifugal sedimentation, PAGE, LC-MS	26
Amorphous silica NPs	Serum albumin, apolipoprotein A-I, prothrombin, apolipoprotein B-100	Centrifugation	Label-free LC-MS, 1D-PAGE, 2D-PAGE, immunoblotting	72

The influence of different plasma concentrations on the composition of the protein corona around silica and sulfonated polystyrene NPs was studied by Monopoli *et al.*[26] Using differential centrifugal sedimentation (DCS), dynamic light scattering (DLS), and zeta potential measurements, they analyzed protein corona formation both *in situ* (i.e., in the presence of plasma) as well as after spinning down, separating, and washing of the NP–protein complexes. They reported that the structures of the NP protein complexes from *in situ* studies were almost identical to the structures found *in vitro* after isolation from excess plasma, indicating a high biological relevance of the *in vitro* findings.

4. Mechanistic Aspects of Protein Corona Formation

Understanding both causes and effects of protein corona formation requires a physicochemical picture on the molecular level. In recent years, numerous studies have been performed trying to shed light on the mechanisms responsible for forming and shaping the protein corona. The biophysical aspects of corona formation have been reviewed and the biological implications have been discussed.[78]

The physical and chemical characteristics of NPs determine their interactions with the surrounding medium by promoting the adsorption of ions, proteins, natural organic materials, and detergents, particle dissolution, or even by allowing the free surface energy to be minimized by surface restructuring.[25,79,80] The effect of degradation on these parameters should always be considered whenever the biological action of the NPs is being studied. Changes in these characteristics can occur upon protein adsorption and may then affect their internalization by biological systems. In addition, even though the as-synthesized NPs may be rather insoluble in aqueous solvents, interactions with proteins may dramatically change this behavior. Detailed knowledge of the mechanistic steps involved in these processes is, however, still scarce.[81]

To obtain a picture of the interaction process at the molecular level, Shang *et al.*[82] studied the structure, thermodynamic and kinetic stabilities, and the activity of cytochrome c (cyt c) on silica NPs. They found evidence that the structure, function, and stability of cyt c adsorbed onto silica NPs strongly depended on NP size.

Nienhaus *et al.*[21] quantitatively analyzed the adsorption of human serum albumin (HSA) onto small (10–20 nm in diameter) polymer-coated, fluorescently labeled FePt and CdSe/ZnS NPs displaying carboxyl functions on the surface. By using fluorescence correlation spectroscopy (FCS), they measured the thickness of the HSA corona formed around these NPs with sub-nanometer precision. Their result, 3.3 nm, confirmed that a single protein monolayer formed around the NP. From the response of the corona thickness to varying

HSA concentrations, apparent binding affinities associated with the NP–protein interaction were determined. Recently, the Nienhaus group has extended their FCS studies to various other important plasma proteins. For transferrin adsorption to FePt NPs, they reported a monolayer with a thickness of 7 nm.[83] This value can also be related to the molecular dimensions of the protein molecule as inferred from the crystal structure.

In a subsequent study, Maffre *et al.*[84] used dual-focus fluorescence correlation spectroscopy (2fFCS) to measure the adsorption of two human blood serum lipoproteins (apolipoprotein A-I and apolipoprotein E4) and for reference, again HSA, onto negatively charged FePt NPs carrying carboxyl groups on their surface. The formation of a protein corona was observed for all three proteins. An intriguing novel aspect of this work was the attempt to explain both the binding affinities and the monolayer thicknesses on the basis of known structural properties of the proteins (Figure 2). The absolute increase in hydrodynamic radius due to corona formation was found to be well correlated with the molecular shapes of the proteins. The electrostatic properties of the proteins were computed by structure-based calculations of the surface potentials. The detailed analysis showed that the binding affinities were not determined by the net charge of the protein, but rather appeared to be related to the presence of positively charged patches on the protein surface. The specific orientations of the corona proteins for optimal electrostatic interaction between the positively charged patches on the proteins and the negatively charged, carboxylated surfaces of the NPs were also revealed. Binding in these orientations led to a protein corona with a thickness compatible with the observed NP size increase. The intriguing revelation of a close structure–affinity relation by this work opened a promising perspective toward a better understanding of NP–protein interactions at the molecular level and may ultimately even lead to a predictability of biological responses.

The effect of the chemical composition of the NP surface on protein corona formation is an important issue. Particularly for nanomedical and biotechnological applications, it is of great interest to identify surface properties that can enhance or diminish the ability of NP surfaces to interact with proteins and other biomolecules.

To identify the influence of polymer coatings on the corona formation, Treuel *et al.*[23] determined apparent affinities for the interaction of serum albumin with citrate-coated silver and gold NPs as well as PVP-coated silver NPs and polystyrene NPs. Using circular dichroism (CD) spectroscopy, they studied the thermodynamic and structural aspects of NP–protein interactions. A fundamental aspect of this work was the elucidation of the effect of the NP surface composition on its interaction with proteins. Their data underlined the importance of the original surface functionality of the NPs, which determines the interaction with the

(a)

(b)

(c)

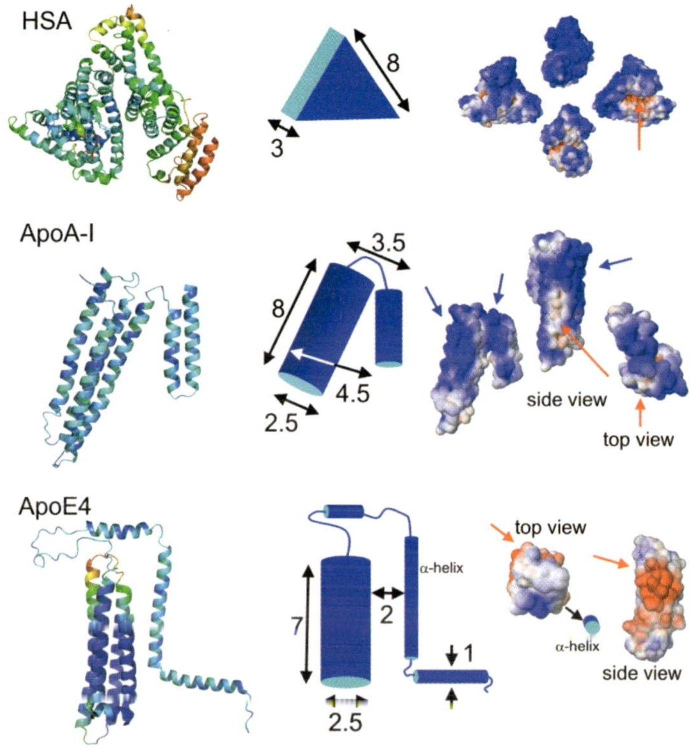

Figure 2. Structural representations of proteins. (a) HSA, (b) apoA-I, and (c) apoE4. Left column: Cartoon representations of HSA (protein data bank accession (pdb) code 1AO6), apoE4 (pdb code 1GS9), and apoA-I (pdb code 2A01). For apoE4, only the X-ray structure of the 22-K domain (four-helix bundle) is known. Center column: Simplified protein structures including approximate dimensions (in nm). Right column: Space-filling models colored according to surface electrostatics at pH 7.4 (blue: negative potential, red: positive potential range –5 kT/e – +5 kT/e; calculated online using the web site http://kryptonite.nbcr.net/pdb2pqr/.) (Figure reproduced with kind permission from Maffre *et al.*[84])

adsorbed proteins. Moreover, the persistence of the surface functionality was found to be an important parameter affecting protein corona formation in the physiological context. Evidently, only a molecular surface functionality on the NP surface that is sufficiently stable under physiological conditions will be able to play a significant role in shaping the *in vivo* protein corona.

5. Effect of Protein Binding on Cellular Uptake and Biodistribution

In principle, NPs can be taken up by different pinocytotic mechanisms — macropinocytosis, clathrin- or caveolae-mediated endocytosis, and mechanisms independent of clathrin and caveolin.[85] In addition, it has been reported that particles may also enter cells via other routes.[86] The cellular uptake of NPs has been shown to depend on NP size,[87,88] shape,[87] surface charge,[89–91] and surface chemistry.[89,92,93]

It was pointed out that the surface properties of NPs can influence their cytotoxicity as well as their intracellular fate.[92] The dependence of the biological action of NPs on size[94] and surface coating was described.[5,6,95,96] It was also observed that NPs can translocate from the tissue where they have originally been absorbed to other target tissues.[5,96] This issue adds considerable complexity to the biodistribution problem.

Correlations between protein corona formation and biological behavior have been established. For example, it has been reported that immunoglobulin binding caused particle opsonization, thereby promoting receptor-mediated phagocytosis.[97] Decreased protein absorption on injected polyethylene glycol (PEG)-coated particles led to longer circulation times and altered biodistribution.[97,98] It was also suggested that proteins constituting the corona may be transported across membranes alongside the NPs, bringing them into biological entities which they would not normally reach.[22]

Jiang *et al.*[83] quantitatively analyzed the adsorption of human transferrin onto small (radius approximately 5 nm) polymer-coated FePt NPs using FCS. They demonstrated the formation of a protein corona consisting of a single transferrin monolayer around the NP surfaces. Using confocal fluorescence microscopy, they also observed the uptake of these NPs by human cancer (HeLa) cells. Cellular uptake was found to be suppressed by this corona compared with the bare NPs.

Moreover, Jiang *et al.*[99] studied the uptake and intracellular transport of D-penicillamine coated quantum dots (DPA-QDs) of 4 nm radius by live HeLa cells using spinning disk and 4Pi confocal microscopies. Unlike larger NPs, these small DPA-QDs were observed to accumulate at the plasma membrane prior to internalization, and the uptake efficiency scaled nonlinearly with the NP concentration. Both observations indicated that a critical threshold density has to be exceeded for triggering the internalization process. Using specific inhibitors, they showed that DPA-QDs were predominantly internalized by clathrin-mediated endocytosis and to a smaller extent, macropinocytosis. They found clusters of DPA-QDs in endosomes, which were actively transported toward the perinuclear region. Subsequently, a significant fraction of endocytosed DPA-QDs was localized in lysosomes, while others were actively transported to the cell periphery and exocytosed.

These detailed studies illustrated the complexity of the situation and emphasized the need for comprehensive mechanistic studies in this field.

Protein corona formation and its biological effects cannot be considered independently from other physical and chemical aspects of the NPs. Some characteristics affecting cellular uptake might be altered by corona formation (e.g., charge, chemical surface composition, hydrophobicity/hydrophilicity, colloidal stability), whereas others may change very little or remain unchanged (e.g., size, chemical, and physical properties of the core material, optical properties). A deeper understanding of protein binding remains a critical challenge for assessing transport and effects of NPs in biological systems.

6. The Protein Corona in Medical Applications

Numerous efforts have been made to modify the surfaces of NPs for the directed transport of drugs or vaccines,[8,35,100–104] however, largely without any detailed mechanistic knowledge. In particular, polymeric materials have been used to modulate the physicochemical characteristics of NPs (e.g., hydrophobicity, zeta potential, drug release properties, biological behavior). These include synthetic poly(lactic acids) (PLA), PEG, poly(lactic-coglycolic acids) (PLGA), poly(ε-caprolactone) (PCL),[105,106] poly(methyl methacrylates), and poly(alkyl cyanoacrylates), or natural polymers (e.g., albumin, gelatin,[107] alginate,[108] collagen, chitosan).[103] Polyesters and combinations of polyesters with other polymers[35] such as PLGA[109–113] and PLA[114–116] were reported to be highly biocompatible and biodegradable[117,118] and have been used for numerous *in vivo* applications such as implants and controlled drug release.[119–122]

More recently, formulations based on natural polymers have been developed and some of their medical implications were studied.[123–126] However, interactions of these surfaces with proteins are currently elusive and mechanistic information remains largely unknown. Specifically, the physical and chemical characteristics of the surface and their mechanistic influence on interactions with proteins and cells are still unresolved.

PEG was described to be amphiphilic,[127] a property resulting in an enhanced solubility of the coated NPs in hydrophilic and lipophilic solvents. Consequently, PEG coatings were frequently investigated for biomedical applications.[128] They were shown to be biocompatible[129] and non-immunogenic,[130] and it was reported that PEG coatings can be used to increase blood circulation time after intravenous injection by reducing opsonization by proteins and hindering uptake by macrophages.[98] The coating of NPs with PEG also resulted in a decrease in renal clearance,[131] especially when using PEG chains larger than 20 kDa.[127] Again, mechanistic details, likely determined by PEG–protein interactions, are still unresolved.

Nanomedical approaches using gold NPs are also of broad interest in many studies[132–135] and (potential) applications include photo-diagnostics and photothermal therapy of cancers and other main diseases.[136] The use of gold NPs in drug (and DNA) delivery is a very promising future application.[135] DNA delivery vectors based on gold NPs were developed and showed a promising potential for the creation of transfection vectors.[137] Oligodeoxynucleotide-conjugated gold NPs were investigated for gene therapy[138] and moreover, gold NPs were successfully employed in the cryosurgery of tumor tissues.[139] A study of corona formation around gold NPs revealed a strong loss of protein structure in the corona around the metallic surface, which was decreased when a polymer coating was applied.[23] Nevertheless, biological implications of these results again remain elusive.

7. Final Remarks

Information about the mechanistic aspects of NP–protein interactions and on the importance of this interaction for cellular uptake and intracellular transport remains scarce.[16,35,37] Optimal NP surface properties for NP applications are being fiercely debated; yet, it is fully accepted that NP surface properties are of utmost importance.[35] In recent years, researchers have become increasingly interested in a better understanding of NP–protein interactions and the biological implications associated herewith.[16,19]

The use of NPs in targeted delivery applications in nanomedicine needs a reliable foundation on the molecular scale. Only if the impact of physical and chemical NP characteristics on protein adsorption are well understood, the protein corona can then be systematically utilized. The circumstance that both desired and undesired biological effects are affected by the nature of the protein corona further underlines the importance of advancing our knowledge about its formation and biological implications.

Acknowledgments

This work was supported by the Deutsche Forschungsgemeinschaft (DFG) through the Center for Functional Nanostructures (CFN) and the Priority Program SPP1313.

References

1. Zhang YQ, Wang YF, Jiang XD. The application of nanoparticles in biochips. *Recent Pat Biotechnol* 2008;2:55–59.

2. Roco MC. The journal of nanoparticle research at 10 years. *J Nanopart Res* 2008;10: 1–2.
3. Aitken RJ, Chaudhry MQ, Boxall ABA, Hull M. Manufacture and use of nanomaterials: Current status in the UK and global trends. *Occup Med* 2006;56:300–306.
4. Anselmann R. Nanoparticles and nanolayers in commercial applications. *J Nanopart Res* 2001;3:329–336.
5. Abbas K, Cydzik I, Del Torchio R, Farina M, Forti E, Gibson N, Holzwarth U, Simonelli F, Kreyling W. Radiolabelling of TiO_2 nanoparticles for radiotracer studies. *J Nanopart Res* 2010;12:2435–2443.
6. Maynard AD, Aitken RJ, Butz T, Colvin V, Donaldson K, Oberdörster G, Philbert MA, Ryan J, Seaton A, Stone V, *et al.* Safe handling of nanotechnology. *Nature* 2006;444:267–269.
7. Barnard AS. Nanohazards: Knowledge is our first defence. *Nat Mater* 2006;5:245–248.
8. Nel A, Xia T, Mädler L, Li N. Toxic potential of materials at the nanolevel. *Science* 2006;311:622–627.
9. Linse S, Cabaleiro-Lago C, Xue W-F, Lynch I, Lindman S, Thulin E, Radford SE, Dawson KA. Nucleation of protein fibrillation by nanoparticles. *Proc Natl Acad Sci U S A* 2007;104:8691–8696.
10. Lunov O, Zablotskii V, Syrovets T, Röcker C, Tron K, Nienhaus GU, Simmet T. Modeling receptor-mediated endocytosis of polymer-functionalized iron oxide nanoparticles by human macrophages. *Biomaterials* 2011;32:547–555.
11. Lunov O, Syrovets T, Röcker C, Tron K, Nienhaus GU, Rasche V, Mailänder V, Landfester K, Simmet T. Lysosomal degradation of the carboxydextran shell of coated superparamagnetic iron oxide nanoparticles and the fate of professional phagocytes. *Biomaterials* 2010;31:9015–9022.
12. Jain R, Stylianopoulus T. Delivering nanomedicine to solid tumors. *Nat Rev Clin Oncol* 2010;7:653–664.
13. Zhang L, Gu FX, Chan JM, Wang AZ, Langer RS, Farokhzad OC. Nanoparticles in medicine: Therapeutic applications and developments. *Clin Pharmacol Ther* 2007;83:761–769.
14. Davis ME, Chen Z, Shin DM. Nanoparticle therapeutics: An emerging treatment modality for cancer. *Nat Rev Drug Discovery* 2008;7:771–782.
15. Asha Rani PV, Low Kah Mun G, Hande MP, Valiyaveettil S. Cytotoxicity and genotoxicity of silver nanoparticles in human cells. *ACS Nano* 2008;3:279–290.
16. De Paoli Lacerda SH, Park JJ, Meuse C, Pristinski D, Becker ML, Karim A, Douglas JF. Interaction of gold nanoparticles with common human blood proteins. *ACS Nano* 2010;4:365–379.
17. Cedervall T, Lynch I, Foy M, Berggård T, Donnelly SC, Cagney G, Linse S, Dawson KA. Detailed identification of plasma proteins adsorbed on copolymer nanoparticles. *Angew Chem Int Ed* 2007;46:5754–5756.
18. Cedervall T, Lynch I, Lindman S, Berggård T, Thulin E, Nilsson H, Dawson KA, Linse S. Understanding the nanoparticle-protein corona using methods to quantify exchange rates and affinities of proteins for nanoparticles. *Proc Natl Acad Sci U S A* 2007;104: 2050–2055.
19. Lundqvist M, Stigler J, Elia G, Lynch I, Cedervall T, Dawson KA. Nanoparticle size and surface properties determine the protein corona with possible implications for biological impacts. *Proc Natl Acad Sci U S A* 2008;105:14265–14270.

20. Lynch I. Are there generic mechanisms governing interactions between nanoparticles and cells? Random epitope mapping for the outer layer of the protein-material interface. *Phys A* 2007;373:511–520.
21. Röcker C, Pötzl M, Zhang F, Parak WJ, Nienhaus GU. A quantitative fluoresence study of protein monolayer formation on colloidal nanoparticles. *Nat Nanotechnol* 2009;4: 577–580.
22. Klein J. Probing the interactions of proteins and nanoparticles. *Proc Natl Acad Sci USA* 2007;104:2029–2030.
23. Treuel L, Malissek M, Gebauer JS, Zellner R. The influence of surface composition of nanoparticles on their interactions with serum albumin. *Chem Phys Chem* 2010;11:3093–3099.
24. Leszczynski J. Bionanoscience: Nano meets bio at the interface. *Nat Nanotechnol* 2010;5:633–634.
25. Nel AE, Mädler L, Velegol D, Xia T, Hoek EM, Somasundaran P, Klaessig F, Castranova V, Thompson M. Understanding biophysicochemical interactions at the nano-bio interface. *Nat Mater* 2009;8:543–557.
26. Monopoli MP, Walczyk D, Campbell A, Elia G, Lynch I, Bombelli FB, Dawson KA. Physical-chemical aspects of protein corona: Relevance to *in vitro* and *in vivo* biological impacts of nanoparticles. *J Am Chem Soc* 2011;133:2525–2534.
27. Aubin-Tam M-E, Hamad-Schifferli K. Gold nanoparticle-cytochrome c complexes: The effect of nanoparticle ligand charge on protein structure. *Langmuir* 2005;21: 12080–12084.
28. Roach P, Farrar D, Perry CC. Surface tailoring for controlled protein adsorption: Effect of topography at the nanometer scale and chemistry. *J Am Chem Soc* 2006;128: 3939–3945.
29. Medintz IL, Konnert JH, Clapp AR, Stanish I, Twing ME, Mattoussi H, Mauro JM, Deschamps JR. A fluorescence resonance energy transfer-derived structure of a quantum dot-protein bioconjugate nanoassembly. *Proc Natl Acad Sci U S A* 2004;101:9612–9617.
30. Vertegel AA, Siegel RW, Dordic JS. Silica nanoparticle size influences the structure and enzymatic activity of adsorbed lysozyme. *Langmuir* 2004;20:6800–6807.
31. Rodriguez CE, Fukuto JM, Taguchi K, Froines J, Cho AK. The interactions of 9,10-phenantrenequinone with glyceraldehyde-3-phosphatedehydrogenase (GAPDH), a potential site for toxic actions. *Chem Biol Interact* 2005;155:97–110.
32. Lynch I, Dawson KA, Linse S. Detecting cryptic epitopes created by nanoparticles. *Science STKE* 2006;2006:14.
33. Baron MH, Revault M, Servagent-Noinville S, Abadie J, Qui-Quampoix HJ. Chymotrypsin adsorption on montmorillonite: Enzymatic activity and kinetic FTIR structural analysis. *J Coll Interf Sci* 1999;214:319–332.
34. Brandes N, Welzel PB, Werner C, Kroh LW. Adsorption-induced conformational changes of proteins onto ceramic particles: Differential scanning calorimetry and FTIR analysis. *J Colloid Interface Sci* 2006;299:56–69.
35. Des Rieux A, Fievez V, Garinot M, Schneider Y-J, Préat V. Nanoparticles as potential oral delivery systems of proteins and vaccines: A mechanistic approach. *J Controlled Release* 2006;116:1–27.
36. Yan M, Du J, Gu Z, Liang M, Hu Y, Zhang W, Priceman S, Wu L, Hong Zhou Z, Liu H, *et al.* A novel intracellular protein delivery platform based on single-protein nanocapsules. *Nat Nanotechnol* 2009;5:48–53.

37. Liu L, Xu K, Wang H, Tan PKJ, Fan W, Venkatraman SS, Li L, Yang Y-Y. Self-assembled cationic peptide nanoparticles as an efficient antimicrobial agent. *Nat Nanotechnol* 2009; 4:457–463.
38. Jeong SK, Kwon MS, Lee EY, Lee HJ, Cho SY, Kim H, Yoo JS, Omenn GS, Aebersold R, Hanash S, *et al.* BiomarkerDigger: A versatile disease proteome database and analysis platform for the identification of plasma cancer biomarkers. *Proteomics* 2009;9: 3729–3740.
39. Aggarwal P, Hall JB, McLeland CB, Dobrovolskaia MA, McNeil SE. Nanoparticle interaction with plasma proteins as it relates to particle biodistribution, biocompatibility and therapeutic efficacy. *Adv Drug Delivery Rev* 2009;61:428–437.
40. Jansch M, Stumpf P, Graf C, Rühl E, Müller RH. Adsorption kinetics of plasma proteins on ultrasmall superparamagnetic iron oxide (USPIO) nanoparticles. *Int J Parmaceut* 2012;428:125–133. 2012.
41. Thode K, Lück M, Semmler W, Müller R, Kresse M. Determination of plasma protein adsorption on magnetic iron oxides: Sample preparation. *Pharm Res* 1997;14:905–910.
42. Shang L, Wang Y, Jiang J, Dong S. pH-dependent protein conformational changes in albumin — gold nanoparticle bioconjugates: A spectroscopic study. *Langmuir* 2007; 23:2714–2721.
43. Wang T, Bai J, Jiang X, Nienhaus GU. Cellular uptake of nanoparticles by membrane penetration: A study combining confocal microscopy with FTIR spectroelectrochemistry. *ACS Nano* 2012;6:1251–1259.
44. Shao M, Lu L, Wang H, Luo S, Duo Duo Ma D. Microfabrication of a new sensor based on silver and silicon nanomaterials, and its application to the enrichment and detection of bovine serum albumin via surface-enhanced Raman scattering. *Microchim Acta* 2009;164:157–160.
45. Royer CA. Probing protein folding and conformational transitions with fluorescence. *Chem Rev* 2006;106:1769–1784.
46. Mátyus L, Szöllösi J, Jenei A. Steady-state fluorescence quenching applications for studying protein structure and dynamics. *J Photochem Photobiol B: Biol* 2006;83:223–236.
47. Printz M, Friess W. Simultaneous detection and analysis of protein aggregation and protein unfolding by size exclusion chromatography with post column addition of the fluorescent dye BisANS. *J Pharmaceut Sci* 2012;101:826–837.
48. Mori S, Barth HG. *Size Exclusion Chromatography*. Springer, 1999.
49. Carpenter JF, Randolph TW, Jiskoot W, Crommelin DJA, Middaugh CR, Winter G. Potential inaccurate quantitation and sizing of protein aggregates by size exclusion chromatography: Essential need to use orthogonal methods to assure the quality of therapeutic protein products. *J Pharmaceut Sci* 2010;99:2200–2208.
50. Vogt A, D, Angelo C, Oswald F, Denzel A, Mazel CH, Matz MV, Ivanchenko S, Nienhaus GU, Wiedenmann J. A green fluorescent protein with photoswitchable emission from the deep sea. *PLoS ONE* 2008;3:1–8.
51. Wiedenmann J, Ivanchenko S, Oswald F, Nienhaus GU. Identification of GFP-like proteins in nonbioluminescent, azooxanthellate anthozoa opens new perspectives for bioprospecting. *Mar Biotechnol* 2004;6:270–277.
52. Wiedenmann J, Schenk A, Röcker C, Girod A, Spindler KD, Nienhaus GU. A far-red fluorescent protein with fast maturation and reduced oligomerization tendency from Entacmaea quadricolor (Anthozoa, Actinaria). *Proc Natl Acad Sci U S A* 2002;99: 11646–11651.

53. Nienhaus GU. *Protein-Ligand Interactions: Methods and Applications.* Humana Press, 2005.
54. Baier G, Costa C, Zeller A, Baumann D, Sayer C, Araujo PHH, Mailänder V, Musyanovych A, Landfester K. BSA adsorption on differently charged polystyrene nanoparticles using isothermal titration calorimetry and the influence on cellular uptake. *Macromol Biosci* 2011;11:628–638.
55. Cheng Y, Wang M, Borghs G, Chen H. Gold nanoparticle dimers for plasmon sensing. *Langmuir* 2011;27:7884–7891.
56. Kim H, Andrieux K, Delomenie C, Chacun H, Appel M, Desmaële D, Taran F, Georgin D, Couvreur P, Taverna M. Analysis of plasma protein adsorption onto PEGylated nanoparticles by complementary methods: 2-DE, CE and Protein Lab-on-chip® system. *Electrophoresis* 2007;28:2252–2261.
57. Göppert TM, Müller RH. Protein adsorption patterns on poloxamer- and poloxamine-stabilized solid lipid nanoparticles (SLN). *Eur J Pharm Biopharm* 2005;60:361–372.
58. Göppert TM, Müller RH. Polysorbate-stabilized solid lipid nanoparticles as colloidal carriers for intravenous targeting of drugs to the brain: Comparison of plasma protein adsorption patterns. *J Drug Target* 2005;13:179–187.
59. Lück M, Paulke BR, Schröder W, Blunk T, Müller RH. Analysis of plasma protein adsorption on polymeric nanoparticles with different surface characteristics. *J Biomed Mater Res* 1998;39:478–485.
60. Seehof K, Kresse M, Mäder K, Müller RH. Interactions of nanoparticles with body proteins — improvement of 2D-PAGE-analysis by internal standard. *Int J Pharmaceut* 2000;196:231–234.
61. Diederichs JE. Plasma protein adsorption patterns on liposomes: Establishment of analytical procedure. *Electrophoresis* 1996;17:607–611.
62. Gref R, Lück M, Quellec P, Marchand M, Dellacherie E, Harnisch S, Blunk T, Müller RH. 'Stealth' corona-core nanoparticles surface modified by polyethylene glycol (PEG): Influences of the corona (PEG chain length and surface density) and of the core composition on phagocytic uptake and plasma protein adsorption. *Colloid Surf B* 2000;18:301–313.
63. Lemarchand C, Gref R, Passirani C, Garcion E, Petri B, Müller RH, Costantini D, Couvreur P. Influence of polysaccharide coating on the interactions of nanoparticles with biological systems. *Biomaterials* 2006;27:108–118.
64. Lück M, Pistel K-F, Li YX, Blunk T, Müller RH, Kissel T. Plasma protein adsorption on biodegradable microspheres consisting of poly(d,l-lactide-co-glycolide), poly(l-lactide) or ABA triblock copolymers containing poly(oxyethylene): Influence of production method and polymer composition. *J Controlled Release* 1998;55:107–120.
65. Dutta D, Sundaram SK, Teeguarden JG, Riley BJ, Fifield LS, Jacobs JM, Addleman SR, Kaysen GA, Moudgil BM, Weber TJ. Adsorbed proteins influence the biological activity and molecular targeting of nanomaterials. *Toxicol Sci* 2007;100:303–315.
66. Berggård T, Arrigoni G, Olsson O, Fex M, Linse S, James P. 140 mouse brain proteins identified by Ca^{2+}-calmodulin affinity chromatography and tandem mass spectrometry. *J Proteome Res* 2006;5:669–687.
67. Ehrenberg M, McGrath JL. Binding between particles and proteins in extracts: Implications for microrheology and toxicity. *Acta Biomater* 2005;1:305–315.
68. Kim HR, Andrieux K, Gil S, Taverna M, Chacun H, Desmaële D, Taran F, Georgin D, Couvreur P. Translocation of poly(ethylene glycol-co-hexadecyl)cyanoacrylate nanoparticles

into rat brain endothelial cells: Role of apolipoproteins in receptor-mediated endocytosis. *Biomacromolecules* 2007;8:793–799.

69. Norman ME, Williams P, Illum L. Influence of block copolymers on the adsorption of plasma proteins to microspheres. *Biomaterials* 1993;14:193–202.
70. Stolnik S, Daudali B, Arien A, Whetstone J, Heald C, Garnett M, Davis S, Illum L. The effect of surface coverage and conformation of poly(ethylene oxide) (PEO) chains of poloxamer 407 on the biological fate of model colloidal drug carriers. *Biochim Biophys Acta* 2001;1514:261–279.
71. Nagayama S, Ogawara K, Fukuoka Y, Higaki K, Kimura T. Time-dependent changes in opsonin amount associated on nanoparticles alter their hepatic uptake characteristics. *Int J Parmaceut* 2007;342:215–221.
72. Tenzer S, Docter D, Rosfa S, Wlodarski A, Kuharev J, Rekik A, Knauer SK, Bantz C, Nawroth T, Bier C, *et al.* Nanoparticle size is a critical physicochemical determinant of the human blood plasma corona: A comprehensive quantitative proteomic analysis. *ACS Nano* 2011;5:7155–7167.
73. Göppert T, Müller R. Plasma protein adsorption of Tween 80- and poloxamer 188-stabilized solid lipid nanoparticles. *J Drug Target* 2003;11:225–231.
74. Peracchia M, Harnisch S, Pinto-Alphandary H, Gulik A, Dedieu J, Desmaële D, d'Angelo J, Müller R, Couvreur P. Visualization of *in vitro* protein-rejecting properties of PEGylated stealth polycyanoacrylate nanoparticles. *Biomaterials* 1999;20:1269–1275.
75. Gessner A, Waicz R, Lieske A, Paulke BR, Mäder K, Müller RH. Nanoparticles with decreasing surface hydrophobicities: Influence on plasma protein adsorption. *Int J Pharmaceut* 2000;196:245–249.
76. Labarre D, Vauthier C, Chauvierre C, Petri B, Müller RH, Chehimi MM. Interactions of blood proteins with poly(isobutylcyanoacrylate) nanoparticles decorated with a polysaccharidic brush. *Biomaterials* 2005;26:5075–5084.
77. Salvador-Morales C, Flahaut E, Sim E, Sloan J, H. Green ML, Sim RB. Complement activation and protein adsorption by carbon nanotubes. *Mol Immunol* 2006;43:193–201.
78. Treuel L, Nienhaus GU. Toward a molecular understanding of nanoparticle-protein interactions. *Biophys Rev* 2012;4:137–147. 2012.
79. Gilbert B, Huang F, Zhang H, Waychunas GA, Banfield JF. Nanoparticles: Strained and stiff. *Science* 2004;305:651–654.
80. Min Y, Akbulut M, Kristiansen K, Golan Y, Israelachvili J. The role of interparticle and external forces in nanoparticle assembly. *Nat Mater* 2008;7:527–538.
81. Geiser M, Rothen-Rutishauser B, Kapp N, Schürch S, Kreyling W, Schulz H, Semmler M, Im Hof V, Heyder J, Gehr P. Ultrafine particles cross cellular membranes by nonphagocytic mechanisms in lungs and in cultured cells. *Environ Health Perspect* 2005;113:1555–1560.
82. Shang W, Nuffer JH, Muñiz-Papandrea VA, Colón W, Siegel RW, Dordick JS. Cytochrome c on silica nanoparticles: Influence of nanoparticle size on protein structure, stability, and activity. *Small* 2009;5:470–476.
83. Jiang X, Weise S, Hafner M, Röcker C, Zhang F, Parak WJ, Nienhaus GU. Quantitative analysis of the protein corona on FePt nanoparticles formed by transferrin binding. *J R Soc Interface* 2010; 7: S5–S13.
84. Maffre P, Nienhaus K, Amin F, Parak WJ, Nienhaus GU. Characterization of protein adsorption onto FePt nanoparticles using dual-focus fluorescence correlation spectroscopy. *Beilstein J Nanotechnol* 2011;2:374–383.

85. Conner S, Schmid SL. Regulated portals of entry into the cell. *Nature* 2003; 422:37–44.
86. Rothen-Rutishauser B, Schurch S, Haenni B, Kapp N, Gehr P. Interaction of fine particles and nanoparticles with red blood cells visualized with advanced microscopic techniques. *Environ Sci Technol* 2006;40:4353–4359.
87. Chithrani BD, Ghazani AA, Chan WCW. Determining the size and shape dependence of gold nanoparticle uptake into mammalian cells. *Nano Lett* 2006;6:662–668.
88. Rejman J, Oberle V, Zuhorn I, Hoekstra D. Size-dependent internalization of particles via the pathways of clathrin-and caveolae-mediated endocytosis. *Biochem J* 2004;377: 159–169.
89. Labhasetwar V, Song C, Humphrey W, Shebuski R, Levy RJ. Arterial uptake of biodegradable nanoparticles: Effect of surface modifications. *J Pharmaceut Sci* 1998;87: 1229–1234.
90. Arbab AS, Bashaw LA, Miller BR, Jordan EK, Lewis BK, Kalish H, Frank JA. Characterization of biophysical and metabolic properties of cells labeled with superparamagnetic iron oxide nanoparticles and transfection agent for cellular MR imaging. *Radiology* 2003;229:838–846.
91. Sun XK, Rossin R, Turner JL, Becker ML, Joralemon MJ, Welch MJ, Wooley KJ. An assessment of the effects of shell cross-linked nanoparticle size, core composition, and surface PEGylation on *in vivo* biodistribution. *Biomacromolecules* 2005;6:2541–2554.
92. Nativo P, Prior IA, Brust M. Uptake and intracellular fate of surface-modified gold nanoparticles. *ACS Nano* 2008;2:1639–1644.
93. Holzapfel V, Lorenz M, Weiss CK, Schrezenmeier H, Landfester K, Mailänder V. Synthesis and biomedical applications of functionalized fluorescent and magnetic dual reporter nanoparticles as obtained in the miniemulsion process. *J Phys Condens Matter* 2006;18:S2581–S2594.
94. Pan Y, Neuss S, Leifert A, Fischler M, Wen F, Simon U. Size-dependent cytotoxicity of gold nanoparticles. *Small* 2007;3:1941–1949.
95. Poland CA, Duffin R, KinlochI I, Maynard A, Wallace WAH, Seaton A, Stone V, Brown S, Macnee W, Donaldson K. Carbon nanotubes introduced into the abdominal cavity of mice show asbestos-like pathogenicity in a pilot study. *Nature* 2008;3:123–128.
96. Oberdörster G, Maynard A, Donaldson K, Castranova V, Fitzpatrick J, Ausman K, Carter J, Karn B, Kreyling W, Lai D, *et al.* Principles for characterizing the potential human health effects from exposure to nanomaterials: Elements of a screening strategy. *Part Fibre Toxicol* 2005;2:1–35.
97. Owens DE, Peppas NA. Opsonization, biodistribution, and pharmacokinetics of polymeric nanoparticles. *Int J Pharm* 2006;307:93–102.
98. Kah JCY, Wong KY, Neoh KG, Song JH, Fu JWP, Mhaisalkar S, Olivo M, Sheppard CJR. Critical parameters in the PEGylation of gold nanoshells for biomedical applications: An *in vitro* macrophage study. *J Drug Target* 2009;17:181–193.
99. Jiang X, Röcker C, Hafner M, Brandholt S, Dörlich RM, Nienhaus GU. Endo- and exocytosis of zwiterionic quantum dot nanoparticles by live HeLa cells. *ACS Nano* 2010;4:6787–6797.
100. Kostarelos K, Bianco A, Prato M. Promises, facts and challenges for carbon nanotubes in imaging and therapeutics. *Nat Nanotechnol* 2009;4:627–633.
101. Tasis D, Tagmatarchis N, Bianco A, Prato M. Chemistry of carbon nanotubes. *Chem Rev* 2006;106:1105–1136.

102. Hamman JH, Enslin GM, Kotze AF. Oral delivery of peptide drugs: Barriers and developments. *BioDrugs* 2005;19:165–177.
103. Galindo-Rodriguez SA, Allemann E, Fessi H, Doelker E. Polymeric nanoparticles for oral delivery of drugs and vaccines: A critical evaluation of *in vivo* studies. *Crit Rev Ther Drug Carr Syst* 2005;22:419–464.
104. Ghosh P, Han G, De M, Kim CK, Rotello VM. Gold nanoparticles in delivery applications. *Adv Drug Delivery Rev* 2008;60:1307–1315.
105. Gan ZH, Yu DH, Zhong ZY, Liang QZ, Jing XB. Enzymatic degradation of poly(epsilon-caprolactone)/poly(DL-lactide) blends in phosphate buffer solution. *Polymer* 1999; 40:2859–2862.
106. Leroueil-Le VM, Fluckiger L, Kim YI, Hoffmann M, Maincent P. Preparation and characterisation of nanoparticles containing an antihypertensive agent. *Eur J Pharm Biopharm* 1998;46:137–143.
107. Balthasar S, Michaelis K, Dinauer N, von Kreuter J, Langer K. Preparation and characterisation of antibody modified gelatin nanoparticles as drug carrier system for uptake in lymphocytes. *Biomaterials* 2005;26:2723–2732.
108. Johnson FA, Craig DQ, Mercer AD. Characterization of the block structure and molecular weight of sodium alginates. *J Pharm Pharmacol* 1997;49:639–643.
109. Fonseca C, Simoes S, Gaspar R. Paclitaxel-loaded PLGA nanoparticles: Preperation, physicochemical charactertization and *in vitro* anti-tumoral activity. *J Controlled Release* 2002;83:273–286.
110. Kumar Ravi MN, Bakowsky U, Lehr CM. Preparation and characterization of cationic PLGA nanospheres as DNA carriers. *Biomaterials* 2004;25:1771–1777.
111. Dillen K, Vandervoort J, van den MG, Verheyden L, Ludwig A. Factorial design, physicochemical characterisation and acitivity of ciprofloxacin-PLGA nanoparticles. *Int J Pharm* 2004;275:171–187.
112. Govender T, Stolnik S, Garnett MC, Illum L, Davis SS. PLGA nanoparticles prepared by nanoprecipitation: Drug loading and release studies of a water soluble drug. *J Controlled Release* 1999;57:171–185.
113. Lemoine D, Preat V. Polymeric nanoparticles as delivery system for influenza virus glycoproteins. *J Controlled Release* 1998;54:15–27.
114. Krause HJ, Schwarz A, Rohdewald P. Polylactic acid nanoparticles, a colloid drug delivery system for lipophilic drugs. *Int J Pharm* 1985;27:145–155.
115. Zambaux MF, Bonneaux F, Gref R, Dellacherie E, Vigneron C. Preparation and characterization of protein C-loaded PLA nanoparticles. *J Controlled Release* 1999;60:179–188.
116. Leo B, Brina B, Forni F, Vandelli MA. *In vitro* evaluation of PLA nanoparticles containing a lipophilic drug in water-soluble or insoluble form. *Int J Pharm* 2004;278:133–141.
117. Langer RS, Peppas NA. Present and future applications of biomaterials in controlled drug delivery systems. *Biomaterials* 1981;2:201–214.
118. Miller JD, Anderson MG. Therapeutic effects of leuprorelin microspheres on endometriosis and uterine leiomyomata. *Adv Drug Delivery Rev* 1997;60:179–155.
119. Ayalasomayajula SP, Kompella UB. Subconjunctivally administered celecoxib-PLGA microparticles sustain retinal drug levels and alleviate diabetes-induced oxidative stress in a rat model. *Eur J Pharm Biopharm* 2005;511:191.
120. Kou JH, Emmett C, Shen P, Aswani S, Iwamoto T, Vaghefi F, Cain G, Sanders L. Bioerosion and biocompatibility of poly(*d,l*-lactic-co-glycolic acid). *J Controlled Release* 1997;43:123–130.

121. Lee SY, Oh JH, Kim YH, Kim SH, Choi JW. *In vivo* conjunctival reconstruction using modified PLGA grafts for decreased scar formation and contraction. *Biomaterials* 2003;24:5049–5059.
122. Zhou T, Lewis H, Foster RE, Schwendman SP. Development of a multiple-drug delivery implant for intraocular management of proliferative vitreoretinopathy. *J Controlled Release* 1998;55:281–295.
123. Janes KA, Fresneau MP, Marazuela A, Fabra A, Alonso MJ. Chitosan nanoparticles as delivery systems for doxorubicin. *J Controlled Release* 2001;73:255–267.
124. van der Lubben I, Verhoef JC, Borchard G, Junginger HE. Chitosan for mucosal vaccination. *Adv Drug Delivery Rev* 2001;73:139–144.
125. Prego C, Garcia M, Torres D, Alonso MJ. Transmucosal macromolecular drug delivery. *J Controlled Release* 2005;101:151–162.
126. Vila A, Sanchez A, Janes K, Behrens I, Kisel T, Vila Jato JL, Alonso MJ. Low molecular weight chitosan nanoparticles as new carriers for nasal vaccine delivery in mice. *Eur J Pharm Biopharm* 2004;57:123–131.
127. Bailon P, Won CY. PEG-modified biopharmaceuticals. *Expert Opin Drug Delivery* 2009; 6:1–16.
128. Lipka J, Semmler-Behnke M, Sperling RA, Wenk A, Takenaka S, Schleh C, Kissel T, Parak WJ, Kreyling WG. Biodistribution of PEG-modified gold nanoparticles following intratracheal instillation and intravenous injection. *Biomaterials* 2010;31:6574–6581.
129. Lai WC, Liau WB. Thermo-oxidative degradation of poly(ethylene glycol)/poly (L-lactic acid) blends. *Polymer* 2003;44:8103–8109.
130. Vonarbourg A, Passirani C, Saulnier P, Benoit JP. Parameters influencing the stealthiness of colloidal drug delivery systems. *Biomaterials* 2006;27:4356–4373.
131. Harris JM, Martin NE, Modi M. PEGylation: A novel process for modifying pharmacokinetics. *Clin Pharmacokinet* 2001;40:539–551.
132. Huang X, El-Sayed IH, Qian W, El-Sayse MA. Cancer cell imaging and photothermal therapy in the near-infrared region by using gold nanorods. *J Am Chem Soc* 2006;128:2115–2120.
133. Villalonga R, Cao R, Gragoso A. Supramolecular chemistry of cyclodextrin in enzyme technology. *Chem Rev* 2007;107:3088–3116.
134. Yeh HC, Ho YP, Wang TH. Quantum dot mediated biosensing assays for specific nucleic acid detection. *Nanomedicine* 2005;1:115–121.
135. Boisselier E, Astruc D. Gold nanoparticles in nanomedicine: Preparations, imaging, diagnostics, therapeutics and toxicity. *Chem Soc Rev* 2009;38:1759–1782.
136. Ferrari M. Cancer nanotechnology: Opportunities and challenges. *Nat Rev Cancer* 2005;5:161–171.
137. Han G, Chari NS, Verma A, Hong R, Martin CT, Rotello VM. Controlled recovery of the transcription of nanoparticle-bound DNA by intracellular concentrations of glutathione. *Bioconjugate Chem* 2005;16:1356–1359.
138. Rosi NL, Giljohann DA, Thaxton CS, Lytton-Jean AKR, Han MS, Mirkin CA. Oligonucleotide-modified gold nanoparticles for intracellular gene regulation. *Science* 2006;312:1027–1030.
139. Goel R, Swanlund D, Coad J, Paciotti GF, Bischof JC. TNF-α based accentuation in cryoinjury — dose, delivery, and response. *Mol Cancer Ther* 2007;6:2039–2047.

Chapter 7

The Effects of Engineered Nanomaterials on Erythrocytes

*Bridget Wildt, Richard A. Malinauskas, and Ronald P. Brown**

Office of Science and Engineering Laboratories
Center for Devices and Radiological Health
Food and Drug Administration
Silver Spring, MD 20993, USA

Both engineered and naturally occurring nanomaterials (NMs) may come in contact with blood through different pathways, either directly (e.g., as nanoparticles carrying chemotherapy agents injected intravenously, imbedded in a coating on a vascular graft, or as wear debris released from an orthopedic device) or indirectly (e.g., through inhalation, oral, or topical exposure routes). Due to the ubiquitous importance of red blood cells in sustaining the different tissues of the body, in this chapter, we review the possible adverse effects of NMs on erythrocytes and the experimental techniques which have been applied to evaluate and better understand their mechanisms of interaction. This chapter provides a comprehensive overview of the current literature, outlines practical approaches to study nanoparticle interaction with erythrocytes, describes challenges, and suggests solutions to overcome them.

Disclaimer:

The mention of commercial products, their sources, or their use in connection with material reported herein is not to be construed as either an actual or implied endorsement of such products by the Department of Health and Human Services. The conclusions and opinions expressed in this article are those of the authors and do not constitute FDA policy, and the mention of any products and/or manufacturers does not imply FDA endorsement.

* Corresponding author: E-mail: ronald.brown@fda.hhs.gov

1. Introduction

Engineered nanomaterials (NMs) differ from naturally occurring materials of the same scale in that they are typically designed with the intent of using their nanoscale properties to promote a particular physical or biological response. Naturally occurring NMs, on the other hand, may simply be particulates that have dislodged from bulk materials or by-products arising from the combustion of materials. Both engineered and naturally occurring NMs may come in contact with blood through different pathways, either directly (e.g., as nanoparticles carrying chemotherapy agents injected intravenously, imbedded in a coating on a vascular graft, or as wear debris released from an orthopedic device) or indirectly (e.g., through inhalation, oral, or topical exposure routes). Due to the ubiquitous importance of red blood cells (RBCs) in sustaining the different tissues of the body, in this chapter, we review the possible adverse effects of NMs on erythrocytes and the experimental techniques which have been applied to evaluate and better understand their mechanisms of interaction.

2. Erythrocyte Structure and Function

Erythrocytes, or RBCs, are the most numerous blood cell type and comprise approximately 45% of human blood by volume; white blood cells and platelets constitute only about 1% of the blood volume, and plasma makes up the remaining 54%. RBCs are primarily responsible for oxygen transport to tissues and the removal of carbon dioxide, and they also contribute to the maintenance of acid–base balance.[1] RBCs are discoid in shape, about 6–8 μm in diameter, and have strong, flexible plasma membranes, which allow the cells to squeeze through narrow capillaries as small as 3 μm in diameter.[2–4] The cell membranes are composed of 50% proteins, 40% lipids, and 10% carbohydrates, and are semi-permeable to water.[5] A mature RBC is devoid of a nucleus, ribosomes, and mitochondria, and relies on the diffusion of glucose into the cell as a fuel source. Glucose metabolism provides the energy to sustain intracellular antioxidant function, preserves cytoplasmic balance against osmotic lysis by operating a sodium–potassium ion membrane pump, and maintains the integrity of the RBC membrane.[5] The RBC cytoplasm consists almost entirely of the iron-carrying protein hemoglobin for transporting oxygen. Due to the continuous intracellular production of reactive oxygen species (ROS), an RBC utilizes enzymatic and non-enzymatic pathways to provide antioxidant protection throughout its lifetime.[5]

A typical RBC has a life cycle of approximately 120 days. As it nears the terminal stage of its lifespan, an RBC rapidly undergoes changes shortly before it is removed from the circulation;[6] the RBC volume and hemoglobin content decrease through the release of membrane vesicles, the density of the cell may increase, structural alterations occur to the Band 3 membrane protein, the negative surface charge decreases (which weakens electrostatic repulsion to other cells), the capacity against ROS decreases, binding of autologous immunoglobulin G and the expression of phosphatidylserine on the outer plasma membrane mark the cell for removal, and it becomes more fragile to chemical and mechanical stress.[1,6–11] To recycle iron from hemoglobin for use in new RBCs, the majority of aged RBCs are destroyed extravascularly when captured and ingested by macrophages of the mononuclear phagocytic system of the spleen and liver.[1]

3. *In Vitro* Techniques Used to Investigate NM Interactions with RBCs

A critical aspect in determining the biocompatibility of NMs is the evaluation of their effects on the structure, function, and lifespan of RBCs. Current *in vitro* methods to assess whether NMs adversely affect RBCs include techniques for evaluating acute hemolysis, oxidative stress, RBC aggregation or agglutination, membrane interaction and intracellular uptake, as well as changes in RBC morphology and mechanical properties. Published studies illustrating the use of these specific techniques to evaluate RBC interaction with a wide range of NMs are summarized in Table 1, and discussed in the following sections.

3.1. *Hemolysis*

In a healthy individual, hemoglobin is almost completely confined within the RBCs, with very little cell-free hemoglobin found in the blood plasma. Hemolysis occurs when hemoglobin is released from a compromised or ruptured erythrocyte plasma membrane. Acute massive intravascular hemolysis is the worst effect a material can have on RBCs as it results in a condition called hemolytic anemia, characterized by a decreased RBC count followed by severely oxygen-deprived tissues and ultimately, multiple organ failure.[109,110] Extracellular hemoglobin is itself toxic and although there are multiple hemoglobin clearance mechanisms for preventing buildup in the blood plasma, it can cause clinical sequelae by affecting vascular, myocardial, renal, and central nervous system tissues.[111,112] For example, hemoglobin filtration by renal tubules can result in tubular obstruction due to heme protein precipitation, vasoconstriction, and extensive oxidative damage, which may lead to renal

Table 1. A summary of published studies using different techniques to evaluate the effects of NMs on RBC-related *in vitro* endpoints. The NMs listed in this table are categorized by the primary component of their base material. *NMs are often complex and may consist of composite base materials functionalized using a variety of surface coatings.

Endpoint	**Assay or technique**	**Studies of RBC interaction with the following NMs***
Hemolysis	*In vitro* hemolysis – analysis of released hemoglobin	Metallic: Au, Ag, Pt[12–19] Dendrimer[20–54] Quantum dots[55,56] Magnetic: Fe_2O_3, Fe_3O_4[57–63] Silica[64–67] Titanium dioxide[68–70] Carbon nanotubes[71–74] Polymeric: poly(aminoacid), poly(ethylene glycol)[75–86] Lipid[87–90] Emulsion[91–93]
Oxidative stress	Dichlorofluorescein assay – determination of intracellular ROS generation Thiobarbituric acid assay – estimation of lipid peroxidation	Fullerene[94,95] Titanium dioxide[69] Polyaniline nanofibers[84]
RBC aggregation or agglutination	Observation of RBC aggregation, settling (erythrocyte sedimentation rate), in vial or by optical microscopy	Titanium dioxide[69] Polymeric: poly(N-vinylacetamide), poly(ethylenimine)[96–98] Self-assembling peptide nanofibers[99] Dendrimer[40,53,100,101] Metallic: Au, Ag, Pt[12–19]
NM cellular uptake	TEM/EDS, SERS, CLSM, ICP-MS, AFM	Silica[67] Titanium dioxide[69] Metallic: Ag, Au[102,103] Magnetic: Fe_2O_3[104,105] Metallofullerene[106]
Morphological and mechanical property changes	SEM/optical microscopy, membrane filtration, SNFUH, MSAFM	Carbon nanohorns[107,108] Silica[67,107,108] Metallic: Au, Pt, Ag[12] Dendrimer[40,101]

Abbreviations

AFM	Atomic force microscopy
CLSM	Confocal laser scanning microscopy
EDS	Energy dispersive spectroscopy
ICP-MS	Inductively coupled plasma mass spectroscopy
MSAFM	Mode synthesizing atomic force microscopy
SEM	Scanning electron microscopy
SERS	Surface enhanced Raman scattering
SNFUH	Scanning probe microscale ultrasound holography
TEM	Transmission electron microscopy

failure.[5,111,113] Recent studies have shown that cell-free oxyhemoglobin binds nitric oxide, a critical regulator of vasodilation and vascular homeostasis. Insufficient nitric oxide levels can lead to smooth muscle dystonias, including hypertension, gastrointestinal contractions, erectile dysfunction, and clot formation.[112] Hemolysis can also release ATP from RBCs, which is a potent platelet activator.[109]

The determination of an NM's tendency to induce acute hemolysis is by far the most commonly used method to assess direct erythrocyte damage as a result of NM exposure (Table 1). An *in vitro* hemolysis assay typically involves the incubation of diluted whole blood or washed RBCs with a known concentration of NMs for 30 min up to 4 h. The samples are then centrifuged to separate the RBCs from the supernatant. If significant hemolysis has occurred, the supernatant will appear pink or reddish due to released hemoglobin. The amount of hemolysis can be quantified by measuring the spectrophotometric absorbance at a particular wavelength of the sample supernatant (relative to a 100% hemolysis positive control). In some versions of the assay, a chemical reagent is added to reduce most forms of hemoglobin to stable cyanmethemoglobin (CMH), which exhibits an absorption peak at 540 nm. In other versions, such reagents are omitted and the amount of hemolysis is quantified by measuring oxyhemoglobin at one of its primary absorbance peaks (i.e., 415, 541, or 577 nm).[114] Figure 1 shows a schematic

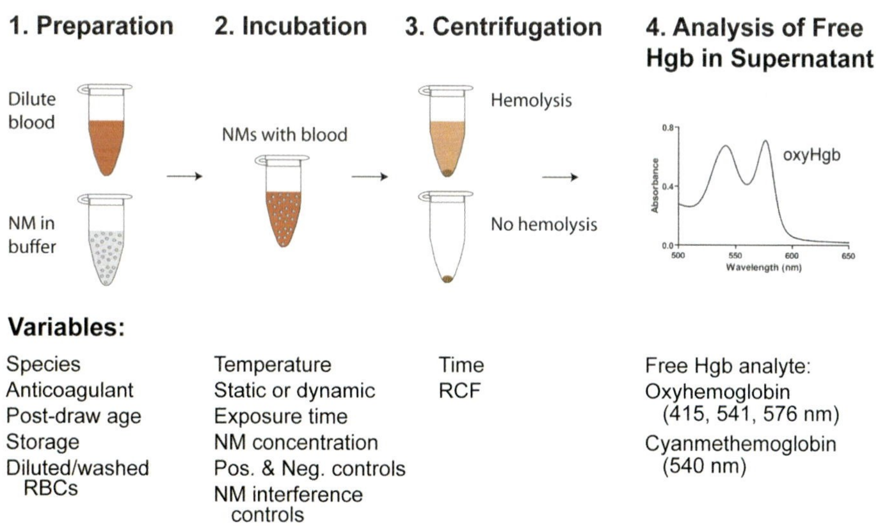

Figure 1. Schematic of the hemolysis assay summarizing the important procedural variables.

Hgb: hemoglobin, oxyHgb: oxyhemoglobin, RCF: relative centrifugal force.

diagram of the hemolysis assay, highlighting the important procedural variables.

It is difficult to quantitatively compare the results of *in vitro* hemolysis evaluations of NMs among studies since the experimental parameters for conducting the testing found in the literature vary greatly (i.e., in terms of the blood source, blood age, hematocrit, anticoagulant, use of whole blood or pre-washed RBCs, buffer, incubation time and temperature, static or dynamic incubation, centrifugation conditions, and the assay to measure the released hemoglobin), which may affect the results.[31,115] A standardized consensus testing protocol, ASTM E2524-08 ("Standard test method for analysis of hemolytic properties of nanoparticles"),[116] was recently developed to guide researchers and to instil some uniformity to the way in which the acute hemolysis assay can be conducted on unbound nanoparticles.[115] ASTM E2524-08 was adapted from the widely used testing standard for evaluating hemolysis caused by materials used in medical devices (i.e., ASTM F756-08, "Standard practice for the assessment of hemolytic properties of materials"[117]). The procedures in ASTM E2524-08 describe the use of reduced sample volumes appropriate for testing nanoparticles, diluted whole human blood containing plasma, positive and negative testing controls, a percent hemolysis evaluation criteria when using the test protocol, and techniques for determining assay interference by the nanoparticles. However, as NMs are extremely diverse, the recommendations in the ASTM E2524-08 standard may not be appropriate for all materials. For instance, the use of phosphate buffered saline as a dilution buffer in the standard protocol causes silver nanoparticles to agglomerate and may form insoluble silver chloride.[13]

As changes in the size and toxicity of NMs can quickly occur due to their interaction with plasma proteins,[13,18,29,68,71,72,94,118,119] one of the most critical aspects of any hemolysis test may be the use of either buffer-diluted whole blood, in which plasma proteins are retained, or washed RBCs, in which the plasma proteins have been removed. The preparation of the cells is important as *in vitro* hemolytic activity for NM samples tested with washed RBCs is often greater than when NMs are incubated with diluted whole blood.[13,18,29,53,68,71,72,94,118,119] The reduction in the hemolysis level when plasma is present is most likely due to protein interactions that affect the NM agglomeration state and cause an increased NM size with decreased effective surface area.[18,53] However, conducting the hemolysis test with diluted whole blood is more representative of the biological environment that a NM is likely to encounter *in vivo*.

Another critical experimental detail in conducting *in vitro* hemolysis testing with NMs is the identification of possible assay interferences which might result in false-positive or false-negative results. The unique properties of NMs, such as their ability to scatter, absorb, or emit light at specific wavelengths, can lead to measurement interference in common *in vitro* assays.[115,120–122] Unfortunately, many *in vitro* hemolysis studies using NMs in the literature do not mention procedures to assess or to compensate for assay interference. For example, the surface plasmon band of 50-nm gold nanoparticles has an absorption peak around 533 nm,[123] which can result in an increased signal if the particles are not sufficiently removed by centrifugation prior to measuring at the 540–541 nm CMH or oxyhemoglobin absorbance peaks.[115] Conversely, the amount of measurable hemolysis may be artificially decreased if coagulation around the NMs preclude the RBCs from interacting with them, or if the test materials absorb the released hemoglobin.[13] As a result, it is recommended to visually observe the samples throughout the testing and to include positive, negative, and cell-free controls (to assess whether NMs cause interference in the assay or aggregate when added to buffer solutions with and without proteins) in any hemolysis test of NMs. While the hemolysis assay is the most widely used test for evaluating RBC interaction with NMs, interlaboratory validation of the testing technique has not been conducted. Furthermore, the criteria that a nanoparticle is biologically acceptable if it does not cause more than 5% hemolysis in the ASTM E2524-08 procedure has not been validated to *in vivo* performance.

3.1.1. *Factors associated with NM-induced hemolysis*

As NMs are often composed of complex base materials with a variety of functionalized surface coatings and testing protocols differ significantly, only limited comparisons can be made among NMs in different *in vitro* hemolysis studies. Of the numerous hemolysis studies conducted to evaluate the interaction between NMs and RBCs (Table 1), most are in general agreement that primary surface properties, such as hydrophobicity, charge/charge density, and functionalization, are critical factors in a material's ability to induce hemolysis.[118] Cationic moieties, in particular unprotected primary amines, present on the surface of many types of NMs have been found to be especially hemolytic.[40,53,118,124–126] As a possible mechanism for inducing hemolysis, cationic NMs have been shown to interact with negatively charged biological membranes, causing nanoholes, membrane thinning, and erosion leading to leakage of cytosolic material.[126] NM surfaces containing surfactant-like properties (e.g., water-soluble fullerene derivatives[125]) or NM formulations which contain surfactants [e.g., lauric acid stabilized poly(lactide-co-glycolide) acid nanoparticles[86] or Triton X-100 stabilized polystyrene particles[115]] have also been shown to increase hemolytic activity.[118] Conversely, the addition of polyethylene glycol (PEG) to the NM surface has been shown to decrease the degree of hemolysis.[26,51,86] As noted earlier, the degree to which plasma proteins are present during the testing, and the extent of their interaction with the NM surface, can also act to suppress hemolysis.[18,53,68] Although NMs may be prepared in buffered saline solutions, they may still cause significant alterations in pH that can contribute to *in vitro* hemolysis.[53]

In general, NMs tend to produce a greater degree of hemolysis than micron-sized particles of the same material when compared on an equivalent mass basis.[13,18,64,69,82] The reason for this is not well understood, but it has been proposed that smaller particles would be present in larger numbers and therefore, have larger total surface areas for reacting with cells.[13,127] For example, the effect of NM size on RBC lysis was investigated through *in vitro* hemolysis testing of various sizes of polystyrene particles (20 nm to 1 μm) functionalized with either carboxyl or amidine chemical moieties.[82] All of the tested particles showed extensive hemolytic activity at very high doses. However, smaller carboxylated particles tended to have higher hemolytic activity than larger particles at similar doses.[82] The authors attributed the high hemolytic activity of the smaller carboxyl polystyrene particles to direct RBC membrane disruption.

A comparison of the hemolytic activity of silver nanoparticle (AgNP) agglomerates ranging from 43 nm to 1.4 μm in size revealed that the smallest AgNP agglomerates generated significantly more hemolytic activity than the larger AgNP agglomerates.[18] Nano-sized, uncoated AgNPs were also shown to be more hemolytic than micron-sized Ag particles.[13] This result may be due to the higher dissolution of Ag^+ ions by smaller AgNPs, since nano-sized particles present a higher total surface area on a per mass basis compared to micron-sized particles.[13,18] Another study showed that nano-titanium dioxide (nano-TiO_2) particles caused increased hemolysis as well as abnormal RBC sedimentation and agglutination, when compared to micron-sized TiO_2 particles at equivalent mass concentrations.[69] Similar size-dependent effects were seen in hemolysis studies using mesoporous silica particles of sizes ranging from 24 nm (causing ~90% hemolysis) to 263 nm (causing ~ 40% hemolysis) at a particle concentration of 200 μg/mL.[64]

In addition to size and surface coatings, NM interactions with RBCs are further affected by material porosity and crystal structure. One study showed that ~100-nm MCM-41–type mesoporous silica NMs were found to absorb to the surface of RBCs without disrupting the membrane, while interactions of ~600-nm SBA-15–type mesoporous silica NMs with RBCs resulted in significant localized membrane deformation, which altered RBC morphology and led to NM internalization and hemolysis.[64,67] Furthermore, micron-sized (~5 μm) TiO_2 with anatase crystal structure was found to be 73 times more likely to induce 50% hemolysis than nano-sized (~50 nm) amorphous TiO_2, and 11 times more likely to cause significant hemolysis than micron-sized (~5 μm) TiO_2 in the rutile form.[68] These examples demonstrate that the disruption of the RBC membrane is dependent on various NM physicochemical factors.

3.2. *Oxidative stress*

ROS are highly interactive molecules containing oxygen derivatives and are important chemical mediators in signal transduction pathways.[128] However, ROS can damage cell membranes by affecting the structure and function of proteins and lipids.[129] RBCs undergo a high level of oxidative stress during their 120-day life cycle due to their extensive oxygen transport activities and the generation of ROS. RBCs contain several enzymatic (superoxide dismutase, glutathione peroxidase) and non-enzymatic (vitamin E, vitamin C, glutathione) antioxidants[5] to combat the high levels of reactive oxidants, which could otherwise damage the plasma membrane and induce hemolysis.[130–132] *In vitro* assays to determine the amount of generated intracellular ROS are common and typically involve the incubation of the cells and test material with cell-permeable 2′,7′-dichlorodihydrofluorescein diacetate, which is oxidized by ROS to highly fluorescent dichlorofluorescein (DCF).[121,133] The fluorescence intensity of DCF is proportional to the amount of reactive species generated.

Although many studies have reported increased oxidative stress in RBCs exposed to chemical oxidants, there are currently only a few published studies which used assays to evaluate ROS generation in RBCs incubated with NMs. One study compared the potential of C_{60} fullerene nanoparticles (chemically non-modified) prepared by different methods to generate oxidative stress-mediated hemolysis.[94] Flow cytometry was used to quantify ROS by measuring the green fluorescence intensity emitted by the resulting DCF. The study found that C_{60} prepared by tetrahydrofuran solvent exchange produced oxidative stress-mediated damage to the erythrocyte membrane, whereas C_{60} prepared by other methods did not induce RBC membrane damage.

The peroxidation of membrane lipids is another common marker of oxidative stress. Lipid peroxides are unstable and decompose to a variety of compounds, the most abundant of which is malondialdehyde (MDA). Cross-linking of RBC membrane proteins and phospholipids by MDA can decrease RBC survival.[5] Li *et al.* used a lipid peroxidation assay to show that 20-nm

TiO_2 particles elicited a dose-dependent increase in MDA production when incubated with rabbit RBCs.[69] C_{60} fullerene derivatives were also shown to cause photo-induced lipid peroxidation of erythrocyte membranes using the MDA assay.[95] Conversely, polyaniline nanofibers demonstrated antioxidant capabilities as they prevented oxidative damage and hemolysis by inactivating hydrogen peroxide when it was added to RBCs.[84]

3.3. *RBC aggregation and agglutination*

An abundance of RBC membrane proteins with sialic acid residues create a highly negative surface charge that repels the adhesion of RBCs to other blood cells and to the vascular endothelium.[134] However, under low *in vivo* shear force or stagnant flow conditions, normal human RBCs form weak, reversible aggregates[a] with one another that break apart when exposed to fluid forces.[135] Increases in RBC aggregation (commonly assessed by measuring the erythrocyte sedimentation rate) are associated with pathologic disease states, such as diabetes and hypertension.[136] NM interactions with RBCs may potentially exacerbate the formation of RBC aggregates. Although the terms appear to be used interchangeably in NM testing, RBC agglutination differs from aggregation in that it usually refers to irreversible processes (e.g., strong antigen–antibody binding).[134] The mechanisms of RBC aggregation are not completely understood, but may be a result of alterations in plasma membrane proteins that include changes in surface charge density and bridging between cells involving the plasma protein fibrinogen.[137,138] The determination of an NM's tendency to induce RBC aggregation is critical because RBC aggregates may lead to elevated blood viscosity, which increases blood flow resistance particularly in low flow conditions, and increased binding with other cell types (e.g., monocytes, macrophages, endothelial cells).[136] Close examination of vials containing RBCs and NMs can indicate abnormal RBC aggregation (or agglutination) when compared side by side to an identical tube without NMs. Li *et al.* used this technique to show that 20 nm TiO_2 particles induced increased RBC agglutination and sedimentation when compared to less reactive 200-nm TiO_2 particles.[69]

Aggregation of RBCs can be more closely observed using optical microscopy where a droplet containing an NM–RBC mixture is placed on a glass slide and observed at high magnification. Asharani *et al.* used bright

[a] In biology, weak, reversible RBC clumping is commonly referred to as aggregation,[135] whereas in nanotechnology the term "aggregation" refers to the irreversible clumping of primary nanomaterials.[18]

field and dark field optical microscopy to study the agglutination of human RBCs treated with 100 μg/ml polyvinyl alcohol-capped metallic gold (Au), platinum (Pt), and AgNPs. They observed that AgNPs induced significant agglutination while AuNPs and PtNPs did not.[12] RBC agglutination was also shown to increase with increasing peptide concentration and nanofiber formation in studies of peptide nanofibers self-assembling in the presence of RBCs.[99] Han *et al.* reported that high concentrations (90 nM) of cationic polyamidoamine dendrimers caused rapid RBC lysis, but at lower levels (0.9 nM and 9 nM), RBC aggregation was observed along with a slower rate of hemolysis.[53] Based on similar results in different studies, it has been postulated that the electrostatic attraction between positively charged dendrimers and negatively charged RBCs leads to agglutination through cell–dendrimer cross-linking, whereas anionic dendrimers have less impact on the RBC membrane.[40,101] These studies demonstrated that NM composition, size, and concentration may contribute to alterations in RBC aggregation.

3.4. *NM membrane interaction and cellular uptake*

Unlike most other cell types which have been observed to internalize NMs by mechanisms such as phagocytosis and endocytosis,[139,140] RBCs are thought to have limited phagocytic ability to engulf particulates. Despite this, a number of studies have shown that NMs can rapidly embed in RBC membranes, penetrate into the RBC cytoplasm, alter cell morphology, and cause agglutination and hemolysis.[67,69,102,103] The mechanism for the uptake of NMs by RBCs has not been elucidated, but is likely due to electrostatic-dependent interactions (e.g., binding of NMs to the cell membrane), hydrophobic forces, and particle size and shape effects on the bending energy of the membrane.[67,106,141]

Zhao *et al.*[67] used transmission electron microscopy (TEM) and scanning electron microscopy (SEM) to show that 600-nm mesoporous silica particles embedded into the surface of RBC membranes, penetrated into the cells, and resulted in spiculation (spiky protrusions) of the RBCs and hemolysis. In contrast, smaller (100 nm) mesoporous silica NMs were found to adsorb to the surface of RBCs without causing damage to the RBC membranes or being engulfed by the RBCs.[67] The authors contended that the RBC membrane wrapped around and engulfed only the larger silica particle because the bending energy required for the RBC–NM interaction is lower than the energy required for membrane deformation around the smaller NMs (i.e., due to steeper bending angles of the membrane over a smaller area).[67]

Rothen-Rutishauser *et al.*[103] studied the interaction of RBCs with fluorescently labeled polystyrene, TiO_2, and AuNPs using TEM and confocal laser scanning microscopy (CLSM). They found that cellular uptake depended on particle size (i.e., only particles less than or equal to 200 nm in diameter entered the RBCs), but not on material type or surface charge.[103] Zhang *et al.* used atomic force microscopy (AFM) to visualize membrane penetration by ~22-nm aggregated metallofullerene particles ($[Gd@C_{82}OH_{22}]_n$), which they attributed to strong hydrophobic interactions between the nanoparticles and RBCs.[106] In their study, they incubated the particles with RBCs for 10 min and washed the suspension five times before separating the RBC cytoplasm from the RBC membrane fragments by centrifugation. Inductively coupled plasma mass spectroscopy (ICP-MS) was used to conduct a time series study of metallofullerene cellular interaction and uptake. They reported metallofullerene adherence to RBC membranes after 3 min and intracellular uptake after only 5 min.

Li *et al.* presented TEM images showing ~20-nm TiO_2 particles outlining the membranes of RBCs and devoid of intracellular hemoglobin (i.e., ghost cells). The authors suggested that the adsorption of the particles altered the membrane surface repulsion and deformed the RBCs, which led to agglutination and eventual hemolysis.[69]

Xu *et al.* used TEM combined with energy dispersive spectrometry (EDS) to compare the uptake of superparamagnetic iron oxide nanoparticles (SPIONs) with a hydrodynamic radius of 60–80 nm in cell types with various phagocytic capacities (macrophages, fibroblasts, and RBCs). They reported a high degree of SPION internalization by macrophages, less internalization by fibroblasts, and no internalization by RBCs.[105] Importantly, they point out that their findings do not exclude the possible internalization of SPIONs by RBCs, but note that the relative degree of internalization was much less compared to the other cell types. As NMs are often very difficult to identify optically in TEM images, this study is a good example of how an analytical technique, such as EDS, can be combined with imaging to provide sufficient evidence to positively identify NMs. Further details on the physicochemical characterization of NMs can be found in Chapter 2.

Besides using advanced microscopic techniques to visualize NM penetration into RBCs, other novel approaches used to study the interactions of NMs and RBCs include surface enhanced Raman scattering (SERS), in which RBC properties such as hemoglobin distribution and structure can be measured in intact RBCs.[102]

3.5. NM-induced changes in RBC morphology and mechanical properties

A change in RBC morphology from that of a smooth-surfaced discoidal shape is indicative of pathological conditions affecting the metabolism, hemoglobin structure, or the plasma membrane of the cell.[142] The highly flexible RBC membrane enables cells (approximately 6–8 μm in diameter) to traverse capillaries as small as 3 μm in diameter.[2–4] Changes in the cell's morphology adversely affect its ability to deform in the circulation, its adhesiveness, and its lifespan.[134,143] Thus, to better understand how NMs alter erythrocytes, there have been a few studies which explored the mechanisms of NM binding to RBC membranes and their effects on cell morphology and deformability. For example, Asharani *et al.* used SEM to visualize RBCs which had spiky protrusions present on their surfaces (i.e., echinocytes) after being exposed to AgNPs capped with polyvinyl alcohol.[12] No such protrusions were seen in control RBCs or RBCs exposed to similarly capped AuNPs or PtNPs. They concluded that AgNPs severely altered the RBC morphology, while AuNPs and PtNPs did not. Cationic dendrimers have also been reported to alter the shape and contour of RBCs, as characterized by the formation of echinocytes and spherocytes, even at NM concentrations lower than those which cause hemolysis.[101]

Scanning near field ultrasonic holography[144] (SNFUH) is an emerging technique which combines scanning probe technology with microscale ultrasound and holography to acquire images of sub-surface structural features with nanoscale resolution.[145] One of the original applications of this technique was to image malaria parasites in RBCs under ambient conditions.[144] More recently, SNFUH was used to verify the location of single-walled carbon nanohorns (SWCNs) in RBCs obtained from mice exposed to 31 μg of SWCNs.[108] A related technique, mode synthesizing atomic force microscopy (MSAFM), has the potential to provide sub-surface nanoscale information on the mechanical properties of nanoparticles below the membrane surface of the RBC.[107]

A simple method to assess RBC deformability is to force cells under controlled pressure through 3–5-μm diameter pores in a polycarbonate membrane filter.[146] Zhao *et al.* used this technique to show that the RBC deformability index, assessed as the volume of RBCs filtered per minute, decreased with increasing concentration of mesoporous silica nanoparticles.[67] Moreover, they also observed that particle size affected cell deformability; larger mesoporous silica particles (531 nm) reduced RBC deformability more significantly than smaller mesoporous silica particles (~122 nm). They concluded that the attachment of nanoparticles to the RBC membrane, particularly the larger silica nanoparticles, restricted cell flexibility and impaired deformability. The RBC deformability index was closer to normal values when PEG and aminopropyl moieties were added to the surface of the silica particles, as the binding of the surface-modified nanoparticles to RBCs was lower compared to the uncoated nanoparticles.[67] Increased RBC membrane interaction with the larger uncoated nanoparticles was also correlated to higher hemolytic activity compared to the smaller silica nanoparticles.[67] While limited, these studies showed that NM size and surface properties can influence RBC morphology and mechanical properties.

4. *In Vivo* Studies of RBC and NM Interactions

In contrast to the relatively large number of literature studies using *in vitro* techniques to identify detrimental interactions between NMs and RBCs, a much smaller number of studies that investigated the effects of NMs on RBCs *in vivo* are available. A summary of these studies is provided in Table 2.

A limited number of studies have been conducted to assess the effects of non-metallic NMs on hematological indices in rodents.[21,54,126] In one such study, Agashe *et al.* examined the *in vivo* effect of polycationic dendrimers in Sprague–Dawley rats.[21] Intravenous injection of the parent poly(propylene imine) dendrimer resulted in reduced RBC count (RBCC), hemoglobin (Hb), and hematocrit level (HCT) after seven days, whereas functional

Table 2. A summary of studies used to evaluate the effects of NMs on RBC-related *in vivo* endpoints. *Endpoints typically reported as part of a routine laboratory complete blood cell count.

Endpoint	Assay or technique	Studies of RBC interaction with the following NMs
Hematology profile (RBCC, HCT, Hb, MCV, MCH, MCHC, RDW)*	Automated hematology analyzer or manual analysis	Metallic: Au, Ag[147–153] Dendrimers[21,54,126]
Cellular uptake	TEM, ICP-OES, AAS	Metallic: Au[151]

Abbreviations

AAS	Atomic absorption spectroscopy
Hb	Hemoglobin concentration
HCT	Hematocrit level
ICP-OES	Inductively coupled plasma optical emission spectroscopy
MCH	Mean corpuscular hemoglobin
MCHC	Mean corpuscular hemoglobin concentration
MCV	Mean corpuscular volume
RBCC	Red blood cell count
RDW	Red-cell distribution width
TEM	Transmission electron microscopy

coating of the dendrimer (e.g., using mannose to replace the primary amine peripheral groups with carbohydrate–OH groups) prevented these effects.[21]

The effect of AgNPs on hematological parameters has been investigated in experimental animals after inhalation and oral exposure to the particles. Ji *et al.* and Sung *et al.* reported no effects on RBCC, Hb levels, or HCT following inhalation of AgNPs in rats (size 16 ± 1.4 nm) for 28 and 90 days, respectively, at airborne concentrations of nanoparticles sufficient to result in elevated levels of silver in the blood and organs such as the liver.[149,152] It is interesting to note that adverse hematological effects did not occur in the 90-day inhalation study conducted by Sung *et al.* at high-dose exposure concentrations of AgNPs, even though there was an increased incidence of liver effects in exposed animals.[149] These findings suggested that the animals were exposed to a sufficient dose of the nanoparticles to produce a biological adverse effect, although no hematological effects occurred at this dose.[149] Similarly, Kim *et al.* reported no adverse hematological effects in a 90-day oral exposure study of rats to AgNPs (mean diameter 56 nm).[153] In contrast, this same laboratory reported slightly increased levels of RBCs, Hb, and HCT in female rats following oral exposure to AgNPs (60 nm) for 28 days.[148] In addition, Tiwari *et al.* reported dose-dependent hematological effects of AgNPs (15–40 nm) in a Wistar rat model.[147] AgNPs were injected intravenously every day for 32 days and all three parameters (RBCs, Hb, and HCT) were measured once a week for five weeks. All but the lowest dose of AgNPs (4 mg/kg) resulted in a statistical decrease in all three of the RBC study endpoints for the first three weeks of the study. Interestingly, the animals' RBCs appeared to adapt to the treatment over time, possibly due to the accumulation of most of the nanoparticles in organs or the removal of nanoparticles by renal clearance, as no significant changes in these parameters were detected in the final two weeks of the study.[147]

In addition to AgNPs, at least one other study examined the *in vivo* hematological effects of AuNPs (13.5 nm) administered to mice orally, intraperitoneally, and through intravenous tail vein injection.[151] After 14 days, there were no dose-dependent changes to the HCT, Hb, or RBCC. In addition, the HCT and Hb levels were found to be independent of the administration route. However, the RBCC values were found to decrease after oral administration and intraperitoneal injection, but were not affected in mice administered nanoparticles through tail vein injection.[151] The authors concluded that the decrease in RBCC can be attributed to changes in the hematopoietic system and, when this decrease is combined with the abnormal responses in the animals' weight and spleen index, the results suggest that the oral administration route was more toxic than the intraperitoneal and tail vein injection routes.

5. *In Vitro–In Vivo* Correlation

In addition to investigating the mechanisms of NM interactions with RBCs, *in vitro* assays may be used to screen NM formulations prior to animal and clinical testing. When evaluating materials in a toxicity testing paradigm, the need exists to not only compare the response of the NMs in the *in vitro* and *in vivo* assays, but also to consider dosimetry issues so that the concentration of NMs that produce adverse effects *in vitro* can be compared to levels of NMs in the blood or other target tissues *in vivo*. In addition, species-related differences in the sensitivity of RBCs to different NMs should also be taken into account. While sometimes ignored during *in vitro* testing, the effect of plasma proteins on NMs in the *in vivo* environment is also important as they may have a significant effect on NM self-association,[b] toxicity, and biodistribution.[18,154]

Dobrovolskaia *et al.* addressed the relevance of *in vitro* NM hemolysis testing to *in vivo* results by performing the hemolysis assay on nanotechnology-containing formulations approved for use in clinical applications.[115] The study was conducted using a liposomal formulation of doxorubicin (Doxil®), albumin-bound paclitaxel nanoparticles (Abraxane®), and a nanoemulsion-based anesthetic agent (Propofol®).[115] The determination that none of these formulations caused significant *in vitro* hemolysis agrees with previous *in vivo* studies, providing some support for the relevance of the *in vitro* assay.[115] Additional data, supporting a correlation between results from an *in vitro* hemolysis test and *in vivo* hematological toxicity testing of several engineered NMs, is described in Chapter 17 of this book. Li *et al.* similarly reported a good correlation between the lack of effects produced by composite iron–gold NMs in an *in vitro* hemolysis assay and hematological effects in acute toxicity studies of these NMs in mice and dogs.[155] In contrast, Jain *et al.* summarized studies which demonstrated that uncoated polycationic dendrimers induce *in vitro* hemolysis and also cause decreases in hematological parameters in animals.[126] In a study which correlated the results of a battery of *in vitro* assays to broncho-alveolar lavage assessment using Wistar rats, the *in vitro* hemolysis assay was identified as the best screening tool to predict the pro-inflammatory potential of metal oxide nanoparticles.[156]

Extrapolation of the results of *in vitro* hemolysis studies to *in vivo* conditions ideally requires an understanding of the concentration of the NMs

[b] As used here, self-association of NMs includes both NM aggregation (strong, irreversible) and agglomeration (weak, reversible) clumping, both of which may occur due to instability of the NMs in solution.[18]

in vitro and the internal dosimetry of the NMs in the body after exposure. Teeguarden *et al.* have outlined many of the challenges associated with assessing the actual dose of NMs that interacts with cells in *in vitro* test systems, as dosage is dependent on particle diffusion, settling, and agglomeration in the suspension media.[157] Implementation of these principles will be essential for comparing the results of *in vitro* and *in vivo* studies.

An additional challenge when comparing the *in vitro* and *in vivo* effects of NMs on the erythrocytes are the potential differences in sensitivity when blood from different species is used for the *in vitro* and *in vivo* studies. Typically, rabbit, cow, or human blood is used in the *in vitro* hemolysis assay, but *in vivo* studies are most often conducted in rodents. RBCs are known to show remarkable species variations in hemolysis sensitivity,[158–160] RBC morphology,[161] hematological parameters, deformability, and aggregation.[162] Hence, the selection of donor species is not a trivial variable in the design of experiments, especially when correlating *in vitro* results with *in vivo* findings.

As a result of these challenges, NM interactions with RBCs must be evaluated on a case-by-case basis in studies designed to provide the most relevant data through careful consideration of comparative dosimetry and blood source. The limited studies available comparing the *in vitro* and *in vivo* response of RBCs to NMs suggest that *in vitro* studies can be used with some success to predict *in vivo* response.[126]

6. Case Study: Interaction of AgNPs with RBCs

As AgNPs have been widely studied, we present a case study to demonstrate their complex interaction modes and effects on RBCs. Since ancient times, silver has been used for its anti-microbial properties.[163,164] More recently, AgNPs have been shown to exhibit even greater anti-bacterial potency than Ag^+ ions alone[165] and also the potential to act as an anti-fungal agent.[166] AgNPs are currently being investigated for use in a wide variety of more than 200 consumer products.[167] This includes applications in a number of medical devices, such as cardiovascular implants, catheters, bone cement, and wound dressings.[168] Understanding the effects of AgNPs on RBCs is critical for addressing their safe use in medical devices, as their physicochemical properties (e.g., Ag^+ ion generation, reactivity with plasma proteins) may depend on how they are used (e.g., as unbound nanoparticles or imbedded in polymer device coatings). Under *in vitro* test conditions, AgNPs have been shown to cause toxic effects, including hemolysis.[12,13,127,169]

In vitro studies revealed that AgNPs tend to have much greater hemolytic activity than similarly sized Au or PtNPs,[12] and that nano-sized AgNPs result in greater hemolysis than larger micron-sized Ag particles.[13] Similarly, aggregated/agglomerated AgNPs result in less hemolysis than well-dispersed AgNPs.[18] The choice of particle dispersant is critical when evaluating AgNP hemolytic activity in these studies since Ag^+ ions will react with chloride (Cl^-) ions in media containing saline to produce insoluble AgCl, thereby preventing Ag^+ ion interactions with RBCs.[13] Moreover, proteins in suspending media will coat the particles and affect NM self-association and toxicity.[13,18]

As mentioned previously, *in vivo* studies measuring the toxicity of AgNPs revealed no long-term effects on RBC hematology parameters when animals were exposed to small (<35 nm) AgNPs either through inhalation[149] or intravenous injection.[147] However, small but significant increases were detected in RBCC, HCT, and Hb for orally administered particles in female rats using 60-nm AgNPs.[148] Elemental Ag was detected in whole blood for both the oral and inhalation administration routes, suggesting that some form of Ag could be interacting with RBCs even when AgNPs were not placed in direct contact with blood. The results of these studies were likely dependent on nanoparticle size, delivery method, exposure concentration, and animal model. More *in vivo* studies are needed to properly assess how these experimental design parameters affect RBCs and the potential to induce hemolysis.

The mechanisms by which AgNPs may damage erythrocytes are complex and have not been fully elucidated. Cell damage may be due to the rate of release of free Ag^+ ions, the overall Ag^+ ion concentration, direct cellular interaction with particles, or a combination of these events.[13,127,170–171] In contact with body fluids, metallic Ag releases ions in proportion to particle surface area.[170] This is especially important for nanoparticles as the surface-to-volume ratio increases as the particle diameter decreases. Low concentrations of Ag^+ ions *in vitro* are reported to cause RBC hemolysis.[172] Ag^+ ions have also been shown to incite eryptosis (programmed RBC death) as characterized by increased phosphatidylserine exposure on the RBC surface and decreased intracellular ATP content.[169] In physiological media, Ag^+ ions quickly bind to Cl^- ions, inorganic sulfide, and thiols (–SH),[173] which is thought to decrease the bioavailability of Ag^+ ions for damaging cells.[170] However, high-affinity binding of Ag^+ ions to thiol groups in proteins (e.g., albumin) and important antioxidants (e.g., glutathione, superoxide dismutase) may cause structural and functional changes.[127] Garner *et al.* observed a large depletion of the RBC intracellular antioxidant glutathione when erythrocytes were exposed to 30-nm Ag colloids; this may have been due to binding with Ag^+ ions or glutathione depletion as a result of an increase in intracellular ROS levels.[127,174] Even in the presence of Cl^- ions, Ag^+ ions were shown to remain bioavailable and cause significant hemolysis by using sulphadimidine as a carrier.[172]

As the amount of available unbound Ag^+ ions may be decreased in biological media (due to binding, decreased reactive surface area from particle agglomeration, and the formation of a protein corona on the particles), direct particle interaction with the RBC membrane should also be considered as a potential damage mechanism. It is possible that AgNPs in direct contact with RBCs could continuously release ions locally that may not be detoxified by binding to plasma chloride or proteins.[13] Asharani *et al.* observed *in vitro* that AgNPs attached to RBC membranes, including RBC ghost cells devoid of hemoglobin.[12] It was also reported that Ag-based nanoparticles (5–30 nm in diameter) caused an increase in RBC membrane lipid peroxidation and hemolysis, while exposure to Ag^+ ions alone did not cause cell damage.[12] Recent studies have shown that several different types of nanoparticles can bind to RBC membranes and enter the cells through an unknown non-specific pathway.[12,67,69,103] Hence, AgNP-induced alterations to RBCs leading to hemolysis may be due to many mechanisms including Ag^+ ion generation, direct particle interaction with the RBC membrane, particle uptake by RBCs, and oxidative stress.

7. Points to Consider Regarding NM–RBC Interaction

In addition to the earlier precautions concerning potential NM-induced interference confounding the results of *in vitro* hemolysis tests, there are other factors that should be considered as well when conducting biocompatibility tests of NMs and interpreting the results.

The variety of NMs is quite profound and ever increasing. Not only are NMs synthesized from an extensive array of materials,[175] they can also be engineered to have unique geometries,[176,177] structures,[178] mechanical properties,[179] and surface functionalities.[82] Due to these complexities, in which a small change in a particle construct may significantly affect its behavior, it is necessary to evaluate the toxicity of each NM on an individual basis. This is especially important as the size, shape, stability, reactivity, and toxicity of a NM may be very different depending on its functionalized surface, its fabrication source (and batch), and its exposure to different media, including storage dispersants and physiological solutions containing plasma proteins.[154] Until thorough physicochemical characterization and biocompatibility testing protocols are established, it will be difficult to compare toxicological results among laboratories.

The patient population that a NM-based treatment is applied to may also prove to be important. *In vitro* testing is most often performed using blood from healthy donors whose RBCs might be less sensitive to NM interactions than the RBCs of compromised individuals. This is a concern given the large number of nanotherapies in which NMs may come into contact with blood from unhealthy patients during treatment. Various disease states, genetic defects, or medical treatments can increase the susceptibility of RBCs to hemolysis and oxidative stress. For example, reduced intracellular antioxidant levels in erythrocytes from critically ill patients,[180] neonates with elevated levels of bilirubin in the blood,[181] or patients with chronic diseases such as rheumatoid arthritis[182] and cancer[183] could make the RBCs from these patients more susceptible to the effects of oxidative stress produced by NMs. In addition, medical procedures such as apheresis,[184] hemodialysis,[185] and cardiopulmonary bypass can produce subhemolytic damage to the RBC membrane from shear stress and can increase intracellular oxidative stress and RBC aggregation while causing a decrease in RBC deformability.[186] Hence, RBCs which are altered due to patient disease states or by being exposed to foreign surfaces and shear stresses in medical devices may be more susceptible to oxidative stress and hemolysis caused by subsequent exposure to NMs.

Glucose-6-phosphate dehydrogenase (G6PD) deficiency is the most common enzyme defect in humans, affecting over 400 million people worldwide.[189] This defect results in an increased susceptibility of erythrocytes to oxidative stress-induced hemolysis. Given the widespread incidence of G6PD deficiency in the human population and the ability of some NMs to produce oxidative stress in RBCs, it would be useful to determine whether RBCs from individuals with G6PD deficiency are more susceptible to oxidative stress and hemolysis produced by NMs than RBCs from individuals without the enzyme defect.

Hemolysis can also occur *in vitro* during routine laboratory testing procedures due to poor venipuncture technique, exposure to extreme temperatures, or the use of certain anticoagulants or prolonged blood storage.[188,189] Thus, careful blood handling and preparation is critical when evaluating RBC hemolysis *in vitro* and *in vivo*.

8. Summary

Understanding the conditions in which RBCs and NMs come in contact with each other and their mechanisms of interaction, is critical for safely and effectively exploiting the great potential of NMs in applications such as cellular therapies, drug delivery, and medical devices.[190] This chapter reviewed the current *in vitro* techniques reported in the literature for investigating the effects of NMs on RBCs using various endpoints (Table 1). Based mainly on these *in vitro* studies, the possible adverse effects of NMs on RBCs are schematically shown in Figure 2.

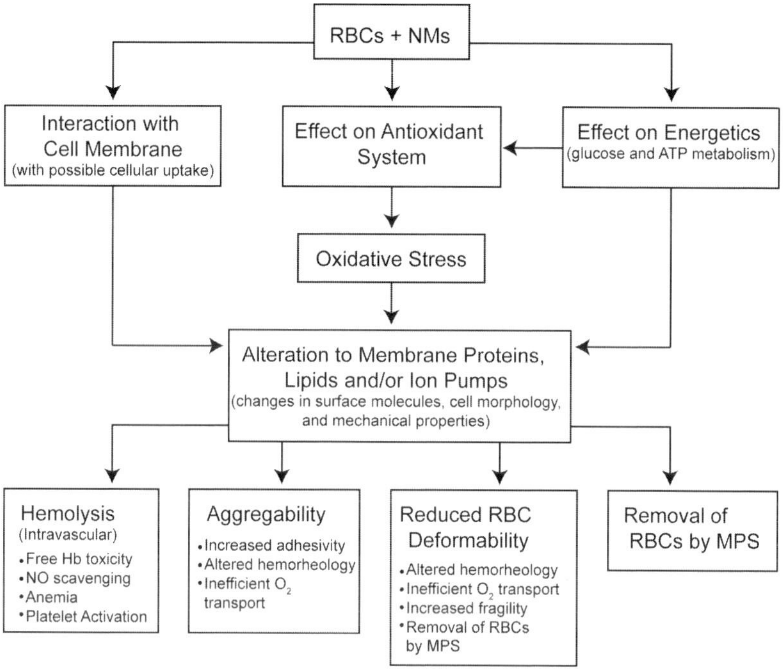

Figure 2. Summary of the possible effects of NMs on RBCs.
MPS: mononuclear phagocytic system, NO: nitric oxide.

Physical properties such as surface charge, shape, size, and chemical surface functionality can vary between manufactured lots of NMs and greatly affect the way NMs interact with RBCs. Moreover, the important effects that different media and plasma proteins play in altering the physicochemical properties and toxicity of NMs (e.g., hemolysis) require further study. While the *in vitro* hemolysis assay is the most widely used test for evaluating RBC interaction with NMs, interlaboratory validation of this testing technique has not been conducted. Of great importance in conducting these tests is that researchers must be aware of the possibility that NMs can interfere with the assays and that it is critical to include appropriate controls in the experimental design. In summary, techniques such as the hemolysis assay are useful for screening materials, but only as part of a battery of biocompatibility tests. Furthermore, important changes to RBCs may occur at NM concentrations that do not cause hemolysis. The field of nanotoxicology is still relatively young and many more *in vitro* studies are needed to properly assess NM interaction with RBCs. New techniques and assays, possibly based on the molecular cell markers associated with the normal removal of RBCs from the circulation at the end of their lifetimes, will certainly be developed.

The latter half of this chapter described the first *in vivo* studies that investigated the biodistribution and effects of NMs on RBCs. These initial studies suggest that Au and Ag nanomaterials will come into contact with RBCs when animals are exposed through the intravenous, oral, or inhalation routes. As evaluated using typical hematology endpoints, these studies did not uncover any substantial long-term changes to RBCs in rodent models. However, cationic dendrimers, which caused hemolysis during *in vitro* testing, also led to a decrease in RBC hematological values when injected intravenously in rats.[21,54] Further *in vivo* studies are needed to properly assess the interaction potential as well as the effects of NMs on RBCs in different animal models. These studies are also important to properly interpret the applicability of comparative *in vitro* assessments to predict *in vivo* results.

A case study using AgNPs was presented to demonstrate the complex mechanisms of action by which a NM may induce RBC damage. The literature suggested that AgNP-induced hemolysis may be related to Ag^{+} ion generation, direct particle interaction with the RBC membrane, particle uptake by RBCs, and oxidative stress.

9. Suggestions for Future Directions

In vitro testing protocols, validated through interlaboratory comparisons (i.e., round robin testing) using NMs with markedly different compositions, need to be developed along with appropriate NM-based controls for the assays listed in Table 1. Many of these assays are straightforward in concept but become more involved with the incorporation of complex NMs, the use of different media, and blood samples from different species.

The inability of an NM to induce a measurable change in a particular RBC-related endpoint using current evaluation techniques does not imply that the NM–RBC interaction is harmless. As noted by Zhao *et al.*, "The lack of hemolysis, however, does not necessarily warrant the absence of interactions between the particles and red blood cells, which could lead to subtle side effects."[67] Indeed, alterations to RBC morphology at low concentrations of NMs that do not induce hemolysis have been reported.[101] Hence, a wide range of subtle yet important biochemical and mechanical endpoints may be assessed when evaluating the effect of NMs on erythrocytes. These include effects on the rheological properties of the RBC, cellular energetics, and biomarkers of aging (e.g., phosphatidylserine expression on the outer RBC membrane[7,8,11]). Consequently, future studies on the effects of NMs on erythrocytes may also explore endpoints such as deformability, adhesivity, oxygen delivery, membrane proteomics,[191] and ATP production and metabolism.

More in-depth, systematic studies need to be conducted in order to clearly elucidate the mechanisms through which NMs and RBCs interact in physiological environments. Special attention must be given to protein interactions with NMs and the potential changes which may affect physico-chemical properties and toxicity of NMs.

Lastly, to ensure that the results of *in vitro* NM evaluation techniques are relevant and can be extrapolated to patients, comparative studies also need to be performed *in vivo* using the same NMs at different concentrations. Moreover, many investigators currently studying the toxicity of nanoparticles *in vivo* fail to include hematology endpoints, such as the tests listed in Table 2, in their studies. These endpoints are well defined, relatively easy to collect, and provide insight into the effects of NMs on the health of test subjects participating in *in vivo* studies.

Acknowledgements

The authors would like to thank the following for review of the chapter and helpful comments — David Pudwill, Molly Ghosh, Jennifer Goode, Martha Betz, Girish Kumar, Peter Goering (Center for Devices and Radiological Health, FDA), Katherine Tyner (Center for Drug Evaluation and Research, FDA), and Marina Dobrovolskaia (Nanotechnology Characterization Laboratory, NCI).

References

1. Lichtman MA, Kipps TJ, Seligsohn U, Kaushansky K, Prchal JT. *Williams Hematology*, 8th edn., McGraw-Hill, 2010.
2. Lingard PS. Capillary pore rheology of erythrocytes. I. Hydroelastic behaviour of human erythrocytes. *Microvasc Res* 1974;8:53–63.
3. Mchedlishvili G, Maeda N. Blood flow structure related to red cell flow: Determinant of blood fluidity in narrow microvessels. *Jpn J Physiol* 2001;51:19–30.
4. Secomb TW. Flow-dependent rheological properties of blood in capillaries. *Microvasc Res* 1987;34:46–58.
5. Cimen MYB. Free radical metabolism in human erythrocytes. *Clin Chim Acta* 2008;390:1–11.
6. Antonelou MH, Kriebardis AG, Papassideri IS. Aging and death signalling in mature red cells: From basic science to transfusion practice. *Blood Transfus* 2010;8(Suppl 3):s39–s47.
7. Boas FE, Forman L, Beutler E. Phosphatidylserine exposure and red cell viability in red cell aging and in hemolytic anemia. *Proc Natl Acad Sci U S A* 1998;95:3077–3081.
8. Connor J, Pak CC, Schroit AJ. Exposure of phosphatidylserine in the outer leaflet of human red blood cells. Relationship to cell density, cell age, and clearance by mononuclear cells. *J Biol Chem* 1994;269:2399–2404.

9. Bosman GJ, Werre JM, Willekens FLA, Novotný VMJ. Erythrocyte ageing *in vivo* and *in vitro*: Structural aspects and implications for transfusion. *Transfus Med* 2008;18:335–347.
10. Huang, YX, Wu Z J, Mehrishi J, Huang BT, Chen XY, Zheng XJ, Liu WJ, Luo M. Human red blood cell aging: Correlative changes in surface charge and cell properties. *J Cell Mol Med* 2011;15:2634–2642.
11. Lang F, Lang KS, Lang PA, Huber SM, Wieder T. Mechanisms and significance of eryptosis. *Antioxid Redox Signal* 2006;8:1183–1192.
12. Asharani PV, Sethu S, Vadukumpully S, Zhong S, Lim CT, Hande MP, Valiyaveettil S. Investigations on the structural damage in human erythrocytes exposed to silver, gold, and platinum nanoparticles. *Adv Funct Mater* 2010;20:1233–1242.
13. Choi J, Reipa V, Hitchins VM, Goering PL, Malinauskas RA. Physicochemical characterization and *in vitro* hemolysis evaluation of silver nanoparticles. *Toxicol Sci* 2011; 123:133–143.
14. Golubeva O, Shamova O, Orlov D, Pazina T, Boldina A, Kokryakov V. Study of antimicrobial and hemolytic activities of silver nanoparticles prepared by chemical reduction. *Glass Phys Chem* 2010;36:628–634.
15. Goodman CM, McCusker CD, Yilmaz T, Rotello VM. Toxicity of gold nanoparticles functionalized with cationic and anionic side chains. *Bioconjug Chem* 2004;15:897–900.
16. Lin IC, Liang M, Liu TY, Ziora ZM, Monteiro MJ, Toth I. Interaction of densely polymer-coated gold nanoparticles with epithelial Caco-2 monolayers. *Biomacromolecules* 2011;12:1339–1348.
17. Tai Y, Tran NT, Tsai YC, Fang JY, Chang LW. One-step synthesis of highly biocompatible multi-shaped gold nanostructures with fruit extract. *IET Nanobiotechnol* 2011;5:52.
18. Zook JM, MacCuspie RI, Locascio LE, Halter MD, Elliott JT. Stable nanoparticle aggregates/agglomerates of different sizes and the effect of their size on hemolytic cytotoxicity. *Nanotoxicology* 2011;5:517–30.
19. Chanda N, Kan P, Watkinson LD, Shukla R, Zambre A, Carmack TL, Engelbrecht H, Lever JR, Katti K, Fent GM, Casteel SW, Smith CJ, Miller WH, Jurisson S, Boote E, Robertson JD, Cutler C, Dobrovolskaia MA, Kannan R, Katti KV. Radioactive gold nanoparticles in cancer therapy: Therapeutic efficacy studies of GA-198AuNP nanoconstruct in prostate tumor-bearing mice. *Nanomedicine* 2010;6:201–209.
20. Agarwal A, Gupta U, Asthana A, Jain NK. Dextran conjugated dendritic nanoconstructs as potential vectors for anti-cancer agent. *Biomaterials* 2009;30:3588–3596.
21. Agashe HB, Dutta T, Garg M, Jain NK. Investigations on the toxicological profile of functionalized fifth-generation poly (propylene imine) dendrimer. *J Pharm Pharmacol* 2006;58:1491–1498.
22. Alavidjeh MS, Haririan I, Khorramizadeh MR, Ghane ZZ, Ardestani MS, Namazi H. Anionic linear-globular dendrimers: Biocompatible hybrid materials with potential uses in nanomedicine. *J Mater Sci Mater Med* 2010;21:1121–1133.
23. Anantharaman A, Rizvi MS, Sahal D. Synergy with rifampin and kanamycin enhances potency, kill kinetics, and selectivity of de novo-designed antimicrobial peptides. *Antimicrob Agents Chemother* 2010;54:1693–1699.
24. Bansal KK, Kakde D, Gupta U, Jain NK. Development and characterization of triazine based dendrimers for delivery of antitumor agent. *J Nanosci Nanotechnol* 2010;10: 8395–8404.

25. Bhadra D, Bhadra S, Jain S, Jain NK. A PEGylated dendritic nanoparticulate carrier of fluorouracil. *Int J Pharm* 2003;257:111–124.
26. Chen HT, Neerman MF, Parrish AR, Simanek EE. Cytotoxicity, hemolysis, and acute *in vivo* toxicity of dendrimers based on melamine, candidate vehicles for drug delivery. *J Am Chem Soc* 2004;126:10044–10048.
27. Duncan R, Izzo L. Dendrimer biocompatibility and toxicity. *Adv Drug Delivery Rev* 2005;57:2215–2237.
28. Dutta T, Agashe HB, Garg M, Balakrishnan P, Kabra M, Jain NK. Poly(propyleneimine) dendrimer based nanocontainers for targeting of efavirenz to human monocytes/macrophages *in vitro*. *J Drug Target* 2007;15:89–98.
29. Eliyahu H, Servel N, Domb AJ, Barenholz Y. Lipoplex-induced hemagglutination: potential involvement in intravenous gene delivery. *Gene Ther* 2002;9:850–858.
30. Fischer D, Li Y, Ahlemeyer B, Krieglstein J, Kissel T. *In vitro* cytotoxicity testing of polycations: Influence of polymer structure on cell viability and hemolysis. *Biomaterials* 2003;24:1121–1131.
31. Fitzgerald KT, Holladay CA, McCarthy C, Power KA, Pandit A, Gallagher WM. Standardization of models and methods used to assess nanoparticles in cardiovascular applications. *Small* 2011;7:705–717.
32. Gonzalo T, Clemente MI, Chonco L, Weber ND, Diaz L, Serramia MJ, Gras R, Ortega P, de la Mata FJ, Gomez R, Lopez-Fernandez LA, Munoz-Fernandez MA, Jimenez JL. Gene therapy in HIV-infected cells to decrease viral impact by using an alternative delivery method. *ChemMedChem* 2010;5:921–929.
33. Gupta U, Dwivedi SK, Bid HK, Konwar R, Jain NK. Ligand anchored dendrimers based nanoconstructs for effective targeting to cancer cells. *Int J Pharm* 2010;393:185–196.
34. Han L, Li J, Huang S, Huang R, Liu S, Hu X, Yi P, Shan D, Wang X, Lei H, Jiang C. Peptide-conjugated polyamidoamine dendrimer as a nanoscale tumor-targeted T1 magnetic resonance imaging contrast agent. *Biomaterials* 2011;32:2989–2998.
35. Haririan I, Alavidjeh MS, Khorramizadeh MR, Ardestani MS, Ghane ZZ, Namazi H. Anionic linear-globular dendrimer-*cis*-platinum (II) conjugates promote cytotoxicity *in vitro* against different cancer cell lines. *Int J Nanomed* 2010;5:63–75.
36. Jiang YY, Tang GT, Zhang LH, Kong SY, Zhu SJ, Pei YY. PEGylated PAMAM dendrimers as a potential drug delivery carrier: *In vitro* and *in vivo* comparative evaluation of covalently conjugated drug and noncovalent drug inclusion complex. *J Drug Target* 2010;18:389–403.
37. Klajnert B, Janiszewska J, Urbanczyk-Lipkowska Z, Bryszewska M, Shcharbin D, Labieniec M. Biological properties of low molecular mass peptide dendrimers. *Int J Pharm* 2006;309:208–217.
38. Klajnert B, Appelhans D, Komber H, Morgner N, Schwarz S, Richter S, Brutschy B, Ionov M, Tonkikh AK, Bryszewska M, Voit B. The influence of densely organized maltose shells on the biological properties of poly(propylene imine) dendrimers: New effects dependent on hydrogen bonding. *Chemistry* 2008;14:7030–7041.
39. Kumar PV, Asthana A, Dutta T, Jain NK. Intracellular macrophage uptake of rifampicin loaded mannosylated dendrimers. *J Drug Target* 2006;14:546–556.
40. Malik N, Wiwattanapatapee R, Klopsch R, Lorenz K, Frey H, Weener JW, Meijer EW, Paulus W, Duncan R. Dendrimers: Relationship between structure and biocompatibility *in vitro*, and preliminary studies on the biodistribution of 125I-labelled polyamidoamine dendrimers *in vivo*. *J Controlled Release* 2000;65:133–148.

41. Mishra V, Gupta U, Jain NK. Influence of different generations of poly(propylene imine) dendrimers on human erythrocytes. *Pharmazie* 2010;65:891–895.
42. Navath RS, Menjoge AR, Wang B, Romero R, Kannan S, Kannan RM. Amino acid-functionalized dendrimers with heterobifunctional chemoselective peripheral groups for drug delivery applications. *Biomacromolecules* 2010;11:1544–1563.
43. Prajapati RN, Tekade RK, Gupta U, Gajbhiye V, Jain NK. Dendimer-mediated solubilization, formulation development and *in vitro-in vivo* assessment of piroxicam. *Mol Pharm* 2009;6:940–950.
44. Prieto MJ, Temprana CF, del Rio Zabala NE, Marotta CH, Alonso SV. Optimization and *in vitro* toxicity evaluation of G4 PAMAM dendrimer-risperidone complexes. *Eur J Med Chem* 2011;46:845–850.
45. Qi R, Gao Y, Tang Y, He RR, Liu TL, He Y, Sun S, Li BY, Li YB, Liu G. PEG-conjugated PAMAM dendrimers mediate efficient intramuscular gene expression. *AAPS J* 2009; 11:395–405.
46. Russ V, Gunther M, Halama A, Ogris M, Wagner E. Oligoethylenimine-grafted polypropylenimine dendrimers as degradable and biocompatible synthetic vectors for gene delivery. *J Controlled Release* 2008;132:131–140.
47. Singh P, Gupta U, Asthana A, Jain NK. Folate and folate-PEG-PAMAM dendrimers: Synthesis, characterization, and targeted anticancer drug delivery potential in tumor bearing mice. *Bioconjug Chem* 2008;19:2239–2252.
48. Tam JP, Lu YA, Yang JL. Antimicrobial dendrimeric peptides. *Eur J Biochem* 2002; 269:923–932.
49. Uchegbu IF, Sadiq L, Pardakhty A, El-Hammadi M, Gray AI, Tetley L, Wang W, Zinselmeyer BH, Schatzlein AG. Gene transfer with three amphiphilic glycol chitosans-the degree of polymerisation is the main controller of transfection efficiency. *J Drug Target* 2004;12:527–539.
50. Wang P, Zhao XH, Wang ZY, Meng M, Li X, Ning Q. Generation 4 polyamidoamine dendrimers is a novel candidate of nano-carrier for gene delivery agents in breast cancer treatment. *Cancer Lett* 2010;298:34–49.
51. Wang W, Xiong W, Zhu Y, Xu H, Yang X. Protective effect of PEGylation against poly(amidoamine) dendrimer-induced hemolysis of human red blood cells. *J Biomed Mater Res B Appl Biomater* 2010;93:59–64.
52. Yellepeddi VK, Kumar A, Palakurthi S. Surface modified poly(amido)amine dendrimers as diverse nanomolecules for biomedical applications. *Expert Opin Drug Delivery* 2009;6:835–850.
53. Han MH, Chen J, Wang J, Chen SL, Wang XT. Blood compatibility of polyamidoamine dendrimers and erythrocyte protection. *J Biomed Nanotechol* 2010;6:82–92.
54. Bhadra D, Yadav AK, Bhadra S, Jain NK. Glycodendrimeric nanoparticulate carriers of primaquine phosphate for liver targeting. *Int J Pharm* 2005;295:221–233.
55. Liu YF, Yu JS. *In situ* synthesis of highly luminescent glutathione-capped CdTe/ZnS quantum dots with biocompatibility. *J Colloid Interface Sci* 2010;351:1–9.
56. Liu YF, Xie B, Yin ZG, Fang SM, Zhao JB. Synthesis of highly stable CdTe/CdS quantum dots with biocompatibility. *Eur J Inorg Chem* 2010;2010:1501–1506.
57. Chouly C, Bordenave L, Bareille R, Guerin V, Baquey A, Pouliquen D, Baquey C, Jallet P. *In vitro* study of the hemocompatibility of superparamagnetic contrast agent for magnetic resonance imaging. *Clin Mater* 1994;15:293–301.

58. Creangă DE, Culea M, Nădejde C, Odncea S, Curecheriu L, Racuciu M. Magnetic nanoparticle effects on the red blood cells. *J Phys Condens Matter* 2009;170:012019.
59. Guowei D, Adriane K, Chen X, Jie C, Yinfeng L. PVP magnetic nanospheres: Biocompatibility, *in vitro* and *in vivo* bleomycin release. *Int J Pharm* 2007;328:78–85.
60. Kempe M, Kempe H, Snowball I, Wallen R, Arza CR, Gotberg M, Olsson T. The use of magnetite nanoparticles for implant-assisted magnetic drug targeting in thrombolytic therapy. *Biomaterials* 2010;31:9499–9510.
61. Ma HL, Qi XR, Ding WX, Maitani Y, Nagai T. Magnetic targeting after femoral artery administration and biocompatibility assessment of superparamagnetic iron oxide nanoparticles. *J Biomed Mater Res A* 2008;84:598–606.
62. Wang ZY, Song J, Zhang DS. Nanosized As2O3/Fe2O3 complexes combined with magnetic fluid hyperthermia selectively target liver cancer cells. *World J Gastroenterol* 2009;15:2995–3002.
63. Wu W, Chen B, Cheng J, Wang J, Xu W, Liu L, Xia G, Wei H, Wang X, Yang M, Yang L, Zhang Y, Xu C, Li J. Biocompatibility of Fe3O4/DNR magnetic nanoparticles in the treatment of hematologic malignancies. *Int J Nanomed* 2010;5:1079–1084.
64. Lin YS, Haynes CL. Impacts of mesoporous silica nanoparticle size, pore ordering, and pore integrity on hemolytic activity. *J Am Chem Soc* 2010;132:4834–4842.
65. Park HS, Kim CW, Lee HJ, Choi JH, Lee SG, Yun YP, Kwon IC, Lee SJ, Jeong SY, Lee SC. A mesoporous silica nanoparticle with charge-convertible pore walls for efficient intracellular protein delivery. *Nanotechnology* 2010;21:225101.
66. Yu T, Malugin A, Ghandehari H. Impact of silica nanoparticle design on cellular toxicity and hemolytic activity. *ACS Nano* 2011;5:5717–5728.
67. Zhao Y, Sun X, Zhang G, Trewyn BG, Slowing II, Lin VS. Interaction of mesoporous silica nanoparticles with human red blood cell membranes: Size and surface effects. *ACS Nano* 2011;5:1366–1375.
68. Aisaka Y, Kawaguchi R, Watanabe S, Ikeda M, Igisu H. Hemolysis caused by titanium dioxide particles. *Inhal Toxicol* 2008;20:891–893.
69. Li SQ, Zhu RR, Zhu H, Xue M, Sun XY, Yao SD, Wang SL. Nanotoxicity of TiO(2) nanoparticles to erythrocyte *in vitro*. *Food Chem Toxicol* 2008;46:3626–3631.
70. Yang Y, Lai Y, Zhang Q, Wu K, Zhang L, Lin C, Tang P. A novel electrochemical strategy for improving blood compatibility of titanium-based biomaterials. *Colloids Surf B Biointerfaces* 2010;79:309–313.
71. Crouzier T, Nimmagadda A, Nollert MU, McFetridge PS. Modification of single walled carbon nanotube surface chemistry to improve aqueous solubility and enhance cellular interactions. *Langmuir* 2008;24:13173–13181.
72. Salvador-Morales C, Flahaut E, Sim E, Sloan J, Green ML, Sim RB. Complement activation and protein adsorption by carbon nanotubes. *Mol Immunol* 2006;43:193–201.
73. Wu H, Liu G, Wang X, Zhang J, Chen Y, Shi J, Yang H, Hu H, Yang S. Solvothermal synthesis of cobalt ferrite nanoparticles loaded on multiwalled carbon nanotubes for magnetic resonance imaging and drug delivery. *Acta Biomater* 2011;7:3496–3504.
74. Wu H, Liu G, Zhuang Y, Wu D, Zhang H, Yang H, Hu H, Yang S. The behavior after intravenous injection in mice of multiwalled carbon nanotube/Fe(3)O(4) hybrid MRI contrast agents. *Biomaterials* 2011;32:4867–4876.
75. Akagi T, Kim H, Akashi M. pH-dependent disruption of erythrocyte membrane by amphiphilic poly(amino acid) nanoparticles. *J Biomater Sci Polym Ed* 2010;21:315–328.

76. Cavadas M, González-Fernández Á, Franco R. Pathogen-mimetic stealth nanocarriers for drug delivery: A future possibility. *Nanomedicine* 2011;7:730–743.
77. Chen C, Cheng YC, Yu CH, Chan SW, Cheung MK, Yu PH. *In vitro* cytotoxicity, hemolysis assay, and biodegradation behavior of biodegradable poly(3-hydroxybutyrate)-poly(ethylene glycol)-poly(3-hydroxybutyrate) nanoparticles as potential drug carriers. *J Biomed Mater Res A* 2008;87:290–298.
78. Chouhan R, Bajpai AK. Release dynamics of ciprofloxacin from swellable nanocarriers of poly(2-hydroxyethyl methacrylate): An *in vitro* study. *Nanomedicine* 2010;6:453–462.
79. Falamarzian A, Lavasanifar A. Optimization of the hydrophobic domain in poly(ethylene oxide)-poly(varepsilon-caprolactone) based nano-carriers for the solubilization and delivery of amphotericin B. *Colloids Surf B Biointerfaces* 2010;81:313–320.
80. Falamarzian A, Lavasanifar A. Chemical modification of hydrophobic block in poly(ethylene oxide) poly(caprolactone) based nanocarriers: Effect on the solubilization and hemolytic activity of amphotericin B. *Macromol Biosci* 2010;10:648–656.
81. Gou M, Zheng L, Peng X, Men K, Zheng X, Zeng S, Guo G, Luo F, Zhao X, Chen L, Wei Y, Qian Z. Poly(epsilon-caprolactone)-poly(ethylene glycol)-poly(epsilon-caprolactone) (PCL-PEG-PCL) nanoparticles for honokiol delivery *in vitro*. *Int J Pharm* 2009;375:170–176.
82. Mayer A, Vadon M, Rinner B, Novak A, Wintersteiger R, Frohlich E. The role of nanoparticle size in hemocompatibility. *Toxicology* 2009;258:139–147.
83. Pignatello R, Cenni E, Micieli D, Fotia C, Salerno M, Granchi D, Avnet S, Sarpietro MG, Castelli F, Baldini N. A novel biomaterial for osteotropic drug nanocarriers: Synthesis and biocompatibility evaluation of a PLGA-ALE conjugate. *Nanomedicine (Lond)* 2009;4:161–175.
84. Banerjee S, Saikia JP, Kumar A, Konwar BK. Antioxidant activity and haemolysis prevention efficiency of polyaniline nanofibers. *Nanotechnology* 2010;21:045101.
85. Tamura, M, Ichinohe S, Tamura A, Ikeda Y, Nagasaki Y. *In vitro* and *in vivo* characteristics of core-shell type nanogel particles: Optimization of core cross-linking density and surface poly(ethylene glycol) density in PEGylated nanogels. *Acta Biomater* 2011;7:3354–3361.
86. Kim D, El-Shall H, Dennis D, Morey T. Interaction of PLGA nanoparticles with human blood constituents. *Colloids Surf B Biointerfaces* 2005;40:83–91.
87. Barauskas J, Cervin C, Jankunec M, Spandyreva M, Ribokaite K, Tiberg F, Johnsson M. Interactions of lipid-based liquid crystalline nanoparticles with model and cell membranes. *Int J Pharm* 2010;391:284–291.
88. Estella-Hermoso de MA, Rayo M, Mollinedo F, Blanco-Prieto MJ. Lipid nanoparticles for alkyl lysophospholipid edelfosine encapsulation: Development and *in vitro* characterization. *Eur J Pharm Biopharm* 2008;68:207–213.
89. Huang ZR, Hua SC, Yang YL, Fang JY. Development and evaluation of lipid nanoparticles for camptothecin delivery: A comparison of solid lipid nanoparticles, nanostructured lipid carriers, and lipid emulsion. *Acta Pharmacol Sin* 2008;29:1094–1102.
90. Schubert MA, Muller-Goymann CC. Characterisation of surface-modified solid lipid nanoparticles (SLN): Influence of lecithin and nonionic emulsifier. *Eur J Pharm Biopharm* 2005;61:77–86.
91. Darole PS, Hegde DD, Nair HA. Formulation and evaluation of microemulsion based delivery system for amphotericin B. *AAPS PharmSciTech* 2008;9:122–128.

92. Pietkiewicz J, Zielinska K, Saczko J, Kulbacka J, Majkowski M, Wilk KA. New approach to hydrophobic cyanine-type photosensitizer delivery using polymeric oil-cored nanocarriers: Hemolytic activity, *in vitro* cytotoxicity and localization in cancer cells. *Eur J Pharm Sci* 2010;39:322–335.
93. Yanasarn N, Sloat BR, Cui Z. Nanoparticles engineered from lecithin-in-water emulsions as a potential delivery system for docetaxel. *Int J Pharm* 2009;379:174–180.
94. Trpkovic A, Todororic-Markoriz B, Klent D, Misirkic M, Janjetoric K, Vuciceviz L, *et al.* Oxidative stress-mediated hemolytic activity of solvent exchange-prepared fullerene (C60) nanoparticles. *Nanotechnology* 2010;21:375102.
95. Yang XL, Huang C, Qiao XG, Yao L, Zhao DX, Tan X. Photo-induced lipid peroxidation of erythrocyte membranes by a bis-methanophosphonate fullerene. *Toxicol in Vitro* 2007;21:1493–1498.
96. Kurosaki T, Morishita T, Kodama Y, Sato K, Nakagawa H, Higuchi N, Nakamura T, Hamamoto T, Sasaki H, Kitahara T. Nanoparticles electrostatically coated with folic acid for effective gene therapy. *Mol Pharm* 2011;8:913–919.
97. Chung YC, Hsieh WY, Young TH. Polycation/DNA complexes coated with oligonucleotides for gene delivery. *Biomaterials* 2010;31:4194–4203.
98. Hiwatari K, Sakuma S, Iwata K, Masaoka Y, Kataoka M, Tachikawa H, Shoji Y, Yamashita S. Poly(N-vinylacetamide) chains enhance lectin-induced biorecognition through the reduction of nonspecific interactions with nontargets. *Eur J Pharm Biopharm* 2008; 70:453–461.
99. Luo Z, Wang S, Zhang S. Fabrication of self-assembling d-form peptide nanofiber scaffold d-EAK16 for rapid hemostasis. *Biomaterials* 2011;32:2013–2020.
100. Oka H, Onaga T, Koyama T, Guo CT, Suzuki Y, Esumi Y, Hatano K, Terunuma D, Matsuoka K. Syntheses and biological evaluations of carbosilane dendrimers uniformly functionalized with sialyl alpha(2–3) lactose moieties as inhibitors for human influenza viruses. *Bioorg Med Chem* 2009;17:5465–5475.
101. Ziemba B, Matuszko G, Bryszewska M, Klajnert B. Influence of dendrimers on red blood cells. *Cell Mol Biol Lett.* 2011.
102. Brazhe NA, Abdali S, Brazhe AR, Luneva OG, Bryzgalova NY, Parshina EY. Sosnovtseva OV. Maksimov GV. New insight into erythrocyte through *in vivo* surface-enhanced Raman spectroscopy. *Biophys J* 2009;97:3206–3214.
103. Rothen-Rutishauser BM, Schürch S, Haenni B, Kapp N, Gehr P. Interaction of fine particles and nanoparticles with red blood cells visualized with advanced microscopic techniques. *Environ Sci Technol* 2006;40:4353–4359.
104. Soler MA, Bao SN, Alcantara GB, Tiburcio VH, Paludo GR, Santana JF, Guedes MH, Lima EC, Lacava ZG, Morais PC. Interaction of erythrocytes with magnetic nanoparticles. *J Nanosci Nanotechnol* 2007;7:1069–1071.
105. Xu H, Dai W, Han Y, Hao W, Xiong F, Zhang Y, Cao JM. Differential internalization of superparamagnetic iron oxide nanoparticles in different types of cells. *J Nanosci Nanotechnol* 2010;10:7406–7410.
106. Zhang M, Xing G, Yuan H, Chang X, Jing L, Zhao Y, Zhu C, Fang X. Transmembrane delivery of aggregated [GdC82(OH)22]n nanoparticles. *J Nanosci Nanotechnol* 2010;10:8556–8561.
107. Tetard L, Passian A, Farahi RH, Thundat T. Atomic force microscopy of silica nanoparticles and carbon nanohorns in macrophages and red blood cells. *Ultramicroscopy* 2010;110:586–591.

108. Tetard L, Passian A, Venmar KT, Lynch RM, Voy BH, Shekhawat G, Dravid VP, Thundat T. Imaging nanoparticles in cells by nanomechanical holography. *Nat Nanotechnol* 2008;3:501–505.
109. Simak J. Nanotoxicity in blood: Effects of engineered nanomaterials on platelets. In: Sahu S, Casciano D, eds. *Nanotoxicity: From In Vivo and In Vitro Models to Health Risks.* John Wiley & Sons, 2009, pp. 191–225.
110. Scharte M, Fink MP. Red blood cell physiology in critical illness. *Crit Care Med* 2003;31:541–544.
111. Buehler PW, D'Agnillo F. Toxicological consequences of extracellular hemoglobin: Biochemical and physiological perspectives. *Antioxid Redox Signal* 2009;12:275–291.
112. Rother RP, Bell L, Hillmen P, Gladwin MT. The clinical sequelae of intravascular hemolysis and extracellular plasma hemoglobin. *JAMA* 2005;293:1653–1662.
113. Minetti M, Agati L, Malorni W. The microenvironment can shift erythrocytes from a friendly to a harmful behavior: Pathogenetic implications for vascular diseases. *Cardiovasc Res* 2007;75:21–28.
114. Malinauskas RA. Plasma hemoglobin measurement techniques for the *in vitro* evaluation of blood damage caused by medical devices. *Artif Organs* 1997;21:1255–1267.
115. Dobrovolskaia MA, Clogston JD, Neun BW, Hall JB, Patri AK, McNeil SE. Method for analysis of nanoparticle hemolytic properties *in vitro. Nano Lett* 2008;8:2180–2187.
116. ASTM E2524-08. Standard Test Method for the Analysis of Hemolytic Properties of Nanoparticles. ASTM International, 2008.
117. ASTM F756-08. Standard Practice for the Assessment of Hemolytic Properties of Materials. ASTM International, 2008.
118. Dobrovolskaia MA, Aggarwal P, Hall JB, McNeil SE. Preclinical studies to understand nanoparticle interaction with the immune system and its potential effects on nanoparticle biodistribution. *Mol Pharm* 2008;5:487–495.
119. Trpkovic A, Todorovic-Markovic B, Kleut D, Misirkic M, Janjetovic K, Vucicevic L, Pantovic A, Jovanovic S, Dramicanin M, Markovic Z, Trajkovic V. Oxidative stress-mediated hemolytic activity of solvent exchange-prepared fullerene (C60) nanoparticles. *Nanotechnology* 2010;21:375102.
120. Dobrovolskaia MA, Germolec DR, Weaver JL. Evaluation of nanoparticle immunotoxicity. *Nat Nano* 2009;4:411–414.
121. Zhang Q, Hitchins VM, Schrand AM, Hussain SM, Goering PL. Uptake of gold nanoparticles in murine macrophage cells without cytotoxicity or production of pro-inflammatory mediators. *Nanotoxicology* 2011;5:284–295.
122. McNeil SE. Challenges for nanoparticle characterization. In: McNeil SE. *Characterization of Nanoparticles Intended for Drug Delivery*, Springer, 2009, pp. 9–15.
123. Link S, El-Sayed MA. Size and temperature dependence of the plasmon absorption of colloidal gold nanoparticles. *J Phys Chem B* 1999;103:4212–4217.
124. Boas U, Heegaard PMH. Dendrimers in drug research. *Chem Soc Rev* 2004;33:43–63.
125. Bosi S, Feruglio L, Da RT, Spalluto G, Gregoretti B, Terdoslavich M, Decorti G, Passamonti S, Moro S, Prato M. Hemolytic effects of water-soluble fullerene derivatives. *J Med Chem* 2004;47:6711–6715.
126. Jain K, Kesharwani P, Gupta U, Jain NK. Dendrimer toxicity: Let's meet the challenge. *Int J Pharm* 2010;394:122–142.
127. Johnston HJ, Hutchison G, Christensen FM, Peters S, Hankin S, Stone V. A review of the *in vivo* and *in vitro* toxicity of silver and gold particulates: Particle attributes and

biological mechanisms responsible for the observed toxicity. *Crit Rev Toxicol* 2010;40:328–346.
128. Fruehauf JP, Meyskens FL. Reactive oxygen species: A breath of life or death? *Clin Cancer Res* 2007;13:789–794.
129. Finkel T, Holbrook NJ. Oxidants, oxidative stress and the biology of ageing. *Nature* 2000;408:239–247.
130. McMullin MF. The molecular basis of disorders of red cell enzymes. *J Clin Pathol* 1999;52:241–244.
131. Hattangadi SM, Lodish HF. Regulation of erythrocyte lifespan: Do reactive oxygen species set the clock? *J Clin Invest* 2007;117:2075–2077.
132. Chang JC, van der Hoeven LH, Haddox CH. Glutathione reductase in the red blood cells. *Ann Clin Lab Sci* 1978;8:23–29.
133. Wang H, Joseph JA. Quantifying cellular oxidative stress by dichlorofluorescein assay using microplate reader. *Free Radic Biol Med* 1999;27:612–616.
134. Cooke BM, Lim CT. Mechanical and adhesive properties of healthy and diseased red blood cells. In: Baskurt OK, Hardeman MR, Rampling MW, Meiselman MJ (eds.), *Handbook of Hemorheology and Hemodynamics*, Vol. 69. Biomedical and Health Research, IOS Press, 2007.
135. Neu B, Meiselman HJ. Red blood cell aggregation. In: Baskurt OK, Hardeman MR, Rampling MW, Meiselman MJ (eds.), *Handbook of Hemorheology and Hemodynamics*, Vol. 69. Biomedical and Health Research, IOS Press, 2007.
136. Letcher RL, Chien S, Pickering TG, Laragh JH. Elevated blood viscosity in patients with borderline essential hypertension. *Hypertension* 1983;5:757–762.
137. Lominadze D, Dean WL. Involvement of fibrinogen specific binding in erythrocyte aggregation. *FEBS Lett* 2002;517:41–44.
138. Weng X, Cloutier G, Beaulieu R, Roederer GO. Influence of acute-phase proteins on erythrocyte aggregation. *Am J Physiol* 1996;271:H2346–H2352.
139. Thurn, K, Brown E, Wu A, Vogt S, Lai B, Maser J, Paunesku T, Woloschak G. Nanoparticles for applications in cellular imaging. *Nanoscale Res Lett* 2007;2:430–441.
140. Zhao F, Zhao Y, Liu Y, Chang X, Chen C, Zhao Y. Cellular uptake, intracellular trafficking, and cytotoxicity of nanomaterials. *Small* 2011;7:1322–1337.
141. Leroueil PR, Berry SA, Duthie K, Han G, Rotello VM, McNerny DQ, James R, Orr BG, Banaszak Holl MM. Wide varieties of cationic nanoparticles induce defects in supported lipid bilayers. *Nano Lett* 2008;8:420–424.
142. An X, Mohandas N. Disorders of red cell membrane. *Br J Haematol* 2008;141:367–375.
143. Park Y, Best C.A, Badizadegan K, Dasari R.R, Feld M.S, Kuriabova T, Henle ML, Levine AJ, Popescu G. Measurement of red blood cell mechanics during morphological changes. *Proc Natl Acad Sci U S A* 2010;107:6731–6736.
144. Shekhawat GS, Dravid VP. Nanoscale imaging of buried structures via scanning near-field ultrasound holography. *Science* 2005;310:89–92.
145. Shekhawat G, Dravid V. Seeing the invisible: Scanning near-field ultrasound holography (SNFUH) for high resolution buried imaging and pattern recognition. *Microsc Microanal* 2007;13:1220–1221.
146. Reid HL, Barnes AJ, Lock PJ, Dormandy JA, Dormandy TL. A simple method for measuring erythrocyte deformability. *J Clin Pathol* 1976;29:855–858.
147. Tiwari DK, Jin T, Behari J. Dose-dependent *in-vivo* toxicity assessment of silver nanoparticle in Wistar rats. *Toxicol Mech Methods* 2011;21:13–24.

148. Kim YS, Kim JS, Cho HS, Rha DS, Kim JM, Park JD, Choi BS, Lim R, Chang HK, Chung YH, Kwon IH, Jeong J, Han BS, Yu IJ. Twenty-eight-day oral toxicity, genotoxicity, and gender-related tissue distribution of silver nanoparticles in Sprague-Dawley rats. *Inhal Toxicol* 2008;20:575–583.
149. Sung JH, Ji JH, Park JD, Yoon JU, Kim DS, Jeon KS, Song MY, Jeong J, Han BS, Han JH, Chung YH, Chang HK, Lee JH, Cho MH, Kelman BJ, Yu IJ. Subchronic inhalation toxicity of silver nanoparticles. *Toxicol Sci* 2009;108:452–461.
150. Simpson CA, Agrawal AC, Balinski A, Harkness KM, Cliffel DE. Short-chain PEG mixed monolayer protected gold clusters increase clearance and red blood cell counts. *ACS Nano* 2011;5:3577–3584.
151. Zhang XD, Wu HY, Wu D, Wang YY, Chang JH, Zhai ZB, Meng AM, Liu PX, Zhang LA, Fan FY. Toxicologic effects of gold nanoparticles *in vivo* by different administration routes. *Int J Nanomed* 2010;5:771–781.
152. Ji JH, Jung JH, Kim SS, Yoon JU, Park JD, Choi BS, Chung YH, Kwon IH, Jeong J, Han BS, Shin JH, Sung JH, Song KS, Yu IJ. Twenty-eight-day inhalation toxicity study of silver nanoparticles in sprague-dawley rats. *Inhal Toxicol* 2007;19:857–871.
153. Kim Y, Song M, Park J, Song K, Ryu H, Chung Y, Chang H, Lee J, Oh K, Kelman B, Hwang I, Yu I. Subchronic oral toxicity of silver nanoparticles. *Part Fibre Toxicol* 2010;7:20.
154. Murdock RC, Braydich-Stolle L, Schrand AM, Schlager JJ, Hussain SM. Characterization of nanomaterial dispersion in solution prior to *in vitro* exposure using dynamic light scattering technique. *Toxicol Sci* 2008;101:239–253.
155. Li YT, Liu J, Zhong YJ, Zhang J, Wang ZY, Wang, An Y, Lin M, Gao ZQ, Zhang DS. Biocompatibility of Fe_3O_4@Au composite magnetic nanoparticles *in vitro* and *in vivo*. *Int J Nanomed* 2011;6:2805–2819.
156. Lu S, Duffin R, Poland C, Daly P, Murphy F, Drost E, MacNee W, Stone V, Donaldson K. Efficacy of simple short-term *in vitro* assays for predicting the potential of metal oxide nanoparticles to cause pulmonary inflammation. *Environ Health Perspect* 2008;117.
157. Teeguarden JG, Hinderliter PM, Orr G, Thrall BD, Pounds JG. Particokinetics *in vitro*: Dosimetry considerations for *in vitro* nanoparticle toxicity assessments. *Toxicol Sci* 2007;95:300–312.
158. Amin K, Dannenfelser RM. *In vitro* hemolysis: Guidance for the pharmaceutical scientist. *J Pharm Sci* 2006;95:1173–1176.
159. Kumar S, Joshi PC. Haemolysis by ultraviolet B of red blood cells from different animal species. *Toxicol In Vitro* 1992;6:345–347.
160. Wennberg A, Hensten-Pettersen A. Sensitivity of erythrocytes from various species to *in vitro* hemolyzation. *J Biomed Mater Res* 1981;15:433–435.
161. Barger MA. Erythrocyte morphology. *Schalm's Vet Hematol* 2011;5:144–151
162. Plasenzotti R, Stoiber B, Posch M, Windberger U. Red blood cell deformability and aggregation behaviour in different animal species. *Clin Hemorheol Microcirc* 2004;31:105–111.
163. Slawson RM, Van Dyke MI, Lee H, Trevors JT. Germanium and silver resistance, accumulation, and toxicity in microorganisms. *Plasmid* 1992;27:72–79.
164. Klasen HJ. A historical review of the use of silver in the treatment of burns. II. Renewed interest for silver. *Burns* 2000;26:131–138.
165. Choi O, Deng KK, Kim NJ, Ross J, Surampalli RY, Hu Z. The inhibitory effects of silver nanoparticles, silver ions, and silver chloride colloids on microbial growth. *Water Res* 2008;42:3066–3074.

166. Juneyoung L, Keuk-Jun K, Woo SS, Jong GK, Dong GL. The silver nanoparticle (nano-Ag): A new model for antifungal agents, Retrieved July 1, 2011, from http://www.intechopen.com/articles/show/title/the-silver-nanoparticle-nano-ag-a-new-model-for-antifungal-agents.
167. Christensen FM, Johnston HJ, Stone V, Aitken RJ, Hankin S, Peters S, Aschberger K. Nano-silver — feasibility and challenges for human health risk assessment based on open literature. *Nanotoxicology* 2010;4:284–295.
168. Chaloupka K, Malam Y, Seifalian AM. Nanosilver as a new generation of nanoproduct in biomedical applications. *Trends Biotechnol* 2010;28:580–588.
169. Sopjani M, Föller M, Haendeler J, Götz F, Lang F. Silver ion-induced suicidal erythrocyte death. *J Appl Toxicol* 2009;29:531–536.
170. Lansdown, AGB. A pharmacological and toxicological profile of silver as an antimicrobial agent in medical devices. *Adv Pharmacol Sci* 2010.
171. Wijnhoven SWP, Peijnenburg WJGM, Herberts CA, Hagens WI, Oomen AG, Heugens EHW, Roszek B, Bisschops J, Gosens I, *et al.* Nano-silver — a review of available data and knowledge gaps in human and environmental risk assessment. *Nanotoxicology* 2009;3:109–138.
172. Ballinger PM, Brown, BS, Griffin MM, Steven FS. Evidence for carriage of silver by sulphadimidine haemolysis of human erythrocytes. *Br J Pharmacol* 1982;77:141–145.
173. Liu X, Lee PY, Ho CM, Lui V, Chen Y, Che CM, Tam P, Wong K. Silver nanoparticles mediate differential responses in keratinocytes and fibroblasts during skin wound healing. *ChemMedChem* 2010;5:468–475.
174. Garner M, Reglinski J, Smith WE, Stewart MJ. The interaction of colloidal metals with erythrocytes. *J Inorg Biochem* 1994;56:283–290.
175. Cozzoli PD, Pellegrino T, Manna L. Synthesis, properties and perspectives of hybrid nanocrystal structures. *Chem Soc Rev* 2006;35:1195–1208.
176. Grzelczak M, Perez-Juste J, Mulvaney P, Liz-Marzan LM. Shape control in gold nanoparticle synthesis. *Chem Soc Rev* 2008;37:1783–1791.
177 Tao AR, Habas S, Yang P. Shape control of colloidal metal nanocrystals. *Small* 2008;4:310–325.
178. Jun YW, Choi JS, Cheon J. Heterostructured magnetic nanoparticles: Their versatility and high performance capabilities. *Chem Commun (Camb)* 2007;1203–1214.
179. Wang J, Byrne JD, Napier ME, DeSimone JM. More effective nanomedicines through particle design. *Small* 2011;7:1919–1931.
180. Metnitz PGH, Bartens C, Fischer M, Fridrich P, Steltzer H, Druml W. Antioxidant status in patients with acute respiratory distress syndrome. *Intensive Care Med* 1999;25:180–185.
181. Davutoglu M, Guler E, Olgar S, Kurutas E, Karabiber H, Mesut G, Ekerbicer H. Oxidative stress and antioxidant status in neonatal hyperbilirubinemia. *Saudi Med J* 2008;29:1743–1748.
182. Staron A, Mąkosa G, Koter-Michalak M. Oxidative stress in erythrocytes from patients with rheumatoid arthritis. *Rheumatol Int* 2010;1–4.
183. Pandey KB, Rizvi SI. Biomarkers of oxidative stress in red blood cells. *Biomed Pap Med Fac Univ Palacky Olomouc Czech Repub* 2011;155:131–136.
184. Amer J, Frankenburg S, Fibach E. Apheresis induces oxidative stress in blood cells. *Ther Apher Dial* 2010;14:166–171.

185. Antonelou MH, Kriebardis AG, Velentzas AD, Kokkalis AC, Georgakopoulou SC, Papassideri IS. Oxidative stress-associated shape transformation and membrane proteome remodeling in erythrocytes of end stage renal disease patients on hemodialysis. *J Proteomics* 2011;74:2441–2452.
186. Kameneva MV, Antaki JF. Mechanical trauma to blood. In: Baskurt OK, Hardeman MR, Rampling MW, Meiselman MJ (eds.), *Handbook of Hemorheology and Hemodynamics* — Vol. 69. Biomedical and Health Research, IOS Press, 2007.
187. Cappellini MD, Fiorelli G. Glucose-6-phosphate dehydrogenase deficiency. *Lancet* 2008;371:64–74.
188. Kirschbaumweg LT. The Journal of the International Federation of Clinical Chemistry and Laboratory Medicine, Retrieved September 12, 2011, from http://www.ifcc.org.
189. Lippi G, Blanckaert N, Bonini P, Green S, Kitchen S, Palicka V, Vassault A, Plebani M. Haemolysis: An overview of the leading causes of unsuitable specimens in clinical labortories. *Clin Chem Lab Med* 2008;46:764–772.
190. Salata OV. Applications of nanoparticles in biology and medicine. *J Nanobiotechnol* 2004;2:3.
191. Bosman GJCG, Lasonder E, Groenen-Döpp YAM, Willekens FLA, Werre JM, Novotný VMJ. Comparative proteomics of erythrocyte aging *in vivo* and *in vitro*. *J Proteomics* 2010;73:396–402.

Chapter 8

The Effects of Engineered Nanomaterials on Cultured Endothelial Cells

Jan Simak

Center for Biologics Evaluation and Research
Food and Drug Administration
1401 Rockville Pike, HFM-335
Rockville, MD 20852-1448, USA
jan.simak@fda.hhs.gov

The vascular endothelium forms the inner lining of the vasculature and is an essential system in multicellular organisms that delivers nutrients and oxygen, removes CO_2 and waste products, and facilitates interorgan communication. Endothelial dysfunctions may lead to life-threatening vascular disorders, including hypoperfusion and organ damage, and thrombotic and/or bleeding pathologies. As various engineered nanomaterials are being designed for biomedical applications for intravascular use and other nanomaterials may reach the vasculature as a result of occupational, environmental, or other types of exposure, it is necessary to evaluate the effects of engineered nanomaterials on the vascular endothelium. In the introduction of this chapter, the physiology of endothelial cells (ECs) and the pathophysiology of the possible adverse effects of nanomaterials on ECs are briefly described. Next, the methods that are used for the *in vitro* evaluation of the effects of nanomaterials on ECs are reviewed. Investigations of the effects of nanomaterials on ECs are primarily based on the observation of structural and functional changes in the different types of cultured ECs upon (usually) short-term exposure (up to 72 h) to various doses of nanomaterials. Numerous microscopic methods are employed to study morphological and structural changes of ECs and nanomaterial uptake. Different cell biology assays are used to assess the cytotoxicity of nanomaterials and their effects on proliferation and induction of apoptosis, necrosis, and autophagy. In addition, assays of various EC activation markers, nitric oxide (NO) and reactive oxygen species (ROS) production or functional EC mobility and morphogenesis, monolayer permeability, and vasoactivity assays also provide valuable information on nanomaterial effects on the vascular intima. Lastly, a summary of the published results regarding the *in vitro* effects of various nanomaterials on ECs, as well as selected assay protocols used in our laboratory are presented.

Disclaimer:

The findings and conclusions in this article have not been formally disseminated by the Food and Drug Administration and should not be construed to represent any agency determination or policy.

1. Introduction

The interaction of biomaterials with the endothelium has been a focus of extensive research.[1] Newly engineered nanomaterials hold great promise for applications in several fields, including nanomedicine. In biomedical applications, different types of nanomaterials are under development for use in diagnostic biosensors, drug delivery nanosystems, imaging nanoprobes for intravascular use, and other devices that come into contact with the blood and vasculature. The general categories of recommended testing for the different types of medical devices in contact with the blood and vascular system have been specified by the ISO standard 10993-4.[2] Because nanomaterials may interfere with certain cytotoxicity or biocompatibility assays and because their evaluation requires specific approaches (e.g., the characterization and quantification of nanomaterials and the preparation of samples for testing), several new international standard assays need to be developed.[3]

The evaluation of the effects of nanomaterials on vascular endothelial cells (ECs) is not yet a routine part of toxicological screening, even though thrombotic diseases stemming from endothelial dysfunction are a leading cause of morbidity and mortality in the developed world.[4] Therefore, the purpose of this chapter is to highlight the adverse effects of nanomaterials on ECs and show the availability of feasible methodological approaches for *in vitro* studies using cultured ECs.

2. The Pathophysiology of Possible Adverse Effects of Nanomaterials on Endothelial Cells

The vascular endothelium forms the inner lining of the vasculature and is an essential system in multicellular organisms that delivers nutrients and oxygen, removes CO_2 and waste products, and facilitates interorgan communication. In the following section on the physiology and pathophysiology of ECs, references are mostly limited to comprehensive review articles that should serve as sources for a more complete coverage of the literature.[5–13] The vascular endothelium is composed of a single layer of ECs and the underlying extracellular matrix (ECM). In adults, the endothelium consists of 1–6 × 10^{13} ECs weighing approximately 1 kg in total and covers a surface area of approximately 4,000 to 7,000 m^2.[14] ECs perform a unique role in the regulation of a variety of processes such as vascular tone, microvascular permeability and tissue–fluid balance, lumen patency and hemostasis, inflammatory response, and angiogenesis. The vascular endothelium is very heterogeneous — the ECs from different types of vasculature and organs differ in their structure, organization of intercellular junctions, and functional properties.[15,16] Under physiological conditions, ECs play a key role in maintaining the antithrombotic and fibrinolytic potential of the vascular wall, participate in the regulation of vascular tone and microvascular perfusion, and control microvascular permeability. Furthermore, healthy ECs have antioxidant and anti-inflammatory properties. Endothelial functions are facilitated by the release of a variety of biologically active molecules (mediators) and by the surface expression of various membrane receptors. The endothelial plasma membrane, depending on the type of EC, contains a number of different membrane-associated microdomains, such as plasmalemmal vesicles, transendothelial channels, fenestrae, coated pits, and coated vesicles.[17] Among these microdomains, the caveolae are the most characteristic of ECs. ECs may also change their exposure of membrane phospholipids and upon specific stimulation, release different types of membrane phospholipid microvesicles (membrane microparticles).

Nanomaterials may exert direct toxic effects on ECs. Such activities would be either analogous to the general cytotoxicity observed in other cell types or may be EC-specific. ECs are very delicate cells; concentrations of nanomaterials that are not toxic in standard cytotoxicity assays may still cause endothelial perturbation and dysfunction. In addition to the possible direct toxicity to ECs, any foreign material given intravascularly, unless optimally biocompatible, induces a complex cooperative defense response involving several cell types and other blood components, the vessel wall, and other exposed tissues (Figure 1). The foreign material is recognized as a pathogenic factor (insult) and elicits the defense process at the tissular level, which is known as the acute inflammatory response.[18] In response to a local or systemic pathogenic insult, ECs play an essential part in the defense process, including changes in the vascular tone and microvascular permeability, intercellular and transcellular transport, redistribution of blood flow, hemostasis, and a complex inflammatory response. The principles and mechanisms of the inflammatory response have been extensively reviewed.[9,19–23] In general, the acute inflammatory response has the following functions: 1) eliminate the

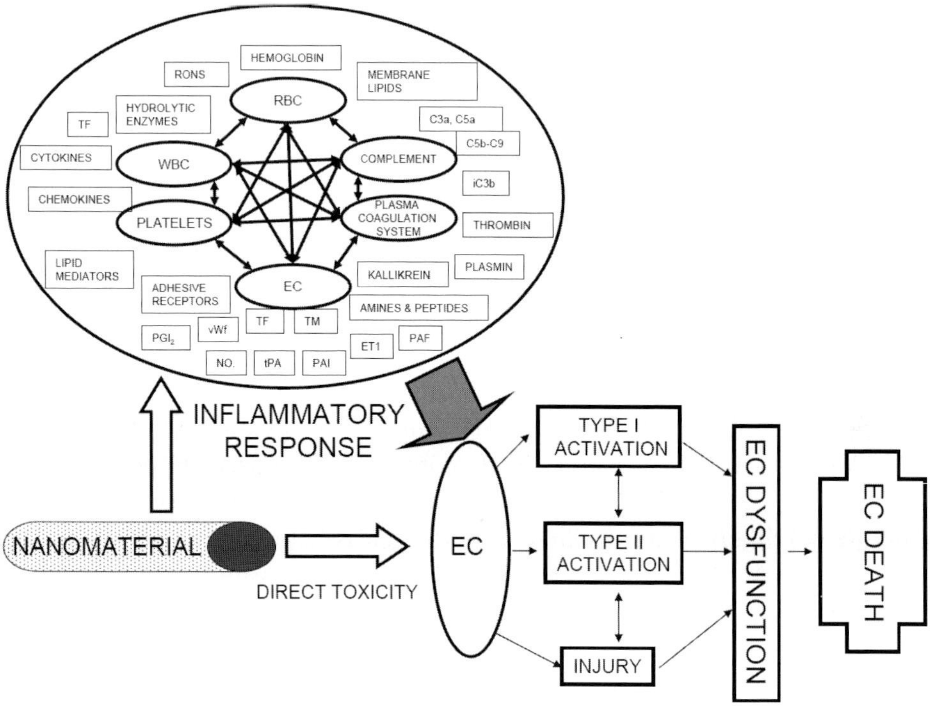

Figure 1. Adverse effects of nanomaterials of endothelial cells.

EC: endothelial cells, ET1: endothelin-1, NO: nitric oxide, PAF: platelet activating factor, PAI: plasminogen activator inhibitor, PGI_2: prostacyclin, RBC: red blood cells, RONS: reactive oxygen and nitrogen species, TF: tissue factor, TM: thrombomodulin, tPA: tissue plasminogen activator, vWf: von Willebrand factor, WBC: white blood cells.

pathogenic factor(s), 2) remove irreversibly destroyed tissues, 3) repair damaged tissues and restore their physiological functions, and 4) resolve the inflammatory process and achieve homeostasis. In the case of extensive tissue damage and/or when a pathogenic factor cannot be eliminated, the acute inflammatory response progresses to the chronic adaptive response, which includes the replacement of damaged tissues with fibroblasts and other cells that are not fully functional. Although the classic concept of inflammation is based on infectious triggers, the inflammatory response can also be initiated by several types of exogenous or endogenous "sterile" pathogenic factors of a physical or chemical nature or by metabolic malfunctions. Thus, we can expect that the general mechanisms of the inflammatory response to new types of nanomaterials will be similar to the response to previously studied pathogenic factors. The acute inflammatory response involves the coordinated delivery of the participating cellular and non-cellular components to the site of injury. ECs are one of the key systems involved in the inflammatory response and they interact with blood platelets, white blood cells (particularly neutrophils, monocytes, tissue residential macrophages, mast cells, and lymphocytes), red blood cells, the plasma coagulation system, and the complement system.

ECs are equipped with a number of potent vasoactive mediators that modulate the vascular hemodynamic response under physiological conditions and in the host response to pathogenic factors.[7] Upon stimulation with a variety of stimuli such as shear stress, epinephrine, serotonin, adenosine diphosphate (ADP), vascular endothelial growth factor (VEGF), histamine, bradykinin, or thrombin, ECs release nitric oxide (NO), which relaxes the vascular smooth muscle cells (SMCs) that surround the ECs.[11,24,25] NO is generated in ECs from L-arginine, predominantly by the constitutive Ca^{2+}-dependent isoform of NO synthase (eNOS), in the presence of cofactors such as tetrahydrobiopterin. NO activates guanylate cyclase in SMCs, thus inducing cGMP-mediated vasodilation. NO not only has a direct vasodilating effect, but also inhibits the aggregation of platelets, the expression of adhesion molecules at the surface of ECs, and the action of the vasoconstricting and the mitogenic peptide endothelin-1 (ET-1), which is also released from ECs. Shear stress is a key activator of eNOS under physiological conditions, whereas the other mediators take over during the inflammatory response. During the inflammatory response, there is a fundamental switch in EC signaling from the NO-mediated silencing of cellular processes toward their activation by redox signaling.[11,25] The formation of reactive oxygen and nitrogen species (RONS) is increased, which alters the functions of many EC proteins. eNOS can switch to generate RONS during EC activation.[24] In endothelial injury and dysfunction, NO production may be insufficient. In contrast, upon EC activation with pro-inflammatory cytokines, ECs and some other cell types express inducible NO synthase (iNOS), which may lead to the overproduction of NO, thereby causing refractory hypotension. In addition to NO, ECs may also induce vasodilation by causing vascular SMCs to hyperpolarize via different pathways.[26] Another vasodilator released by ECs is prostacyclin PGI_2, which is a product of the metabolism of arachidonic acid by the cyclooxygenase pathway in ECs. Although PGI_2 may contribute to other EC regulatory roles, it appears to have a more limited role in the maintenance of vasodilator tone in humans.[24] Several prostanoids can be synthesized in ECs, including vasoconstricting thromboxane A_2, PGH_2, $PGF_{2\alpha}$, and PGE_2. Paradoxically, even PGI_2 may act as a vasoconstricting factor.[27] The vasoconstricting phospholipid platelet-activating factor (PAF) is also produced from arachidonic acid after EC stimulation. PAF remains associated with the cell membrane and promotes leukocyte adhesion. ECs produce the most potent vasoconstrictor, the peptide ET-1, which also induces the proliferation of smooth muscles. Angiotensin-converting enzyme (ACE) produced by ECs is part of another vasoconstricting pathway. ACE converts angiotensin I to the potent vasoconstrictor angiotensin II and also degrades the vasodilating peptide bradykinin.

Keeping the antithrombotic surface of the vascular wall intact is a critical role of ECs for maintaining tissue perfusion. Circulating platelets do not adhere to ECs under physiological conditions due to the endothelial production of PGI_2 and NO, and the expression of ecto-ADPases. In addition, the balance of blood coagulation is maintained by a specific battery of endothelial mediators.[7,28] The main task of these mediators is to control the activity of thrombin, which is the key proteinase of the plasma coagulation system (see Chapter 9 for more details). ECs synthesize heparan sulfates and other glycosaminoglycans, which enhance the activity of the plasma proteinase inhibitor antithrombin (AT), whereas dermatan sulfates promote the activity of another plasma proteinase inhibitor, heparin cofactor II. The expression of the receptor thrombomodulin (TM) on ECs plays a major role in modulating thrombin activity. The thrombin bound to the TM on the EC surface is unable to activate platelets, does not convert fibrinogen to fibrin, and does not activate the plasma coagulation factors V (FV) or XIII (FXIII); instead, it activates the protein C present in the plasma to activated protein C (APC). APC and its cofactor protein S inactivate the plasma coagulation factors Va and VIIIa. The protein C system is a vitally important negative feedback loop in the generation of thrombin. Hence, the dysfunction or deficiency of the protein C system leads to catastrophic thromboembolic events. The release of the tissue factor pathway inhibitor (TFPI) represents another antithrombotic potential of ECs. Finally, stimulated ECs activate fibrinolysis. ECs release tissue plasminogen activator (tPA), which activates the plasminogen present in the plasma to the fibrinolytic enzyme plasmin. In the case of vascular injury with the loss of vascular integrity, the procoagulant potential of the endothelium is crucial to achieve hemostasis and prevent bleeding. The main trigger of the generation of thrombin activity in the plasma coagulation system is the membrane receptor tissue factor (TF). TF, in a complex with FVIIa, activates FX and FIX and thus initiates the TF pathway, the most efficient pathway for thrombin generation.[4,7,28]

TF is not expressed in its active form on ECs under physiological conditions, whereas it is present on subendothelial tissues and may be rapidly exposed after EC retraction or desquammation. Dramatic changes in the antithrombotic to prothrombotic phenotypes of ECs are the result of the induction of TF expression on ECs by pro-inflammatory cytokines, thrombin, and other pro-inflammatory stimuli. Although this process is easy to demonstrate in cultured ECs *in vitro*, it has not been fully elucidated in different organs *in vivo*. It also appears that TF has functions independent of the thrombin generation pathway. The EC receptors for thrombin and other plasma coagulation factors play additional regulatory roles in the plasma coagulation system. Stimulated ECs release von Willebrand factor (vWf), a large multimeric protein that facilitates the binding of platelets to ECs and subendothelial tissues. After pro-inflammatory stimulation, ECs produce plasminogen activator inhibitor (PAI-1), which inhibits fibrinolysis. In addition, thrombin-activatable fibrinolysis inhibitor (TAFI) is activated on the surface of ECs by the thrombin–TM complex. The cell adhesion receptors expressed on activated ECs also facilitate the recruitment of leukocytes and the adhesion of platelets, thereby potentiating the procoagulant activity of the vascular wall.[7,28] The preceding cases are examples of the major potential of the EC mediators involved in vasoregulation and hemostasis. Although slightly oversimplified, vasodilating mediators mostly have antithrombotic and anti-inflammatory properties, whereas vasoconstricting mediators generally have prothrombotic and pro-inflammatory effects.

According to the classic concept of EC activation, there are two stages of this process.[29,30] The first, called "EC stimulation" or "activation type I," is a rapid response that does not require *de novo* protein synthesis or gene upregulation. This process includes the activation of the cytoskeleton, resulting in the retraction of ECs on the basal membrane and the release of the EC secretory organelles, termed Weibel–Palade (WP) bodies. The major components released in this response are the adhesive receptor P-selectin, which is subsequently expressed on the EC surface membrane, and the hemostatic protein vWf, which is released into the plasma. Other mediators released from the WP bodies include tPA, interleukin-8 (IL-8), angiopoietin-2, and ET-1. Several agonists induce the rapid type I activation of ECs via different signaling pathways, including thrombin, histamine, epinephrine, and vasopressin. The interaction between cell surface receptors and the cross-talk between signaling pathways can fine-tune and specialize the endothelial secretory response.[31]

The acute response of ECs to systemic insult may have dramatic clinical manifestations, including hypotension, systemic and pulmonary edema, hyperfibrinolysis or acute disseminate intravascular coagulation (DIC), and the rapid onset of the dysfunction or failure of vital organs. However, if the pathogenic factor is removed and homeostasis is recovered to prevent secondary complications such as hypoxia-reperfusion injury, EC dysfunction may be rapidly reversible. If the pathogenic factor is not removed and a complex inflammatory response continues to develop, the pro-inflammatory cytokines from stimulated white blood cells, such as IL-1 and tumor necrosis factor-α (TNF-α), have critical effects on ECs. EC activation, driven by pro-inflammatory cytokines, is characterized by the increased or *de novo* expression of leukocyte adhesion molecules, a change in phenotype from antithrombotic to prothrombotic, cytokine production, and the upregulation of HLA (human leukocyte antigen) molecules.[30] This "activation type II" response is the second stage of EC activation and requires time for the stimulating agent to cause an effect via gene transcription and protein synthesis.[32,33] The genes involved include those that encode adhesion molecules, cytokines, and TF. The upregulation of leukocyte adhesion molecules, such as E-selectin, P-selectin, intercellular adhesion molecule-1 (ICAM-1), and vascular cell adhesion molecule (VCAM-1), facilitates leukocyte adherence to the endothelium and subsequent relocation of leukocytes into the tissues. The prothrombotic effects of EC activation include the loss of the surface anticoagulant molecules TM and heparan sulfate, reduced fibrinolytic potential due to the enhanced release of PAI-1, the loss of platelet anti-aggregatory effects of ecto-ADPases and PGE_2, the production of PAF, and the expression of TF.[30] Cytokines are then synthesized, including IL-6 which regulates the acute phase response, and chemoattractants such as IL-8 and monocyte chemoattractant protein-1 (MCP-1). The expression of class II HLA molecules allows ECs to act as antigen-presenting cells, permitting cross-talk with T lymphocytes in antigen clearance.[34] Of note, EC activation is not an all-or-nothing phenomenon, but rather a spectrum of responses.[35] Depending on the level of biocompatibility, the interactions of a biomaterial with blood components and the vessel wall may or may not be clinically significant. While the controlled inflammatory response is essential for the physiological mechanism of tissue defense, it can become detrimental if dysregulated. To eliminate a pathogenic factor and to remove damaged cells, highly potent mediators such as RONS and proteinases are released, which induce collateral damage at the site of inflammation. To minimize autoaggressive effects, the inflammatory processes must be localized at the site of injury and be well controlled. When the action of the pathogenic factor is systemic and not locally restricted, such as with the non-targeted intravascular administration of a nanoparticulate material, the host defense becomes a systemic inflammatory response. The delocalization and dysregulation of the inflammatory response leads to autoaggressive inflammation, in which inflammatory cells and mediators damage host tissues.[5,36–40] Whereas EC activation is associated with the host defense, EC dysfunction is the result of injury caused by the direct toxic effect of a pathogenic factor and/or by the autoaggressive inflammatory response.[41] At a certain level of intensity, the systemic inflammatory response and sustained endothelial activation and/or injury cause the failure of the essential constitutive physiological functions of ECs and organ perfusion is not maintained. This failure leads to micro- and/or macrovascular occlusion, hypoxia, and organ dysfunction progressing to multiple organ dysfunction syndrome (MODS).[5] MODS is associated with the failure of adequate perfusion of vital organs due to microcirculation disorders, including edema, microthromboses, and other thrombohemorrhagic complications. EC death by necrosis, apoptosis, and autophagy ends this process.[41] Thus, removing the pathogenic insult and regenerating the physiological functions of ECs is the limiting factor for the survival of organ tissues on a local scale and the whole organism on a systemic scale. Therefore, the level of EC injury and the degree to which it can be reversed in vital organs is a critical factor for determining the mortality, morbidity, and reversibility of MODS, and the prognosis of the patient.

The special cellular properties of ECs have led to their use in evaluating the adverse effects of biologics, drugs, and devices such as engineered nanomaterials. Several studies of the effects of different nanomaterials on ECs cultured *in vitro* are reviewed in this chapter. It is important to note that these studies do not provide a model for the possible indirect effects of the nanomaterials on other cell types or other components of the inflammatory response present in the vascular compartment. Although *in vitro* studies on cultured ECs have several limitations, the introduction of endothelial biocompatibility assays into the preclinical safety evaluation of nanomaterials would greatly contribute to their development for medical use.

3. Methods for Evaluating the *In Vitro* Effects of Nanomaterials on Endothelial Cells

A large variety of methods and assays have been used to study the effects of different nanomaterials on cultured ECs. Table 1 summarizes the methodologies of selected studies published to date and related references. The common objective of these studies is to identify the structural and functional changes in the different types of cultured ECs (usually) upon short-term exposure (up to 72 h) to various doses of different nanomaterials. One great challenge in these studies is the lack of standard protocols for the characterization of nanomaterials and analysis of impurities in the published reports. For

Table 1. Methods used for the evaluation of *in vitro* effects of nanomaterials on ECs.

Parameter	Assay	Instrument	NP	EC	Ref.
NP uptake	TEM	TEM	SiNP	RAEC	70
			IONP	HMVEC	53
			PGNP	HUVEC	42
			SPION	PAEC	63
			MNP	HAEC	51
			PSONP	HAEC	52
	AFM	AFM	SiNP	HUVEC	43
	CIFM	CIFM	SiNP	HUVEC	43
	ICP-MS	ICP-MS	SiNP	HUVEC	44
			IONP	HAEC	92
	RMS	RMS	SWCNT	HUVEC	45
	FC	FC	PGNP	HUVEC	42
Morphology & structural changes	SEM	SEM	SWCNT	HUVEC	45
	SEM	SEM	TiNTS	BAEC	60
	TEM	TEM			
	SEM	SEM	FUL	RCMEC	66
	AFM	AFM			
	VE-cadherin	CIFM, WBT	IONP	HMVEC	53
	acetyl-α-tubulin				
	vinculin	CIFM	TiNTS	BAEC	60
	actin				
	actin	CFM-RhoPha	FUL	RCMEC	66
	actin	CFM-RhodPha	TiNS	RAEC	64
	actin	CFM-RhoPha	MWCNT/PU	HUVEC	73
	tubulin	FITC-SAV			
	vinculin	ELISA			
	collagen IV				
	laminin				
Viability/ Cytotoxicity Proliferation	MTT	COL/MPR	SWCNT	HUVEC	45
			SiNP	HUVEC	43
			SNP	BREC	61
			GNP	HRMEC	56
			SiNP	EAHY926	58
			FUL	HUVEC	46
	MTS	COL/MPR	LIP	HUVEC	47
			FUL/CNT	HUVEC	48
			SiNP	ISOHAS1	59
			SiNP	HUVEC	44
			TiNS	RAEC	64
	XTT	COL/MPR	SNP	RCMEC	67
	WST-8	COL/MPR	PSONP	HAEC	52

(*Continued*)

Table 1. (*Continued*)

Parameter	Assay	Instrument	NP	EC	Ref.
			SWCNT	HUVEC	77
			FUL	CMEC	68
				RCMEC	66
	LDH	COL/MPR	IONP	HMVEC	53
			SiNP	HUVEC	43
				ISOHAS1	59
				EAHY926	58
				HUVEC	44
			FUL	RCMEC	68
	NRU	COL/MPR	SWCNT	HUVEC	45
	FDAC	COL/MPR	TiNTS	BAEC	60
	Trypan blue	COL/MPR	MWCNT	HUVEC	50
	Alamar blue	FL/MPR	FUL	HUVEC	46
	Crystal violet	COL/MPR	SWCNT	HUVEC	45
	CyQuant	COL/MPR	LIP	HUVEC	47
Apoptosis/Necrosis	Annexin V/PI	FC	SWCNT	HUVEC	45
			FUL/CNT	HUVEC	48
			FUL	RCMEC	66
			SiNP	HUVEC	44
			FUL	HUVEC	46
	Caspase Glo 3/7	LU/MPR	SiNP	HUVEC	43
	Caspase 3 Ac-DEVD-AMC	FL/MPR	SWCNT	RAEC	78
	Ac-DEVD-pNA	COL/MPR	SNP	BREC	61
	Caspase 3	WBT	FUL	RCMEC	66
	Caspase 3, PARP		FUL	HUVEC	80
DNA damage	TUNEL	FC	FUL	HUVEC	79
	γH2AX phosphorylation	IFM	MWCNT	HUVEC	50
Cell cycle analysis	PI/RNAse	FC	FUL/CNT	HUVEC	48
			FUL	HUVEC	79
Lysosome detection	LysoTracker	FM	SWCNT	HUVEC	45
Autophagy	LC3-II	WBT	FUL	HUVEC	80
Membrane potentials	JC-1	FL/MPR	SWCNT	RAEC	78
		FC	SiNP	HUVEC	44
NO_X production	GRIESS	COL/MPR	TiNTS	BAEC	60
			FUL	CMEC	68
			SNP	RCOEC	65

(*Continued*)

Table 1. (*Continued*)

Parameter	Assay	Instrument	NP	EC	Ref.
ROS production	CM-H2DCFDA	FC	IONP	HMVEC	53
			SiNP	HUVEC	44
		CFM	SPION	PAEC	63
		FL/MPR	FUL	HUVEC	46
	DCFH-DA	FL/MPR	MWCNT	HUVEC	50
		FC	FUL	RCMEC	68
			FUL	RCMEC	66
Peroxynitrite	DAR-4M AM	FM/CCD	PSONP	HAEC	52
	DHR-123	FL/MPR	FUL	HUVEC	46
	HPF				
Redox status	GSH	MBCL/FM	SWCNT	RAEC	78
	GSH-Px		MWCNT	HUVEC	50
	SOD				
	MDA				
	GSH	GR/M2VP	ZnNP	HUVEC	88
Activation status/ Proinflammatory stimulation	$[Ca^{2+}]_i$	FURA-2AM FM/CCD	FUL	HUVEC	79
	vWf	CIFM	SiNP	HUVEC	43
	ET-1	ELISA	TiNTS	BAEC	60
			FUL	CMEC	68
	PGI_2	EIA	TiNT	BAEC	101
	ICAM-1	FC	FUL	HUVEC	79
	TF				
	TF	ELISA	MWCNT/PU	HUVEC	73
	PAI-1				
	ICAM-1	WBT, IFM,	AlNP	PAEC	106
	ELAM-1	qPCR			
	GAPDH				
	ICAM-1	qPCR, WBT	MNP	HAEC	51
	MCP-1	ELISA			
	IL-8				
	sICAM-1	ELISA	SiNP	ISOHAS1	59
	IL-6				
	IL-8				
	eNOS	IFM	PUGNP	BCAEC	85
	eNOS				
	caverolin	IFM	PSONP	HAEC	52
	eNOS	WBT			
	TNF-α, IL-1β, IL-2, PGE_2	ELISA	SNP	RCMEC	67

(*Continued*)

Table 1. (*Continued*)

Parameter	Assay	Instrument	NP	EC	Ref.
Signaling pathways	GSK-3β/Akt	WBT, CIFM	IONP	HMVEC	53
	Akt/FAK	WBT, CIFM FAK-GFP	PUGNP	BCAEC	85
	PI3K/Akt	WBT, Transfection	SNP	BREC	61
	Phospho-Src (Y419)	ELISA	GNP	BREC	96
	ERK 1/2	WBT	GNP	HRMEC	56
	t-JNK p-JNK c-Jun PARP p-ERK	WBT	FUL	RCMEC	66
	ERK p-JNK JNK p-c-Jun c Jun p-p53 Bcl-2 NF-κB caspase-3	WBT	SiNP	HUVEC	44
	Rho, Rac/Cdc42	WBT	MWCNT/PU	HUVEC	73
	Src	Transfection	SNP	PREC	87
	NFκB IκBα, p-IκBα	Transfection WBT	ZnNP	HUVEC	88
Gene expression profiling	qPCR microarrays		PGNP TiNT FUL	HUVEC HAEC HUVEC	42 84 80
Chemotaxis/ Cell mobility	Boyden chambers	Densitometr. analysis	SWCNT	HUVEC	45
	Cell mobility	FDAC/FM	TiNTS	BAEC	60
	In vitro wound healing	PCM	GNP SNP GNP	BREC BREC HRMEC	96 61 56

(*Continued*)

Table 1. (*Continued*)

Parameter	Assay	Instrument	NP	EC	Ref.
Morphogenesis/ Tubulogenesis	Matrigel	LM	SWCNT	HUVEC	45
			FUL/CNT	HUVEC	48
			SNP	BREC	61
			GNP	HRMEC	56
Elastic modulus	AFM	AFM	SPION	PAEC	63
Vasoactivity	Rat aortic rings	Original chamber	SiNP	RAEC	70
		Transducer/ polygraph	SNP	RAEC	65
	Guinea pig aortic rings		FUL	GAEC	71
Permeability	TER	ECIS	IONP	HMVEC	53
			SiNP	ISOHAS1	59
	RITC-dextran	Transwell inserts	GNP	BREC	96
			SNP	PREC	87
	fluorescein		SNP	RCMEC	67

Abbreviations

Ac-DEVD-AMC	N-Acetyl-Asp-Glu-Val-Asp-7-amino-4-methylcoumarin
Ac-DEVD-pNA	N-Acetyl-Asp-Glu-Val-Asp-p-nitroanilide
AFM	atomic force microscopy
Akt	protein kinase B
AlNP	alumina NP
BAEC	bovine aortal EC
Bcl-2	B cell lymphoma 2 apoptosis regulator protein
BREC	bovine retinal EC
CCD	charge-coupled device
CFM	confocal fluorescence microscopy
CIFM	confocal immunofluorescence microscopy
CM-H2DCFDA	5-(and-6)-chloromethyl-2′,7′-dichlorodihydrofluorescein diacetate, acetyl ester
COL	colorimetry
DAR-4M AM	diaminorhodamine-4M acetoxymethyl ester
DCFH-DA	2′,7′-dichlorfluorescein-diacetate
DHR 123	dihydrorhodamine 123
EAHY926	human umbilical vein EC line
ECIS	electric cell-substrate impedance sensing
EIA	enzyme immunoassay
ELAM-1	endothelial-leukocyte adhesion molecule 1
ELISA	enzyme-linked immunosorbent assay
eNOS	endothelial NO synthase
ERK	extracellular signal-regulated kinase
ET-1	endothelin 1
FAK	focal adhesion kinase
FC	flow cytometry
FDAC	fluorescein di-O-acetate
FITC	fluorescein isothiocyanate
FL	fluorescence
FM	fluorescence microscopy
FUL	fullerenes
FURA-2AM	fura-2-acetoxymethyl ester
GAEC	guinea pig aortic EC
GAPDH	glyceraldehyde 3-phosphate dehydrogenase
GFP	green fluorescent protein
GNP	gold NP
GR	glutathione reductase
GRIESS	Griess reagent
GSH	glutathione
GSH-Px	glutathione peroxidase
GSK-3β	glycogen synthase kinase-3β
HAEC	human aortal EC
HMVEC	human microvascular EC
HPF	hydroxyphenyl fluorescein

HRMEC	human retinal microvascular EC
HUVEC	human umbilical vein EC
ICAM-1	intercellular adhesion molecule 1
ICP-MS	inductively coupled plasma mass spectrometry
IFM	immunofluorescence microscopy
IκBα	nuclear factor of kappa light polypeptide gene enhancer in B cells inhibitor, alpha
IL-2	interleukin-2
IL-6	interleukin-6
IL-8	interleukin-8
IONP	iron oxide NP
ISOHAS1	human haemangiosarcoma derived EC line
JC-1	5,5′,6,6′-tetrachloro-1,1′, 3,3′ tetraethylbenzimidazolyl-carbocyanine iodide
JNK	c-Jun N-terminal kinase
LC3-II	microtubule-associated protein 1 light chain 3 II
LDH	lactate dehydrogenase
LIP	liposomes
LM	light microscopy
LU	luminescence
M2VP	1-methyl-2-vinyl-pyridinium trifluoromethane sulfonate
MBCL	monochlorobimane
MCP-1	monocyte chemotactic protein 1
MDA	malondialdehyde
MNP	metal NP
MPR	microplate reader
MTS	3-(4,5-dimethylthiazol-2-yl)-5-(3 carboxymethoxyphenyl)-2-(4-sulfophenyl)-2H-tetrazolium, inner salt
MTT	3-(4,5-dimethyl-2-thiazolyl)--2,5-diphenyl-2H-tetrazolium bromide
MWCNT	multi-walled CNT
NFκB	nuclear factor kappa-light-chain enhancer of activated B cells
NP	nanoparticles
NRU	neutral red uptake
PAEC	porcine aortic EC
PAI-1	plasminogen activator inhibitor 1
PARP	poly(ADP-ribose) polymerase
PCM	phase contrast microscopy
PGE_2	prostaglandin E2
PGI_2	prostacyclin
PGNP	peptide conjugated GNP
PI3K	phosphatidylinositol 3-kinase
PI	propidium iodide
POLNP	polymer NP
PREC	porcine retinal EC
PSONP	polysiloxane NP
PU	polyurethane
qPCR	quantitative polymerase chain reaction
RAEC	rat aortic EC
RCMEC	rat cerebral microvessel EC
RCOEC	rat coronary EC
RhodPha	rhodamine phalloidin
RITC	Rhodamine B isothiocyanate
RMS	raman spectroscopy
SAV	streptavidin
SEM	scanning electron microscopy
sICAM-1	soluble ICAM-1
SiNP	silica NP
SNP	silver NP
SOD	superoxide dismutase
SPION	super paramagnetic iron oxide NP
Src	proto-oncogene tyrosine protein kinase
SWCNT	single-walled CNT
TEM	transmission electron microscopy
TER	transendothelial electrical resistance
TF	tissue factor
TiNS	titanium nanostructures
TiNTS	titanium nanotube surfaces
TNF-α	tumor necrosis factor-α
TUNEL	terminal deoxynucleotidyl transferase dUTP nick end labeling
vWf	von Willebrand factor
WBT	Western blotting
WST-8	2-(2-methoxy-4-nitrophenyl)-3-(4-nitrophenyl)-5-(2,4-disulfophenyl)-2H-tetrazolium, monosodium salt
XTT	sodium 2,3-bis(2-methoxy-4 nitro-5-sulfophenyl)-5-[(phenylamino)-carbonyl]-2H-tetrazolium inner salt
ZnNP	zinc NP

nanomaterials that are insoluble in aqueous solutions, an understanding of how the nanomaterial suspensions in buffer are prepared and how nanomaterial agglomeration is controlled and monitored during *in vitro* or *in vivo* biological experiments is critical. The standardization of such protocols would allow the interlaboratory comparison of the results of the biocompatibility and toxicity studies. The implementation of a consensus on the standard nanomaterials that are used as negative ("non-toxic") and positive ("toxic") nanomaterial controls would help in the development of a relative scale of biocompatibility/toxicity.

3.1 *Endothelial cell cultures*

The *in vitro* effects of nanomaterials on ECs have been studied in various cultured EC types. A great number of studies have been performed in human umbilical vein ECs (HUVECs).[42–50] Other studies have used different types of human ECs, such as human aortic ECs (HAECs),[51,52] human microvascular ECs (HMVECs),[53] human cardiac microvascular ECs (HCAMECs),[54] pulmonary microvascular ECs (HPMVECs), human dermal microvascular ECs (HDMVECs),[55] and human retinal microvascular ECs (HRMECs).[56] Some studies have been performed in human cell lines, such as ECV304,[57] EAHY926,[58] and ISOHAS1,[59] which have certain EC-like properties. The most common studies from other species utilized bovine ECs, such as bovine aortic ECs (BAECs),[60] bovine carotid artery ECs (BCAECs), bovine retinal ECs (BRECs),[61] and bovine lung microvascular ECs (BLMVECs).[62] In addition to bovine ECs, porcine aortic ECs (PAECs),[63] porcine retinal ECs (PREC), rat aortic ECs (RAECs),[64] rat coronary ECs (RCOECs),[65] rat cerebral microvessel ECs (RCMECs),[66–68] and mouse pulmonary microvascular ECs (MPMVECs)[69] have also been used. A few *ex vivo* studies have been performed on aortic rings isolated from rats[70] or guinea pigs.[71] It is important to note that the ECs from different species, organs, and passages have different properties.[1,15,16,72] Additionally, the characteristics of cells from individual donors may vary; therefore, pooled ECs from a number of donors are commonly used. Thus, the translation of the observed results from one EC type to another or the generalization of results should be performed with great caution. In our laboratory, we use at maximum a second passage of pooled HUVECs for the initial basic endothelial toxicity screening studies, with the intention to expand testing to other cell types, such as HPMVECs and HAECs, in more specific studies. The quality of the tissue culture techniques also has a critical impact on the quality of the studies.

3.2 *Nanomaterial uptake*

Transmission electron microscopy (TEM) is frequently used in the studies of nanomaterial uptake by ECs.[42,51,52,53,63,70] The uptake of fluorescent nanoparticles has been studied using flow cytometry[42] or fluorescence microscopy.[43] Studies on nanomaterial uptake have also utilized atomic force microscopy (AFM),[43] Raman spectroscopy,[45] or inductively coupled plasma mass spectrometry (ICP-MS).[44]

3.3 *Morphological and structural changes*

Observing the changes in EC culture morphology and structure using different microscopic techniques has been very valuable. Even simple phase contrast pictures can clearly document EC activation or cytotoxic effects. Immunofluorescence microscopy used to detect specific proteins, such as VE-cadherin,[53] vinculin,[60] and other proteins involved in EC junctions, or staining of actin with phalloidin,[64,66,73] provides a deeper look into at the structural changes of ECs. Finally, scanning electron microscopy (SEM),[45,60,66] TEM,[60] and AFM[66] are used to visualize the interaction of ECs with nanomaterials.

3.4 *Viability/cytotoxicity and proliferation*

Several methods have been used to assess the cytotoxic effects of nanomaterials. The commonly used viability/cytotoxicity or proliferation assays are based on the extracellular reduction of largely cell-impermeable second generation tetrazolium salts by the NADH produced in the mitochondria of live cells via the trans-plasma membrane electron transport and an electron mediator.[74] The resulting soluble formazans are measured by colorimetry using a microplate reader. The most commonly used tetrazolium salts are as follows: 3-(4,5-dimethylthiazol-2-yl)-5-(3-carboxymethoxyphenyl)-2-(4-sulfophenyl)-2H-tetrazolium inner salt (MTS),[43,47,48,59,64] sodium 2,3-bis(2-methoxy-4-nitro-5-sulfophenyl)-5-[(phenylamino)-carbonyl]-2H-tetrazolium inner salt (XTT),[75] 2-(4-iodophenyl)-3-(4-nitrophenyl)-5-(2,4-disulfophenyl)-2H-tetrazolium (WST-1),[76] and 2-(2-methoxy-4-nitrophenyl)-3-(4-nitrophenyl)-5-(2,4-disulfophenyl)-2H-tetrazolium monosodium salt (WST-8).[52,66,68,78] The first generation of tetrazolium salt derivatives is represented by 3-(4,5-di methyl-2-thiazolyl)-2,5-diphenyl-2H-tetrazolium bromide (MTT).[43,45,46,56,58,61] MTT is cell-permeable; its reduction is associated not only with the mitochondria, but also with the cytoplasm and non-mitochondrial membranes, including the endosomal/lysosomal compartment and the plasma membrane, and

MTT-formazan forms insoluble crystals. Single-walled carbon nanotubes (SWCNTs) interact with MTT-formazan crystals, resulting in false-positive results of SWCNT cytotoxicity, whereas the WST-1 assay was unaffected.[76] Other nanomaterials may interfere with these assays in different ways. Therefore, performing the cytotoxicity analysis using more than one assay based on different principles is recommended.

Another frequently used cytotoxicity assay is based on the change of the integrity of cellular membranes. In damaged or dead cells, the cytoplasmic enzyme lactate dehydrogenase (LDH) leaks into the supernatant.[43,44,53,58,59,66] This enzyme can be detected by measuring its catalytic activity through an enzymatic coupling reaction — LDH oxidizes lactate to generate NADH, which then reacts with a tetrazolium salt to generate a formazan that can be assayed by photometry. Formazan products are therefore analyzed using both the cytotoxicity assays described above. Even though these types of assays are generally used as complementary approaches, an additional cytotoxicity assay based on a completely different principle could be useful. For example, Trypan blue[50] is a diazo dye commonly used in light microscopy to stain dead cells. The neutral red uptake (NRU)[45] cytotoxicity assay is based on the ability of viable cells to incorporate neutral red, a weak cationic supravital dye that readily penetrates cell membranes by non-ionic diffusion and predominantly accumulates intracellularly in lysosomes. Changes in the cell surface, lysosomal fragility, and other changes produced by toxic substances decrease the uptake and binding of NR. Thus, it is possible to distinguish between viable, damaged, or dead cells via spectrophotometric measurements.

The fluorescein di-O-acetate (FDAC)[60] cytotoxicity assay is based on the capability of viable cells to take up the nonpolar, non-fluorescent compound FDAC, which is then hydrolyzed by acetyl esterases to fluorescein, a polar, fluorescent compound that is retained within the cell. Non-viable cells do not have esterase activity and will not exhibit fluorescein accumulation. Counterstaining of dead cells with propidium iodide (PI) is usually performed.

Although classic proliferation assays such as ^{3}H-thymidine incorporation assays are less commonly used in the latest studies, methods based on nucleic acid quantification by fluorescent dyes, such as the CyQuant-GR dye[47] whose fluorescence increases when bound to cellular nucleic acids, are becoming popular.

3.5 *Apoptosis, necrosis, and autophagy*

Flow cytometric analysis of cells by annexin V/PI staining is widely used in nanomaterial cytotoxicity studies.[43–46,48,66] Annexin V binds the accessible phosphatidylserine (PS) on the cell surface in a Ca^{2+}-dependent manner. PI is an intercalating agent that binds to nucleic acids; it is membrane-impermeable and generally excluded from viable cells. Thus, PI stains nucleic acids only in dead cells. The different labeling patterns in the annexin V/PI analysis identify the following different cell populations: annexin V-negative and PI-negative viable cells, annexin V-positive and PI-negative early apoptotic cells, annexin V-positive and PI-positive late apoptotic cells or necrotic cells, and annexin V-negative and PI-positive necrotic cells. Theoretically, cationic nanomaterials may bind to the cell surface PS and thus interfere with annexin V binding. Another caveat is the PI staining. In our hands, a prolonged incubation with PI increases the permeability of ECs for PI, thereby giving false-positive results. Therefore, we add PI to the annexin V-labeled cells immediately prior to flow cytometric analysis. Cell apoptosis should be confirmed by an assay based on the terminal deoxynucleotidyl transferase dUTP nick end labeling (TUNEL) principle.[79] Flow cytometric cell cycle analysis is usually performed in parallel with TUNEL.[48,79] Assays of the activities of execution caspases, such as caspases-3 and -7, are also commonly used to confirm the activation of apoptotic pathways.[61,66,78,80]

The autophagy of nanomaterial-treated ECs has been assayed by Western blotting and/or immunofluorescence microscopy of microtubule-associated protein light chain 3 (LC3).[80,81] LC3 is located on the autophagosomal membrane. LC3-II (16 kDa), the product of LC3 conversion, is associated with the autophagosomal membrane and migrates faster than LC3-I (18 kDa) on SDS-PAGE. Since the amount of LC3-II correlates with the number of autophagosomes, an immunoblot analysis of LC3 is a valuable indicator of the autophagic activity in ECs and other cells.[82] Immunofluorescence microscopy of the cellular distribution of LC3 also helps to document the autophagy process. The LysoTracker Red dye uptake assay has also been used in autophagy studies.[45] LysoTracker Red DND-99 is a cationic fluorescent dye that preferentially accumulates in the acidic lysosomal compartments. The amount of dye taken up by cells can be used as an indicator of the lysosomal content and an indirect measure of the autophagolysosomes. CellTracker Green is deacetylated within viable cells to a thiol-reactive dye that remains in the cytosol and is used to normalize the LysoTracker signal to viable cells. Finally, TEM analysis provides valuable supporting data on the apoptosis and autophagy of ECs.

3.6 *NO and ROS production and redox status*

The NO produced by ECs is not stable in tissue culture or *in vivo* and is rapidly scavenged ($t_{1/2}$ = 4 s). Therefore, instead of directly measuring NO,

NO production can be quantified by the determination of the total sum of nitrite (NO_2^-) and nitrate (NO_3^-) concentration in tissue culture. Nitrite/nitrate assays are based on a two-step method. In the first step, the nitrates are reduced to nitrites by nitrate reductase. In the second step, the Griess reagent converts nitrites into a deep purple azo compound, assayed by colorimetry.[60,65,68] Alternatively, the nitrite concentration is determined by a fluorometric assay, e.g., using 2,3-diaminonaphtalene (DAN).

ROS production has been monitored in several studies using the oxidation-sensitive fluorescent dye 5-(and-6)-carboxy-2′,7’-dichlorodihydrofluorescein diacetate (carboxy-H_2DCFDA) or other derivatives of reduced fluorescein. These compounds are non-fluorescent until the acetate groups are removed by intracellular esterases and the product is oxidized by ROS. The fluorescence is usually monitored by fluorometry using a microplate reader,[46,50] flow cytometry,[44,53,66,68] or fluorescence microscopy.[63]

L-7-glutamyl-L-cysteinylglycine (glutathione, GSH) is an important reducing agent and antioxidant involved in the maintenance of the cellular oxidation–reduction balance. The enzymatic recycling method is the most common GSH assay and uses glutathione reductase and 5,5′-dithio-bis-2-(nitrobenzoic acid) (DTNB). GSH reacts with DTNB to produce 5-thio-2-nitrobenzoic acid (TNB), which is assayed photometrically. Glutathione reductase is used to reduce the mixed disulfide GSTNB to recycle GSH and replenish TNB. Thus, the rate of TNB production is proportional to the concentration of GSH in the sample. An alternative GSH assay is based on flow cytometry or fluorescence microscopy following staining with monochlorobimane (MBCL).[78]

Malondialdehyde (MDA), a natural product of lipid peroxidation,[50] is usually assayed as a component of thiobarbituric acid (TBA) reactive substances. The MDA–TBA adducts are measured by fluorometry or photometry. More specific assays for MDA are based on high-performance liquid chromatography (HPLC) analysis. Superoxide dismutase (SOD) is an essential component of the cellular antioxidant defense system,[50] which catalyzes the dismutation of the superoxide radical to molecular oxygen and hydrogen peroxide. SOD enzymatic assays are based on its ability to dismutate superoxide radicals produced by xanthine oxidase. These radicals are then detected with a tetrazolium salt and measured by photometry.

Glutathione peroxidase (GSH-Px) is an important antioxidant enzyme that catalyzes the reduction of hydrogen peroxide and hydroperoxides formed from fatty acids.[50] It is assayed enzymatically or by the enzyme-linked immunosorbent assay (ELISA).

3.7 *EC activation markers*

Monitoring the changes in the concentration of intracellular free Ca^{2+} using a Ca^{2+}-sensitive fluorescent probe such as FURA-2AM provides valuable data on the activation status of ECs.[79]

Markers of the release of WP bodies, particularly tPA, vWf, ET-1, PAI-1, IL-6, and IL-8, may be detected in the EC culture supernatant using ELISA,[51,59,60,67,73] whereas the redistribution of P-selectin (CD62P) from WP bodies to the cell surface can be monitored by flow cytometry or immunofluorescence microscopy. In our experience, the surface localization of P-selectin on the activated ECs is usually not increased in multiply-passaged HUVECs. Thus, primary cultures are useful for this type of experiment. The pro-inflammatory (type II) activation of ECs can be documented by the analysis of the surface expression of ICAM-1 (CD54), TF (CD142), E-selectin (CD62E), or VCAM-1 (CD106). The soluble forms of the adhesive receptors like sICAM-1, sE-selectin, and sVCAM receptors detected in the cell supernatant may also serve as markers of pro-inflammatory activation, whereas sTM is considered to be a marker of EC damage.[12,13] Flow cytometric analysis of the different characteristics of the membrane microparticles released from ECs may also document EC activation or apoptosis, although the interpretation of these results is still problematic.[83] Gene expression profiling using quantitative polymerase chain reaction (PCR) or microarrays has been performed in some studies,[42,80,84] while other studies investigated specific signaling pathways (e.g., GSK-3β/Akt, Akt/FAK) using immunoblotting and immunofluorescence microscopy.[44,53,56,61,66,73,85–88]

3.8 *EC mobility and morphogenesis, monolayer permeability, and vasoactivity assays*

EC migration into *in vitro* scratched "wounds" has been studied by phase contrast microscopy[56,61,86] or fluorescence microscopy of FDAC-stained cells.[60] Alternatively, cell mobility was monitored by SEM. A different EC chemotaxis migration assay is to use a Boyden chamber with two compartments separated by polycarbonate filters coated with collagen IV. The relative migration of ECs was measured by the densitometric analysis of cells stained with toluidine blue.[45] EC morphogenesis has also been studied using Matrigel with microscopic observation.[45,48,56,61] The transendothelial electrical resistance (TER) was measured using the electrical cell-substrate impedance sensing system (ECIS).[53,59] Another method to analyze permeability of ECs is an *in vitro* RITC–dextran assay. To measure solute flux across ECs, the cells were seeded onto 12 mm diameter transwell filter inserts with a 0.4 μm pore size, and fluorometric analysis of the RITC–dextran[86,87] or fluorescein[67] transported across the EC monolayer from the apical to basolateral chamber was performed.

Finally, the effects of nanomaterials on EC vasodilator potency has been studied *ex vivo* in rat[65,70] or guinea pig[71] aortic rings mounted between two fine steel wires in an organ bath setup. An isometric strain transducer was used to measure the degree of tension, which was recorded. The EC-dependent and -independent vasodilator responses were induced by acetylcholine (ACh) and sodium nitropruside, respectively.

4. Studies on the *In Vitro* Interactions of Nanomaterials with Endothelial Cells

The interactions of nanomaterials with ECs have been studied for different purposes, including the endothelial targeting and transport of nanoparticles (NPs) designed for drug delivery.[89] Different types of affinity tools, including antibodies or peptides, have been used to specifically target nanoparticles to ECs. The ability of NPs to be internalized and transcytosed are critical factors for NP drug delivery to be successful. The internalization of NPs may occur via several endocytic pathways in ECs, including clathrin- and caveolin-mediated endocytosis and receptor mediated endocytosis (cell adhesion molecule-mediated). Phagocytosis and macropinocytosis are less characteristic for ECs.[89] A different mechanism of drug delivery is via passive targeting. NPs accumulate and are entrapped in tumors by the enhanced permeation and retention effect caused by leaky angiogenic vessels and poor lymphatic drainage. However, although NPs have been observed accumulating in tumors, the cellular uptake and intracellular drug release have been questioned.[90] In addition to drug delivery, similar principles have been used to design NPs to target ECs for vascular imaging.[91] Another group of studies deals with the pro- and anti-angiogenic effects of NPs for use in vascular engineering and cancer treatment, respectively. The antioxidant and other cytoprotective effects of NPs have also been studied for potential biomedical applications. Finally, the group of studies highlighted in this chapter concerns the endothelial toxicity of nanomaterials and their vascular biocompatibility and potential vascular adverse events. Below, I will summarize the most recent findings on the interactions between ECs and various classes of nanoparticles.

4.1 *Metal nanomaterials*

4.1.1 *Iron-containing nanomaterials*

Iron oxide nanoparticles (IONPs) have been extensively studied for various biomedical uses; however, their cytotoxic effects are an ongoing concern. The two main forms are magnetite, ferriferous oxide (Fe_3O_4), and its oxidized form maghemite, ferric oxide (γ-Fe_2O_3). Zhu *et al.*[57] demonstrated that both Fe_2O_3 and Fe_3O_4 nanoparticles at a concentration range of tens of μg/mL can generate oxidative stress by stimulating the superoxide anion (O_2^-) and increase NO in the ECV304 EC line. The loss of mitochondrial membrane potential and the condensation of apoptotic chromatin in the nucleus were observed as early signs of apoptosis.[57] In a follow-up study, the authors investigated the direct effects of IONPs on HAECs and the possible effects mediated by the phagocytosis and activation of monocytes (U937 cells). HAECs and U937 cells were exposed to 2, 20, and 100 μg/mL of 22 nm Fe_2O_3 and 43 nm Fe_3O_4 particles. Both of the IONPs induced cytoplasmic vacuolation, mitochondrial swelling, and cell death in the HAECs. There was also a significant increase in the production of NO, which coincided with the elevation of NOS activity in the HAECs. The adhesion of monocytes to the HAECs was significantly enhanced as a consequence of the upregulation of ICAM-1 and IL-8 expression, both of which are considered to be markers of EC pro-inflammatory activation. The phagocytosis and dissolution of NPs by monocytes simultaneously provoked oxidative stress and mediated severe endothelial toxicity. The authors concluded that IONPs may induce endothelial system inflammation and dysfunction in three ways: 1) NPs may escape phagocytosis by interacting directly with the endothelial monolayer, 2) NPs are phagocytosed by monocytes and then dissolved, thus impacting the ECs as free iron ions, or 3) NPs are phagocytosed by monocytes, thereby inducing oxidative stress responses.[92] Similarly, Buyukhatipoglu *et al.*[63] showed that 20–40 nm IONPs at a high concentration of 500 μg/mL increases intracellular ROS in PAECs by more than 800% after a 3 h exposure. ECs elongated to more than twice their initial length after 12 h, and actin stress fibers formed within the cells. This change in the actin cytoskeleton increased the cell elastic modulus by 50%. When ROS formation was blocked using scavengers, the initial cell morphology and the actin cytoskeleton remained intact, and cell viability increased. These studies indicated that IONPs induce ROS formation, which disrupts the actin cytoskeleton and alters EC morphology and mechanics.[63] Apopa *et al.*[53] showed that the exposure of HMVECs to Fe_2O_3 IONPs of mean size <10 nm at a concentration of 50 μg/mL for 30 min induces an increase in the cell permeability of HMVECs through the production of ROS and the stabilization of microtubules. The study showed that the Akt/GSK-3β signaling pathways are involved in the IONP-induced cell permeability, with ROS playing a major regulating role.[53]

As bare IONPs are highly unstable in saline solutions and form optically visible aggregates, dextran and citric acid are commonly used as coating materials to stabilize and functionalize IONPs for biomedical applications. Wu *et al.*[93] demonstrated the cytotoxicity of dextran- and citric acid-coated Fe_3O_4 IONPs on HUVECs. Both citric acid- and dextran-coated IONPs, with an approximate core size of 38 ± 8.14 nm and 31 ± 6.58 nm, respectively, appeared to be largely internalized by the HUVECs via endocytosis and showed significant cytotoxicity toward the HUVECs even at nanomolar concentrations. The cytoskeletal structures were greatly disrupted, as evidenced by diminished vinculin spots, and the actin fiber and tubulin networks were disorganized. The capacity of HUVECs to form a vascular network diminished after exposure to IONPs, which also significantly inhibited and cell migration and invasion.[93]

A variety of colloidal dispersions of NPs with an iron oxide core coated with organic or inorganic materials are being developed. These IONPs are called superparamagnetic iron oxide nanoparticles (SPIONs). The magnetic properties of SPIONs, based on their inducible magnetization, may lead to a range of new biomedical applications. For example, SPIONs can be used to non-invasively visualize labeled cells by magnetic resonance imaging (MRI) after transplantation, enhance the efficiency of drug or gene delivery, or be used as tools for magnetic hyperthermia for cancer treatment. Cengelli *et al.*[94] investigated the interaction of the different types of SPIONs with brain-derived ECs. The SPIONs tested consisted of an iron oxide core (9–10 nm in diameter) coated with either dextran (Sinerem and Endorem) or various functionalized polyvinyl alcohols (PVAs) (PVA–SPIONs). None of these nanoparticles were cytotoxic or induced the production of the inflammatory mediator NO, which was used as a reporter for cell activation. Amino-PVA–SPIONs were taken up by isolated brain-derived endothelial and microglial cells at a much higher level than the other SPIONs. Based on NO analysis, inflammatory activation of these cells was not observed.[94] Soenen *et al.*[95] studied four different types of NPs: dextran-coated Endorem® (80–150 nm; IONP core, 4.2–4.8 nm), carboxydextran-coated Resovist® (62 nm; IONP core, 4.2 nm), lipid-coated magnetoliposomes (MLs) (40 nm; IONP core, 14 nm), and citrate-coated very small iron oxide particles (VSOPs) (8.6 nm; IONP core, 4 nm) using a variety of cell types including human blood outgrowth ECs. The cellular cytoskeleton and morphology were affected only at very high NP concentrations (600 μg/mL).

Although VSOPs did not affect the actin cytoskeleton and microtubule network, these particles led to the highest production of ROS. Thus, the high intracellular concentrations of SPIONs and other IONPs required for efficient MRI pose a serious cytotoxicity risk, which needs to be evaluated.[95]

4.1.2 *Gold*

Trickler *et al.*[75] studied the cytotoxic effects of larger (10–60 nm) and smaller (3–7 nm) gold NPs (GNPs) on primary RCMECs *in vitro*. The small GNPs led to higher RCMEC accumulation compared to larger GNPs, whereas only a moderate decrease in cell viability was observed with smaller GNPs at concentrations higher than approximately 8 μg/cm^2. The basal levels of the various pro-inflammatory mediators remained unchanged. RCMEC morphology appeared to be unaffected 24 h after exposure to GNPs, with only mild changes in fluorescein permeability.[75] Kim *et al.*[56] showed that the VEGF-induced proliferation of HRMECs was significantly suppressed by the co-treatment of GNPs in a concentration-dependent manner (1 and 5 μM GNPs). GNP treatment did not affect cell viability, but significantly inhibited the VEGF induced HMREC tube formation on a 2D Matrigel matrix. GNPs blocked the VEGF-induced auto-phosphorylation of VEGFR-2 to consequently inhibit ERK 1/2 activation.[56] Similarly, Kalishwaralal *et al.*[96] demonstrated that 50 nm GNPs (500 nM) significantly inhibited the VEGF-induced proliferation, migration, and tube formation of BRECs. They found that GNPs can block the VEGF activation of Src in BRECs; therefore, modulation of these pathways may contribute to the ability of GNPs to block VEGF-induced angiogenesis.[96]

Another example of using GNPs to modulate angiogenesis involves colloidal peptide-capped GNPs. Bartczak *et al.*[42] showed the specific binding of these peptide-capped particles to two endothelial-expressed receptors that control angiogenesis — vascular endothelial growth factor type 1 receptor (VEGFR-1) and the neuropilin receptor (NRP-1). The selective binding and the biological activity of the peptide-capped GNPs could induce the stimulation or suppression of desired genes.[42]

Mao *et al.*[97] demonstrated that polycaprolactone (PCL)-coating of 100 nm GNPs may reduce their cytotoxicity in ECV304 cells. The subcellular distribution of the GNPs showed that they are located in various cellular compartments (endosomes or lysosomes) — the cytoplasm, nuclear envelope, and nucleus, regardless of the presence of the PCL coating. However, a significant change in the cell morphology and cytoskeleton associated with higher cytotoxicity was caused by bare GNPs. While a GNP concentration of 50 μg/mL did not affect cell proliferation, viability decreased significantly when the concentration exceeded 250 μg/mL. In contrast, no significant differences in the cell morphology or cytoskeleton were caused by the PCL-coated GNPs.[97]

4.1.3 *Silver*

Ragaseema *et al.*[98] investigated effects of polyethylene glycol (PEG)-protected 20 nm silver nanoparticles (SNPs) immobilized in a fibrin matrix on cultured HUVECs. Actin staining was carried out to compare EC attachment and spreading on PEG–SNP-incorporated fibrin surfaces as compared to controls. After 24 h in culture on 0.1 w/w% PEG–SNP fibrin matrix, ECs were well attached and spread, and were comparable with the control without PEG–SNPs. In contrast, at 0.2 w/w% PEG–SNPs, cell spreading was poor. Accordingly, at 0.1 w/w% PEG–SNPs, live, dead and apoptotic cell populations were quantitatively similar to the control group without PEG–SNPs. At 0.2 w/w% PEG–SNPs, the apoptotic/dead population was significantly higher when compared to the control. Real-time PCR showed that the expression of vWf was upregulated, whereas expression of tPA was downregulated at 0.2 w/w% PEG–SNPs, indicating the prothrombotic phenotype of HUVECs grown at higher PEG–SNP concentrations.

Rosas-Hernandez *et al.*[65] showed that 45 nm SNPs had dual effects on the growth of RCOECs. Low SNP concentrations of 1.0–10 μg/mL inhibited cell proliferation compared to cultures grown in the absence of SNPs; however, high SNP concentrations of 50–100 μg/mL induced a significant proliferative effect and stimulated NO production. The antiproliferative effects of the low concentrations of SNPs were associated with an increase in LDH release, a marker of membrane disruption. High concentrations of SNPs, which stimulated cell growth, did not induce toxicity in RCOECs. The authors showed that high concentrations of SNPs induced NO-dependent proliferation through the activation of eNOS by phosphorylation of serine 1177. In isolated rat aortic rings, a low SNP concentration induced vasoconstriction and a high concentration stimulated vasodilation. Both effects were endothelium-dependent. The physiological effects induced by a low concentration of SNPs inhibited ACh-induced NO-mediated relaxation.[65] To understand the underlying mechanism of the inhibitory effect of SNPs on angiogenesis, Gurunathan *et al.*[61] showed that 50 nm SNPs at a concentration of 500 nM inhibited VEGF-induced cell proliferation, migration, and capillary-like tube formation of BRECs. The cells showed a clear increase in caspase-3 activity and the formation of DNA ladders, which provided evidence of the induction of apoptosis. The authors reported that SNPs likely inhibited cell survival by targeting the activation of the PI3K/Akt signaling pathways.[61] Trickler *et al.*[67] studied the effects of different-sized SNPs on primary RCMECs. SNPs of 25 nm and 40 nm in size significantly decreased cell viability at both concentrations of 25 and 50 μg/mL. In contrast, 80 nm SNPs significantly decreased cell viability only at 50 μg/mL. RCMEC monolayers treated with 25 nm SNPs experienced significantly more cellular damage, with large perforations in the monolayers, when compared with control monolayers. Treatment with 40 and 80 nm SNPs also produced some alterations in cell morphology; however, the perforations were significantly smaller and less frequent. The exposure to 25 nm SNPs significantly increased PGE_2 release after 4 h. The larger SNPs (40 and 80 nm) had no observable effect on PGE_2 release from RCMECs. In contrast to 80 nm SNPs, the exposure to 25 and 40 nm SNPs induced a significant release of TNF after 4 h. In addition, 25 and 40 nm SNPs induced a significant release of IL-1. The release of IL-2 from the RCMECs after SNP exposure was insignificant. The increased fluorescein transport observed in the SNP-treated RCMECs clearly indicated a size-dependent increase in RCMEC monolayer permeability, which correlated with the severity of cytotoxicity.[67]

Sheikpranbabu *et al.*[87] studied changes in the permeability of PRECs by measuring the flux of RITC–dextran across the PREC monolayer. The 50 nm SNPs showed a dose-dependent increase in cytotoxicity; exposure of cells to more than 500 nM SNPs caused significant cell death. Studies of the effect of SNPs on vascular permeability were carried out using a low, non-toxic concentration of 100 nM. SNPs blocked the solute flux induced by both VEGF and IL-1. They further elucidated that VEGF and IL-1 stimulated endothelial permeability via a Src-dependent pathway by increasing Src phosphorylation and that SNPs blocked the VEGF- and IL-1–induced Src phosphorylation at Y419. These results suggested that SNPs may inhibit VEGF- and IL-1–induced permeability through the inhibition of the Src kinase pathway.[87]

Liu *et al.*[99] showed that embedding SNPs in a polymer matrix, such as polyetherurethane (PEU), may reduce their cytotoxic effects. The proliferation of ECs on PEU–SNP nanocomposites was enhanced as compared to bare SNPs. The expression of the eNOS gene was upregulated on PEU–SNPs containing 3–4 nm or 5–7 nm SNPs. This effect was not as remarkable in nanocomposites of 10–40 nm SNPs. Overall, nanocomposites made from PEU and 60 ppm of 5–7 nm SNPs showed the best biocompatibility.[99]

Madhavan *et al.*[100] studied the effects of 20 nm SNPs immobilized on a PCL scaffold on HUVECs. Cell morphology, survival, necrosis, apoptosis, and the expression levels of vWf, tPA, and eNOS were not different on the 0.1% (w/w) SNP–PCL scaffold compared to the control (PCL-only) scaffold. However, the cells exposed to higher concentrations of SNPs had increased toxicity, elevated vWf mRNA levels, and downregulated tPA and eNOS mRNA. Thus, there was an optimum SNP concentration at which maintenance of the normal EC phenotype was established, whereas an excessive SNP concentration adversely affected the cells at the molecular level. A total of 0.1% (w/w) SNP on the PCL scaffold was antimicrobial, but nontoxic to ECs at the cellular and molecular levels.[100]

4.1.4 *Titanium*

Peng *et al.*[101] investigated the effects of nanotubular titanium oxide (TiO_2) surfaces on BAECs. They found that the TiO_2 surface significantly enhanced EC proliferation and secretion of PGI_2.[101] In a follow-up study,[84] the response of primary HAECs and vascular SMCs to TiO_2 nanotube arrays 30 and 100 nm in size was studied by gene expression analysis. Microarrays revealed that the nanotubes enhanced EC proliferation and motility, decreased vascular SMC proliferation, and decreased the expression of molecules involved in inflammation and coagulation in both cell types. The networks generated from the significantly affected genes suggest that cells may sense nanotopographical cues via pathways previously implicated in sensing shear stress.[84] Lu *et al.*[64] studied the influence of different titanium surfaces, including submicron rough (lateral feature sizes larger than 100 nm), nanometer rough (lateral and vertical feature sizes smaller than 50 nm), and flat, on vascular cells (ECs and SMCs). The submicron and nanometer rough titanium surfaces promoted RAEC proliferation and long-term intracellular collagen and elastin synthesis, while simultaneously decreasing SMC function.[64] In another study, Brammer *et al.*[60] investigated primary BAECs on a flat titanium surface versus a nanostructured TiO_2 nanotube surface with an average ~100 nm outer diameter with a ~15 nm wall thickness and ~250 nm height. BAECs were able to probe the surface and interlock cellular extensions into the nanotubes for advanced sensing, elongation, and migration. The nanotubes also stimulated an increase in ECM deposition and induced a more natural "vascular bed." The actin filaments formed unidirectional cytoskeletons and more organized lamellipodia on the nanotube surface. A small puncta instead of a large dash of vinculin staining proved that motile cells were observed on the nanotube surface. Finally, the nanostructures increased the functionality of the cells by substantially increasing the NO_x/ET-1 ratio in the medium. These beneficial effects suggested that nanotube structures with enhanced endothelialization, increase ECM formation, and a substantially higher level of NO_x/ET-1 ratio may be useful as a material for vascular stents.[60]

4.1.5 *Quantum dots*

Chen *et al.*[102] demonstrated that the uptake of 3-mercaptopropionic acid-capped CdTe quantum dots (QDs) into HUVECs is associated with a dose-dependent decrease in cellular viability even at a QD concentration of 5 μg/mL.[102] Yan *et al.*[103] further investigated the endothelial toxicity of 4 nm mercaptosuccinic acid (MSA)-capped CdTe QDs. The CdTe QDs (0.1–100 μg/mL) dose-dependently induced significant endothelial toxicity. At a concentration of 10 μg/mL, CdTe QDs induced significant oxidative stress, mitochondrial network fragmentation, and the disruption of mitochondrial membrane potential; accordingly, ROS scavengers could protect HUVECs from QD-induced mitochondrial dysfunction. Moreover, upon a 24 h exposure to 10 μg/mL CdTe QD, apoptosis in HUVECs dramatically increased, accompanied with the upregulation of Bax, downregulation of Bcl-2, release of mitochondrial cytochrome c, and cleavage of caspase-9/caspase-3. These results suggested that CdTe QDs not only could impair mitochondria, but also induce endothelial toxicity through the activation of the mitochondrial death pathway and the induction of endothelial apoptosis.[103] Similarly, Wang *et al.*[104] showed that CdTe QDs had a dose-dependent inhibitory effect on HUVECs. After a 12 h treatment, QDs at 1, 10, and 50 μg/mL induced the formation of γH2AX foci, which were indicative of DNA damage. Moreover, QD treatment clearly induced the generation of reactive oxygen species (ROS). Pre-treatment with N-acetyl-cysteine (NAC), an ROS scavenger, inhibited the QD-induced generation of ROS and the formation of γH2AX foci. These data suggest that CdTe QDs have cytotoxic and genotoxic effects on HUVECs and that ROS generation may be involved in QD-induced DNA damage.[104]

4.1.6 *Other metals*

Gojova *et al.*[51] studied the effects of different metal nanoparticles (0.001–50 μg/mL) on HAECs, including iron oxide (Fe_2O_3, 45 nm and 5 nm mixture), yttrium oxide (Y_2O_3, 20–60 nm), and zinc oxide (ZnO, rod shape 20–70 nm × 100–200 nm) particles. The study showed that all three types of nanoparticles were internalized into HAECs and were often found within intracellular vesicles. The Fe_2O_3 nanoparticles failed to provoke an inflammatory response in HAECs at any of the concentrations tested; however, the Y_2O_3 and ZnO nanoparticles elicited a pronounced inflammatory response above the threshold concentration of 10 μg/mL. At the highest concentration, the ZnO nanoparticles were markedly cytotoxic.[51] In a follow-up study with the same design, the authors showed that cerium oxide (CeO_2) nanoparticles (35–40 nm) caused a very muted inflammatory response in HAECs, even at the highest dose.[105]

In a similar study, Sun *et al.*[54] investigated the cytotoxicity of the following six metal oxide nanoparticles on HCAMECs *in vitro*: zinc oxide (ZnO), iron(III) oxide (Fe_2O_3), iron(II,III) oxide (Fe_3O_4), magnesium oxide (MgO), aluminum oxide (Al_2O_3), and copper(II) oxide (CuO). The mean NP diameter was 39–46 nm. HCMECs were exposed for 12–24 h to NP concentrations of 0.001–100 μg/mL. The results showed that the Fe_2O_3, Fe_3O_4, and Al_2O_3 NPs did not have significant effects on cytotoxicity, permeability, and the inflammation response in HCMECs at any of the concentrations tested. The ZnO, CuO, and MgO NPs were cytotoxic in a concentration- and time-dependent manner, and elicited a permeability and inflammation response in HCMECs. The ZnO and CuO NPs significantly induced the mRNA expression of VCAM-1, ICAM-1, MCP-1, and IL-8 at concentrations of ≤5 μg/mL, whereas the MgO NPs increased the levels of these inflammatory markers only at higher concentrations.[54]

Tsou *et al.*[88] studied the inflammatory effects of 50 and 100 nm ZnO NPs on HUVECs. Interestingly, ZnO NP concentrations of ≤3 μg/mL resulted in increased cell proliferation, whereas higher doses caused increases in oxidized GSH levels. A marked drop in survival rate was observed when the ZnO NP concentration was increased to 45 μg/mL. In addition, the ZnO NPs induced a dose-dependent increase in ICAM-1 expression and marked increases in NF-κB reporter activity. The overexpression of IκBα completely inhibited the ZnO-induced ICAM-1 expression, suggesting that NF-κB plays a pivotal role in the regulation of ZnO NP-induced inflammation in HUVECs.[88]

Yu *et al.*[69] compared the alteration of PAI-1 expression in MPMVECs exposed to non-toxic doses (<5 μg/mL) of 42 nm nano-sized CuO NPs and 28 nm nano-sized TiO_2 NPs. The results showed that the CuO NPs caused a dose- and time-dependent increase in PAI-1 expression. Moreover, the exposure of MPMVECs to CuO NPs, but not TiO_2 NPs, at 0.625 μg/mL for 12 h induced the generation of ROS. The exposure of MPMVECs to CuO NPs also caused a dose- and time-dependent increase in p38 phosphorylation, as detected by Western blotting. In addition, the overexpression of PAI-1 was attenuated by pre-treatment with a p38 inhibitor. The results suggested that the CuO NP upregulation of PAI-1 in ECs is mediated by p38 phosphorylation due to oxidative stress.[69] Thus, CuO NPs may increase the anti-fibrinolytic potential of ECs via PAI-1 upregulation.

Peters *et al.*[55] studied the effects of different types of NPs on human dermal MVECs, including TiO_2 (70 nm), Co (120 nm), Ni (50 nm), SiO_2 (14 nm), and PVC NPs (130 nm). The Co NPs at 5 and 50 μg/mL, but not the other NPs, decreased cell viability. Exposure to the Co NPs and to a lesser extent, the SiO_2 NPs, impaired the proliferative activity and the pro-inflammatory stimulation of ECs. The detection of the cell cycle-associated antigen Ki67 showed a 50% decrease in cell proliferation after treatment with 50 μg/mL Co NPs. This effect was associated with the release of IL-8 and actin reorganization.[55] Thus, Co NPs should be considered to be cytotoxic to ECs.

Ge *et al.*[49] studied the effects of 100 nm MgO NPs on HUVECs. Although concentrations below 200 μg/mL showed no cytotoxicity, MgO NP concentrations higher than 500 μg/mL caused a reduction in the relative growth rate relative to the control. In addition, the MgO NPs significantly enhanced NO release in HUVECs.[49]

Oesterling *et al.*[106] showed that alumina nanoparticles (10–20 nm in size at 100 μg/mL) induced a significant increase in the mRNA and protein expression of VCAM-1, ICAM-1, and ELAM-1 in both porcine pulmonary artery ECs and HUVECs. Furthermore, human ECs treated with alumina NPs showed an increased adhesion of activated monocytes. These data show that alumina NPs can elicit a pro-inflammatory response.[106]

4.2 *Carbon nanomaterials*

4.2.1 *Carbon nanotubes*

There are controversial reports about the effects of carbon nanomaterials on cultured ECs, particularly with regard to carbon nanotubes (CNTs). A recent study reported that SWCNTs at a very high concentration (200 μg/mL) induced apoptosis in RAECs. The generation of ROS was involved in the activation of the mitochondria-dependent apoptotic pathway.[78] The discovery of apoptosis was supported by a number of morphological and biochemical hallmarks, including chromatin condensation, internucleosomal DNA fragmentation, and caspase-3 activation. A problem with the interpretation of the results when using very high concentrations of nanomaterial, such as 200 μg/mL, is that cells may simply suffocate under the heavy load of nanomaterial sediment. Therefore, an appropriate control material should always be used for comparison. In contrast, Walker *et al.*[107] showed endothelial toxicity of CNTs at very low concentrations. Marked actin filament and VE-cadherin disruption, cytotoxicity, and reduced tubule formation occurred consistently following 24 h exposure to the highest concentrations (50–150 μg/10^6 cells, 1.5–4.5 μg/mL) of both SWCNTs and MWCNTs. These effects were not observed with carbon black exposure or CNT exposure at lower concentrations (1–10 μg/10^6 cells, 0.04–0.4 μg/mL). Zhiqing *et al.*[108] demonstrated that the exposure of RAECs to SWCNTs induces oxidative stress, thereby increasing ICAM-1 and VCAM-1 expression. SWCNT-induced nuclear NF-κB/p65 translocation was inhibited by N-acetylcysteine. Thus, oxidative stress plays an important role in the SWCNT-induced vascular endothelium damage. Similarly, Guo *et al.*[50] showed that exposure to MWCNTs at ≥10 μg/mL reduced the viability and induced apoptosis of HUVECs. Furthermore, MWCNTs could cause DNA damage. MWCNTs were shown to affect the cellular redox status, e.g., increasing intracellular ROS, MDA, and GSH-Px levels and altering SOD activity. These results demonstrated that MWCNTs could induce cytotoxic and genotoxic effects in ECs, probably through oxidative damage pathways. In contrast, Albini *et al.*[45] showed that pristine and oxidized SWCNTs, free of large aggregates, did not show any significant toxicity in HUVECs at concentrations up to 50 μg/mL, as determined by growth, migration morphogenesis, and survival assays. The ECs transiently took up the SWCNTs, which were associated with an enhanced acidic vesicle compartment within the cells.[45] The controversy in the results from different laboratories may stem from impurities in the studied CNT material. There is an urgent need for standards of different well-characterized, maximally pure CNT materials.

The effects of carbon nanomaterials on angiogenesis are the focus of several studies. While their anti-angiogenic potential may be used to treat tumors, their pro-angiogenic activities are useful in designing EC scaffolds for vascular engineering. Chaudhuri *et al.*[48] reported that PEGylated SWCNTs exerted pro-angiogenic effects *in vitro* and *in vivo*. In contrast, fullerenol significantly inhibited angiogenesis. PEGylated CNTs increased EC proliferation and tubulogenesis. The treatment of HUVECs with PEGylated SWCNTs resulted in the upregulation and clustering of $\alpha_v\beta_3$ integrin, a major adhesive receptor complex implicated in angiogenesis. The phosphorylation of focal adhesion kinase (FAK) and activation of the downstream PI3K-Akt pathway were observed, suggesting a mechanism for the pro-angiogenic effect of CNTs.[48]

Another area of investigation on the effects of nanomaterials on ECs is the design of EC scaffolds for vascular engineering. Han *et al.*[109] showed that the aligned nanofibrous scaffold of the MWCNT/polyurethane (PU) composite promoted EC proliferation by supporting its migration and aggregation along the axis of the aligned nanofibers. The ECs exhibited a non-thrombogenic phenotype, implying a promising potential for vascular engineering.[109] Similarly, Bianco *et al.*[110] showed that fibrous composites based on PCL and SWCNTs increased the viability of RCMECs.[110] Tan *et al.*[111] demonstrated that composites of covalently functionalized SWCNTs with collagen showed significantly higher endothelial coverage with pulmonary arterial ECs at low concentration of SWCNTs (1%), compared to cells seeded on polyethylene–, polytetrafluoroethylene–, and carboxylated SWCNT–collagen composites, or collagen alone.[111] By studying the mechanism of these CNT effects, Meng *et al.*[73] reported that the aligned nanofibrous structure of the MWCNT/PU composite could function as an extracellular signal to stimulate cell growth, proliferation, and extracellular collagen secretion, and preserve the anticoagulant function of HUVECs. The nanofibrous structures play important roles in the activation of Rac and Cdc42, whereas CNTs regulate the activation of Rho. Both of these activities synergistically activate Rho GTPases, which transmit cell–substrate signals to the cytoplasm. These signals are then relayed to the nucleus by the MAP kinase pathway to direct cytoskeletal arrangement and cell orientation.[73] The interaction of the CNTs in the blood modulates their effects on ECs. Ge *et al.*[77] demonstrated that coating SWCNTs with plasma proteins may decrease their endothelial toxicity. Transferrin-, gamma globulin-, bovine serum albumin (BSA)-, and bovine fibrinogen-coated SWCNTs induced less cytotoxicity in HUVECs than in uncoated SWCNTs. The fibrinogen coating had the highest cytoprotective effect.[77]

4.2.2 *Fullerenes*

Yamawaki and Iwai[80] showed that chronic treatment with low-dose $C_{60}(OH)_{24}$ (10 μg/mL for eight days) inhibited cell attachment and delayed EC growth. Although fullerenes altered cell morphology in a dose-dependent manner, only the maximal dose (100 μg/mL) caused cytotoxic injury and/or death and inhibited cell growth. EC death seemed to be caused by the activation of ubiquitin-autophagy cell death pathways. In our laboratory, we performed a pilot study on the effects of fullerenes on HUVECs.[79] Our study showed that fullerenol $C_{60}(OH)_{24}$ at a dose of 100 μg/mL had both pro-inflammatory and pro-apoptotic effects on HUVECs. We found that a 24 h treatment of

HUVECs with $C_{60}(OH)_{24}$ at 100 μg/mL significantly increased the cell surface expression of ICAM-1 (CD54). In addition, this treatment induced the expression of TF (CD142) in HUVECs and increased exposure of PS. Analysis of the cell cycle and DNA fragmentation by TUNEL showed that both nC_{60} and $C_{60}(OH)_{24}$ caused G1 arrest of HUVECs, and $C_{60}(OH)_{24}$ induced significant apoptosis. We also demonstrated that both nC_{60} and $C_{60}(OH)_{24}$ induced a rapid concentration-dependent elevation of intracellular free Ca^{2+}. This increase could be inhibited by EGTA, suggesting that the source of intracellular Ca^{2+} in the fullerene-stimulated Ca^{2+} flux is predominantly from the extracellular environment.

In contrast, some fullerene derivatives demonstrated potent antioxidant properties, and several studies have indicated their usefulness as cytoprotective agents.[112] The cytoprotective effects of a water-soluble fullerene vesicle-based phenyl penta-substituted fullerene cyclopentadienide ($Ph_5C_{60}K$, PhK) markedly improved viability and decreased apoptosis, ROS production, and lipid peroxidation in angiotensin II-treated HUVECs.[46] The carboxylated fullerene derivative $C_{60}[C(COOH)_2]_2$ selectively entered oxidation-damaged RCMECs instead of normal cells *in vitro*, and maintained CMEC integrity by attenuating H_2O_2-induced F-actin depolymerization. Additionally, $C_{60}[C(COOH)_2]_2$ NPs greatly inhibited the H_2O_2-induced apoptosis of RCMECs. $C_{60}[C(COOH)_2]_2$ NPs likely regulated several downstream signaling events related to the c-Jun N-terminal kinase (JNK) pathway, including the reduction of JNK phosphorylation, activation of activator protein 1 (AP-1) and caspase-3, and inhibition of polyADP-ribose polymerase (PARP) cleavage and mitochondrial cytochrome c release.[66]

In addition, the scavenging activity of NO was investigated with the hydroxylated fullerene derivative $C_{60}(OH)_{22}$ and malonic acid derivative $C_{60}[C(COOH)_2]_2$ using primary RCMECs. The fullerene derivatives protected against NO-induced apoptosis in sodium nitroprusside-treated cells. The fullerene derivatives could also prevent the NO-induced depolymerization of the cytoskeleton and nuclear damage and accelerate EC repair. The increase of intercellular ROS induced by NO was significantly attenuated by post-treatment with the fullerene derivatives. These results implied a potential application of fullerene derivatives in the treatment of NO-related disorders.[68] Another study demonstrated the EC-protective ROS scavenging properties of fullerene derivatives. The endohedral metallofullerenol, $[Gd@C_{82}(OH)_{22}]_n$, has been extensively investigated for its anti-tumor properties and the reduction of H_2O_2-induced ROS formation and mitochondrial damage in rat brain capillary ECs.[113]

4.3 *Silica-based nanomaterials*

Bauer *et al.*[43] showed that amorphous silica NPs (ASiNPs) induced necrotic cytotoxicity in HUVECs. Exposing HUVECs to ASiNPs with diameters of 304 nm and 310 nm led to a significant increase in WP body exocytosis, which was associated with the release of vWf and the formation of ultra-large fibers. The internalization and perinuclear localization of 310 nm perylene-labeled ASiNPs not only affected viability, but also cell migration and proliferation.[43]

Napierka *et al.*[114] investigated the effect of monodisperse ASiNPs of different sizes on the viability of the EC line EAHY926. The concentrations leading to a 50% reduction in cell viability (TC_{50}) for the smallest particles tested (14, 15, and 16 nm in diameter), ranged from 33 to 47 μg/cm^2, which was significantly different from the TC_{50} values for the bigger nanoparticles — 89 and 254 μg/cm^2 for 19 and 60 nm ASiNPs, respectively. Two fine Si particles with diameters of 104 and 335 nm showed a very low cytotoxic response compared to nanometer-sized particles, with TC_{50} values of 1,095 and 1,087 μg/cm^2, respectively. The smaller particles also appeared to affect the exposed cells faster, with necrotic cell death observed within just a few hours. The surface area of the tested particles is likely a key parameter in determining the EC toxicity of ASiNPs.[114] Due to their small size and large specific surface area, insoluble NPs are not greatly affected by the gravitational force and are generally formulated in stable suspensions or solutions.[58] However, this formulation creates a potential difficulty in *in vitro* assay systems in that cells adhering to the bottom of a culture vessel may not be exposed to the majority of the NPs in the suspension. Lison *et al.*[58] hypothesized that because of the convection forces that usually develop in solutions, the majority of the particles may reach the target cells and exert their potential toxicity. To address this issue, they exposed different cell lines, including the EAHY926 cell line, to a monodisperse suspension of Stober ASiNPs in three different laboratories. Four different endpoints (LDH release, LDH cell content, tetrazolium salt [MTT], and crystal violet staining) were used to assess the cellular response to NPs. They found that in all cell lines and for all endpoints, the cellular response was determined by the total mass/number/surface area of the particles and their concentration. The authors concluded that the minimal dose remains the most appropriate metric for the *in vitro* toxicity testing of insoluble ASiNPs dispersed in an aqueous medium.[58]

In a study on the molecular mechanism of ASiNP cytotoxicity, Liu *et al.*[44] showed that 20 nm ASiNPs at concentrations ranging from 50 to 200 μg/mL markedly induced ROS production, mitochondrial depolarization, and apoptosis in HUVECs. At the highest ASiNP concentration, the necrotic rate, LDH leakage, expression of CD54 and CD62E, and release of TF, IL-6, IL-8, and MCP-1 were significantly increased. ASiNPs also activated JNK, c-Jun, p53, caspase-3, and NF-κB, increased Bax expression, and suppressed Bcl-2 protein expression. Moreover, the inhibition of ROS attenuated SiNP-induced apoptosis and inflammation and the activation of JNK, c-Jun, p53, and NF-κB.[44]

Tavano *et al.*[115] investigated the effects of synthetic ASiNPs, including organically modified silica (ORMOSIL) NPs approximately 50 nm in size and PLGA NPs with a mean diameter of 282 nm, on HUVECs. All tested NPs, including non-PEGylated and PEGylated ORMOSIL NPs, did not significantly affect TF gene transcription in HUVECs. Consistently, none of the tested NPs caused an increase in cell procoagulant activity. Similarly, the incubation of HUVECs with NPs did not increase the expression of E-selectin or other EC activation markers (e.g., CD31, CD106, CD54). Interestingly, PLGA and small unilamellar vesicle (SUV) NPs were endocytosed by ECs, regardless of their PEGylation state, whereas the PEGylation of ORMOSIL NPs resulted in a virtually complete reduction of NP cell association and engulfment.[115]

Kasper *et al.*[59] examined the inflammatory and cytotoxic responses of monodisperse 30 nm ASiNPs in an *in vitro* co-culture model mimicking the alveolar–capillary barrier and compared the results with those from conventional monocultures. The epithelial cell line H441 and the EC line ISOHAS1 were used in monoculture and in co-culture on opposite sides of a filter membrane. The study showed that the *in vitro* co-culture model of the alveolar–capillary barrier is clearly advantageous compared to conventional monocultures. At concentrations that were 10–100-fold below the toxic concentrations, the apically exposed co-culture released IL-6 and IL-8 to the basolateral side, which may mimic the early inflammatory events that take place in the pulmonary alveoli after ASiNP inhalation. Furthermore, a number of markers of the intrinsic apoptotic pathway were upregulated in the co-culture following ASiNP treatment. Analysis of the individual markers indicated that the cells suffered from DNA damage, hypoxia, and endoplasmic reticulum stress.[59]

Akbar *et al.*[70] investigated the direct influence of ASiNP uptake on the vasodilator responses of rat aortic vessels *in vitro* using fabricated ASiNPs of defined sizes (97 ± 7.60 and 197 ± 7.50 nm) and charges (positive and non-modified). The dilator responses to a series of doses of endothelial-dependent (ACh) and endothelial-independent (sodium nitroprusside) agonists were determined before and 30 min after incubation with SiNPs (1.1×10^{11} NPs/mL). The acute exposure to SiNPs led to their rapid uptake by the ECs of the rat aorta, which had no significant influence on dilator responses, although a greater degree of attenuation was evident after the uptake of positively, charged SiNPs.[70]

The interactions of cells with nanostructures are of great interest due to their application in controlling the tissue response to biomedical implants. Lee *et al.*[116] showed that dense and upright SiO_2-coated nanorods nearly eliminated EC adhesion. The lack of adhesion was not due to a decrease in the matrix protein adsorption of the nanostructures, but rather an inability of the cells to assemble focal adhesions. Using spatially patterned nanorods, they showed that the cells displayed a preference for flat regions. The results supported a model in which interfering with the nanoscale spacing of ligated integrins leads to reduced cell adhesion and subsequent cell death.[116]

4.4 *Dendrimer nanomaterials*

Stasko *et al.*[117] reported that cationic, primary amine-containing, and fluorescently labeled derivatives of the fifth generation of polypropylenimine dendrimers caused drastic time-dependent changes in the permeability of the plasma membrane and prominent cytotoxicity in HUVECs. The complete removal of the primary amines or the masking of the cationic surface via PEGylation decreased dendrimer cytotoxicity. The authors suggested that preventing electrostatic interactions between the dendrimers and the cellular membrane is apparently a necessary step toward minimizing the toxicity of the delivery vehicles to the endothelium.[117]

Nam *et al.*[118] investigated the endothelial (HUVEC) toxicity of biodegradable polycationic polyamidoamine (PAMAM) esters (e-PAM-R, e-PAM-K) that are linked to arginines or lysines at the peripheral ends of the PAMAM–OH dendrimer through ester bonds. The HUVEC viability in the presence of these NPs was

$$\text{e-PAM-K} > \text{e-PAM-R} > \text{PAM-K} > \text{PAM-R} \gg \text{polyethylenimine (PEI)}.$$

As expected, the biodegradable ester-bonded PAMAM dendrimers, which had a non-toxic PAMAM–OH backbone, allowed high cell viability. The shorter half-life of e-PAM-K was reflected in its lower cytotoxicity compared to e-PAM-R. In addition, the high cell viability observed with e-PAM-R and e-PAM-K may be due to their non-toxic building blocks, PAMAM–OH and amino acids, which are released by the hydrolysis of ester bonds.[118]

The early generation PAMAM dendrimers (up to G4.0) have better biocompatibility compared to later generations.[119] The modification of early generation dendrimers with PEG to create stealth dendrimers is still necessary. However, to efficiently use early generation dendrimer-based stealth dendrimers for drug loading, PEGylation needs to be optimized. Yang *et al.*[119] showed that the degree of PEGylation on early generation dendrimers could be dramatically reduced to leave as many unoccupied surface groups as possible for drug loading, while maintaining the drug carrier cytocompatibility, as tested in bovine brain microvascular ECs. The study showed that G2.5 and G3.5 at 0.2 and 2 μM had negligible cytotoxic effects in ECs for up to 24 h.

Not surprisingly, polycationic G3.0 and G4.0 PAMAM dendrimers showed time-, concentration-, and generation-dependent cytotoxicity in ECs. A high concentration of G3.0 and G4.0 (20 μM) and a long incubation period (24 h) caused a significant decrease in cell viability, with G4 being more cytotoxic than G3. The PEGylation of G3 and G4 significantly improved their cytocompatibility compared to non-PEGylated controls. The PEGylated products 3PEGs–G3.0 (3PEG per G3; PEGylation degree 9%) and 10PEGs–G4.0 (PEGylation degree 16%) were considered to be optimized stealth dendrimers for EC compatibility and possible drug load.[119]

Dendrimeric compounds have also been investigated as potential modulators of the EC inflammatory response. Dernedde *et al.*[120] studied synthetic dendrimeric polyglycerol sulfates (dPGSs) as macromolecular inhibitors of inflammation that operate via a multivalent binding mechanism that mimics naturally occurring ligands. dPGSs inhibited both leukocytic L-selectin and endothelial P-selectin in HUVECs with high efficacy. The size and degree of the sulfation of the polymer core determined the selectin binding affinity.[120]

4.5 *Polymer nanomaterials*

Ehrenberg *et al.*[121] studied the binding of different surface-modified polystyrene (PST) NPs (100–120 nm in size) to HUVECs and the influence of serum proteins on this binding. The results showed a rapid coating of particles by serum proteins, and a positive correlation was observed between the protein adsorption to the particle surfaces and cellular binding. Covalent modifications of the negatively charged COOH base reduced protein binding capacity and similarly reduced cell association. In rank order for both protein binding and cell association, they found the lowest effect was observed with the positively charged lysine, then the neutral CH_3 and cysteine, and was the greatest with PEG. They also showed that the identity of proteins on the NP surfaces was not a critical determining factor for particle–cell association and therefore unlikely to require specific binding to any specific cellular receptors.[121]

Xia *et al.*[122] showed the cell-specific toxicity of 60 nm NH_2-labeled PST nanospheres. These cationic NPs, which are highly toxic in a macrophage cell line (RAW 264.7), were taken up into HMVECs by the same mechanism as LAMP-1$^+$ endosomes. In contrast to the macrophage cell line, the NP uptake in HMVECs did not lead to lysosomal permeabilization, increased Ca^{2+} flux, or mitochondrial damage. The results of this study demonstrated the importance of cell-specific uptake mechanisms and pathways that could lead to sensitivity or resistance to cationic particle toxicity.[122]

Wang *et al.*[62] used BSA-coated carboxylate-modified polymer NPs varying from 20 to 100 nm in diameter in BLMVECs to demonstrate that caveolae-mediated NP uptake is dependent on the albumin coating of the particles. They also showed that caveolae could accommodate up to 100 nm NPs, a size larger than the diameter of typical caveolae, suggesting a compliant property of caveolae.[62]

Hung *et al.*[123] investigated the response of ECs to PU that is characteristic of surface micelles (approximately 89 nm in diameter) and two novel nanocomposites, PU–GNPs and PU SNPs, which contain smaller surface micelles (approximately 14–22 nm in diameter). Both nanocomposites increased cell migration rate, which was accompanied by the upregulation of eNOS and phosphorylated Akt (p-Akt) expression. The induced eNOS and p-Akt expression was inhibited by LY294002, indicating that the nanocomposites acted through the PI3K pathway. An elevation of intracellular Ca^{2+} concentration was also observed. In addition, actin fibers were induced by PU–GNPs and PU–SNPs. A reduction in actin expression upon the addition of Y-27632 (an inhibitor of Rho-GTPase) and SU-1498 (an inhibitor of VEGFR 2) was observed. Thus, the nanometric micelles on the PU surface may interact with ECs and accelerate their migration by increasing cytoplasmic Ca^{2+} and stimulating the PI3K/Akt/eNOS signaling pathway.[123]

In a follow-up study, the authors investigated nanocomposites from PU incorporated with various low concentrations (17.4–174 ppm) of ~5 nm GNPs. ECs had the highest migration rate on the nanocomposite containing 43.5 ppm of gold (PU–GNPs 43.5 ppm). The high EC migration rate was associated with increased levels of eNOS and p-Akt expressed by ECs cultured on PU–GNP. The induction of both eNOS and p-Akt on PU–GNPs was abolished by the addition of the PI3K inhibitor LY294002, suggesting that these cellular events may be regulated through the PI3K signaling pathway. An increase in the protein expression of FAK on PU–GNPs (43.5 ppm) was observed in FAK–GFP-transfected ECs. Thus,

PU–GNP nanocomposites activated FAK and the PI3K/Akt signaling pathway in ECs, leading to EC proliferation and migration on these surfaces.[85]

Similarly, Nishikawa *et al.*[52] studied the cellular uptake of 40–400 nm polysiloxane NPs and the concomitant influence on NO release in HAECs. They found that polysiloxane NPs were endocytosed via caveolae in HAECs. NP uptake was associated with enhanced NO release due to the activation of eNOS.[52]

4.6 *Liposomes*

Targeting the tumor vasculature with liposome-formulated drugs is increasingly recognized as an effective strategy to obtain superior therapeutic efficacy with limited host toxicity when compared with conventional treatments.[124] The use of cationic liposomes for the delivery of cytotoxic molecules to the tumor microvasculature is currently being developed for treating cancer and other angiogenesis-related diseases.[125]

Coating liposomes with PEG by incorporating PEG-derived lipids into the liposomal bilayer results in the inhibition of liposomal uptake by the reticuloendothelial system and a significant increase of liposomal residence time in the bloodstream.[126] Dabbas *et al.*[127] prepared different PEG-modified cationic liposomes with varying percentages of cationic lipid content and lipid type, including 1,2-dioleoyl-3-trimethylammonium propane (DOTAP), 1,2-dioleoyl-3-dimethylammonium propane (DODAP), 1,2-dimethyldioctadecylammonium bromide (DDAB), 1,2-distearoyl-3-trimethylammonium propane (DSTAP), and 1,2-dimyristoyl-3-trimethylammonium propane (DMTAP). The diameters of each type were between 111 and 161 nm. DMTAP and DDAB formed significantly smaller liposomes compared to DOTAP, DODAP, and DDAB preparations. The authors tested the cytotoxicity of these liposomal preparations in different EC lines, including HUVECs, lung microvascular ECs (HMVEC-L and HPVE-26), coronary microvascular ECs (HMVEC-C), dermal microvascular ECs (HMVEC-D), and immortalized dermal microvascular ECs (HMEC-1). Some of the cationic liposomes (reported without the drug payload) were considerably more toxic to cells compared to other liposomes. Moreover, human ECs varied in their sensitivity level to the toxic effects of the different PEG-modified cationic liposomes. For example, the relative toxic effects were

$$\text{DMTAP} = \text{DSTAP} > \text{DOTAP} = \text{DODAP} = \text{DDAB}.$$

Interestingly, the quaternary amine of DMTAP, DSTAP, and DOTAP was a common feature that separated the toxic from the relatively less toxic varieties. The physicochemical properties of cationic liposomes and their ability to interact with ECs are important factors to consider during the early stages of developing formulations for the treatment of cancer and other angiogenesis-dependent diseases.[127]

Another example of research into the interactions between liposomes and ECs is in the area of oxygen carriers. PEGylated liposomes (~200 nm in diameter) encapsulating human hemoglobin as oxygen carriers were prepared from a purified carbonylhemoglobin (HbCO) solution and a lipid mixture composed of 1,2-dipalmitoyl-sn-glycero-3-phosphatidylcholine (DPPC), cholesterol, 1,2-dimyristoyl-sn-glycero-3-phosphoethanolamine-N-polyethylene glycol (2000) [DMPE-PEG(2000)], and palmitic acid. The liposomes showed increasing cytotoxicity in HUVECs at a high concentration range of 1–5 mg/mL.[47]

4.7 *Other nanomaterials*

Fine *et al.*[128] demonstrated that rosette nanotubes in a complex with lysine (RNT-K), a new class of biomimetic nanotubes that self-assemble from DNA base analogs, can be used to coat conventional stent materials such as titanium. Even at low concentrations, RNT-K significantly increased the EC density compared to uncoated titanium. Specifically, titanium coated with 0.01 mg/mL RNT-K increased EC density by 37% and 52% after 4 h and three days, respectively, compared to uncoated titanium. The excellent cytocompatibility properties of RNTs suggest the need for further studies of these novel nanostructured materials for vascular stent applications.[128]

In the area of nanomaterial applications in osteogenesis, Pezzatini *et al.*[129] showed that nanostructured hydroxyapatite (HA) at concentrations ranging from 2 to 10 μg/mL promoted endothelium survival and proliferation, preserved $\alpha_v\beta_3$-integrin localization, and stimulated β-actin reorganization and Akt phosphorylation. HA increased the expression of eNOS and fibroblast growth factor-2 (FGF-2). HA enhanced cell responsiveness to VEGF in terms of NOS activity, thereby increasing the ability of the microvascular endothelium to differentiate into capillary-like structures when grown in a 3D fibrin gel.[129]

In cell imaging research, van Kasteren *et al.*[130] described the design, construction, and *in vivo* application of carbohydrate-functionalized NPs that allowed the direct detection of the endothelial markers E-/P-selectin (CD62E/CD62P) during acute inflammation. These first examples of

MRI-visible glyconanoparticles displayed multiple copies of the natural glycan ligand complex of selectins.[130]

5. Selected Protocols for *In Vitro* Endothelial Cell Studies Used in Our Laboratory

Below, I provide a description of the reagents and procedures we use in our laboratory to study nanomaterial toxicity to ECs. The information on vendors for reagents and manufacturers of the instruments and data processing softwares is given for informational purposes only and not for endorsement. In all cases, equivalent reagents, instruments, and softwares can be used.

5.1 *Culture of human umbilical vein endothelial cells*[79]

HUVECs (second passage) from Lonza (Walkersville, MD) are seeded in either six-well (5,000 cells/well) or 24-well (1,250 cells/well) plates (Becton Dickinson Labware, Franklin Lakes, NJ) with EGM-2 media containing 2% fetal bovine serum and supplements (Lonza). Cells are used for experiments when they reached 80–90% confluence (~48 h in culture). HUVECs are washed and incubated for 24–72 h at 37°C/5% CO_2 with a nanomaterial suspension in water diluted 1:5 with EGM-2 medium. Unless specified otherwise, after incubation, the medium containing the nanomaterials is removed, and the cells are washed three times with Hanks' balanced salt solution with 0.35% BSA (HBSS/BSA) and either processed in culture for different assays or harvested with 20 mM HEPES buffer with 100 mM NaCl, 0.5% BSA, and 10 mM EDTA, pH 7.4. After harvesting, the cells are centrifuged (300 *g* for 5 min), washed with HBSS/BSA, and used for analysis.

5.2 *Cell Counting Kit-8 viability assay*[66,68]

The Cell Counting Kit-8 (CCK-8) assay (Dojindo Molecular Technologies, Inc., Rockville, MD) is based on a water-soluble tetrazolium salt, 2-(2-methoxy-4-nitrophenyl)-3-(4-nitrophenyl)-5-(2,4-disulfophenyl)-2H-tetrazolium monosodium salt (WST-8). CCK-8 is used to determine the effect of different nanomaterials on cell viability. HUVECs at a density of 5×10^3 cells are seeded into each well of a six-well plate and grown for two days. After reaching 80% confluence, cells are treated in triplicate with different concentrations of nanomaterial controls (negative control: nanomaterial vehicle, H_2O; positive control: 2 mM H_2O_2) in the culture medium for 24–72 h. After incubation, phase contrast microscopy is performed to observe changes in cell culture morphology. The medium containing nanomaterials is then discarded, and cells are washed three times with phosphate buffered saline (PBS). Fifty mL of the CKK-8 solution are then added into each well of the plate with 500 μL of culture medium, and the cells are incubated for 4 h at 37°C and 5% CO_2. The absorbance is measured at 450 nm using a microplate reader.

5.3 *Lactate dehydrogenase assay*[68,107,114]

The effect of different nanomaterials on LDH release by HUVECs is detected by an LDH Cytotoxicity Detection Kit (Takara Bio, Inc., Otsu, Japan). HUVECs at a density of 5 × 10^3 cells are seeded into each well of a six-well plate and grown for two days. After reaching 80% confluence, cells are treated in triplicate with different concentrations of nanomaterials or controls (negative control: nanomaterial vehicle, H_2O; positive control: 2 mM H_2O_2) in the culture medium for 24–72 h. At the end of the incubation, an additional set of negative control cells is used to determine the maximum amount of releasable LDH enzyme by permeabilizing the cells with 1% Triton X-100. After the incubation, supernatants are removed and centrifuged for 5 min at 3,000 rpm; 100 μL of the prepared supernatant are transferred in triplicate into 96-well plates and incubated with 100 μL reaction solution for 30 min protected from light. The absorbance is measured at 490 nm using a microplate reader.

5.4 *Analysis of ROS production*[78,131]

ROS production is monitored using the oxidation-sensitive fluorescent dye 5-(and 6)-carboxy-2′,7′-dichlorodihydro-fluorescein diacetate (carboxy-H_2DCFDA; Molecular Probes, Inc., Eugene, OR). HUVECs at a density of 1.25 × 10^3 cells/well are seeded in 24-well plates and grown for two days. After reaching 80% confluence, cells are treated in triplicate with different concentrations of nanomaterials or controls in the culture medium for 24–72 h. After incubation, the cell supernatant is removed, and the cells are washed three times with PBS and incubated with 20 μM carboxy-H_2DCFDA in PBS at 37°C for 30 min protected from light. H_2O_2 (300 μM) is used as a positive control for ROS production. The increase in fluorescence is measured at 495 nm (excitation)/529 nm (emission) by a fluorescence multi-well plate reader.

5.5 *Flow cytometry annexin V/PI and TUNEL apoptosis assays*[44,66,78,79]

The annexin V/PI apoptosis detection kit (BD Biosciences, San Diego, CA) or equivalent reagents can be used. Briefly, HUVECs are exposed to different concentrations of nanomaterials for 24 h, washed three times with PBS, and trypsinized. After centrifugation at 2,700 rpm for 10 min, the cell pellet is washed with cold PBS and resuspended in the appropriate volume of binding buffer. The cell suspension (100 μL) is then transferred into 5 mL flow tubes and incubated with 5 μL annexin V-FITC for 15 min in the dark at room temperate, followed by a short incubation with 5 μL PI. The samples are then diluted with 400 μL binding buffer and analyzed with the LSR II flow cytometer (Becton Dickinson, San Diego, CA) equipped with FACSDiva (Becton Dickinson) software. At least 10^4 cells are counted per sample. The cell population of interest is gated based on the forward and side scatter properties. The different labeling patterns in the annexin V/PI analysis identify the different cell populations: FITC-negative and PI-negative cells are viable cells, FITC-positive and PI-negative are early apoptotic cells, FITC-positive and PI-positive are late apoptotic cells or necrotic cells, and FITC-negative and PI-positive are necrotic cells. The data analysis is performed using FlowJo software (TreeStar, Inc., Ashland, OR).

The TUNEL apoptosis assay is carried out using the APO-BRDU kit following the manufacturer's instructions exactly (Becton Dickinson, San Diego, CA). In addition, cell cycle analysis is performed using ModFit software (Verity Software House, Topsham, ME).

5.6 *Flow cytometric assay of EC activation markers*[44,79]

Annexin V-FITC– and Cy-Chrome–conjugated anti-human monoclonal antibody (mAB) to CD54 (ICAM-1, clone HA58), phycoerythrin-conjugated (PE) anti-human mAB to CD142 (TF, clone HTF-1), and isotype-matched controls can be purchased from Becton Dickinson (San Diego, CA) or an equivalent source. All antibodies need to be titrated to ensure saturating concentrations. Approximately 10^5 HUVECs are resuspended in 50 μL HBSS/BSA and incubated with a saturating concentration of the mABs or annexin V. Non-labeled cells and cells incubated with either relevant isotype controls or with annexin V in the presence of 20 mM EDTA are prepared as controls. After a 20 min incubation at room temperature, the suspension of labeled cells is diluted with 2 mL HBSS/BSA and centrifuged at 300 *g* for 5 min.

The sedimented cells are resuspended in 0.5 mL HBSS/BSA and analyzed using flow cytometry.

The analysis of cell samples has been described previously.[132,133] Briefly, a FACSCalibur flow cytometer (Becton Dickinson, San Diego, CA) equipped with CELLQuest software with forward and side scatter in the linear mode is used. Alternatively, in more recent studies, cells are analyzed using the LSR II flow cytometer equipped with FACSDiva software (Becton Dickinson). The populations of intact cells are gated according to their light-scattering characteristics to exclude debris, and 10,000 gated cells are analyzed for each sample. The data analysis can be performed performed using FlowJo software (TreeStar). The total percentage (%) of $CD54^+$, $CD142^+$, and PS^+ (annexin V-binding) cells is evaluated.

5.7 *Free intracellular Ca^{2+} assay*[79]

The acute effect of nanoparticles on free intracellular Ca^{2+} concentration is studied in HUVECs, GT1-7, and PC12 cells loaded with a Ca^{2+}-sensitive probe (FURA-2AM). The changes in fluorescence in individual cells (n = 100) are monitored at 340 nm and 380 nm excitation (the rate of data capture was 170/min) using a Nikon inverted epi-fluorescence/phase microscope equipped with a low-light level integrating CCD camera with a microphotometer assembly (InCyt I/P-2 TM Imaging & Photometry System, Intracellular Imaging, Inc., Cincinnati, OH). The free intracellular Ca^{2+} concentration in real-time was calculated from the ratio of the emission detected at 510 nm from two excitation wavelengths (340 nm and 380 nm) and by comparison to a standard curve established for these settings using buffers of known free intracellular Ca^{2+} concentration with the InCyt Im2 software.

5.8 *Autophagy assay: Western blot analysis of LC3-II*[78,81,134,135]

HUVECs are plated in a six-well plate, grown to 80% confluence and either pre-treated in triplicate with 1 mM of the autophagy inhibitor 3-methyladenine (3-MA) for 2 h prior to the addition of different concentrations of nanomaterials for a further 24 h or directly treated in triplicate with nanomaterials, 100 μM chloroquine as a positive control, or negative control medium for 24 h. After washing once with ice-cold PBS, cells are lysed in an ice-cold cell extraction buffer (10 mM Tris, 100 mM NaCl, 1 mM EDTA, 1 mM EGTA, 1 mM NaF, 20 mM $Na_4P_2O_7$, 2 mM Na_3VO_4, 1% Triton X-100, 10% glycerol, 0.1% SDS, 0.5% deoxycholate) (Invitrogen Corp., Carlsbad, CA) containing 1 mM phenylmethylsulphonyl fluoride (PMSF)

(Sigma-Aldrich Corp., St. Louis, MO) and a phosphatase inhibitor cocktail (Sigma) for 30 min. After centrifuging the lysates at 15,000 rpm at 4°C for 10 min, the supernatants are collected and stored at –80°C until use. The protein concentrations of the extracts are then determined by the BCA protein assay (Pierce Protein Research Products, Thermo Scientific, Inc., Rockford, IL). Equal amounts of lysate proteins (30 μg) are loaded onto 4–20% Tris-Glycine gels and electrophoretically transferred to nitrocellulose membranes (Invitrogen). After blocking with a Block Buffer (Pierce) containing 0.01% Tween-20 for 1 h at room temperature, the membrane is incubated with primary anti-LC3B rabbit polyclonal antibody (1:1,000, Invitrogen) overnight at 4°C, washed with Tris buffered saline (TBS), and incubated with alkaline phosphatase-conjugated goat anti-rabbit secondary antibody (1:5,000; Jackson ImmunoResearch Laboratories, Inc., West Grove, PA) for 1 h at 37°C. The membrane is washed again and developed using BCIP/NBT Single Reagent Alkaline Phosphatase Substrate (Invitrogen). Following an incubation of 10–15 min, the reaction is terminated by distilled water rinses. The intensity of each band has to be normalized to the β-actin loading control (rabbit polyclonal anti-β-actin; 1:1,000; Abcam, Cambridge, MA). The band densities are calculated using ImageJ software (NIH, Bethesda, MD).

5.9 *Autophagy assay: Immunofluorescence microscopy of LC3-II*[81,136]

HUVEC cells are seeded in a two-well glass chamber and after attachment, are treated with different concentrations of nanomaterials with or without the autophagy inhibitor 3-MA. For nanomaterial and 3-MA co-treatment, cells are pre-treated with 3-MA for 2 h. The final 3-MA concentration is 1 mM. Chloroquine (100 μM) is used as a positive control for autophagy. After incubation for 24 h, the cells are fixed with 2% paraformaldehyde and permeabilized with 0.1% Triton X-100. The permeabilization buffer is then removed and cells are washed three times with PBS. The LC3B rabbit polyclonal primary antibody (Cell Signaling Technology, Inc., Danvers, MA) is added to cells at a final concentration of 0.5 μg/mL, and the cells are incubated for 1 h at room temperature. After additional washes and an incubation for 30 min with the anti-rabbit secondary antibody, the cells are washed and stained for actin (Phalloidin 555, Invitrogen) and nuclei (TO-PRO, Invitrogen). The cells are mounted with gold antifade mounting media (Invitrogen) for laser confocal microscopy performed with a Zeiss 710 LSM 710 NLO with a Plan Apochromat 60× oil objective.

5.10 *Premo Autophagy Sensor*

To verify the results obtained with the standard immunofluorescent methods, we seed 40,000 cells in two-well glass chamber and analyze HUVEC autophagy with the Premo Autophagy Sensor (Invitrogen) according to the manufacturer's instructions.

The Premo Autophagy Sensor is another approach for evaluating autophagy in cells. It combines the selectivity of an LC3B-fluorescent protein chimera with the transduction efficiency of the BacMam (insect baculovirus with a mammalian promoter) 2.0 technology. Following an overnight incubation of HUVECs with LC3B-FP (MOI of 50), we follow the standard fixation and permeabilization procedures for the immunofluorescent analysis of autophagy in cells.

5.11 *LysoTracker Red dye uptake assay*[135,137]

LysoTracker Red DND-99 is a cationic fluorescent dye that preferentially accumulates in acidic lysosomal compartments. The amount of dye taken up by cells in culture can be used as an indicator of lysosome content and an indirect measure of autophagolysosomes. CellTracker Green is deacetylated in viable cells to a thiol-reactive dye that remains in the cytosol and is used to normalize the LysoTracker signal to viable cells. HUVECs are plated in six-well plates and processed according to Stern *et al.*[135,137] Briefly, following cell attachment, cells are treated with 20 or 100 μg/mL CNTs. The treated cells are then washed and stained with 100 μL of a 50 nM LysoTracker Red/10 μM CellTracker Green co-staining solution prepared in phenol-free RPMI-1640 for 1 h at 37°C. Following dye uptake, the co-staining solution was removed, the plates are rinsed with RPMI, and 200 μL phenol-free RPMI are added to each well. LysoTracker Red fluorescence (λ_{max} = 590 nm) and CellTracker Green fluorescence (λ_{max} = 517 nm) are measured using a microplate reader. LysoTracker Red uptake for treated cells is expressed as the ratio percent of the control normalized to CellTracker Green (percent control LysoTracker Red fluorescence/percent control CellTracker Green fluorescence).

6. Conclusion

In vitro testing of the interactions of nanomaterials with ECs is an important part of vascular biocompatibility screening and should be performed on all nanomaterials intended for intravascular use, as well as with nanomaterials with an identified risk of environmental or occupational exposure. Feasible assays generally used for other cell types are available for application on different types of cultured ECs. The development of a standard panel of assays for the evaluation of effects on nanomaterials on ECs is needed.

The oxidative stress, pro-inflammatory activation, induction of apoptosis of ECs, and/or inhibition of proliferation, migration, and tube formation are the most common adverse effects of nanomaterials on ECs. More comprehensive studies are needed to make generalized conlusions on material, size, surface, and structure/function (i.e., activity) relationships with regard to endothelial adverse effects.

References

1. McGuigan AP, Sefton MV. The influence of biomaterials on endothelial cell thrombogenicity. *Biomaterials* 2007;28:2547–2571.
2. ANSI/AAMI/ISO. Biological evaluation of medical devices — Part 4: Selection of tests for interaction with blood. American National Standards Institute, Vol. 10993-4, 2002.
3. Dobrovolskaia MA, Aggarwal P, Hall JB, McNeil SE. Preclinical studies to understand nanoparticle interaction with the immune system and its potential effects on nanoparticle biodistribution. *Mol Pharm* 2008;5:487–495.
4. Owens AP III, Mackman N. Tissue factor and thrombosis: The clot starts here. *Thromb Haemost* 2010;104:432–439.
5. Aird WC. The role of the endothelium in severe sepsis and multiple organ dysfunction syndrome. *Blood* 2003;101:3765–3777.
6. Aird WC. Endothelium as an organ system. *Crit Care Med* 2004;32:S271–279.
7. Cines DB, Pollak ES, Buck CA, *et al.* Endothelial cells in physiology and in the pathophysiology of vascular disorders. *Blood* 1998;91:3527–3561.
8. Nag S, Kapadia A, Stewart DJ. Review: Molecular pathogenesis of blood-brain barrier breakdown in acute brain injury. *Neuropathol Appl Neurobiol* 2011;37:3–23.
9. Pober JS, Sessa WC. Evolving functions of endothelial cells in inflammation. *Nat Rev Immunol* 2007;7:803–815.
10. Rabelink TJ, de Boer HC, van Zonneveld AJ. Endothelial activation and circulating markers of endothelial activation in kidney disease. *Nat Rev Nephrol* 2010;6:404–414.
11. Vanhoutte PM, Shimokawa H, Tang EH, Feletou M. Endothelial dysfunction and vascular disease. *Acta Physiol (Oxf)* 2009;196:193–222.
12. Zhang J, Defelice AF, Hanig JP, Colatsky T. Biomarkers of endothelial cell activation serve as potential surrogate markers for drug-induced vascular injury. *Toxicol Pathol* 2010;38:856–871.
13. Zhang J, Hanig JP, de Felice AF. Biomarkers of endothelial cell activation: Candidate markers for drug-induced vasculitis in patients or drug-induced vascular injury in animals. *Vascul Pharmacol* 2012;56:14–25.
14. Ramcharan KS, Lip GY, Stonelake PS, Blann AD. The endotheliome: A new concept in vascular biology. *Thromb Res* 2011;128:1–7.
15. Aird WC. Phenotypic heterogeneity of the endothelium: II. Representative vascular beds. *Circ Res* 2007;100:174–190.

16. Aird WC. Phenotypic heterogeneity of the endothelium: I. Structure, function, and mechanisms. *Circ Res* 2007;100:158–173.
17. Simionescu M, Antohe F. Functional ultrastructure of the vascular endothelium: Changes in various pathologies. In: Moncada S, Higgs A (eds.), *The Vascular Endothelium I*, Vol. 176/I, Springer, 2006, pp. 41–69.
18. Simak J. Nanotoxicity in blood: Effects of engineered nanomaterials on platelets. In: Sahu S, Casciano D (eds.), *Nanotoxicity: From In Vivo and In Vitro Models to Health Risks*, John Wiley & Sons, 2009, pp. 191–225.
19. Medzhitov R. Origin and physiological roles of inflammation. *Nature* 2008; 454:428–435.
20. Nathan C. Neutrophils and immunity: Challenges and opportunities. *Nat Rev Immunol* 2006;6:173–182.
21. Nathan C. Points of control in inflammation. *Nature* 2002;420:846–852.
22. Barton GM. A calculated response: Control of inflammation by the innate immune system. *J Clin Invest* 2008;118:413–420.
23. Serhan CN, Chiang N, Van Dyke TE. Resolving inflammation: Dual anti-inflammatory and pro-resolution lipid mediators. *Nat Rev Immunol* 2008;8:349–361.
24. Deanfield JE, Halcox JP, Rabelink TJ. Endothelial function and dysfunction: Testing and clinical relevance. *Circulation* 2007;115:1285–1295.
25. Feletou M, Kohler R, Vanhoutte PM. Nitric oxide: Orchestrator of endothelium-dependent responses. *Ann Med* 2011; Early online: 1–23.
26. Edwards G, Feletou M, Weston AH. Endothelium-derived hyperpolarising factors and associated pathways: A synopsis. *Pflugers Arch* 2011;459:863–879.
27. Feletou M, Huang Y, Vanhoutte PM. Endothelium-mediated control of vascular tone: COX-1 and COX-2 products. *Br J Pharmacol* 2011;164:894–912.
28. Arnout J, Hoylaerts MF, Lijnen HR. Haemostasis. In: Moncada S, Higgs A (eds.), *The Vascular Endothelium II*, Vol. 176/II, Springer, 2006, pp. 1–41.
29. Hunt BJ, Jurd KM. Endothelial cell activation. A central pathophysiological process. *Br Med J* 1998;316:1328–1329.
30. Blann AD. Endothelial cell activation, injury, damage and dysfunction: Separate entities or mutual terms? *Blood Coagul Fibrinolysis* 2000;11:623–630.
31. Valentijn KM, Sadler JE, Valentijn JA, Voorberg J, Eikenboom J. Functional architecture of Weibel-Palade bodies. *Blood* 2011;117:5033–5043.
32. Bach FH, Hancock WW, Ferran C. Protective genes expressed in endothelial cells: A regulatory response to injury. *Immunol Today* 1997;18:483–486.
33. Charreau B. Molecular regulation of endothelial cell activation: Novel mechanisms and emerging targets. *Curr Opin Organ Transplant* 2011;16:207–213.
34. Pober JS. Immunobiology of human vascular endothelium. *Immunol Res* 1999;19: 225–232.
35. Aird WC. Endothelium in health and disease. *Pharmacol Rep* 2008;60:139–143.
36. Matsuda N, Hattori Y. Systemic inflammatory response syndrome (SIRS): Molecular pathophysiology and gene therapy. *J Pharmacol Sci* 2006;101:189–198.
37. van Eeden SF, Hogg JC. Systemic inflammatory response induced by particulate matter air pollution: The importance of bone-marrow stimulation. *J Toxicol Environ Health A* 2002;65:1597–1613.
38. Bone RC. Toward an epidemiology and natural history of SIRS (systemic inflammatory response syndrome). *JAMA* 1992;268:3452–3455.

39. Bone RC. Toward a theory regarding the pathogenesis of the systemic inflammatory response syndrome: What we do and do not know about cytokine regulation. *Crit Care Med* 1996;24:163–172.
40. Davies MG, Hagen PO. Systemic inflammatory response syndrome. *Br J Surg* 1997;84: 920–935.
41. Pober JS, Min W. Endothelial cell dysfunction, injury and death. In: Moncada S, Higgs A (eds.), *The Vascular Endothelium II*, Vol. 176/II, Springer, 2006, pp. 135–156.
42. Bartczak D, Sanchez-Elsner T, Louafi F, Millar TM, Kanaras AG. Receptor-mediated interactions between colloidal gold nanoparticles and human umbilical vein endothelial cells. *Small* 2011;7:388–394.
43. Bauer AT, Strozyk EA, Gorzelanny C, *et al.* Cytotoxicity of silica nanoparticles through exocytosis of von Willebrand factor and necrotic cell death in primary human endothelial cells. *Biomaterials* 2011;32:8385–8393.
44. Liu X, Sun J. Endothelial cells dysfunction induced by silica nanoparticles through oxidative stress via JNK/P53 and NF-kappaB pathways. *Biomaterials* 2010;31:8198–8209.
45. Albini A, Mussi V, Parodi A, *et al.* Interactions of single-wall carbon nanotubes with endothelial cells. *Nanomedicine* 2010;6:277–288.
46. Maeda R, Noiri E, Isobe H, *et al.* A water-soluble fullerene vesicle alleviates angiotensin II-induced oxidative stress in human umbilical venous endothelial cells. *Hypertens Res* 2008;31:141–151.
47. Centis V, Vermette P. Physico-chemical properties and cytotoxicity assessment of PEG-modified liposomes containing human hemoglobin. *Colloids Surf B Biointerfaces* 2008; 65:239–246.
48. Chaudhuri P, Harfouche R, Soni S, Hentschel DM, Sengupta S. Shape effect of carbon nanovectors on angiogenesis. *ACS Nano* 2010;4:574–582.
49. Ge S, Wang G, Shen Y, *et al.* Cytotoxic effects of MgO nanoparticles on human umbilical vein endothelial cells *in vitro*. *IET Nanobiotechnol* 2011;5:36.
50. Guo YY, Zhang J, Zheng YF, Yang J, Zhu XQ. Cytotoxic and genotoxic effects of multiwall carbon nanotubes on human umbilical vein endothelial cells *in vitro*. *Mutat Res* 2011;721:184–191.
51. Gojova A, Guo B, Kota RS, Rutledge JC, Kennedy IM, Barakat AI. Induction of inflammation in vascular endothelial cells by metal oxide nanoparticles: Effect of particle composition. *Environ Health Perspect* 2007;115:403–409.
52. Nishikawa T, Iwakiri N, Kaneko Y, *et al.* Nitric oxide release in human aortic endothelial cells mediated by delivery of amphiphilic polysiloxane nanoparticles to caveolae. *Biomacromolecules* 2009;10:2074–2085.
53. Apopa PL, Qian Y, Shao R, *et al.* Iron oxide nanoparticles induce human microvascular endothelial cell permeability through reactive oxygen species production and microtubule remodeling. *Part Fibre Toxicol* 2009;6:1.
54. Sun J, Wang S, Zhao D, Hun FH, Weng L, Liu H. Cytotoxicity, permeability, and inflammation of metal oxide nanoparticles in human cardiac microvascular endothelial cells: Cytotoxicity, permeability, and inflammation of metal oxide nanoparticles. *Cell Biol Toxicol* 2011;27:333–342.
55. Peters K, Unger RE, Kirkpatrick CJ, Gatti AM, Monari E. Effects of nano-scaled particles on endothelial cell function *in vitro*: Studies on viability, proliferation and inflammation. *J Mater Sci Mater Med* 2004;15:321–325.

56. Kim JH, Kim MH, Jo DH, Yu YS, Lee TG. The inhibition of retinal neovascularization by gold nanoparticles via suppression of VEGFR-2 activation. *Biomaterials* 2011;32: 1865–1871.
57. Zhu MT, Wang Y, Feng WY, *et al.* Oxidative stress and apoptosis induced by iron oxide nanoparticles in cultured human umbilical endothelial cells. *J Nanosci Nanotechnol* 2010;10:8584–8590.
58. Lison D, Thomassen LC, Rabolli V, *et al.* Nominal and effective dosimetry of silica nanoparticles in cytotoxicity assays. *Toxicol Sci* 2008;104:155–162.
59. Kasper J, Hermanns MI, Bantz C, *et al.* Inflammatory and cytotoxic responses of an alveolar-capillary coculture model to silica nanoparticles: Comparison with conventional monocultures. *Part Fibre Toxicol* 2011;8:6.
60. Brammer KS, Oh S, Gallagher JO, Jin S. Enhanced cellular mobility guided by TiO2 nanotube surfaces. *Nano Lett* 2008;8:786–793.
61. Gurunathan S, Lee KJ, Kalishwaralal K, Sheikpranbabu S, Vaidyanathan R, Eom SH. Antiangiogenic properties of silver nanoparticles. *Biomaterials* 2009;30:6341–6350.
62. Wang Z, Tiruppathi C, Minshall RD, Malik AB. Size and dynamics of caveolae studied using nanoparticles in living endothelial cells. *ACS Nano* 2009;3:4110–4116.
63. Buyukhatipoglu K, Clyne AM. Superparamagnetic iron oxide nanoparticles change endothelial cell morphology and mechanics via reactive oxygen species formation. *J Biomed Mater Res A* 2011;96:186–195.
64. Lu J, Khang D, Webster TJ. Greater endothelial cell responses on submicron and nanometer rough titanium surfaces. *J Biomed Mater Res A* 2010;94:1042–1049.
65. Rosas-Hernandez H, Jimenez-Badillo S, Martinez-Cuevas PP, *et al.* Effects of 45-nm silver nanoparticles on coronary endothelial cells and isolated rat aortic rings. *Toxicol Lett* 2009;191:305–313.
66. Lao F, Chen L, Li W, *et al.* Fullerene nanoparticles selectively enter oxidation-damaged cerebral microvessel endothelial cells and inhibit JNK-related apoptosis. *ACS Nano* 2009;3:3358–3368.
67. Trickler WJ, Lantz SM, Murdock RC, *et al.* Silver nanoparticle induced blood-brain barrier inflammation and increased permeability in primary rat brain microvessel endothelial cells. *Toxicol Sci* 2010;118:160–170.
68. Lao F, Li W, Han D, *et al.* Fullerene derivatives protect endothelial cells against NO-induced damage. *Nanotechnology* 2009;20:225103.
69. Yu M, Mo Y, Wan R, Chien S, Zhang X, Zhang Q. Regulation of plasminogen activator inhibitor-1 expression in endothelial cells with exposure to metal nanoparticles. *Toxicol Lett* 2010;195:82–89.
70. Akbar N, Mohamed T, Whitehead D, Azzawi M. Biocompatibility of amorphous silica nanoparticles: Size and charge effect on vascular function, *in vitro*. *Biotechnol Appl Biochem* 2011;58:353–362.
71. Huang SS, Mashino T, Mochizuki M, *et al.* Effect of hexasulfobutylated C(60) on the isolated aortic ring of guinea pig. *Pharmacology* 2002;64:91–97.
72. Watson CA, Camera-Benson L, Palmer-Crocker R, Pober JS. Variability among human umbilical vein endothelial cultures. *Science* 1995;268:447–448.
73. Meng J, Han Z, Kong H, *et al.* Electrospun aligned nanofibrous composite of MWCNT/polyurethane to enhance vascular endothelium cells proliferation and function. *J Biomed Mater Res A* 2010;95:312–320.

74. Berridge MV, Herst PM, Tan AS. Tetrazolium dyes as tools in cell biology: New insights into their cellular reduction. *Biotechnol Annu Rev* 2005;11:127–152.
75. Trickler WJ, Lantz SM, Murdock RC, *et al.* Brain microvessel endothelial cells responses to gold nanoparticles: *In vitro* pro-inflammatory mediators and permeability. *Nanotoxicology* 2011;5:479–492.
76. Worle-Knirsch JM, Pulskamp K, Krug HF. Oops they did it again! Carbon nanotubes hoax scientists in viability assays. *Nano Lett* 2006;6:1261–1268.
77. Ge C, Du J, Zhao L, *et al.* Binding of blood proteins to carbon nanotubes reduces cytotoxicity. *Proc Natl Acad Sci U S A* 2011;108:16968–16973.
78. Cheng WW, Lin ZQ, Wei BF, *et al.* Single-walled carbon nanotube induction of rat aortic endothelial cell apoptosis: Reactive oxygen species are involved in the mitochondrial pathway. *Int J Biochem Cell Biol* 2011;43:564–572.
79. Gelderman MP, Simakova O, Clogston JD, *et al.* Adverse effects of fullerenes on endothelial cells: Fullerenol $C_{60}(OH)_{24}$ induced tissue factor and ICAM-I membrane expression and apoptosis *in vitro*. *Int J Nanomed* 2008;3:59–68.
80 Yamawaki H, Iwai N. Cytotoxicity of water-soluble fullerene in vascular endothelial cells. *Am J Physiol Cell Physiol* 2006;290:C1495–1502.
81. Xie Y, You SJ, Zhang YL, *et al.* Protective role of autophagy in AGE-induced early injury of human vascular endothelial cells. *Mol Med Report* 2011;4:459–464.
82. Patschan S, Chen J, Polotskaia A, *et al.* Lipid mediators of autophagy in stress-induced premature senescence of endothelial cells. *Am J Physiol Heart Circ Physiol* 2008;294: H1119–1129.
83. Simak J, Gelderman MP. Cell membrane microparticles in blood and blood products: Potentially pathogenic agents and diagnostic markers. *Transfus Med Rev* 2006;20: 1–26.
84. Peng L, Barczak AJ, Barbeau RA, *et al.* Whole genome expression analysis reveals differential effects of TiO2 nanotubes on vascular cells. *Nano Lett* 2010;10:143–148.
85. Hung HS, Wu CC, Chien S, Hsu SH. The behavior of endothelial cells on polyurethane nanocomposites and the associated signaling pathways. *Biomaterials* 2009;30: 1502–1511.
86. Kalishwaralal K, Banumathi E, Ram Kumar Pandian S, *et al.* Silver nanoparticles inhibit VEGF induced cell proliferation and migration in bovine retinal endothelial cells. *Colloids Surf B Biointerfaces* 2009;73:51–57.
87. Sheikpranbabu S, Kalishwaralal K, Venkataraman D, Eom SH, Park J, Gurunathan S. Silver nanoparticles inhibit VEGF-and IL-1beta-induced vascular permeability via Src dependent pathway in porcine retinal endothelial cells. *J Nanobiotechnol* 2009;7:8.
88. Tsou TC, Yeh SC, Tsai FY, *et al.* Zinc oxide particles induce inflammatory responses in vascular endothelial cells via NF-kappaB signaling. *J Hazard Mater* 2010;183: 182–188.
89. Muro S, Koval M, Muzykantov V. Endothelial endocytic pathways: Gates for vascular drug delivery. *Curr Vasc Pharmacol* 2004;2:281–299.
90. Wang M, Thanou M. Targeting nanoparticles to cancer. *Pharmacol Res* 2010;62: 90–99.
91. Saravanakumar G, Kim K, Park JH, Rhee K, Kwon IC. Current status of nanoparticle-based imaging agents for early diagnosis of cancer and atherosclerosis. *J Biomed Nanotechnol* 2009;5:20–35.

92. Zhu MT, Wang B, Wang Y, *et al.* Endothelial dysfunction and inflammation induced by iron oxide nanoparticle exposure: Risk factors for early atherosclerosis. *Toxicol Lett* 2011;203:162–171.
93. Wu X, Tan Y, Mao H, Zhang M. Toxic effects of iron oxide nanoparticles on human umbilical vein endothelial cells. *Int J Nanomed* 2010;5:385–399.
94. Cengelli F, Maysinger D, Tschudi-Monnet F, *et al.* Interaction of functionalized super-paramagnetic iron oxide nanoparticles with brain structures. *J Pharmacol Exp Ther* 2006; 318:108–116.
95. Soenen SJ, Himmelreich U, Nuytten N, De Cuyper M. Cytotoxic effects of iron oxide nanoparticles and implications for safety in cell labelling. *Biomaterials* 2011;32: 195–205.
96. Kalishwaralal K, Sheikpranbabu S, Barathmanikanth S, Haribalaganesh R, Ramkumarpandian S, Gurunathan S. Gold nanoparticles inhibit vascular endothelial growth factor-induced angiogenesis and vascular permeability via Src dependent pathway in retinal endothelial cells. *Angiogenesis* 2011;14:29–45.
97. Mao Z, Wang B, Ma L, Gao C, Shen J. The influence of polycaprolactone coating on the internalization and cytotoxicity of gold nanoparticles. *Nanomedicine* 2007;3:215–223.
98. Ragaseema VM, Unnikrishnan S, Krishnan VK, Krishnan LK. The antithrombote and antimicrobial properties of PEG-projected silver nanoparticle-coated surfaces. *Biomaterials* 2012;33:3083–3092.
99. Liu HL, Dai SA, Fu KY, Hsu SH. Antibacterial properties of silver nanoparticles in three different sizes and their nanocomposites with a new waterborne polyurethane. *Int J Nanomed* 2010;5:1017–1028.
100. Madhavan RV, Rosemary MJ, Nandkumar MA, Krishnan KV, Krishnan LK. Silver nanoparticle impregnated poly (varepsilon-Caprolactone) scaffolds: Optimization of antimicrobial and noncytotoxic concentrations. *Tissue Eng Part A* 2010;17:439–449.
101. Peng L, Eltgroth ML, LaTempa TJ, Grimes CA, Desai TA. The effect of TiO2 nanotubes on endothelial function and smooth muscle proliferation. *Biomaterials* 2009;30: 1268–1272.
102. Chen W, Liu P, Zhang A, Ren J, Xu LX. Quantification of quantum dots in HUVECs by confocal laser scanning microscopy. *Conf Proc IEEE Eng Med Biol Soc* 2006;1: 1478–1481.
103. Yan M, Zhang Y, Xu K, Fu T, Qin H, Zheng X. An *in vitro* study of vascular endothelial toxicity of CdTe quantum dots. *Toxicology* 2011;282:94–103.
104. Wang L, Zhang J, Zheng Y, Yang J, Zhang Q, Zhu X. Bioeffects of CdTe quantum dots on human umbilical vein endothelial cells. *J Nanosci Nanotechnol* 2010;10: 8591–8596.
105. Gojova A, Lee JT, Jung HS, Guo B, Barakat AI, Kennedy IM. Effect of cerium oxide nanoparticles on inflammation in vascular endothelial cells. *Inhal Toxicol* 2009;21(Suppl 1):123–130.
106. Oesterling E, Chopra N, Gavalas V, *et al.* Alumina nanoparticles induce expression of endothelial cell adhesion molecules. *Toxicol Lett* 2008;178:160–166.
107. Walker VG, Li Z, Hulderman T, Schwegler-Berry D, Kashon ML, Simeonova PP. Potential *in vitro* effects of carbon nanotubes on human aortic endothelial cells. *Toxicol Appl Pharmacol* 2009;236:319–328.

108. Zhiqing L, Zhuge X, Fuhuan C, *et al.* ICAM-1 and VCAM-1 expression in rat aortic endothelial cells after single-walled carbon nanotube exposure. *J Nanosci Nanotechnol* 2010;10:8562–8574.
109. Han Z, Kong H, Meng J, Wang C, Xie S, Xu H. Electrospun aligned nanofibrous scaffold of carbon nanotubes-polyurethane composite for endothelial cells. *J Nanosci Nanotechnol* 2009;9:1400–1402.
110. Bianco A, Del Gaudio C, Baiguera S, *et al.* Microstructure and cytocompatibility of electrospun nanocomposites based on poly(epsilon-caprolactone) and carbon nanostructures. *Int J Artif Organs* 2010;33:271–282.
111. Tan W, Twomey J, Guo D, Madhavan K, Li M. Evaluation of nanostructural, mechanical, and biological properties of collagen-nanotube composites. *IEEE Trans Nanobiosci* 2010;9:111–120.
112. Satoh M, Takayanagi I. Pharmacological studies on fullerene (C60), a novel carbon allotrope, and its derivatives. *J Pharmacol Sci* 2006;100:513–518.
113. Yin JJ, Lao F, Meng J, *et al.* Inhibition of tumor growth by endohedral metallofullerenol nanoparticles optimized as reactive oxygen species scavenger. *Mol Pharmacol* 2008;74: 1132–1140.
114. Napierska D, Thomassen LC, Rabolli V, *et al.* Size-dependent cytotoxicity of monodisperse silica nanoparticles in human endothelial cells. *Small* 2009;5:846–853.
115. Tavano R, Segat D, Reddi E, *et al.* Procoagulant properties of bare and highly PEGylated vinyl-modified silica nanoparticles. *Nanomedicine (Lond)* 2010;5:881–896.
116. Lee J, Chu BH, Chen KH, Ren F, Lele TP. Randomly oriented, upright SiO_2 coated nanorods for reduced adhesion of mammalian cells. *Biomaterials* 2009;30:4488–4493.
117. Stasko NA, Johnson CB, Schoenfisch MH, Johnson TA, Holmuhamedov EL. Cytotoxicity of polypropylenimine dendrimer conjugates on cultured endothelial cells. *Biomacromolecules* 2007;8:3853–3859.
118. Nam HY, Nam K, Hahn HJ, *et al.* Biodegradable PAMAM ester for enhanced transfection efficiency with low cytotoxicity. *Biomaterials* 2009;30:665–673.
119. Yang H, Lopina ST, DiPersio LP, Schmidt SP. Stealth dendrimers for drug delivery: Correlation between PEGylation, cytocompatibility, and drug payload. *J Mater Sci Mater Med* 2008;19:1991–1997.
120. Dernedde J, Rausch A, Weinhart M, *et al.* Dendritic polyglycerol sulfates as multivalent inhibitors of inflammation. *Proc Natl Acad Sci USA* 2010;107:19679–19684.
121. Ehrenberg MS, Friedman AE, Finkelstein JN, Oberdorster G, McGrath JL. The influence of protein adsorption on nanoparticle association with cultured endothelial cells. *Biomaterials* 2009;30:603–610.
122. Xia T, Kovochich M, Liong M, Zink JI, Nel AE. Cationic polystyrene nanosphere toxicity depends on cell-specific endocytic and mitochondrial injury pathways. *ACS Nano* 2008;2:85–96.
123. Hung HS, Hsu SH. The response of endothelial cells to polymer surface composed of nanometric micelles. *N Biotechnol* 2009;25:235–243.
124. Abu Lila AS, Ishida T, Kiwada H. Recent advances in tumor vasculature targeting using liposomal drug delivery systems. *Expert Opin Drug Delivery* 2009;6:1297–1309.
125. Abu Lila AS, Ishida T, Kiwada H. Targeting anticancer drugs to tumor vasculature using cationic liposomes. *Pharm Res* 2010;27:1171–1183.
126. Gabizon AA, Shmeeda H, Zalipsky S. Pros and cons of the liposome platform in cancer drug targeting. *J Liposome Res* 2006;16:175–183.

127. Dabbas S, Kaushik RR, Dandamudi S, Kuesters GM, Campbell RB. Importance of the liposomal cationic lipid content and type in tumor vascular targeting: Physicochemical characterization and *in vitro* studies using human primary and transformed endothelial cells. *Endothelium* 2008;15:189–201.
128. Fine E, Zhang L, Fenniri H, Webster TJ. Enhanced endothelial cell functions on rosette nanotube-coated titanium vascular stents. *Int J Nanomed* 2009;4:91–97.
129. Pezzatini S, Morbidelli L, Solito R, *et al.* Nanostructured HA crystals up-regulate FGF-2 expression and activity in microvascular endothelium promoting angiogenesis. *Bone* 2007;41:523–534.
130. van Kasteren SI, Campbell SJ, Serres S, Anthony DC, Sibson NR, Davis BG. Glyconanoparticles allow pre-symptomatic *in vivo* imaging of brain disease. *Proc Natl Acad Sci U S A* 2009;106:18–23.
131. Sharma CS, Sarkar S, Periyakaruppan A, *et al.* Single-walled carbon nanotubes induces oxidative stress in rat lung epithelial cells. *J Nanosci Nanotechnol* 2007;7:2466–2472.
132. Simak J, Holada K, D'Agnillo F, Janota J, Vostal JG. Cellular prion protein is expressed on endothelial cells and is released during apoptosis on membrane microparticles found in human plasma. *Transfusion* 2002;42:334–342.
133. Simak J, Holada K, Vostal JG. Release of annexin V-binding membrane microparticles from cultured human umbilical vein endothelial cells after treatment with camptothecin. *BMC Cell Biol* 2002;3:11.
134. Ren Y, Huang F, Liu Y, Yang Y, Jiang Q, Xu C. Autophagy inhibition through PI3K/Akt increases apoptosis by sodium selenite in NB4 cells. *BMB Rep* 2009;42:599–604.
135. Johnson-Lyles DN, Peifley K, Lockett S, *et al.* Fullerenol cytotoxicity in kidney cells is associated with cytoskeleton disruption, autophagic vacuole accumulation, and mitochondrial dysfunction. *Toxicol Appl Pharmacol* 2010;248:249–258.
136. Lee CM, Huang ST, Huang SH, *et al.* C60 fullerene-pentoxifylline dyad nanoparticles enhance autophagy to avoid cytotoxic effects caused by the beta-amyloid peptide. *Nanomedicine* 2011;7:107–114.
137. Stern ST, Zolnik BS, McLeland CB, Clogston J, Zheng J, McNeil SE. Induction of autophagy in porcine kidney cells by quantum dots: A common cellular response to nanomaterials? *Toxicol Sci* 2008;106:140–152.

Chapter 9

The Effects of Engineered Nanomaterials on the Plasma Coagulation System

Jan Simak

Center for Biologics Evaluation and Research
Food and Drug Administration
1401 Rockville Pike, HFM-335
Rockville, MD 20852-1448, USA
jan.simak@fda.hhs.gov

The plasma coagulation system (PCS) consists of plasma proteins and other factors that, together with platelets and vascular endothelial cells, maintain hemocoagulation balance. Under physiological conditions, these systems prevent blood clotting and, in cases of vascular injury, facilitate hemostasis to prevent blood loss. The dysregulation of hemocoagulation may lead to life-threatening thrombotic and/or bleeding pathologies. Since 1) various engineered nanomaterials are being designed for biomedical applications that will come into contact with the blood and 2) other nanomaterials may reach the circulation as a result of occupational, environmental, or other types of exposure, it is necessary to evaluate the effects of engineered nanomaterials on the PCS. In the introduction of this chapter, the components and mechanisms of the kallikrein–kinin system (KKS), the PCS, fibrinolysis, and the pathophysiology of the possible adverse effects of nanomaterials on these systems are briefly described. Next, the methods that are used for the *in vitro* evaluation of the effects of nanomaterials on the KKS, PCS, and fibrinolysis are reviewed. Screenings for the effects of nanomaterials are primarily based on their preincubation with citrated platelet-poor plasma (PPP). The PPP is subsequently subjected to clotting tests, including measurements of the activated partial thromboplastin time (APTT), prothrombin time (PT), thrombin time (TT), and recalcification time. In certain studies, plasmas that are deficient in specific plasma coagulation factors are used for comparison purposes. The activities of certain specific coagulation factors can be assayed photometrically using chromogenic or fluorogenic substrates, while antigens can be detected using ELISAs. Thromboelastography is another technique that is used for the complex analysis of the effects of nanomaterials on blood clotting and fibrinolysis. Gravimetric *in vitro* thrombolytic assays have also been used to this end. Lastly, a summary of the published results regarding the *in vitro* effects of various nanomaterials on the KKS, the PCS, and fibrinolysis is presented.

Disclaimer:

The findings and conclusions in this article have not been formally disseminated by the Food and Drug Administration and should not be construed to represent any agency determination or policy.

1. Introduction

The evaluation of the effects of nanomaterials on the plasma coagulation system (PCS) is critical for the assessment of these materials' thrombogenicities and other possible adverse effects on hemostasis. After decades of material hemocompatibility research, thrombosis remains a significant barrier to the development and implementation of advanced blood-contacting medical devices.[1] Thrombogenicity is defined as the ability of a material to induce or promote the formation of thromboemboli. This definition of thrombogenicity is based on *in vivo* adverse effects and does not provide complete information as to which type of material activities should be studied *in vitro*.[2] The general categories of the recommended tests for different types of blood-contacting medical devices are specified in ISO 10993-4.[3] In principle, thrombotic and hemorrhagic symptoms are caused by a dysregulation of the hemostatic balance of the organism. This balance is maintained by an interplay between the vascular endothelium, platelets, and the PCS.[4] As explained in the detail in the chapter that deals with endothelial cells (ECs), the hemostatic balance is also influenced by the systems that generate inflammatory responses. Thus, hemostasis is an integral part of inflammation and must be understood in this context. The dysregulation of hemostasis leads to a disorder in which either thrombotic or hemorrhagic symptoms prevail. It is important to note that in certain types of hemocoagulation disorders, thrombosis and bleeding can be observed simultaneously, i.e., thromboembolic symptoms may be associated with a failure of post-traumatic hemostasis. Therefore, an understanding of the mechanisms that underlie the dysregulation of the hemostatic balance is essential.

2. The Pathophysiology of the Possible Adverse Effects of Nanomaterials on the Kallikrein–Kinin System, the Plasma Coagulation System, and Fibrinolysis

2.1 *The kallikrein–kinin system*

The plasma kallikrein–kinin system (KKS) consists of two serine protease zymogens, factor XII (FXII) and prekallikrein (PK), and a substrate/cofactor, high molecular weight kininogen (HK). The KKS is reviewed in several of the references listed at the end of this chapter.[1,5–7] These proteins influence the results of surface-activated *in vitro* coagulation assays, including the activated partial thromboplastin time (APTT) assay, and are referred to as "contact factors." This nomenclature arose because, for many years, the only known pathway of KKS activation was FXII autoactivation on contact surfaces. *In vivo*, FXII activation occurs where the thrombus develops and contributes to the size of the thrombus. The FXI/HK complex is also present at the contact surface, which leads to the activation of FXI by FXIIa. FXI, however, can also be activated by thrombin in an FXII-independent manner. Therefore, the KKS is not essential to the intrinsic pathway of thrombin generation, and a deficient KKS does not cause bleeding conditions. Various artificial negatively charged surfaces, including glass, kaolin, and celite have been demonstrated to promote FXII autoactivation, as have several (patho) physiological substances, including cartilage, skin, heparins, chondroitin sulfates, endotoxin, bacteria, sodium urate crystals, and amyloid β-protein.[6] Material biocompatibility studies indicate that anionic hydrophilic surfaces are the most efficient with respect to plasma contact activation, but FXII has also been demonstrated to be activated by hydrophobic surfaces. The surface activation of FXII is moderated by the protein composition of the fluid phase via protein adsorption competition.[1] FXII autoactivation leads to PK activation, initiating a positive feedback loop of both FXII and PK activation. In addition to their activation of the KKS at contact surfaces, HK, FXII, and PK bind to ECs, platelets, and granulocytes, where the KKS is constitutively active in an FXII- and contact-independent manner.[5,6] Endothelial prolylcarboxypeptidase has been identified as an FXII-independent PK activator. Moreover, under physiological conditions, the binding of PK/HK to ECs predominates over the binding of FXI/HK. Thus, the constitutively activated KKS on ECs does not likely promote intrinsic thrombin generation. Kallikrein (KAL) releases the nonapeptide bradykinin (BK) from HK. BK stimulates the release of NO, PGI_2, SMC (smooth muscle cell) hyperpolarization factor, and superoxides from ECs. BK also potently stimulates the release of tissue plasminogen activator (tPA) from ECs, thereby activating fibrinolysis. An excessive release of BK contributes to hypotension. The serpin C1 esterase inhibitor (C1INH) is the most important physiologic inhibitor of FXIIa and KAL. The uncontrolled activation of the KKS in C1INH deficiency leads to angioedema due to the release of BK. BK is proteolyzed in the circulation primarily by the angiotensin-converting enzyme (ACE). HK, low molecular weight kininogens, BK, and its degradation product have certain anti-thrombin (AT) activities. Thus, the KKS may influence the risk of thrombosis independently of the intrinsic pathway of the PCS. In addition, the KKS exhibits complex effects on the cellular inflammatory response, including effects on angiogenesis and EC apoptosis.[6]

In conclusion, the KKS can be activated by artificial surfaces, and it is expected that certain engineered nanomaterials may induce an uncontrolled systemic activation of this system. The resulting adverse *in vivo* effects include hypotension, angioedema, hyperfibrinolysis, and the potentiation of the complex inflammatory response. Thrombotic complications may also occur, but a role for KKS activation in mediating the thrombogenicity of intravascularly exposed materials has not been fully elucidated.

2.2 *The plasma coagulation system and fibrinolysis*

The mechanisms that underlie the PCS are being extensively investigated, and several of its pathways are not yet fully understood. This brief overview is based on expert review articles,[8–10] which may serve as sources of primary references. The principal task of the PCS is to generate and control thrombin activity. The serine protease thrombin, a key enzyme in hemostasis, is formed by the activation of the zymogen prothrombin, which is present in the plasma. The final prothrombin activation product α-thrombin (FIIa) activates platelets and converts fibrinogen (FBG) to fibrin. During this conversion, thrombin cleaves fibrinopeptides A and B from FBG, with the cleavage of fibrinopeptide B occuring at a slower rate. The kinetics of fibrinopeptide release likely differ for FBG in solution compared to when it is adhered to various surfaces. Following cleavage, the fibrin monomers polymerize into a fibrin network. Thrombin also activates FXIIIa, a transamidase that covalently cross-links fibrin monomers and thus stabilizes the fibrin network.[8,9]

Fibrin, together with activated platelets, forms a structural basis for a blood clot. By forming a positive feedback loop that favors its own activation, thrombin activates FXI, FIX, FV, and FVIII. As part of a negative feedback loop, thrombin that is bound to the endothelial receptor thrombomodulin (TM) loses its procoagulant ability and activates protein C (PC). Activated PC

(APC) and its cofactor protein S subsequently proteolytically inactivate FVa and FVIIIa, which are key to accelerating the thrombin generation pathway. Thrombin has additional roles in the complex inflammatory response, including the activation of the complement system and the stimulation of ECs, white blood cells (WBCs), SMCs, and other cell types.[8,9]

Prothrombin (FII) activation involves two bond cleavages, with the predominant pathway proceeding through meizothrombin, which is subsequently converted to α-thrombin (FIIa). In contrast to α-thrombin, meizothrombin preferentially interacts with the TM/PC system; meizothrombin exhibits very low activity toward platelets and FBG, and also exhibits low affinity for the serpin-type inhibitor AT. Under flow conditions *in vivo*, meizothrombin accumulates during prothrombin conversion and thus, a significant fraction of thrombin is primarily anticoagulant in nature.

The generation of thrombin following the activation of hemostasis can be mechanistically divided into three phases: initiation, propagation, and termination. However, hemostasis and thrombosis are dynamic processes *in vivo*, where these phases overlap and proceed simultaneously. In *in vitro* tests using both whole blood and plasma, FBG clotting occurs at the transition between the initiation and the propagation phases, where ≤5% of the prothrombin is converted to thrombin. Thus, clot-based assays exclude 95% of the thrombin generation reaction.[8,11]

The primary trigger of PCS activation is tissue factor (TF), a transmembrane receptor (Figure 1). TF is constitutively expressed on fibroblasts and pericytes in the adventitia and in the medial SMCs of blood vessel walls. TF is also constitutively expressed on the various extravascular tissues that come into contact with the blood in cases of vascular perforation. Following stimulation with pro-inflammatory cytokines, TF is expressed intravascularly on stimulated mononuclear cells and ECs. Moreover, it is hypothesized that TF is expressed intravascularly in an inactive encrypted conformation. TF in this conformation would be activated by protein disulfide isomerases, which are released from activated platelets and ECs. The active form of TF binds FVIIa to activate FX and to a lesser extent, FIX. Moreover, active TF is exposed on the cell membrane microparticles that are released during various pathologies by stimulated monocytes, and likely also by ECs, tumor cells, and certain other cell types.[8,9,12,13] Once activated, FXa potentiates the activation of FIX via the TF–FVIIa complex. The activity of the TF–FVIIa–FXa complex is controlled by the tissue factor pathway inhibitor (TFPI).[9]

In addition to its role in the PCS, TF has multiple PCS-independent signaling functions in angiogenesis, neoplasia, and other (patho)physiological pathways.

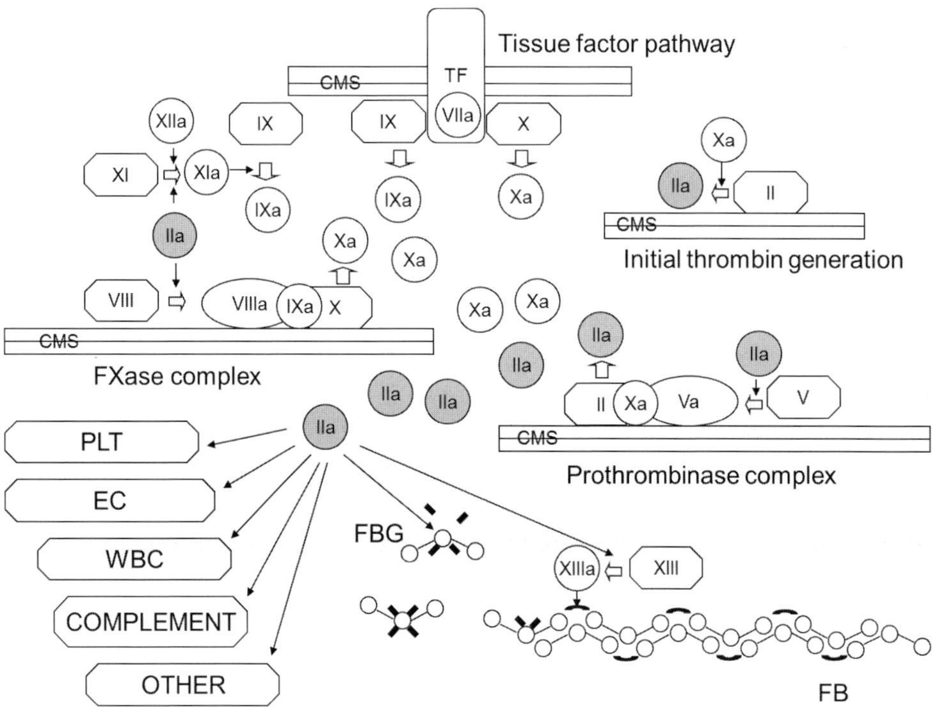

Figure 1. Schematic of plasma coagulation cascade. The pathways of thrombin (FIIa) generation in the plasma coagulation system (PCS) during hemostasis and thrombosis (modified from Mann *et al.* and Furie *et al.*[8,9]).

CMS: cell membrane surfaces, EC: endothelial cells, FBG: fibrinogen, FB: fibrin, PLT: platelets, WBC: white blood cells.

Initially, a small amount of thrombin (FIIa) is produced by FXa from the prothrombin (FII) that is bound to the surface of activated cells. This thrombin further activates platelets and initiates the activation of the accelerating cofactors FV and FVIII. Platelet-bound FVa binds FXa, resulting in the formation of the prothrombinase complex FXa–FVa, which greatly accelerates the generation of thrombin activity. Similarly, FIIa activates FVIII, releasing it from the plasma transporter vWf (von Willebrand factor). Platelet membrane-bound FVIIIa binds FIXa, resulting in the formation of a tenase complex of FIXa–FVIIIa, which greatly accelerates FX activation. The membrane-bound prothrombinase and tenase complexes are protected from AT inhibition and other plasma inhibitors acting within a major amplification loop in the thrombin generation pathway. The assembly of serine proteases, protein cofactors, and substrates on a suitable membrane surface is essential to most PCS processes; however, the molecular mechanism of the binding of the plasma coagulation factors to the membrane surfaces is not fully understood. The exposure of anionic phospholipids, particularly phosphatidylserine (PS), on the phospholipid membrane dramatically stimulates clotting reactions. Thus, it has been proposed that suitable membrane surfaces must present PS, similar to activated platelets or certain damaged cells.[10] However, fibrin generation in Par4-null mice, whose platelets cannot be activated by thrombin, is normal, suggesting the importance of other membrane surfaces *in vivo*.[9] The complex catalysts are 10^5–10^7-fold more active than their constituent serine proteases. The complex therefore facilitates the propagation phase of thrombin generation as long as the stimulus is sufficient to overcome the synergistic inhibition of plasma protease inhibitors and the simultaneously active negative feedback loops. The membrane binding of the tenase and prothrombinase complexes localizes their reactivity to the site of vascular injury. Thrombin also activates FXI. Active FXI (FXIa) in turn activates FIX, potentiating the thrombin generation pathway.[8]

The propagation of thrombin generation is independent of the initially presented TF as long as blood flow is continuous. Thus, the dynamics of the coagulation process *in vivo* is very different from that *in vitro*, where there is no flow or influx of new reactants.

Major differences in the dynamics of thrombin generation also exist within the vasculature, particularly between venous and arterial vessels. The generation of thrombin by immobilized prothrombinase under flow conditions is under dilutional control. Under high flow rate conditions, the local thrombin concentration at the site of generation is diluted by the blood supply, and the diluted thrombin is inhibited by a continuous supply of inhibitors in the plasma. Under conditions of high shear in arterial vessels, platelets play a more critical role in hemostasis and thrombosis compared to venous circulation. The procoagulant surface and the obstruction of flow afforded by activated platelets are critical to maintain a high local thrombin concentration.[8]

Different types of plasma protease inhibitors, including TFPI, AT, heparin cofactor II, α_2-macroglobulin, α_1-antitrypsin, and others, participate in regulating the PCS. Stoichiometric inhibitors in the plasma, i.e., Kunitz-type TFPI and the serpin AT, effectively inhibit all of the active serine proteases. The AT cofactors are the heparan sulfate proteoglycans that are present on ECs. The free serine proteases (IIa, IXa, and Xa) are rapidly inhibited by AT to localize and control the generation of thrombin.[8]

The dysregulation of thrombin activity may lead to life-threatening arterial or venous thrombosis and/or thromboembolic events. In addition to large vessel thrombosis, microcirculation disorders that result in the dysfunction of vital organs during stroke, heart attack, or lung or renal failure may also occur. These conditions can lead to multiple organ dysfunction syndrome (MODS) and death. Severe systemic dysregulation of thrombin activity manifests as a thrombohemorrhagic syndrome known as disseminated intravascular coagulation (DIC), which is associated with MODS and high mortality.[14,15]

To induce blood clot lysis, thrombin activates fibrinolysis by stimulating the release of tPA and urokinase (uPA) from ECs. tPA and uPA activate the fibrin-bound zymogen plasminogen to generate the fibrinolytic enzyme plasmin. The proteolytic activity of these two plasminogen activators is inhibited by plasminogen activator inhibitors PAI-1 and PAI-2. During the pro-inflammatory activation of ECs by tumor necrosis factor (TNF), PAI-1 expression increases and tPA expression is downregulated,[16] resulting in the suppression of fibrinolysis. The plasmin enzyme cleaves fibrin and FBG to generate fibrin and FBG degradation products (FDP). Specific immunoassays targeted to the neoepitopes of cross-linked fibrin are able to detect D-dimers, which are markers of the fibrinolysis of stabilized fibrin. Plasmin, however, is not a fibrin-specific enzyme and its dysregulation may lead to hyperfibrino(geno)lysis, which in turn leads to afibrinogenemia and the proteolysis of other plasma proteins and cell membrane receptors. Therefore, plasmin activity in the circulation is strictly controlled by α_2-antiplasmin (α_2-AP). The plasmin that is bound to a fibrin clot is protected from α_2-AP inhibition, facilitating fibrinolysis. Moreover, plasmin and thrombin/TM activate the thrombin activatable fibrinolysis inhibitor (TAFI). Active TAFI (TAFIa) is a metallocarboxypeptidase that cleaves C-terminal lysines from partially degraded fibrin, thereby inhibiting fibrin cofactor function in plasminogen activation.[17]

3. Methods for Evaluating the *In Vitro* Effects of Nanomaterials on the Kallikrein–Kinin System, the Plasma Coagulation System, and Fibrinolysis

The evaluation of the effects of biomaterials on hemocoagulation aims to determine how the tested materials interact with or affect the functions of ECs, the PCS, and platelets. Most *in vitro* screening tests of the effects of a biomaterial on hemostasis focus on the PCS and platelet function. To evaluate the interaction of a given material with the PCS for the purpose of assessing the material's biocompatibility, the first question that is addressed is how the material affects the activities of the three most important effector enzymes of the PCS — thrombin, plasmin, and KAL. The manner in which the biomaterial interferes with the conversion of soluble FBG to fibrin polymer — a structural base for the plasmatic clot — is also assessed. Table 1 summarizes different assays used in published studies on the interaction of nanomaterials with the KKS, the PCS, and fibrinolysis. Thrombin activity is the most important factor in terms of hemostasis and also the most dangerous when

Table 1. Methods used for the evaluation of *in vitro* effects of nanomaterials on the KKS, PCS, and fibrinolysis.

Assay	Instrument	Blood/Plasma	NP	Ref.
APTT, PT, TT	CGMT	PPP	GNP	37
APTT, PT, TT	CGMT	PPP	POLNP	38
APTT, PT, TT	CGMT	PPP	ACDNP	39
APTT, PT, TT	CGMT	PPP	SFNF	40
APTT, PT, TT	CGMT	PPP	LPNP	41
APTT, PT	CGMT	PPP	POLNP	42
APTT, PT	CGMT	PPP	SNP	43
TT, RT	MAN	FBG	SNP	44
RCT	MAN	PPP	POLNP	45
FBG	CGMT	PPP	POLNP	38
	ELISA	PPP		
FII, FV, FVII+X	CGMT	Factor def.	POLNP	38
FVIII, FIX, FXI, FXII		PPP		
FXII	TURB	Factor def.	SiNP	46
		PPP		
KAL	COL/CHS	PPP	SPION	47
TEG	TEG	WB	PBRS	28
TEG	TEG	WB	POLNP	29
ELT	MAN	PPP	PEPNP	34
TLA	GM	Rat CA	PEPNP	34
TLA	GM	WB w/o AC	LPNP	35
TLA	GM	Rat WB w/o AC	PEPNP	36

Abbreviations

ACDNP	arginine–chitosan DNA NP
APTT	activated partial thromboplastin time
CA	carotid artery
CGMT	coagulometer
CHS	chromogenic substrate
COL	colorimetry
ELISA	enzyme-linked immunosorbent assay
ELT	euglobulin lysis time
FVII	plasma coagulation factor VII
FBG	fibrinogen (human, if not specified otherwise)
GM	gravimetry
GNP	gold NP
KAL	kallikrein
LPNP	lipid NP
MAN	manual
NP	nanoparticles
PEPNP	peptide NP
POLNP	polymer NP
PPP	citrate-anticoagulated platelet poor plasma (human, if not specified otherwise)
PBRS	polymer brush surfaces
PT	prothrombin time
RCT	recalcification time
RT	reptilase time
SFNF	silk fibrin nanofibers
SiNP	silica NP
SNP	silver NP
SPION	super paramagnetic iron oxide NP
TEG	thromboelastography
TLA	thrombolytic activity
TT	thrombin time
TURB	turbidimetry
w/o	without
WB	citrate-anticoagulated whole blood (human, if not otherwise specified)

dysregulated. Thrombin activity is not restricted to the PCS. Thrombin is a strong activator of platelets and the complement system. Thrombin also stimulates WBCs and ECs to enter a pro-inflammatory state. Therefore, the evaluation of the effects of biomaterials on thrombin activity is critical. Similarly, plasmin is a major player in the inflammatory response. In addition to its role in fibrinolysis, plasmin interacts with all of the inflammatory response systems. Furthermore, it has been suggested that plasminogen and plasmin play key roles in the inflammatory response to biomaterials.[18] In most cases, intravascular biomaterials indirectly affect the generation of plasmin activity by stimulating the release of plasminogen activators, including tPA, or inhibitors (PAI) from WBCs and ECs. Therefore, the usefulness of the evaluation of fibrinolytic activity in plasma *in vitro* is very limited. *In vitro* whole blood experiments and standardized endothelial and macrophage tissue culture assays may be more useful.

PK activation has historically been associated with the FXII-dependent contact phase of plasma coagulation. However, it has recently been demonstrated that a more favorable proteolytic pathway that leads to PK activation occurs *in vivo* on ECs and certain other cell types. Moreover, this pathway is independent of both contact activation and FXII.[6] KAL plays several roles in the inflammatory response. Its most recognized activity is the cleavage of kininogens to liberate kinin peptides, particularly BK. BK is a potent stimulator of ECs and induces the release of vasodilating mediators (NO, PGI_2) and tPA. Thus, the KKS is involved in the regulation of blood pressure, blood flow, microvascular permeability, and other inflammatory response processes. The methods that are used to assess the effects of biomaterials on the KKS are not well established because the clinical importance of this system with respect to the adverse effects of intravascular biomaterials is unclear. Nevertheless, the evaluation of recently developed nanomaterials for intravascular nanobiomaterials regarding their effects on the KKS should not be overlooked.

3.1 *The preparation of a nanomaterial sample for testing*

The preparation of a nanomaterial sample for testing and characterization is a critical part of the analysis. Materials that are insoluble in aquatic solvents and unstable in suspension must be particularly well characterized. For example, certain nanomaterials that are provided as a powder from the manufacturer require resuspension in a buffer. This resuspension is generally achieved by sonication, which causes dramatic changes in the material's characteristics. Certain studies are performed with nanomaterials that are resuspended in the presence of a detergent. Every additive or impurity affects the experiment differently. Another important issue concerning the preparation of nanomaterials is its resuspended stability, as well as its stability following oxidative and other chemical changes. Lastly, the agglomeration of nanomaterials and the time course of this process in the blood, plasma, or in protein solutions should be monitored. Involving a material scientist early in the project and working with a well-characterized nanomaterial are strongly advised. Otherwise, the interpretation of results will be very problematic and the value of the study will be compromised.

3.2 *The plasma and blood samples for testing*

The testing of biomaterials for their effects on the PCS is performed with venous blood from normal healthy individuals who are not taking any medications that might interfere with the PCS (e.g., oral anticoagulants). For most plasma coagulation studies, blood (9 vol.) is collected in a buffered 0.109 M trisodium citrate anticoagulant (1 vol.). Platelet-poor plasma (PPP) supernatant is harvested following centrifugation at 2,500*g* for 20 min. The tests should be performed with minimal delay, generally within 4 h. For studies with fresh PPP samples, small pools from at least three normal donors are used.[19] For certain assays, snap-frozen samples can be used. In these cases, larger pools of PPP from at least 20 normal donors should be used. Standard procedures for blood collection, transport, and processing have been developed for coagulation diagnostics[20] and are largely applicable to material studies as well. According to the ANSI/AAMI/ISO 10993–4:2002 standard, the variables that should be considered when using *in vitro* test methods include the hematocrit, presence of anticoagulants, sample collection, age of the sample, conditions of sample storage, aeration and pH, temperature, sequence of the test versus that of the control studies, surface-to-volume ratio, and fluid dynamic conditions (especially the wall shear rate).[3]

3.3 *Clot-based plasma coagulation tests*

For screening purposes, the effects of biomaterials on the PCS are routinely evaluated using established plasma clotting assays. The protocol for screening the effects of nanoparticles (NPs) on plasma coagulation has been developed and consists of four clotting assays: APTT, prothrombin time (PT), thrombin time (TT), and reptilase time (RT).[21]

The tested material is preincubated with citrate-anticoagulated plasma, and PCS activation is subsequently triggered via different pathways. The time to clot formation is monitored either manually or using a coagulometer.

In the APTT test, PCS activation is triggered via the contact pathway (i.e., the intrinsic pathway, via FXII). In this test, a contact activator (e.g., ellagic acid) and phospholipids (cephalin) are used. In the less common partial thromboplastin time (PTT) test, only phospholipids are mixed with the PPP. Following the incubation period in both the APTT and PTT tests, Ca^{2+} is added, and the time to clot is monitored. The addition of negatively charged phospholipids with Ca^{2+} allows for the formation of the activation complexes that lead to thrombin activity. It was previously believed that materials that are exposed to blood induce thrombin activity via the contact pathway. *In vivo*, however, it is more likely that biomaterials induce the expression of TF on monocytes. Thus, the biomaterial-induced thrombin activity in the blood is predominantly mediated by the TF pathway (formerly extrinsic) and not by contact activation.[22] The TF pathway is activated in the PT test. In this test, plasma coagulation is triggered by an excess dose of TF with phospholipids (i.e., tissue thromboplastin). Following the incubation period, Ca^{2+} is added, and the time to clot formation is monitored. In a modified PT test that is known as the dilute PT (dPT) test, a lower concentration of the thromboplastin reagent is used to increase sensitivity.[23]

Generally, the results of tests that use plasma to determine the effect of a biomaterial on clotting time, such as the APTT assay, are very difficult to interpret. A biomaterial may interact with coagulation factors and inhibitors at different stages of PCS activation, e.g., during coagulation and during the activation of the fibrinolytic pathway. Furthermore, these interactions may be antagonistic. In addition, all of the clotting time-based assays are dependent on the conversion of FBG to fibrin. The clotting time marks the moment when sufficient levels of thrombin have had sufficient time to induce FBG clotting.[24] Thus, if the biomaterial interferes with the conversion of FBG to fibrin, thrombin generation cannot be monitored by any clotting based test. The direct effects of biomaterials on the conversion of FBG to fibrin can be evaluated using the TT and RT tests. Both of these enzymes convert FBG to fibrin. In contrast to thrombin, reptilase, a fibrinopeptide A-releasing enzyme from the *Bothrops atrox* venom, is insensitive to heparin–AT inhibition. In addition to interfering with the conversion of FBG to fibrin, the lengthening of the RT and TT may indicate activated fibrinolysis, i.e., plasmin activity. Plasmin cleaves fibrin and, when dysregulated, FBG to generate their degradation products. High FDP levels inhibit the conversion of FBG to fibrin and thus, prolong both TT and RT. Therefore, clot-based assays are not sufficiently specific or sensitive to investigate the specific effects of nanomaterials on the PCS and should be used for initial screenings only.

Other clotting-based tests are used for more specific applications. To evaluate a material's direct inhibitory effects on the activities of meizothrombin and thrombin that are formed in the plasma, the ecarin clotting time (ECT) assay may be used. The prothrombin-converting enzyme, ecarin from the venom of the saw-scaled viper (*Echis carinatus*),[25] is used in this assay to convert prothrombin to meizothrombin. Meizothrombin is not inhibited by AT or heparin but can be inhibited by other direct thrombin inhibitors, including hirudin. Thus, the ECT is useful in the evaluation of the anticoagulant activity of certain thrombin inhibitors that are used in clinical practice.[22] Another clotting assay is the prothrombinase-induced clotting time (PiCT) test, which is sensitive to FXa and thrombin inhibitors.[23,26] In this assay, PPP is added to a reagent containing FXa, phospholipids, and the FV activator from Russell's viper (*Daboia russelii*) venom. Following an incubation to allow for FV activation, the recalcification clotting time is measured. This assay can be used to measure the anticoagulant effects of unfractionated and low molecular weight heparins (LMWH).

3.4 *Thromboelastography*

Thromboelastography is a global test of blood coagulation that quantitatively measures the elasticity of the blood from the initiation of coagulation through the end of fibrinolysis.[27] Two types of systems have been developed for this assay, the thromboelastogram (TEG) and the rotational thromboelastometer (ROTEM). Both of these systems record viscoelastic changes during the blood coagulation process. The principal difference between these two assays is that the TEG hardware includes a rotating cap, whereas the ROTEM hardware includes an immobile cap and an oscillating pin/wire transduction system.[27] TEG evaluates the following parameters during the coagulation process: the reaction time (R), the coagulation time (K), the clot polymerization rate (α angle), the maximum amplitude or maximum clot strength (MA), the time to reach MA (TMA), the degree of fibrinolysis at 30 min, or the rate of amplitude reduction 30 min following the TMA (LY30). Similar parameters are obtained with ROTEM, but the results are system-specific. Therefore, the data from the TEG and ROTEM cannot be used interchangeably.[27] The TEG assay has been used in a few nanomaterial studies,[28,29] and the application of this assay to biomaterials testing has been comprehensively reviewed.[27] Although the TEG assay is simple and versatile and can be performed with various activators to investigate the effects of biomaterials on the different blood coagulation pathways, the TEG has not been widely used for this purpose because neither standard protocols nor reference standards have been established. It is a very complex whole blood assay, and platelets, the PCS, and other blood components can influence the results. Therefore, the ability of thromboelastegraphy to determine a biomaterial's effect on a specific PCS mechanism is limited. Thromboelastegraphy should be considered a useful complementary method to screen the effects of nanomaterials on blood coagulation and fibrinolysis.

3.5 *Synthetic substrate-based assays*

Various assays and commercially available kits have been developed for the evaluation of the activities of different components of the KKS, the PCS, and fibrinolysis. These tests make use of synthetic substrates, which are generally chromogenic or fluorogenic.[30] Synthetic substrate-based assays are used to determine the activity of endogenous proenzymes, enzymes, cofactors, and inhibitors. These assays are also used to monitor therapeutic substances and for purposes of quality control.[30] Synthetic substrates can be used in plasma assays as well as in studies of purified systems. The most widely used substrates are based on the chromogenic peptide para-nitroanilide (pNA), the release of which is assayed using colorimetry. Fluorogenic substrates are used for assays that require higher sensitivity.

Thrombin generation tests were developed to mimic physiological coagulation and used synthetic substrates that allow for the direct measurement of plasma thrombin activity.[23,31] These assays measure the ability of the plasma sample to generate thrombin following *in vitro* plasma activation, which is achieved with TF or another trigger. A widely used thrombin generation assay is calibrated automated thrombography (CAT). CAT employs a low affinity fluorogenic substrate (Z–Gly–Gly–Arg–AMC) to continuously monitor plasma thrombin activity. The test can be performed in PPP or platelet-rich plasma (PRP) to mimic physiological conditions. CAT is performed in a microtiter plate, and the appropriately calibrated fluorescence readings are converted into thrombin generation curves using a dedicated software.[31] The parameters that are frequently measured in this assay include the lag time, the time to peak, and the endogenous thrombin potential (ETP).

Other assays include FXa chromogenic assays, which are used to measure the plasma activities of LMWH and certain newly developed anticoagulants.[23] Chromogenic assays to determine the activity of the KKS factors, including FXIIa and KAL, would be very useful for nanomaterial hemocompatibility studies.

3.6 *ELISA and other immunoassays*

In certain biocompatibility studies, surrogate markers of thrombin generation are examined. Such markers include the thrombin–AT (TAT) complex and prothrombin fragment 1.2 (PF1.2), which are analyzed using ELISA.[32] The immunodetection of fibrinopeptide A, FDP, and D-dimers may be useful for investigating the effects of nanomaterials on the conversion and degradation of FBG.

3.7 *Fibrinolytic and thrombolytic activity assays*

Various *in vitro* assays can be used for testing the effects of biomaterials on the fibrinolytic system. Although none of these assays have been standardized, they may be useful for screening studies. In addition to simple assays that measure the timing of clot lysis in a tube or fibrinolysis using thromboelastegraphy, the euglobulin lysis time (ELT) test is historically used to determine the overall degree of intrinsic fibrinolytic activity in the plasma. In the ELT test, euglobulin is isoelectrically precipitated from diluted plasma using acetic acid, such that fibrinolysis inhibitors remain in the supernatant. The euglobulin fraction contains FBG, PAI-1, tPA, plasminogen, and to a lesser extent, α_2-AP. The elimination of the plasmin inhibitors α_2-AP and α_2-macroglobulin allows for the measurement of the lysis time of the euglobulin clot. The activities of both tPA and PAI-1 in the euglobulin precipitate appear to be the major factors in determining the ELT results.[33] The euglobulin precipitate is dissolved in a borate buffer and clotted by recalcification, and the lysis time is monitored either manually or spectrophotometrically. Alternatively, the euglobulin fraction in the borate buffer is recalcified on the plate, forming a layer. A sample of the tested material is then pipetted onto the euglobulin layer and, following incubation, the diameter of the lysis ring is measured.[34] Similarly, a simple classical method for the screening of fibrinolytic activity is through a fibrin plate assay, in which the diameter of the lysis zone produced by the tested material on a fibrin plate is measured. *Ex vivo* or *in vitro* thrombolytic assays are often based on gravimetric methods that measure a decrease in the weight of the whole blood thrombus following incubation.[34–36]

4. Studies of the *In Vitro* Interactions of Nanomaterials with the Plasma Coagulation System, the Kallikrein–Kinin System, and Fibrinolysis

There are different types of studies that examine the interaction of nanomaterials with the PCS.

Nanomaterials are employed in the development of delivery systems of various drugs that affect the PCS. These drugs are primarily antithrombotics and anticoagulants, including heparins,[40,48–55] synthetic thrombin inhibitors,[56] and PC.[57] Other drugs include thrombolytic agents, including tPA,[58–60] streptokinase,[61–63] uPA,[64] uPA with anti-fibrin antibodies,[65] and thrombolytic peptides.[34,36] Hemostatic agents, including thrombin,[66] FVIII,[67] or FVIII DNA,[68] are also being examined. Moreover, the nanomaterials themselves may be designed for antithrombotic,[29,69] thrombolytic,[35] or hemostatic[70,71] use. Furthermore, NPs that are made from FBG have been proposed for use in cancer drug delivery.[72–74]

Several studies have examined the adhesion of nanomaterials to plasma proteins, including components of the PCS, particularly FBG. Protein adhesion is generally important in the inflammatory response to nanomaterials and may have a specific effect on hemostasis. To understand the interactions of a biomaterial with the components of the inflammatory response, and to predict the clearance and biodistribution of a particular biomaterial, it is essential to evaluate the adsorption of the plasma proteins onto the material's surface, a process known as opsonization.[21,75] Opsonization is a critical issue with respect to the blood biocompatibility of a biomaterial, which will likely be instantly coated with various plasma proteins upon intravascular exposure. Depending on the surface characteristics, immunoglobulins, albumin, complement proteins, kininogens, and FBG generally comprise the major coating proteins. The binding of FBG is particularly important. When soluble, this protein is not pro-inflammatory. However, upon adsorption to material surfaces, FBG prompts a complex inflammatory response, primarily by recruiting WBCs to the material surface.[76] For example, it has been demonstrated that negatively charged poly(acrylic acid)-conjugated gold NPs (GNPs) with diameters of 5 and 20 nm bind to and induce the unfolding of FBG. This event promotes an interaction of FBG with the integrin receptor Mac-1, which is expressed on THP-1 cells. Mac-1 activation stimulates the NF-κB signaling pathway, resulting in the release of inflammatory cytokines.[77] Another study reported the specific self-assembly of FBG into nanofibrils on a gold Au(1,1,1) surface.[78] This observed FBG polymerization was based on the interaction of the central αC domains of FBG and thus, was completely unlike thrombin-induced polymerization. The nanofibril formation was specific to the gold surface and was not observed on the other materials that were tested (i.e., mica and highly ordered pyrolytic graphite). The authors suggested that the interaction of the gold surface with the FBG cysteinyl residues plays a key role in the FBG assembly process.

Apart from the effects of NPs on protein conformation, the interaction of the NP with plasma proteins may significantly impact NP agglomeration. Therefore, it is imperative to evaluate a time course of NP agglomeration in all biological assays.

For the screening of plasma protein adhesion to different types of nanomaterials, a protocol using two-dimensional gel electrophoresis and mass spectrometry has been developed.[21,75] Certain previous studies used liquid chromatography-tandem mass spectrometry,[79,80] spectroscopic ellipsometry,[81] or surface plasmon resonance techniques to examine NP agglomeration.[82] The adsorption of different plasma proteins, including FBG, has been examined for various nanomaterials, including TiO_2 NPs,[83–85] titania nanotubes,[86] nanoroughness gradients,[87] iron oxide NPs,[84] zinc oxide NPs,[83] nanoporous alumina,[88] GNPs,[37,89,90] nano-rough platinum surfaces,[91] ZrN–Ag nanocomposite films,[92] silica NPs,[80,84,93–95] carbon nanotubes,[96–98] various polymer and dendrimer NPs,[99–102] nanostructured surfaces[103–111] and solid lipid NPs.[112] Adsorption studies are performed using either purified proteins or complex samples, including human plasma. The NP material, size, and surface properties have been suggested to play a role in determining the composition of the protein corona of the NP in the plasma; however, the underlying mechanism has not been fully elucidated.[80,82,113,114] Furthermore, protein adsorption onto solid surfaces is time-dependent. Initially, the more concentrated plasma proteins, including FBG, are adsorbed. However, these proteins are displaced within seconds by less abundant plasma proteins with higher affinity for the NP's surface, forming a more stabile protein corona around the engineered particles. This transient protein adsorption is referred to as the Vroman effect.[81] By studying PCS proteins, Vroman first observed that the initial adsorption of FBG is displaced by HK and to a lesser extent, by FXII. In studies of the adsorption of the PCS proteins onto NPs in the plasma, the quality of the plasma sample and the use of different types of anticoagulants (Ca^{2+} chelation) may highly impact the study results. Comprehensive comparative studies are needed in this respect.

Very few *in vitro* studies (Table 1) have performed detailed investigations on the activatory or inhibitory effects of nanomaterials on the KKS, PCS, or fibrinolysis. These studies would be able to predict the possible thrombogenic or hemorrhagic activities and other adverse effects of a given nanomaterial on the hemocoagulation balance *in vivo*.

Another area of research is the optimization of nanostructured procoagulant surfaces, with the aim of developing of contact hemostatic agents. Porous silica materials are attractive in this area because of their procoagulant surface chemistries. Baker *et al.*[115] investigated the effect of the nanostructure of silicate mesocellular foams (MCFs) on the contact activation of PCS. This study demonstrated that the conventional description of the surface area as defined by the gas sorption assay (i.e., the BET area) does not necessarily correlate with the plasma contact-activating potency. This is because for materials with small pore window diameters, the area that is accessible to the proteins is much smaller than the BET surface area. Thromboelastographic experiments demonstrated that the MCF procoagulant activity markedly increased when the pore window size was increased from 6 to 33 nm. The sharp increase in the MCF procoagulant activity with window sizes greater that 11 nm likely corresponds to the increased access of FXII, which has an estimated hydrodynamic diameter of 7.5 nm. Thus, the window size and volume that determine the surface area that is actually accessible to KKS proteins are critical factors with respect to plasma contact activation.

In contrast to studies of nanomaterial-based hemostatic agents, most studies of the interactions of nanomaterials with the KKS are focused on the optimization of these materials for intravascular use, with the aim of achieving minimal interference with the KKS and the PCS.

Simberg *et al.*[116] reported that plasma PK and HK attach to the surface of dextran-coated superparamagnetic iron oxide NPs (SPIONs) at uncovered sites of the iron oxide core. In a follow-up study, the authors demonstrated that SPIONs also activate the KKS in both human and mouse plasma.[47] The serine protease activity of KAL was observed to be stably associated with the SPIONs and was detected on the NP even after extensive washing steps. The KAL enzymatic activity was not detectable in either HK- or FXII-deficient plasma. Interestingly, while non- and low-aminated SPIONs were demonstrated to activate PK in mouse plasma, highly aminated SPIONs did not. PK was bound to the highly aminated SPIONs but was uncleaved, likely because of the inactivation of KAL, FXIIa, or both by the highly basic surface of the SPIONs. Because PK binds to SPIONs via the histidine-rich domain 5 (D5) of HK, competitive inhibition experiments were performed. PK activation was blocked by precoating the SPION with D5, indicating an essential role of HK in SPION-induced KAL activity. Importantly, KAL activity was detected in the plasma of SPION-injected mice, but not in the plasma of D5/SPION-injected mice. When injected into HK-deficient and control mice, the tumor-targeted SPIONs produced high levels of vascular clotting in tumors, suggesting that it is unlikely that PK activation plays a major role in SPION-induced thrombosis.

Tavano *et al.*[46] compared the effects of synthetic amorphous silica (SAS) NPs (35 nm), organically modified silica (ORMOSIL) NPs (45 nm), poly(D,L-lactide-co-glycolide) (PLGA) NPs (280 nm), and small (150 nm) unilamellar vesicles (SUV) NPs made from dipalmitoylphosphatidylcholine and dipalmitoylphosphatidyl glycerol on whole blood and plasma recalcification time. The SAS NPs markedly accelerated the clotting of recalcified whole blood, whereas the SUV NPs caused only a moderate acceleration. The ORMOSIL and PLGA NPs triggered no acceleration of the clotting of recalcified whole blood. Similarly, in recalcified plasma, the SAS NPs markedly accelerated clotting, and ORMOSIL NPs weakly accelerated clotting. The PLGA and SUV NPs both promoted clotting to a greater extent than did the ORMOSIL NPs. Experiments in FXII-deficient plasma demonstrated that the NP-induced acceleration of clotting in recalcified plasma was FXII-dependent. Notably, polyethylene glycol (PEG)ylation prevented both the whole blood and plasma coagulation activity in all of the tested NPs.

Kannan *et al.*[29] investigated the anticoagulant properties of nanocomposites formed from silica nanocomposite polyhedral oligomeric silsesquioxane (POSS) and poly(carbonate-urea)urethane (PCU) for their potential use in cardiovascular bypass grafts and as the microvascular components of artificial capillary beds. Thromboelastographic analyses demonstrated a significant decrease in clot strength and an increase in clot lysis on the nanocomposites when compared to both poly(tetrafluoroethylene) (PTFE) and PCU. Interestingly, increasing the concentration of the POSS nanocages within these polymers proportionally inhibited FX activity. Thus, POSS–PCU represents a promising type of composite material with anticoagulant and antithrombogenic properties.

Martinez-Gutierez *et al.*[43] demonstrated that 24 nm silver NPs (SNPs) prolonged APTT, suggesting a moderate anticoagulant effect. SNPs did not affect PT, nor did they induce coagulation in citrated plasma. Shrivastava *et al.*[44] also reported that SNPs with a mean diameter of approximately 10–15 nm can significantly retard fibrin polymerization kinetics in both pure and plasma-incorporated systems, and can therefore impede thrombus formation. SNP-treated FBG exhibited delayed polymerization in response to both thrombin and atroxin (reptilase). This result suggests a reduced interaction between the "GPR knob" on the fibrin α-chain and the "distal hole" of γ-chain carboxyl domains in the presence of SNPs. Although a much higher concentration of SNPs was required to inhibit fibrin polymerization in the plasma compared to the concentration required for the purified system, SNPs may be used effectively in certain blood contact applications, where their potential anticoagulant activity may complement the properties of antibacterials. Interestingly, Dobrovolskaia *et al.*[37] demonstrated no effect of 30 and 50 nm GNPs on APTT, PT, or TT.

Lai *et al.*[28] showed that high graft density hydrophilic non-charged poly(N,N-dimethyl acrylamide) (PDMA) brushes prevent the initiation of blood coagulation on synthetic surfaces. Performing a multifaceted analysis, they observed that PDMA brushes with graft densities of greater than 0.27 chains/nm^2 exhibited this highly desirable property. Non-specific protein adsorption is greatly reduced on high density brushes compared to bare surfaces. Thromboelastographic assays demonstrated that only the high graft density brushes are inert with respect to blood coagulation. As the graft density decreased, the PDMA brushes exhibited signs of surface-initiated coagulation in whole blood and PRP as evidenced by a significant decrease in the R-value of the TEG. This result indicated more rapid blood clot generation. As polymer brush synthesis can be adapted to most biomedical materials that are currently in use, these results have a significant impact on the design of highly hemocompatible surfaces.

Studies of the hemocompatibility of amphiphilic block copolymers are important for the development of hydrophobic compound delivery systems. Letchford *et al.*[42] synthesized and characterized two methoxy PEG–block–poly(caprolactone) (MePEG–b–PCL) amphiphilic diblock copolymers, both based on MePEG with 114 repeat units and PCL block lengths of either 19 or 104 repeat units. The NPs were formed from these copolymers using a nanoprecipitation and dialysis technique. The MePEG(114)–b–PCL(19) copolymer was water-soluble and formed micelles with hydrodynamic diameters of 40 nm. The practically water-insoluble MePEG(114)–b–PCL(104) copolymer formed NPs with a larger hydrodynamic diameter, the magnitude of which was dependent on the copolymer concentration. The micelles solubilized the hydrophobic anticancer agent paclitaxel (PTX) and did not affect either the APTT or the PT.

Salvador-Morales *et al.*[41] investigated the coagulation effects of methoxyl functionalized lipid–polymer hybrid NPs that were composed of a PLGA core, a PEG shell, and a lipid monolayer at the core–shell interface. The NPs with a mean diameter of approximately 100 nm did not exhibit any significant effect on the PT, APTT, or TT, and thus may be promising candidates for drug delivery applications.

Sahli *et al.*[38] investigated the effects of sodium cholate-stabilized suspensions of poly(lactic acid) (PLA–Ch) and poly(lactic acid-co-ethylene oxide) (PLA–PEO–Ch) NPs (150 nm) on the PCS. The measured APTT, PT, and TT remained similar to those of the controls in the presence of PLA–PEO–Ch NPs, whereas these times were markedly prolonged in the presence of PLA–Ch NPs, which also exhibited a slightly prolonged RT. The PLA–PEO–Ch NP suspensions were remarkably inert with respect to the PCS. The steric repulsion caused by the high surface density of PEO is sufficient to avoid both strong interactions with proteins and the formation of aggregates between particles.[38]

Song *et al.*[39] demonstrated that both arginine–chitosan/DNA NPs (153 nm) and chitosan/DNA NPs (163 nm) prolonged the APTT, but did not influence the PT or the TT *in vitro*. The introduction of arginine moieties into chitosan slightly enhanced its anticoagulant properties, an effect that would be beneficial to applications that involve direct contact with the blood.[39]

Dash *et al.*[45] investigated the influence of the size and the charge of chitosan/polyglutamic acid hollow spheres on plasma recalcification clotting time. When compared to the control, the recalcification times were significantly shorter for all of the samples, i.e., neutrally charged, PEGylated, and negatively charged 100, 300, 500, and 1,000 nm NPs. The negatively charged spheres resulted in a significant decrease in the clotting time when compared with the 100 and 300 nm spheres, which have other surface charges. The size of the PEGylated spheres did not significantly influence the associated clotting time, whereas the 300 nm negatively charged spheres resulted in significantly reduced the clotting time when compared to the 500 and 100 nm spheres.[45]

Another area is of study is the development of nanomaterial-based thrombolytic agents. Yang *et al.*[34] investigated the activity of Cu(II) nanocomplexes of P6A peptide analogs. Ala–Arg–Pro–Ala–Lys, a pentapeptide known as P6A, was isolated from the products of the plasmin-induced degradation of the fibrin B β-chain. P6A was demonstrated to possess various vasodilating and thrombolytic effects, and this peptide and its PAK-containing analogs likely enhance plasminogen activation. The precise mechanism of this effect remains to be elucidated. The complexing of Cu(II) with previously synthesized thrombolytic peptides, including Pro–Ala–Lys (6), Arg–Pro–Ala–Lys (7), Ala–Arg–Pro–Ala–Lys (8), Gly–Arg–Pro–Ala–Lys (9), and Gln–Arg–Pro–Ala–Lys (10), resulted in the formation of the following complexes: Cu(II–(Pro–Ala–Lys) (6-Cu), Cu(II)–(Arg–Pro–Ala–Lys) (7-Cu), Cu(II)–(Ala–Arg–Pro–Ala–Lys) (8-Cu), Cu(II)–(Gly–Arg–Pro–Ala–Lys) (9-Cu), and Cu(II)–(Gln–Arg–Pro–Ala–Lys) (10-Cu). The mean diameter of the aggregates was approximately 200–300 nm. *In vitro* euglobulin lysis and thrombolytic assays indicated that the binding of Cu(II) resulted in a significant increase in the activities of peptides 6–10. The increase in the thrombolytic potency was also demonstrated *in vivo* in rats. These findings suggested that the self-assembly of the Cu(II)-peptide complexes into nanoscale aggregates improved the thrombolytic activity of the peptides.

Similarly, a thrombolytic agent was designed based on the nanoscale assembly of poly-α,β-DL-aspartyl-L-alanine.[36] This compound formed particles of various sizes, depending on the pH and temperature. At the physiological pH (7.4), particles of 19–43 nm and nanosticks of up to 150 × 522 nm in size were observed. These particles exhibited significant thrombolytic activity *in vitro* as assessed by a gravimetric assay using rat whole blood clots; the thrombolytic activity was also confirmed in rats *in vivo*. The optimization of particle size distribution for maximum thrombolytic potency is a challenge for future investigation.

Becker *et al.*[35] investigated the ultrasound-induced thrombolytic effects of various liposomal formulations using a gravimetric method and human blood clots *in vitro*. The human blood clots were incubated with various lipid dispersions: 1,2-dipalmitoyl-sn-glycero-3-phosphocholine (DPPC)/cholesterol (CH), DPPC/PEG (40) stearate (PEG40S), 1,2-distearoyl-sn-glycero-3-phosphocholine (DSPC)/PEG40S, and the commercially available ultrasound contrast agent SonoVue®. The PEG40S model liposomes exhibited high polydispersity indices and mean particle sizes of approximately 200 nm. The weight of each clot was determined prior to and following continuous insonation of 1 h at 1.4 MHz. All of the nanoscale formulations, as well as the SonoVue® contrast agent, reduced the thrombus weight compared to the weight loss of the clots that were not insonated. A highly significant weight reduction was observed with DSPC/PEG40S compared to SonoVue®. The nanoscale DSPC/PEG40S dispersion may be a promising formulation in ultrasound-enhanced thrombolysis, even without thrombolytic drugs.

5. Conclusion

The testing of the interactions between nanomaterials and the PCS is essential to hemocompatibility screening that is performed during the development of biocompatible nanomaterials. Simple screening assays that use plasma or whole blood clotting tests (e.g., APTT, PT, TT, TEG) are available; however, their sensitivity and specificity are questionable. To perform a complete investigation of the specific effects of nanomaterials on the KKS, the PCS, and fibrinolysis, comprehensive studies using both whole blood/plasma and purified systems, clot-based assays, enzyme activity assays, and immunoassays are required. Based on the present review of the published studies that investigated the effects of nanomaterials on the KKS, the PCS, and fibrinolysis, it is evident that limited information concerning a small number of types of nanomaterials is available. Therefore, no generalized conclusions can be made at this time. Systematic structure/function (i.e., activity) studies of different types of nanomaterials, including particles of various sizes and with various surface modifications, are required. The development of standard protocols for coagulation testing is also essential to allow comparisons among the results of different laboratories.

References

1. Vogler EA, Siedlecki CA. Contact activation of blood-plasma coagulation. *Biomaterials* 2009;30:1857–1869.
2. Sefton MV, Gemmell CH, Gorbet MB. What really is blood compatibility? *J Biomater Sci Polym Ed* 2000;11:1165–1182.
3. AAMI. Biological evaluation of medical devices — Part 4: Selection of tests for interaction with blood. ANSI/AAMI/ISO 2002;Vol. 10993-4.
4. McGuigan AP, Sefton MV. The influence of biomaterials on endothelial cell thrombogenicity. *Biomaterials* 2007;28:2547–2571.
5. Sainz IM, Pixley RA, Colman RW. Fifty years of research on the plasma kallikrein-kinin system: from protein structure and function to cell biology and *in-vivo* pathophysiology. *Thromb Haemost* 2007;98:77–83.
6. Schmaier AH, McCrae KR. The plasma kallikrein-kinin system: Its evolution from contact activation. *J Thromb Haemost* 2007;5:2323–2329.
7. Schmaier AH. Assembly, activation, and physiologic influence of the plasma kallikrein/kinin system. *Int Immunopharmacol* 2008;8:161–165.
8. Mann KG. Thrombin generation in hemorrhage control and vascular occlusion. *Circulation* 2011;124:225 235.
9. Furie B, Furie BC. Mechanisms of thrombus formation. *N Engl J Med* 2008;359:938–949.
10. Morrissey JH, Davis-Harrison RL, Tavoosi N, *et al.* Protein-phospholipid interactions in blood clotting. *Thromb Res* 2010;125(Suppl 1):S23–25.
11. Mann KG, Orfeo T, Butenas S, Undas A, Brummel-Ziedins K. Blood coagulation dynamics in haemostasis. *Hamostaseologie* 2009;29:7–16.
12. Simak J, Holada K, Vostal JG. Release of annexin V-binding membrane microparticles from cultured human umbilical vein endothelial cells after treatment with camptothecin. *BMC Cell Biol* 2002;3:11.
13. Simak J, Gelderman MP. Cell membrane microparticles in blood and blood products: potentially pathogenic agents and diagnostic markers. *Transfus Med Rev* 2006; 20:1–26.
14. Levi M. Disseminated intravascular coagulation. *Crit Care Med* 2007;35:2191–2195.
15. Levi M, Toh CH, Thachil J, Watson HG. Guidelines for the diagnosis and management of disseminated intravascular coagulation. British Committee for Standards in Haematology. *Br J Haematol* 2009;145:24–33.
16. Medcalf RL. Fibrinolysis, inflammation, and regulation of the plasminogen activating system. *J Thromb Haemost* 2007;5(Suppl 1):132–142.
17. Declerck PJ. Thrombin activatable fibrinolysis inhibitor. *Hamostaseologie* 2011;31: 165–166, 168–173.
18. Busuttil SJ, Ploplis VA, Castellino FJ, Tang L, Eaton JW, Plow EF. A central role for plasminogen in the inflammatory response to biomaterials. *J Thromb Haemost* 2004;2:1798–1805.
19. Neun BW, Dobrovolskaia MA. Method for *in vitro* analysis of nanoparticle thrombogenic properties. *Methods Mol Biol* 2011;697:225–235.

20. Adcock D, Hoefner D, Kottke-Marchant K, Marlar R, Szamosi D, Warunek D. Collection, transport, and processing of blood specimens for testing plasma-based coagulation assays and molecular hemostasis assays; Approved guidline — 5th edn. *Clinical and Laboratory Standards Institute* 2008; H21–A5:Vol. 28.
21. Dobrovolskaia MA, Aggarwal P, Hall JB, McNeil SE. Preclinical studies to understand nanoparticle interaction with the immune system and its potential effects on nanoparticle biodistribution. *Mol Pharm* 2008;5:487–495.
22. Gorbet MB, Sefton MV. Biomaterial-associated thrombosis: roles of coagulation factors, complement, platelets and leukocytes. *Biomaterials* 2004;25:5681–5703.
23. Samama MM, Guinet C. Laboratory assessment of new anticoagulants. *Clin Chem Lab Med* 2011;49:761–772.
24. Hemker HC. The initiation phase — a review of old (clotting-) times. *Thromb Haemost* 2007;98:20–23.
25. Kornalik F, Schieck A, Habermann E. Isolation, biochemical and pharmacologic characterization of a prothrombin-activating principle from *Echis carinatus* venom. *Naunyn Schmiedebergs Arch Pharmakol* 1969;264:259–260.
26. Calatzis A, Peetz D, Haas S, Spannagl M, Rudin K, Wilmer M. Prothrombinase-induced clotting time assay for determination of the anticoagulant effects of unfractionated and low-molecular-weight heparins, fondaparinux, and thrombin inhibitors. *Am J Clin Pathol* 2008;130:446–454.
27. Peng HT. Thromboelastographic study of biomaterials. *J Biomed Mater Res B Appl Biomater* 2010;94:469–485.
28. Lai BF, Creagh AL, Janzen J, Haynes CA, Brooks DE, Kizhakkedathu JN. The induction of thrombus generation on nanostructured neutral polymer brush surfaces. *Biomaterials* 2010;31:6710–6718.
29. Kannan RY, Salacinski HJ, De Groot J, *et al.* The antithrombogenic potential of a polyhedral oligomeric silsesquioxane (POSS) nanocomposite. *Biomacromolecules* 2006;7: 215–223.
30. Rosen S. Chromogenic methods in coagulation diagnostics. *Hamostaseologie* 2005; 25:259–266.
31. Castoldi E, Rosing J. Thrombin generation tests. *Thromb Res* 2011;127(Suppl 3): S21–S25.
32. Seyfert UT, Biehl V, Schenk J. *In vitro* hemocompatibility testing of biomaterials according to the ISO 10993-4. *Biomol Eng* 2002;19:91–96.
33. Urano T, Nishikawa T, Nagai N, Takada Y, Takada A. Amounts of tPA and PAI-1 in the euglobulin fraction obtained at different pH: Their relation to the euglobulin clot lysis time. *Thromb Res* 1997;88:75–80.
34. Yang M, Cui G, Zhao M, *et al.* The effect of complexation of Cu(II) with P6A peptide and its analogs on their thrombolytic activities. *Int J Pharm* 2008;362:81–87.
35. Becker A, Marxer E, Brussler J, *et al.* Ultrasound active nanoscaled lipid formulations for thrombus lysis. *Eur J Pharm Biopharm* 2011;77:424–429.
36. Gui L, Zhao M, Wang Y, Qin Y, Liu J, Peng S. Synthesis, nanofeatures, *in vitro* thrombus lysis activity and *in vivo* thrombolytic activity of poly-alpha,beta-aspartyl-L-alanine. *Nanomedicine (Lond)* 2010;5:703–714.
37. Dobrovolskaia MA, Patri AK, Zheng J, *et al.* Interaction of colloidal gold nanoparticles with human blood: Effects on particle size and analysis of plasma protein binding profiles. *Nanomedicine* 2009;5:106–117.

38. Sahli H, Tapon-Bretaudiere J, Fischer AM, *et al.* Interactions of poly(lactic acid) and poly(lactic acid-co-ethylene oxide) nanoparticles with the plasma factors of the coagulation system. *Biomaterials* 1997;18:281–288.
39. Song L, Zhu D, Liu L, Dong X, Zhang H, Leng X. Evaluation of the coagulation properties of arginine-chitosan/DNA nanoparticles. *J Biomed Mater Res B Appl Biomater* 2010;95:374–379.
40. Wang S, Zhang Y, Wang H, Dong Z. Preparation, characterization and biocompatibility of electrospinning heparin-modified silk fibroin nanofibers. *Int J Biol Macromol* 2011;48:345–353.
41. Salvador-Morales C, Zhang L, Langer R, Farokhzad OC. Immunocompatibility properties of lipid-polymer hybrid nanoparticles with heterogeneous surface functional groups. *Biomaterials* 2009;30:2231–2240.
42. Letchford K, Liggins R, Wasan KM, Burt H. *In vitro* human plasma distribution of nanoparticulate paclitaxel is dependent on the physicochemical properties of poly (ethylene glycol)-block-poly(caprolactone) nanoparticles. *Eur J Pharm Biopharm* 2009; 71:196–206.
43. Martinez-Gutierrez F, Thi EP, Silverman JM, *et al.* Antibacterial activity, inflammatory response, coagulation and cytotoxicity effects of silver nanoparticles. *Nanomedicine* 2012;8:328–326.
44. Shrivastava S, Singh SK, Mukhopadhyay A, Sinha AS, Mandal RK, Dash D. Negative regulation of fibrin polymerization and clot formation by nanoparticles of silver. *Colloids Surf B Biointerfaces* 2011;82:241–246.
45. Dash BC, Rethore G, Monaghan M, Fitzgerald K, Gallagher W, Pandit A. The influence of size and charge of chitosan/polyglutamic acid hollow spheres on cellular internalization, viability and blood compatibility. *Biomaterials* 2010;31:8188–8197.
46. Tavano R, Segat D, Reddi E, *et al.* Procoagulant properties of bare and highly PEGylated vinyl-modified silica nanoparticles. *Nanomedicine (Lond)* 2010;5:881–896.
47. Simberg D, Zhang WM, Merkulov S, *et al.* Contact activation of kallikrein-kinin system by superparamagnetic iron oxide nanoparticles *in vitro* and *in vivo*. *J Controlled Release* 2009;140:301–305.
48. Bai S, Thomas C, Ahsan F. Dendrimers as a carrier for pulmonary delivery of enoxaparin, a low-molecular weight heparin. *J Pharm Sci* 2007;96:2090–2106.
49. Bai S, Ahsan F. Synthesis and evaluation of PEGylated dendrimeric nanocarrier for pulmonary delivery of low molecular weight heparin. *Pharm Res* 2009;26:539–548.
50. Feng X, Cheng Y, Yang K, Zhang J, Wu Q, Xu T. Host-guest chemistry of dendrimer-drug complexes. 5. Insights into the design of formulations for noninvasive delivery of heparin revealed by isothermal titration calorimetry and NMR studies. *J Phys Chem B* 2012;114:11017–11026.
51. Hoffart V, Lamprecht A, Maincent P, Lecompte T, Vigneron C, Ubrich N. Oral bioavailability of a low molecular weight heparin using a polymeric delivery system. *J Controlled Release* 2006;113:38–42.
52. Jeon O, Kang SW, Lim HW, Hyung Chung J, Kim BS. Long-term and zero-order release of basic fibroblast growth factor from heparin-conjugated poly(L-lactide-co-glycolide) nanospheres and fibrin gel. *Biomaterials* 2006;27:1598–1607.
53. Meng N, Zhang SQ, Zhou NL, Shen J. Biopolymer-modified graphite oxide nanocomposite films based on benzalkonium chloride-heparin intercalated in graphite oxide. *Nanotechnology* 2010;21:185101.

54. Shan L. Glycol chitosan/heparin-immobilized gold-deposited iron oxide nanoparticles. In: *Molecular Imaging and Contrast Agent Database* (MICAD) [internet]. Bethesda (MD), National Center for Biotechnology Information (US), 2011.
55. Stevens KN, Croes S, Boersma RS, *et al.* Hydrophilic surface coatings with embedded biocidal silver nanoparticles and sodium heparin for central venous catheters. *Biomaterials* 2011;32:1264–1269.
56. Myerson J, He L, Lanza G, Tollefsen D, Wickline S. Thrombin-inhibiting perfluorocarbon nanoparticles provide a novel strategy for the treatment and magnetic resonance imaging of acute thrombosis. *J Thromb Haemost* 2011;9:1292–1300.
57. Zambaux MF, Bonneaux F, Gref R, Dellacherie E, Vigneron C. Protein C-loaded monomethoxypoly (ethylene oxide)-poly(lactic acid) nanoparticles. *Int J Pharm* 2001; 212:1–9.
58. Ma YH, Wu SY, Wu T, Chang YJ, Hua MY, Chen JP. Magnetically targeted thrombolysis with recombinant tissue plasminogen activator bound to polyacrylic acid-coated nanoparticles. *Biomaterials* 2009;30:3343–3351.
59. Wang SS, Chou NK, Chung TW. The t-PA-encapsulated PLGA nanoparticles shelled with CS or CS-GRGD alter both permeation through and dissolving patterns of blood clots compared with t-PA solution: An *in vitro* thrombolysis study. *J Biomed Mater Res A* 2009;91:753–761.
60. Uesugi Y, Kawata H, Jo J, Saito Y, Tabata Y. An ultrasound-responsive nano delivery system of tissue-type plasminogen activator for thrombolytic therapy. *J Controlled Release* 2010;147:269–277.
61. Fernandes EG, de Queiroz AA, Abraham GA, San Roman J. Antithrombogenic properties of bioconjugate streptokinase-polyglycerol dendrimers. *J Mater Sci Mater Med* 2006;17:105–111.
62. Marsh JN, Senpan A, Hu G, *et al.* Fibrin-targeted perfluorocarbon nanoparticles for targeted thrombolysis. *Nanomedicine (Lond)* 2007;2:533–543.
63. Wang X, Inapagolla R, Kannan S, Lieh-Lai M, Kannan RM. Synthesis, characterization, and *in vitro* activity of dendrimer-streptokinase conjugates. *Bioconjug Chem* 2007;18:791–799.
64. Bi F, Zhang J, Su Y, Tang YC, Liu JN. Chemical conjugation of urokinase to magnetic nanoparticles for targeted thrombolysis. *Biomaterials* 2009;30:5125–5130.
65. Marsh JN, Hu G, Scott MJ, *et al.* A fibrin-specific thrombolytic nanomedicine approach to acute ischemic stroke. *Nanomedicine (Lond)* 2011;6:605–615.
66. Ziv-Polat O, Topaz M, Brosh T, Margel S. Enhancement of incisional wound healing by thrombin conjugated iron oxide nanoparticles. *Biomaterials* 2010;31:741–747.
67. Dhadwar SS, Kiernan J, Wen J, Hortelano G. Repeated oral administration of chitosan/DNA nanoparticles delivers functional FVIII with the absence of antibodies in hemophilia A mice. *J Thromb Haemost* 2010;8:2743–2750.
68. Bowman K, Sarkar R, Raut S, Leong KW. Gene transfer to hemophilia A mice via oral delivery of FVIII-chitosan nanoparticles. *J Controlled Release* 2008;132:252–259.
69. Rangnekar A, Zhang AM, Li SS, *et al.* Increased anticoagulant activity of thrombin-binding DNA aptamers by nanoscale organization on DNA nanostructures. *Nanomedicine* 2011; 8:673–681.
70. Okamura Y, Handa M, Suzuki H, Ikeda Y, Takeoka S. New strategy of platelet substitutes for enhancing platelet aggregation at high shear rates: Cooperative effects of a

mixed system of fibrinogen gamma-chain dodecapeptide- or glycoprotein Ibalpha-conjugated latex beads under flow conditions. *J Artif Organs* 2006;9:251–258.
71. Okamura Y, Takeoka S, Eto K, *et al.* Development of fibrinogen gamma-chain peptide-coated, adenosine diphosphate-encapsulated liposomes as a synthetic platelet substitute. *J Thromb Haemost* 2009;7:470–477.
72. Rejinold NS, Muthunarayanan M, Chennazhi KP, Nair SV, Jayakumar R. Curcumin loaded fibrinogen nanoparticles for cancer drug delivery. *J Biomed Nanotechnol* 2011;7:521–534.
73. Rejinold NS, Muthunarayanan M, Chennazhi KP, Nair SV, Jayakumar R. 5-fluorouracil loaded fibrinogen nanoparticles for cancer drug delivery applications. *Int J Biol Macromol* 2011;48:98–105.
74. Rejinold NS, Muthunarayanan M, Deepa N, Chennazhi KP, Nair SV, Jayakumar R. Development of novel fibrinogen nanoparticles by two-step co-acervation method. *Int J Biol Macromol* 2010;47:37–43.
75. Owens DE, 3rd, Peppas NA. Opsonization, biodistribution, and pharmacokinetics of polymeric nanoparticles. *Int J Pharm* 2006;307:93–102.
76. Hu WJ, Eaton JW, Ugarova TP, Tang L. Molecular basis of biomaterial-mediated foreign body reactions. *Blood* 2001;98:1231–1238.
77. Deng ZJ, Liang M, Monteiro M, Toth I, Minchin RF. Nanoparticle-induced unfolding of fibrinogen promotes Mac-1 receptor activation and inflammation. *Nat Nanotechnol* 2011;6:39–44.
78. Chen G, Ni N, Wang B, Xu B. Fibrinogen nanofibril growth and self-assembly on Au (1,1,1) surface in the absence of thrombin. *Chemphyschem* 2010;11:565–568.
79. Sund J, Alenius H, Vippola M, Savolainen K, Puustinen A. Proteomic characterization of engineered nanomaterial-protein interactions in relation to surface reactivity. *ACS Nano* 2011;5:4300–4309.
80. Tenzer S, Docter D, Rosfa S, *et al.* Nanoparticle size is a critical physicochemical determinant of the human blood plasma corona: A comprehensive quantitative proteomic analysis. *ACS Nano* 2011;5:7155–7167.
81. Vroman L, Adams AL, Fischer GC, Munoz PC. Interaction of high molecular weight kininogen, factor XII, and fibrinogen in plasma at interfaces. *Blood* 1980;55: 156–159.
82. Cedervall T, Lynch I, Lindman S, *et al.* Understanding the nanoparticle-protein corona using methods to quantify exchange rates and affinities of proteins for nanoparticles. *Proc Natl Acad Sci U S A* 2007;104:2050–2055.
83. Deng ZJ, Mortimer G, Schiller T, Musumeci A, Martin D, Minchin RF. Differential plasma protein binding to metal oxide nanoparticles. *Nanotechnology* 2009;20:455101.
84. Ruh H, Kuhl B, Brenner-Weiss G, Hopf C, Diabate S, Weiss C. Identification of serum proteins bound to industrial nanomaterials. *Toxicol Lett* 2012;208:41–50.
85. Yongli C, Xiufang Z, Yandao G, Nanming Z, Tingying Z, Xinqi S. Conformational changes of fibrinogen adsorption onto hydroxyapatite and titanium oxide nanoparticles. *J Colloid Interface Sci* 1999;214:38–45.
86. Smith BS, Yoriya S, Grissom L, Grimes CA, Popat KC. Hemocompatibility of titania nanotube arrays. *J Biomed Mater Res A* 2010;95:350–360.
87. Rockwell GP, Lohstreter LB, Dahn JR. Fibrinogen and albumin adsorption on titanium nanoroughness gradients. *Colloids Surf B Biointerfaces* 2012;91:90–96.

88. Karlsson M, Tang L. Surface morphology and adsorbed proteins affect phagocyte responses to nano-porous alumina. *J Mater Sci Mater Med* 2006;17:1101–1111.
89. Chen G, Ni N, Zhou J, *et al.* Fibrinogen clot induced by gold-nanoparticle *in vitro*. *J Nanosci Nanotechnol* 2011;11:74–81.
90. Lacerda SH, Park JJ, Meuse C, *et al.* Interaction of gold nanoparticles with common human blood proteins. *ACS Nano* 2010;4:365–379.
91. Dolatshahi-Pirouz A, Pennisi CP, Skeldal S, *et al.* The influence of glancing angle deposited nano-rough platinum surfaces on the adsorption of fibrinogen and the proliferation of primary human fibroblasts. *Nanotechnology* 2009;20:095101.
92. Kertzman Z, Marchal J, Suarez M, *et al.* Mechanical, tribological, and biocompatibility properties of ZrN-Ag nanocomposite films. *J Biomed Mater Res A* 2008;84:1061–1067.
93. Roach P, Farrar D, Perry CC. Surface tailoring for controlled protein adsorption: Effect of topography at the nanometer scale and chemistry. *J Am Chem Soc* 2006;128:3939–3945.
94. Kendall M, Ding P, Kendall K. Particle and nanoparticle interactions with fibrinogen: The importance of aggregation in nanotoxicology. *Nanotoxicology* 2011;5:55–65.
95. Lord MS, Cousins BG, Doherty PJ, *et al.* The effect of silica nanoparticulate coatings on serum protein adsorption and cellular response. *Biomaterials* 2006;27:4856–4862.
96. Song L, Meng J, Zhong J, *et al.* Human fibrinogen adsorption onto single-walled carbon nanotube films. *Colloids Surf B Biointerfaces* 2006;49:66–70.
97. Salvador-Morales C, Flahaut E, Sim E, Sloan J, Green ML, Sim RB. Complement activation and protein adsorption by carbon nanotubes. *Mol Immunol* 2006;43:193–201.
98. Li D, Yuan L, Yang Y, *et al.* Adsorption and adhesion of blood proteins and fibroblasts on multi-wall carbon nanotubes. *Sci China C Life Sci* 2009;52:479–482.
99. Vauthier C, Persson B, Lindner P, Cabane B. Protein adsorption and complement activation for di-block copolymer nanoparticles. *Biomaterials* 2011;32:1646–1656.
100. Peracchia MT, Harnisch S, Pinto-Alphandary H, *et al.* Visualization of *in vitro* protein-rejecting properties of PEGylated stealth polycyanoacrylate nanoparticles. *Biomaterials* 1999;20:1269–1275.
101. Minelli C, Kikuta A, Tsud N, Ball MD, Yamamoto A. A micro-fluidic study of whole blood behaviour on PMMA topographical nanostructures. *J Nanobiotechnol* 2008;6:3.
102. Matsumoto E, Fukuda T, Miura Y. Bioinert surface to protein adsorption with higher generation of dendrimer SAMs. *Colloids Surf B Biointerfaces* 2011;84:280–284.
103. Luck M, Paulke BR, Schroder W, Blunk T, Muller RH. Analysis of plasma protein adsorption on polymeric nanoparticles with different surface characteristics. *J Biomed Mater Res* 1998;39:478–485.
104. Luck M, Schroder W, Harnisch S, *et al.* Identification of plasma proteins facilitated by enrichment on particulate surfaces: Analysis by two-dimensional electrophoresis and N-terminal microsequencing. *Electrophoresis* 1997;18:2961–2967.
105. Labarre D, Vauthier C, Chauvierre C, Petri B, Muller R, Chehimi MM. Interactions of blood proteins with poly(isobutylcyanoacrylate) nanoparticles decorated with a polysaccharidic brush. *Biomaterials* 2005;26:5075–5084.
106. Koh LB, Rodriguez I, Venkatraman SS. A novel nanostructured poly(lactic-co-glycolic-acid)-multi-walled carbon nanotube composite for blood-contacting applications: thrombogenicity studies. *Acta Biomater* 2009;5:3411–3422.

107. Kawakami H, Okuyama Y, Matsumoto N, Nagaoka S. Plasma protein adsorption to nanofabricated fluorinated polyimide surface. *J Biomater Sci Polym Ed* 2005;16: 1023–1034.
108. Gref R, Luck M, Quellec P, *et al.* 'Stealth' corona-core nanoparticles surface modified by polyethylene glycol (PEG): Influences of the corona (PEG chain length and surface density) and of the core composition on phagocytic uptake and plasma protein adsorption. *Colloids Surf B Biointerfaces* 2000;18:301–313.
109. Besheer A, Vogel J, Glanz D, Kressler J, Groth T, Mader K. Characterization of PLGA nanospheres stabilized with amphiphilic polymers: Hydrophobically modified hydroxyethyl starch *vs.* pluronics. *Mol Pharm* 2009;6:407–415.
110. Dell'Orco D, Lundqvist M, Oslakovic C, Cedervall T, Linse S. Modeling the time evolution of the nanoparticle-protein corona in a body fluid. *PLoS One* 2010;5:e10949.
111. Wyszogrodzka M, Haag R. Synthesis and characterization of glycerol dendrons, self-assembled monolayers on gold: A detailed study of their protein resistance. *Biomacromolecules* 2009;10:1043–1054.
112. Goppert TM, Muller RH. Adsorption kinetics of plasma proteins on solid lipid nanoparticles for drug targeting. *Int J Pharm* 2005;302:172–186.
113. Lynch I, Cedervall T, Lundqvist M, Cabaleiro-Lago C, Linse S, Dawson KA. The nanoparticle-protein complex as a biological entity; a complex fluids and surface science challenge for the 21st century. *Adv Colloid Interface Sci* 2007;134–135:167–174.
114. Lundqvist M, Stigler J, Elia G, Lynch I, Cedervall T, Dawson KA. Nanoparticle size and surface properties determine the protein corona with possible implications for biological impacts. *Proc Natl Acad Sci U S A* 2008;105:14265–14270.
115. Baker SE, Sawvel AM, Fan J, Shi Q, Strandwitz N, Stucky GD. Blood clot initiation by mesocellular foams: Dependence on nanopore size and enzyme immobilization. *Langmuir* 2008;24:14254–14260.
116. Simberg D, Park JH, Karmali PP, *et al.* Differential proteomics analysis of the surface heterogeneity of dextran iron oxide nanoparticles and the implications for their *in vivo* clearance. *Biomaterials* 2009;30:3926–3933.

Chapter 10

The Effects of Engineered Nanomaterials on Platelets

Jan Simak

Center for Biologics Evaluation and Research
Food and Drug Administration
1401 Rockville Pike, HFM-335
Rockville, MD 20852–1448, USA
jan.simak@fda.hhs.gov

Platelets (PLTs) are small discoidal cells that are anucleated in mammals and circulate in the blood, where they play a critical role in hemostasis and thrombosis. PLT adhesion at the site of vascular injury and their activation and aggregation result in the formation of the primary hemostatic plug. This plug is enforced by a fibrin network structure as a result of the activation of the plasma coagulation system (PCS). Thus, PLTs, together with the PCS and vascular endothelial cells, maintain the hemocoagulation balance. Under physiological conditions, these systems prevent blood clotting, while in cases of vascular injury, they facilitate hemostasis to prevent blood loss. The dysregulation of hemocoagulation can lead to life-threatening thrombotic and/or bleeding pathologies. As various engineered nanomaterials that will come into contact with the blood are being designed for biomedical applications and other nanomaterials may reach the circulation as a result of occupational, environmental, or other routes of exposure, it is necessary to evaluate the effects of engineered nanomaterials on PLTs. In the introduction of this chapter, both the physiology of PLTs and the pathophysiology of the possible adverse effects of nanomaterials on PLTs are briefly described. Next, the methods that are used for the *in vitro* evaluation of the effects of nanomaterials on PLTs are reviewed. Screenings for the effects of nanomaterials on PLTs are generally based on PLT aggregation assays. These tests include cell counter-based assays, light transmission aggregometry, and impedance aggregometry. The adhesion of PLTs to nanomaterial surfaces is frequently monitored by scanning electron microscopy or light microscopy. Ultrastructural changes are examined using transmission electron microscopy. PLT membrane activation markers (CD62P, CD63, PAC-1) and the release of PLT membrane microparticles are analyzed by flow cytometry or via immunomicroscopic techniques. Soluble activation markers that are released by PLTs (e.g., βTG, PF4, and TSP-1) are analyzed using ELISA. PLT integrity is assayed using a colorimetric lactate dehydrogenase (LDH) assay in the PLT supernatant. Other techniques are used to investigate the effects of nanomaterials on specific signaling pathways. Lastly, a summary of the published results regarding the *in vitro* effects of various nanomaterials on PLTs is presented.

Disclaimer:

1. Introduction

Different types of engineered nanomaterials have had profound impacts on the development of diagnostic biosensors, drug delivery nanosystems, imaging nanoprobes for intravascular use, and other devices that come into contact with blood.[1–10] Thus, designing nontoxic nanomaterials with optimum biocompatibility is a critical task. Biocompatibility is defined as the ability of a biomaterial to perform its desired function with respect to a medical therapy without eliciting any undesirable local or systemic effects in the recipient or beneficiary of said therapy. A further characteristic of a biocompatible material is that it generates the most appropriate beneficial cellular or tissular response in a specific context, thereby optimizing the clinically relevant performance of the given therapy.[11] The biocompatibility of engineered nanomaterials in the context of blood and the vasculature is based on the material's effects on a complex inflammatory response, including its impact on the hemocoagulation balance.[12] The hemocoagulation balance is maintained by the collaboration of vascular endothelial cells (ECs), the plasma coagulation system (PCS), and blood platelets (PLTs). This review chapter is complementary to the chapters on ECs and the PCS (Chapters 8 and 9, respectively) and is focused on the *in vitro* investigation of nanomaterial interactions with blood PLTs.

2. The Pathophysiology of the Possible Adverse Effects of Nanomaterials on Platelets

2.1. *Blood platelets*

PLT physiology and pathophysiology have been extensively reviewed;[13–23] only a brief introduction is presented here to provide basic information on PLTs and their hemostatic functions. PLTs are small discoidal cells that are anucleated in mammals. They are approximately 2–5 μm × 0.5 μm in size, with a volume of 6–10 fL. PLTs are derived from megakaryocytes in the bone marrow,[24] and one megakaryocyte generates approximately 5,000–10,000 PLTs. Under physiological conditions, the PLT count in peripheral blood is 150–450 × 10^9/L, and PLTs circulate for approximately 7–10 days. In addition, a pool of PLTs is stored in the spleen, and these PLTs can be rapidly mobilized into the circulation. Old PLTs are removed by macrophages (as part of the reticuloendothelial system, RES) in both the liver and spleen.

PLTs play a key role in hemostasis. In the bloodstream, PLTs circulate near the vessel wall and maintain the integrity of the vasculature. At least 5 × 10^9 functional PLTs/L are required to prevent spontaneous bleeding in non-traumatized thrombocytopenic patients. The clinical PLT count threshold for prophylactic PLT transfusion is 10–20 × 10^9/L. Trauma or surgery requires a substantially higher functional PLT count to prevent bleeding.[25] Furthermore, PLTs are very reactive cells and are equipped with multiple surface receptors that bind various activatory and inhibitory ligands.[26] The activation of different signaling pathways eventually leads to the increase of the intracellular calcium (Ca^{2+}) concentration and subsequent changes in cytoskeletal organization, resulting in 1) alterations in the PLT's shape from discoidal to spherical, 2) the formation of pseudopodia, 3) adhesion and aggregation, 4) the exposure of the procoagulant surface to the PCS activation complexes, 5) the release of procoagulant membrane microparticles (MPs), and 6) the release of mediators from different types of PLT granules. Thus, PLT activation results in the formation of a primary hemostatic plug that is enforced with a fibrin network as a structural result of PCS activation. Clot retraction is an additional function of PLTs that is dependent on a functional PLT cytoskeleton. PLT activation signals are generated during any type of injury of the endothelial intima of blood vessels and by the exposure of PLTs to subendothelial matrix proteins such as collagen, or subendothelial tissues.

In addition to vessel wall injury, there are several other types of natural and artificial PLT activation agonists. For example, PLTs can be activated by immunoglobulins, microorganisms (bacteria, parasites) and their by-products, drugs, and artificial surfaces and particles such as engineered nanomaterials.

The activatory or inhibitory effects of biomaterials are generally evaluated based on the material's interference with the PLTs' abilities to perform their hemostatic functions, namely adhesion, aggregation, and the release reaction.

Under physiological conditions within an intact healthy vasculature, most PLTs never significantly interact with the endothelial surface.[15] The vascular endothelium actively releases antithrombotic mediators, such as prostacyclin (PGI_2) and nitric oxide (NO), to prevent undesired adhesion and activation of PLTs at the intact vessel wall (for details, see Chapter 8). At sites of vascular injury, however, the subendothelial matrix is exposed. This matrix contains several adhesive proteins, such as collagen, von Willebrand factor (vWf), laminin, fibronectin, and thrombospondin, which causes PLTs to adhere to the site of injury, thereby limiting the degree of the hemorrhage and promoting tissue healing. Among the subendothelial matrix proteins, type I and type III fibrillar collagens are the strongest promoters of PLT adhesion due to their strong PLT-activating potential and affinity to vWf. The PLT receptors that directly bind collagen are glycoprotein VI (GPVI), a member of the immunoglobulin super-family, and integrin $\alpha_2\beta_1$ (also known as GPIa-IIa, VLA-2, or CD49b/CD29). However, despite extensive research, the precise roles of these receptors in the context of PLT adhesion and activation remain unclear.[15] PLT adhesion to the extracellular matrix involves tethering, rolling, activation, and firm adhesion.[15] At low shear rates (<1,000 s^{-1}), such as in veins and larger arteries, PLTs primarily adhere to collagen, fibronectin, and laminin. At higher shear rates (>1,000 s^{-1}), such as in smaller arteries, the microvasculature, or stenotic vessels, the interaction between the PLT surface glycoprotein Ibα (GPIbα, CD42b) and vWf becomes critically important for slowing the fast-moving PLTs. GPIbα is part of the GPIb/IX/V complex, which consists of the leucine-rich repeat glycoproteins GPIbα, GPIbβ, GPIX, and GPV. vWf is a large multimeric adhesive protein that is synthesized in ECs and megakaryocytes. Circulating vWf binds to exposed collagen. vWf immobilization and/or high fluid shear stress reveal the cryptic GPIbα binding site on vWf, thus facilitating the deceleration of PLTs and maintaining their close contact with the subendothelial matrix. The firm adhesion of PLTs is then mediated by integrin receptors and requires PLT activation, shifting several β_1 and β_3 integrins into their ligand-binding conformation. The most abundant integrin receptor on PLTs is $\alpha_{IIb}\beta_3$ (GPIIb-IIIa, CD41a). This receptor facilitates strong adhesion to vWf and immobilized fibrinogen and is also the primary aggregation receptor. The precise physiological importance of other integrins, such as $\alpha_V\beta_3$ (a vitronectin receptor), $\alpha_5\beta_1$ (a fibronectin receptor), and $\alpha_6\beta_1$ (a laminin receptor), has not been elucidated.[15]

PLTs can be activated by different signal transduction pathways that primarily involve the activation of tyrosine receptor kinase cascades or transmembrane G protein-coupled receptors (GPCRs).[15] Certain receptors, such as integrin $\alpha_{IIb}\beta_3$, are involved in both outside-in and inside-out signaling. Inside-out signaling is caused by stimulation-induced increases in free intracellular Ca^{2+} and diacylglycerol levels. This type of signaling is associated with cytoskeletal rearrangement and conformational changes in the integrin receptors, which result in the exposure of their fibrinogen-binding sites.

PLT adhesion and activation are followed by PLT aggregation, which in turn leads to the formation of the primary hemostatic plug at the site of vascular injury. PLT aggregation is a complex and dynamic process that involves many receptors, such as GPIbα and $\alpha_{IIb}\beta_3$, and several ligands, such as fibrinogen, fibronectin, and vWf. PLT aggregation also involves PLTs in different states of activation.[15] Three distinct mechanisms that initiate PLT aggregation have been identified. At low shear rates, PLT aggregation is primarily mediated by an $\alpha_{IIb}\beta_3$–fibrinogen interaction. At high shear rates, i.e., between 1,000 and 10,000 s^{-1}, the initial aggregation step depends on the adhesive functions of GPIbα and $\alpha_{IIb}\beta_3$ and is mediated by the formation of reversible PLT aggregates. The second step toward the formation of stable aggregates is dependent on the generation of PLT agonists and the irreversible activation of $\alpha_{IIb}\beta_3$. At very high shear stress rates, i.e., >10,000 s^{-1}, PLT aggregation is exclusively mediated by GPIbα–vWf interactions.[15]

Biologically active PLT-derived mediators are released either newly synthesized or from pre-existing stores in different granules. The former group of mediators is represented by PLT cyclooxygenase (COX) products.[14,17] PLTs constitutively express COX-1 and can also express COX-2 during inflammation. The predominant COX product in PLTs is thromboxane A_2 (TXA_2), which has pro-inflammatory, prothrombotic, and vasoconstricting activities. PLT-activating factor (PAF) is another PLT-derived pro-inflammatory lipid mediator. Sphingosin 1-phosphate, lysophosphatidic acid (LPA), and 2-arachidonyl glycerol are other examples of bioactive mediators that are synthesized and released by PLTs.[17]

Following the binding of PLTs to the subendothelial matrix and the subsequent transmission of signaling into the PLT, a controlled release reaction takes place. The granule contents are secreted when the granule membrane fuses with the intracellular membrane systems that connect with the surface, such as the open canalicular system (OCS) or the plasma membrane itself. This membrane fusion involves two categories of SNARE proteins, namely vesicular SNARE proteins in the membranes of secretory organelles (v-SNARES) and SNARE proteins that are present on the target membrane (t-SNARES).[17]

PLTs contain three types of granules that are released in a regulated manner upon stimulation.[14,17] PLT granules contain a variety of mediators that facilitate the functions of PLTs in both hemostasis and inflammation. Dense granules store small molecules such as ATP, ADP, serotonin, glutamate, and Ca^{2+}. In addition, dense PLT granules contain significant amounts of inorganic polyphosphates. Polyphosphates regulate plasma coagulation and fibrinolysis and activate factor XII, indicating that PLTs release a plasma contact system activator.[27] Alpha granules contain an extensive variety of proteins, including adhesive proteins such as vWf, fibronectin, fibrinogen, thrombospondin (TSP-1, TSP-2), laminin, and vitronectin. Alpha granules also contain different plasma coagulation and fibrinolytic factors, protease inhibitors, and growth and mitogenic factors (e.g., PDGF, VEGF). Important components of alpha granules also include chemokines and cytokines, including PLT factor 4 (PF4/CXCL4), β-thromboglobulin (βTG and

NAP-2/CXCL7 as its active breakdown products), RANTES (CCL5), IL-1α, IL-1β, TGF-β, TNF-α, membrane glycoproteins, such as GPIb (CD42b), GPIIb-IIIa (CD41a), PECAM-1 (CD36), P-selectin (CD62P), and CD40L, and other proteins. Lysosomes are the third type of PLT granules. They contain proteases (cathepsins, elastases) and other hydrolases (phosphatases, glycosidases) that are responsible for protein and matrix degradation. CD63, a member of the tetraspanin family, is present on the lysosomal membrane and together with alpha-granular CD62P, is frequently used as a PLT activation marker.[17]

PLTs also release different types of exocytic vesicles of approximately 100–1,000 nm in diameter; these vesicles are commonly referred to as PLT membrane MPs. MP populations of different sizes have been identified and are likely released by distinct mechanisms from the plasma membranes of stimulated PLTs. The majority of PLT MPs expose phosphatidylserine, indicating their procoagulant nature. PLT MPs contain a number of membrane receptors and other protein and lipid mediators and have been implicated in various pathophysiological processes, including tumor metastasis, stem cell engraftment, and inflammatory and neurodegenerative diseases.[17,28] PLT MPs are distinct from the smaller exosomes (40–100 nm in diameter) that are released from multivesicular bodies and alpha granules.

In addition to their critical role in hemostasis, PLTs play multiple roles in other inflammation-related processes. PLTs are involved in the host defense against pathogens, the regulation of vascular tone, and wound healing by participating in the regeneration and reparation of tissues. PLTs also play important roles in tumor growth and metastasis. PLTs themselves release different growth factors and other mediators. As was discussed in Chapter 8, PLTs interact with all of the cellular and protein systems of the inflammatory response that have potentially adverse effects when delocalized or dysregulated. Moreover, PLTs are involved in the recruitment of endothelial progenitor cells; PLTs interact with monocytes and neutrophils in the circulation and at the site of injury. They also induce the activation of monocytes and neutrophils leading to the production of cytokines, proteolysis, oxidative bursts, and other processes.[29,30] Under physiological conditions, the healthy vascular endothelium attempts to prevent PLT activation and maintain PLTs in a resting state in the circulation by means of different inhibitory mediators. The interference of biomaterials with these processes can have dramatic clinical consequences.

2.2. *Possible adverse effects of nanomaterial interactions with platelets*

To evaluate the possible clinical effects of nanomaterials on PLTs, we must consider the different categories of nanomaterials that come into contact with them: 1) nanoparticles (NPs, e.g., fullerenes, dendrimers, liposomes), 2) macrosurfaces that are composed of nanomaterials or NP-containing composites (e.g., plastics with NP additives), and 3) nanotextured macrosurfaces (e.g., filtration membranes, tissue scaffolds). The type of exposure is also a critical issue. The basic types of nanomaterial intravascular contact and blood exposure are as follows. First, NPs may be administered directly to the intravascular compartment via either injection or an implanted intravascular or extravascular device. Other types of direct exposure involve the use of nanomaterials in surgery or trauma treatment or in different extracorporeal circulation techniques (e.g., hemodialysis, apheresis, cardiopulmonary bypass, extracorporeal membrane oxygenation). Lastly, nanomaterials can be directly exposed to the blood during the collection, processing, and storage of blood transfusion products. In addition, the direct exposure of nanomaterials to blood and the localization of NPs to either the intravascular compartment or the bone marrow following other types or exposure (e.g., oral, inhalation, dermal) should also be considered. The type of exposure and the pharmacokinetic behavior (i.e., the ADME characteristics) of each specific nanomaterial for its intended application are critically important with respect to the clinical effects of the material on PLTs. The possible adverse effects of nanomaterials on both PLT count and functions and the clinical manifestation of such effects are summarized in Figure 1. In the worst-case scenario, the interaction of the PLT with the nanomaterial may lead to the development of irreversible multiple organ dysfunction and death. In most cases, the preclinical evaluation of the adverse effects of biomaterials, particularly with respect to *in vivo* animal studies, is based on a single insult model in which a healthy organism is exposed to the tested biomaterial. The same principle is also applied in human studies, which test materials on healthy volunteers. In clinical practice, however, the majority of patients in whom nanomaterials will be used for diagnostic or therapeutic purposes will already be severely ill. These patients will exhibit a certain level of dysregulation of both hemostasis and the inflammatory response and a certain level of dysfunction of multiple vital organs. It is likely that even minor effects of the nanomaterial on PLTs may have critical consequences in these patients compared with the effects of these materials on healthy individuals. A strong example of the differential role of PLTs in healthy versus injured tissue can be observed in the context of lung injury. Whereas intact PLTs in healthy humans are not sequestered or cleared in the lungs, activated or damaged PLTs adhere to the pulmonary vasculature in healthy lungs. Moreover, PLTs strongly stimulate the recruitment of neutrophils and other leukocytes in the injured pulmonary vasculature. Thus, PLTs may play important roles in the progression and possibly in the initiation of inflammatory lung diseases, including acute lung injury.[31] Therefore,

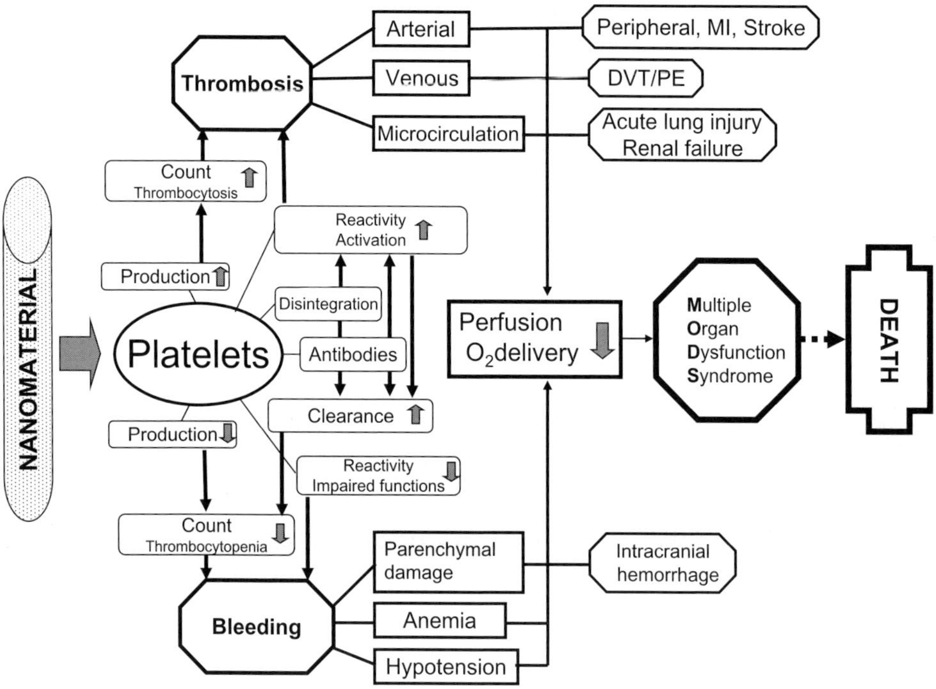

Figure 1. The potentially serious *in vivo* adverse effects resulting from nanomaterial interactions with platelets (PLTs).[12]

DVT: deep venous thrombosis, MI: myocardial infarction, PE: pulmonary embolism.

potential pulmonary adverse events should always be carefully monitored when testing the interactions between nanomaterials and PLTs *in vivo*.

3. Methods for Evaluating the *In Vitro* Effects of Nanomaterials on Platelets

Several types of assays can be used to evaluate PLT hemostatic functions, including adhesion, aggregation, and release reaction assays, which are based on various principles. The methods that are used for *in vitro* PLT testing have been reviewed extensively.[32–36] For practical reasons, the assays that are described in the following review are grouped according to the required instrumentation. No standard panel of PLT assays for the evaluation of nanomaterials is available. The protocols for the assays used in our laboratory are included at the end of this chapter. Based on a review of published results, a list of assays that are used to evaluate the effects of nanomaterials on PLTs is summarized in Table 1. The usefulness of the different assays for this purpose is discussed below.

Table 1. Methods that are used to study PLT interactions with engineered nanomaterials.

Method	PLT sample	Assay	Nanomaterial	References
Cell counter	PRP	PLT–LIP aggregation	LIP	39
	PRP	PLT–LIP aggregation	LIP	40
	PRP	PLT adhesion; counting of adhered PLTs following trypsin-mediated release	GNP–PU Film	41
	PRP	PLT aggregation	DENNP	65
	Rat WB	PLT aggregation	TiNP	136
Light microscopy				
Phase contrast	PRP	PLT–LIP aggregation	LIP	44
	PRP	PLT–LIP aggregation	LIP	40
	PLS	PLT–LIP aggregation	LIP	45
	PLS	PLT–POLNP aggregation	POLNP	46
	PRP	PLT–POLNP aggregation	POLNP	47
	PLS	PLT aggregation single cell counting assay	SNP	49
	PRP	CD62P, CD41 immunostaining peroxidase detection	Nanopores	48
Fluorescence	PLS	Immunodetection of MMP-9	CNP	57
	PLS	Binding of NBD-labeled liposomes to activated PLTs	LIP	59
	PRP	PLT adhesion: acridine orange staining	DEN–SK film	58
	PLS	Binding of Alexa488 labeled NPs to PLTs	DENNP	60
	PRP	CD62P	CNP	55
	PLS	Intracellular free Ca^{2+} (FURA-2AM)	CNP	61
	PRP	CD61	LIP	54
	PLS	CD41a	LIP	51

(*Continued*)

Table 1. (*Continued*)

Method	PLT sample	Assay	Nanomaterial	References
	PRP	CD41a, LIP	LIP	52
	PRP	STIM1, ORAI1	CNP	56
	PLS	CD41a	LIP	53
	PRP	Calcein-AM	TiNP	78
TEM		PLT ultrastructure	CNP	57
		PLT ultrastructure	LIP	45
		PLT ultrastructure	SiNP	62
		PLT ultrastructure	IONP	63
		PLT ultrastructure	SNP	49
		STIM1, DTS	CNP	56
		CD41a	LIP	64
SEM	PLS	PLT–LIP interaction	LIP	53
	PLS	PLT–LIP interaction	LIP	51
	PRP	PLT morphology	CNP–PU film	66
	PRP	PLT aggregation	CNP	56
	PLS	PLT aggregation	POLNP	46
	PRP	PLT aggregation	DENNP	65
	PRP	PLT adhesion	GNP–PU film	41
	PRP	PLT adhesion	CNP	68
	PRP	PLT adhesion	Scaffold	67
	PRP	PLT adhesion	Nanopores	48
	PRP	PLT adhesion	POLNP	69
	PRP	PLT adhesion	POLNP	70
	PLS	PLT adhesion	GNP	71
	PRP	PLT adhesion	POLNP	72
	PRP	PLT adhesion	Al surface	157
	PRP	PLT adhesion	POL surface	135
	PRP	PLT adhesion	TiNP	74
	PRP	PLT adhesion	POLNP	75
	PRP	PLT adhesion	SNP	139
	PRP	PLT adhesion	AgBrNP	77
	PRP	PLT adhesion	TiNP	78
AFM	PRP	PLT adhesion	CNP	79
Aggregometry— impedance	WB	PLT aggregation + ATP release luminescence assay	CNP	57
	WB	PLT aggregation + ATP release luminescence assay	CNP	86

(*Continued*)

Table 1. (*Continued*)

Method	PLT sample	Assay	Nanomaterial	References
	WB	PLT aggregation	LIP	85
	PRP	PLT aggregation	SiNP	62
	PRP	PLT aggregation	POLNP	87
Aggregometry— light transmission	PRP	PLT aggregation	GNP	71
	PRP	PLT aggregation	CNP	61
	PRP	PLT aggregation	LIP	44
	PRP, PLS	PLT aggregation	LIP	81
	PLS	PLT aggregation	DENNP	60
	PLS	PLT aggregation	IONP	63
	PRP	PLT aggregation	DENNP	65
	PRP	PLT aggregation	POLNP	82
	PRP	PLT aggregation	LIP	145
	PRP	PLT aggregation	SiNP	84
Aggregometry— screen filtration pressure	Rab WB	PLT aggregation	CNP	89
Flow cytometry	WB	CD62P, PAC-1, CD42b, CD61	CNP	57
	WB	CD62P, CD41a	LIP	53
	WB	CD62P, CD41a,	LIP	52
	PRP	CD62P, CD61	LIP	97
	PRP	CD62P, CD63, CD41a, MPs	CNP	61
	PRP	CD62P, CD63	DEN-SK film	58
	PRP	CD62P, PAC-1, AnV binding	POLNP	100
	PRP	CD62P, PAC-1	SiNP	62
	PRP	CD62P, PAC-1	POLNP	87
	PRP	CD62P, PAC-1	POLNP	75
	PLS	CD62P, CD41a,	LIP	51
	PRP	CD62P, CD42b, FITC–IONP	IONP	63
	WB	CD62P, CD42b,	POLNP	98
	PRP	CD62P, CD42b,	SiNP	84
	WB	CD62P, PLT–WBC complexes	CNP	86
	WB	CD62P	CNP	55
	PLS	CD62P	LIP	76
	PRP	PAC-1	CNP	68
	PLS	PAC-1	GNP	71

(*Continued*)

Table 1. (*Continued*)

Method	PLT sample	Assay	Nanomaterial	References
	WB	PAC-1	LIP	85
	PRP, WB	CD42b, FITC–LIP	LIP	95
	PRP	CD41/CD61^{+} MP	Scaffold	67
	PRP	CD42b, CD61	LIP	54
	PLS	LIP–PLT	LIP	64
	PRP	LIP–PLT	LIP	148
ELISA	PRP	PF4	POLNP	47
	WB	βTG, PF4	POLNP	98
	PRP	βTG	Nanopores	48
	PRP	βTG	AgBrNP	77
	WB	TSP-1	Al membrane	73
Photometry	PRP	PLT aggregation	GdNP	104
	PRP	PLT adhesion — alamar blue	POL membrane	105
	PRP	$[Ca^{2+}]_i$ –Fura-FF/AM	CNP	56
	PRP	LDH release	SNP	139
	PRP	LDH release	AgBrNP	77
Luminometry	PLS	ATP release: Luciferin–luciferase assay	POLNP	46
Scintillation	PRP	[3H] and [14C] LIP Phagocytosis, exocytosis, intracellular localization	LIP	45
	PLS	14C 5HT	SNP	49
Perfusion chamber	WB	PLT aggregation	LIP	54
	WB	PLT adhesion	POLNP	83
	PRP	PLT–LIP interaction	LIP	52
QCM-D	PRP	PLT adhesion	SiNP, GNP POLNP	115
Thromboelastography	WB	Blood clotting	POLNP	158
SDS-PAGE zymography		Gelatinase MMP-2, MMP-9 activities	CNP	57
Immunoblotting		Phosphotyrosine	GNP	71

Abbreviations

AFM	atomic force microscopy
βTG	β-thromboglobulin
CNP	carbon NP
DENNP	dendrimer NP
FITC	fluorescein isothiocyanate
FURA-2AM	fura-2-acetoxymethyl ester
GdNP	gadolinium-containing NP
GNP	gold NP
IONP	iron oxide NP
LIP	liposome
MMP	matrix metalloproteinase
MP	membrane microparticle
NP	nanoparticle
NBD	nitrobenzodiazole
PF4	platelet factor 4
PLT	platelet
PLS	platelet suspension in buffer (Tyrode's)
PRP	platelet-rich plasma
POLNP	polymer NP
PU	polyurethane
QCM-D	quartz crystal microbalance with dissipation
SEM	scanning electron microscopy
SiNP	silica NP
SNP	silver NP
SK	streptokinase
TiNP	titanium-containing NP
TEM	transmission electron microscopy
TSP-1	thrombospondin 1
WB	anticoagulated whole blood
WBC	white blood cell

3.1. *The preparation of a nanomaterial sample for testing*

The standard preparation of a nanomaterial sample for its characterization and testing is a critical aspect of the given study.[12] Particularly, materials that are insoluble in aqueous solvents and unstable in suspension must be well characterized. For example, carbon nanotubes that are provided as a powder from the manufacturer must be resuspended in a buffer, which is generally performed with sonication. This process dramatically alters the material's characteristics. Certain studies are performed with nanomaterials that are resuspended in the presence of a detergent. Every additive or impurity in the nanomaterial significantly alters the interpretation of the results of the experiment. Another important issue is the stability of the resuspended nanomaterial, including possible oxidative and other chemical changes. It is strongly recommended to involve a material scientist early in the project and to work with a well-characterized nanomaterial. Otherwise, the interpretation of results will be very problematic, and the value of the study may be compromised.

3.2. *Platelets for testing*

To achieve the maximum robustness of PLT assays, it is critical to collect and process blood samples in a standardized manner, with minimal impact on the PLTs.[12] Human blood should be used if possible, unless *ex vivo* studies are being performed on PLTs that were exposed to nanomaterials in an *in vivo* animal model. The reactivity of human PLTs from individual donors is highly variable for a number of reasons. Various drugs, including widely used nonsteroidal anti-inflammatory agents, impact PLT function for several days.[37] Therefore, the donor screening process is essential. Shear stress, inappropriate anticoagulation, temperature changes, and metabolic stress due to long-term storage are the primary factors to consider during blood processing and PLT preparation.[32] For PLT assays, citrate-anticoagulated or acid–citrate–dextrose (ACD)-anticoagulated blood is used. With respect to *in vitro* testing, nanomaterials are generally exposed to anticoagulated whole blood, PLT-rich plasma (PRP), or PLT suspension in Tyrode's buffer. Nanomaterials should be tested using PLTs from different donors. Sample size is determined by a variability of results of the used assays. As an example, in robust flow cytometry or aggregation assays, $n = 6$ may provide sufficient power ($\alpha = 0.05$, $\beta = 0.8$) to detect 20% difference from the control.

3.3. *Cell counter-based assays*

Cell counters that are based on the Coulter principle of the electrical sensing zone method are routinely used to assay PLT count, size, volume, and the percentage of aggregation.[38] Therefore, cell counters are useful in any PLT assay for the characterization of whole blood and PRP or PLT suspensions. Cell counters are also used to assess both PLT–NP aggregation[39,40] and PLT adhesion to macrosurfaces. In this latter type of study, adherent PLTs were cleaved from the material surface by trypsin and counted.[41] Additionally, a simple screening method for NP-induced PLT aggregation can be performed using cell counters.[42,43] Briefly, PRP is incubated with the material sample for 15 min at 37°C under static conditions and analyzed using a Coulter analyzer. As PLT aggregates are not counted in the single PLT window, the decrease in the PLT count allows for the calculation of the percentage of aggregation. The primary drawback of this assay is the absence of any agitation, which would provide homogeneous contact of PLTs with the tested NPs under low shear conditions. The introduction of stirring, however, would be difficult to standardize and would require additional instrumentation. This assay is simple to perform and may be useful for the screening of nanomaterials during product development to identify NPs with strong aggregation activity on PLTs. However, this assay only evaluates decreases in the single PLT count and does not distinguish between the aggregation and disintegration of PLTs. Therefore, other more specific methods are essential for a full investigation of the effects of a nanomaterial on PLTs.

3.4. *Microscopy*

Different light and electron microscopic techniques have been widely used in studies of PLT–nanomaterial interactions. Phase contrast light microscopy has been employed in several studies to observe PLT–NP aggregates.[40,44–49] PLT counts and basic morphology are other useful parameters that can be determined by phase contrast microscopy. PLT morphology can be evaluated using different scoring systems. For example, Kunicki's morphology score is frequently used to evaluate PLT status in transfusion products.[50] Immunomicroscopy with peroxidase staining has been used for the detection of both constitutively expressed PLT receptors and activation markers on PLTs that are adhered to a nanoporous material.[48] Immunodetection can be performed with more advanced fluorescence microscopy techniques. Various PLT antigens, including CD41a,[51–53] CD61,[54] CD62P,[55] STIM1, and ORAI1,[56] or the gelatinase MMP-9[57] have been detected using this method in nanomaterial studies. Fluorescence microscopy has various applications and has been used to detect adherent PLTs that are labeled with nucleic acid-specific dyes, such as acridine orange.[58] This type of microscopy has also been used to examine the adhesion of fluorescent dye-labeled NPs to PLTs.[59,60] With respect to other applications, our laboratory have used fluorescence microscopy to examine PLTs that are loaded with the calcium probe FURA-2AM to monitor the intracellular free Ca^{2+} concentration following the exposure of PLTs to NPs.[61] Transmission electron microscopy (TEM) has been used to evaluate the detailed morphology of NP-exposed PLTs,[45,49,56,57,62–64] and immunogold detection was employed in certain studies.[56,64] Scanning electron microscopy (SEM) has been used to visualize PLT–NP aggregation[46,51,53,56,65] or PLT adhesion to nanocomposite surfaces, scaffolds, or nanoporous materials.[41,48,66–78] Atomic force microscopy (AFM) has also been used to image the interactions between nanomaterials and PLTs.[79] Generally, microscopy techniques require well-trained and experienced professionals. From a statistical perspective, low numbers of PLTs are generally evaluated by microscopy, and sample processing may lead to artifacts. Although microscopic techniques are extremely valuable tools, these approaches should be complementary to other methods such as flow cytometry.

3.5. *Aggregometry*

PLT aggregation can be detected using different physical principles. Light transmission aggregometry (LTA) is performed using PRP that is stirred in an aggregometer cuvette. The transmission of light through the PRP sample is set as 0% aggregation, whereas the transmission through PLT-poor plasma (PPP) is set as 100%. Thus, the aggregation response following the addition of an agonist can be monitored in real-time. A typical panel of the agonists that are used in PLT aggregation studies includes thrombin receptor-activating peptide (TRAP), collagen, ADP, epinephrine, arachidonic acid, and ristocetin, which facilitates the binding of vWf to the PLT receptor GPIb.[16]

PLT aggregation can also be examined using whole blood aggregometry (WBA). In this assay, whole blood is stirred in a cuvette, and changes in the impedance between two electrodes are measured. The increase in impedance is proportional to PLT aggregation because PLT aggregates adhere to the electrodes. Impedance aggregometry may be performed with PRP, but whole blood is primarily used for this assay. Whole blood impedance aggregometry has several advantages over LTA, including the fact that there is no need for PRP preparation, the blood is hence minimally manipulated prior to testing. Moreover, smaller volumes of whole blood are required for testing, as compared to LTA. However, the presence of blood cells makes the analysis more complex. LTA and WBA are considered the gold standard assays of PLT function for the diagnosis of hereditary and several acquired PLT disorders. Along with these aggregation assays, the release reaction can be monitored using a lumi-aggregometer. In this assay, the ADP released from dense granules is converted to ATP, which reacts with a luciferin–luciferase complex to produce adenyl-luciferon. The luminescence that is generated is proportional to the ATP content when adenyl-luciferon is oxidized. Both LTA[44,60,61,63,65,71,80–84] and impedance WBA[44,57,85–87] have been used in different NP–PLT interaction studies. A different principle, i.e., WBA based on the screen filtration method, can also be used. This technique measures the resistance of flow of a whole blood sample through a microsieve with 30 μm^2 openings and returns the pressure rate as an index of PLT aggregation.[88] This method is not widely used; nevertheless, it was used in one NP study.[89] PLT aggregometry assays are strong candidates for the development of a standard assay for characterizing the effects of nanomaterials on PLTs. First, a possible interference of different nanomaterials on the results of LTA, impedance aggregometry, or other aggregometry methods must be evaluated. The quality of the PLTs that are used for the aggregation experiment is also a critical issue. Different drugs influence the ability of PLTs to aggregate. Furthermore, different methods of blood collection and manipulation, including shear stress or temperature changes, have a strong impact on PLT function in aggregation assays. Additionally, PLTs must be used for aggregation experiments within 4 h following blood collection because their reactivity changes over time. Appropriate positive and negative controls in aggregation experiments are required for each experimental run. When characterizing the effects of nanomaterials on PLTs, the following types of information can be obtained: 1) the direct PLT-aggregating activity of PRP and washed PLTs, 2) the effects of different inhibitors, such as ASA (acetylsalicyclic acid), prostacyclin, NO donors, calcium channel blockers, kinase inhibitors, or ion chelators (e.g., EDTA), 3) ADP/ATP release, and 4) the potentiating or inhibitory effects of nanomaterials on PLT aggregation induced by different concentrations of standard agonists (ADP, TRAP, collagen).

Simple assays that are recommended for the *in vitro* evaluation of PLT transfusion products include tests of the hypotonic stress response and the extent of shape change (ESC).[90] These assays are generally performed using LTA, although a photometer that is able to stir the suspension in the cuvette can also be used. The hypotonic stress response (HSR) is based on the ability of PLTs to extrude water and electrolytes after rapid swelling, which occurs when the PLTs are placed in a hypotonic solution. The ESC is determined in ADP-activated PLTs in the presence of EDTA to prevent PLT aggregation. While HSR indicates the metabolic status of the PLTs and correlates this status with the ATP level, the ESC is correlated with the morphology score. Both the HSR and ESC correlate with *in vivo* PLT recovery. In addition to their use in evaluating PLT transfusion products, ESC and HSR measurements may also be explored as simple complementary tests for the evaluation of nanomaterial-treated PLTs. No data, however, are available in this respect.

3.6. *Flow cytometry*

Flow cytometry is a multi-purpose tool that allows for a wide variety of PLT assays.[91–94] With regard to studies of PLT–nanomaterial interactions, PLT flow cytometry may be particularly useful for the following purposes: 1) to measure the activation state of circulating PLTs and their reactivity based on activation-dependent changes in PLT surface antigens, leukocyte–PLT aggregation, or the release of PLT-derived MPs, and 2) to assay intracellular PLT changes following the activation and induction of apoptosis, necrosis, or other types of stimulation (intracellular free Ca^{2+} concentration, mitochondrial potential). Flow cytometry is also valuable for the evaluation of PLT–NP interactions *in vivo*, assaying the binding of immunoglobulins to PLTs for the diagnosis of immune thrombocytopenias, or monitoring thrombopoiesis by counting young "reticulated" PLTs. When using fluorophore-conjugated NPs or naturally fluorescent NPs, the binding of NPs to PLTs can also be evaluated.[95] In addition, counts of both PLTs and PLT aggregates can be assayed. Flow cytometry allows for the analysis of several thousands of PLTs in one experimental run. In this technique, PLTs and other cells or processes are counted or assayed in the flow of a hydrodynamically focused stream of fluid. These counts or assays are performed using one or more laser beams with specific wavelengths. A detector is aligned with the light beam and determines a forward scatter (FSC) that is characteristic for each passing particle. The FSC is correlated with particle volume or size. A detector that is situated perpendicularly to the light beam records side scatter (SSC) characteristics, which depend on the inner complexity of the cell, such as its granularity, nucleus shape, and membrane roughness. Thus, PLTs and PLT membrane MPs form characteristic populations on an FSC/SSC plot. Additional fluorescence detectors recognize the fluorescence signal in certain contexts, e.g., when examining PLTs that are labeled with different monoclonal antibodies (mABs), each of which can be conjugated to a specific fluorophore that emits in a different fluorescence band. Results of flow cytometry studies are usually presented in both precentage of PLTs positive for the detected marker and mean of fluorescence intensity of analyzed populations of PLTs in arbitrary units. For the immunodetection of PLTs, mABs are generally used against major PLT surface receptors GPIIb-IIIa (CD41/CD61, CD41a) and GP Ib (CD42b).[51–53,57,61,63,65,67,80,84,95–98] Laboratory markers of PLT activation include activation-dependent conformational changes in the GPIIb-IIIa complex, exposure of granule membrane proteins, PLT surface binding of secreted PLT proteins, PLT and PLT–leukocyte aggregates, and the development of a procoagulant surface.[92] To detect activation-dependent conformational changes in the GPIIb-IIIa complex, the mAB PAC-1 is used, which is directed against the GPIIb-IIIa fibrinogen-binding site. This site is only exposed following PLT activation. The most widely used PLT surface activation marker is P-selectin (CD62P). P-selectin is expressed on the membrane of PLT alpha granules and is exposed on the PLT surface following alpha granule secretion. The activation-dependent increase in PLT surface P-selectin exposure may change over time *in vitro*, e.g., P-selectin may be released from PLT surface in MPs. Moreover, it has been documented that *in vivo* circulating degranulated PLTs rapidly lose their surface P-selectin expression. Therefore, P-selectin is not a reliable marker for detecting circulating degranulated PLTs.[92] Another example of a degranulation-dependent PLT surface marker is CD63, which resides on the membranes of lysosomes and dense granules in resting PLTs.[99] Flow cytometry analysis of the PLT surface expression of CD62P and CD63 and analysis of PLT binding of the PAC-1 antibody, have been performed in several studies on nanomaterial–PLT interactions.[51–53,55,57,58,61–63,65,68,71,75,76,80,84–87,97,98,100] Changes in PLT reactivity can also be investigated by the analysis of surface activation markers after the stimulation of PLTs with a range of concentrations of standard agonists, e.g., 1 and 20 μmol/L of ADP or TRAP,[94] or after PLT inhibition by NO using an NO donor like MAHMA NONOate.[101] The analysis of leukocyte–PLT aggregates appears to be a useful technique to monitor PLT activation *in vivo*.[92,93] Circulating monocyte ($CD14^+$)–PLT aggregates are more sensitive markers of *in vivo* PLT activation than surface CD62P expression.[92] Flow cytometry can also be used to analyze PLT membrane MPs, where MPs are phospholipid vesicles of approximately 0.1–1 μm in size that are released from the plasma membrane of stimulated PLTs and other cell types.[28,102] Thus far, the release of MPs from PLTs has been assayed only in a small number of nanomaterial studies,[61,65,67] including studies from our laboratory. PLT MPs expose various PLT membrane antigens. Furthermore, the majority of PLT MPs also expose accessible phosphatidylserine, which can be detected using fluorophore-conjugated annexin V in the presence of Ca^{2+} or by lactadherin. In addition, PLT MPs are procoagulant by promoting the activation of the PCS and may be prothrombotic *in vivo*. Therefore, the assessment of whether a nanomaterial induces MP release from PLTs is a useful complementary assay to both *in vitro* and *in vivo* testing.

3.7. *Solid phase assays (ELISA)*

The activation markers that are released into the PLT supernatant (either true soluble markers or those markers that are associated with different types of membrane vesicles) can be assayed using solid phase assays, such as ELISA.[103] Different proteins (e.g., PF4, βTG, TSP-1) or "soluble" forms of surface antigens or activation markers (e.g., sCD40L, sCD62P, GPV) can be analyzed to evaluate PLT activation. These assays have been used to study *in vivo* and *in vitro* PLT activation, including studies of nanomaterials.[47,48,73,77,98] Since these methods are prone to giving artifactual results, the variability of the results, particularly with respect to *in vivo* experiments, is generally very high. With respect to non-protein mediators, stable thromboxane metabolite thromboxane B_2 (TXB_2) and its metabolite 11-dehydro-TXB_2 can be assayed in serum and in urine, respectively, to detect *in vivo* PLT activation.

3.8. *Photometry, luminometry, and scintillation assays*

There are anecdotal reports of the use of a microplate reader for photometry-based PLT aggregation[104] or adhesion[105] assays. In certain studies, the enzymatic colorimetric assay of lactate dehydrogenase (LDH) release is employed to assess the effects of nanomaterials on the integrity of the PLT plasma membrane.[77,139] Similarly, the determination of the degree of LDH release in cell lines is part of the standard assays for the evaluation of the *in vitro* cytotoxicity of nanomaterials.[106] This assay is useful and simple and should be included when screening the effects of nanomaterials on PLTs. Luminometric assays of ATP release from PLTs have been used in certain nanomaterial studies, either alone or as a complementary approach to WBA.[46,57,86]

Scintillation assays have been employed for the [3H] or [14C] labeling of liposomes[45] and studies of 14C-5HT release from PLTs.[49]

3.9. *Biochemical status: pH, pO_2, pCO_2, HCO_3^-, lactate, glucose, and ATP*

The biochemical parameters of PLTs are generally assayed using clinical analyzers and are used to evaluate PLT transfusion products.[36] These assays will be useful for testing PLTs over long periods of exposure to nanomaterials, e.g., in devices that are used to process and store transfusion products. The biochemical status of PRP or PLT suspensions may be assayed to characterize PLTs prior to any testing to ensure that the PLTs are not under metabolic stress.

3.10. *Dynamic light scattering*

The analysis of the hydrodynamic size of PLTs and the temperature dependence of this parameter using dynamic light scattering (DLS) was proposed, together with a DLS analysis of the number of MPs, as a quality assurance test for PLT transfusion products.[107] Pilot studies appear promising; however, few researchers have applied DLS for PLT analysis. It would be very helpful to establish a DLS assay to evaluate the effects of nanomaterials on PLTs. DLS instrumentation is generally available in nanomaterial laboratories and is simple to use. Nevertheless, the primary complication with DLS is that it is not ideal for polydisperse particle suspensions or precipitating suspensions, which is precisely what PLTs with MPs and plasma proteins are. Therefore, the interpretation of DLS results for a complex mixture, such as for PLTs in plasma or other solutions, should be performed with great caution.

3.11. *Platelet function analyzers and thromboelastography*

Several types of analyzers are used, generally at the point of care, for the clinical evaluation of PLT dysfunctions.[103] These analyzers are based on different principles. Some are based on high shear PLT adhesion or aggregation, such as the PFA-100 or cone-and-platelet analyzer, whereas others are global hemostasis analyzers, such as the thromboelastograph analyzer (see Chapter 9). PFA-100 and thromboelastography have thus far been used to a very limited extent in the evaluation of the effects of nanomaterials in the blood.[108,109] The cone-and-platelet analyzer may be useful for investigation of nanomaterial effects on PLT adhesion to vWf under high shear rates.[110–112] Certain PLT funtion analyzers may eventually be demonstrated to be useful as screening or complementary tests.

3.12. *Perfusion chambers*

Different assays that are based on perfusion chambers have been used to evaluate PLT adhesion to biomaterial surfaces under specific shear stress conditions.[113] PLT adhesion and aggregation under dynamic flow conditions in perfusion chambers have been examined in a small number of nanomaterial studies.[52,54,83] The advantage of these methods is the possibility of modeling specific shear stresses, thereby mimicking the environment of different regions of *in vivo* circulation. Perfusion may be very useful for evaluating the impact of shear stress on PLT adhesion to nanomaterial-containing composites or nanotextured surfaces. However, a standard perfusion chamber assay is necessary for an interlaboratory comparison of the results. In a classical Baumgartner perfusion chamber, the PLTs flow along the subendothelial surface of a vessel segment.[114] Such chambers or modifications thereof, e.g., those chambers that use a collagen surface or even cultured ECs, may be useful to assess the adhesive functions of PLTs that are treated with NPs prior to or during perfusion.

3.13. *Quartz crystal microbalance with dissipation*

A new method for measuring PLT microaggregation using a commercially available quartz crystal microbalance with dissipation (QCM-D) under flow conditions has recently been developed.[115] The principle of QCM analysis is based on determining the resonance frequency of a quartz crystal that is induced to vibrate by applying an alternating electric field across the crystal. The deposition of mass on the quartz surface decreases the crystal's oscillation frequency (i.e., a negative frequency shift). For thin, rigid, and uniformly distributed layers, the resonance frequency is proportional to the mass. However, when a soft or thick layer is bound to the crystal, a high dissipation shift results. In this case, the mass can be underestimated when only the frequency is measured. The combined information from changes in frequency and dissipation is superior to frequency measurements alone. Both of these parameters can be monitored simultaneously in real-time using a QCM-D.[115] The Q-Sense® E4 QCM-D system has four temperature- and flow-controlled modules that are arranged in a parallel configuration. The samples can be perfused using a peristaltic microflow system. Using polystyrene-coated quartz crystals that were treated with fibrinogen, perfusion with PRP that was incubated with either 50 nm silica NPs (SiNPs) or with 60 nm polystyrene NPs (up to 200 μg/mL) induced frequency and dissipation changes. Both 10 nm SiNPs (100 μg/mL) and 18 nm gold NPs (GNPs) (5 μg/mL) induced significant changes in dissipation compared to PRP alone. These effects were confirmed using phase contrast microscopy and helium ion microscopy. These techniques demonstrated the presence of large PLT aggregates on the surface of the crystals following the perfusion of PRP in the presence of NPs.

In contrast, at the same concentrations, no PLT aggregation effects were observed for 50 nm SiNPs, 60 nm polystyrene NPs, or 18 nm GNPs based on the results of LTA. The 10 nm SiNPs only induced PLT aggregation at 200 μg/mL and not at lower concentrations. Similarly, an elevation in surface CD62P exposure was only observed following 10 nm SiNP treatment at 200 μg/mL, but not following treatment with the other tested NPs.[115] Based on the results of this pilot study, the QCM-D may be a promising sensitive method for evaluating the effects of NPs on PLTs. Further studies are required to conclusively demonstrate the usefulness of this method.

3.14. *Platelet proteomic, genomic, and transcriptomic analyses*

Recent advances in proteomic and genomic analysis technologies make these powerful tools widely available and routinely performed in several PLT research laboratories.[116] Although the role of proteomic analysis for the testing of the effects of materials on PLTs has not been demonstrated, it is likely that a comparative proteomic analysis of treated and untreated PLTs may lead to the identification of a panel of proteomic markers that will be useful for evaluating material biocompatibility or toxicity. As PLTs are anucleated, they have been assumed to have limited or no capacity for protein synthesis. However, recent discoveries demonstrated that PLTs synthesize proteins with important biological activities in response to stimulation. This protein synthesis is termed signal-dependent translation and involves the use of a constitutive transcriptome and specialized pathways. Such synthesis can alter PLT phenotypes and functions.[117] Genomic and transcriptomic analyses may therefore be useful for determining different nanomaterial interaction-induced changes in PLT status. Extensive controls for these assays are absolutely critical to exclude the possibility of the contamination of the PLT sample with blood cells, thereby avoiding misleading results. We expect that proteomic, genomic, and transcriptomic analyses will open a new chapter in our understanding of nanomaterial–PLT interactions and in the general principles of nanomaterial, blood, and vascular biocompatibility.

3.15. *In vivo animal models and clinical trials*

Observations from *in vitro* and tissue culture studies regarding the toxicity and other effects of nanomaterials may not correspond to *in vivo* results.[118] Therefore, *in vivo* animal studies are essential for the preclinical investigation of the effects of nanomaterials on PLTs. General toxicity studies in animals should be performed, including careful histological examination for thrombosis in both large vessels and the microcirculation and for signs of hemorrhage. Additionally, the effect of nanomaterials on PLTs *in vivo* can be evaluated in specific animal models that are designed according to the intended use of the material. Several animal models of arterial, venous, or microvascular thrombosis have been developed.[119–123] The potentiating or inhibitory effects of NPs can be examined in these models.[57] Similarly, the effects of nanomaterials on the hemostatic efficacy of PLTs can be evaluated in different hemorrhage or thrombocytopenia models.[124–126] PLT activation can also be examined in animal models of complement-mediated hypersensitivity reactions to nanomaterials and in other *in vivo* models of nanomaterial-induced inflammatory responses.[127] In addition, as was discussed in the section above focused on the clinical adverse effects of nanomaterials, evaluating nanomaterial interactions with PLTs by administration to healthy animals is not generally pathophysiologically relevant to the intended clinical use of the nanomaterial. To mimic the clinical situation in the preclinical testing of nanomaterials, "two-hit" animal models should be developed. In such models, nanomaterials could be evaluated in animals that are preconditioned with acute lung injury, sepsis, circulatory shock with hypoxia-reperfusion injury, or other types of multiple organ dysfunctions.[124–126,128–130] This approach would provide preclinical *in vivo* results that are more relevant for clinical use. The correct selection of the animal species is also important because there are considerable differences among various animal species and humans with respect to PLTs, other systems that participate in hemostasis, and the inflammatory responses.

Lastly, the evaluation of a nanomaterial's safety for specific clinical indications, including the investigation of clinical adverse effects, must be performed in appropriately designed clinical trials. PLT–nanomaterial interactions should be investigated not only in the context of bleeding and thromboembolic complications, but also in relation to other adverse events, such as acute lung injury, renal failure, or other microcirculation disorder-induced organ dysfunctions.

4. Studies on the *In Vitro* Interactions of Nanomaterials with Platelets

Only a limited number of studies have investigated the effects of novel nanomaterials on PLTs. More extensive literature is available for liposomes, which are often nanosized, but have a long history of medical use. Therefore, liposomes are not typical representatives of novel types of engineered NPs. The primary focus of the following review is to highlight *in vitro* preclinical testing of the effects of nanomaterials on PLTs. The absence of any standard assay makes it difficult to compare results from different studies and to draw general conclusions. Unless an animal species is specified, the following studies evaluated the effects of nanomaterials on human PLTs.

4.1. *Carbon nanomaterials*

Radomski *et al.*[57] reported a key study on the effects of carbon nanomaterials on PLTs. Multi-walled carbon nanotubes (MWCNTs), single-walled carbon nanotubes (SWCNTs), C_{60} fullerenes (nC_{60}), and mixed carbon NPs (MCNs), which are a mixture of amorphous carbon with approximately 7% of C_{60}, were investigated at concentrations 0.2–300 μg/mL. The NPs were compared with standard urban particulate matter (SRM1648 with an average size of 1.4 μm). This study is a good example of a comprehensive methodological approach for the evaluation of the effects of nanomaterials on PLTs. Since not all tested materials are soluble in aqueous solvents, it is critical to understand how the suspensions were prepared for testing. Here, the particles were suspended in Tyrode's solution, sonicated, and vortexed prior to testing. Under these conditions, fullerene C_{60} likely formed a nanocrystalline suspension of stable aggregates (nC_{60}). Therefore, the material that is described as C60CS in the original report is referred to as nC_{60} in this review. nC_{60} refers to an unknown number "n" of C_{60} molecules that agglomerate to form the suspension. The clusters contain unmodified C_{60} at their centers, which are surrounded by partially hydroxylated C_{60} on the outside.[131] Radomski *et al.* did not report the size distribution of the NPs in the tested materials. For example, in our study of C_{60} fullerenes,[132] we prepared nC_{60} nanocrystalline suspensions by stirring C_{60} in water for two weeks. This step was followed by a final spinning of the suspension at 4,000 *g* to remove large agglomerates. Based on the DLS size distributions of the particles that were present in the nC_{60} water suspension, the major water-"soluble" component was approximately 220 nm in hydrodynamic diameter. In the study that was performed by Radomski *et al.*, the size of the agglomerate was likely larger and the surface was likely less hydroxylated because the suspension was not stirred for such a long period of time.

Radomski *et al.* also performed aggregation studies using washed PLTs that were resuspended in Tyrode's solution and pre-incubated with various inhibitors (prostacyclin, S-nitrosoglutathione, aspirin, 2-methylthio-AMP, phenanthroline, EDTA, and Gö6976) prior to the addition of the nanomaterials. PLT aggregation was assayed in a whole blood ionized calcium lumi-aggregometer. ADP release was measured using a luciferin–luciferase assay. For the microscopic analysis, PLT aggregation was terminated at 20% of the maximal response as determined by the aggregometer, and the samples were fixed and evaluated by phase contrast microscopy or further processed for immunofluorescence microscopy or TEM. Flow cytometry of the PLT surface expression of activated GPIIb-IIIa (PAC-1), GPIb, and CD62P or the expression of β_3 integrin (GPIIIa, CD61) was also performed using unfixed diluted samples. Zymographic analyses of the gelatinase activities in the PLT releasate were performed using SDS-PAGE with copolymerized gelatin. The activities of the PLT matrix metalloproteinases MMP-2 and MMP-9 were identified by their molecular weights and quantified using reference standards.

The study reported that MCNs, SWCNTs, MWCNTs, and SRM1648 induced activation and aggregation of human PLTs (MCN ≥ SWCNTs > MWCNTs > SRM1648). Interestingly, nC_{60} fullerenes did not activate the PLTs. Compared to a strong PLT response to MCNs, CNTs caused partial aggregation with little or no granular release. SRM1648, MCNs, MWCNTs, and SWCNTs all induced GPIIb-IIIa activation, and treatment with EDTA completely inhibited aggregation, indicating the essential role of this integrin complex in the carbon NP-induced aggregation response. MCNs caused a maximum aggregation response and PLT degranulation that was associated with ADP release. In agreement with this result, MCN-induced aggregation was inhibited by the P2Y12 receptor antagonist 1-MeSAMP. In addition, MCNs, but not other carbon nanomaterials, triggered increased surface expression of CD62P and decreased the expression of GPIb. All carbon NP-induced aggregation was inhibited by prostacyclin. Aggregation was inhibited with less efficacy by S-nitrosoglutathione (an endogenous NO donor), but not by aspirin. The classical protein kinase C (PKC) isoform inhibitor Gö6976 did not affect MCN-, SWCNT-, or MWCNT-induced aggregation. Phenanthroline, which inhibits the MMP gelatinase-dependent aggregation pathway, significantly reduced the aggregation response to carbon NPs. Moreover, all of the carbon NPs tested, including non-aggregating nC_{60}, released cytosolic matrix matalloproteinases (gelatinases), proaggregatory MMP-2 and antiaggregatory MMP-9 from the PLTs. The authors suggested that the release of MMP-9 may be a sensitive index of cellular exposure to nanoparticles.

Radomski *et al.*[57] discussed different characteristics of carbon NPs that may impact their ability to activate and aggregate PLTs. The size, shape, and surface characteristics of particles are the primary features to be considered. Previously tested carbon nanomaterials form polydisperse suspensions in aquatic solvents and can form agglomerates up to tens of micrometers in size. In this study, the authors did not identify a clear relationship between the size of agglomerates and PLT aggregating potency. With respect to the shape of the NPs, nanotubes may form inter-PLT bridges, thereby promoting aggregation, whereas spherical fullerenes are unlikely to do so. However, based on the assumption that agglomerates will form, it would be interesting to compare the shape of the agglomerates that are formed by CNTs and nC_{60}. Moreover, surface characteristics, including surface charge, are different between CNTs and nC_{60}. NP surface charges may play an important role in the interaction of NPs with PLT receptors, such as GPIIb-IIIa.

Another important factor with respect to carbon nanomaterial toxicity is the level of contaminants, particularly transition metals, that are present in different carbon NP preparations. Transition metals on the surface of NPs can increase the generation of reactive oxygen species (ROS).

Soluble contaminants are unlikely to affect NP aggregatory potency because the aggregating ability of particle suspension was greatly reduced following filtration through a 100 nm filter. The hypothesis that NPs containing amorphous carbon have a higher potency to activate and aggregate PLTs compared to crystal-like carbon NPs is unlikely to be true because our experiments demonstrated that several CNT materials induced higher PLT aggregation responses compared to amorphous carbon black (CB) NPs (Sigma Aldrich).[61] The activating effects of carbon NPs on PLTs that were observed by Radomski *et al. in vitro* were confirmed by the authors in a rat thrombosis model *in vivo*. The PLT-activating activity of carbon NPs may be important in humans at the systemic levels of NPs that are reached in therapeutic and diagnostic applications and as a result of environmental pollution or other exposures. Although a maximum aggregation response was also observed *in vitro* at concentrations >100 μg/mL, activated GPIIb-IIIa was observed and zymographic analyses indicated MMP-9 activity at concentrations that were four- to 100-fold lower.

Furthermore, Meng *et al.*[68] investigated PLT interactions with non-woven SWCNTs, which are novel carbon materials with a nanotopography macroscopic surface. Interestingly, the study demonstrated that non-woven SWCNTs caused significantly lower PLT adhesion and lower PAC-1 expression compared to a polyurethane (PU) surface.

Our study investigated the PLT aggregating activitities of different SWCNTs and MWCNTs.[61] We further demonstrated that MWCNTs activate PLTs by inducing extracellular Ca^{2+} influx, which was inhibited by the calcium channel blockers SKF 96365 and 2-APB. In addition, we showed that CNT-induced PLT activation is associated with a marked release of PLT membrane MPs that are positive for the granular secretion markers CD62P and CD63.[61]

In a follow-up study,[56] we showed that pristine MWCNTs penetrate the PLT plasma membrane without any discernible membrane damage, but interact with the dense tubular system (DTS), causing a depletion of the PLT's intracellular Ca^{2+} stores. This process was accompanied by the clustering of stromal interaction molecule 1 (STIM1), which colocalized with ORAI1 and thus indicated the activation of store-operated Ca^{2+} entry (SOCE). Our findings therefore revealed the molecular mechanism of CNT-induced PLT activation.

In agreement with our results, Bihary *et al.* [86] demonstrated that SWCNTs induced PLT aggregation and increased CD62P surface expression and the formation of PLT granulocyte complexes *in vitro*. In contrast, diesel emission NPs, carboxylated polystyrene NPs (60 nm), and titanium dioxide NPs did not. Amine-conjugated polystyrene NPs triggered significant CD62P expression and the formation of PLT–granulocyte complexes, but did not induce PLT aggregation. Similarly, Burke *et al.*[55] demonstrated that MWCNTs activate PLTs *in vitro*, with amidated MWCNTs exhibiting a greater PLT activation potential than carboxylated or pristine MWCNTs.

Studies of soluble C_{60} derivatives could further elucidate the effects of fullerenes on PLTs, eliminating problems that are related to the formation of different types of agglomerates. When fullerene molecules are hydroxylated, they become soluble in aquatic solvents. Niwa and Iwai[89] compared the effects of the polyhydroxylated soluble derivative of C_{60} fullerenol $C_{60}(OH)_{24}$ with those of amorphous CB on rabbit PLTs. The authors used a type of WBA based on the screen filtration method for these analyses.[88] The effects of CB or $C_{60}(OH)_{24}$ on ADP-induced PLT aggregation in whole blood were evaluated. Neither CB nor $C_{60}(OH)_{24}$ alone induced PLT aggregation. However, when the whole blood was pre-treated with $C_{60}(OH)_{24}$,

ADP-induced aggregation threshold index values were elevated in a dose-dependent manner. In contrast, collagen- and thrombin-induced PLT aggregation was not affected by $C_{60}(OH)_{24}$. Although these results indicated that $C_{60}(OH)_{24}$ specifically facilitates ADP-induced PLT aggregation, we were unable to confirm this conclusion using LTA on human PRP. In contrast, in our experience, the pre-treatment of human PRP with $C_{60}(OH)_{24}$ significantly inhibits ADP-induced PLT aggregation (unpublished results). These controversial results emphasized the role of experimental settings and possible differences in the functions of human and rabbit PLTs.

Graphene oxide (GO), a novel 2D carbon nanomaterial, has been extensively investigated for potential biomedical applications. Singh *et al.*[133] demonstrated a strong PLT aggregating activity of GO. At 2 μg/mL, the aggregation induced by GO was even stronger than that elicited by thrombin (1 U/mL), one of the most potent PLT agonists. Aspirin had no effect on GO-induced PLT aggregation, whereas apyrase only slightly attenuated GO-induced PLT aggregation. On the contrary, U73122 and Ro-31-8425, which are specific inhibitors of phospholipase C (PLC) and PKC, respectively, significantly attenuated GO-induced PLT aggregation, thus implicating the PLC–PKC axis in GO-mediated signaling. Interestingly, higher concentrations of GO (5 and 20 μg/mL) induced only a small increase in CD62P expression, whereas 2 μg/mL of GO significantly augmented PAC-1 binding to PLTs. Accordingly, a near twofold increase in the number of GO-treated PLTs adhered to a fibrinogen matrix. The level of Src pTyr-529 significantly decreased in GO-treated PLTs, which suggested that Src is stimulated in the presence of GO. GO (2 μg/mL) evoked an initial increase in intracellular Ca^{2+} levels, which was followed by a concentration plateau. This effect was similar to that observed for thrombin on PLT intracellular Ca^{2+} flux. Inhibitors of PLC (U73122) and PKC (Ro-31-8425) and the IP3 receptor antagonist 2-aminoethoxydiphenyl borate (2-APB) significantly suppressed the GO-induced increase in PLT intracellular Ca^{2+} levels, thus implicating the PLC–IP3/diacylglycerol–Ca^{2+}/PKC axis in the GO-triggered signaling cascade. GO (5–20 μg/mL) also induced intracellular ROS formation and membrane alterations, but did not elicit a significant release of LDH from PLT cytosol, thus ruling out a breach in membrane integrity. In summary, GO-induced PLT activation may be attributed to the release of intracellular free Ca^{2+} from cytosolic stores and the activation of Src family non-receptor protein tyrosine kinases in PLTs. The prothrombotic capabilities of GO are dependent on the surface charge distribution of the NP, as reduced GO (RGO) is significantly less effective in inducing PLT aggregation.

Karagkiozaki *et al.*[134] investigated the thrombogenic potential of amorphous hydrogenated carbon thin films (a-C:H) that were developed by magnetron sputtering. These films had an approximately 42% sp^3 content, and the deposition was performed under hydrogen partial pressure. *In situ* observation of the PLTs' activation and their subsequent adhesion was performed using AFM. The results demonstrated that the film's surface morphology and the spatial electric field distribution influenced the thrombogenicity of a-C:H and that the PLTs are negatively charged. Moreover, variations in the deposition parameters can lead to nanomaterials with different surface charges. Such materials could be used as coatings for cardiovascular applications and as nanosensors for PLTs that are prone to clotting. Carbon films that are grown without a bias voltage with 20% H_2 in the plasma exhibit less thrombogenicity compared to those films formulated in the presence of a biased voltage. Electrostatic force microscopic (EFM) measurements of the electrostatic force distribution during the adhesion of PLTs to biomaterials provide information regarding the contribution of positive surface charges of artificial materials in attracting PLTs. Thus, negatively charged biomaterial surfaces exhibit lower thrombogenicity. Additionally, more negatively charged thin films, such as floating a-C:H films, when compared to the examined biased films, may be used to sense less negatively charged PLTs, which are prone to form aggregates. In a follow-up study,[79] the authors demonstrated that PLT activation can be affected by altering the surface topography of nanocoatings. Carbon nanocoatings, which exhibited higher surface roughness, were less thrombogenic in terms of PLT adhesion. Carbon nanocoating with higher surface roughness may be a solution for improving biocompatibility stent coating fabrications.

Carbon nanomaterials are very attractive candidates as plastic composite components for improving the mechanical properties of plastics that can be used for various biomedical applications. Lin and Wu[66] examined PLT adhesion to a C_{60}-grafted PU surface. Samples of C_{60}-grafted PU film that was preconditioned in a buffer were incubated with PRP. Following incubation, the PRP was removed, and the samples were incubated with buffer, then fixed and processed for SEM analysis. The results demonstrated significantly higher PLT adhesion on the C_{60}-grafted PU than on the untreated PU control. In addition, the degree of PLT activation on the C_{60}-grafted surfaces was higher than in the control conditions. The authors concluded that this effect may be attributed to the synergistic effect of the grafted C_{60} molecules and the few residual amine functional groups that remain following the C_{60} grafting reaction. Testing PLT adhesion on CNT-containing plastic composite surfaces is essential to avoid potential adverse effects of these materials in clinical use. The standardization of a perfusion chamber assay would be very useful in this regard.

4.2. *Silica nanomaterials*

Tavano *et al.*[84] compared the effects on PLTs of synthetic amorphous SiNPs (35 nm), organically modified silica (ORMOSIL) NPs (45 nm), poly(lactic-*co*-glycolic acid) (PLGA) NPs (280 nm), and small unilamellar vesicles (SUV) NPs (150 nm) that were fabricated from dipalmitoylphosphatidylcholine (DPPC) and dipalmitoylphosphatidylglycerol (DPPG). The incubation experiments with PRP demonstrated that CD62P expression is not increased by any of the NPs, indicating a lack of PLT activation. PLT aggregation in suspension and under stirring, a more physiologically relevant PLT activation test, was also performed to assess NP prothrombotic activity. LTA revealed that amorphous SiNPs and bare and polyethyleneglycol(PEG)ylated ORMOSIL NPs did not induce a significant PLT aggregation effect, in agreement with CD62P upregulation analysis. Interestingly, bare PLGA NPs induced PLT aggregation, whereas PEGylation abolished such activity. Bare SUV NPs demonstrated a weak PLT aggregating effect, which was further diminished by PEGylation. Despite the lack of activation effects, all of the NPs rapidly associated with the PLTs. The binding of NPs to the PLTs was saturable and reached its maximum at a concentration of approximately 50 μg/mL for bare ORMOSIL NPs and bare or PEGylated PLGA NPs. Bare SUV binding was less efficient and did not saturate at concentrations of up to 200 μg/mL.

Corbalan *et al.*[62] examined the biocompatibility of amorphous SiNPs (10 nm, 10SiNPs; 50 nm, 50SiNPs; 150 nm, 150SiNPs; 500 nm, 500SiNPs) with human PLTs. When interacting with PLTs, SiNPs can stimulate the release of NO and to a greater extent, peroxynitrite ($ONOO^-$). These effects led to a low NO to $ONOO^-$ concentration ratio that strongly correlated with SiNP-induced PLT aggregation. SiNP stimulated GPIIb-IIIa activation and led to PLT aggregation via ADP- and MMP-2–dependent mechanisms. These effects were augmented with increasing NP concentration, but were inversely proportional to the size of SiNPs. In fact, the smallest SiNPs (10 nm) induced the most severe effects on the PLTs.

Liu *et al.*[135] investigated porous SiO_2/pHEMA biocomposites that were synthesized *in situ* by incorporating SiNPs with a hydroxyethyl methacrylate (HEMA) monomer following UV-induced photopolymerization. The authors observed that the addition of 35% or 50% H_2O during pHEMA polymerization decreased the degree of PLT adhesion to the polymer surface. The addition of 9% SiNPs enhanced the mechanical properties of the composite without influencing PLT adhesion. Thus, the resulting SiO_2/pHEMA biocomposite with optimized hemocompatibility may be promising for drug delivery applications.

4.3. *Metal nanomaterials*

4.3.1. *Titanium*

Smith *et al.*[78] demonstrated significant PLT activation (adhesion and aggregation) on arrays of titania nanotubes with diameters of 70–90 nm and lengths of 1 mm when compared to smooth titanium surfaces. Interestingly, Yang *et al.*[74] demonstrated that the controllable fabrication of a superhydrophobic surface on titanium biomedical implants improved their blood biocompatibility, particularly in terms of decreased PLT adhesion. Electrochemical anodization was used to form a layer of TiO_2 nanotubes on the titanium substrate. A hydrophobic monolayer subsequently self-assembled on the nanotube surface. Based on the *in vitro* blood compatibility evaluation, very few PLTs attached onto the super hydrophobic surface, and those few attached PLTs were not activated in this condition. Comparatively, a large number of PLTs adhered and spread out on both a bare titanium substrate and the superhydrophilic surface. This latter surface was obtained by exposing the TiO_2 nanotubes to UV irradiation. These results indicated that the superhydrophobic TiO_2 nanotube layers exhibit promising blood compatibility and performance in preventing PLTs from adhering to the implant surface. In contrast, Nemmar *et al.*[136] demonstrated that rutile TiO_2 nanorods with primary diameters of approximately 4–6 nm induced PLT aggregation in whole blood at very low concentrations (0.4–10 μg/mL) in a dose-dependent manner. Thus, similar to other nanomaterials, the nanostructural design of titanium and the physicochemical characteristics of the surface dictate the nature of the material's effects on PLTs.

4.3.2. *Iron*

Oldenburg *et al.*[107] demonstrated that rehydrated lyophilized PLTs can be labeled with superparamagnetic iron oxide NPs (SPIONs) (5 nm core) to facilitate their imaging by magnetomotive optical coherence tomography (OCT). Based on the results of this technique, the SPIONs were internalized by rehydrated lyophilized PLTs and formed rod-like structures in the PLTs. This technique may enable the development of novel technologies for *in vivo* monitoring of the adherence of SPION-loaded rehydrated lyophilized PLTs to sites of bleeding and vascular damage. This advance would allow for the investigation of the hemostatic functions and potential adverse thrombotic effects of these materials.

Magnetic labeling of PLTs may offer a new tool in diagnostics and the study of transfusion and cardiovascular medicine. Aurich *et al.*[63]

used ferucarbotran superparamagnetic NPs of approximately 30–90 nm in diameter for PLT labeling. PLTs endocytosed these NPs without the use of linkers or binding agents. Flow cytometry using FITC-conjugated magnetic NPs demonstrated *ex vivo* labeling of approximately 98% of the PLTs; the NPs were predominantly located inside the PLT granules. Labeling, however, led to the significant activation of PLTs, as demonstrated by a marked increase in CD62P surface expression. Therefore, further optimization of this fabrication technique is required.

4.3.3. *Silver*

Shrivastava *et al.*[138] demonstrated that 10–15 nm silver NPs (SNPs) have intrinsic anti-PLT activation properties and effectively prevent integrin-mediated PLT responses in a concentration-dependent manner. SNPs effectively suppressed integrin-mediated PLT functional responses, such as aggregation and alpha and dense granule secretion, in a dose-dependent manner. Tyrosine phosphorylation in thrombin-stimulated PLTs, adhesion to immobilized fibrinogen and collagen, and fibrin clot retraction were similarly inhibited, irrespective of the nature of agonists used. SNPs perturbed the PLT membrane microenvironment. The particles occupied PLT granules and vacuolar spaces and prevented hyaloplasmic extensions, reducing inter-PLT proximity. PLTs did not release intracellular LDH even after exposure to a considerably high concentration of SNPs, reflecting a lack of cell lysis and the maintenance of membrane integrity. The authors suggested that the observed inhibition may be due to a conformational modulation of the PLT surface integrin $\alpha_{IIb}\beta_3$, thus preventing the interaction of these surface molecules with fibrinogen.

In contrast, Jun *et al.*[49] demonstrated that SNPs (10–100 nm) induced PLT aggregation and procoagulant activation. These effects were evidenced by increased phosphatidylserine exposure and thrombin generation. Interestingly, sub-threshold levels of thrombin significantly enhanced the observed SNP-induced PLT activation, indicating that the prothrombotic effects of SNPs may be further potentiated in activated PLTs. SNPs induced increases in intracellular Ca^{2+} levels and were associated with PLT activation and CD62P expression. SNPs also enhanced serotonin release from PLTs. Similarly, Stevens *et al.*[139] demonstrated that a silicon tubing coated with a hydrophilic polymer containing either 8% or 15% of 20 nm SNPs caused significant PLT activation when compared to a hydrophilic polymer coating without SNPs or other biocompatible polymer tubings. The same research group[77] examined the PLT biocompatibility of a highly antimicrobial composite of poly(4-vinylpyridine)-*co*-poly(4-vinyl-*N*-hexylpyridinium bromide) (NPVP) and AgBr

NPs as a thin adherent surface coating on Tygon elastomer tubes. The study demonstrated that PLTs that come into contact with this material are rapidly and highly activated, after which they become substantially disrupted. PLT activation and disintegration were most evident from measurements of released βTG. These results implied that NPVP/AgBr is by no means blood-compatible and may be more useful outside of the biomedical field.

Wang *et al.*[140] suggested the use of silver-coated magnetic NPs for bacterial inactivation in PLT transfusion products. While the bacterial inactivation potency of silver-coated NPs may be promising, the primary concern is the effect of these NPs on PLTs, which was not investigated in the study.

Although the data regarding the interactions of SNPs with PLTs are controversial, a greater number of studies have demonstrated the PLT-activating effects of SNPs, which is likely size-dependent.

4.3.4. *Gold*

Deb *et al.*[71] investigated the effects of GNPs on human PLTs. The PLT aggregation response decreased with increasing size and decreasing concentration of GNPs. The smaller GNPs (about 20 nm in diameter), at a final concentration of 40 μM, exerted maximal effect on PLTs. On the other hand, the larger GNPs (about 70 nm in diameter) had almost no effect on PLTs, even at higher concentrations. Accordingly, the expression of surface CD62P was elevated after treatment with smaller GNPs. The GNP effect was associated with an enhanced release reaction, tyrosine phosphorylation, and CD62P expression levels. The uptake study, which examined the quenching of quinacrine hydrochloride fluorescence by GNPs, indicated that the accumulation of 20 nm GNPs is severalfold higher than for the 70 nm GNPs.

GNPs have been examined for their capacity to improve the mechanical properties and biostability of PU polymers. Hsu *et al.*[41] prepared nanocomposites that were composed of a polyether-type waterborne PU. Different amounts (17.4–174 ppm) of GNPs (approximately 5 nm in diameter) were incorporated into the PU. Biocompatibility of the nanocomposite that contained a certain amount (43.5 ppm) of gold (PU–GNP 43.5 ppm) was assessed. For the evaluation of the effects of these nanocomposites on PLTs, the material samples were incubated in culture plates with PRP. Following incubation, the plates were washed and the adherent PLTs were either detached with trypsin and counted using a cell counter or fixed on the surface and processed for SEM. A morphology score was evaluated, and the average degree of PLT activation was calculated. The results demonstrated that PLT adherence on PU–GNP nanocomposites was not significantly different compared to the original PU; however, the adhered PLTs were less activated in all of the PU–GNP nanocomposites when compared to PU alone. Interestingly, the concentration of gold in these NPs appears to affect PLT compatibility. Nanocomposite PU–GNPs at a concentration of 43.5 ppm resulted in lower PLT adhesion and a less activated PLT morphology when compared to nanocomposites with higher gold contents (174 ppm PU–GNP). The counting of adhered PLTs following trypsin detachment is not a widely used approach; it would be interesting to determine the robustness and reproducibility of this method.

Davies *et al.*[141] investigated the pH-controlled delivery of luminescent europium-coated NPs into PLTs. Water-soluble luminescent GNPs were delivered into human PLTs via a rapid, pH-controlled mechanism using the low pH insertion peptide pHLIP. This approach included a co-coating of GNPs with a luminescent europium complex, EuL, and the pHLIP peptide, resulting in the formation of pHLIP*EuL*GNPs. The 13 nm GNPs acted as a scaffold for the attachment of both the luminescent probe and the peptide for targeting the NP delivery. The internalization of pHLIP*EuL*GNPs into the PLTs, which took only minutes, was examined using a variety of imaging modalities, including luminescence, confocal reflection, and TEM. The results indicated that pHLIP*EuL*GNPs only enter PLTs in low pH conditions (pH 6.5), but not at pH 7.4. This internalization was mediated by pHLIP translocation across the membrane. Thus, this study demonstrated the potential of this design to translocate multimodal NP probes into cells in a pH-dependent manner.

4.3.5. *Aluminum*

Ferraz *et al.*[73] investigated the PLT activation potential of alumina membranes with pore size ranging from 20 to 200 nm in diameter. The PLT activation characteristics and time profiles differed between the two membranes. PLT adhesion increased over time for the 20 nm surface, whereas the clusters of PLT MPs on the 200 nm pore surface did not appreciably change during the course of the experiment. Very few PLTs were observed on the 200 nm pore membrane as compared to the 20 nm pore membrane. The PLTs observed on the 20 nm pore membrane exhibited signs of activation, such as a spread morphology, protruding filopodia, and CD62P expression; however, no MPs were detected on this surface. Despite the fact that very few PLTs were observed on the 200 nm pore alumina compared to the 20 nm pore membrane, many MPs were detected on the 200 nm pore surface.

Interestingly, all of the MPs were observed within circular areas of approximately 3 μm in diameter. As this is the approximate size of a PLT, the authors speculated that this result is evidence of transient, non-adherent PLT contact with the surface, which triggered the generation of PLT MPs. This study demonstrated how nanotexture can influence PLT MP generation. The release of TSP-1 increased with time for both membrane types; however, this release occurred much later for the 200 nm pore alumina (240 min) than for the 20 nm pore membrane (60 min). The surface topography of the alumina most likely influences the protein transition rate, in turn altering the effects of the material on PLT activation kinetics.

4.3.6. *Gadolinium*

Gadolinium-containing NPs (GdNPs) have great importance as contrast agents in NMR (nuclear magnetic resonance) imaging. In addition, GdNPs have also been examined for their potential use in tumor-specific neutron capture therapy (NCT). Folate receptors are effective targets for the tumor-specific targeting of GdNPs. Oyewumi *et al.*[104] evaluated the effects of folate-coated GdNPs on PLTs using a simple spectrophotometric assay. Various concentrations of folate-coated GdNPs were incubated at 37°C for 10 min with PRP under constant stirring. PLT aggregation in the presence of various NP concentrations was measured by absorbance at 500 nm. The absorbance values of PRP and PPP and NP aqueous suspensions were measured as reference values, and both collagen and epinephrine were used as positive controls. The study demonstrated that, as assayed by this simple screening method of turbidity measurements, GdNPs in concentrations ranging from 50 to 300 μg/mL did not cause PLT aggregation. This study is an example of a simple screening method of a material's effects on PLTs, and this technique could be very useful during product development. More comprehensive studies are required to obtain conclusive results.

4.4. *Polymer nanomaterials*

McGuinnes *et al.*[100] examined the interaction of PLTs with polystyrene latex NPs. All of the NP formulations had a mean size of 60–80 nm, but were designed with different surface charges, i.e., unmodified, aminated, or carboxylated. Both the carboxylated and aminated NPs induced PLT aggregation, whereas the unmodified polystyrene latex NPs did not. While carboxylated NPs caused aggregation that was associated with the upregulation of the activation markers CD62P and PAC-1, aminated NPs did not upregulate activation markers and appeared to act by perturbing the PLT membrane, thereby revealing anionic phospholipids. Neither oxidative stress generation by particles nor metal contamination was responsible for these effects, which instead were the result of differential surface derivatization. This study revealed that NPs that are composed of insoluble low toxicity materials exhibit significantly altered potency with respect to the induction of PLT aggregation when their surface chemistry is modified. Furthermore, the two surface modifications that caused aggregation (i.e., aminated and carboxylated) did so via different mechanisms. This study highlighted the fundamental role of surface chemistry on the bioactivity of NPs in a PLT activation model. Similarly, Mayer *et al.*[98] demonstrated that both carboxylated polystyrene NPs with a mean diameter of 26–160 nm and 220 nm amidine-conjugated beads at very high concentrations (0.5–2 mg/mL) activated PLTs as determined by elevated CD62P surface exposure and PF4 and βTG release.

Functionalized latex NPs have also been used to examine the structural requirements of collagen-related peptides (CRPs) with respect to their PLT-activating activity.[142] Normally, CRPs, even those as long as 30 amino acids [10 Gly–Pro–Hyp (GPO) repeats], cannot effectively induce collagen's PLT-activating behavior. Two short CRPs, AcHN–(Gly–Pro–Hyp)$_n$Gly–OH, with $n = 5$ and $n = 10$, were attached via the C-terminus to amino-functionalized latex NPs and resulted in the creation of a multimeric display of triple helical motifs. The authors demonstrated that that a 31-mer CRP can mimic the function of collagen when this peptide is displayed in a multivalent fashion by linking its C-terminus to an NP scaffold. These results indicated the importance of multiple triple-helical motifs for the robust stimulation of PLTs by CRPs via GPVI receptors.

Li *et al.*[87] investigated the effects on PLTs of negatively charged PLGA and PLGA–chitosan (CS) NPs with sizes ranging from 580 ± 230 nm to 640 ± 150 nm (CS–PLGA) and positively charged CS NPs with sizes ranging from 430 ± 70 nm to 580 ± 50 nm. The examined NPs did not induce PLT activation (CD62P or GPIIb-IIIa/PAC-1) or aggregation. PLGA, PLGA–CS, and CS NPs weakly inhibited collagen-induced PLT aggregation. The inhibition of PLT aggregation may be due to an NP-mediated decrease in collagen binding and/or PLT–PLT contact. Similarly, Ramtoola *et al.*[82] investigated PLT interactions with biodegradable MPs and NPs based on PLGA, PLGA–macrogol, and CS. The formulated MPs had a median diameter of 2–9 μm, whereas the NPs had an average diameter of 100–500 nm. The surface morphologies of the NPs ranged from smooth and spherical to irregular depending on the polymer and preparation method used. The particles were reconstituted at a concentration range of 0.1–500 μg/mL and were tested for their ability to induce or inhibit PLT aggregation. No effects on either the induction of PLT activity or the inhibition of ADP-induced aggregation were detected. In another study, Cenni *et al.*[47] evaluated the PLT compatibility of NPs of approximately 200 nm in diameter that were composed of a PLGA–alendronate conjugate. The examined NPs did not exhibit any effect on the PLT count (aggregation) or the PF4 concentration in the supernatant (release reaction). However, no positive control was used in these experiments. Positive controls are essential because the reactivities of PLTs in individual experiments may be extremely varied due to multiple factors, as was discussed previously.

Koh *et al.*[72] investigated the interaction of PLTs with composite films of PLGA with either randomly aligned or vertically oriented MWCNTs. The results of this study demonstrated that PLGA–MWCNT composites with vertically aligned nanotubes exhibit very low levels of fibrinogen adsorption and PLT adhesion, which can be attributed to both chemical and topographical effects. These results suggested that in addition to chemical alterations, nanotopographical modification may be effective in the development of low thrombogenic and hemocompatible materials.

Specific nanostructures with polymer surfaces may significantly affect PLT biocompatibility. Poly(carbonate urethane)s (PCUs) have been widely used in various biomedical applications due to their long-term biostability and moderate blood compatibility. Fluorinated alkyl side chains may aid in lowering the surface free energy and improving polymer properties. These types of side chains have been successfully used in the development of novel biomaterials. Sun *et al.*[75] investigated two types of PCUs with different ratios of fluorinated alkyl side chains (20%, referred to as FPCU20, and 50%, referred to as FPCU50). Aligned CNT films were used as the nanostructured templates. These films are composed of densely packed MWCNTs with an average diameter of approximately 39.7 nm and a length of approximately 20 μm. The tubes are aligned vertically relative to the substrate.

After typical dip-coating and a subsequent drying process, the polymers formed a thin coating on the outer wall of the CNTs, thus forming nanostructured polymer films. The aligned nanostructure of the CNTs remained well intact following this process. Moreover, the upper surfaces of the films that were prepared in this manner were quite rough and were composed of the tail ends of the polymer-coated CNTs. The study demonstrated that compared to the significant PLT adhesion of the smooth PFCU20 or PFCU50 surfaces, nearly no PLT adhesion occurs on the nanostructured superhydrophobic PFCU20 or PFCU50 films. Additionally, the PLT activation markers

PAC-1 and CD62P exhibited a marked decrease following the introduction of the special nanostructures to both polymers. The obtained nanostructured superhydrophobic films are highly compatible with PLTs. The superior properties of the obtained nanostructured materials, including excellent anti-adhesion to PLTs, outstanding blood compatibility, and superhydrophobicity, would be highly advantageous in intra- or extracorporeal medical devices. These materials are expected to play important roles in future medical applications.

The porosity of composite materials is another factor that affects their interaction with PLTs. Ghanbari *et al.*[105] examined the surface properties of the novel nanocomposite polymeric material polyhedral oligomeric silsesquioxane (POSS–PCU). The results demonstrated that fewer PLTs adhered to POSS–PCU than to PU. In a follow-up study, Ahmed *et al.*[69] investigated a small diameter (<5 mm) vascular graft that was composed of POSS–PCU, which was itself produced using an extrusion/phase inversion method. Through the dispersion of controlled concentrations (0–55%) of the porogen sodium bicarbonate ($NaHCO_3$), the authors produced grafts with well-defined pore morphologies. The study demonstrated that the material prepared without $NaHCO_3$ caused the highest PLT adhesion and that the adhered PLTs exhibited an activated phenotype. In contrast, materials with a 35–55% $NaHCO_3$ composition caused significantly lower PLT adhesion in a $NaHCO_3$ concentration-dependent manner, and the adhered PLTs exhibited minimal pseudopodia formation.

Chen *et al.*[70] investigated the nanoscale topography of a thermally responsive poly(N-isopropylacrylamide) (PNIPAAm) surface by grafting the polymer from silicon nanowire arrays via surface-initiated atom transfer radical polymerization. The prepared surface exhibited largely reduced PLT adhesion *in vitro* both below and above the lower critical solution temperature (LCST) of PNIPAAm (approximately 32°C). In contrast, a smooth PNIPAAm surface exhibited anti-adhesive properties for PLTs only below the LCST. The contact angle and adhesive force measurements on oil droplets (1,2-dichloroethane) in water demonstrated that the nanoscale topography maintained a relatively high water content ratio on the as-prepared surface and played a key role in largely reducing PLT adhesion. However, this PLT anti-adhesion effect was not observed on the smooth PNIPAAm surface. These results can be used to extend the applications of PNIPAAm to the fields of biomaterials and biomedicine under physiological temperatures, providing a novel strategy for the fabrication of other blood-compatible materials.

4.5. *Dendrimer nanomaterials*

In a collaborative study with our laboratory, Dobrovolskaia *et al.*[65] evaluated 12 formulations of poly(amidoamine) (PAMAM) dendrimers that varied in both size and surface charge. Using a cell counting-based method, LTA, and SEM, we demonstrated that only large cationic dendrimers, but not anionic, neutral, or small cationic dendrimers, induce the aggregation of human PLTs in plasma *in vitro*. The aggregation that was induced by large cationic dendrimers was proportional to the number of surface amines. The observed aggregation was not associated with membrane MP release and was insensitive to a variety of chemical and biological inhibitors that interfere with various pathways of PLT activation. Taken in context with previously reported studies, our data suggested that large cationic PAMAM dendrimers induce PLT aggregation by disrupting PLT membrane integrity.

Dendrimer chemistry allows for the synthesis of a wide variety of NPs with specific properties that are suited for their intended use. Fernandes *et al.*[58] examined the effects of a bioconjugate of a thrombolytic agent streptokinase (SK) and a polyglycerol dendrimer (PGLD, generation 5) on PLTs. In this study, polystyrene microplates that were coated with the PGLD–SK bioconjugate were incubated with PRP, and the surface expressions of CD62P and CD63 were analyzed on unfixed PLTs using flow cytometry. This experiment demonstrated that PLT contact with PGLD–SK plates induced significantly less CD62P expression than contact with plates that were coated with only PGLD or uncoated polystyrene plates. In addition, the authors examined PLT adhesion using microscopy. Following incubation with citrate-anticoagulated whole blood, the plates were washed and the adhered PLTs were fixed, stained with acridine orange, and evaluated using epifluorescence microscopy. The authors demonstrated that uncoated polystyrene plates exhibited PLT adhesion and thrombus formation. In contrast, PGDL–SK coated plates were nearly free of both PLT adhesion and thrombus formation.

A promising approach is a targeted modification of NP surfaces with PLT-inhibiting ligands. For example, the activation of the GPCR A2A by extracellular adenosine is anti-aggregatory in PLTs and anti-inflammatory. This PLT inhibition pathway was employed by Kim *et al.*[60] Multiple copies of the nucleoside CGS21680, an A2A agonist, were covalently coupled to PAMAM dendrimers. Treatment with Alexa 488 fluorescent PAMAM–CGS21680 "conjugate 5" inhibited the aggregation of washed human PLTs, and the conjugate was effectively internalized. Thus, multivalent dendrimer conjugates may improve the overall pharmacological profile compared to monovalent GPCR ligands.

4.6. *Liposomes*

Liposomes rapidly interact with cellular components in the blood. Constantinescu *et al.*[95] examined the effects of various liposomes on PLTs. Liposomes that were labeled with carboxy-fluorescein, membrane-sequestering R18, or FITC-labeled phospholipids were mixed with washed PLTs in buffer, PRP, or whole blood. The PLTs were labeled with CD42b mAB and analyzed using flow cytometry. Irrespective of the liposomal composition, and with or without PEG, all of the liposomes interacted rapidly and dose-dependently with PLTs and other blood cells. These interactions took place equally in the presence and absence of plasma proteins. These results suggested that the interaction of liposomes with PLTs is opsonization-independent.

Do liposomes induce PLT activation or inhibition? This important question has been addressed in several studies. Juliano *et al.*[81] investigated the effects of liposomes on *in vitro* PLT aggregation using LTA analyses of human PRP and PLT suspensions in buffer. SUVs and multilamellar vesicles (MLVs) that were prepared from various conventional lipids and from a photopolymerizable phosphatidylcholine derivative (bis[1,2-(methacryloyloxy) dodecanoyl]-L-alpha-phosphatidylcholine, a dipolymerizable lipid, DPL) were used. None of the liposomal preparations that were examined caused marked PLT aggregation in either plasma or the buffer solution. However, positively charged vesicles that contained stearylamine impaired the ability of PLTs in plasma to aggregate in response to treatment with ADP. In contrast, negatively charged vesicles that contained phosphatidylglycerol (PG) impaired the ability of the PLTs in the buffer to aggregate in response to thrombin stimulation. DPL vesicles exhibited only modest effects on PLTs in plasma or buffer. This study demonstrated that the surface charge of liposomes and the availability of plasma components are critical factors for the interactions between liposomes and PLTs.

Similar results demonstrating that certain liposomes can inhibit ADP-induced PLT aggregation were also observed by Bonte *et al.*[143] The authors examined the biocompatibility properties of polymerizable phosphatidylcholine bilayer membranes in the form of liposomes. This analysis was performed with the aim of eventually utilizing such polymerized lipid assemblies in drug carrier systems or as surface coatings for biomaterials. The SH-based polymerizable lipid 1,2-bis[1,2-(lipoyl)dodecanoyl]-sn-glycero-3-phosphocholine (a dilipoyl lipid, DLL) and the methacryl-based lipid 1,2-bis[(methacryloyloxy)-dodecanoyl]-sn-glycero-3-phosphocholine (DPLA) were examined in comparison to "conventional" zwitterionic or charged phospholipids. The authors then examined the effects of liposomes on PLT aggregation. Polymerized

DPLA liposomes and DLL in polymerized or non-polymerized liposomes exhibited no substantial effects on PLT aggregation. However, while non-polymerized DPL vesicles did not induce aggregation, they impaired ADP-induced PLT aggregation.

A comprehensive study of the *in vitro* PLT compatibility of liposomal NPs with promising potential for clinical use was performed by Koziara *et al.*[85] The effects of cetyl alcohol/polysorbate NPs (E78 NPs) and PEG-coated E78 NPs (PEG–E78 NPs) on the activation and aggregation of human PLTs were investigated using flow cytometry and whole blood impedance aggregometry. Neither the NP E78 nor the PEG–E78 formulations resulted in PLT activation as assayed by PAC-1 mAB binding (which is specific for activated GPIIb-IIIa), nor did these NPs induce PLT aggregation. Interestingly, both of the NP formulations very rapidly inhibited agonist-induced PLT activation and aggregation in a dose-dependent manner. Apparently, these liposomes interact rapidly with PLTs without causing their activation. When a high NP-to-PLT ratio is achieved, liposomes interfere with PLT activation irrespective of the presence of agonists. The authors estimated that E78 NPs should not interfere with PLT functions *in vivo* at clinically relevant doses.

Although this chapter is focused on *in vitro* studies, specific interactions of certain liposomes with PLTs that are observed *in vivo* in animal studies also warrant brief descriptions. Reinish *et al.*[39] examined the interaction of liposomes with PLTs in rats. The rats were injected intravenously with liposomes of various compositions and sizes, and the blood PLT count was measured using a Coulter counter. Negatively charged liposomes induced a transient reduction in the PLT count in the first 5 min following the injection. This decrease recovered by 60 min post-injection. This effect was most striking for MLVs that contained PG. The authors demonstrated that negatively charged liposomes temporarily associate with PLTs both *in vitro* and *in vivo*. *In vivo*, PLTs are very likely transiently sequestered with liposomes in the lungs and the liver. An *in vivo* study of radiolabeled PLTs indicated that the transiently sequestrated PLTs reappear in the circulation 60 min following liposomal administration. Thus, PLT count recovery is not due to the mobilization of stored PLTs from the spleen. It was not possible to prevent the transient liposome–PLT interaction with ASA, heparin, or ancrod, indicating that this liposome–PLT interaction is not associated with the induction of irreversible PLT aggregation. Interestingly, the observed transient thrombocytopenia was prevented by liposomal pre-treatment, which likely saturated the binding sites on the endothelium. Compared to negatively charged liposomes, positively charged liposomes produced a less pronounced transient reduction in PLT count. Neutral liposomes induced only a mild transient PLT decline. In addition to rats, this study was also performed in rabbits, in which similar results were observed.[144] A similar study was also performed in a different laboratory using negatively charged liposomes (phosphatidylcholine/phosphatidyl acid) in guinea pigs.[44]

In an *in vitro* study, Loughrey *et al.*[40] demonstrated that the transient interaction of negatively charged PG liposomes with rat PLTs is mediated by a complement factor, likely to be C3b. The aggregation of PLTs with PG liposomes was monitored using phase contrast microscopy. Aggregation was also monitored in terms of a reduction in PLT count using a Coulter counter. The interaction of PLTs with PG liposomes did not occur in C3b-depleted plasma or in plasma that was treated with a cobra venom factor that specifically degrades C3 (confirmed by Western blotting). The interaction of PLTs with the PG liposomes was also inhibited by EDTA. This study therefore demonstrated that the transient interaction of liposomes with rat PLTs requires the presence of negatively charged lipids in the liposomal composition, divalent cations, and intact complement factors. The authors suggested that the binding of C3b to rat PLTs likely occurred due to CR1 receptor expression on the rodent PLTs. In contrast, human PLTs do not express CR1 receptors. Accordingly, the authors did not observe an interaction between human PLTs and PG liposomes in the presence of human plasma. This study is a strong illustration of how the results of rodent studies do not necessarily translate to human studies.

Following the initial adhesion of an NP, such as a liposome, to the PLT membrane, the next question would be the mechanism that is further employed. Among the several possibilities are energy-independent sequestration in the OCS, fusion with the plasma membrane, or active phagocytosis. The actual mechanism is very likely dependent on the size and surface characteristics of the NPs. Male *et al.*[45] demonstrated that PLTs can phagocytose liposomes rather than simply sequestering particles. The authors used ^{3}H-cholesteryl hexadecyl ether as the lipid label and ^{3}H-inulin as a label of the aqueous phase. The incubation of human PLTs with small neutral unilamellar liposomes (approximately 74 nm in size) resulted in liposomal uptake. While the liposomal lipids were retained, the PLTs rapidly released the components of the liposome's aqueous phase. The uptake of the liposomes was proportional to both the number of liposomes that were added and the incubation time. Approximately 250 liposomes per PLT were taken up over 5 h of incubation. Thin-section electron microscopy confirmed that the uptake of the liposomes occurred through the OCS. Changes in the fluorescence of a pH-sensitive probe and the hydrolysis of the cholesteryl [14C]oleate membrane marker indicated that the liposomes accumulated and were degraded in acid- and esterase-containing vesicles. This uptake was inhibited by the addition of EDTA, indicating that divalent cations are involved in the phagocytosis process. The inhibitory effect of cytochalasin B indicated that microfilaments and membrane plasticity play important roles in this process. The energy dependence of liposomal uptake has been confirmed by the inhibitory effects of 2,4-dinitrophenol and iodoacetate, which interfere with glycolysis and oxidative phosphorylation. Neither PLT function nor morphology was affected by liposomal loading or phagocytosis. These results were confirmed by a serotonin release assay, a micro-aggregation assay, a fluorescein diacetate membrane integrity assay, and electron microscopy.

Liposomal constructs have also been investigated as possible PLT substitutes. The dodecapeptide HHLGGAKQAGDV (H12) within the carboxy-terminus of the fibrinogen gamma-chain is a specific binding site of the PLT GPIIb-IIIa complex ligand. Suzuki *et al.*[64] evaluated liposomes of 200–400 nm in diameter that carried H12 as a synthetic PLT substitute. The study demonstrated that H12–liposomes bind to thrombin-activated PLTs and accumulate between adherent PLTs in a manner similar to fibrinogen, leading to large-scale aggregation. Abundant H12–liposomes were localized not only on the surface membrane of thrombin-activated PLTs, but also in the lumen of the large, swollen OCS, the formation of which was inhibited by pre-incubation with unbound H12, EDTA, or an anti-GPIIb–IIIa antibody. The authors observed electron-transparent areas between adherent PLTs in thrombin-induced PLT aggregates. Abundant H12–liposomes were distributed in these areas. Preclinical studies are required to evaluate the hemostatic potency of H12–liposomes and their potential adverse effects. To strengthen the hemostatic ability of H12-coated particles as PLT substitutes, Okamura *et al.*[145] exploited a drug delivery function by encapsulating ADP into H12–(ADP)–liposomes, which had a mean diameter of 250 nm. Via selective interaction of activated PLTs with GPIIb-IIIa, the H12–(ADP)–liposomes could augment agonist-induced PLT aggregation by releasing ADP in an aggregation-dependent manner. A follow-up study[83] assessed the possibility of a novel PLT substitute that consisted of disk-shaped nanosheets. These sheets had a large contact area for the targeting site rather than the conventional small contact area of the spherical carriers. The H12 peptide was conjugated to the surface of the free-standing nanosheets, which were constructed from biodegradable PLGA. These H12–PLGA nanosheets were fabricated onto 3 μm disk-shaped patterned hydrophobic octadecyl regions on an SiO_2 substrate. For comparison, spherical H12–PLGA MPs with the same surface area and H12 conjugation number were also prepared. The resulting H12–PLGA nanosheets specifically interacted with the activated PLTs that were adhered to the collagen surface at twice the rate of the H12–PLGA MPs under flow conditions. Moreover, the H12–PLGA nanosheets exhibited PLT thrombus formation in a 2D spreading manner. Thus, H12–PLGA nanosheets may be suitable candidates as a novel PLT substitute.

Tokutomi *et al.*[146] investigated the relationship between H12 density on ADP-encapsulated H12–liposomes [H12–(ADP)Lipo] and their ability to bind to PLTs. The authors also evaluated the inhibitory effect of fibrinogen on the binding of H12–(ADP)Lipo to PLTs. The binding ability of the liposomes to PLTs steeply increased with increasing H12 density until saturation was reached. The 50% inhibition concentration of fibrinogen on the binding of H12–(ADP)Lipo to PLTs was approximately 25-fold higher than the concentration of H12 molecules on the liposome. Moreover, nearly no inhibition by fibrinogen was observed at physiological fibrinogen concentrations. This result suggested that the ability of H12 to bind to GPIIb-IIIa is overwhelmingly increased by anchoring H12 to the liposome, thereby enabling the cooperative binding of H12 peptides to the PLTs.

Another example of a biomedical application of liposomes is the development of liposome-based red blood cell substitutes. Agashe *et al.*[97] demonstrated that hemoglobin-encapsulating liposomes of 200–300 nm in size containing the novel anionic non-phospholipid 2-carboxyheptadecanoyl heptadecylamide (CHHDA) did not activate PLTs. This effect was evidenced by the absence of an increase in CD62P expression.

Finally, we describe examples of a drug delivery strategy using liposomes, including the strategy of PLT-targeting. Biomimetic glycoliposomes that target CD62P appear to be promising in this respect. Zhu *et al.*[59] reported on glycocalyx-mimicking liposomes that were prepared by incorporating the glycolipid ligand 3'-sulfo-Lewis a, resulting in SuLe(a)–PEG–DSPE. This compound consisted of a SuLe(a) headgroup and a PEG spacer that was linked to two hydrophobic tails. This PEG-spaced structure was used to mimic the extended structure of the CD62P glycoprotein ligand 1 (PSGL-1) on activated leukocytes, thereby facilitating the specific binding of liposomes to the CD62P receptor on activated PLTs. The binding ability of liposomes to recombinant rat CD62P-coated plates was assayed by ELISA. In addition, the binding of nitrobenzodiazole-modified phosphatidylcholine-labeled liposomes to activated PLTs was evaluated by fluorescence microscopy. In this experiment, the PLTs themselves were adsorbed to collagen-coated glass cover slips. The results demonstrated that SuLe(a)–PEG–DSPE can form stable, narrowly size-distributed liposomes with 1,2-distearoyl-sn-glycero-3-phosphocholine (DSPC) and cholesterol, with a vesicle size of approximately 113 nm. The resultant

SuLe(a)–PEG–liposomes bound to CD62P at 22-fold higher affinity than did SuLe(a)–liposomes without a PEG spacer. Fluorescence microscopy studies demonstrated that SuLe(a)–PEG–liposomes can effectively bind to activated PLTs *in vitro*. These results suggested that biomimetic SuLe(a)–PEG–liposomes may be used as nanocarriers to target activated PLTs for drug delivery.

In addition, PEG-grafted phosphatidylcholine liposomes are used as drug carriers due to their low immunogenicity and prolonged circulation time. Heger *et al.*[54] investigated the interaction between resting and convulxin-activated hamster and human PLTs and calcein- or 5,6-carboxyfluorescein–encapsulating PEGylated liposomes. The liposomes themselves were composed of distearoylphosphatidylcholine, DSPC, and PEG-derivatized distearoylphosphatidylethanolamine. Neither resting nor activated hamster or human PLTs interacted with liposomes at a 0.70 mM lipid concentration. Flow cytometry assays demonstrated that human PLTs interacted with liposomes at lipid concentrations of ≥1.35 mM. These interactions were stronger for activated PLTs than for resting PLTs. The authors concluded that the use of PEGylated lecithin liposomes at lipid concentrations of <1.35 mM has no detrimental impact on liposomal drug delivery formulations that are based on these types of liposomes. However, the authors also stated that these drug carriers may be associated with a reduced targeting efficacy or compromised primary hemostatic system when used at concentrations of ≥1.35 mM.

Vaidya *et al.*[147] developed and characterized target-sensitive (TS) liposomes that encapsulated SK. The developed TS liposomes were composed of dioleylphosphatidyl ethanolamine (DOPE) and dipalmityl-c(RGDfK) (10:1 mol/mol). Dipalmityl-c(RGDfK) was synthesized using typical carbodiimide chemistry with palmitic acid and cyclic pentapeptide c(RGDfK), and lysine was used as a spacer. The liposomes were 100–120 nm in size. An *in vitro* drug release study demonstrated that nearly 40% of the entrapped drug was released in 12 h in phosphate buffered saline (PBS) (pH 7.4); however, on incubation with activated PLTs, approximately 90% of the drug was released within 45 min. The results suggested the target sensitivity of the liposomes. In a follow-up study,[148] the authors demonstrated that RGD–liposomes for SK delivery bound to thrombin-activated but not resting PLTs. The study also revealed that RGD–liposomes did not cause PLT aggregation or influence thrombin-induced PLT aggregation.

Paclitaxel that was encapsulated in cationic liposomes (EndoTAG-1), which had a mean size of 180–200 nm, strongly impaired tumor growth by significantly reducing functional tumor microcirculation and inducing EC apoptosis. Strieth *et al.*[76] demonstrated that EndoTAG-1 liposomes activate

PLTs. The authors suggested that significantly unbalancing the coagulation system within tumors by the targeted induction of microthromboses within the tumor microcirculation appears to be an important mechanism of EndoTAG-1 therapy. How selective the effect of EndoTAG-1 is with respect to tumor circulation and the associated risk of adverse thromboembolic effects remain unknown.

PLT-targeted NPs hold the promise of vascular site-selective delivery of drugs and imaging probes. Shrinivasan *et al.*[53] targeted activated PLTs using liposomes with cyclic RGD peptide (cRGD) surface modifications. RGD peptides specifically bind PLT integrins, particularly GPIIb-IIIa. This integrin is upregulated and stimulated to acquire a ligand-binding conformation on the surface of activated PLTs. The peptide-modified liposomes in this study were labeled either with a lipophilic fluorophore or lipid-tethered Nanogold®. All of the microscopy results indicated successful PLT targeting by the peptide-modified liposomes. The *in vitro* SEM results also enabled the visualization of the nanoscopic liposomes that were attached to the activated PLTs. The results validated this cRGD NP design for site-selective vascular delivery.

Modery *et al.*[52] described liposomal nanoconstructs that were surface-modified with two types of peptides. The use of two peptides allowed for the simultaneous binding of integrin $\alpha_{IIb}\beta_3$ and CD62P on activated PLTs. This formulation potentially induces synergistic mechanisms to enhance the targeting of drugs to vascular disease sites. The study used fluorescently labeled liposomes of 150 nm in mean diameter that were surface-modified by an RGD peptide. This peptide was targeted to active $\alpha_{IIb}\beta_3$, and an EWVDV peptide was targeted to CD62P. The liposomes were examined for their PLT-specific interactions inside a parallel plate flow chamber at low-to-high shear stresses. The results indicated that the liposomes that were surface-modified with both RGD and EWVDV exhibit higher selectivity to and retention on activated PLTs under flow conditions than liposomes with either single peptide. These results established the potential of these nanoconstructs for enhanced site-selective drug delivery in vascular diseases. Furthermore, Huang *et al.*[51] reported on liposomal surface modifications with a conformationally constrained high affinity cRGD motif to modulate the GPIIb-IIIa binding capability of the liposomes. The study demonstrated that cRGD liposomes bind activated PLTs with significantly higher affinity than linear RGD liposomes. Hence, these results established the feasibility of modulating the PLT-targeting and binding abilities of vascularly targeted liposomes by manipulating the affinity of surface-modifying ligands.

4.7. *Platelet coating with nanoparticles*

Another area of interest is the development of specific methods for the assembly of nanostructures on the PLT surface. The coating of PLTs with different engineered NPs has been examined.[149] For example, PLTs have been coated with 78 nm SiNPs, 45 nm fluorescent nanospheres, or bovine IgG. This coating was performed using layer-by-layer assembly by alternate adsorption with oppositely charged linear polyions. The sequential deposition on the PLT surfaces of cationic poly(dimethyldiallylammonium chloride) and anionic poly(styrene sulfonate) was followed by the adsorption of NPs or immunoglobulins. This technique may be used to modify the PLT surface for various purposes or modulate PLT receptor–agonist interactions, including aggregation or adhesion responses. For example, antibody-coated PLTs may be targeted to specific *in vivo* sites for diagnostic or therapeutic purposes.

4.8. *Tissue scaffolds*

Progress in the field of tissue scaffolds for tissue engineering has been greatly enhanced by the development of nanotechnology. Because many of these products are intended for intravascular use, investigating the effects of these materials on PLTs is important. Interfacial polyelectrolyte complexation (PEC) fiber has been investigated as a biostructural unit and as a biological construct for tissue engineering applications. These uses have been proposed given the ability of PEC to incorporate proteins, molecular drugs, DNA NPs, and cells. PEC is a process of self-assembly that occurs when two oppositely charged polyelectrolytes meet. The interaction between two naturally derived polyelectrolytes, CS and alginate, has been exploited to produce different biomaterials, including tissue scaffolds. Yim *et al.*[67] evaluated the biocompatibility of a PEC fiber to assess its potential for *in vivo* applications in tissue engineering. Although the CS–alginate PEC fibrous scaffold was thrombogenic, the blood compatibility of the scaffold could be significantly improved by incorporating a small amount of heparin in the polyelectrolyte solution during fiber formation. To test the effects of the scaffold on PLTs, the different samples were incubated with rabbit PRP. The PLT MP production and PLT adhesion on the CS–alginate–heparin fibrous scaffold were comparable to the results on the resting control. Only PLT compatibility was determined from these assays. The analysis of MP release is a valuable complementary assay for evaluating PLT-activating effects. This assay, however, should always be accompanied by the evaluation of other activation markers, such as the activation of GPIIb-IIIa (PAC-1), CD62P, and CD63, on the PLT surface.

Kushwaha *et al.*[150] developed a native endothelial extracellular matrix that mimicked a self-assembled nanofibrous matrix to serve as a coating for implantable cardiovascular devices. Nanofibrous matrices self-assembled from peptide amphiphiles (PAs) that contained NO-donating residues, endothelial cell-adhesive ligands composed of YIGSR peptide sequences, and enzyme-mediated degradable sites. NO was successfully and rapidly released from the nanofibrous matrix within 48 h. This rapid release was followed by a sustained release over a period of 30 days. The NO-releasing nanofibrous matrix resulted in significantly decreased PLT adhesion when compared to collagen-coated or stainless steel surfaces. Nanofibrous matrices have the potential to be applied to various cardiovascular implants as a self-assembled coating, thereby providing a native mimicking environment.

4.9. *In vitro interactions of nanomaterials with platelets: Conclusions*

Conclusions from the presented studies should be made with caution because the available data are very limited. In most cases, nanomaterials either induced PLT aggregation or did not interfere with this process. With the exception of anecdotal reports,[138] there are no studies on nanomaterial inhibition of PLT aggregation. Transient PLT agglomeration without PLT activation is also possible and has been described in certain liposome studies.[40] Carbon nanomaterials, GO,[133] and pristine, amino-conjugated, and carboxy-conjugated CNTs[57,61,86] activate PLTs, whereas fullerene C_{60} and fullerenol $C_{60}(OH)_{24}$[61] do not. Surface nanotexture/roughness and hydrophobicity dictate the PLT adhesion properties of titanium oxide surfaces. Superhydrophobic TiO_2 prevents PLT adhesion, whereas bare or superhydrophilic TiO_2 surfaces do not.[74] Both SNPs[49] and small GNPs (20 nm)[71] trigger PLT activation, whereas larger GNPs (70 nm) remain nearly inert with respect to PLT activation.[71] Unmodified polystyrene NPs do not activate PLTs[100] except at very high concentrations (0.5–2 mg/mL).[98] In contrast, carboxylated and amino-conjugated polystyrene NPs activate PLTs, each via a different mechanism.[100] Whereas carboxylated polystyrene NPs induce PLT aggregation and upregulate PLT membrane activation markers, amino-conjugated polystyrene NPs likely disintegrate the PLT plasma membrane; large (generation 6) cationic amino-conjugated PAMAM dendrimers also apparently have this effect.[65] Among polymer nanomaterials, PLGA demonstrated favorable PLT biocompatibility.[47,72] Furthermore, several nanocomposite nanomaterials that are based on PCUs and other polymers with promising PLT biocompatibility have been developed.[75,105]

Comprehensive studies are required to assess the effects of the type of material and the size, surface characteristics, and nanotexture of these materials on PLT biocompatibility.

5. Selected Protocols for *In Vitro* Platelet Studies Used in Our Laboratory

5.1. *Basic assays*

5.1.1. *Cell counter-based method*[65,151]

To study particle effects on PLT aggregation, whole blood was centrifuged 8 min at 200 *g* in order to obtain PRP. The PRP was treated with NPs, PBS (negative control), or collagen (positive control) for 15 min at 37°C. After that, single PLT count was conducted using Z2 counter and size analyzer (Beckman Coulter). The difference in single PLT counts between the negative control and test samples was used to calculate percent PLT aggregation. Additional controls included the incubation of the PPP and PBS with particles and analyzing these samples on the instrument. These controls were used to monitor potential particle aggregation in the presence of plasma proteins to avoid false-negative results. The detailed protocol is available at http://ncl.cancer.gov/NCL_Method_ITA-2.pdf.

5.1.2. *LTA in PRP*[61]

To further study PLT aggregation activity of nanomaterials, we employed LTA (PAP-8E aggregometer, Bio/Data Corp., Horsham, PA). PRP was prepared from ACD-anticoagulated blood of healthy donors. Fifty mL of whole blood were centrifuged at 150 *g* for 10 min at room temperature and the PRP was collected. The sediment was then centrifuged at 1,000 *g* for 15 min at room temperature to obtain PPP, which was used as a blank. The PRP PLT count was assayed (ABX Pentra 60, Horiba ABX, Inc., Irvine, CA) and diluted with PPP to 250×10^3 PLTs/μL. Aggregation experiments were completed within 4 h after blood collection to ensure normal PLT responsiveness. TRAP-6 or collagen were used as positive controls. All PLT samples were preincubated for 2 min before adding the tested nanomaterials or controls in the stirring wells of the aggregometer, and a 20 min reading was obtained for every sample set. Each experiment was performed using at least three blood samples from different donors. Means of maximum aggregation responses ± SEM were presented.

5.1.3. *Flow cytometry of PLT activation markers and membrane MPs*[61]

Analysis of PLT surface activation markers and PLT membrane MPs was performed by flow cytometry. PRP was prepared as described above and then diluted to 250×10^3 PLTs/μL with PPP for MP experiments and to 30×10^3 PLTs/μL with Tyrode's–HEPES buffer (THB: 130 mM NaCl, 2.6 mM KCl, 0.42 mM NaH_2PO_4, 5.5 mM glucose, 10 mM HEPES, 0.3% bovine serum albumin, BSA) for assessing PLT surface activation markers. PLTs were equilibrated for 30 min after dilution and were incubated with tested nanomaterials (100 μg/mL) or a positive/negative control (20 μM of TRAP-6 and PLTs only, respectively) for 15 min on a rocking platform at 37°C. PLT surface markers and MP assays were run in parallel, using the same blood specimen. For the analysis of PLT surface markers, PLT samples were spun briefly for 1 min at 150 *g* at room temperature to avoid the presence of large nanomaterial aggregates. Fifty μL of supernatant were stained with saturating concentration of mABs against CD41a (FITC-labeled) and CD62P (PE-labeled). Matching isotype controls and non-labeled samples were used as controls. After 15 min of incubation in the dark at room temperature, samples were diluted 10-fold with THB and immediately acquired using a FACSCalibur flow cytometer (Becton Dickinson, San Diego, CA, USA) equipped with CELLQuest software, with FSC and SSC in logarithmic mode and subsequently analyzed using FlowJo (Tree Star, Inc., Ashland, OR). Representative histograms and FSC/SSC plots of three independent experiments were presented.

MP samples, after incubation with nanomaterials, were spun for 10 min at 10°C at 10,000 *g* to obtain PLT-free plasma (PFP) for the MP assay. Samples were then processed and analyzed as described previously.[102,152] Aliquots of 50 μL PFP were incubated for 20 min at room temperature in the dark with saturating concentrations of FITC- and PE-conjugated antibodies. In parallel, non-labeled samples and samples labeled with relevant isotype controls were analyzed. After incubation and washing with 1 mL HBSS/Ca^{2+}/BSA, samples were resuspended in 500 μL HBSS/Ca^{2+}/BSA (Hanks' balanced salt solution, Gibco, Grand Island, NY, was supplemented with $CaCl_2$ to 2.5 mM Ca^{2+} and with 0.3% BSA) and analyzed by flow cytometry. Data were acquired using the flow cytometer and software as described above. MPs were analyzed using both FSC and SSC in logarithmic mode. The flow rate was evaluated using TruCount beads from Becton Dickinson (San Diego, CA) analyzed as separate samples in parallel. MPs were defined as particles ≤1 μm based on size comparison on FSC with the size standard polystyrene beads of 1 μm in diameter. Counts of $CD41a^+CD62P^+$ MPs in the

PLT supernatant were evaluated using double fluorescence plots acquired for 60 s at the standard flow rate. The relative increase in MP count in the PLT supernatant after different treatments was presented as mean ± SEM of three independent experiments.

5.1.4. *SEM*[56]

PRP was untreated or treated with dendrimers at a final concentration of 100 μg/mL for 7 min, as described in the cell counter-based screening method above. At the end of incubation, PLTs were fixed with 2% paraformaldehyde (PA) for 15 min, centrifuged at 1,000 *g* for 10 min, and resuspended in 500 μL of PBS; 100 μL of the PLT suspension was deposited on a poly-L-lysine–coated cover slip and incubated for 20 min. The cover slips were rinsed thrice with PBS and incubated for 30 min in 2% glutaraldehyde (GTA) solution in 0.1 M PBS buffer (pH 7.4). The cover slips were then washed three times with 0.1 M cacodylate buffer. The samples were immersed in a series of graded ethanol solution (20%, 50%, 70%, and 90% in water, and 100%) for 10 min and two changes were performed for 100% ethanol. At the last step of dehydration, tetramethylsilane (TMS) was added and kept for 10 min. Three changes were done to remove any residue of ethanol. The TMS was removed and the samples were dried in a desiccator under vacuum. The cover slips were attached to SEM aluminum stub and carbon, and a thin layer of gold was evaporated on the cover slips. The SEM imaging was performed using Hitachi S4700 field emission SEM (FESEM).

5.2. *Special investigational assays*

5.2.1. *PLT–NP interaction studies by FESEM and TEM*[56]

PRP was prepared from the ACD-anticoagulated blood of healthy donors. Fifty mL of whole blood were centrifuged at 150 *g* for 10 min at room temperature, and the PRP was collected. The sediment was then centrifuged at 1,000 *g* for 15 min to obtain the PPP. PRP PLT count was performed using ABX Pentra 60 (Horiba ABX, Inc., Irvine, CA). A stock solution was prepared by diluting the PRP with PPP to 250×10^3 PLTs/μL, after which the PLT suspension was incubated at 37°C for 30 min. As described previously,[61] the interaction of nC_{60} and M60 CNTs with human blood PLTs was analyzed by mixing 450 μL PLT stock suspension with 50 μL NP suspension (freshly prepared suspension, 1 mg/mL), TRAP, or thapsigargin (TPG; final concentrations were 20 and 6 μmol/L, respectively) and incubating at 37°C for 15 min under gentle agitation. The ultrastructural analysis of the carbon NP–PLT interaction was performed by FESEM and TEM. For FESEM, PLTs were fixed with 4% PA for 30 min, carefully deposited onto clean glass cover slips by cytospin, washed twice with PBS, fixed with 2% GTA, washed extensively with cacodylate buffer, dehydrated in an ethanol series, washed with TMS, and dried at room temperature under vacuum. Some samples were sputter-coated with a thin film of gold at 13.3 Pa and 45 mA for 90 s. For TEM, PLTs were fixed in 4% PA and 1% GTA for 1 h, post-fixed with 4% osmium tetroxide (OsO_4) solution in cacodylate buffer, dehydrated in an ethanol series, embedded with Spur, and polymerized for two days at 65°C. Ultrathin sections (approximately 60 nm) were cut on a Leica ultramicrotome and collected onto 600 mesh copper grids.

5.2.1.1. DTS visualization by peroxidase labeling with DAB[153]

PLTs were fixed with 0.1% GTA in PBS for 15 min, washed twice with PBS to remove the excess plasma proteins, fixed with 2.5% PA and 0.5% GTA for 30 min, washed extensively with 0.05 mol/L Tris buffer, and incubated overnight with DAB solution (2.5 mg/mL in 0.05 mol/L Tris buffer) at 4°C. On the following day, the DAB solution was replaced with freshly prepared DAB solution containing 0.003 % hydrogen peroxide and incubated for 2 h at room temperature. The PLTs were then extensively washed with cacodylate buffer, post-fixed with 4% OsO_4 for 1 h at room temperature in the dark, dehydrated in an ethanol series, embedded in Spur, and sectioned using a microtome for TEM.

5.2.1.2. STIM1 gold immunolabeling

SDS-PAGE and Western blot analysis of STIM1 from PLTs to confirm the TS of the anti-STIM1 antibody used for immunolabeling was performed as previously described.[154] STIM1 in PLTs was labeled for TEM as follows: after PLTs were incubated with carbon NPs or TPG as described above, samples were fixed with 1% PA for 15 min, washed three times with PBS, and permeabilized with 0.1% Triton X-100 for 15 min. After extensive washes with PBS, the PLTs were blocked with Image-iT FX signal enhancer for 30 min. The PLTs were then incubated with 150 µL mouse anti-GOK/STIM1 antibody (10 µg/mL) for 2 h, washed four times with PBS for 5 min each time, and incubated with 5 nm immunogold-conjugated goat anti-mouse IgG (150 µL, 10 µg/mL) for 2 h at room temperature. After five washes with PBS for

5 min each, the samples were fixed with 2.0% PA and 2.5% GTA for 30 min at room temperature to avoid artifacts during the embedding process required for the TEM, as described above. All of the above steps were performed at room temperature.

5.2.2. *Confocal immunomicroscopy*[56]

STIM1 and ORAI1 in PLTs were labeled for laser scanning confocal microscopy (LSCM) using the protocol described above, except the immunogold conjugate was replaced with Alexa Fluor 488-conjugated goat anti-mouse IgG to detect STIM1, and Alexa Fluor 633-conjugated donkey anti-goat IgG to detect ORAI1. After incubation with the antibodies, the samples were extensively washed with PBS. The PLTs were then deposited onto a clean glass slide by cytospin, mounted with VectaShield and imaged with a Zeiss 710 LSM 710 NLO with a Plan Apochromat 100×/1.40 oil objective. Once adjusted, the laser configuration and pinhole aperture were kept the same for all samples analyzed. MatLab and Image J software were used for image deconvolution.

5.2.3. *Intracellular Ca^{2+} assay*[56,61]

5.2.3.1. Assay 1

The acute effect of CNTs on intracellular free Ca^{2+} concentration in individual PLTs loaded with a Ca^{2+}-sensitive probe FURA-2AM was investigated by employing ratio fluorometry as described previously,[155] with modifications for PLTs. PRP was prepared from heparinized blood and recalcified as described above, diluted with PPP to a PLT count of 250×10^3 PLTs/μL and incubated with 100 μM ASA for 30 min in 37°C to inhibit PLT activation during the procedure. PRP was then incubated with 2 μM FURA-2AM for 1 h at 37°C. After subsequent washing, PLTs were resuspended in 300 μL of THB and incubated for 20 min on poly-L-lysine–coated cover glass. Changes in the fluorescence of individual PLTs ($n = 100$) were monitored at 340 and 380 nm excitation wavelenghts (the rate of data capture was 170/min) using a Nikon inverted epi-fluorescence/phase microscope equipped with a low-light level integrating CCD camera with a microphotometer assembly (InCyt I/P-2 TM Imaging and Photometry System, Intracellular Imaging, Inc., Cincinnati, OH). Real-time intracellular Ca^{2+} concentration was calculated from the ratio of emission detected at 510 nm at two excitation wavelengths (340 and

380 nm) and by comparison to a standard curve established for these settings using buffers of known free Ca^{2+}.

5.2.3.2. Assay 2

Qualitative analysis of the Ca^{2+} content in PLTs was performed as per Lopez *et al.*[156] with some modifications. PRP (10^8 cells/mL) was incubated at 37°C with 5 μmol/L Fura-FF/AM and Pluronic F-127 (0.025%) for 45 min, after which 1 μM prostaglandin E1 was added. After 10 min, the cells were collected by centrifugation at 350 *g* for 10 min and resuspended in a modified intracellular solution containing 20 mmol/L NaCl, 10 mmol/L HEPES–KOH, 120 mmol/L KCl, and 1.13 mmol/L $MgCl_2$, at pH 7.2, and supplemented with 40 μg/mL apyrase; EGTA (100 μmol/L) was then added 10 min before the experiment started. Cell membranes were permeabilized, allowing cytosolic Ca^{2+} to be chelated by mixing 225 μL Fura-FF/AM-loaded PLTs placed in a 96-well plate with 0.5 U/mL streptolysin. After 1 min, 25 μL M60, nC_{60}, or TPG (6 μmol/L) solution was added to begin the experiment. Changes in the Ca^{2+} content in PLT stores were monitored by measuring the excitation at 380 nm and emission at 510 nm.

6. Materials

3,3'-diaminobenzidine tetrahydrochloride hydrate (DAB), EGTA, apyrase, prostaglandin E1, NaCl, HEPES-KOH, KCl, $MgCl_2$, and streptolysin were purchased from Sigma-Aldrich (St. Louis, MO). Sodium cacodylate, Spur resin, glutaraldehyde, paraformaldehyde and osmium tetroxide were purchased from Electron Microscopy Science. Thrombin receptor-activating peptide SFLLRN (TRAP-6) was from AnaSpec, Inc. (Fremont, CA). Alkaline phosphatase-conjugated AffiniPure F(ab')s fragment donkey anti-mouse IgG was from Jackson ImmunoResearch (West Grove, PA). Image-iT FX signal enhancer, Fura-FF/AM, Pluronic F-127, TPG were purchased from Invitrogen (Grand Island, NY).

Other chemicals: Collagen (Bio/Data Corp., Horsham, PA), adenosine diphosphate (ADP; Bio/Data Corp., Horsham, PA), EDTA (Gibco, Grand Island, NY), SKF-96365 (Alexis Biochemicals, Axxora LLC, San Diego, CA), MRS 2500 tetraammonium salt (Tocris Bioscience, Ellisville, MO), NF449 (Tocris Bioscience, Ellisville, MO), DM BAPTA, AM (Invitrogen Corp., Carlsbad, CA), 2-aminoethoxydiphenyl borate (2-APB) (EMD-Calbiochem Brand, San Diego, CA), t-butylhydroquinone (TBHQ) (Alexis Biochemicals, Axxora LLC, San Diego, CA), calcium chloride

(Sigma-Aldrich Inc., St. Louis, MO), FURA-2AM (Invitrogen Corp., Carlsbad, CA), acetylsalicylic acid (ASA; Sigma-Aldrich, Inc., St. Louis, MO), Heparin Sodium Inj. 1000 USP units/mL (Baxter Healthcare Corp., Deerfield, IL), dimethyl sulfoxide (DMSO; Sigma-Aldrich, Inc., St. Louis, MO), phosphate buffered saline (PBS; Gibco, Grand Island, NY), poly-L-lysine (Sigma-Aldrich, Inc., St. Louis, MO).

Tyrode's–HEPES buffer (THB) was prepared as follows: 130 mM NaCl, 2.6 mM KCl, 0.42 mM NaH_2PO_4, 5.5 mM glucose, 10 mM HEPES, 0.3% bovine serum albumin (BSA), with or without 2.5 mM Ca^{2+}; all chemicals were from Sigma-Aldrich, Inc., St. Louis, MO. Hanks' balanced salt solution (HBSS; Gibco, Grand Island, NY) was supplemented with $CaCl_2$ to 2.5 mM Ca^{2+} and with 0.3% BSA (HBSS/BSA).

Antibodies: CD41a FITC (clone HIP8), CD62P PE (clone AC1.2), CD63PE (clone H5C6), and corresponding isotype controls (mouse IgG1 kappa, clone X40), and mouse anti-GOK/STIM1 were all from BD Bioscience (San Jose, CA). Anti-ORAI1 antibody, STIM1 peptide and anti-STIM1 were purchased from ProSci (Poway, CA). Immunogold conjugates (5 nm) goat anti-mouse IgG, Alexa Fluor 633-conjugated donkey anti-goat IgG, Alexa Fluor 488-conjugated goat anti-mouse IgG, and F(ab')2 fragment of alkaline phosphatase-conjugated goat anti-rabbit IgG were purchased from Invitrogen (Grand Island, NY).

7. Conclusion

Thrombotic and hemorrhagic complications have a major impact on the mortality and morbidity in both cancer patients and in patients with vascular disorders and other diseases. Thus, understanding the effects of nanomaterials on PLTs, other blood systems, and the vasculature is a critical safety issue. There are few available studies on the effects of different engineered nanomaterials on PLTs. The results of these studies have demonstrated that certain nanomaterials induce PLT activation, adhesion, aggregation, and the release reaction, whereas other nanomaterials may result in an impairment of PLT function and PLT damage. A panel of *in vitro* standard assays for the screening and characterization of the effects of NPs and macrodevices with nanostructured surfaces that come into contact with the blood on PLTs is needed. Presented reviews of methods, nanomaterial studies, and assay protocols may provide a good ground for the development of such a panel of *in vitro* assays. In addition, standard animal models are needed for the *in vivo* evaluation of the effects of nanomaterials on PLTs. These models should include "two-hit" models in which the animals are pre-conditioned prior to nanomaterial exposure. Lastly, well-designed clinical trials should carefully monitor the possible adverse effects of nanomaterials on PLTs with respect to consequent thrombotic or bleeding complications, or other PLT-mediated organ dysfunctions.

References

1. Banerjee HN, Verma M. Application of nanotechnology in cancer. *Technol Cancer Res Treat* 2008;7:149–154.
2. Buxton DB. Nanotechnology in the diagnosis and management of heart, lung and blood diseases. *Expert Rev Mol Diagn* 2007;7:149–160.
3. Liu Y, Miyoshi H, Nakamura M. Nanomedicine for drug delivery and imaging: A promising avenue for cancer therapy and diagnosis using targeted functional nanoparticles. *Int J Cancer* 2007;120:2527–2537.
4. Moghimi SM, Hunter AC, Murray JC. Nanomedicine: Current status and future prospects. *Faseb J* 2005;19:311–330.
5. Morrow KJ, Jr, Bawa R, Wei C. Recent advances in basic and clinical nanomedicine. *Med Clin North Am* 2007;91:805–843.
6. Nie S, Xing Y, Kim GJ, Simons JW. Nanotechnology applications in cancer. *Annu Rev Biomed Eng* 2007;9:257–288.
7. Patel DN, Bailey SR. Nanotechnology in cardiovascular medicine. *Catheter Cardiovasc Interv* 2007;69:643–654.
8. Pope-Harman A, Cheng MM, Robertson F, Sakamoto J, Ferrari M. Biomedical nanotechnology for cancer. *Med Clin North Am* 2007;91:899–927.
9. Wei C, Wei W, Morris M, Kondo E, Gorbounov M, Tomalia DA. Nanomedicine and drug delivery. *Med Clin North Am* 2007;91:863–870.
10. Zuo L, Wei W, Morris M, Wei J, Gorbounov M, Wei C. New technology and clinical applications of nanomedicine. *Med Clin North Am* 2007;91:845–862.
11. Williams DF. On the mechanisms of biocompatibility. *Biomaterials* 2008;29:2941–2953.
12. Simak J. Nanotoxicity in blood: Effects of engineered nanomaterials on platelets. In: Sahu S, Casciano D (eds.), *Nanotoxicity. From In Vivo and In Vitro Models to Health Risks*, John Wiley & Sons, 2009, pp. 191–255.
13. Michelson AD (ed.), *Platelets*, Elsevier, Inc., 2007.
14. Shi G, Morrell CN. Platelets as initiators and mediators of inflammation at the vessel wall. *Thromb Res* 2010;127:387–390.
15. Broos K, Feys HB, De Meyer SF, Vanhoorelbeke K, Deckmyn H. Platelets at work in primary hemostasis. *Blood Rev* 2011;25:155–167.
16. Jennings LK, McCabe-White M. Platelet aggregation. In: Michelson AD (ed.), *Platelets*, Elsevier, Inc., 2007, pp. 495–507.
17. Nurden AT. Platelets, inflammation and tissue regeneration. *Thromb Haemost* 2011; 105(Suppl 1):S13–33.
18. Cox D, Kerrigan SW, Watson SP. Platelets and the innate immune system: Mechanisms of bacterial-induced PLT activation. *J Thromb Haemost* 2011;9:1097–1107.
19. Vieira-de-Abreu A, Campbell RA, Weyrich AS, Zimmerman GA. Platelets: Versatile effector cells in hemostasis, inflammation, and the immune continuum. *Semin Immunopathol* 2011;34:5–30.

20. Semple JW, Freedman J. Platelets and innate immunity. *Cell Mol Life Sci* 2010;67:499–511.
21. Lievens D, von Hundelshausen P. Platelets in atherosclerosis. *Thromb Haemost* 2011;106:827–838.
22. Smith TL, Weyrich AS. Platelets as central mediators of systemic inflammatory responses. *Thromb Res* 2011;127:391–394.
23. Li Z, Yang F, Dunn S, Gross AK, Smyth SS. Platelets as immune mediators: Their role in host defense responses and sepsis. *Thromb Res* 2011;127:184–188.
24. Kaushansky K. The molecular mechanisms that control thrombopoiesis. *J Clin Invest* 2005;115:3339–3347.
25. Heal JM, Blumberg N. Optimizing platelet transfusion therapy. *Blood Rev* 2004;18: 149–165.
26. Varga-Szabo D, Pleines I, Nieswandt B. Cell adhesion mechanisms in platelets. *Arterioscler Thromb Vasc Biol* 2008;28:403–412.
27. Caen J, Wu Q. Hageman factor, platelets and polyphosphates: Early history and recent connection. *J Thromb Haemost* 8:1670–1674.
28. Simak J, Gelderman MP. Cell membrane microparticles in blood and blood products: Potentially pathogenic agents and diagnostic markers. *Transfus Med Rev* 2006; 20:1–26.
29. Freedman JE. Oxidative stress and platelets. *Arterioscler Thromb Vasc Biol* 2008; 28:S11–16.
30. May AE, Seizer P, Gawaz M. Platelets: Inflammatory firebugs of vascular walls. *Arterioscler Thromb Vasc Biol* 2008;28:S5–10.
31. Tabuchi A, Kuebler WM. Endothelium-platelet interactions in inflammatory lung disease. *Vascul Pharmacol* 2008;49:141–150.
32. Harrison P. Platelet function analysis. *Blood Rev* 2005;19:111–123.
33. Michelson AD, Frelinger AL III, Furman MI. Current options in platelet function testing. *Am J Cardiol* 2006;98:4N–10N.
34. Harle CC. Point-of-care platelet function testing. *Semin Cardiothorac Vasc Anesth* 2007;11: 247–251.
35. Gurbel PA, Becker RC, Mann KG, Steinhubl SR, Michelson AD. Platelet function monitoring in patients with coronary artery disease. *J Am Coll Cardiol* 2007;50:1822–1834.
36. Cardigan R, Turner C, Harrison P. Current methods of assessing Platelet function: Relevance to transfusion medicine. *Vox Sang* 2005;88:153–163.
37. Shen YM, Frenkel EP. Acquired platelet dysfunction. *Hematol Oncol Clin North Am* 2007;21:647–661, vi.
38. Briggs C, Harrison P, Machin SJ. Platelet counting. In: Michelson AD (ed.), *Platelets*, Elsevier, Inc., 2007, pp. 475–483.
39. Reinish LW, Bally MB, Loughrey HC, Cullis PR. Interactions of liposomes and platelets. *Thromb Haemost* 1988;60:518–523.
40. Loughrey HC, Bally MB, Reinish LW, Cullis PR. The binding of phosphatidylglycerol liposomes to rat platelets is mediated by complement. *Thromb Haemost* 1990;64:172–176.
41. Hsu SH, Tang CM, Tseng HJ. Biocompatibility of poly(ether)urethane-gold nanocomposites. *J Biomed Mater Res A* 2006;79:759–770.
42. Dobrovolskaia MA, Aggarwal P, Hall JB, McNeil SE. Preclinical studies to understand nanoparticle interaction with the immune system and its potential effects on nanoparticle biodistribution. *Mol Pharm* 2008;5:487–495.

43. Neun BW, Dobrovolskaia MA. Method for *in vitro* analysis of nanoparticle thrombogenic properties. *Methods Mol Biol* 2011;697:225–235.
44. Zbinden G, Wunderli-Allenspach H, Grimm L. Assessment of thrombogenic potential of liposomes. *Toxicology* 1989;54:273–280.
45. Male R, Vannier WE, Baldeschwieler JD. Phagocytosis of liposomes by human platelets. *Proc Natl Acad Sci U S A* 1992;89:9191–9195.
46. Miyamoto M, Sasakawa S, Ozawa T, Kawaguchi H, Ohtsuka Y. Platelet aggregation induced by latex particles. I. Effects of size, surface potential and hydrophobicity of particles. *Biomaterials* 1989;10:251–257.
47. Cenni E, Granchi D, Avnet S, *et al.* Biocompatibility of poly(D,L-lactide-co-glycolide) nanoparticles conjugated with alendronate. *Biomaterials* 2008;29:1400–1411.
48. Ferraz N, Carlsson J, Hong J, Ott MK. Influence of nanoporesize on platelet adhesion and activation *J Mater Sci Mater Med.* 2008;19:3115–3121.
49. Jun EA, Lim KM, Kim K, *et al.* Silver nanoparticles enhance thrombus formation through increased platelet aggregation and procoagulant activity. *Nanotoxicology.* 2011;5:157–167.
50. Kunicki TJ, Tuccelli M, Becker GA, Aster RH. A study of variables affecting the quality of platelets stored at "room temperature". *Transfusion* 1975;15:414–421.
51. Huang G, Zhou Z, Srinivasan R, *et al.* Affinity manipulation of surface-conjugated RGD peptide to modulate binding of liposomes to activated platelets. *Biomaterials* 2008;29:1676–1685.
52. Modery CL, Ravikumar M, Wong TL, Dzuricky MJ, Durongkaveroj N, Sen Gupta A. Heteromultivalent liposomal nanoconstructs for enhanced targeting and shear-stable binding to active platelets for site-selective vascular drug delivery. *Biomaterials* 2011;32:9504–9514.
53. Srinivasan R, Marchant RE, Gupta AS. *In vitro* and *in vivo* platelet targeting by cyclic RGD-modified liposomes. *J Biomed Mater Res A* 2009;93:1004–1015.
54. Heger M, Salles, II, van Vuure W, *et al.* On the interaction of fluorophore-encapsulating PEGylated lecithin liposomes with hamster and human platelets. *Microvasc Res* 2009;78:57–66.
55. Burke AR, Singh RN, Carroll DL, *et al.* Determinants of the thrombogenic potential of multiwalled carbon nanotubes. *Biomaterials* 2011;32:5970–5978.
56. Lacerda SH, Semberova J, Holada K, Simakova O, Hudson SD, Simak J. Carbon nanotubes activate store-operated calcium entry in human blood platelets. *ACS Nano* 2011;5:5808–5813.
57. Radomski A, Jurasz P, Alonso-Escolano D, *et al.* Nanoparticle-induced platelet aggregation and vascular thrombosis. *Br J Pharmacol* 2005;146:882–893.
58. Fernandes EG, de Queiroz AA, Abraham GA, San Roman J. Antithrombogenic properties of bioconjugate streptokinase-polyglycerol dendrimers. *J Mater Sci Mater Med* 2006;17:105–111.
59. Zhu J, Xue J, Guo Z, Zhang L, Marchant RE. Biomimetic glycoliposomes as nanocarriers for targeting P-selectin on activated platelets. *Bioconjug Chem* 2007;18: 1366–1369.
60. Kim Y, Hechler B, Klutz AM, Gachet C, Jacobson KA. Toward multivalent signaling across G protein-coupled receptors from poly(amidoamine) dendrimers. *Bioconjug Chem* 2008;19:406–411.

61. Semberova J, De Paoli Lacerda SH, Simakova O, Holada K, Gelderman MP, Simak J. Carbon nanotubes activate blood platelets by inducing extracellular Ca^{2+} influx sensitive to calcium entry inhibitors. *Nano Lett* 2009;9:3312–3317.
62. Corbalan JJ, Medina C, Jacoby A, Malinski T, Radomski MW. Amorphous silica nanoparticles aggregate human platelets: Potential implications for vascular homeostasis. *Int J Nanomed* 2012;7:631–639.
63. Aurich K, Spoerl MC, Furll B, *et al.* Development of a method for magnetic labeling of platelets. *Nanomedicine* 2012;8:537–544.
64. Suzuki H, Okamura Y, Ikeda Y, Takeoka S, Handa M. Ultrastructural analysis of thrombin-induced interaction between human platelet and liposomes carrying fibrinogen gamma-chain dodecapeptide as a synthetic platelet substitute. *Thromb Res* 2011;128:552–559.
65. Dobrovolskaia MA, Patri AK, Simak J, *et al.* Nanoparticle size and surface charge determine effects of PAMAM dendrimers on human platelets *in vitro*. *Mol Pharm* 2011;9:382–393.
66. Lin JC, Wu CH. Surface characterization and platelet adhesion studies on polyurethane surface immobilized with C60. *Biomaterials* 1999;20:1613–1620.
67. Yim EK, Liao IC, Leong KW. Tissue compatibility of interfacial polyelectrolyte complexation fibrous scaffold: Evaluation of blood compatibility and biocompatibility. *Tissue Eng* 2007;13:423–433.
68. Meng J, Song L, Xu H, *et al.* Effects of single-walled carbon nanotubes on the functions of plasma proteins and potentials in vascular prostheses. *Nanomedicine* 2005;1:136–142.
69. Ahmed M, Ghanbari H, Cousins BG, Hamilton G, Seifalian AM. Small calibre polyhedral oligomeric silsesquioxane nanocomposite cardiovascular grafts: Influence of porosity on the structure, haemocompatibility and mechanical properties. *Acta Biomater* 2011;7:3857–3867.
70. Chen L, Liu M, Bai H, *et al.* Antiplatelet and thermally responsive poly(N-isopropylacrylamide) surface with nanoscale topography. *J Am Chem Soc* 2009;131:10467–10472.
71. Deb S, Patra HK, Lahiri P, Dasgupta AK, Chakrabarti K, Chaudhuri U. Multistability in platelets and their response to gold nanoparticles. *Nanomedicine* 2011;7:376–384.
72. Koh LB, Rodriguez I, Venkatraman SS. A novel nanostructured poly(lactic-co-glycolic-acid)-multi-walled carbon nanotube composite for blood-contacting applications: Thrombogenicity studies. *Acta Biomater* 2009;5:3411–3422.
73. Ferraz N, Ott MK, Hong J. Time sequence of blood activation by nanoporous alumina: Studies on platelets and complement system. *Microsc Res Tech* 2010;73:1101–1109.
74. Yang Y, Lai Y, Zhang Q, *et al.* A novel electrochemical strategy for improving blood compatibility of titanium-based biomaterials. *Colloids Surf B Biointerfaces* 2010;79:309–313.
75. Sun T, Tan H, Han D, Fu Q, Jiang L. No platelet can adhere — largely improved blood compatibility on nanostructured superhydrophobic surfaces. *Small* 2005;1:959–963.
76. Strieth S, Nussbaum CF, Eichhorn ME, *et al.* Tumor-selective vessel occlusions by platelets after vascular targeting chemotherapy using paclitaxel encapsulated in cationic liposomes. *Int J Cancer* 2008;122:452–460.
77. Stevens KN, Knetsch ML, Sen A, Sambhy V, Koole LH. Disruption and activation of blood platelets in contact with an antimicrobial composite coating consisting of a pyridinium polymer and AgBr nanoparticles. *ACS Appl Mater Interfaces* 2009;1:2049–2054.

78. Smith BS, Yoriya S, Grissom L, Grimes CA, Popat KC. Hemocompatibility of titania nanotube arrays. *J Biomed Mater Res A* 2010;95:350–360.
79. Karagkiozaki VC, Logothetidis SD, Kassavetis SN, Giannoglou GD. Nanomedicine for the reduction of the thrombogenicity of stent coatings. *Int J Nanomed* 2010;5:239–248.
80. Semberova J, Holada K, Simakova O, Gelderman-Fuhrmann M, Simak J. Carbon nanotubes activate platelets by facilitating extracellular Ca^{2+} influx. *Blood* 2008;112:365–366 (abs).
81. Juliano RL, Hsu MJ, Peterson D, Regen SL, Singh A. Interactions of conventional or photopolymerized liposomes with platelets *in vitro*. *Exp Cell Res* 1983;146:422–427.
82. Ramtoola Z, Lyons P, Keohane K, Kerrigan SW, Kirby BP, Kelly JG. Investigation of the interaction of biodegradable micro- and nanoparticulate drug delivery systems with platelets. *J Pharm Pharmacol* 2011;63:26–32.
83. Okamura Y, Fukui Y, Kabata K, *et al.* Novel platelet substitutes: Disk-shaped biodegradable nanosheets and their enhanced effects on platelet aggregation. *Bioconjug Chem* 2009;20: 1958–1965.
84. Tavano R, Segat D, Reddi E, *et al.* Procoagulant properties of bare and highly PEGylated vinyl-modified silica nanoparticles. *Nanomedicine (Lond)* 2010;5:881–896.
85. Koziara JM, Oh JJ, Akers WS, Ferraris SP, Mumper RJ. Blood compatibility of cetyl alcohol/polysorbate-based nanoparticles. *Pharm Res* 2005;22:1821–1828.
86. Bihari P, Holzer M, Praetner M, *et al.* Single-walled carbon nanotubes activate platelets and accelerate thrombus formation in the microcirculation. *Toxicology* 2010;269:148–154.
87. Li X, Radomski A, Corrigan OI, *et al.* Platelet compatibility of PLGA, chitosan and PLGA-chitosan nanoparticles. *Nanomedicine (Lond)* 2009;4:735–746.
88. Ozeki Y, Sudo T, Toga K, *et al.* Characterization of whole blood aggregation with a new type of aggregometer by a screen filtration pressure method. *Thromb Res* 2001;101: 65–72.
89. Niwa Y, Iwai N. Nanomaterials induce oxidized low-density lipoprotein cellular uptake in macrophages and platelet aggregation. *Circ J* 2007;71:437–444.
90. Holme S, Moroff G, Murphy S. A multi-laboratory evaluation of *in vitro* platelet assays: The tests for extent of shape change and response to hypotonic shock. Biomedical Excellence for Safer Transfusion Working Party of the International Society of Blood Transfusion. *Transfusion* 1998;38:31–40.
91. Michelson AD, Linden MD, Barnard MR, Furman MI, Frelinger AL III. Flow cytometry. In: Michelson AD (ed.), *Platelets*, Elsevier, Inc., 2007, pp. 545–563.
92. Michelson AD. Evaluation of platelet function by flow cytometry. *Pathophysiol Haemost Thromb* 2006;35:67–82.
93. Linden MD, Furman MI, Frelinger AL III, *et al.* Indices of platelet activation and the stability of coronary artery disease. *J Thromb Haemost* 2007;5:761–765.
94. Frelinger AL III, Barnard MR, Fox ML, Michelson AD. The platelet activity after clopidogrel termination (PACT) study. *Circ Cardiovasc Interv* 2010;3:442–449.
95. Constantinescu I, Levin E, Gyongyossy-Issa M. Liposomes and blood cells: A flow cytometric study. *Artif Cells Blood Substit Immobil Biotechnol* 2003;31:395–424.
96. Shin SH, Ye MK, Kim HS, Kang HS. The effects of nano-silver on the proliferation and cytokine expression by peripheral blood mononuclear cells. *Int Immunopharmacol* 2007;7:1813–1818.
97. Agashe H, Lagisetty P, Awasthi S, Awasthi V. Improved formulation of liposome-encapsulated hemoglobin with an anionic non-phospholipid. *Colloids Surf B Biointerfaces* 2010;75:573–583.

98. Mayer A, Vadon M, Rinner B, Novak A, Wintersteiger R, Frohlich E. The role of nanoparticle size in hemocompatibility. *Toxicology* 2009;258:139–147.
99. Israels SJ, McMillan-Ward EM. Platelet tetraspanin complexes and their association with lipid rafts. *Thromb Haemost* 2007;98:1081–1087.
100. McGuinnes C, Duffin R, Brown S, *et al.* Surface derivatization state of polystyrene latex nanoparticles determines both their potency and their mechanism of causing human platelet aggregation *in vitro*. *Toxicol Sci* 2011;119:359–368.
101. Villagra J, Shiva S, Hunter LA, Machado RF, Gladwin MT, Kato GJ. Platelet activation in patients with sickle disease, hemolysis-associated pulmonary hypertension, and nitric oxide scavenging by cell-free hemoglobin. *Blood* 2007;110:2166–2172.
102. Gelderman MP, Simak J. Flow cytometric analysis of cell membrane microparticles. *Methods Mol Biol* 2008;484:79–93.
103. Harrison P, Keeling D. Clinical tests of platelet functions. In: Michelson AD (ed.), *Platelets*, Esevier, Inc., 2007, pp. 445–474.
104. Oyewumi MO, Yokel RA, Jay M, Coakley T, Mumper RJ. Comparison of cell uptake, biodistribution and tumor retention of folate-coated and PEG-coated gadolinium nanoparticles in tumor-bearing mice. *J Controlled Release* 2004;95:613–626.
105. Ghanbari H, Kidane AG, Burriesci G, Ramesh B, Darbyshire A, Seifalian AM. The anticalcification potential of a silsesquioxane nanocomposite polymer under *in vitro* conditions: Potential material for synthetic leaflet heart valve. *Acta Biomater* 2010;6: 4249–4260.
106. ASTM. ASTM Standard E 2526–08, Standard text method for evaluation of cytotoxicity of nanoparticulate materials in pocine kidney cells and human hepatocarcinoma cells. ASTM International 2008;Vol. E: 2526–08.
107. Maurer-Spurej E, Chipperfield K. Past and future approaches to assess the quality of platelets for transfusion. *Transfus Med Rev* 2007;21:295–306.
108. Jovanovic AV, Flint JA, Varshney M, Morey TE, Dennis DM, Duran RS. Surface modification of silica core-shell nanocapsules: Biomedical implications. *Biomacromolecules* 2006;7:945–949.
109. Murugesan S, Park TJ, Yang H, Mousa S, Linhardt RJ. Blood compatible carbon nanotubes — nano-based neoproteoglycans. *Langmuir* 2006;22:3461–3463.
110. Varon D, Dardik R, Shenkman B, *et al.* A new method for quantitative analysis of whole blood platelet interaction with extracellular matrix under flow conditions. *Thromb Res* 1997;85:283–294.
111. Matetzky S, Shenkman B, Guetta V, *et al.* Clopidogrel resistance is associated with increased risk of recurrent atherothrombotic events in patients with acute myocardial infarction. *Circulation* 2004;109:3171–3175.
112. Spectre G, Brill A, Gural A, *et al.* A new point-of-care method for monitoring antiplatelet therapy: Application of the cone and plate(let) analyzer. *Platelets* 2005;16: 293–299.
113. Sakakibara M, Goto S, Eto K, Tamura N, Isshiki T, Handa S. Application of *ex vivo* flow chamber system for assessment of stent thrombosis. *Arterioscler Thromb Vasc Biol* 2002;22:1360–1364.
114. Baumgartner HR, Turitto V, Weiss HJ. Effect of shear rate on platelet interaction with subendothelium in citrated and native blood. II. Relationships among platelet adhesion, thrombus dimensions, and fibrin formation. *J Lab Clin Med* 1980;95:208–221.

115. Santos-Martinez MJ, Inkielewicz-Stepniak I, Medina C, *et al.* The use of quartz crystal microbalance with dissipation (QCM-D) for studying nanoparticle-induced platelet aggregation. *Int J Nanomed* 2012;7:243–255.
116. Macaulay IC, Carr P, Gusnanto A, Ouwehand WH, Fitzgerald D, Watkins NA. Platelet genomics and proteomics in human health and disease. *J Clin Invest* 2005;115: 3370–3377.
117. Zimmerman GA, Weyrich AS. Signal-dependent protein synthesis by activated platelets: New pathways to altered phenotype and function. *Arterioscler Thromb Vasc Biol* 2008; 28:S17–24.
118. Fischer HC, Chan WC. Nanotoxicity: The growing need for *in vivo* study. *Curr Opin Biotechnol* 2007;18:565–571.
119. Verbeuren TJ. Experimental models of thrombosis and atherosclerosis. *Therapie* 2006; 61:379–387.
120. Harker LA, Kelly AB, Hanson SR. Experimental arterial thrombosis in nonhuman primates. *Circulation* 1991;83:IV41–55.
121. Folts J. An *in vivo* model of experimental arterial stenosis, intimal damage, and periodic thrombosis. *Circulation* 1991;83:IV3–14.
122. Badimon L. Atherosclerosis and thrombosis: Lessons from animal models. *Thromb Haemost* 2001;86:356–365.
123. Rumbaut RE, Slaff DW, Burns AR. Microvascular thrombosis models in venules and arterioles *in vivo*. *Microcirculation* 2005;12:259–274.
124. Blajchman MA, Lee DH. The thrombocytopenic rabbit bleeding time model to evaluate the *in vivo* hemostatic efficacy of platelets and platelet substitutes. *Transfus Med Rev* 1997; 11:95–105.
125. Majde JA. Animal models for hemorrhage and resuscitation research. *J Trauma* 2003;54:S100–105.
126. James ML, Warner DS, Laskowitz DT. Preclinical models of intracerebral hemorrhage: A translational perspective. *Neurocrit Care* 2008;9:139–152.
127. Szebeni J, Alving CR, Rosivall L, *et al.* Animal models of complement-mediated hypersensitivity reactions to liposomes and other lipid-based nanoparticles. *J Liposome Res* 2007;17:107–117.
128. Matute-Bello G, Frevert CW, Martin TR. Animal models of acute lung injury. *Am J Physiol Lung Cell Mol Physiol* 2008;295:L379–399.
129. Lomas-Niera JL, Perl M, Chung CS, Ayala A. Shock and hemorrhage: An overview of animal models. *Shock* 2005;24(Suppl 1):33–39.
130. Parker SJ, Watkins PE. Experimental models of Gram-negative sepsis. *Br J Surg* 2001;88:22–30.
131. Fortner JD, Lyon DY, Sayes CM, *et al.* C_{60} in water: Nanocrystal formation and microbial response. *Environ Sci Technol* 2005;39:4307–4316.
132. Gelderman MP, Simakova O, Clogston JD, *et al.* Adverse effects of fullerenes on endothelial cells: Fullerenol $C_{60}(OH)_{24}$ induced tissue factor and ICAM-I membrane expression and apoptosis *in vitro*. *Int J Nanomed* 2008;3:59–68.
133. Singh SK, Singh MK, Nayak MK, *et al.* Thrombus inducing property of atomically thin graphene oxide sheets. *ACS Nano* 2011;5:4987–4996.
134. Karagkiozaki V, Logothetidis S, Lousinian S, Giannoglou G. Impact of surface electric properties of carbon-based thin films on platelet activation for nano-medical and nanosensing applications. *Int J Nanomed* 2008;3:461–469.

135. Liu YY, Liu TY, Chen SY, Liu DM. Synthesis and characterization of nanoporous SiO(2)/pHEMA biocomposites. *J Mater Sci Mater Med* 2008;19:2903–2911.
136. Nemmar A, Melghit K, Ali BH. The acute proinflammatory and prothrombotic effects of pulmonary exposure to rutile TiO_2 nanorods in rats. *Exp Biol Med (Maywood)* 2008;233:610–619.
137. Oldenburg AL, Gallippi CM, Tsui F, *et al.* Magnetic and contrast properties of labeled platelets for magnetomotive optical coherence tomography. *Biophys J* 2010;99: 2374–2383.
138. Shrivastava S, Bera T, Singh SK, Singh G, Ramachandrarao P, Dash D. Characterization of novel anti-platelet properties of silver nanoparticles. *ACS Nano* 2009;3: 1357–1364.
139. Stevens KN, Crespo-Biel O, van den Bosch EE, Dias AA, Knetsch ML, Aldenhoff YB, van der Veen FH, Maessen JG, Stobberingh EE, Koole LH. The relationship between the antimicrobial effect of catheter coatings containing silver nanoparticles and the coagulation of contacting blood. *Biomaterials* 2009;30:3682–3690.
140. Wang L, Luo J, Shan S, *et al.* Bacterial inactivation using silver-coated magnetic nanoparticles as functional antimicrobial agents. *Anal Chem* 2011;83:8688–8695.
141. Davies A, Lewis DJ, Watson SP, Thomas SG, Pikramenou Z. pH-controlled delivery of luminescent europium coated nanoparticles into platelets. *Proc Natl Acad Sci U S A* 2012;109:1862–1867.
142. Cejas MA, Chen C, Kinney WA, Maryanoff BE. Nanoparticles that display short collagen-related peptides. Potent stimulation of human platelet aggregation by triple helical motifs. *Bioconjug Chem* 2007;18:1025–1027.
143. Bonte F, Hsu MJ, Papp A, Wu K, Regen SL, Juliano RL. Interactions of polymerizable phosphatidylcholine vesicles with blood components: Relevance to biocompatibility. *Biochim Biophys Acta* 1987;900:1–9.
144. Doerschuk CM, Gie RP, Bally MB, Cullis PR, Reinish LW. Platelet distribution in rabbits following infusion of liposomes. *Thromb Haemost* 1989;61:392–396.
145. Okamura Y, Takeoka S, Eto K, *et al.* Development of fibrinogen gamma-chain peptide-coated, adenosine diphosphate-encapsulated liposomes as a synthetic platelet substitute. *J Thromb Haemost* 2009;7:470–477.
146. Tokutomi K, Tagawa T, Korenaga M, *et al.* Decoration of fibrinogen gamma-chain peptide on adenosine diphosphate-encapsulated liposomes enhances binding of the liposomes to activated platelets. *Int J Pharm* 2011;407:151–157.
147. Vaidya B, Nayak MK, Dash D, Agrawal GP, Vyas SP. Development and characterization of site specific target sensitive liposomes for the delivery of thrombolytic agents. *Int J Pharm* 2011;403:254–261.
148. Vaidya B, Agrawal GP, Vyas SP. Platelets directed liposomes for the delivery of streptokinase: Development and characterization. *Eur J Pharm Sci* 2011;44:589–594.
149. Ai H, Fang M, Jones SA, Lvov YM. Electrostatic layer-by-layer nanoassembly on biological microtemplates: Platelets. *Biomacromolecules* 2002;3:560–564.
150. Kushwaha M, Anderson JM, Bosworth CA, *et al.* A nitric oxide releasing, self assembled peptide amphiphile matrix that mimics native endothelium for coating implantable cardiovascular devices. *Biomaterials* 2010;31:1502–1508.
151. Dobrovolskaia MA, Clogston JD, Neun BW, Hall JB, Patri AK, McNeil SE. Method for analysis of nanoparticle hemolytic properties *in vitro*. *Nano Lett* 2008;8:2180–2187.

152. Simak J, Gelderman MP, Yu H, Wright V, Baird AE. Circulating endothelial microparticles in acute ischemic stroke: A link to severity, lesion volume and outcome. *J Thromb Haemost* 2006;4:1296–1302.
153. White JG. Electron microscopy methods for studying platelet structure and function. *Methods Mol Biol* 2004;272:47–63.
154. Brouckova A, Holada K. Cellular prion protein in blood platelets associates with both lipid rafts and the cytoskeleton. *Thromb Haemost* 2009;102:966–974.
155. Simakova O, Arispe NJ. The cell-selective neurotoxicity of the Alzheimer's Abeta peptide is determined by surface phosphatidylserine and cytosolic ATP levels. Membrane binding is required for Abeta toxicity. *J Neurosci* 2007;27:13719–13729.
156. Lopez JJ, Redondo PC, Salido GM, Pariente JA, Rosado JA. Two distinct Ca^{2+} compartments show differential sensitivity to thrombin, ADP and vasopressin in human platelets. *Cell Signal* 2006;18:373–381.
157. Ferraz N, Hong J, Karlsson Ott M. Procoagulant behavior and platelet microparticle generation on nanoporous alumina. *J Biomater Appl* 2010;24:675–692.
158. Fischer TH, Hays WE, Valeri CR. Poly-N-acetyl glucosamine fibers accelerate hemostasis in patients treated with antiplatelet drugs. *J Trauma* 2011;71:S176–182.

Chapter 11

Complement Activation

Carolina Salvador-Morales[*,†,§] *and Robert B. Sim*[‡]

[*]*Bioengineering Department, George Mason University*
4400 University Drive, MS 1G5 Fairfax
VA 22030, USA
[†]*Krasnow Institute for Advanced Study*
George Mason University, 4400 University Drive
MS 2A1 Fairfax, VA 22030, USA
[‡]*Department of Pharmacology, University of Oxford*
Mansfield Road, OX1 3QT, UK

The complement system is the most important biochemical cascade in the blood for the recognition, opsonization, and elimination of foreign materials. To date, the leading causes of death in the United States include cancer, cardiovascular and neurodegenerative diseases, and diabetes. New treatments are urgently needed to treat these devastating diseases and nanotechnology potentially provides new avenues to fight such illnesses. These avenues include the development of novel nanocarriers that deliver drugs in a specific and controlled manner, while minimizing secondary effects. The success of bioengineering effective nanocarriers for drug delivery purposes requires a deep understanding of the interaction between the complement system and the nanocarriers. This review focuses on reporting the current state of complement activation by different nanomaterials. Here, we assess various important parameters that influence the activation of the complement system, which include the physicochemical characteristics of both nanocarriers and complement proteins. We next evaluate the most recent engineering approaches to prevent or reduce complement activation. Finally, we discuss different *in vitro* and *in vivo* procedures to assess complement activation.

[§]Corresponding author. E-mail: csalvado@gmu.edu

1. The Complement System

The complement system is a group of about 35 soluble and cell surface proteins in blood and other body fluids which interact to recognize, opsonize, and clear or kill invading microorganisms, altered host cells, and other foreign materials, including many synthetic materials.[1] A simplified representation of the complement system is shown in Figure 1. The activation of the complement system can occur by any of three pathways, termed the classical, lectin, and alternative pathways.

1.1. *Pathways of complement activation*

1.1.1. *The classical pathway*

In the classical pathway (Figure 1), the recognition protein C1q binds to targets (activators), and the binding causes two proteases, C1r and C1s, attached to C1q, to become active. When C1s is activated, it cleaves and activates the next two proteins of the system, which are called C4 and C2. Domains of these proteins then form a complex called C4b2a, which is itself a protease that cleaves and activates the most abundant complement protein in blood, C3. C3 is activated to form C3b, which binds back on to the surface

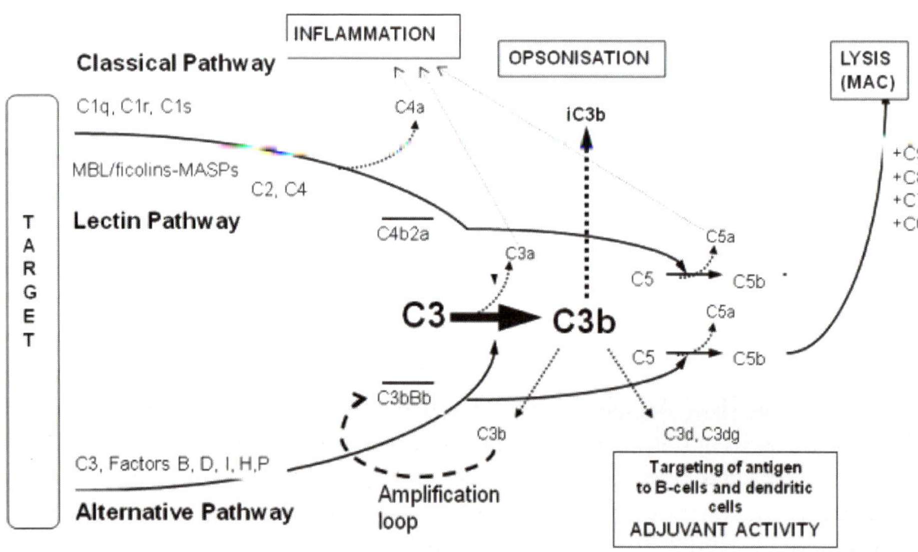

Figure 1. The pathways of the complement system. The complement system can be activated via three pathways, namely the classical, lectin, and alternative pathways.

of the target. The target becomes coated with clusters of hundreds of C3b molecules, which gradually are cleaved by other proteases to forms of C3 called iC3b or C3d/C3dg.

C3b is recognised by CR1 (complement receptor 1) which is on red blood cells and some white blood cells, and this causes the target to bind to the cells. A C3b-coated target, bound to red blood cells, can circulate in the blood, with the C3b gradually being converted to iC3b. As the red blood cell and target pass through the liver or spleen, they come into contact with macrophages, which have on their surface receptors for iC3b (named CR3 and CR4). The target, with its attached iC3b, is stripped off of the red blood cell and transferred to the macrophage, which ingests the target and destroys it intracellularly. This phenomenon, by which targets are coated with C3b or iC3b that promote their uptake and destruction by macrophages, is called opsonization (Figure 1). When iC3b is further broken down to C3d/C3dg, it interacts with another receptor, CR2, which is present on B lymphocytes. This interaction can stimulate the synthesis of antibodies against the target. Dendritic cells, which capture foreign materials and present them to the adaptive immune system as antigens, also have complement receptors, thus coating of the target with C3 fragments also helps to develop the adaptive immune response (antibodies and cytotoxic T cells) against the target. These activities that promote the antibody or T cell response are called the adjuvant activities of complement (Figure 1).

Once C3 has been activated, the protease C4b2a can then activate the next complement protein C5, forming the fragments C5a and C5b. C5b binds to C6, C7, C8, and C9, and this large protein complex called the MAC (membrane attack complex) can insert itself into lipid bilayers (cell membranes). If the target has a cell membrane, the MAC will effectively make holes in the membrane, killing the target ("lysis") (Figure 1).

During the activation of C4, C3, and C5, small peptides C4a, C3a, and C5a, collectively called anaphylotoxins, are released, and these potentially have inflammatory effects. They affect the smooth muscle of blood vessels and cause fluid leakage from the blood into the tissue spaces. They have effects on cytokine and chemokine release, and C5a is also a chemotactic factor, i.e., it attracts cells (i.e., granulocytes). At the site of a wound, these activities would cause the leakage of fluid into the wound, releasing more complement proteins from the blood into the site, to opsonize infectious microorganisms and also cause granulocytes to migrate to the site, where they ingest and kill bacteria.

1.1.2. *The lectin pathway*

The lectin pathway is initiated by the binding of mannan-binding lectin (MBL) or ficolins to carbohydrate structures present on a wide range of microorganisms including bacteria, viruses, fungi, and parasites. MBL has also been reported to bind to IgA and to agalactosyl IgG (IgG-Go). MBL and ficolins circulate in serum complexed with serine protease proenzymes called MBL-associated serine proteases (MASPs) and a small 19-kDa related protein (Map 19). MASPs are structurally similar to C1r and C1s with identical domain composition. The mechanisms for MASP activation have yet to be fully determined but upon MBL of ficolin binding to targets, MASP-2 is auto-activated and cleaves C4 and C2 to form the C3 convertase C4b2a, similar to that in the classical pathway.

1.1.3. *The alternative pathway*

The alternative complement pathway is initiated differently from the classical and lectin pathways. To trigger this pathway, C3b has to be deposited on the surface of a target. C3b may be derived from the activation of the other pathways, arise from a non-enzymic slow turnover of C3, or it may arise because other non-complement proteases can activate C3 to a minor extent. Once one molecule of C3b has bound to the target surface, factor B can bind to it, and is then cleaved by factor D to form C3bBb, which is a protease complex homologous to C4b2a in the other pathways. Once this C3-cleaving enzyme has formed, events occur as per the classical and lectin pathways. Since C3bBb can form more molecules of C3b, the alternative pathway acts as an amplification loop (Figure 1), causing more C3b to be produced and deposited on the target surface.

The different pathways of complement respond to different targets. C1q in the classical pathway binds to a very wide range of targets. These include antibody–antigen complexes formed with IgG or IgM, Gram-negative bacteria, some viruses, polyanionic molecules like DNA/RNA, or anionic phospholipid micelles, altered host proteins such as clots or amyloids, and many synthetic materials such as carbon nanotubes (Perspex). C1q recognises mainly charged clusters, but probably also hydrophobic patches on surfaces.

1.2. *Physicochemical characteristics of key complement proteins*

Among 3,700 different proteins in human serum,[2] only a handful play a role in the activation of the complement system. C1q is the recognition molecule for the classical pathway, MBL or the ficolins act for the lectin pathway, while C3 and C3b are key molecules for the alternative pathway. The direct or indirect binding of C1q[3] and C3b[3] to a carbon nanotube surface activates the complement system. The extent of the complement activation induced by a nanomaterial will depend on the physicochemical characteristics of the nanomaterial and its interaction with complement proteins. Here, we discuss the most important physicochemical characteristics of C1q, C3, and C3b that activate the complement system when interacting with foreign materials.

1.2.1. *Morphology of C1q*

C1q is a 460-kDa protein composed of six heterotrimeric collagen-like triple helices that converge in their N-terminal half to form a stalk, then diverge to form individual stems, each terminating in a C-terminal heterotrimeric globular domain[4] (Figure 2).

C1q binds to target ligands via the globular domains, or heads, triggering the activation of C1r and C1s, the proteases associated with C1q.[5] One of the key features of C1q morphology in initiating the complement system via the classical pathway is that its hexameric structure allows binding by multiple heads. Each head has three lobes, made up of homologous domains (called A, B, and C lobes). Each has distinct but overlapping binding specificity. C1q binds to targets by multiple weak binding interactions. Recognition of one motif (a charge cluster or hydrophobic patch) by one lobe of a head is a weak interaction, but since there are six heads and 18 lobes, multiple interactions

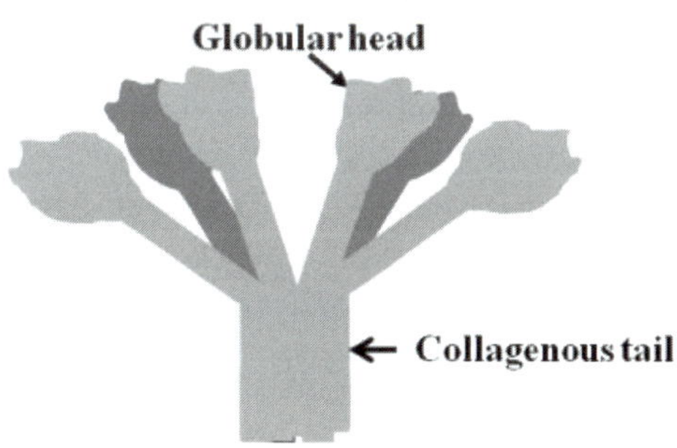

Figure 2. The structure of C1q. C1q is made up of three types of polypeptide chains — A, B, and C — which are homologous to each other. One of each type of chain interact with one another and intertwine to form a collagen triple-helical stalk and a three-lobed globular head. Six of these subunits assemble to form the umbel shape shown here. MBL and the ficolins have similar subunit structures, and form final assemblies with three–six subunits (i.e., three–six heads).

can form, and this results in high avidity binding. A requirement for high avidity binding is that the motifs recognized are in a certain form of regularly spaced array on the target surface. The spacing between motifs must correspond to the spacing between lobes (about 5 nm) or between heads (variable because of the flexibility of the whole molecule) of about 10–40 nm. C1q globular heads are approximately 6 nm wide and have multivalent charged groups for target recognition.[6] The whole C1q molecule measures approximately 40 nm in diameter. C1q morphology and its dimensions are relevant features in understanding its interaction with invaders and foreign materials. We have earlier reported the direct binding of C1q to double-walled carbon nanotubes (DWNTs).[3] C1q binds to DWNTs presumably through its globular heads, activating the complement system via the classical pathway. High-pressure monoxide single-walled carbon nanotubes (HIPco SWNTs) and multi-walled carbon nanotubes (MWNTs) also activate the complement system via the classical pathway. Although the direct binding of C1q to HIPco SWNTs and MWNTs has not been shown, it is likely that C1q binds to them since these carbon nanotubes are not chemically modified, and therefore have a large hydrophobic surface area. Ling *et al.*[7] in a series of transmission electron microscope studies, reported the binding of C1q and C1s–C1r–C1r–C1s to MWNTs, although no binding of C1q to either DWNTs or SWNTs was observed in their study. They observed binding of C1s–C1r–C1r–C1s to DWNTs, but not to SWNTs. In summary, Ling's team did not observe binding of C1q to SWNTs and DWNTs in contrast to our previous studies.[3] Although these two separate findings disagree with each other, we provide strong evidence of the binding of C1q to DWNTs.[3] For example, we measured the direct binding of C1q to DWNTs in the presence and absence of human serum proteins. C1q binds to DWNTs in both conditions. This result addressed the argument given by Ling and co-workers to explain the discrepancies in the findings. Ling's team believes that the SWNT and DWNT studies reported by us activated the complement system via the classical pathway because of the formation of a serum protein layer on DWNTs. We, however, clearly showed that C1q binds to DWNTs even in the absence of serum proteins. We believe that these differences are mainly due to the surface chemistry of these carbon nanotubes. A careful characterization of the SWNTs and DWNTs used in Ling's study would be extremely useful to further explain the disparate results.

There is a vacuum of knowledge with respect to the direct binding of C1 to core-shell nanoparticles — the preferred choice of nanoparticle design for drug delivery purposes. These particles are often coated with a layer of polyethylene glycol (PEG), which is an uncharged hydrophilic polymer that helps reduce protein adsorption. The dimensions and the morphology of C1q become relevant for its interaction with core-shell nanoparticles since the whole C1q molecule is about 40 nm wide (Figure 2). Core-shell nanoparticles are mainly covered with PEG units in which the distance between each unit can be around 5 nm. Polymeric nanoparticles synthesized with poly(lactic-co-glycolic acid) (PLGA) and polylactic acid (PLA) are usually coated with PEG. Given these dimensions, it will be difficult for the C1 complex or C1q to reach the core even when PEG units are flexible. Nevertheless, if the surface of the core-shell particles is partially covered with PEG, C1 and C1q will most likely interact with the core via hydrophobic interactions and will therefore initiate the complement system activation via the classical pathway. C3 is another protein of the complement system that plays a key role in the complement activation cascade and will be discussed in detail in the next section.

1.2.2. *Morphology and chemistry of C3 and C4*

C3 is the central and most abundant protein of the complement system,[8] and it is present in the plasma at a concentration of 1.3 g/L.[8] It is a protein of 185 kDa with dimensions 15.2 nm × 9 nm × 8.4 nm[9] and contains an internal thioester bond. When C3 is activated to C3b, the thioester is exposed and is highly reactive, and can bind covalently to surfaces (e.g., the target) bearing NH_2 and OH groups, forming amide and ester bonds, respectively. It is generally agreed that this covalent binding reaction is the main route by which C3b becomes attached to biological targets. The same mechanism applies to C4b, which is a homolog of C3b.[10] For synthetic materials, however, there may not be suitable surface groups to which C3b and C4b can form covalent bonds. It is possible that other plasma proteins which bind to synthetic materials would provide such chemical groups. However, for carbon nanotubes, C3 and C4 fragments do bind to them, but we found no evidence that they were covalently bound to other proteins; instead, it appears that C3b and C4b bind by hydrophobic, non-covalent interactions.[11]

Recent findings report that although C3 is a large protein, it is able to interact with the diffuse shell of core-diffuse shell-structured nanoparticles.[12] In this case, complement activation takes place by the complex formed between C3 and bovine serum albumin (BSA), where the latter is able to reach the core of the particle. Other studies report that if C3b binds to or becomes trapped between surface PEG "brushes," C3 hydration and conformational changes (C3 tickover) may become accelerated, leading to the assembly of fluid phase C3 convertase.[13]

The assessment of the complement system via the alternative pathway might be more relevant from a clinical point of view than the assessment of the classical pathway since there are more reports which discuss the initiation of the alternative pathway via the binding of C3b and C3 convertase on the surface of artificial materials.[14–16] Surface plasmon resonance (SPR) is a quantitative technique that can measure the amount of C3b bound to the nanomaterials as reported by Toda *et al.*[17]

2. Physicochemical Characteristics of Nanoparticles

The physicochemical characteristics of different nanocarriers influence the adsorption of human serum and plasma proteins which in turn, might include those which activate the complement system cascade. The coating of different nanomaterials such as liposomes, carbon nanotubes, and polymeric nanoparticles with PEG is very well known as an effective way to prevent protein adsorption. The success of this technical strategy depends on the density of the coating (e.g., maximizing the surface coverage of the core), the length, and the configuration of the PEG (e.g., mushroom, brush, or mushroom–brush). These parameters will be examined carefully in the next section.

2.1. *PEG chain density*

As mentioned above, the decoration of a particle's surface by covalently grafting, entrapping, or adsorbing PEG chains diminishes protein adsorption. Each of these methods has its advantages and disadvantages. For instance, one of the biggest advantages of the covalent binding of the PEG chain to the nanoparticle surface is that it prolongs the nanoparticle's circulation half-life. However, the biggest drawback of this method is that it does not ensure complete surface coverage. In contrast, the adsorption method overcomes this disadvantage but poses new challenges which include the desorption of PEG chains from the nanoparticle's surface. The desorption of PEG will eventually cause the precipitation of the nanoparticles, preventing further use of these particles.

The spatial distribution of PEG on a nanoparticle surface is also relevant to protein adsorption and consequently important to diminishing or preventing complement activation. Research has demonstrated that PEG chains can take on two main different spatial arrangements on a nanoparticle surface, referred to as the mushroom and brush configurations.[18] In the mushroom configuration, particles present very low surface coverage of PEG chains. On the contrary, particles holding a brush configuration have a high surface coverage of PEG chains. The prevention or diminution of protein adsorption will not only depend on balancing the PEG chain density on the particle surface, but also on other factors. For instance, high PEG chain density ensures the complete coverage of the nanoparticle surface, but decreases the mobility of the PEG chains, thus diminishing the steric hindrance properties of the PEG layer.[19] To date, there is no rule of thumb for obtaining this delicate balance. Some scientists believe that the optimal thickness of hydrodynamic layer effectively shielding the particle surface from protein adsorption is 5% of the particle's hydrodynamic size. According to another view, the optimal thickness of hydrodynamic layer should be at least twofold greater than the radius of the polymer coil at the polymer conformation in its diluted solution.[19]

The physicochemical characteristics of the proteins are factors that also influence protein adsorption. New studies report that the shell of core-diffuse shell structured nanoparticles synthesized with PIBCA [poly (isobutylcyanoacrylate)]–dextran block copolymers by a self-assembly process[12] do not prevent the adsorption of small flexible proteins such as BSA, even when the density of chains within the diffuse shell is quite high. PIBCA is a bioerodible and bioeliminable polymer. Large proteins such as fibrinogen (340 kDa) and C3 (185 kDa) interact with these nanoparticles in a specific manner. Fibrinogen not only reaches the diffuse shell of the nanoparticles, but also reaches their hydrophobic core. On the other hand, C3b penetrates the diffuse shell, but does not reach the core of the nanoparticle.[20] The authors observed complement activation on nanoparticles that trapped C3b in their diffuse shell. They hypothesized that complement activation in this case is due to a complex formed between C3b and BSA.

2.2. *PEG chain length*

PEG chain length is another important parameter that plays a crucial role in both protein adsorption and complement activation. Mosqueira *et al.*[21] demonstrated that polymeric particles covered with a high density of 20-kDa PEG chains show the lowest protein adsorption and lowest complement activation among the materials they tested. This is presumably due to steric barriers surrounding these particles. Mosqueira's team hypothesized that the long PEG chains act through creating a PEG-hydrated cloud "shielding" negatively charged groups located underneath it. In addition, they reported that particles coated with 5-kDa PEG with high density are more effective in reducing complement activation than at low density.

Recent studies on PEG–PLGA nanoparticles further corroborate that long chain lengths reduce protein adsorption and complement activation.[22] Research on the *in vivo* effects of short and long chains of monolayer-protected gold nanoparticles indicated that the former presented short circulation half-life, whereas longer half-life was observed with long PEG chains.[23]

2.3. *Morphology of the nanocarrier*

Activation of the complement system has been assessed on nanomaterials that are either spherical or tubular, including liposomes, micelles, lipid-polymer nanoparticles, and carbon nanotubes. Although these studies are signs of progress on this topic, there still lacks a deep understanding of the influence of the nanocarrier's shape on complement activation. To the best of our knowledge, the only insight on this topic comes from a study that suggested that the higher the curvature, the lower the human plasma protein surface coverage, thus rendering activation of the complement system less likely.[24] For spherical nanocarriers, the curvature is directly related to the size of the particles.[24] Small nanoparticles (hydrodynamic diameter of ~70 nm) have higher curvature than large nanoparticles (hydrodynamic diameter of ~300 nm). Despite this finding, the morphology of the nanocarrier itself seems to be irrelevant for the activation of the complement system; rather, it is the surface chemistry of the nanocarrier that is the key to complement activation. As a matter of fact, it has been demonstrated that lipid PEGylated carbon nanotubes activate the complement system regardless of the terminal functional group and the PEG brush-like surface structure on HIPco SWNTs.[25] The activation of the complement system by lipid PEGylated HIPco SWNTs might be due to the incomplete coverage of PEG on their surface. Since the un-PEGylated surface area of HIPco SWNTs is hydrophobic, it is likely that C1q molecules bind to that surface via hydrophobic interactions. Previous studies on non-chemically modified HIPCo SWNTs showed the activation of the complement system via the classical pathway, which is initiated by the binding of C1q to the activator surface.[3]

In the field of polymeric nanoparticles, Gref *et al.*[26] reported the influence of the core composition on protein adsorption. Protein binding studies were conducted with nanoparticles synthesized with three different polymeric cores made of PLGA, PLA, and poly(ε-caprolactone) (PCL) polymers. During this synthesis, the length and density of the PEG were kept constant. The results showed slight differences in the pattern of protein binding which suggested that the core has a part in determining the proteins that bind to the nanoparticle surface.

3. Recent Engineering Approaches to Avoid or Reduce Complement Activation

Understanding the factors that activate the complement system is very important to engineering biocompatible nanocarriers that can be used for a wide range of applications in medicine. In the last two decades, there has been great progress in understanding the complement system cascade and its interaction with different organic and inorganic materials. The complement system views liposomes, carbon nanotubes, polymeric nanoparticles, and micelles as invaders. Synthetic strategies such as the use of PEG to coat the surface of these nanocarriers to evade the complement system cascade have been employed.[18,25,27] This chemical approach has worked only to a certain extent for immune evasion. For instance, studies have demonstrated that PEGylation does not necessarily suppress complement opsonization.[27] Complement activation induces the opsonization of PEGylated liposomes because of the covalent deposition of C3b on the liposomal surfaces.

In previous sections, we mentioned that PEG can present two different spatial arrangements known as the mushroom- and brush-like configurations. These configurations influence protein adsorption which might or might not lead to complement activation. Lately, it has been found that the transition from one configuration to another influences the activation of the complement system.[28]

The reduction or prevention of complement activation is relevant for several reasons. Preventing the activation of the complement system can help to extend the circulation half-life of the nanocarrier in the bloodstream, improve the nanocarrier's biodistribution profile, and avoid a set of allergies known as CARPA (complement activation-related pseudoallergy).[29] For all these reasons, it is urgent to develop new synthetic strategies for diminishing or preventing complement system activation. We next discuss the most recent advances in this area.

3.1. *Influence of PEG mushroom- and brush-like configurations on complement activation*

Surface PEGylation not only helps diminish protein adsorption, but can also cause alterations of the pathway for complement activation. Moghimi *et al.*[28] reported that the transition of copolymer architecture on nanoparticles with polyethylene oxide (PEO) chains from the mushroom–brush to brush-like configuration not only switches complement activation from the C1q-dependent classical to lectin pathway, but also reduces the level of generated complement activation products C4d, Bb, C5a, and SC5b-9.

According to the authors, changes in adsorbed polymer configuration trigger alternative pathway activation differently and through different initiators. This was the first study to demonstrate the importance of configurational mobility of surface-projected PEG chains in the modulation of complement activation with a spectrum of model nanoparticles that exhibited different pharmacokinetic profiles.

The mushroom–brush and brush-like PEG configurations are observed on the surface of different nanocarriers, including liposomes,[29] polymeric nanoparticles,[26] gold nanoshells[30] and carbon nanotubes.[25] In all these cases, the brush-like configuration has been identified as the surface structure that reduces protein adsorption, though it does not necessarily diminish complement activation. PEGylated liposomes are a good example of this case. Bradley *et al.*[31] found that the incorporation of PE–mPEG [phosphatidylethanolamine–monomethoxylpoly(ethylene glycol)2000] from 5–7.5 mol% into the liposomal bilayer was not enough to prevent complement activation. Their results revealed that the inhibitory effect of mPEG–lipid on complement activation is highly dependent on the liposomal concentration used in the complement assay.[31] Other studies have reported that the concentration of PEG in liposomes causes the transition from the mushroom- to brush-like regime.[32] Liposomes prepared with up to 4 mol% of grafted PEG exhibited the mushroom configuration because the neighboring coils did not interact laterally. On the other hand, liposomes synthesized with PEG concentration above 4 mol% showed a brush regime since the neighboring PEG chains pushed against each other, extending farther out from the surface on which they were grafted.[32] This study did not report on complement system activation under these circumstances.

The spatial configuration of PEG on polymeric nanoparticles has been studied for more than a decade. It is well known that PEG can exhibit either the mushroom- or brush-like configuration on a polymeric nanoparticle surface. PEG with high molecular weight and at high density is expected to exhibit the brush-like configuration.[33] Mosqueira *et al.*[21] demonstrated that 20-kDa PEG chains are more effective in preventing C3 cleavage than 5-kDa PEG chains because of the steric barrier created by the PEG surrounding the particles.

Core-shell nanoparticles synthesized with di-block copolymer of methoxyPEG–PCL (MePEG–PCL) and tri-block copolymer of PCL–PEG–PCL present mushroom- and brush-like surface structures, respectively. The mushroom-like structure was thought to be more efficient in preventing opsonization because it could form a more effective conformational cloud.[34] Recently, a quantitative and nondestructive assay based on surface enhanced Raman scattering (SERS) spectroscopy has been reported to determine the number of PEG molecules bound to gold nanoshell surfaces.[30] From this estimate, one can obtain the pack density of PEG units on nanoshell surfaces, which helps to infer their configuration.[30]

Heterogeneity is a phenomenon that sooner or later will have impact on complement activation, and therefore must be taken into account when designing new nanocarriers. The concept of surface heterogeneity in essence describes the incomplete coverage of the nanoparticle's surface with PEG.[35] Moghimi and Szebeni discussed the direct effect of the partial coverage of PEG on the nanoparticles' circulation half-life in the bloodstream.[35] The less the surface coverage, the poorer the steric shielding, and the shorter the circulation life, which is presumably caused by the binding of opsonic molecules to the unshielded area.[4] The higher the surface coverage, the greater the resistance to protein binding. For instance, microsphere populations covered with a high density of mPEG with the mushroom–brush intermediate and/or brush-like configurations were most resistant to phagocytosis and activated the complement system poorly.[36]

To date, there has been little discussion of the influence that factors such as temperature,[37] autoxidation catalyzed by transient metals,[38] and salt concentration[39] might have on the configuration and stability of PEG. Changes in the physicochemical characteristics of PEG might affect PEG configuration. There is evidence that the aforementioned parameters have a direct effect on PEG. As a matter of fact, it has been observed that increasing the salt concentration to 3 M and raising the temperature to 37°C resulted in the rapid aggregation of PEGylated nanoparticles.[40] This is due to the fact that high temperature causes the dissolution of the PEG hydration layer, leading to particle precipitation. Although the assessment of complement activation and protein binding studies were not conducted at such high salt concentrations, these parameters should be kept in mind until new studies report the temperature and salt concentration at which mushroom- and brush-like configurations are not only formed, but are also stable.

The identification of these parameters is very important for the efficient design of nanocarriers, especially when application is envisioned in the medical field. Depending on the application, the experimental conditions in which the nanocarriers will be immersed might be subject to high temperature and high salt concentration. Temperature is the one of the parameters that might have a strong influence in complement activation depending on the method used. For example, hemolytic assays are conducted at 37°C and a great number of papers has described the assessment of complement activation using hemolytic assays (which can only be done between about 20°C and 37°C).

New synthetic approaches are already available as an alternative to PEG, which is traditionally the workhorse for preventing protein adsorption.[40] These novel polymers are not temperature-sensitive and the surface structure does not change at high salt concentration.[41] However, they are not as well studied and characterized as PEG since they have emerged only recently.

3.2. *Modulation of complement system activation via functional groups*

For more than 30 years, scientists have been trying to prevent or diminish complement activation to achieve blood biocompatibility. In the 1980s and 1990s, there were reports of various methods to diminish complement activation.[41–45] Carreno *et al.*[45] reported that the substitution of hydroxyl groups of Sephadex® (an activator of the alternative pathway) by carboxymethyl (CM), groups can reduce the activating capacity of the resulting polymer (CMSeph). Complement activation by CMSeph can be abolished when an average of one CM group is present per glycosyl unit. However, this method is a multistep procedure and involves complicated organic synthesis.

Another example of reducing activation of the complement system via direct chemical modification of the nanomaterial is in the field of carbon nanotubes. In previous studies, we reported that the chemical modification of pristine MWNTs reduces complement activation.[46] In this study, four different types of chemically modified MWNTs were tested for complement activation via the classical and alternative pathways using hemolytic assays. It was found that MWNTs functionalized with e-caprolactan or L-alanine showed, respectively, >90% and >75% reduction in classical pathway activation compared with unmodified MWNTs. The reduced level of complement activation via the classical pathway that is likely to increase biocompatibility is directly correlated with the amount of C1q protein bound to chemically modified carbon nanotubes. These results demonstrated for the first time that these types of chemical modifications are able to considerably alter the level of specific complement proteins bound by pristine MWNTs.

Lately, new synthetic methods have been developed not only to diminish complement system activation, but most importantly, to manipulate it at will. In this method, the manipulation of the complement system is due to the modulation of the nanoparticles' surface charge. For some nanocarriers, the surface charge dictates the activated pathway. This is the case with liposomes.

Negatively charged liposomes containing phosphatidyl glycerol, phosphatidic acid, cardiolipin, and/or phosphatidylinositol activate the complement system via the classical pathway.[47] Positively charged liposomes containing stearylamine activate the alternative pathway.[47] Neutral liposomes hardly activate the complement system.[47]

For other potential drug delivery systems such as PEGgylated lipid carbon nanotubes and polymeric nanoparticles, the role of surface charge is still relevant for complement activation, but differences among liposomes and in charges in carbon nanotubes and polymeric nanoparticles do not determine which pathway will be activated. Results reported by Hamad *et al.*[25] demonstrated that PEGylated lipid HIPco SWNTs activate the whole complement system independently of the terminal end moiety of the projected PEG chain. The authors hypothesized that HIPco SWNTs most likely activate the complement system via the lectin pathway. In addition, PEGylated lipid polymeric nanoparticles functionalized with NH_2, COOH, and CH_3 activated the alternative pathway, but not the classical pathway.[48] This scientific evidence suggested that the presence of functional groups in the PEG chain does not necessarily determine the pathway of the complement system that is activated.

So far, scientists have not yet determined the set of parameters that absolutely dictate the pathway activated. In an effort to shed light on this matter, Toda *et al.*[49] conducted studies assessing complement activation of PEGylated self-assembly monolayers (SAMs) functionalized with a mixture of NH_2–COOH and NH_2–CH_3. Briefly, functionalized SAMs were prepared with various molar ratios of a pair of NH_2–COOH or NH_2–CH_3 on gold-coated glass plates. These coated glass plates were immersed in reaction mixtures for 24 h followed by several washes. Their results showed that the NH_2–COOH mixture activated the complement system, whereas the NH_2–CH_3 pair did not activate it. The authors believed that the NH_2–CH_3 mixture did not activate the complement system because of the numerous serum proteins adsorbed onto those SAMs, including albumin, that formed a protein layer which inhibited access of C1q or C3b to the surface. On the contrary, a high amount of C3b or C3 convertase was found to be deposited on the NH_2–COOH SAM.

We showed a direct correlation between the zeta potential, levels of complement activation, and the amount of C3b deposited on lipid–polymeric nanoparticles functionalized with a mixture of NH_2–COOH and NH_2–CH_3 and synthesized via the nanoprecipitation method.[48] We observed that as the presence of NH_2 on PEGylated lipid–polymer nanoparticles reduces, the level of complement activation diminishes and the zeta potential becomes more negative. The C3b binds to NH_2 and OCH_3 as well as the different molar ratio mixtures of these two functional groups. The molar ratio mixture of

COOH–NH_2 shows complement activation, which agrees with Toda's findings, even when those studies were conducted on SAMs and not in polymeric nanoparticles. This pair of functional groups mixtures also binds the C3 β chain, which might explain the mechanism involved in the activation of the complement system. The main difference between Toda's findings and our results is that we found that lipid–polymeric nanoparticles functionalized with several molar ratio mixtures of NH_2–CH_3 activate the complement system.[48] This could probably be due to the partial coverage of NH_2–CH_3 on the nanoparticle surface. Our findings clearly demonstrated a direct correlation of the surface charge of the particles and the activation of the complement system. In addition, our team's method was a simple, practical, and inexpensive way to modulate the complement system. Using this method, we can make these particles act as adjuvants by functionalizing their polymeric core with DSPE–PEG–NH_2 or make them function as biocompatible nanocarriers for drug delivery purposes by functionalizing their surfaces with CH_3.

Hydroxyl groups are another terminating group that has been used to functionalize SAMs to study their effects on complement activation. Sperling and co-workers reported that these surfaces strongly activate the complement system.[50] As the amount of surface OH increases, the amount of C5a generated also increases.[50] Other studies also reported the activation of the complement system induced by terminal hydroxyl group of tri(ethylene glycol)-terminated alkanethiol (HS-TEGOH).[51] Again, the cause of such activation was the deposition of C3b on these surfaces.

4. Methods for the *In Vitro* Study of Complement Activation by Different Nanomaterials

4.1. *Hemolytic assay (CH_{50})*

For several decades, hemolytic assays have been used as the standard procedure to assess complement activation. CH_{50}, or total hemolytic complement assay, measures the ability of the classical pathway-activated MAC to lyse sheep red blood cells (SRBCs) coated with an antibody.[52] The alternative pathway hemolytic assay (APH_{50}) measures the ability of the MAC generated by this pathway to lyse rabbit red blood cells. Both assays indicate a deficiency of a complement component by the absence of lysis. Hemolytic assays have several advantages over other complement tests. For example, the biggest advantage of the CH_{50} assay is the possibility of evaluating the activation of the complement system in most mammalian species' sera. Also, this method is inexpensive. However, limitations include their labor intensity (particularly the pipetting of small volumes), the short shelf-life of SRBCs, the variable inter-lot performance of SRBCs, and the low sensitivity of the assay relative to measuring C scission products such as C5a and SC5b-9.

Complement hemolytic assays can vary widely in sensitivity from day to day, so it is possible to compare only results obtained in the same assay and not results obtained on different days. This is mainly because of metabolic changes in the cells, which make them more sensitive to lysis as they age. The most common assay for complement consumption is done in a buffer containing Ca^{2+} and Mg^{2+} ions. This measures classical pathway activation. If Ca^{2+} ions are removed by the chelator EGTA, the classical and lectin pathways are completely inhibited, and only the alternative pathway is measured. There are numerous versions of the hemolytic CH_{50} assay, which is arbitrarily standardized with respect to the concentration of sensitized SBRCs (10^8/1.5 mL), and the concentration and type of sensitizing antibody (heterophilic Forssmann, i.e., rabbit anti-RBCs).

Hemolytic assays have also been used to evaluate the activation of the complement system by different nanomaterials such as liposomes[47] and pristine and chemically modified carbon nanotubes.[3,46] It is worth noting that this simple complement consumption assay can only be used if the activator is particulate and can be separated from the serum by filtration or centrifugation before complement activity is assayed in the serum. If the activator is soluble and cannot be easily separated from the serum, more sophisticated assays are required to distinguish between complement consumption and inhibition of the complement assay.

4.2. *ELISA kits*

Lately, ELISA-like assays have been used to assess *in vitro* complement activation.[25,48–49] In these studies, complement activation was assessed using Quidel kit SC5b-9. This ELISA-based assay measures the cleavage of C5 and subsequent terminal pathway activation. Specifically, SC5b-9 is generated by the assembly of C5–C9 as a consequence of complement activation via all three pathways and subsequent binding to the naturally occurring regulatory serum protein, the S protein (vitronectin). SC5b-9 forms when the MAC (Figure 1) fails to insert into a lipid bilayer, but instead reacts with S protein. C9 within this complex expresses a neo-epitope (i.e., an epitope not present in C9 which is not incorporated in the SC5b-9 complex), so C9 itself does not interfere in the assay. ELISA-like assays have several advantages over hemolytic assays. For instance, the former is a faster and easier method than the latter. These methods do not involve the use of SRBCs so they eliminate

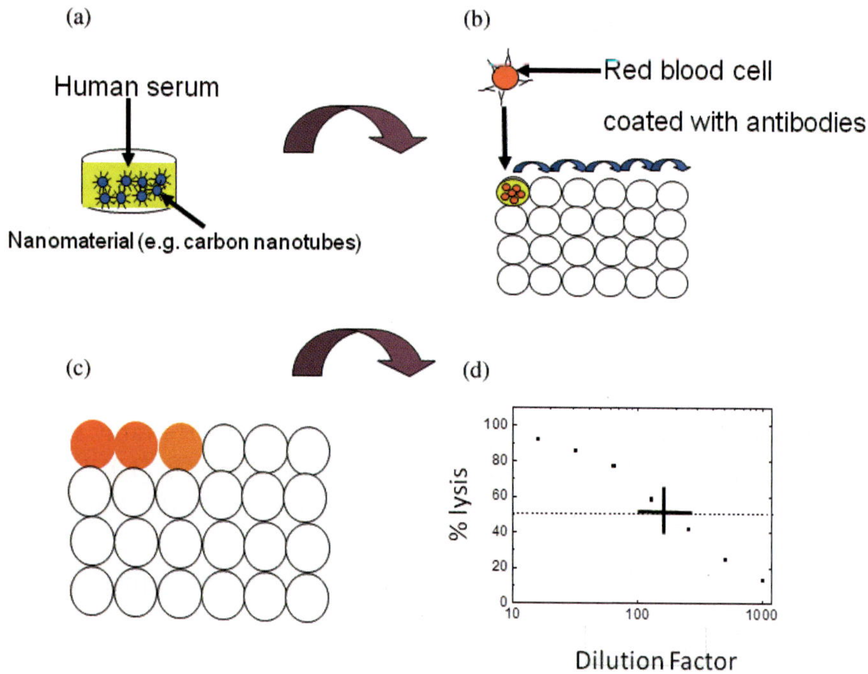

Figure 3. Schematic diagram showing the main steps in hemolytic assays. This diagram illustrates the assessment of the complement system via the classical pathway. This type of assay measures the ability of the classical pathway-activated membrane attack complex (MAC) to lyse sheep red blood cells (SRBCs) coated with an antibody. Likewise, the functional activity of the alternative pathway and the terminal components can be measured by the lysis of rabbit erythrocytes in human serum. (a) Assessing the activation of the complement system is done by simply incubating the nanomaterial (e.g., carbon nanotubes) with serum. Samples are incubated at 37°C for 1 h followed by centrifugation. (b) The supernatant of the sample is serially diluted and placed in a microtiter plate. One hundred microliters of each dilution are incubated with 100 μL of antibody-sensitized SRBCs (EA) (10^8 cells/mL in veronal buffer). SRBCs sensitized with an antibody (EA) are used as an immune complex to activate the classical pathway. When EA cells are added to serum or plasma, they are lysed as a result of their activation of the complement system and the intercalation of C5b-9 MAC into their cell membranes. (c) After incubation, cells are spun down (2500 rpm, 10 min, room temperature) and hemoglobin is measured at 405 nm in the supernatant. (d) Calculating complement activation is done by plotting the percent lysis against the dilution of human serum (logarithmic scale). (Adapted from Ref. 52.)

problems such as short shelf-life and inter-lot variability and sensitivity. For these reasons, the ELISA-like assay method is superior to hemolytic assays. In addition, this method can be adapted to assess complement activation in a high-throughput screening fashion. On the other hand, disadvantages of ELISA-based assays include the evaluation of the complement system only in human sera and not in other mammalian species (although a few reagents, e.g., antibodies, are available for mice or rats). Another disadvantage is that this assay only indicates whether the complement system is activated or not. Further experiments are needed to identify the activated complement system pathway. For this purpose, there are similar kits that can be used. For instance, Quidel C4d and Bb ELISA kits are specific for classical/lectin and alternative pathway complement activators, respectively. Both analytes are by-products of complement activation; C4d is a scission derivative of C4b, whereas Bb rises in blood as a consequence of spontaneous dissociation of the alternative pathway C3 convertase. Other assay kits which measure C3a or C5a generation are available from Enzo Life Sciences, BD Biosciences, R& D systems, and others. All of these kits are relatively expensive and they expire in less than a year.

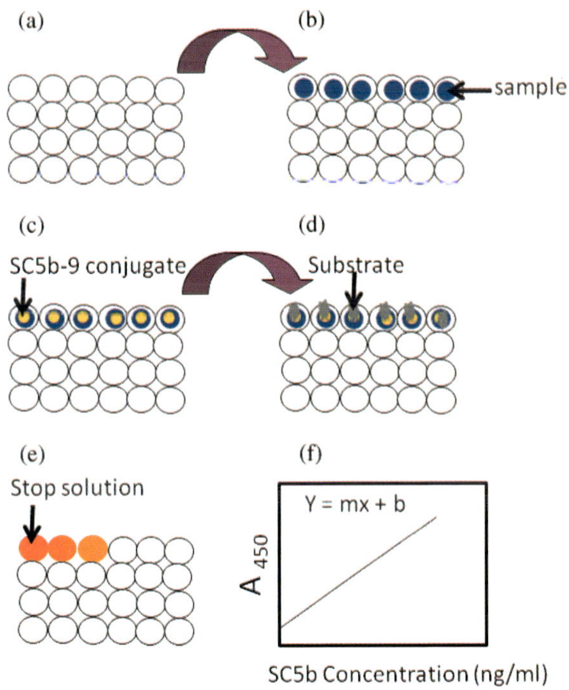

Figure 4. Schematic diagram of the Quidel-kit as an example of Elisa-like assays. (a) Washing assay wells with wash solution as indicated in Quidel kit SC5b-9. (b) Pipetting specimen diluents (black), standards, controls, and specimens into assay wells followed by incubation and several washes. (c) Pipetting SC5b-9 with the conjugate into assay wells followed by incubation and several washes. (d) Pipetting substrate into assay wells. (e) Adding stop solution into assay wells. (f) Reading the optical density at 450 nm and analyzing the assay results using a linear curve fit ($y = mX + b$). (Adapted from Ref. 48.)

4.3. *Wieslab diagnostics kits*

Wieslab diagnostics kits are also ELISA-like assays for the qualitative determination of functional classical, alternative, and lectin pathways in human serum.[53] Wieslab products are available via Eurodiagnostica (http://www.eurodiagnostica.com/). These complement assays combine the principles of the hemolytic assay for complement activation with the use of labeled antibodies specific for the SC5b-9 neoepitope produced as a result of complement activation. The amount of SC5b-9 generated is proportional to the functional activity of complement pathways. The wells of the microtiter strips are coated with specific activators of the classical, lectin (MBL-only), or the alternative pathway. Patient serum is diluted in diluents containing specific reagents to ensure that only the appropriate pathway is activated. During the incubation of the diluted patient serum in the wells, complement is activated by the specific coating. It is worth noting that the level of complement activity evaluated by functional assays such as the Wieslab complement kits takes into account the rate of synthesis, degradation, and consumption of the components, and provides a measure of the integrity of the pathways as opposed to immunochemical methods, which specifically measure the concentration of various complement components.

4.4. *2D immunoelectrophoresis method*

The 2D immunoelectrophoresis method is another technique that has been used to assess complement system activation.[12] This technique separates and characterizes proteins based on electrophoresis and reaction with antibodies. The cleavage of C3 into breakdown products C3b, iC3b, and C3c alters the electrophoretic mobility of C3. C3 is separated from its breakdown products by electrophoresis on agarose, and the proteins are electrophoresed at right angles into an agarose gel containing anti-C3 antibodies. The proteins react with antibodies and form a visible precipitate over an area proportional to the protein concentration. This is a time-consuming procedure, unsuitable for high-throughput screening, but for a one-time assessment, it could be a good option. Also, this technique is not as sensitive for complement activation assessment as ELISA-like assays. However, this technique could be less expensive than ELISA-Quidel Kits and hemolytic assays as the reagents for the test can be used for other research purposes and be frozen for future use. The long shelf-life of antibodies against complement proteins is one of the biggest advantages of the 2D

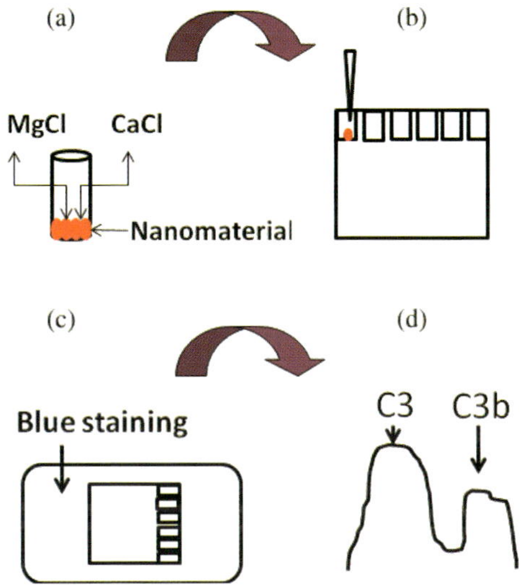

Figure 5. Schematic diagram of activation of the complement system evaluated from 2D immunoelectrophoresis method. (a) Incubation of the nanomaterial (e.g., nanoparticles) with human serum in veronal buffer supplemented with MgCl and CaCl. (b) 2D electrophoresis gel loaded with incubated samples as described in Ref. 12. (c) Coomassie blue staining of 2D electrophoresis gel. (d) Analysis of the immunoelectrophoregrams showing the presence of C3 and C3b proteins after incubation of the serum with the nanomaterial. (Adapted from Ref. 12.)

Table 1. Summary of *in vitro* studies of complement activation on different nanomaterials.

Assay method	Assay source/protocol
Hemolytic assays (CH_{50})	Salvador-Morales, *et al.*[3] Gbadamosi J K. *et al.*[4] Whaley K[54]
ELISA-like kits	Quidel Corp., San Diego, CA
Enzyme immunoassays	Ferraz N *et al.*[55]
Wieslab	http://www.eurodiagnostica.com/
2D immunoelectrophoresis	Vauthier C *et al.*[12] Bertholon I *et al.*[56]

immunoelectrophoresis method as both ELISA-like assays and CH_{50} have short shelf-lives. For example, reagents for hemolytic assays such as SRBCs only last for several days, whereas ELISA kits last for 12 months. Table 1 summarizes all the *in vitro* assays employed to assess complement activation described above.

5. *In Vivo* Studies of Complement Activation by Different Nanomaterials

To date, *in vivo* studies on complement system activation are still limited.[29,57–60] Szebeni and co-workers have made substantial contributions on this topic by assessing complement activation in different animal models including pigs, dogs, and rats.[29] Their findings showed that these animals present a set of allergy symptoms when they are injected with lipid-based nanomaterials. These allergy symptoms include tachypenia, tachycardia, hypotension, hypertension, chest pain, and back pain. These reactions, commonly known as infusion reactions, form the concept of CARPA.[60] Evidence from *in vitro*, *in vivo,* and clinical trials suggest that these reactions are derived from complement system activation and different mammalian species respond differently to the intravenous (i.v.) administration of liposomes. For example, rats are less sensitive than dogs and pigs; pigs are the most sensitive animal model for the i.v. injection of lipid-based materials.[60] Szebeni and co-workers' findings also show that the presence of cholesterol and phospholipids in liposomes are responsible for complement system activation.[61–62] In addition, it is well known that cationic and anionic liposomes activate the complement system via the alternative and classical pathways, respectively.[47] In this case, it is reported that the complement system is activated not only due to the presence of cholesterol and phospholipids, but also cationic and anionic molecules in liposomes.

Other lipid-based materials such as micelles cause infusion reactions in dogs and pigs.[60] However, other studies demonstrated that mPEG–phosholipid in micelles and liposomes coated with methoxyl functional groups did not activate the complement system. These results indicated that charge is the property of the liposomes that causes complement activation. PEGylated HIPco SWNTs with CH_3 and NH_2 terminal ends also induce *in vitro* and *in vivo* complement activation.[25] In this study, the *in vivo* complement activation assessment was measured indirectly by the levels of thromboxane B2 in rat blood.

The differences in complement system activation across mammals are based on species, dose dependence, and the influence of lipid composition. For instance, the mechanism involved in the activation of the complement system in pigs that are injected with Doxil® is the anaphylaxis phenomenon, which is basically the increment of C5a and C3a elements. This increment causes severe cardiac abnormalities in porcine models. On the other hand, a similar amount of undiluted Doxil® can be fatal for men and dogs. Rats are two to three orders of magnitude less sensitive to liposomes, at least to those containing <50% cholesterol,[42] although complement-dependent shock and tissue damage induced by the i.v. injection of cholesterol-enriched liposomes in rats have been observed.

Lipid composition plays an important role in complement activation. Szebeni and co-workers reported that the cholesterol content of liposomes may be an important determinant of high sensitivity reactions (HSR). Pulmonary hypertensive effects of multilayer vesicles in pigs were found to be proportional to the amount of cholesterol in the vesicle in the 20–71% range,[63] and rat murine leukemic virus (MLV) containing 71% cholesterol were significantly more reactogenic compared to liposomes with 45% cholesterol.[42]

Recently, Szebeni and co-workers have reported the activation of the complement system in porcine models because of the presence of polyethylenimine polymers.[64] This type of polymer triggers anaphylactic reactions in low percentages of hypersensitive individuals regardless of PEGylation. The results should be kept in mind when engineering new drug delivery systems, as transient and mild reactions can be fatal for patients who suffer from allergies and heart diseases.[65] More studies are urgently needed to be conducted on this topic to fully elucidate the mechanisms involved in *in vivo* complement activation.

6. Concluding Remarks

Complement system activation can be triggered by the direct binding of key complement proteins such as C1q, MBL, ficolins, and C3b to different nanoparticle surfaces. The binding of other plasma proteins such as fibrinogen and BSA can serve as a bridge for the binding of proteins such as C3b, which in turn activates the complement system. Thus, protein adsorption undoubtedly plays a significant role in complement activation. Therefore, there is a great need to reduce protein adsorption by understanding the parameters of the nanosurfaces that enhance or avoid protein binding.

In this chapter, we have highlighted the physicochemical characteristics of both complement proteins and nanomaterials that cause complement activation. Parameters such as the PEG chain length, density, and conformation strongly influence complement activation. The process of engineering nanocarriers with a long circulation half-life and a good pharmacokinetic profile involves making several trade-offs.

After more than three decades of research in this area, we still have not produced a recipe that guarantees the synthesis of ideal nanocarriers. The existing literature on this topic undoubtedly has helped us to much better understand the parameters that we need to consider to engineer the ideal nanocarrier. We have remarked that future decisions and trade-offs will depend on the severity of the disease and the willingness of patients to take the risk. As shown above, lipid-based nanocarriers activate the complement system *in vitro* and *in vivo.* Most importantly, clinical trials have demonstrated the activation of the complement system. These studies should not be overlooked, but should be seriously considered when Doxil® and other lipid-based drugs are consumed by patients who have heart diseases. Future developments in the area of biomaterials await to abrogate complement activation, minimizing secondary effects. Bacteria have evolved ways of avoiding attack by complement, e.g., by binding to their surface a complement downregulatory protein, factor h, from the host. It has recently been suggested that mimicking bacteria such as by attaching factor h-binding peptides to nanomaterial surfaces, may be a strategy worth exploring.[66–67]

References

1. Carroll MV, Sim RB. Complement in health and disease. *Adv Drug Rev* 2011; 63:965–975.
2. Muthusamy BHG, Suresh S, Rekha B, Srinivas D, Karthick L, Vrushabendra BM, Sharma S, Mishra G, Chatterje P. Plasma proteome database as a resource for proteomics research. *Proteomics* 2005;3531–3536.
3. Salvador-Morales C, Flahaut E, Sim E, Sloan J, Green ML, Sim RB. Complement activation and protein adsorption by carbon nanotubes. *Mol Immunol* 2006;43:193–201.
4. Gaboriaud C, Juanhuix J, Gruez A, *et al.* The crystal structure of the globular head of complement protein C1q provides a basis for its versatile recognition properties. *J Biol Chem* 2003;278:46974–46982.
5. Arlaud GJ, Gaboriaud C, Thielens NM, Budayova-Spano M, Rossi V, Fontecilla-Camps JC. Structural biology of the Cl complex of complement unveils the mechanisms of its activation and proteolytic activity. *Mol Immunol* 2002;39:383–394.
6. Kishore U, Gaboriaud C, Water P, *et al.* C1q and tumor necrosis factor superfamily: Modularity and versatility. *Trends Immunol* 2004;25:551–561.
7. Ling WL, Biro A, Bally I, *et al.* Proteins of the innate immune system crystallize on carbon nanotubes but are not activated. *ACS Nano* 2011; 5: 730–737.
8. Law SKA, Reid KBM. *Complement.* IRL Press, 1995.
9. Sim RB, Perkins SJ. Molecular modelling of C3 and its ligands. *Curr Top Microbiol Immunol* 1990;153:209–222.
10. Sim RB, Sim E. Autolytic fragmentation of complement components C3 and C4 and its relationship to covalent binding activity. *Ann N Y Acad Sci* 1983;421:259–276.
11. Malgorzata J, Rybak-Smith MJ, Pondman KM, Flahaut E, Salvador-Morales C, Sim RB. Recognition of carbon nanotubes by the human innate immune system. In: Klingeler R, Sim RB, eds. *Carbon Nanotubes for Biomedical Applications.* Springer-Verlag, 2011, pp. 183–210.
12. Vauthier C, Persson B, Lindner P, Cabane B. Protein adsorption and complement activation for di-block copolymer nanoparticles. *Biomaterials* 2011;32:1646–1656.
13. Moghimi SM, Andersen AJ, Hashemi SH, *et al.* Complement activation cascade triggered by PEG-PL engineered nanomedicines and carbon nanotubes: The challenges ahead. *J Controlled Release* 2010;146:175–181.

14. Hed J, Johansson M, Lindroth M. Complement activation according to the alternative pathway by glass and plastic surfaces and its role in neutrophile adhesion. *Immunol Lett* 1984;8:295–299.
15. Kazatchkine MD, Carreno MP. Activation of the complement system at the interface between blood and artificial surfaces. *Biomaterials* 1988;9:30–35.
16. Chenoweth DE. Complement activation in extracorporeal circuits. *Ann N Y Acad Sci* 1987;516:306–313.
17. Toda M, Arima Y, Iwata H. Complement activation on degraded polyethylene glycol-covered surface. *Acta Biomater* 2010;6:2642–2649.
18. Peracchia MT, Harnisch S, Pinto-Alphandary H, *et al.* Visualization of *in vitro* protein-rejecting properties of PEGylated stealth (R) polycyanoacrylate nanoparticles. *Biomaterials* 1999;20:1269–1275.
19. Stolnik S, Illum L, Davis SS. Long circulation microparticle drug carriers. *Adv Drug Delivery Rev* 1995;16:195–214.
20. Thomas SN, van der vlies AJ, O' Neil CP, *et al.* Engineering complement activation on polypropylene sulfide vaccine nanoparticles. *Biomaterials* 2011;32:2194–2203.
21. Mosqueira VC, Legrand P, Gulik A, *et al.* Relationship between complement activation, cellular uptake and surface physicochemical aspects of novel PEG modified nanocapsules. *Biomaterials* 2001;22:2967–2979.
22. Yang A, Liu W, Li Z, Jiang L, Xu H, Yang X. Influence of polyethyleneglycol modification on phagocytic uptake of polymeric nanoparticles mediated by immunoglobulin G and complement activation. *J Nanosci Nanotechnol* 2010;10:622–628.
23. Simpson CA, Agrawal AC, Balinski A, Harkness KM, Cliffel DE. Short-chain PEG mixed monolayer protected gold clusters increase clearance and red blood cell counts. *ACS Nano* 2011;5:3577–3584.
24. Cedervall T, Lynch I, Lindman S, *et al.* Understanding the nanoparticle–protein corona using methods to quantify exchange rates and affinities of proteins for nanoparticles. *Proc Natl Acad Sci* 2007;104:2050–2055.
25. Hamad I, Hunter C, Rutt KJ, Liu Z, Dai H. Complement activation by PEGylated single-walled carbon nanotubes is independent of C1q and alternative pathway turnover. *Mol Immunol* 2008;45:3797–3803.
26. Gref R, Luck M, Quellec P, *et al.* Stealth corona-core nanoparticles surface modified by polyethylene glycol (PEG): Influences of the corona (PEG chain length and surface density) and of the core composition on phagocytic uptake and plasma protein adsorption. *Colloids Surf B Biointerfaces* 2000;18:301–313.
27. Laverman P, Brouwers AH, Dams ET, Oyen WJ, Storm G, van Rooijen N. Preclinical and clinical evidence for disappearance of long-circulating characteristics of polyethylene glycol liposomes at low lipid dose. *J Pharmcol Exp Ther* 2000;293:996–1001.
28. Hamad I, Al-Hanbali O, Hunter AC, Rutt KJ, Andresen TL, Moghimi SM. Distinct polymer architecture mediates switching of complement activation pathways at the nanosphere-serum interface: Implications for stealth nanoparticle engineering. *ACS Nano* 2010;4:6629–6638.
29. Szebeni J, Alving CR, Rosivall L, *et al.* Animal models of complement-mediated hypersensitivity reactions to liposomes and other lipid-based nanoparticles. *J Liposome Res* 2007;17:107–117.
30. Levin CS, Bishnoi SW, Grady NK, Halas NJ. Determining the conformation of thiolated poly (ethylene glycol) on Au nanoshells by surface-enhanced raman scattering spectroscopy assay. *Anal Chem* 2006;78:3277–3281.

31. Bradley AJ, Devine DV, Ansell AM, Janzen J, Brooks DE. Inhibition of liposome-induced complement activation by incorporated poly(ethylene glycol)-lipids. *Arch Biochem Biophys* 1998;357:185–194.
32. Garbuzenko O, Barenholz Y, Priev A. Effect of grafted PEG on liposome size and on compressibility and packing of lipid bilayer. *Chem Phys Lipids* 2005;135:117–129.
33. Buscall R. The stability of polymer latices. In: Buscall R, Corner T, Stagema JF, eds. *Polymer Colloids.* Elsevier Applied Science, 1986, pp. 141–217.
34. Hu Y, Xie J, Tong YW, Wang CH. Effect of PEG conformation and particle size on the cellular uptake efficiency of nanoparticles with the HepG2 cells. *J Controlled Release* 2007;118:7–17.
35. Moghimi SM, Szebeni J. Stealth liposomes and long circulating nanoparticles: Critical issues in pharmacokinetics, opsonization and protein-binding properties. *Prog Lipid Res* 2003;42:463–478.
36. Gbadamosi JK, Hunter AC, Moghimi SM. PEGylation of microspheres generates a heterogeneous population of particles with differential surface characteristics and biological performance. *FEBS* 2002;532:338–344.
37. Efremova NV, Sheth SR, Leckband DE. Protein-induced changes in poly(ethylene glycol) brushes molecular weight and temperature dependence. *Langmuir* 2011;17: 7628–7636.
38. Luk YY, Kato M, Mrksich M. Langmuir. Self-assembled monolayer of alkane thiolates presenting mannitol groups are inert to protein adsorption and cell attachment. *Langmuir* 2000;16:9604–9608.
39. Ataman M. Properties of aqueous salt solutions of poly(ethylene oxide). Cloud points, θ temperatures. *Colloid Polym Sci* 1987;265:19–25.
40. Estephan ZG, Schlenoff PS, Schlenoff JB. Zwitteration as an alternative to PEGylation. *Langmuir* 2011;27:6794–6800.
41. Crepon B, Maillet F, Kazatchkine MD, Jozefonvicz J. Molecular weight dependency of the acquired anticomplementary and anticoagulant activities of specifically substituted dextrans. *Biomaterials* 1987;8:248–253.
42. Baranyi L, Szebeni J, Savay S, *et al.* Complement-dependent shock and tissue damage induced by intravenous injection of cholesterol-enriched liposome in rats. *J Appl Res Clin Exp Ther* 2003,3:3.
43. Montdargent B, Maillet F, Carreno MP, Jozefowicz M, Kazatchkine M, Labarre D. Regulation by sulphonate groups of complement activation induced by hydroxymethyl groups on polystyrene surfaces. *Biomaterials* 1993;14:203–208.
44. Montdargent B, Toufik J, Carreno MP, Labarre D, Jozefowicz M. Complement activation and adsorption of protein fragments by functionalized polymer surfaces in human serum. *Biomaterials* 1992;13:571–576.
45. Carreno MP, Labarre D, Jozefowicz M, Kazatchkine MD. The ability of Sephadex to activate human complement is suppressed in specifically substituted functional Sephadex derivatives. *Mol Immunol* 1988;25:165–171.
46. Salvador-Morales C, Basiuk EV, Basiuk VA, Green ML, Sim R. Effect of covalent functionalization on the biocompatibility characteristic of multi-walled carbon nanotubes. *J Nanosci Nanotechnol* 2007;8:2347–2356.
47. Chonn A, Cullis PR, Devine DV. The role of surface charge in the activation of the classical and alternative pathways of complement by liposomes. *J Immunol* 1991;146:4234–4241.

48. Salvador-Morales C, Zhang L, Langer R, Farokhzad OC. Immunocompatibility properties of lipid-polymer hybrid nanoparticles with heterogeneous surface functional groups. *Biomaterials* 2009;30:2231–2240.
49. Toda M, Hirata I. Effects of hydrophobicity and electrostatic charge on complement activation by amino groups. *ACS Appl Mater Interfaces* 2010;2:1107–1113.
50. Sperling C, Schweiss RB, Streller U, Werner C. *In vitro* hemocompatibility of self-assembled monolayers displaying various functional groups. *Biomaterials* 2005;26: 6547–6557.
51. Arima Y, Toda M, Iwata M. Complement activation on surfaces modified with ethylene glycol units. *Biomaterials* 2008;29:551–560.
52. Whaley K, North, J. Haemolytic assays for whole complement activity and individual components. In: Dodds AW, Sim RB (eds.), *Complement: A Practical Approach*, IRL Press, Oxford University Press, 1997, pp. 19–47.
53. Eelen MA, Roos A, Wieslander J, *et al.* Functional analysis of the classical, alternative, and MBL Pathways of the complement system: Standardization and validation of a simple ELISA. *J Immunol Methods* 2005;296:187–198.
54. Whaley K. Methods in complement for clinical immunologists. In: Whaley, D (ed.), *Measurement of Complement*, Churchill Livingstone, 1985, pp. 77–140.
55. Ferraz N, Karlsson-Ott M, Hong J. Time sequence of blood activation by nanoporous alumina: studies on platelets and complement activation. *Microsc Res Tech* 2010;73: 1101–1109.
56. Bertholon I, Vauthier C, Labarre C. Complement activation by core-shell poly (isobutylcyanoacrylate)-polysaccharide nanoparticles: Influences of surface morphology, length, and type of polysaccharide. *Pharm Res* 2006;23:1313–1323.
57 Szebeni J, Wassef NM, Spielberg H, Rudolph AS, Alving CR. Complement activation in rats by liposomes and liposome-encapsulated hemoglobin: Evidence for anti-lipid and antibodies and alternative pathway activation. *Biochem Biophys Res Commun* 1994;205:255–263.
58. Uetrecht J. Role of animal models in the study of drug-induced hypersensitivity reactions. *AAPS J* 2006;7:E914-E921.
59. Szebeni J, Fontana JL, Wassef NM, *et al.* Hemodynamic changes induced by liposomes and liposome- encapsulated hemoglobin in pigs: A model for pseudoallergic cardiopulmonary reactions to liposomes: Role of complement and inhibition by soluble CR1 and anti-C5a antibody. *Circulation* 1999;99:2302–2309.
60. Szebeni J. Complement activation-related pseudoallergy caused by amphiphilic drug carriers: the role of lipoproteins. *Curr Drug Delivery* 2005;2:443–449.
61. Moghimi MM, Hamad I, Bunger R, *et al.* Activation of the human complement system by cholesterol-rich and PEGylated liposomes-modulation of cholesterol-rich liposome-mediated complement activation by elevated serum LDL and HDL levels. *J Liposome Res* 2006;16:167–174.
62. Szebeni J, Wassef NM, Spielberg H, Rudolph AS, Alving CR. Complement activation in rats by liposomes and liposome-encapsulated hemoglobulin: Evidence for anti-lipid antibodies and alternative pathway activation. *Biochem Biophys Res Commun* 1994;205:255–263.
63. Szebeni J, Baranyi L, Savay S, *et al.* Complement activation-related cardiac anaphylaxis in pigs: Role of C5a anaphylatoxin and adenosine in liposome-induced abnormalities in ECG and heart function. *Am J Physiol* 2006;290:H1050-H1058.

64. Merkel OM, Urbanics R, Bedocs P, *et al.* *In vitro* and *in vivo* complement activation and related anaphylactic effects associated with polyethylenimine and polyethylenimine-graft-poly(ethylene glycol) block copolymers. *Biomaterials* 2011;32:4936–4942.
65. Szebeni J. Complement activation-related pseudoallergy: A new class of drug-induced immune toxicity. *Toxicology* 2005;216:106–121.
66. Sim RB, Willis R. Surface properties: Immune attack on nanoparticles. *Nat Nanotechnol* 2011;6:80–81.
67. Wu YQ, Qu H, Sfyroera G, *et al.* Protection of nonself surfaces from complement attack by factor H-binding peptides: Implications for therapeutic medicine. *J Immunol* 2011;186:4269–4277.

Chapter 12

Bidirectional Interaction Between Nanoparticles and Cells of the Mononuclear Phagocytic System

Whitney P. Caron[*], *Sumit Rawal*[*], *Gina Song*[*], *Parag Kumar*[*], *John C. Lay*[†], *and William C. Zamboni*[*,‡,§,¶,||,**]

[*] *Division of Pharmacotherapy and Experimental Therapeutics*
UNC Eshelman School of Pharmacy
[†] *UNC School of Medicine*
[†] *UNC Institute for Pharmacogenomics and Individualized Therapy*
[§] *UNC Lineberger Comprehensive Cancer Center*
[¶] *Carolina Center of Cancer Nanotechnology Excellence*
University of North Carolina at Chapel Hill
Chapel Hill, NC 27599, USA
[||] *North Carolina Biomedical Innovation Network*
The Research Triangle Park
Durham, NC 27709, USA

Nanoparticle (NP) uptake by the mononuclear phagocytic system (MPS) has been described through several *in vitro* and *in vivo* studies, and is a field of research that is constantly evolving and continuing to grow. While many NP agents contain traditional small molecules that have been used therapeutically for decades, the unique delivery system of NPs allows for greater exposure and efficacy of the active chemical entity in the body. Once an NP enters the body, they encounter a much different host response than that observed with small molecule administration. The network of opsonins and circulating and tissue phagocytes is very complex, and their interaction with NPs often depends on the biological atmosphere as well as the physicochemical properties of the NP. This chapter will discuss NP uptake by the MPS

** Corresponding author. E-mail: zamboni@email.unc.edu

and the resulting clinical manifestations. More specifically, key concepts will include differences in NP physicochemical properties, cell lines and/or animal models used, and the bidirectional interaction between the MPS and NPs in patients.

1. Introduction

Elie Metchnikoff was the first to demonstrate the role of certain cells in the host immunity in his studies during the 1880s, in which he showed that the cellular ingestion of foreign matter was paramount to host defense. Observing leukocytes ingesting and destroying foreign matter, or antigens, formed the groundwork for the now widely known concept of "phagocytosis." It is a literal translation of Greek origin, meaning "devouring cell." Metchinkoff then went on to describe that these "devouring cells" or phagocytes acted after being stimulated by serum substances called "stimulins." Denys and Leclef in 1895 clarified that it was actually the antigen or bacteria that these "stimulins" acted on, and not the leukocyte. In 1903, "stimulins" would receive their present-day name of "opsonins" from Wright and Douglas. "Opsonin," also of Greek origin, means "I cater for, I prepare victuals for." It was discovered that the presence of serum opsonins can make a microorganism more susceptible to phagocytic ingestion.[1]

While Metchinkoff's work has been fundamental to our understanding of phagocytosis, it was Ribbert who years later built upon this concept when he showed the selective distribution and exposure of intravenous (i.v.)-injected colloid dyes in certain cells of the liver, spleen, and bone marrow. In 1924, Aschoff introduced to the scientific community the term "reticuloendothelial system" or RES to describe these specific phagocytic cells. The term RES refers to the collective body of mesenchymal-derived cells that forms a system of resident mononuclear phagocytes, situated in reticular organs including the liver, spleen, lymph nodes, and other organs, all with the collective property of rapidly ingesting and accumulating colloidal and particulate matter. As not all endothelial and reticular cells are phagocytic, the term RES has largely been replaced with the more preferred nomenclature "mononuclear phagocytic system," or MPS. While polymorphonuclear granulocytes (PMNs or neutrophils and eosinophils) are also phagocytic cells, they are not included as part of the MPS.[1]

2. Description of Nanoparticles: Terminology and Basic Concepts as They Relate to MPS uptake

2.1 *Nanoparticle terminology*

The colloidal dyes that Aschoff used in his studies were effectively what we refer to today as nanoparticles (NPs) or nanotechnology. In recent years, there has been a surge in the field of NP research and development, with thousands of different agents being developed for medical applications, including the delivery of chemotherapeutic agents. This has generated a new lexicon of terminology associated with the field of NP research and description of NPs. Examples of various types of NPs include liposomes, fullerenes, carbon tubes, quantum dots, nanoshells, polymers, dendrimers, and conjugates.[2] The theoretical advantages of NPs include increased solubility, prolonged duration of exposure, selective delivery of entrapped drug to the site of action, increased efficacy, decreased toxicity, and potentially overcoming resistance associated with the small molecule form of the active agent.[2] Many NPs have been developed specifically for the purpose of evading rapid clearance from the bloodstream, thereby extending systemic circulation time. A common alteration of the physiochemical properties of NPs is accomplished via conjugation of polymers such as polyethylene glycol (PEG) to the outer surface of the NP, allowing it to evade recognition by the immune system and prolong the duration of exposure in plasma. Although this allows for a greater probability of NP delivery to a target site, an increase in circulation time results in a proportional increase of the duration of contact with the components of the immune system.[3]

Pharmacokinetics (PK) is the study of a drug and/or its metabolite in various compartments (such as plasma or tissue) of the body over time. PK encompasses several unique processes as a drug is being absorbed, distributed throughout the body, metabolized, and/or excreted (ADME). Traditionally, PK considers that the drug has been eliminated from a compartment when any biotransformation of the compound takes place and is no longer in its original chemical structure, even if the resulting metabolites or components of the compound remain in the body.[4] The PK disposition of NP agents is dependent on the carrier enclosing the parent drug and not the small molecule drug contained within the carrier. This remains true until the small molecule drug is released from the carrier (or undergoes "liberation") and resumes its classical PK disposition.[5] The nomenclature used to describe carrier-mediated PK includes encapsulated (the drug within or bound to the carrier), released (active drug that is released from the carrier), and sum total (encapsulated drug plus released drug).[6,7] When the drugs are encapsulated in NPs, they are protected from metabolizing enzymes in liver and other tissues, and from renal clearance before they are released, because of the increased size of their carrier. After the drug is released from its carrier, it is pharmacologically active and subjected to the same routes of metabolism and clearance as the non-carrier form of the drug.[7] Thus, the PK profiles of the parent drug and the drug encapsulated in the NPs are different.

2.2 *Interaction between nanoparticles and the MPS*

The uptake and removal of NPs from the circulation are mediated by both circulating and resident tissue phagocytes of the MPS. Peripheral blood leukocytes capable of NP uptake include circulating monocytes, granulocytes, dendritic cells (DCs), and even platelets.[3,8] In addition, NPs can be taken up in tissues by phagocytes, such as Kupffer cells in the liver, macrophages in the spleen and bone marrow, and by macrophages and DCs in the lymph nodes[3] and lung. Figure 1 illustrates the interaction and clearance of NPs with the MPS. Many NPs have been developed for the purpose of evading rapid clearance from the bloodstream, thereby extending systemic circulation time.

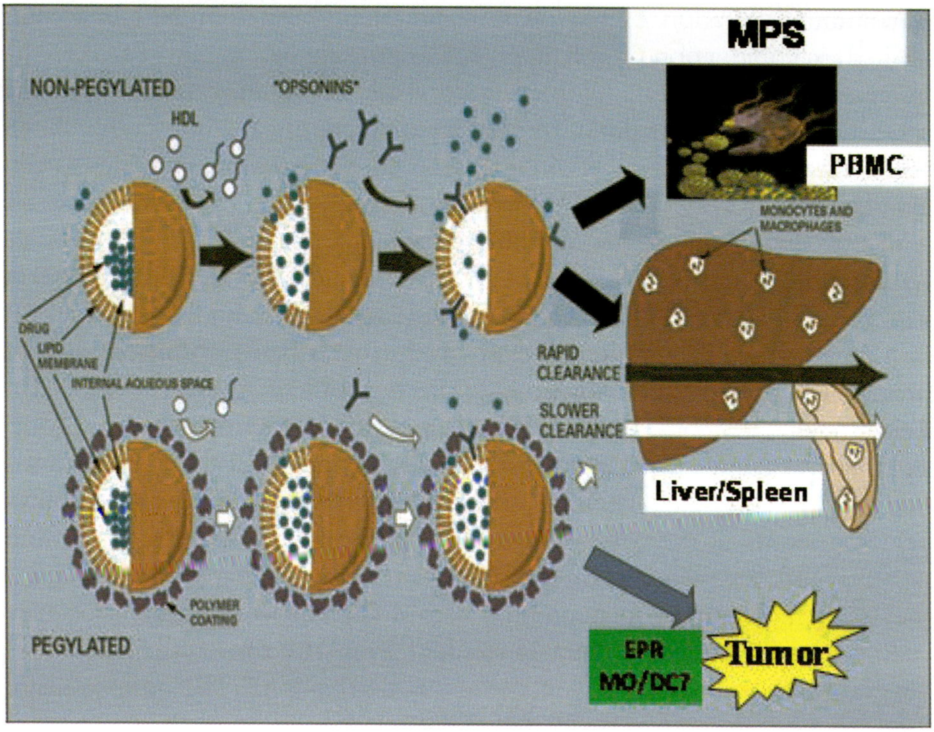

Figure 1. Clearance of PEGylated and non-PEGylated liposomal agents via the MPS in the blood, liver, and spleen. When non-stabilized liposomal agents were first tested, they were found to only minimally increase the circulation time of the small molecule agent encapsulated within the carrier (rapid clearance). However, stabilization with polyethylene glycol (PEG) has helped to reduce uptake of NPs by the MPS (slower clearance). While the clearance of PEGylated liposomes are slower than non-PEGylated liposomes, both are phagocytosed by peripheral blood mononuclear cells (PBMCs), phagocytes of the liver and spleen. Greater tumor exposure is seen after administration of PEGylated liposomes, which is in part due to the enhanced permeability and retention (EPR) effect and possibly, the presence of MPS-related cells in tumors.

While this allows for the greater probability of NP delivery to a target site, an increase in circulation time results in a proportional increase of the duration of contact with the components of the immune system.[3] Variability in the PK and pharmacodynamics (PD) of NP agents seen in patients can be attributed to many different variables, particularly components of the MPS. In this chapter, we will focus on the steps and processes involved in NP uptake, how NPs can be modified to affect uptake, the resulting cytotoxicity to immune cells following uptake, and the clinical manifestations related to the bidirectional interaction between MPS and NP-based chemotherapy.

2.3 *Capture versus hijacking*

There are many studies that document NP interaction with the MPS, and in this chapter, we will discuss both *in vitro* and *in vivo* examples. We will also highlight NP clearance and drug release, which is a direct outcome of the interaction with the MPS. Although the interaction is documented, it is not known whether NP uptake occurs through phagocytes in the tumor, liver, and spleen, "capturing" the NP as it reaches the organ/tissue through the bloodstream, or if the circulating MPS cells "hijack" the NP in the bloodstream and carry it to the site of action, leading to the increased tumor accumulation seen with these agents. Figure 2 shows the phenomenon of "capture" and "hijacking," terms which we have coined to bring attention to this component of NP uptake that has not been fully elucidated.

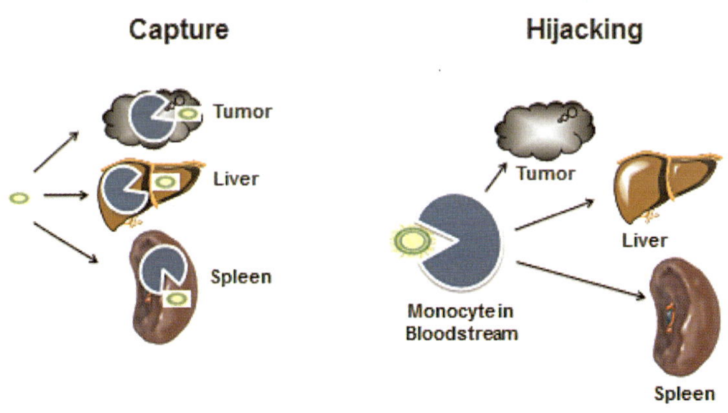

Figure 2. Two theories for how NPs reach organs of the MPS and the tumor when delivering chemotherapeutics. On the left is the concept of "capture," in which the NP is delivered to the tumor and tissues and phagocytosed by MPS cells in the tumor and tissues. On the right is "hijacking," in which circulating phagocytes in the systemic circulation take up the particle in the blood and carry the NP to its ultimate destination, such as the liver, spleen, or tumor.

3. Mechanisms and Factors Involved in Nanoparticle Uptake

3.1 *Opsonization*

Two of the first biological objects NPs encounter upon entering the bloodstream are plasma proteins, namely opsonins, and immune cells. Opsonizing proteins can have a profound effect on hydrodynamic size (diameter) of an NP. Thus, dynamic light scattering (DLS) of NPs is a common method for determining hydrodynamic size in solution and is often used only at baseline, when particles are dispersed in their stock solution. However, coating of NPs by opsonizing plasma proteins can cause significant "growth" in hydrodynamic size (i.e., increased diameter). For this reason, some groups incubate particles in serum to report a "true" hydrodynamic size using DLS, which may be more relevant than the reported core size in terms of interactions with biological systems. For example, Franca *et al.*[9] and Dobrovolskaia *et al.*[10] showed that the hydrodynamic size of 30-nm gold colloids changed markedly before and after opsonization with human serum (34.4 ± 0.2 nm and 94.8 ± 0.2 nm, respectively). In addition, 150-nm gold colloid particles were 149.8 ± 0.7 nm and 263.6 ± 4.7 nm before and after opsonization, respectively. The zeta potential of the 30-nm gold colloids was −38.2 ± 1.2 mV before and −16.4 ± 0.6 mV after incubation with serum. The zeta potential of the 150-nm particles changed from −46.3 ± 0.9 mV to −20.4 ± 1.9 mV before and after opsonization, respectively. The amount of gold in the two formulations was approximately equal, 48.0 ± 3.9 µg/mL and 42.5 ± 3.7 µg/mL in 30- and 150-nm colloids, respectively. Endotoxin concentrations were below the limit of detection, indicating that changes in core size are not affected by contamination. Franca *et al.* demonstrated the profound effects that opsonization may have on NP size and charge, and that these should be taken into account during the early phase of NP development.

3.2 *Endocytosis*

NP drugs are eliminated primarily by the macrophages and monocytes of the MPS. These phagocytic cells normally function in removing pathogens from the circulation and also may migrate to sites of inflammation following recruitment by chemotactic cytokines or complement proteins.[11] The uptake of NPs by macrophages occurs by endocytosis, which includes phagocytosis and pinocytosis. Both endocytic mechanisms are further subdivided into complement receptor-mediated, FCyR-mediated, and scavenger receptor (SR)-mediated for phagocytosis, and macropinocytosis, clathrin-mediated, caveolin-mediated, and clathrin/caveolin-independent endocytosis for pinocytosis. Due to the specificity of active receptors in each of these pathways, different receptors are involved in the uptake of particles and molecules with different physical and chemical properties.[9]

FCyR- and complement receptor (CR)-mediated phagocytosis rely on the opsonization of particles by immunoglobulin G (IgG) and iC3b, respectively, which bind with these specific cell surface receptors to initiate phagocytosis. In contrast, SRs, which are expressed at various levels on a variety of cells including macrophages and some DCs and monocytes, function as opsonin-independent molecular pattern recognition receptors which recognize and bind various molecular structural components. SRs comprise several different classes including SR-A, B, C-type (calcium-dependent) lectin receptors, and others (types C–F). The macrophage SR (CD204, SR-A1) is the proteotypic SR-A receptor and binds a broad array of ligands including proteins, lipoproteins, and oxidized lipoproteins, as well as pathogen-associated molecular patterns (PAMPs) on pathogenic Gram-positive and -negative bacteria.[12,13] Other SR-A receptors include CD163 (SR-A2) and MARCO (macrophage receptor with collagenous structure). The class B receptors (SR-B) comprise SR-B1 and CD36, which function primarily as receptors for low-density lipoproteins (LDL), oxidized lipoproteins, and phospholipids. SR-B receptors are also expressed on endothelial cells and are important players in the pathogenesis of atherosclerosis. Other SRs classified as types C, D, E, or F receptors bind a variety of ligands. The macrophage mannose receptor (CD206, SRCL) is a C-type (calcium-dependent) lectin receptor which binds to carbohydrate moieties. A good overview of the scavenger receptors is provided in reviews by Aereshcoug, Murphy, Kraal, Perez, Pieser, and Valacchi.[14,15]

Receptor-mediated endocytosis is an active and very complex process that occurs via specific or non-specific receptors and is performed by specialized cells.[16,17] An illustration of this process is provided in Figure 3. Chen *et al.* described a film tension theory in which specific or non-specific opsonins adsorbed onto NPs and then bind to receptors localized in the host cell membrane.

This then leads to phagocytosis by the lowering of membrane tension.[18] Spontaneous actin polymerization causes the formation of membrane extensions.[11] The resulting size of the phagocytic vesicles depends on the size of NPs, but is always greater than 250 nm.[19] Receptor-mediated endocytosis is also an active process which involves the uptake of NPs via membrane receptors.[20] The difference between receptor-mediated endocytosis and phagocytosis is explained by the involvement of clathrin-coated pits in endocytosis and intact actin filaments in phagocytosis. Pinocytosis is fluid-phase; it is a

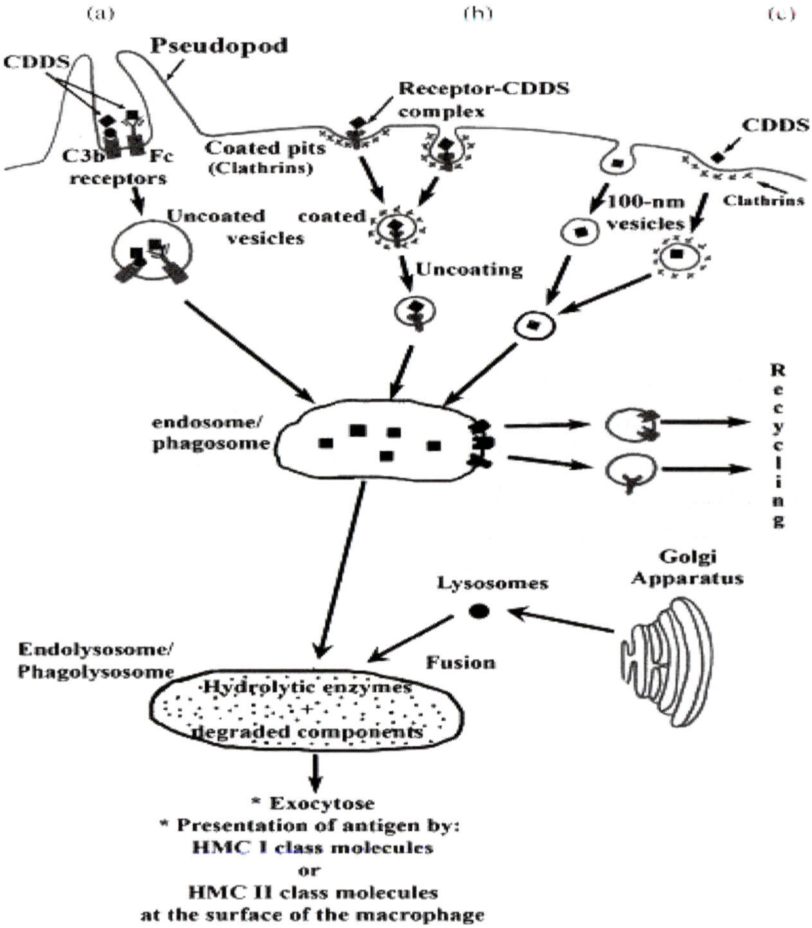

Figure 3. Schematic representation of different capture mechanisms by endocytosis at the surface of macrophages. (a) Phagocytosis, (b) receptor-mediated endocytosis, and (c) pinocytosis via clathrin/non-coated vesicles. (Adapted from Ref. 11.)

receptor-independent, passive uptake of insoluble particles, lipoproteins, immune complexes, enzymes, and low molecular weight solutes.[21]

Both phagocytosis and macropinocytosis utilize active transport and rely on actin polymerization.[9] To test this theory, the actin polymerization inhibitor cytochalasin D was assessed for its ability to affect the uptake of gold colloids by adherent RAW264.7 murine macrophages.[9] When cells were pretreated with cytochalasin D, there was a decreased uptake of both the 30- and 150-nm gold colloids. There was also a similar decrease in the uptake between the two different-sized particles when lower doses of the inhibitor were incubated with the cells. Conversely, when greater concentrations of cytochalasin D were used, the internalization of the larger 150-nm colloids was decreased to a greater degree than the smaller 30-nm colloids (>50%). To distinguish between the two routes of uptake, 5-(N,N-dimethyl) amiloride hydrochloride, an inhibitor of membrane Na^+/H^+ ATPase, and therefore macropinocytosis, was added to the cells prior to gold colloid incubation. Treatment with the inhibitor did not result in a decrease in the uptake of either size of gold colloid.[9] Therefore, the uptake of these particles was due to a phagocytotic pathway, but not pinocytosis.

When examining the clathrin-mediated pathway associated with endocytosis, cells were pretreated with chlorpromazine hydrochloride, a known inhibitor of the pathway. This resulted in a nearly 50% decrease in uptake of the 30-nm NPs, but had negligible effect on the uptake of the 150-nm NPs. The same size effects on uptake were seen in the absence of SRs. Therefore, it was hypothesized that the uptake of 30-nm colloids could be almost completely blocked by co-inhibition of both clathrin-mediated and SR-mediated routes if these were the only two paths involved in uptake. This was tested using an RAW264.7-SR–negative clone in the presence of chlorpromazine hydrochloride, and a similar degree of inhibition of the uptake of 30-nm gold colloids was observed as in the chlorpromazine only experiment, questioning if SRs are truly as involved as the clathrin-mediated pathway in phagocytosis.[9]

Since neither clathrin-mediated pinocytosis nor SR-mediated phagocytosis resulted in complete inhibition of the uptake of gold colloids, additional pathways must be involved, such as caveolin-mediated uptake, another actin independent form of pinocytosis. Filipin III is an inhibitor of caveolae formation, and was therefore used to determine its involvement in the uptake of gold colloids. When filipin III was present, there was a decrease in the uptake of both NPs. The decrease in uptake of the smaller NP was slightly greater than that of the larger NP. However, in both cases, the degree of decrease was significant when compared to the negative control, suggesting that NP uptake is in part due to caveolin-mediated uptake.[9]

3.3 *Surface receptors*

There are several receptors present on the surface of macrophages. The receptors include CRs, SRs, fibronectin, receptors for immunoglobulin fragments (Fc), and receptors for glycoproteins.[22–23] Pattern recognition receptors, which include Toll-like receptors, SR-A (CD204, CD36), and the mannose receptor (CD206), are primarily involved in mediating receptor-mediated endocytosis.[24,25] In particular, CD204 has been shown to contribute to the non-specific uptake of NPs surface-functionalized with carboxylic acids, antibodies, and synthetic polymers.[26]

Class A SRs are specifically expressed on the surface of macrophages and are well known to endocytose a variety of liposomes.[27] Several studies have reported contrasting views on the role of class A SRs in the recognition of liposomes. While some *in vitro* studies excluded the participation of class A SRs in the uptake of liposomes containing 10 mol% phosphatidylserine (PS) by Kupffer cells,[28] others suggested that their role is in the uptake of liposomes containing 30 mol% PS.[29] Therefore, the relative amount of PS may regulate the charge and steric organization of liposomes, which are essential properties in their recognition by class A SRs and subsequent uptake by Kupffer cells.

Most of the macrophage surface receptors are distributed according to the function of phagocytes, tissue origin, degree of activation, and the species studied. Species-dependent liposomal clearance from blood was reported by Liu *et al.*[30,31] In mice, serum opsonins were not involved in liposomal uptake by the liver. Liposomes were captured via direct interactions with macrophage receptors. In contrast, rat liver strongly exhibited serum opsonin-dependent uptake of liposomes. Furthermore, the activity of serum opsonins for a given liposomal composition differed among animal species. Human serum exhibits higher oposonin activities and it is likely that C3 of the human complement system plays an important role in mediating the uptake of liposomes by the liver.[30]

3.4 *The role of the complement system in nanoparticle uptake by the MPS*

The complement (C) system, made up of approximately 30 circulating and membrane-expressed proteins, plays an essential role in both the innate and antibody-mediated acquired immune responses.[32] The three pathways of C system activation are the classical, lectin, and alternative pathways. The initiation of each pathway involves distinct recognition events and components, but the later stages of all three pathways involve the same components. The characteristics of each pathway of C system activation are discussed in detail in Chapter 11.

3.5 *Complement affects nanoparticle stability, biodistribution, and PK*

After C activation, the resulting products including opsonins (C3b/iC3b), the membrane attack complex (MAC), and anaphylatoxins (C3a, C5a), exert their protective functions by opsonization, direct lysis of invader cells, and activation of inflammatory cells, respectively.[32] Opsonization is essential for efficient phagocytosis by signaling cells in ingesting the particles. It is also known that the activation of the C system after the i.v. administration of certain liposomal drugs such as Doxil® (PEGylated liposomal doxorubicin) and AmBisome® (liposomal amphotericin B), can result in hypersensitivity reactions called C activation-related pseudoallergy (CARPA) due to anaphylatoxin production.[33] Opsonization is the major factor that induces and facilitates the recognition and uptake of NP agents by the MPS. Thus, factors influencing C activation such as particle surface characteristics may greatly impact NP drug carrier stability and the biodistribution and PK of NP agents.[34]

3.6 *Accelerated blood clearance phenomenon*

Many groups have shown that a repeated administration of PEGylated liposomes into animals triggers the rapid clearance of the second dose of liposomes via their accumulation in the liver. These results are dependent on the amount of the first dose, time interval between injections, lipid dose, and drug encapsulation.[35] This phenomenon is called accelerated blood clearance (ABC). The ABC phenomenon has been investigated and observed in mice, rats, dogs, and rhesus monkeys.

Ishida *et al.* investigated the effect of lipid dose on a first injection on the induction of the ABC phenomenon by PEGylated liposomes. Rats were treated with PEGylated liposomes at a dose of 0.001, 0.01, 0.1, 1, or 5 μmol/kg of their body weight. It was found that there is a strong inverse relationship between the dose of initially injected PEGylated liposomes and the extent to which the ABC phenomenon was induced, i.e., the higher the dose (more than 1 μmol phospholipids/kg), the smaller the ABC phenomenon.[36] Laverman *et al.* evaluated the impact of enhanced clearance for the commercially available Caelyx® in rats.[37] Three groups of five rats were injected with either Caelyx®, PEG–liposomes without doxorubicin, or phosphate buffered saline (PBS). One week later, all groups were injected with ^{111}In-labeled Caelyx®. It was found that when rats had received Caelyx®, the PK and biodistribution of a second ^{111}In-labeled empty Caelyx® injection one week later was similar to the PK and biodistribution of the control group (after a first injection with PBS). However, the PK and biodistribution of the ^{111}In-labeled Caelyx® was altered dramatically when empty PEG–liposomes were injected one week before.

In addition, Koide *et al.* confirmed the effects of a drug encapsulated in PEGylated liposomes on the ABC phenomenon. Interestingly, the pre-administration of PEGylated liposomes encapsulating doxorubicin (Doxil®) did not trigger the rapid clearance of the second dose of PEGylated liposomes from the bloodstream in contrast to empty PEGylated liposomes and the mixture of free doxorubicin and empty PEGylated liposomes. It has been suggested that T cell-independent B cell response mediated by anti-PEG IgM antibody is responsible for the ABC phenomenon induced by repeated injections of PEGylated liposomes. Thus, B cells responsible for the ABC phenomenon might be selectively damaged by doxorubicin encapsulated in PEGylated liposomes because PEGylated liposomes might be recognized by B cells as a thymus-independent type 2 (TI-2) antigen.[38] However, repeated administrations of Doxil® in patients has led to a significant decrease in Doxil® clearance from cycle 1 to cycle 3 ($p < 0.0001$).[39] Additionally, it has been found that animal models are not always predictive of patients in terms of NP PK, likely due to differences in the MPS.[40] Thus, the ABC phenomenon results in animal models that may not be predictive of PK of NPs after repeated doses in humans.

4. The Effects of Physicochemical Properties on Nanoparticle Uptake by the MPS

4.1 *Modifiable factors*

Designing optimal NPs for drug delivery requires a minimization of the interaction between macrophages and the synthetic NPs. There are several modifiable physicochemical properties of NPs such as size, shape, hydrophobicity, charge, surface chemistry that regulate the interaction with macrophages.[41–42] In addition, the geometry of NPs has been shown to be important in the design of carriers. Particle geometry plays an important role in various biological functions such as phagocytic internalization, transport in the vasculature, and blood circulation half-life.[10] The actual process of phagocytosis is determined by a complex interplay among these physicochemical characteristics.[43]

4.2 *Nanoparticle composition*

The NP components play an important role in their recognition by macrophages. NPs formulated for the purpose of drug delivery mostly consist of synthetic polymers such as PLGA, polystyrene, PLA, and polyalkyl acrylates. While most of the components used are biodegradable, some are not and lead to bioaccumulation in tissues and cytotoxicity. As discussed, the macrophage surface has various receptors or glycoproteins that aid in their recognition and interaction with potential targets.[44] Thus, NP uptake can be mediated by complement proteins or antibodies and their receptors. However, in the presence of certain components like phosphatidylinositol (PI) and PS, the complement-mediated uptake of liposomes can be reduced.[45]

SR-mediated uptake is mostly opsonin-independent and the surface chemistry of the NP may be involved in its recognition by SRs and other molecules. Certain cell surface molecules including the class B SR, CD36 (receptor for oxidized LDL), and other non-scavenger surface molecules including CD14 (lipopolysaccharide receptor) and CD61/51 (integrin cell adhesion molecules) are known to bind PEGylated liposomes. CD14 is a membrane glycoprotein expressed on the surface of mononuclear phagocytes which recognizes PAMPs in the components of bacterial cell walls, specifically, lipopolysaccharide (endotoxin), lipoteichoic acid, and lipoarabinomannin. It has also been implicated as a receptor for PS in the plasma membranes of apoptotic cells.[46] CD61/51 is a cell adhesion molecule expressed by many cell types. Of these three cell surface molecules, only CD36 is officially classified as a scavenger molecule capable of initiating endocytosis upon binding to its ligand. However, it has been shown that CD14 may bind with NPs as reported by Yu *et al.*[47] There is also evidence that CD14 is involved in the removal of PEGylated liposomes. Shibuya-Fujiwara *et al.* reported the clearance of PEGylated liposomal hemoglobin via non-opsonin mediated mechanisms involving not only CD36, but also CD14 and CD61/51.[48] While the exact mechanism of phagocytosis is unclear, the removal of PEGylated liposomal hemoglobin was explained on the basis of the recognition of PEG or some new determinants by these receptors.[48] Analogous to the "tethering" of apoptotic cells following the binding of CD14 to PS, we might speculate that since, unlike CD36, neither CD14 nor CD16/51 are phagocytic receptors, they may function to "tether" the liposomes in proximity to the plasma membrane, allowing subsequent endocytosis via binding to SRs on the membrane.[46]

Among numerous factors influencing C activation, surface charge has been extensively studied. It has been demonstrated that the single negatively charged phosphodiester moiety of the mPEG–phospholipid is responsible for enhanced C activation, at least for mPEGylated dipalmitoylphosphatidylcholine (DPPC) liposomes in mice.[49] In addition, liposomes in which the negatively charged PEG–DSPE with replaced near-neutral PEG–DS (2K-PEG-3methoxy polyethylene glycol-oxycarbonyl 3-amino-1, 2,-propandiol distearoyl ester) showed decreased C activation in pigs.[33] This phenomenon may explain that the negatively charged liposomes demonstrate a significantly higher rate of clearance from the blood than neutral liposomes. The negatively charged liposomes also showed an increased rate of MPS uptake in the liver compared to the neutral liposomes, indicating that enhanced C activation with the negatively charged liposomes induced the uptake by the phagocytic cells and increased rate of clearance of NPs from the blood.[30]

Conventional liposomes, an NP platform without protective polymers or other particle surface components to prevent adsorption of opsonins, were reported to significantly activate the C system and be removed from the blood circulation within seconds by Kupffer cells, with accumulation in the liver.[50,51] It has been known that liposomes can activate the C system via both the classical and the alternative pathways.[51,52] The most commonly used strategy to minimize opsonization is to conjugate PEG polymers onto the surface of NPs. PEG is an inert hydrophilic polymer that forms a water shell and provides good steric hindrance for preventing protein binding to the NP.[53] It has been shown that PEGylation reduces the rate of MPS uptake and increases the circulation half-life for various types of NPs.[54,55] However, Moghimi *et al.* found that PEGylated liposomes can trigger C activation in the absence of anti-PEG antibodies through both C1q-dependent classical and MBL–MASP-dependent alternative pathways using normal and C1q-depleted human serum.[34,56,57] Thus, the presence of surface mPEG molecules does not interfere with the production of opsonins, but may sterically suppress the deposition of C3 convertases and/or C receptor binding, which subsequently results in slower recognition and clearance by the MPS.[58]

Structure–activity relationships have been also demonstrated with other molecules surface-grafted onto NPs. Pham *et al.* measured the C response to lipid-encapsulated NPs that are functionalized on the surface with various lipid-anchored gadolinium chelates.[59] The results showed that surface functionalization of NPs with certain chemical structures can result in swift C activation that is initiated by a natural IgM antibody and propagated with the classical pathway. It was also reported that the intensity of the response is dependent on the chemical structures of the lipid-anchored chelates and not surface charge or zeta potential effects alone.[59] In addition, conformational states of surface-projected polyethyleneoxide (PEO) chains, derived from the block copolymer poloxamine908 adsorption, on polystyrene NPs were found to activate the C system differently.[34] Alteration of copolymer architecture on nanospheres from the mushroom- to brush-like configuration not only switches C activation from the C1a-dependent classical pathway to the lectin pathway, but also decreases the level of generated C activation products C4d, Bb, C5a, and SC5b-9.[34] Interestingly, released doxorubicin from the PEGylated liposomes was shown to bind to the liposomal surface and lead to aggregation of vesicles.[33] Morphological evidence of the presence of low-curvature oval, elongated, or irregular liposomes and aggregates due to doxorubicin binding on the surface was found in a commercial vial of Doxil®, which may significantly change the surface area of vesicles in blood and subsequently lead to C activation.[33] Taken together, these findings suggest that the systemic NP structure–activity relationship may provide the basis for design of immunologically safer surface-engineered NPs with more favorable PK properties.

4.3 *Particle size and dimension*

The size of a NP is a very important parameter in determining its interaction with macrophages and its resulting clearance from the blood. NPs with smaller sizes (< 100 nm) have been shown to be advantageous as they are removed more slowly from blood.[58,60] Several authors have reported the size-dependent activation of the C system for particles less than 1 μm in size.[61,62] Harashima *et al.* showed that the surface area of liposomes recognized by the C system is directly proportional to the number of particles and their diameter.[63] This is due to the greater size, increasing the degree of opsonization, and is correlated with higher NP uptake. In addition, the larger surface area possibly increases the density of recognizable sites on the surface of liposomes. For particles less than 1 μm in size, a good correlation has been observed between the increasing size of liposomes and their uptake by macrophages both *in vitro* and *in vivo*.[64]

Yu *et al.* investigated the effects of PEGylated NP size and charge on the uptake by macrophages.[47] The internalization of NPs into macrophages was investigated *in vitro* by the synthesis of a library of polymer-coated iron oxide micelles, spanning a range of 30–100 nm in diameter and –23 mV to +9 mV. NP size was found to be a dominant predictor of uptake as compared to charge. The effect of size was evaluated using protease-sensitive NPs displaying a matrix metalloproteinase-9 (MMP-9)-cleavable polymeric corona. These NPs can shed 10–20 nm hydrodynamic diameter in response to MMP-9 activity. MMP-9 triggered a reduction in size and reduced NP uptake by macrophages by up to six-fold, as observed by T2-weighted magnetic resonance imaging.

Doshi *et al.* reported a strong dependence between the size and shape of NPs with different geometries and their attachment to macrophages.[42] NPs with the longest dimensions exhibited the strongest attachment with macrophages, which also correlated with the size range of commonly found bacteria in nature. The surface features of macrophages, in particular membrane ruffles, were implicated in this geometry-based target recognition which could have a significant effect on the design of drug delivery carriers.

Champion *et al.* studied phagocytosis by alveolar macrophages using polystyrene particles of various sizes and shapes, and reported that particle shape plays a predominant role in phagocytosis.[43] Whether or not macrophages initiate phagocytosis is dictated by the particle shape when present near the macrophages, as measured by tangent angles at the point of initial contact. The local shape also determines the complex structure of actin required for associated phagocytosis. Failure in creating such a structure results in simple spreading and not internalization of NPs. Gratton *et al.* reported on the internalization of monodisperse hydrogel particles into HeLa cells as a function of size, shape, and surface charge[65] using a top-down particle fabrication technique called PRINT that is able to generate uniform NPs. These findings suggest that HeLa cells readily internalize nonspherical particles with dimensions as large as 3 μm by using several different mechanisms of endocytosis.

4.4 *Surface charge*

The importance of charge in the clearance of NPs has been studied extensively. NP charge, either positive or negative, is determined by different surface structures that in turn, determine the degree of protein adsorption.[66] As the surface of proteins is heterogenous with both positive and negative charges, hydrogen bonding abilities, as well as nonpolar regions, the resulting interaction of the proteins with NPs can be highly complex. Thus, a variety of different interactions such as ionic, hydrophobic, hydrogen bonding, van der Waals, acid–base, and hydration forces can result between NPs and proteins.[67,68] However, it is generally assumed that hydrophobic and ionic forces are the two major interactions responsible for protein adsorption.

5. Nanoparticle Cytotoxicity to Cells of the MPS — *In Vitro*

There are numerous studies, mainly *in vitro* experiments, which have looked at the toxicological effects of NPs on immune cells. Wilhelmi *et al.*[69] used a variety of NPs, including crystalline silica, amorphous silica, titanium dioxide (TiO_2), zinc oxide (ZnO), and magnesium oxide (MgO), and incubated them at various concentrations and times in the macrophage-like murine cell line RAW264.7. After 24 h of treatment, the untreated control and MgO-treated cells displayed similar viability/healthy conditions with regard to their morphology and number, indicating the lack of toxicity for MgO particles. Staurosporine, a known apoptosis inducer, was used as a positive control and showed a reduced macrophage number and alterations of shape that were similar to ZnO-treated cells. The amorphous and crystalline silica also caused distinct damage to cells, but not to the extent of the ZnO-treated cells. TiO_2-treated cells had no effect on macrophage number; however, the medium containing the cells and particles was nearly absent of TiO_2 following incubation, indicating that the cells had a massive uptake of these particles. The investigators showed that toxicity to macrophages is dependent on the specific NP composition.

Additionally, cellular response to conducting polymers (CPs) deposited onto silica NPs (Si NPs; 22 nm in diameter) was assessed in the same mouse macrophage RAW264.7 cell line. The cellular uptake of the NPs by the macrophages was assessed by transmission electron microscopy (TEM). It was determined that internalization resulted from phagocytosis and endocytosis. Subsequently, cytotoxicity was measured by adenosine triphosphate (ATP) production, propidium iodide (PI) staining, reactive oxygen species (ROS) detection, and lactate dehydrogenase (LDH) assays.[70] LDH is released by cells when they are damaged, and specifically serves as an indicator of membrane integrity. The cellular toxicity as measured by the LDH assay was NP concentration-dependent. The production of ROS can indicate external or internal cell damage, or the dysregulation of normal cellular antioxidant mechanisms. In this study, cells were stained with 2',7'-dichlorodihydrofluorescein diacetate (DCF-DA), a non-fluorescent compound that becomes oxidized to fluorescent DCF in the presence of ROS. The NP-treated cells did not have a significantly higher ROS production than the cells of the negative control, suggesting that not all NP compositions cause direct cellular toxicity.[70]

ROS production was also assessed by Bhattacharjee *et al.* after Si NPs were incubated with NR8383 rat alveolar macrophage cells at various concentrations. After 24 h of exposure to various Si NPs, NR8383 cells had a dose-dependent increase in intracellular ROS production (EC_{50} = 22 ng/mL). Altering the surface charge of the NPs was found to play an important role in their interaction with cells and also in the resulting cytotoxicity. Positively charged particles such as Si NP–NH_2 were more cytotoxic toward NR8383 cells than neutral Si NP–N_3. Negatively charged particles such as Si NP–COOH showed very low or no cytotoxicity altogether. The Si NP–NH_2 formulation produced the most intracellular ROS followed by Si NP–N_3, and no ROS was measured after Si NP–COOH incubation. The authors noted that that Si NP–NH_2 and Si NP–COOH are slightly more hydrophilic than Si NP–N_3. However, since the difference in cytotoxicity and ROS production was much larger between Si NP–NH_2 and Si NP–COOH than between Si NP–NH_2 or Si NP–COOH and Si NP–N_3, it can be concluded that surface charge is much more significant than differences in hydrophilicity or hydrophobicity.

The relationship between NP size and cytotoxic properties on different immune cells were once again studied, this time using ZnO NPs. ZnO NPs of 4–20 nm were synthesized by modifying the hydrolysis molar ratio of water to zinc acetate. TEM determined the sizes of the three different particles to be 4 to 8, 13, and 20 nm. Following a 24-h incubation with the particles, the freshly isolated human PBMCs were stained with PI to determine viability. All of the lymphocyte populations displayed similar IC_{50} (half-maximal inhibitory concentration of the NP) values (~5.0 mM), with no significant difference observed between them at any NP concentration evaluated. However, the natural killer (NK) cells were markedly more sensitive to NP-induced cytotoxicity, with an IC_{50} of ~1.0 mM. Statistically significant differences were detected between NK cells and B and T lymphocytes at 1–5 mM NP concentrations (e.g., $p = 0.003$ at 1 mM, $p = 0.0002$ at 2.5 mM, $p = 0.002$ at 5 mM, and $p = 0.05$; NK cells *vs.* $CD3^+$ T cells at 10 mM NPs). The greatest cytotoxicity observed was in monocytes, with >50% of the monocytes killed at the lowest NP concentration tested (0.5 mM). Monocytes experienced significantly more cytotoxicity than NK cells (0.5 mM and 1 mM; $p = 0.0002$ and $p = 0.0008$, respectively), and T and B lymphocytes ($p < 0.0001$ for both cell types at both concentrations tested). Therefore, T and B lymphocytes were found to be more resistant to NP toxicity compared to monocytes and NK cells. As the clearance of NPs is attributed to cells of the MPS (monocytes/macrophages), it is not surprising that lymphocytes were minimally affected in terms of cytotoxicity.[71]

Commensurate with other studies of the effect of NP size on cytotoxicity, Hanley *et al.* demonstrated that significantly greater cytotoxicity was observed in human PBMCs with smaller NPs. The 4-nm NPs (80.0% ± 2.0%), compared to either 13- or 20-nm NPs ($p = 0.04$ and $p = 0.01$, respectively) had significantly greater toxicity. In addition, significantly more cytotoxicity ($p = 0.05$) was observed for 13-nm NPs (70.1% ± 8.0%), compared to 20-nm NPs (44.0% ± 6.8%). To validate the finding that cytotoxicity increases with decreasing NP size, experiments were performed over a range of NP concentrations. A significantly greater toxicity in human PBMCs using 4-nm NPs compared to 20-nm NPs remained at all concentrations tested ($p = 0.03$ at 1 mM, $p = 0.0001$ at 5 mM, and $p = 0.01$ at 10 mM NP concentration).[71]

6. Nanoparticle Cytotoxicity to Cells of the MPS — *In Vivo*

In vivo toxicities of monocytes and macrophages of the MPS in animals and humans have also been observed as a result of the interaction with NPs. Doxil® and Caelyx® are PEGylated liposomal formulations of encapsulated doxorubicin commercially available for the treatment of recurrent ovarian cancer, multiple myeloma, and Kaposi's sarcoma. Interestingly, these liposomal formulations of doxorubicin have been shown to reduce the incidence and severity of acute cardiotoxicity commonly associated with doxorubicin use in both preclinical and clinical studies,[72,73] as well as allow for improved drug accumulation in tumor tissues.[74] Unfortunately, new toxicities appear to occur as a result of the altered PK disposition of encapsulated doxorubicin. It has been demonstrated that the encapsulation of doxorubicin in liposomes can extend blood circulation time and can result in increased delivery of doxorubicin to phagocytic cells of the MPS in the livers of rats following i.v. administration. This in turn leads to significant pharmacologic and cytotoxic effects on these cells, including both decreased specific phagocytic capacity and reduced cell numbers.[75] The toxic effects on monocytes have also been demonstrated with other NP drugs. For example, the administration of liposome-encapsulated clondronate at 390 mg i.v. × 1 given over 1 min to gerbils resulted in striking reductions of spleen and liver macrophages by > 95% and decreased blood monocytes by 55% at five days after injection.[76] The potential clinical implications of these toxicities are emphasized by the results of several preclinical studies demonstrating that the reduction in the phagocytic activity of macrophages is associated with increases in metastatic growth of tumors.[77]

Signs and symptoms of acute hypersensitivity reactions, such as shortness of breath, flushing, chest pain, back pain, and large changes in blood pressure have also been observed to occur after the administration of liposomes. These adverse reactions all resolve within minutes after the termination of liposomal infusions, and the symptoms can be attenuated by treatment or prevented by pretreatment with agents such as epinephrine and/or steroids.[78,79] In one study, hypersensitivity reactions were observed in a surprisingly high rate of 13 out of 29 patients with solid tumors treated with Doxil® for the first time.[80] Interestingly, infusion of radio-labeled (99m-Tc) empty PEGylated liposomes (no active encapsulated drug) also resulted in two of 35 patients experiencing mild flushing and tightness of the chest that subsided when the infusion was temporarily stopped.[81] In another trial with non-radiolabeled PEGylated liposomes, three of nine patients displayed similar symptoms.[82] These reactions with the empty PEGylated liposomes even occurred at relatively low lipid doses (0.5 μmol/kg), suggesting a possible immune-mediated hypersensitivity or pseudoallergic reaction.[82] The role of liposomes in activating the C system and the subsequent opsonization of liposomes and C receptor-mediated endocytosis by macrophages is the predominant proposed mechanism for the occurrence of these NP-mediated hypersensitivity reactions.[83]

ZnO NPs are currently being used in a variety of consumer products and are being explored for their use as anticancer agents and in drug delivery technologies.[84] In a preclinical nanotoxicological study of mice receiving ZnO

NPs orally at sub-acute exposure doses of 300 mg/kg or 50 mg/kg, the 300 mg/kg group showed significantly ($p < 0.05$) higher levels of alanine aminotransferase (ALT) and alkaline phosphatase (ALP) compared to control mice.[85] A significant ($p < 0.05$) induction in lipid peroxidation was also observed at 300 mg/kg dose in the liver and kidney of these mice compared to the control group; however, the brain showed no increase in lipid peroxidation in these mice. The liver and kidney of the mice in the ZnO NPs at 300 mg/kg also showed pathological lesions of hepatocellular necrosis and cystic dilation of tubules, respectively, after 14 consecutive days of treatment. A statistically significant ($p < 0.05$) qualitative and quantitative increase in oxidative DNA damage was observed in the liver at 300 mg/kg dose in mice by the Fpg-modified comet assay; however, was no significant DNA damage was found in the kidney cells of these mice. Lastly, a significant ($p < 0.05$) percentage (35%) of TUNEL-positive cells was found throughout the liver section of mice treated with ZnO NPs at 300 mg/kg compared to the control mice. These results demonstrated that sub-acute oral exposure to ZnO NPs in mice leads to an accumulation of NPs in metabolic organs, with a preference for the liver, a primary organ of the MPS, resulting in oxidative stress-mediated DNA toxicity and apoptosis.[85]

Interestingly, the depletion of monocytes (*in vivo*) due to the administration of NP agents increases the toxicity of subsequently administered NP drugs. This effect was demonstrated in mice administered with liposomal dichloromethylene diphosphonate (L-DMDP) via the tail vein. There was a near-complete depletion of macrophage/PMN cell populations in the liver and spleen at 24 h.[86] Twenty-four hours after the injection of L-DMDP, free amphotericin B (Fungizone®) (0.5–4.0 mg/kg) was administered to half of the animals and liposome intercolated Amphotericin B (L-Amp B) (10–80 mg/kg) to the other half. There was a 2.6-fold increase in the toxicity (as measured by the median lethal dose, LD_{50}) of L-Amp B with the co-administration of L-DMDP treatment, as compared to the animals that did not receive L-DMDP. However, PMN/macrophage depletion via L-DMDP had no effect on the toxicity (LD_{50}) of Fungizone®. Twenty-four hours after the administration of L-Amp B, there were increases in the concentration of Amp B in the lungs (3.3-fold), spleens (3.2-fold), and kidneys (1.4-fold) of macrophage-depleted mice compared to concentrations in organs of control mice not treated with L-DMDP. L-Amp B is primarily known to be taken up by the MPS and accumulates in the lung, liver, and spleen tissues of animals.[87] The increase of L-Amp B in circulation in macrophage-depleted mice, as demonstrated by increased serum concentrations, could result from the depletion of resident macrophages/PMN cells from the liver and the spleen, and may give greater opportunity for the drug to interact with potentially sensitive cells, which could explain the enhanced toxicity observed in these animals.[88]

7. Translational and Clinical Relevance

7.1 *The MPS in tumors*

In studies comparing the disposition of PEGylated liposomal CKD-602 (S-CKD602) and non-liposomal CKD-602 (NL-CKD602) in mice bearing A375 human melanoma xenografts, S-CKD602 provided PK advantages in the plasma, tumor, and tumor ECF at one-thirtieth of the dose of NL-CKD602.[89] This was consistent with the improved antitumor efficacy of S-CKD602 in preclinical studies. Tumor exposure and antitumor activity of liposomal anticancer agents was found to be related to the presence of the MPS in tumors.[89] This was demonstrated in mice bearing SKOV-3 human ovarian and A375 human melanoma xenografts. The ratio of S-CKD602 in tumors to plasma was 1.7-fold higher in mice bearing SKOV-3 compared to A375. The ratio of released CKD-602 to S-CKD602 in tumors was twofold higher in mice bearing SKOV-3 than A375. The staining of MPS cells was 4.5-fold higher in SKOV-3 than in A375 ($p < 0.0001$). In addition, SKOV-3 was four-fold more sensitive to S-CKD602 compared to A375. The increased tumor delivery and release of CKD-602 from S-CKD602 in SKOV-3 ovarian compared to A375 melanoma xenografts was consistent with the increased staining of MPS cells in SKOV-3, suggesting that variability in the MPS may affect tumor disposition and the activity of nanosomal anticancer agents.[89,90]

7.2 *Clinical effects*

A phase I study looked at the relationship between the disposition of the carrier-mediated agent S-CKD602 (PEGylated liposomal CKD-602, a camptothecin analog) and changes in monocytes and absolute neutrophil count (ANC).[91] As a control, changes in monocytes and ANC were also evaluated after patients received non-liposomal CKD-602 in a phase I study of refractory solid tumors and a phase II study of refractory ovarian cancer.[91] The percent decrease in ANC and monocytes from baseline to nadir from the blood of all patients administered S-CKD602 in cycle 1 was measured. For S-CKD602, the mean ± S.D. days of ANC and monocyte nadir were 15.3 ± 5.0 and 8.6 ± 3.3 days, respectively ($p = 0.008$). For S-CKD602, the mean ± S.D. percent decreases in ANC and monocytes at nadir were 42 ± 30% and

58 ± 34%, respectively ($p = 0.001$). For S-CKD602, the ratio of percent decrease in monocytes to ANC at the nadir was 2.1 ± 2.0. The percent decreases in ANC and monocytes at nadir in the blood of patients were also assessed after the administration of non-liposomal CKD-602. For NL-CKD602, the mean ± S.D. days of ANC and monocyte nadir after the administration of non-liposomal CKD-602 were 14.0 ± 3.5 and 9.2 ± 3.0 days, respectively ($p > 0.05$). For NL-CKD602, after administration, the percent decreases in ANC and monocytes were 86 ± 11 and 87 ± 12%, respectively ($p > 0.05$). For NL-CKD602, the ratio of percent decrease in monocytes to ANC was 1.0 ± 0.2. In addition, the exposure (area under the curve, AUC) of encapsulated CKD-602 and percent decrease in monocytes at nadir showed a relationship among patients less than 60 years of age ($R^2 = 0.75$). However, there was no relationship between the AUC of released CKD-602 and percent decrease in monocytes at nadir, or between the AUC of encapsulated or released CKD-602 and percent decrease in ANC at nadir. The results of this first study evaluating the PK and PD relationships between a liposomal anticancer agent and immune cells in patients suggest that monocytes are more sensitive to S-CKD602 compared to neutrophils. The increased sensitivity also appears to be related to the liposomal formulation and not the released drug from the liposome or the small molecule formulation.

To evaluate how the monocytes can affect the PK of S-CKD602, the relationship between percent decrease of monocytes, clearance of encapsulated drug, and the release of CKD-602 from its carrier in plasma was evaluated.[30] In all patients ($R^2 = 0.51$), i.e., those < 60 years of age ($R^2 = 0.54$) and those ≥60 years of age ($R^2 = 0.49$), there was a linear relationship between percent decrease in monocytes and encapsulated CKD-602 clearance [Figure 4(a)]. In addition, there was a correlation between percent decrease in monocytes and the AUC of released CKD-602, as shown in Figure 4(b).[91]

Some of the significant variability in the pharmacokinetic dispositions of NP agents versus small molecules was found to be related to linear and nonlinear (saturable) clearance of S-CKD602 in patients.[92] In this study, despite being treated with the same dose of 1.7 mg/m^2, patients were found to have linear and nonlinear clearance.[31] Figure 5 shows an example of each of these profiles.

Those with linear clearance had a plasma concentration versus time profile of CKD-602 with a distribution phase and an elimination phase. The area under the concentration–time curve $(AUC)_{0-\infty}$ values of sum-total, encapsulated, and released CKD-602 in this patient were 4,933, 4,646, and 43 ng/ml·h, respectively. For patients with nonlinear clearance, the plasma concentration versus time profile of CKD-602 yielded little change from 0 to 24 h.

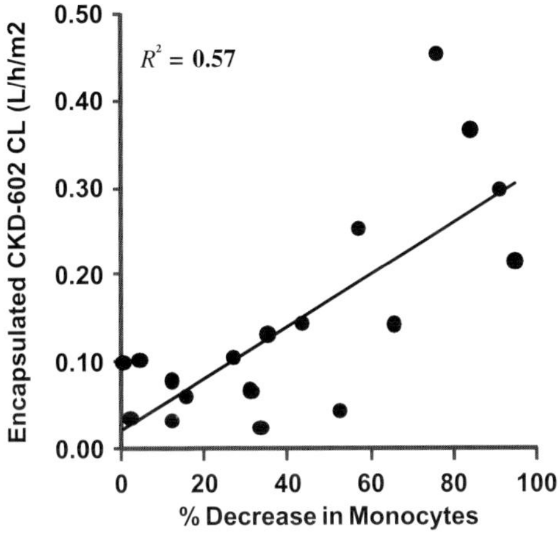

(a)

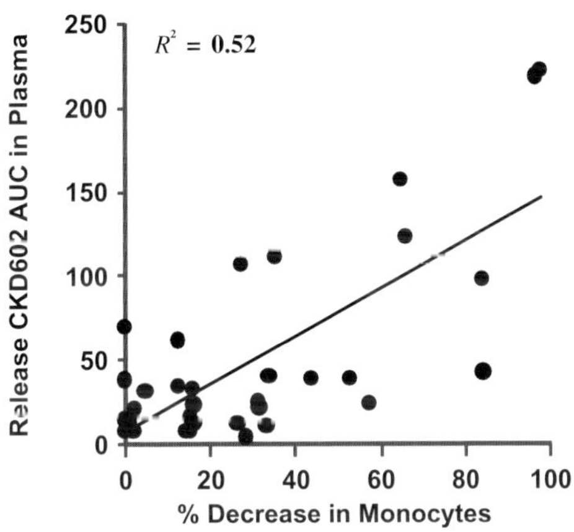

(b)

Figure 4. Graphical representation of the effect of monocytes on the PK of S-CKD602. The relationship between percent decrease in monocytes at the nadir and the clearance of encapsulated CKD-602 in plasma. Linear regression ($R^2 = 0.57$) is represented by the solid line. (b) The relationship between percent decrease in monocytes at the nadir and the AUC of released CKD-602 in monocytes at the nadir for each patient was calculated from using each patient's serial monocyte count obtained during cycle 1 and from using the standard formula: [(Prevalue-nadir)/Prevalue] × 100.

The $AUC_{0-\infty}$ values of sum-total, encapsulated, and released CKD-602 in this patient were 50,700, 52,668, and 152 ng/ml·h, respectively. The incidence of linear or nonlinear clearance was found to be associated with the dose of S-CKD602 administered. At doses from 0.1 to 1.1 mg/m^2, the S-CKD602 sum-total and encapsulated CKD-602 plasma concentrations were best described using a model with linear clearance in all the patients (n = 33). At doses of

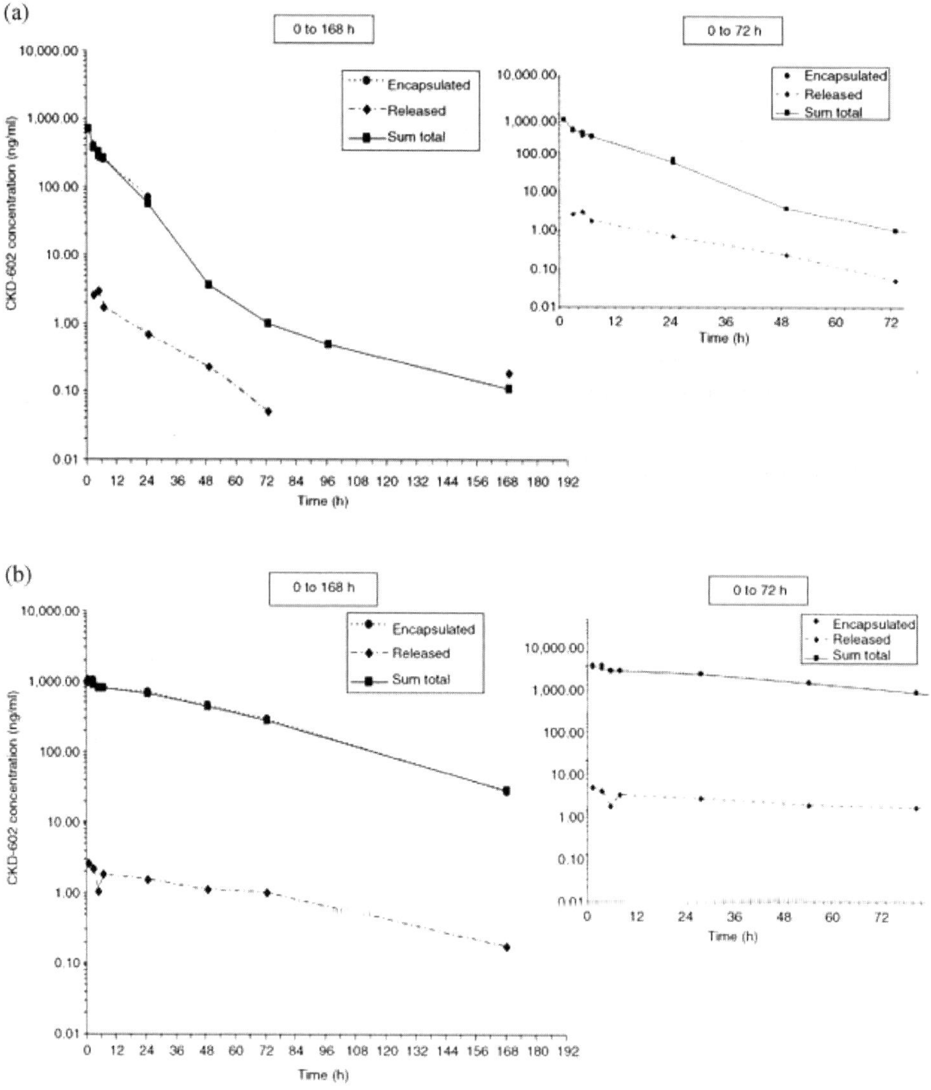

Figure 5. Representative concentration versus time profiles of sum-total, encapsulated, and released CKD-602 in patients with linear and nonlinear clearance. (a) Data from patient (#141) with linear clearance, treated with S-CKD602 at a dose of 1.7 mg/m^2. (b) Data from patient (#142) with nonlinear clearance treated with S-CKD602 at a dose of 1.7 mg/m^2. (a) and (b) represent the concentration *vs.* time profile from 0 to 168 h and 0 to 72 h, respectively. The sum-total, encapsulated, and released CKD-602 AUCs in the patient represented in (a) were 4,933, 4,646, and 43 ng/ml·h, respectively, and the corresponding values in the patient represented in (b) were 50,700, 52,668, and 152 ng/ml·h.

1.7–2.5 mg/m^2, the S-CKD602 sum-total and encapsulated CKD-602 plasma concentrations were best described using a model with linear (n = 2) and nonlinear (n = 10) clearances.[92] The observation of nonlinear PK at higher doses of NP agents may indicate MPS uptake in particle clearance and its subsequent saturation. The occurrence of linear and nonlinear clearance has also been reported for Doxil®.[93]

Some of the interpatient variability in S-CKD602 PK disposition could possibly be explained by the presence of metastatic liver tumors, as the V_{max} in patients with these tumors is 1.5-fold higher than in patients without liver tumors.[94] The presence of liver tumors led to a 35% lower plasma exposure of encapsulated CKD-602, which is in contrast to most studies which showed a decrease in clearance of small molecule drugs in patients with liver tumors compared to those without liver tumors.[95,96]

8. Final Remarks

This chapter discusses just a fraction of studies of NP uptake by the MPS. This field of research is rapidly developing and constantly evolving. It is key to note that many NP agents attach to traditional small molecules and deliver them in a novel way in which we observe therapeutically greater exposure and efficacy in the body. While the advantages to administering an NP agent are numerous, we encounter a much different host response than we do with small molecule administration. The network of opsonins and circulating and tissue phagocytes is very complex and their interaction with NPs often depends on the biological atmosphere as well as the physicochemical properties of the NP.

Some key concepts introduced in this chapter include animal models that are not always equivalent to patient models, and that there is bidirectional interaction between the MPS and NPs. Animal models are not always predictive of patients in terms of NP PK, likely due to differences in the MPS. Differences were seen between animals and humans by the ABC phenomenon, clearance of the NP, and *in vitro* and *in vivo* toxicities resulting from NPs. Additionally, the MPS is responsible for taking up and clearing NPs from the systemic circulation, but this interaction is not unidirectional. Introduced through the concepts of "hijacking" or "capture," as well as uptake and clearance dependent on cell and NP physicochemical properties, this is a complex bidirectional system in which both key players have profound effects on each other. In the future, it is likely that we will see the use of MPS profiling in order to individualize NP therapy in patients, as depicted in Figure 6.

If we can determine the level and extent of MPS interaction with NPs, we could select therapies and doses for patients accordingly. In addition to

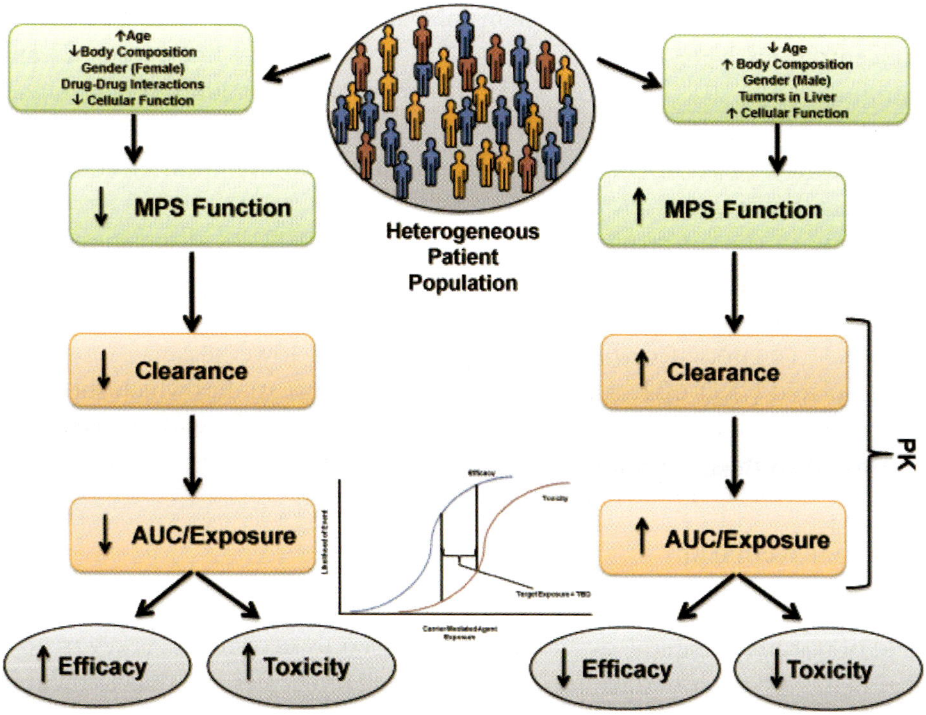

Figure 6. A growing body of evidence indicates that many different factors contribute to the variability in the PK and PD of NPs given to patients. These factors include age, body composition, gender, cellular function, and concomitant medications. They operate on the level of the MPS and have the ability to either up- or downregulate its function, thereby affecting drug clearance, and area under the curve (AUC)/exposure. This will ultimately lead to reduced or greater efficacy and/or toxicity. The target response must balance efficacy and toxicity, creating an ideal NP exposure window that needs to be elucidated through further research.

profiling the MPS through the measurement of cell function, we could seek to identify the various mediators of MPS function. Perhaps we would be able to ultimately observe the intricate network of chemokines, cytokines, complement, and even hormones that all regulate MPS function. Developing probes for MPS function and these mediators will be crucial to predict and individualize NP PK and PD.

References

1. Saba TM. Physiology and physiopathology of the reticuloendothelial system. *Arch Intern Med* 1970;126:1031–1052.
2. Zamboni WC. Concept and clinical evaluation of carrier-mediated anticancer agents. *Oncologist* 2008;13:248–260.

3. Dobrovolskaia MA, Aggarwal P, Hall JB, McNeil SE. Preclinical studies to understand nanoparticle interaction with the immune system and its potential effects on nanoparticle biodistribution. *Mol Pharm* 2008;5:487–495.
4. Ruiz-Garcia A, Bermejo M, Moss A, Casabo VG. Pharmacokinetics in drug discovery. *J Pharm Sci* 2008;97:654–690.
5. Laginha K, Mumbengegwi D, Allen T. Liposomes targeted via two different antibodies: assay, B-cell binding and cytotoxicity. *Biochim Biophys Acta* 2005;1711:25–32.
6. Yurkovetskiy AV, Hiller A, Syed S, *et al.* Synthesis of a macromolecular camptothecin conjugate with dual phase drug release. *Mol Pharm* 2004;1:375–382.
7. Zamboni WC. Liposomal, nanoparticle, and conjugated formulations of anticancer agents. *Clin Cancer Res* 2005;11:8230–8234.
8. Dobrovolskaia MA, McNeil SE. Immunological properties of engineered nanomaterials. *Nat Nanotechnol* 2007;2:469–478.
9. Franca A, Aggarwal P, Barsov EV, Kozlov SV, Dobrovolskaia MA, Gonzalez-Fernandez A. Macrophage scavenger receptor A mediates the uptake of gold colloids by macrophages in vitro. *Nanomedicine (Lond)* 2011;6:1175–1188.
10. Dobrovolskaia MA, Patri AK, Zheng J, *et al.* Interaction of colloidal gold nanoparticles with human blood: Effects on particle size and analysis of plasma protein binding profiles. *Nanomedicine* 2009;5:106–117.
11. Vonarbourg A, Passirani C, Saulnier P, Benoit JP. Parameters influencing the stealthiness of colloidal drug delivery systems. *Biomaterials* 2006;27:4356–4373.
12. Gough PJ, Gordon S. The role of scavenger receptors in the innate immune system. *Microbes Infect* 2000;2:305–311.
13. Guo C, Yi H, Yu X, *et al.* Absence of scavenger receptor A promotes dendritic cell-mediated cross-presentation of cell-associated antigen and antitumor immune response. *Immunol Cell Biol* 2012;90:101–108.
14. Areschoug T, Gordon S. Scavenger receptors: role in innate immunity and microbial pathogenesis. *Cell Microbiol* 2009;11:1160–1169.
15. Valacchi G, Sticozzi C, Lim Y, Pecorelli A. Scavenger receptor class B type I: A multifunctional receptor. *Ann N Y Acad Sci* 2011;1229:E1–E7.
16. Swanson JA. Shaping cups into phagosomes and macropinosomes. *Nat Rev Mol Cell Biol* 2008;9:639–649.
17. Allen LA, Aderem A. Mechanisms of phagocytosis. *Curr Opin Immunol* 1996; 8:36–40.
18. Chen H, Langer R, Edwards DA. A film tension theory of phagocytosis. *J Colloid Interface Sci* 1997;190:118–133.
19. Alberts B, Johnson A, Lewis J, Raff M, Roberts K, Walter P (eds.), *Molecular Biology of the Cell*, Garland Publishing, 2007.
20. Panyam J, Labhasetwar V. Dynamics of endocytosis and exocytosis of poly(D,L-lactide-co-glycolide) nanoparticles in vascular smooth muscle cells. *Pharm Res* 2003;20:212–220.
21. Tabata Y, Ikada Y. Phagocytosis of polymer microspheres by macrophages. *Adv Polym Sci* 1990;94:107–141.
22. Patel H, Moghimi S. Serum-mediated recognition of liposomes by phagocytic cells of the reticuloendothelial system — the concept of tissue specificity. *Adv Drug Deliv Rev* 1998;32:45–60.
23. Kamps JA, Scherphof GL. Receptor versus non-receptor mediated clearance of liposomes. *Adv Drug Deliv Rev* 1998;32:81–97.

24. Brown MS, Goldstein JL. Lipoprotein metabolism in the macrophage: implications for cholesterol deposition in atherosclerosis. *Annu Rev Biochem* 1983;52:223–261.
25. Linehan SA, Martinez-Pomares L, Gordon S. Mannose receptor and scavenger receptor: Two macrophage pattern recognition receptors with diverse functions in tissue homeostasis and host defense. *Adv Exp Med Biol* 2000;479:1–14.
26. Chnari E, Nikitczuk JS, Wang J, Uhrich KE, Moghe PV. Engineered polymeric nanoparticles for receptor-targeted blockage of oxidized low density lipoprotein uptake and atherogenesis in macrophages. *Biomacromolecules* 2006;7:1796–1805.
27. Moghimi SM, Hunter AC. Recognition by macrophages and liver cells of opsonized phospholipid vesicles and phospholipid headgroups. *Pharm Res* 2001;18:1–8.
28. Dijkstra J, van Galen M, Scherphof G. Influence of liposome charge on the association of liposomes with Kupffer cells in vitro. Effects of divalent cations and competition with latex particles. *Biochim Biophys Acta* 1985;813:287–297.
29. Kamps JA, Morselt HW, Scherphof GL. Uptake of liposomes containing phosphatidylserine by liver cells *in vivo* and by sinusoidal liver cells in primary culture: *In vivo-in vitro* differences. *Biochem Biophys Res Commun* 1999;256:57–62.
30. Liu F, Liu D. Serum independent liposome uptake by mouse liver. *Biochim Biophys Acta* 1996;1278:5–11.
31. Liu D, Hu Q, Song YK. Liposome clearance from blood: Different animal species have different mechanisms. *Biochim Biophys Acta* 1995;1240:277–284.
32. Coico R, Sunshine G, Benjamini E (eds.), *Immunology: A short course.* John Wiley & Sons, Inc., 2003.
33. Szebeni J, Bedocs P, Rozsnyay Z, *et al.* Liposome-induced complement activation and related cardiopulmonary distress in pigs: Factors promoting reactogenicity of Doxil and AmBisome. *Nanomedicine* 2012;8:176–184.
34. Hamad I, Al-Hanbali O, Hunter AC, Rutt KJ, Andresen TL, Moghimi SM. Distinct polymer architecture mediates switching of complement activation pathways at the nanosphere-serum interface: Implications for stealth nanoparticle engineering. *ACS Nano* 2010;4:6629–6638.
35. Ishida T, Kiwada H. Accelerated blood clearance (ABC) phenomenon upon repeated injection of PEGylated liposomes. *Int J Pharm* 2008;354:56–62.
36. Ishida T, Harada M, Wang XY, Ichihara M, Irimura K, Kiwada H. Accelerated blood clearance of PEGylated liposomes following preceding liposome injection: Effects of lipid dose and PEG surface-density and chain length of the first-dose liposomes. *J Controlled Release* 2005;105:305–317.
37. Laverman P, Carstens MG, Boerman OC, *et al.* Factors affecting the accelerated blood clearance of polyethylene glycol-liposomes upon repeated injection. *J Pharmacol Exp Ther* 2001;298:607–612.
38. Koide H, Asai T, Hatanaka K, *et al.* T cell-independent B cell response is responsible for ABC phenomenon induced by repeated injection of PEGylated liposomes. *Int J Pharm* 2010;392:218–223.
39. La-Beck NM, Zamboni BA, Gabizon A, *et al.* Factors affecting the pharmacokinetics of PEGylated liposomal doxorubicin in patients. *Cancer Chemother Pharmacol* 2011.
40. Caron WP, Clewell H, Dedrick R, *et al.* Allometric scaling of PEGylated liposomal anticancer drugs. *J Pharmacokinet Pharmacodyn* 2011;38: 653–669.
41. Thomas SN, van der Vlies AJ, O'Neil CP, *et al.* Engineering complement activation on polypropylene sulfide vaccine nanoparticles. *Biomaterials* 2011;32:2194–2203.

42. Doshi N, Mitragotri S. Macrophages recognize size and shape of their targets. *PLoS One* 2010;5:e10051.
43. Champion JA, Mitragotri S. Role of target geometry in phagocytosis. *Proc Natl Acad Sci U S A* 2006;103:4930–4934.
44. Adams DO, Hamilton TA. The cell biology of macrophage activation. *Annu Rev Immunol* 1984;2:283–318.
45. Roerdink F, Wassef NM, Richardson EC, Alving CR. Effects of negatively charged lipids on phagocytosis of liposomes opsonized by complement. *Biochim Biophys Acta* 1983;734:33–39.
46. Maderna P, Godson C. Phagocytosis of apoptotic cells and the resolution of inflammation. *Biochim Biophys Acta* 2003;1639:141–151.
47. Yu B, Hailman E, Wright SD. Lipopolysaccharide binding protein and soluble CD14 catalyze exchange of phospholipids. *J Clin Invest* 1997;99:315–324.
48. Shibuya-Fujiwara N, Hirayama F, Ogata Y, Ikeda H, Ikebuchi K. Phagocytosis *in vitro* of polyethylene glycol-modified liposome-encapsulated hemoglobin by human peripheral blood monocytes plus macrophages through scavenger receptors. *Life Sci* 2001;70:291–300.
49. Li SD, Huang L. Pharmacokinetics and biodistribution of nanoparticles. *Mol Pharm* 2008;5:496–504.
50. Allen TM, Hansen C. Pharmacokinetics of stealth versus conventional liposomes: Effect of dose. *Biochim Biophys Acta* 1991;1068:133–141.
51. Szebeni J, Baranyi L, Savay S, *et al.* The interaction of liposomes with the complement system: *In vitro* and *in vivo* assays. *Methods Enzymol* 2003;373:136–154.
52. Szebeni J. The interaction of liposomes with the complement system. *Crit Rev Ther Drug Carrier Syst* 1998;15:57–88.
53. Torchilin VP, Shtilman MI, Trubetskoy VS, Whiteman K, Milstein AM. Amphiphilic vinyl polymers effectively prolong liposome circulation time *in vivo*. *Biochim Biophys Acta* 1994;1195:181–184.
54. Woodle MC, Lasic DD. Sterically stabilized liposomes. *Biochim Biophys Acta* 1992;1113:171–199.
55. Li SD, Chen YC, Hackett MJ, Huang L. Tumor-targeted delivery of siRNA by self-assembled nanoparticles. *Mol Ther* 2008;16:163–169.
56. Hamad I, Hunter AC, Szebeni J, Moghimi SM. Poly(ethylene glycol)s generate complement activation products in human serum through increased alternative pathway turnover and a MASP-2-dependent process. *Mol Immunol* 2008;46:225–232.
57. Moghimi SM, Hamad I, Andresen TL, Jorgensen K, Szebeni J. Methylation of the phosphate oxygen moiety of phospholipid-methoxy(polyethylene glycol) conjugate prevents PEGylated liposome-mediated complement activation and anaphylatoxin production. *FASEB J* 2006;20:2591–2593.
58. Moghimi SM, Szebeni J. Stealth liposomes and long circulating nanoparticles: Critical issues in pharmacokinetics, opsonization and protein-binding properties. *Prog Lipid Res* 2003;42:463–478.
59. Pham CT, Mitchell LM, Huang JL, *et al.* Variable antibody-dependent activation of complement by functionalized phospholipid nanoparticle surfaces. *J Biol Chem* 2011;286:123–130.
60. Senior JH. Fate and behavior of liposomes *in vivo*: A review of controlling factors. *Crit Rev Ther Drug Carrier Syst* 1987;3:123–193.

61. Harashima H, Sakata K, Funato K, Kiwada H. Enhanced hepatic uptake of liposomes through complement activation depending on the size of liposomes. *Pharm Res* 1994;11:402–406.
62. Devine DV, Wong K, Serrano K, Chonn A, Cullis PR. Liposome-complement interactions in rat serum: Implications for liposome survival studies. *Biochim Biophys Acta* 1994;1191:43–51.
63. Harashima H, Huong TM, Ishida T, Manabe Y, Matsuo H, Kiwada H. Synergistic effect between size and cholesterol content in the enhanced hepatic uptake clearance of liposomes through complement activation in rats. *Pharm Res* 1996;13:1704–1709.
64. Harashima H, Hiraiwa T, Ochi Y, Kiwada H. Size dependent liposome degradation in blood: *In vivo/in vitro* correlation by kinetic modeling. *J Drug Target* 1995;3:253–261.
65. Gratton SE, Ropp PA, Pohlhaus PD, *et al.* The effect of particle design on cellular internalization pathways. *Proc Natl Acad Sci U S A* 2008;105:11613–11618.
66. Luck M, Paulke BR, Schroder W, Blunk T, Muller RH. Analysis of plasma protein adsorption on polymeric nanoparticles with different surface characteristics. *J Biomed Mater Res* 1998;39:478–485.
67. Jeon S, Lee J, Andrade J, De Gennes P. Protein-surface interactions in the presence of polyethylene oxide. *J Colloid Interface Sci* 1991;142:159–66.
68. Nagaoka S, Mori Y, Tanzawa H, Nishiumi S. Interaction between blood components and hydrogelswith poly(oxyethylene) chains. In: Hoffman A, Ratner D, Horbett T (eds.), *Polymers as Biomaterials*, Plenum Press, 1984, pp. 361–374.
69. Wilhelmi V, Fischer U, van Berlo D, Schulze-Osthoff K, Schins RP, Albrecht C. Evaluation of apoptosis induced by nanoparticles and fine particles in RAW 264.7 macrophages: Facts and artefacts. *Toxicol In Vitro* 2012;26:323–334.
70. Jeong YS, Oh WK, Kim S, Jang J. Cellular uptake, cytotoxicity, and ROS generation with silica/conducting polymer core/shell nanospheres. *Biomaterials* 2011;32:7217–7225.
71. Hanley C, Thurber A, Hanna C, Punnoose A, Zhang J, Wingett DG. The influences of cell type and ZnO nanoparticle size on immune cell cytotoxicity and cytokine induction. *Nanoscale Res Lett* 2009; 4:1409–1420.
72. Ewer MS, Martin FJ, Henderson C, Shapiro CL, Benjamin RS, Gabizon AA. Cardiac safety of liposomal anthracyclines. *Semin Oncol* 2004;31:161–181.
73. Safra T. Cardiac safety of liposomal anthracyclines. *Oncologist* 2003;8(Suppl 2):17–24.
74. Gabizon A, Catane R, Uziely B, *et al.* Prolonged circulation time and enhanced accumulation in malignant exudates of doxorubicin encapsulated in polyethylene-glycol coated liposomes. *Cancer Res* 1994;54:987–992.
75. Daemen T, Hofstede G, Ten Kate MT, Bakker-Woudenberg IA, Scherphof GL. Liposomal doxorubicin-induced toxicity: Depletion and impairment of phagocytic activity of liver macrophages. *Int J Cancer* 1995;61:716–721.
76. Zito MA, Koennecke LA, McAuliffe MJ, McNally B, van Rooijen N, Heyes MP. Depletion of systemic macrophages by liposome-encapsulated clodronate attenuates striatal macrophage invasion and neurodegeneration following local endotoxin infusion in gerbils. *Brain Res* 2001;892:13–26.
77. Phillips NC. Kupffer cells and liver metastasis. Optimization and limitation of activation of tumoricidal activity. *Cancer Metastasis Rev* 1989;8:231–252.
78. Laing RB, Milne LJ, Leen CL, Malcolm GP, Steers AJ. Anaphylactic reactions to liposomal amphotericin. *Lancet* 1994;344:682.

79. Huang JY, Wu CH, Shih IH, Lai PC. Complications mimicking lupus flare-up in a uremic patient undergoing PEGylated liposomal doxorubicin therapy for cervical cancer. *Anticancer Drugs* 2004;15:239–241.
80. Chanan-Khan A, Szebeni J, Savay S, *et al.* Complement activation following first exposure to PEGylated liposomal doxorubicin (Doxil): Possible role in hypersensitivity reactions. *Ann Oncol* 2003;14:1430–1437.
81. Dams ET, Oyen WJ, Boerman OC, *et al.* 99mTc-PEG liposomes for the scintigraphic detection of infection and inflammation: Clinical evaluation. *J Nucl Med* 2000;41: 622–630.
82. Szebeni J, Baranyi L, Savay S, *et al.* Liposome-induced pulmonary hypertension: Properties and mechanism of a complement-mediated pseudoallergic reaction. *Am J Physiol Heart Circ Physiol* 2000;279:H1319–28.
83. Ishida T, Kiwada H. Unexpected reactions by *in vivo* applications of PEGylated liposomes. In: Webster T (ed.), *Safety of Nanoparticles: Nanostructure Science and Technology*, Springer Science and Business Media, 2009.
84. Rasmussen JW, Martinez E, Louka P, Wingett DG. Zinc oxide nanoparticles for selective destruction of tumor cells and potential for drug delivery applications. *Expert Opin Drug Deliv* 2010;7:1063–1077.
85. Sharma V, Singh P, Pandey AK, Dhawan A. Induction of oxidative stress, DNA damage and apoptosis in mouse liver after sub-acute oral exposure to zinc oxide nanoparticles. *Mutat Res* 2011;745: 84–91.
86. Moonis M, Ahmad I, Bachhawat BK. Effect of elimination of phagocytic cells by liposomal dichloromethylene diphosphonate on aspergillosis virulence and toxicity of liposomal amphotericin B in mice. *J Antimicrob Chemother* 1994;33:571–583.
87. Lopez-Berestein G, Rosenblum MG, Mehta R. Altered tissue distribution of amphotericin B by liposomal encapsulation: Comparison of normal mice to mice infected with *Candida albicans*. *Cancer Drug Deliv* 1984;1:199–205.
88. Szoka FC. Jr, Milholland D, Barza M. Effect of lipid composition and liposome size on toxicity and *in vitro* fungicidal activity of liposome-intercalated amphotericin B. *Antimicrob Agents Chemother* 1987;31:421–429
89. Zamboni WC, Strychor S, Joseph E, *et al.* Plasma, tumor, and tissue disposition of STEALTH liposomal CKD-602 (S-CKD602) and nonliposomal CKD-602 in mice bearing A375 human melanoma xenografts. *Clin Cancer Res* 2007;13:7217–7223.
90. Zamboni WC, Strychor S, Joseph E, Parise RA, Egorin MJ, Eiseman JL. Tumor, tissue, and plasma pharmacokinetic studies and antitumor response studies of docetaxel in combination with 9-nitrocamptothecin in mice bearing SKOV-3 human ovarian xenografts. *Cancer Chemother Pharmacol* 2008;62:417–426.
91. Zamboni WC, Maruca LJ, Strychor S, *et al.* Bidirectional pharmacodynamic interaction between pegylated liposomal CKD-602 (S-CKD602) and monocytes in patients with refractory solid tumors. *J Liposome Res* 2011;21:158–165.
92. Zamboni WC, Strychor S, Maruca L, *et al.* Pharmacokinetic study of PEGylated liposomal CKD-602 (S-CKD602) in patients with advanced malignancies. *Clin Pharmacol Ther* 2009;86:519–526.
93. Martin FJ. PEGylated liposomal doxorubicin: Scientific rationale and preclinical pharmacology. *Oncology* 1997;11:11–20.
94. Wu H, Ramanathan RK, Zamboni BA, *et al.* Population pharmacokinetics of PEGylated liposomal CKD-602 (S-CKD602) in patients with advanced malignancies. *J Clin Pharmacol* 2011.

95. Robieux I, Sorio R, Borsatti E, *et al.* Pharmacokinetics of vinorelbine in patients with liver metastases. *Clin Pharmacol Ther* 1996;59:32–40.
96. Twelves CJ, O'Reilly SM, Coleman RE, Richards MA, Rubens RD. Weekly epirubicin for breast cancer with liver metastases and abnormal liver biochemistry. *Br J Cancer* 1989;60:938–941.

Chapter 13

The Effects of Nanoparticles on Dendritic Cells

Valentyna Fesenkova

The Wallace H. Coulter
Department of Medical Engineering
Georgia Institute of Technology
315 First Drive, Atlanta
GA 30332, USA
vfesenkova@yahoo.com

Dendritic cells (DCs) are professional antigen-presenting cells (APCs) that induce the primary immune response and are responsible for the activation of T and B cells. The main function of DCs is to serve as a link between the innate and adaptive immunity. Understanding the role of DCs in controlling the immune system can provide powerful tools for the successful manipulation of immune cells to optimize vaccine delivery and to expand possibilities of immunotherapeutic approaches. Nanoparticle (NP) applications for targeted delivery of antigens to the DCs is a promising strategy for vaccine development. Herein, I present an overview of different NPs and approaches of their delivery to DCs, review the interaction between NPs and DCs, NP effects on the maturation, activation, and function of DCs, and outline future directions for NP-enhanced immune control to further foster the development of successful vaccination strategies and cancer immunotherapies.

1. Introduction

Nanoparticles (NPs) are particulate materials with a size of 100 nm or less in at least one dimension. Many conventional materials can change physical properties when made in the nanoscale size and they create a bridge between bulk materials and atomic or molecular structures. NPs present great scientific interest for many applications in the fields of biological sciences and medicine. Some of them are already successfully used in medicine, while many others are being developed for targeted drug delivery, imaging contrast agents, and vaccine development with optimized antigen delivery to antigen-presenting cells (APCs).[1] The success of biomedical applications of NPs depends on their interactions with various components of the immune system. One specific area of interest includes NP interaction with and their effects on APCs called dendritic cells (DCs). A variety of NPs have been designed with different properties aimed at targeting DCs.[2] DCs are potent APCs that initiate and modulate the immune response. They control the function of B and T lymphocytes. While B cells can directly recognize the antigen through the B cell receptor, T cells require the antigen to be processed and presented to them by APCs with molecules of the major histocompatibility complex (MHC) on the cell surface. There are two types of MHC — MHC-I and MHC-II. MHC class I antigens are loaded upon the processing of intracellular pathogens by the APCs through either endogenous pathways or in the case of particulate antigens and immune complexes, through exogenous pathways. Peptides generated in the proteasome are loaded onto MHC class I molecules within the endoplasmic reticulum. Soluble antigens are captured through macropinocytosis and after their degradation in endosomes, resultant polypeptides are transported into MHC class II-rich compartments where they are loaded onto MHC class II molecules (Figure 1).

Immature DCs located at many tissues efficiently capture and process antigens and consequently, display large amounts of MHC–peptide complexes on their surface. During these processes, they undergo maturation and migrate to lymphoid organs such as the spleen and lymph nodes, where they activate T cells. Maturation is accompanied by the upregulation of co-stimulatory molecules. The peptide-binding proteins MHC class I and MHC class II on the cell surface DCs stimulate different subsets of T lymphocytes, including cytotoxic T cells and helper T cells, respectively.

There are two types of DCs — myeloid DCs (mDCs) and plasmacytoid DCs (pDCs). Hematopoietic stem cells ($CD34^+$) differentiate into common myeloid progenitors (CMPs) and common lymphoid progenitors (CLPs), which give rise to mDCs and pDCs, respectively. mDCs share some properties of monocytes, such as IL-12 production and the strong induction of T helper 1 (Th1) and cytotoxic T lymphocytes (CTL) responses. pDCs are responsible for the production of type I interferon (IFN-α/β) and induction of Th2 response. They have plasma cell-like morphology.[3]

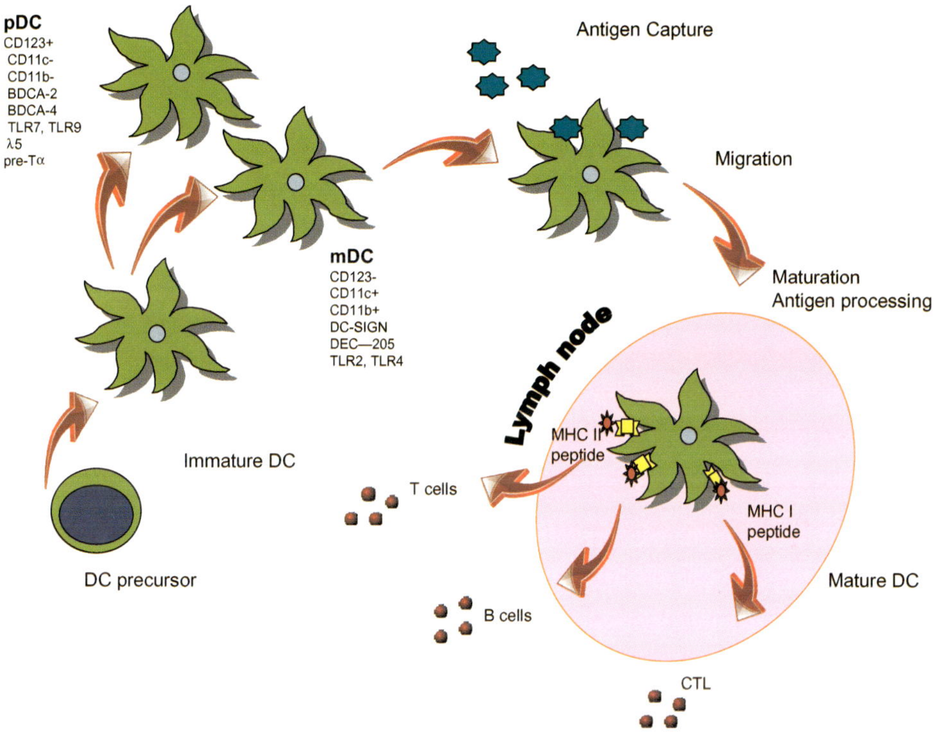

Figure 1. DC properties. The two types of DCs — plasmacytoid DCs (pDCs) and myeloid DCs (mDCs). Steps of DC maturation: 1) Antigen capture, 2) increased expression of co-stimulatory molecules, and 3) the initiation of an immune response.

Not only do mDCs and pDCs have different repertoires of surface molecules (Table 1), migration routes of immature pDCs and mDCs are also different. After capturing pathogens in peripheral tissues, maturing mDCs migrate to T cell zones of lymphatic tissues. pDCs express CD62L and CXCR3 and like the naïve lymphocytes, they migrate to lymphatic tissues through the hematogenous route, where they reside in T cell zones around high endothelial venules.

2. The Role of Dendritic Cells in Linking the Innate and Adaptive Immunity and Their Effector Functions in Antitumor Immune Response

Activation of the innate immune system is often triggered by the immune cell recognition of pathogen-associated molecular patterns (PAMPs). PAMPs are molecules that represent highly conservative motives on the surface of pathogens. DCs possess the pattern recognition receptors (PRRs), such as Toll-like

Table 1. Plasmacytoid and myeloid DCs have different markers expressed on their respective surfaces.

pDC	mDC
CD123$^+$	CD123$^-$
CD11c$^-$	CD11c$^+$
CD11b$^-$	CD11b$^+$
BDCA-2	DEC205
BDCA-4	DC-SIGN
TLR7	TLR2
TLR9	TLR4
λ5	
preTα	

Abbreviations

BDCA	C-type lectin receptor (CLR)
DC-SIGN	dendritic cell-specific inter-cellular adhesion molecule-3-grabbing non-integrin, also known as CD209
DEC	phagocytic receptor-mediating antigen uptake
TLR	Toll-like receptor

receptors (TLRs), which are expressed on immature DCs and can sense and bind antigens. After pathogen recognition, immature DCs undergo maturation. Maturation is a complex process involving multiple changes such as 1) an increase in the surface expression of MHC class I- and class II-peptide complexes, 2) the upregulation of the expression of adhesion and co-stimulatory molecules (CD40, CD54, CD80, CD86), 3) the production of pro-inflammatory cytokines such as IL-1, IL-6, IL-12, IL10, IL-23, and 4) the expression of chemokine receptor CCR7, which is responsible for DC migration to the lymphatic system.[4]

The repertoires of PRRs are different between mDCs and pDCs. mDCs respond to bacterial infection through sensing bacterial peptidoglycans and endotoxin via TLR2 and TLR4, while pDCs express TLR7, TLR8, and TLR9, recognizing viral DNA and RNA.[3] pDCs produce large amounts of IFN-α/β in response to TLR activation. The production of IFN-α/β increases the cytotoxicity of natural killer (NK) cells and CD8$^+$ T cells. The secretion of IL-6 and IL-12 cytokines by pDCs induces IFN-γ production in NK cells, CD8$^+$ T cells, and CD4$^+$ T helper cells. This subset of DCs promotes the differentiation of B cells into plasma cells, which produce anti-viral antibodies (Figure 2).

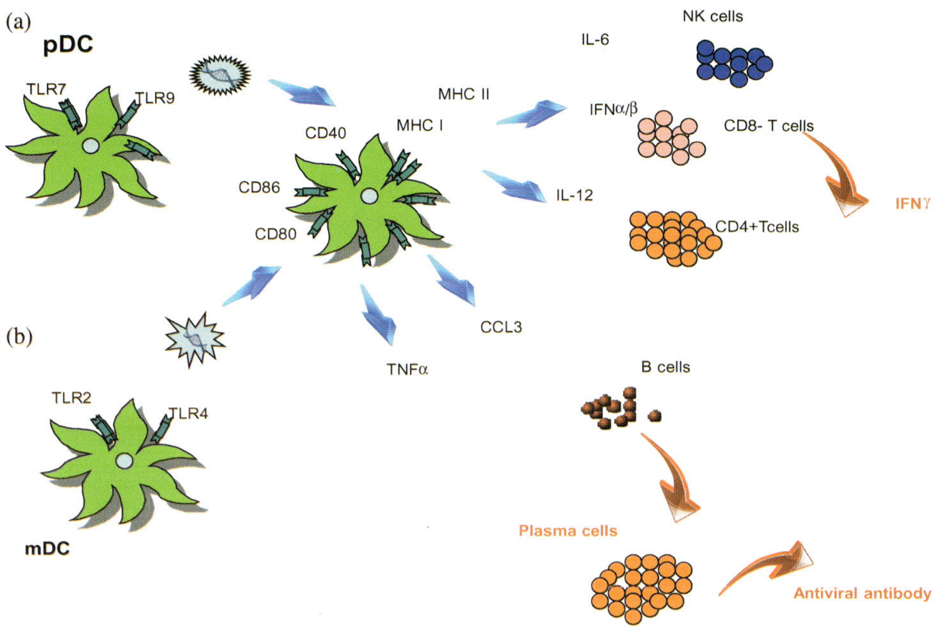

Figure 2. DC immune response. (a) Antiviral activity of pDCs: pDCs are responsible for the maintenance of the innate immunity, which accounts for their antiviral activity. pDCs recognize viral DNA and RNA via TLR7 and TLR9, increase the expression of co-stimulatory molecules, and produce large amounts of IFN-α/β, IL-6, and IL-12. (b) mDCs initiate an adaptive immune response through the activation of TLR2 and TLR4 in response to bacterial infection. Upon encounter with a pathogen, mDCs are stimulated, increasing the expression of co-stimulatory molecules and production of cytokines. After activation, pDCs and mDCs induce a variety of effector cells (NK cells, $CD8^+$ T cells, $CD4^+$ T cells, and B cells), which is followed by the development of a strong immune response against invading pathogens.

Immature DCs can also recognize tumor antigens. They capture cellular apoptotic bodies, followed by maturation, and display tumor antigens to tumor-specific lymphocytes $CD8^+$ T cells.[5,6]

Activated immature DCs also produce IFN-α, which in turn stimulates a variety of effector cells — macrophages, NK cells, and NK T (NKT) cells. These effector cells are responsible for targeting and killing tumor cells. CTLs and NK cells are the first line of cellular defense against invading cancer cells. They are activated through the interaction with tumor antigen-loaded mature DCs and are cytotoxic to target tumor cells.[7]

The host antitumor immune response is an important mechanism in the protection against cancer. However, the response of the immune system to cancer cells is often inadequate because tumor cells use different mechanisms to avoid immune control. Defective DCs and their abnormal response to cancer antigens represent one of such mechanisms. Several groups have reported the defective function of DCs in tumor-bearing mice and in cancer patients.[7–11] Defective DCs also have low expression levels of MHC class II and co-stimulatory molecules CD40, CD80, and CD86, as well as some adhesion molecules (e.g., VCAM-1). These cells are unable to induce effective peptide-specific and antitumor cytotoxic immune responses.[7,12]

In addition, IL-10 can convert the antigen-presenting function of DCs to the induction of antigen-specific anergy, which can lead to a state of tolerance against tumor tissues.[7,13–15] In addition, it has been shown that cancer can evade immunosurveillance. Cancer patients are largely immunosuppressed, which allows for tumor growth and the avoidance of immune control.[16] Furthermore, tumor–host interaction is very complex. The immune system is supposed to recognize and destroy transformed cells; however, it fails to perform this function in cancer. Failure of immune control can be explained by several reasons: 1) tumor cells have low immunogenicity due to the low expression of MHC classes I and II, co-stimulatory (CD40, CD80, CD86) and adhesion (ICAM-1, LFA-3) molecules, and weak antigens, 2) tumor cells produce immunosuppressive factors (VEGF, IL-10, PGF2, TGFβ, IL-6), and 3) tumor cells disable DC differentiation, maturation, and migration, and cause functional defects (e.g., a large number of immature DCs are functionally suppressed). Malignancies are often accompanied by an increased pool of suppressive T cells, such as $CD4^+CD25^+FOXP3^+$ and IL-13, producing $CD4^+$ T and NKT cells.[17,18]

Even though DCs are the most potent APCs that can activate and modulate immune response, these properties of DCs are altered in cancer. It has been shown recently that mDCs express B7-H1 cell surface glycoproteins belonging to the B7 family of co-stimulatory molecules. The upregulation of B7-H1 on mDCs in the tumor downregulates T cell immunity. The blockade of B7-H1 on mDCs also mediates T cell activation, which is accompanied by the downregulation of IL-10 and upregulation of IL-2 and IFN-γ.[19]

Moreover, tumor cells alter the balance between Th1 and Th2 cytokine production, with the prevalence of the production of immunosuppressive cytokines (mainly IL-10) and induction of $CD4^+CD25^+$ regulatory T cells with suppressive potential. Additionally, tumor-derived soluble factors TGFβ and VEGF inhibit DC differentiation and maturation, and downregulate the expression of MHC class I and class II, co-stimulatory (CD40, CD86), and adhesion (ICAM-1, LFA) molecules.[20] Furthermore, tumor cells alter DC differentiation by impacting on host hematopoiesis resulting in the production of large numbers of immature DCs or myeloid suppressor cells.[21] These

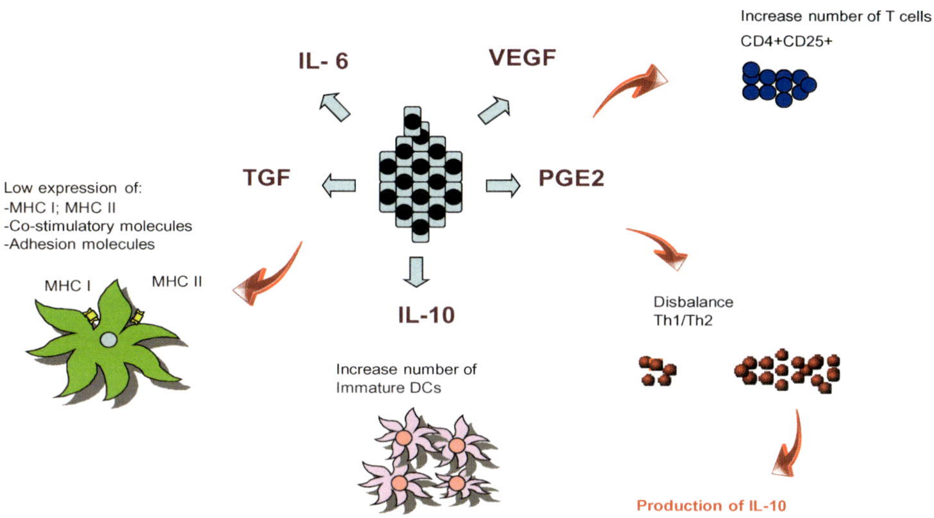

Figure 3. The role of DCs in tumor–host interaction. Defective DCs are unable to develop adequate immune response against tumor cells because of the low expression of MHC class II, co-stimulatory molecules (CD40, CD80, CD86), and adhesion molecules (VCAM-1). In addition, tumor cells suppress the immune response via the production of VEGF, IL-10, PGF2, TGFβ, and IL-6. They also increase the pool of suppressive T cells ($CD4^+CD25^+FOXP3^+$), create a disbalance of Th1/Th2 cells, and increase the number of immature DCs.

cells are unable to efficiently process and present antigens and hence, cannot induce sufficient antitumor response.

The return of immune control to DCs is a big challenge for cancer therapy. Nanotechnology is being increasingly used for drug delivery and the application of NPs as vectors for the targeted delivery of antigens and drugs to APCs, especially DCs, is currently considered as a very promising approach to achieve therapeutic effects on these cells. An immunostimulatory effect of NPs on DCs is important for the purpose of creating a stronger host immune response, and to enhance the cytotoxic T cell response against tumor-associated antigens. The unique ability of professional APCs to internalize foreign antigens and present them along with molecules of MHC class I or MHC class II on cell surfaces provides the opportunity for using antigen-loaded NPs to control immune response. This property is also used to improve vaccine efficacy.

The flip side of the coin includes the undesirable activation or inhibition of DC function by NPs when they are intended for biomedical applications other than vaccine delivery and immunotherapy. Immunotoxicity of engineered nanomaterials is a new and rapidly growing field. Currently, there is a gap in the understanding of the undesirable effects of NPs on DCs due to the lack of comprehensive studies with systematic investigation of the role of NP physicochemical properties on the toxicity to DCs. Due to the lack of knowledge in this area, the toxicity of engineered nanomaterials to DCs, will not be reviewed in this chapter. Below, I will focus on the most recent advances in the rapidly growing field of using DCs for drug and antigen delivery.

3. The Role of Nanoparticle Physicochemical Properties in Their Uptake by Dendritic Cells

The therapeutic effect of targeted drug delivery by DCs depends on the interaction between NPs and APCs. Various NPs have been introduced in medical field — dendrimers, polymeric NPs, magnetic NPs, metallic NPs, and quantum dots. One of the more promising strategies that has emerged in recent years is the loading of magnetic NPs with specific anticancer drugs and targeting them to cells.[14] The interaction of NPs with DCs occurs during receptor-mediated endocytosis, phagocytosis, and macropinocytosis.[2] Here, I will discuss the effects of particle surface characteristics, size, and surface charge on the efficiency of NP uptake by DCs.

It has been reported that size and surface properties of NPs can affect phagocytosis, in that charged and larger NPs have a greater rate of uptake than neutral and smaller ones.[22–25] For example, the phagocytosis rate of 4.5-μm negatively charged particles was greater than that of their 1.0-μm counterparts.[22,25] The uptake of negatively charged carboxylate- and bovine serum albumin-coated particles by macrophages and DCs was low, while positively charged poly-L-lysine (PLL)-coated particles induced a high phagocytosis rate in both cell types.[22] Since these particles not only had different charges, but also different architectures and compositions of surface polymers, it is difficult to determine whether the observed difference in the rate of phagocytosis was due to the charge alone, architecture/composition alone, or a combination of both factors. Nevertheless, the data suggested that phagocytosis may depend not only on the charge of particles, but can also be influenced by their surface coating.[22] Several other studies have also reported that charged cationic particles are more readily taken up by APCs than by anionic particles. For example, internalization was documented for negatively and positively charged and untreated HAP (hydroxyapatite) NPs with similar sizes and shapes; however, greater uptake was detected only when cationic particles were used.[26] One logical explanation for these observations may be due to the electrostatic interactions between the negatively charged cell membrane and positively charged particle surface, and the repulsion between anionic particles and the anionic membrane surface.

4. Challenges and Examples of Nanoparticle Use for Antigen and Drug Delivery to Dendritic Cells

One of the challenges with NP-based antigens and drug delivery is the premature release and inactivation of the payload before it reaches the target site. The optimization of NP synthesis and fine-tuning NP physicochemical parameters are being intensely investigated to overcome this challenge. Another issue would be particle capture by the mononuclear phagocytic system (MPS). Several approaches have been described to avoid the MPS uptake: 1) use of hydrophilic polymer coating (e.g., polyethylene glycol), 2) change in polymer conformation, and 3) change in NP hydrodynamic size[27] (Figure 2).

In addition to modifying NP physicochemical properties, manipulating their route of administration has also been shown to alter their biodistribution and improve uptake by APCs. For example, changing an intravenous (i.v.) route to subcutaneous (s.c.) was shown to protect cells from uptake by the

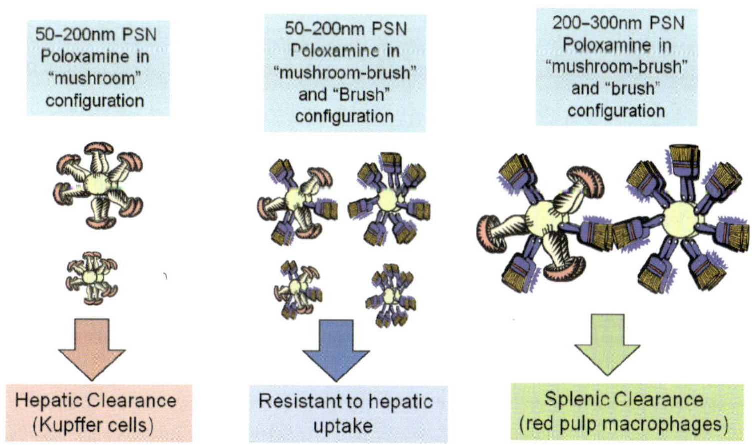

Figure 4. Effects of NP size and surface coating on the NP clearance route. Polymer conformation has been proven to be important for the protection of NP surface from immune recognition. Several studies have demonstrated the protective role of polymer coating, and postulated that loose polymer coating (referred to as "mushroom" configuration) provides little protection from MPS uptake. Increasing the density of polymer coating ("mushroom-brush" and "brush" configurations) helps to ensure more protection. However, an increase in the particle size regardless of the type of coating was shown to either shift particle biodistribution from hepatic (for smaller particles) to splenic uptake (for larger particles), or even completely eliminate protection (for particles larger than 300 nm). The summary is prepared based on Refs. 28–31.

MPS and specifically deliver NPs to lymph nodes, where they accumulate in DCs and stimulate effector cells of the adaptive immunity.[32,33] It is important to note that NP size was a critical factor in determining the lymphatic distribution in that only small particles (25 nm), but not large ones (100 nm), were transported through the lymphatic system and captured by DCs.[32]

Another interesting approach is the loading of DCs with NPs *in vitro* and reinjecting these loaded DCs into the blood. This method is thought to be valuable for drug delivery and diagnostic strategies when maximal uptake by DCs is desired. For example, DCs were used for the delivery of exosomes that were nano-engineered to mediate and modulate immune responses *in vivo* by the semi-direct activation of T cells. The development of these artificial exosomes included the coating of liposomes with an optimized number of MHC class I-associated peptide complexes and attaching a specific range of ligands for adhesion, early and late activation, and survival. When DCs loaded with these engineered exosomes were tested, the activation of antigen-specific T cells and stimulation of IFN-γ secretion were observed.[34]

The use of certain physical properties of NPs was demonstrated to maximize antigen and drug delivery to DCs. For example, magnetic properties of NPs were used to achieve maximal loading, thus making these particles important passive carriers.[35] These properties of magnetic NPs include: 1) the ability to store drugs for extended release by magnetic field stimulation, 2) the absence of toxic effects on the DCs, 3) the absence of effects on cell viability, and 4) internalization into the lysosomal compartment that provides a strong magnetic signal due to optimal pH. To protect an antigen from degradation, it may be encapsulated into and maintained within the NP core.

Successfully engineered NPs for antigen delivery should meet another challenging criterion — the absence of cytotoxicity to DCs. To achieve this, a variety of biomaterials (polymeric NPs, liposomes) were designed and tested to specifically target DCs and enhance their maturation. Dendrimers, for example, were considered for antigen delivery because these hyperbranched polymeric NPs are biodegradable, can be synthesized to carry specific antigens and adjuvants, and their surface properties can be manipulated to reduce toxicity to DCs. It was demonstrated that mannosylated ovalbumin (OVA) linked to polyamidoamine (PAMAM) dendrimers (MDO–NP) is a potent and specific immunostimulant. MDO–NP-induced OVA-specific T cell response *in vitro* demonstrated high binding avidity to DCs. It stimulated DC maturation through enhancing antigen presentation, stimulating $CD4^+$/$CD8^+$ T cells, and upregulating CD40, CD80, CD86, and MHC class II molecules. The antibody response to the OVA antigen was also higher than the response to the same dose of antigen.[36]

Biodegradable NPs are another attractive potential for nanomedicine applications, because the degradation and elimination of the nanocarrier from the body help avoid concerns about bioaccumulation in vital organs such as the spleen, liver, bone marrow, or brain, reducing long-term toxicities. There are many promising results using such nanocarriers for antigen delivery into DCs. For example, when biodegradable poly(γ-glutamic acid) (γ-PGA) NPs were studied in murine spleen-derived DCs, strong activation and stimulation of maturation of DCs were observed.[37] In this study, γ-PGA NPs resulted in an increased expression of co-stimulatory molecules (CD40, CD80, and CD86), induction of cytokines and chemokines (TNF-α, IL-6, RANTES, and MIP-1 α), and MHC class I molecules in splenic DCs. Interestingly, the maturation and activation of splenic DCs by γ-PGA NPs was found to occur through the pathways commonly activated by classical microbial pathogens, i.e., the activation of nuclear-factor kappa B (NF-κB) and mitogen-activated protein kinase (MAPK).[37]

Another example by which engineered NPs were used to stimulate classical inflammatory signaling and DC maturation is the use of layered double hydroxide (LDH) NPs, which consist of positively charged layers with charge-balancing anions between layers to activate DCs. These particles were engineered to provide pH-controlled drug release. Bone marrow-derived DCs (MDDCs) showed enhanced expression of pro- and anti-inflammatory cytokines (TNF-α and IL-12) and co-stimulatory molecules (CD40, CD86, and MHC-II), increased migration capacity to CCL21, and the upregulation of the expression of CCR7 when treated with the LDH NPs. Similar to the previous study, the involvement of the NF-κB pathway in the maturation of DCs stimulated with LDH NPs was also observed.[38]

Polylactic-co-glycolitic acid (PLGA) NPs is another type of biodegradable nanocarrier which can be used for the encapsulation of MHC classes Ia, Ib, and II peptides. DCs loaded with peptide-containing PLGA NPs induced the activation of tumor-specific CTLs.[39]

Another promising form of therapy is the delivery of NP-encapsulated siRNA into linear polyethylenimin (PEI)-based NPs. They were shown to be phagocytosed by regulatory DCs expressing CD11c and programmed cell death 1–ligand 1 (PD–L1) at sites of ovarian cancer in mice. The tumors often overexpress immunosuppressive PD–L1, allowing them to avoid immune control. The uptake of siRNA-loaded PEI NPs transformed these tumor-modified DCs from immunosuppressive cells to efficient APCs with enhanced expression of co-stimulatory molecules CD80 and molecules of MHC-II and MHC-I. siRNA–PEI nanocomplexes also stimulated multiple TLRs expressed by tumor-associated DCs in a dose-dependent manner and induced the maturation of tumor-associated regulatory DCs. This study showed a successful example of DC reprogramming by NPs, leading to efficient T cell-mediated tumor reduction.[40] Although not impossible, the use of NPs for siRNA delivery into DCs is highly challenging and requires the consideration of multiple factors, which has been reviewed in detail by Gao and Huang,[41] and will not be covered in this chapter.

5. Future Directions

The development of new materials and technologies for their application in the biomedical field is often accompanied by concerns regarding their safety.[11] Nanotechnology is not an exception. The properties of nanomaterials that influence their toxicity include chemical composition, shape, architecture, surface charge, aggregation/agglomeration state, and solubility. The key factor in achieving success and spurring the progress in the development of engineered nanomaterials for biomedical applications is the standardization of toxicological tests allowing for the establishment of NP biocompatibility profiles. Nanotoxicological studies require extremely precise characterization of a nanomaterial before its introduction into biomedical applications. Such characterization involves the understanding of the chemical and physical properties of nanomaterials, their compatibility with biological systems, and interactions with components of the immune system.[42,43]

APCs provide a link between the innate and adaptive arms of immunity, which is a highly complex, delicate, and tightly regulated process. The main challenge in studying toxicity of nanomaterials to APCs is the need for the use of immune function tests and unfortunately, there is only a limited number of such tests which can be done *in vitro*. Both immunostimulatory and immunosuppressive properties of an engineered nanomaterial are critical for investigation. When an immunosuppressive or an immunostimulatory property is uncovered and is desirable, it may be successfully developed for clinical applications. For example, the immune stimulation of APCs by NPs is actively pursued in the vaccine field and some of these concepts are already in the preclinical development stage. PDS Biotechnology Corporation has a nanotechnolgy platform (Versamune™), which has shown great potential against various cancers and infectious diseases, and is now being tested in preclinical studies.[44]

Understanding and identifying undesirable effects of NPS on APCs is harder to achieve. Nevertheless, even when the activation of DCs is desired, one has to carefully consider how such activation may impact the host in order to avoid excess activation and potential side effects. This is why the monitoring of DC activation and maturation is needed when we develop novel nanotechnology-based formulations for medical applications.

The effects of NPs on the maturation of DCs can be examined by studying the production of cytokines and chemokines, activation of downstream signaling pathways, and the expression of co-stimulatory molecules and histocompatibility complexes (MHC class I and class II) on the cell surface. Multiple protocols for such studies have been described in the literature and are widely used in traditional immunology.[37–39] Unfortunately, at this point, there is no cytokine, chemokine, or DC functional test which is specifically validated for the screening of engineered nanomaterials. Since NPs often interfere with traditional immunological tests, one has to include additional controls into these traditional protocols.[45] Such controls are prepared by combining NPs with cell culture media and all assay reagents except cells (to screen for false-positive results), and adding NPs into cell culture supernatants collected from cells stimulated with traditional antigens (to screen for false-negative results). Since NPs are so diverse in their physicochemical properties, it is impossible to predict all the possible interactions of these materials with the components of the individual immune function assay. This is why inhibition/enhancement quality controls have to be included in each individual run. Alternatively, the assay has to be validated for a given nanoformulation to rule out interference. It is important to keep in mind that an assay validated for one nanoformulation may still suffer from interference when used to study nanomaterials with different compositions and properties. Hence, is better to conduct validation on a case-by-case basis to ensure accuracy of all test results.

We are at the nascent stage of understanding how DCs are able to orchestrate immune responses to foreign antigens. Ongoing research will uncover information with regard to their potential as vectors for immunotherapy, allowing a deeper understanding of DC heterogeneity, which will be able to explain unique biological functions of DCs, including the existence of different subsets of DCs, their antigen-capturing capacity in the immature stage, the polarization of T cell response toward type 1 or 2 in the mature stage, secretory products of DCs, and regulation of B cell-mediated responses.[5,6]

Future studies will involve the elucidation these features of DCs, the successful manipulation of the immune system, and the broadening of the horizon of DC applications in relation to nanotechnology.

Acknowledgements

I am grateful to Dr. Marina Dobrovolskaia for her help with the preparation of this manuscript.

References

1. Vollhov H, Qin J, Johansson SM, Ahlborg N, Muhammed MA, Scheynius A, Gabrielsson S, *et al.* The importance of an endotoxin-free environment during the production of nanoparticles used in medical applications. *Nano Lett* 2006;6:1682–1686.
2. Klippstein R, Rozo D. Nanotechnology-based manipulation of dendritic cells for enhanced immunotherapy strategies. *Nanomed Nanotechnol Biol Med* 2010;6:523–529.
3. McKenna K, Beignon A-S, Bhardwaj N. Plasmacytoid dendritic cells: Linking innate and adaptive immunity. *J Virol* 2005;70:17–27.
4. Liu Y-J. Dendritic cell subsets and lineages, and their functions in innate and adaptive immunity. *Cell* 2001;106:259–262.
5. Banchereau J, Steinman RM. Dendritic cells and the control of immunity. *Nature* 1998;392:245–252.
6. Banchareau J, Briere F, Caux C, Davoust J, Lebecque S, Liu Y-J, *et al.* Immunobiology of dendritic cells. *Annu Rev Immunol* 2000;18:767–811.
7. Almand B, Resser JR, Lindman B, Nadaf S, Clark JI, Kwon, ED, Carbone DP, *et al.* Clinical significance of defective dendritic cell differentiation in cancer. *Clin Cancer Res* 2000;6:1755–1766.
8. Thurnher M, Radmayar C, Ramoner R, Edner S, Bock G, Klocker H, *et al.* Human renal-cell carcinoma tissue contains dendritic cells. *Int J Cancer* 1996;67:1–7.
9. Nestle FO, Burg G, Fah J, Wrone-Smith T, Nickoloff BJ. Human sunlight-induced basal-cell-carcinoma-associated dendritic cells are deficient in T cell co-stimulatory molecules and are impaired as antigen-presenting cells. *Am J Pathol* 1997;150:641–651.
10. Chaux P, Moutet M, Faivre J, Martin F, Martin M. Inflammatory cells infiltrating human colorectal carcinomas express HLA class II but not B7–1 and B7–2 costimulatory molecules of the T-cell activation. *Lab Invest* 1996;74:975–983.
11. Chaux P, Favre N, Martin M, Martin F. Tumor-infiltrating dendritic cells are defective in their antigen-presenting function and inducible B7 expression in rats. *Int J Cancer* 1997;72:619–624.
12. Gabrilovich D, Ciernik F, Carbon DP. Dendritic cells in anti-tumor immune responses. I. Defective antigen presentation in tumor-bearing hosts. *Cell Immunol* 1996;170:101–110.
13. Enk AH, Jonuleit H, Saloga J, Knop J. Dendritic cells as mediators of tumor-induced tolerance in metastatic melanoma. *Int J Cancer* 1997;73:309–316.
14. Enk AH, Saloga J, Becker D, Mohamadzadeh M, Knop J. Induction of hapten-specific tolerance by interleukin 10 *in vivo. J Exp Med* 1994;179:1397–1402.
15. Steinbrink K, Jonuleit H, Muller G, Schuler G, Knop J, Enk AH. Interleukin-10-treated human dendritic cells induce a melanoma-antigen-specific anergy in CD8(+) T cells resulting in a failure to lyse tumor cells. *Blood* 1999;93:1634–1642.
16. Bell D, Chomarat P, Broyles D, Netto G, Harb GM, Lebecque S, Valladeau J, Davoust J, Palucka KA, Banchereau J. In breast carcinoma tissue, immature dendritic cells reside within the tumor, whereas mature dendritic cells are located in peritumoral areas. *J Exp Med* 1999;190:1417–1426.
17. Steinman RM, Banchereau J. Taking dendritic cells into medicine. *Nature* 2007;449:419–426.
18. Yang L, Carbone DP. Tumor-host immune interactions and dendritic cell dysfunction. *Adv Cancer Res* 2004;92:13–27.

19. Curiel TJ, Wei S, Dong H, Alvarez X, *et al.* Blockage of b7-H1 improves myeloid dendritic cell-mediated antitumor immunity. *Nat Med* 2003;9:562–567.
20. Gray CP, Arosio P, Hersey P. Heavy chain ferritin activates regulatory T cells by induction of changes in dendritic cells. *Blood* 2002;99:3326–3334.
21. Bronte V, Serafini P, Apolloni E, Zanovello P. Tumor-induced immune dysfunctions caused by myeloid suppressor cells. *J Immunother* 2001;24:431–446.
22. Thiele L, Rothen-Rutishauser B, Jilek S, Wunderli-Allenspach H, Merkle HP, Walter E. Evaluation of particle uptake in human blood monocyte-derived cells *in vitro.* Does phagocytosis activity of dendritic cells measure up with macrophages? *J Controlled Release* 2001;76:59–71.
23. Roser M, Fischer D, Kissel T. Surface-modified biodegradable albumin nano- and microspheres. II: Effect of surface charges on *in vitro* phagocytosis and biodistribution in rats. *Eur J Pharm Biopharm* 1998;46:255–263.
24. Tabata Y, Ikada Y. Phagocytosis of polymer microspheres by macrophages. *Adv Polym Sci* 1990;94:107–141.
25. Reddya ST, Rehora A, Schmoekela HG, Huuella JA, Swartza MA. *In vivo* targeting of dendritic cells in lymph nodes with poly(propylene sulfide) nanoparticles. *J Controlled Release* 2006;112:26–34.
26. Chen L, McCrate JM, Lee JC, Li H. The role of surface charge on the uptake and biocompatibility of hydroxyapatite nanoparticles with osteoblast cells. *Nanotechnology* 2011;22:105708.
27. Moghimi SM. Prolonging the circulation time and modifying the body distribution of intravenously injected polystyrene nanospheres by prior intravenous administration of poloxamine -908. A "hepatic-blockage" event or manipulation of nanosphere surface *in vivo*? *Biochim Biophys Acta* 1997;1336:1–6.
28. Moghimi SM, Hedeman H, Muir IS, Illum L, Davis SS. An investigation of the filtration capacity and the fate of large filtered sterically-stabilized microspheres in rat spleen. *Biochim Biophys Acta* 1993;1157:233–240.
29. Moghimi SM, Hedeman H, Christy NM, Ilum L, Davis SS. Enhanced hepatic clearance of intravenously administered sterically stabilized microspheres in zymosan-stimulated rats. *J Leukoc Biol* 1993;54:513–517.
30. Stolnik S, Daudali B, Arien A, Whetstone J, Heald CR, Garnett MC, Davis SS, Illum L. The effect of surface coverage and conformation of poly (ethylene oxide) (PEO) chains of poloxamer 407 on the biological fate of model colloidal drug carriers. *Biomembranes* 2001;1514:261–279.
31. Al-Hanbali Y, Rutt KJ, Sarker DK, Hunter AC, Moghimi SM. Concentration dependent structural ordering of Poloxamine 908 on polystyrene nanoparticles and their modulatory role on complement consumption. *J Nanosci Nanotechnol* 2006;6:1–8.
32. Reddy ST, van der Vlies AJ, Simeoni E, Angeli V, Randolph GJ, O'Neil CP, LeeLK, Swartz MA, Hubbell JA. Exploiting lymphatic transport and complement activation in nanoparticle vaccines. *Nat Biotechnol* 2007;25:1159–1164.
33. Reddy ST, Berk DA, Jain RK, Swartz MA. A sensitive *in vivo* model for quantifying interstitial convective transport of injected macromolecules and nanoparticles. *J Appl Physiol* 2006;101:1162–1169.
34. de la Peña, H, Madrigal JA, Rusakiewicz S, Bencsik M, Cave GWV, Selman A, *et al.* Artificial exosomes as tools for basic and clinical immunology. *J Immunol Methods* 2009;344:121–132.

35. Goya GF, Marcos-Campos I, Fernandez-Pacheco R, Saez B, Godino J, Asin L, Lambea K, Tabuenca P, Mayordomo JI, Larrad L, Ibarra MR, Tres A. Dendritic cell uptake of iron-based magnetic nanoparticles. *Cell Biol Int* 2008;3:1001–1005.
36. Sheng K-C, Kalkanidis M, Pouniotis DS, Esparon S, Tang CK, Apostolopoulos V, Pietersz A. Delivery of antigen using a novel mannosylated dendrimer potentiates immunogenicity *in vitro* and *in vivo*. *Eur J Immunol* 2008;38:424–436.
37. Uto T, Akagi T, Hamasaki T, Akashi M, Baba M. Modulation of innate and adaptive immunity by biodegradable nanoparticles. *Immunology* 2009;125:46–52.
38. Li A, Qin L, Zhu D, Zhu R, Sun J, Wang S. Signaling pathways involved in the activation of dendritic cells by layered double hydroxide nanoparticles. *Biomaterials* 2009;31:748–756.
39. Ma W, Smith T, Bogin V, Zhang Y, Ozkan C, Ozkan M, Hayden M, Schroter S, *et al.* Enhanced presentation of MHC class Ia, Ib and class II-restricted peptides encapsulated in biodegradable nanoparticles: A promising starategy for tumor immunotherapy. *J Transl Med* 2011;9:34.
40. Cubillos-Ruiz JR, Engle X, Scarlett UK, Martinez D, Barber A, Elgueta R, Wang L, *et al.* Polyethylenimine-based siRNA nanocomplexes reprogram tumor-associated dendritic cells via TLR5 to elicit therapeutic antitumor immunity. *J Clin Invest* 2009;119:2231–2244.
41. Guo S, Huang L. Nanoparticles escaping RES and endosome: Challenges for siRNA delivery for cancer therapy. *J Nanomater* 2011.
42. Buzea C, Pacheco I, Robbie K. Nanomaterials and nanoparticles: Sources and toxicity. *Biointerphases* 2007;2:MR17–MR71.
43. NANOSAFE (Safe production and use of nanomaterials). Dissemination reports. Retrieved from www.nanosafe.org/scripts/home/publigen/content/templates/show.asp?P=63&L=EN&ITEMID=13.
44. Vanderson. Nanoparticles could be tiny cancer assassins. Retrieved from Onemedplus http://www.onemedplace.com/blog/archives/5425.
45. Hall JB, Dobrovolskaia MA, Patri AK, McNeil SE. Characterization of nanoparticles for therapeutics. *Nanomedicine* 2007;2:789–803.

Chapter 14

The Effects of Nanoparticles on Bone Marrow Cells

Ekaterina Dadachova

Departments of Radiology and
Microbiology and Immunology
Albert Einstein College of Medicine
1695A Eastchester Rd
Bronx, NY 10461, USA
ekaterina.dadachova@einstein.yu.edu

This chapter reviews the observations made during the last decade in the area of nanoparticle effects on bone marrow cells such as nanoparticle biodistribution in the host including the bone marrow, nanoparticle-assisted imaging of the bone marrow, immunomodulatory effects, toxicity to bone marrow cells, and the use of nanoparticles for the radioprotection of the bone marrow. Depending on the nature of nanoparticles and their surface modifications, their biodistribution can be controlled to some extent, allowing the increase or decrease of their uptake into the bone marrow for the intended application. Magnetic resonance imaging with magnetic iron oxide nanoparticles has been gaining momentum with studies being conducted both in patients and animal models. Nanoparticles have proven to be useful in helping to induce the suppression or stimulation of the bone marrow as well as avoid the immunomodulatory effects of certain drugs altogether. The delivery of radioprotectors to the bone marrow by nanoparticles helps to enhance their radioprotective effect. Since the application of nanomaterials in biomedical research and medicine is a relatively new field, the most thorough evaluation of their potential toxicity, especially to the bone marrow, is paramount. While biodegradable nanoparticles in general lack bone marrow toxicity, metal and metal oxide nanoparticles vary in their effects on the bone marrow depending on the dose and route of administration, with TiO_2 particles having the highest cause for concern. Hence, research in this area should continue for any newly introduced nanomaterial.

1. Introduction

Nanomaterials were used for the imaging of the bone marrow in nuclear medicine field long before the emergence of the term "nanomaterials." The observation that the human body has a self-sieving effect and particles smaller than 1 μm in diameter when injected intravenously (i.v.) will be lodged to significant extent in the bone marrow led to the creation of technitium-99m (^{99m}Tc) sulfur colloid (now often termed "nanocolloid"), which has been in use in nuclear medicine since the 1960s.[1–3]

In this chapter, I will review the observations made during the last decade in the area of nanoparticle effects on bone marrow cells. The topics discussed will include the imaging of the bone marrow with nanoparticles and their biodistribution in the host, immunomodulatory effects, toxicity to bone marrow cells, and nanoparticle delivery for the radioprotection of the bone marrow. This chapter covers only nanomaterials which are truly "nano" in all three dimensions and exclude the so-called high aspect ratio nanoparticles such as single- or multi-walled carbon nanotubes. High aspect ratio materials are "nano" only in one dimension, and because of their size and bioretention, they are proving more and more of a concern from the toxicological and carcinogenic points of view.[4,5]

2. Biodistribution

The biodistribution of novel nanoparticles is usually assessed in small rodents or rabbits and helps to predict their potential usefulness in a variety of applications (Table 1). Banerjee *et al.* performed a study on the biodistribution of ultrafine (30 nm) chitosan nanoparticles after i.v. administration in mice.[6] The experiment demonstrated that particles evade the reticuloendothelial system (RES) and remain in the blood for a prolonged period of time. The follow-up experiments were conducted with ^{99m}Tc-labeled nanoparticles in rabbits and showed a significant percentage of these nanoparticles going into the bone marrow, which allowed the authors to conclude that the chitosan nanoparticles might be suitable for bone imaging and targeting.

Valdivia *et al.* demonstrated the potential of ferro-carbon nanoparticles as drug-carrying vehicles in the localization of drugs in solid organs with the particular purpose of delivering doxorubicin to the kidneys.[7] In contrast to the previous study, they attempted to prevent the deposition of nanoparticles in the RES and thus, they implanted a magnetic microharpoon in the left kidney of New Zealand rabbits before nanoparticle injection. The biodistribution showed that in spite of the microharpoon present, a significant percentage of nanoparticles ended up in the RES including the bone marrow, which led to the conclusion that temporarily shutting down of the RES is needed before administration of these nanoparticles. Likewise, Halder *et al.*[8] attempted to avoid bone marrow uptake to maximize the delivery of chloramphenicol-incorporated polylactide-co-glycolide (PLGA) nanoparticles into the brain. They prepared the nanoparticles via the emulsification solvent evaporation technique, either by using polyvinyl alcohol (PVA) as an emulsion stabilizer or polysorbate-80 (PS-80) as a surfactant, and radiolabeled them with ^{99m}Tc before i.v. injection into mice. It appeared that surface modifications with PS-80 resulted in the redirection of nanoparticles into the brain and minimized bone marrow uptake, in comparison with free chloramphenicol and non-surface modified PLGA nanoparticles. Another nanoparticle carrier capable of localizing in the brain turned out to be magnetic iron oxide (Fe_3O_4) nanoparticles synthesized by Wang *et al.*[9] In the imprinting control region (ICR) of mice, the majority of Fe_3O_4 nanoparticles were found in the liver and spleen in addition to some in the brain. Thus, depending on the nature of nanoparticles and their surface modifications, their biodistribution can be controlled to some extent, allowing the increase or decrease in uptake into the bone marrow for the intended applications.

3. Imaging

Magnetic resonance imaging (MRI) with magnetic Fe_2O_3 nanoparticles has been gaining momentum during the last decade with studies conducted both in patients and in animal models (Table 1). The physicochemical and pharmacokinetic properties of Fe_2O_3 as well as the particle size influence their clinical applications. Superparamagnetic iron oxides (SPIOs) enhance both T1 and T2/T2* relaxation, reduce T2/T2* relaxation time, and improve T1 relaxivity.[10] Specifically for bone marrow imaging, smaller Fe_2O_3 particles with a mean diameter of 20 nm seem to be the most suitable.[10] A variety of conditions have been studied using Fe_2O_3-enhanced MRI in patients such as primary and metastatic cancer, infection, and bone marrow transplantation. Further novel applications are currently being investigated in animal models.

Metz and co-authors investigated the use of ferumoxtran-10–enhanced imaging of the bone marrow before and after conditioning therapy in patients with non-Hodgkin lymphoma.[11] They quantified permeability changes of the blood–bone marrow barrier (BMB) and detected malignant bone marrow infiltrations before and after conditioning therapy for subsequent leukapheresis in 22 patients with malignant non-Hodgkin lymphoma, i.e., nine patients (group A) before and 13 patients (group B) after conditioning therapy. The

Table 1. Brief characteristics of the techniques employed in the evaluation of the interaction of nanoparticles with the bone marrow.

Method	Brief description
Biodistribution	Biodistribution is a method of tracking the route taken by the compounds of interest through an experimental animal or a human subject. The contents of the compound in the tissues are evaluated by its imagable label, usually a radioisotope.
Magnetic resonance imaging (MRI)	MRI is a medical imaging technique utilized to visualize in detail internal structures and organs. MRI often requires the use of enhancing agents such as SPIOs.
Colony forming units for granulocytes and macrophages (CFU-GM)	The CFU-GM method involves the incubation of nanoparticles with bone marrow cells *in vitro* and *in vivo* post-systemic injection, subsequent culturing of GM in a specialized medium for 12 days at 37°C, counting CFUs, and morphology evaluation.
Micronucleus test	The micronucleus test involves the injection of an experimental animal with nanoparticles, isolating the bone marrow, smearing it on a glass slide, staining with H&E, and analyzing for the presence of micronuclei.
Neutral red assay	The neutral red uptake assay provides a quantitative estimation of the number of viable cells in a culture. It is one of the most used cytotoxicity tests and is based on the ability of viable cells to incorporate and bind the supravital dye neutral red in the lysosomes.
Alkaline phosphatase assay	The alkaline phosphotase assay measures the ubiquitous alkaline phosphatase marker in stem cells and is analyzed by immunohistochemistry staining or colorimetric or fluorometric plate readers.
Comet assay	The comet assay (also known as Single Cell Gel Electrophoresis assay) is used for the detection of DNA damage in individual eukaryotic cells. It involves the encapsulation of cells in a low-melting point agarose suspension, lysis of the cells in neutral or alkaline (pH > 13) conditions, and electrophoresis of the suspended lysed cells, followed by visual analysis with the staining of DNA and calculating fluorescence to determine the extent of DNA damage.
Gamma-H2AX immunostaining assay	Gamma-H2AX is emerging as an important marker of ionizing radiation-induced double-strand breaks. Quantification of gamma-H2AX has broad applications in assessing physiological responses to radiation exposure.
8-hydroxy-2′-deoxyguanosine level assay	8-hydroxy-2′-deoxyguanosine (8-OH-dG) is a major form of oxidative DNA damage from heat, chemicals, or various forms of radiation and is analyzed by high-performance liquid chromatography with electrochemical detection (HPLC-ECD).

dynamic ΔSI data were correlated with the quantity of mobilized $CD34^+$ cells, and the number of focal bone marrow lesions was compared before and after ferumoxtran-10 administration. Though the results of the investigation clearly showed that the dynamic ΔSI data were higher in group B than in group A, indicating an increased BMB permeability after conditioning therapy, ΔSI data did not correlate with the quantity of mobilized $CD34^+$ cells. In spite of these observations, the investigators made a conclusion that overall, ferumoxtran-10 significantly ($p < 0.05$) improved the detection of focal bone marrow lesions. In another cancer application of Fe_2O_3-enhanced MRI, Daldrup-Link *et al.* aimed to differentiate normal, hypercellular, and neoplastic bone marrow based on its MR enhancement after i.v. administration of SPIOs in patients with cancer of the hematopoietic system.[12] They recruited 18 patients with that particular cancer who underwent MRI of the spine before and after infusion of ferumoxides ($n = 9$) or ferumoxtran ($n = 9$) using T1- and T2-weighted turbo spin-echo and short tau inversion recovery sequences (STIR). The preliminary diagnosis in all patients was diffuse or multifocal bone marrow infiltration, as shown by iliac crest biopsy, conventional radiographs, MRI, and positron emission tomography. This clinical study demonstrated that SPIOs are taken up by normal and hypercellular reconverted bone marrow, but not by neoplastic bone marrow lesions, which allowed for significantly different enhancement patterns on T2-weighted MR images. The authors concluded that SPIOs are useful for the differentiation between normal and neoplastic bone marrow and for increasing the bone marrow-to-tumor contrast.

The same group carried out another interesting application for SPIO-enhanced MRI by using it for the evaluation of bone marrow before and after total body irradiation (TBI) in an experimental animal model.[13] They performed SPIO-enhanced MRI of the lumbar spine and os sacrum in New Zealand white rabbits without TBI and after 4 and 12 Gy TBI, in an attempt to quantify the permeability of the BMB and phagocytic activity of bone marrow RES cells before and after TBI. The data were correlated with bone marrow histopathology and revealed an irradiation-induced decline of all hematopoetic cell lines. SPIOs were phagocytosed by bone marrow RES cells and caused significant bone marrow signal decline on postcontrast T2-weighted images ($p < 0.05$). Importantly, ΔSI (%) data for T2-weighted images were significantly higher for the irradiated bone marrow as compared to non-irradiated controls ($p < 0.05$). However, dynamic T1-weighted images directly after contrast medium injection were not able to characterize the permeability of the BMB. The most important outcomes of this study were that the hematopoietic bone marrow can be labeled with SPIOs and irradiation does not impair the phagocytic activity of bone marrow RES cells. Their results were corroborated by the study by Azoulay *et al.*, who investigated the ability of ultrasmall SPIO (USPIO)-enhanced MRI to detect RES cells in normal newborn rats[14] and demonstrated that ferumoxtran-10 selectively labeled RES cells after 96 h and produced non-specific labeling at earlier time points.

SPIO MRI is also proving to be useful for the differentiation among various conditions such as bone metastasis and osteomyelitis, or infectious osteomyelitis and aseptic vertebral inflammation. Fukuda *et al.* first established in healthy volunteers that the most appropriate timing to perform SPIO-enhanced MRI in patients with bone metastasis and osteomyelitis would be at 3 h.[15] They evaluated 12 patients, five with bone metastasis and seven with vertebral osteomyelitis by measuring the signal intensities (SIs) of the bone metastases, osteomyelitis, and surrounding normal bone marrow, and calculating the relative enhancement (RE) for each lesion. In patients with bone metastasis, the RE of the bone metastases was –12.2%, which was significantly higher than that in the osteomyelitis (–35.0%; $p < 0.001$) and normal bone marrow (–46.6%; $p < 0.0005$). Such results could provide the basis for performing the procedure routinely in the clinic. To differentiate between infectious osteomyelitis and aseptic vertebral inflammation, Bierry and co-workers employed a rabbit model.[16] They compared gadolinium- and SPIO-enhanced MRI in rabbits with infection and with sterile inflammation. While gadolinium MRI results were not different between the two groups ($p = 0.26$), for SPIO MRI, a significant signal-to-noise ratio (SNR) increase was only observed in the infection group ($p = 0.03$), with a significant difference in SNR between the infection and the sterile inflammation groups ($p = 0.002$). Histological evaluation showed that the bone marrow in infected areas was intensely infiltrated by macrophages, with some of them being iron loaded. In contrast, sterile inflammation showed a replacement of the bone marrow by inflammatory tissues with only rare macrophages, without any Perls blue staining. These results encourage further investigation of the ability of SPIO-enhanced MRI to distinguish infectious osteomyelitis from sterile vertebral inflammation due to different macrophage distributions in the two lesions.

In the most recent examples of the imaging of the bone marrow with the help of nanoparticles, *ex vivo* imaging is utilized as well as "magnetic needle" biopsy. Bodo *et al.* worked on quantum dot immunofluorescence for the *in situ* quantitation of phosphoproteins in fixed samples, as knowing the levels of phosphoproteins is important for the application of protein kinases inhibitors, which are being increasingly used in cancer treatment.[17] The use of quantum dot nanocrystals for the immunofluorescent detection of phosphoproteins was demonstrated in xenografts from experimental animals as well as in bone marrow biopsies from patients with acute myeloid leukemia. Jaetao *et al.* aimed to improve the accuracy of measuring minimal residual disease in patients with acute leukemia, which is critical in determining their prognosis and treatment.[18] They developed a novel apparatus that utilized antibodies to the acute leukemia antigen CD34 conjugated to SPIO nanoparticles and thus, could be detected by the magnetic needle. The method was tested with leukemia cell lines expressing high or minimal CD34. The antibodies to the CD34-conjugated nanoparticles preferentially bind high CD34-expressing cell lines. Importantly, the magnetic needle enabled the identification of both the cell line and patient leukemia cells diluted into normal blood at concentrations below those normally found in remission marrow samples. The authors concluded that bone marrow biopsy using antigen-targeted magnetic nanoparticles and a magnetic needle for the evaluation of minimal residual disease in $CD34^+$ acute leukemias is capable of improving sensitivity in comparison with the current standard of care.

4. Toxicity

Since the application of nanomaterials in biomedical research and medicine is a new area, the most thorough evaluation of their potential toxicity, especially to the bone marrow, is paramount. As mentioned in the beginning of this chapter, high impact ratio nanomaterials such as carbon nanotubes proved to be toxic at least to the same extent as asbestos fibers.[4,5] Here, the results of toxicity investigations of a variety of nanomaterials with an impact ratio of approximately 1 will be discussed.

Chatterjee and colleagues reported the development of polyethyleneimine (PEI)-coated NaYF(4):Yb,Er nanoparticles for use in up-conversion fluorescence imaging with excitation in the near-infrared (NIR) region.[19] The advantages of such imaging include the absence of photo-damage to living organisms, very low auto-fluorescence, high detection sensitivity, and high light penetration depth in biological tissues. The nanoparticles were shown to be stable in phosphate buffered saline (PBS) and non-toxic to bone marrow stem cells and to living animals (rats) post-systemic injection. Shinohara *et al.* also utilized a battery of *in vitro* and *in vivo* tests to clarify conflicting reports on the genotoxicity of fullerene C_{60}.[20] They prepared the suspension of C_{60} in Tween 80 aqueous solution and performed a bone marrow micronucleus test (Table 1) in ICR mice. The test showed that the *in vivo* clastogenic ability of the C_{60} nanoparticles was negative up to the maximum dose of 88 mg/kg when given twice to mice.

While the nanomaterials described above are still being evaluated in the laboratory, silver nanoparticles are already widely utilized in health, electronic, and home products due to their antibacterial properties. For this reason, Kim *et al.* looked at the toxicity and genotoxicity of orally administered silver nanoparticles in Sprague–Dawley rats.[21] The rats were given from 30 to 1,000 mg/kg particles and after 28 days of exposure, their blood biochemistry and hematology were investigated, along with a histopathological examination and a silver distribution study. No statistically significant differences in the micronucleated polychromatic erythrocytes or ratio of polychromatic erythrocytes among the total erythrocytes were detected after silver nanoparticle exposure when compared to the control. The authors concluded that silver nanoparticles do not induce genetic toxicity in both male and female rat bone marrows *in vivo*. However, there was a dose-dependent accumulation of silver content in all examined tissues.

Additionally, biodegradable nanoparticles are gaining acceptance as delivery vehicles for various drugs which by themselves are often toxic to multiple tissues. Cenni and co-workers assessed the biocompatibility of PLGA nanoparticles conjugated with alendronate (ALE).[22] ALE is used to treat and prevent osteoporosis in post-menopausal women. The results of hemocompatibility studies using different concentrations of PLGA–ALE nanoparticles demonstrated no significant effect on hemolysis, leukocyte number, platelet activation, and complement consumption in comparison with the PBS control. The viability of bone marrow stromal cells (BMSCs), evaluated through the neutral red test (Table 1), was also not affected by PLGA–ALE nanoparticles. Furthermore, no differences in cell-associated alkaline phosphatase levels (Table 1) between BMSCs incubated with PLGA–ALE nanoparticles and PBS-added media were detected. The authors concluded that PLGA–ALE nanoparticles had an acceptable degree of blood and bone marrow compatibility and may be considered suitable for i.v. administration. In another application of PLGA-based nanoparticles, they were loaded with amphotericin B (AMB) to be used against paracoccidioidomycosis.[23] The researchers infected BALB/c mice with the yeast *Paracoccidioides brasiliensis* and animals were treated with either encapsulated or non-encapsulated AMB 30 days post-infection. In parallel with efficacy assessment, they also evaluated the hematological toxicity of both forms of the drug. Nano-D–AMB presented results comparable to free D-AMB, with a marked antifungal efficacy. Importantly for bone marrow toxicity, the micronucleus assay showed no significant differences in both the micronucleus frequency and percentage of polychromatic erythrocytes for Nano-D–AMB, indicating the absence of genotoxicity and cytotoxic effects. Based on the efficacy and encouraging lack of toxicity, the authors planned to investigate whether Nano-D–AMB would be useful to treat other systemic fungal infections such as paracoccidioidomycosis, candidiasis, aspergillosis, and cryptococcosis.

Metal or metal oxide-based nanoparticles present endless possibilities in the manipulation of the properties of nanoparticles for a variety of applications. Significant research efforts have already been invested into evaluating their toxicity including that to the bone marrow. Bregoli *et al.* performed a comprehensive study evaluating the toxicity of seven metal or metal oxide-based nanoparticles types — Fe_2O_3, Fe_3O_4, Sb_2O_3, Au, TiO_2, Co, and Ag — on primary cultures of human hematopoietic progenitor cells from the bone marrows of healthy donors.[24] They demonstrated that while antimony oxide (Sb_2O_3) and cobalt (Co) nanoparticles had toxic effects on the human hematopoietic progenitor cells in the concentration range of 5–100 mg/mL, other nanoparticles were not toxic to the primary cells. In particular, Co nanoparticle suspension was toxic to both erythroid and granulocytic/monocytic precursors, and Sb_2O_3 at 5 mg/mL was specifically toxic to erythroid colony development (Table 1), suggesting a highly selective type of toxicity. The authors also cautioned against the use of immortalized cell lines for evaluating the toxicity of nanomaterials as in spite of being toxic to primary cultures, Sb_2O_3 had no toxic effect on immortalized human cell lines of hematopoietic origin, such as K562, HL-60, and Jurkat.

While using primary cells seems to be preferential to cell lines for toxicity evaluation *in vitro,* the ultimate data are produced in *in vivo* experiments. In this regard, TiO_2, which proved to be benign *in vitro* in the above study,[24] has shown a range of toxic effects when given to mice in drinking water.[25] The authors determined the extent of DNA damage in TiO_2 nanoparticle-fed mice by using the comet assay, the micronuclei assay, the gamma-H2AX immunostaining assay (Table 1), and by measuring 8-hydroxy-2'-deoxyguanosine levels (Table 1), and as a genetic instability endpoint, DNA deletions. TiO_2 nanoparticles induced 8-hydroxy-2'-deoxyguanosine, gamma-H2AX foci, micronuclei, and DNA deletions. The formation of gamma-H2AX foci, indicative of DNA double-strand breaks, was the most sensitive parameter. Inflammation was also present as characterized by a moderate inflammatory response. The authors made a justified conclusion that their findings raised concerns about potential health hazards associated with TiO_2 nanoparticle exposure. Other groups demonstrated the toxicity of TiO_2 to bone marrow-derived macrophages which inhibited their nitric oxide (NO) production.[26] These are important and alarming observations as TiO_2 nanoparticles are very widely used in consumer products such as sunscreens and cosmetics. Another type of metal oxide-based nanoparticle which has already entered the market is the ZnO nanoparticle. A very recent research by Heng *et al.* demonstrated that the exposure of primary dendritic cells to ZnO nanoparticles upregulated the expression of CD80 and CD86 (well-known markers of dendritic cell activation and maturation) and stimulated the release of pro-inflammatory cytokines IL-6 and TNF-α, thus pointing to the potential of ZnO nanoparticles in inducing inflammation.[27] This study highlighted the fact that ZnO nanoparticles may potentially trigger detrimental effects on cells even at low doses.

Gold nanoparticles present another potential attractive option for use in biomedical applications because of the inertness of gold chemicals. However, their toxicological effects *in vivo* have not been studied in detail. Zhang *et al.* contributed to closing this gap in knowledge by evaluating the toxicological effects of gold nanoparticles administered to mice via different routes.[28] 13.5-nm gold nanoparticles were given to mice orally, intraperitoneally, or intravenously in the dose range of 137.5–2200 μg/kg. High concentrations of gold nanoparticles induced decreases in body weight, red blood cells, and hematocrit. The researchers also observed that gold nanoparticles administered orally caused significant decreases in body weight, spleen index, and red blood cells. Of the three administration routes, the oral and intraperitoneal routes showed the highest toxicity, while i.v. administration resulted in the lowest toxicity. The authors suggested the latter route for the application of gold nanoparticles for the enhancement of radiotherapy, photothermal therapy, and related medical diagnostic procedures. Overall, the published data summarized above showed that while biodegradable nanoparticles in general lack bone marrow toxicity, metal and metal oxide nanoparticles vary in their effects on the bone marrow depending on the dose and route of administration, with TiO_2 particles causing the most concern. Hence, research in this area should continue for any newly introduced nanomaterial.

5. Immunomodulation

Immunomodulation of the bone marrow plays an important role in the treatment of graft-versus-host disease in patients whose condition requires a bone marrow transplant such as in acute leukemia.[29] Cyclosporine A, a metabolite of the fungus *Tolypocladium inflatum,* has been used for the immunosuppression of the bone marrow for more than 30 years in the form of oral emulsions or capsules. To minimize its toxicity, research is still on-going to deliver it in the form of biodegradable nanoparticles.[29] Interestingly, when not immunosuppression but immunostimulation of bone marrow is required, biodegrable PLGA nanoparticles again seem to the agent of choice. In this regard, Hamdy and colleagues used PLGA nanoparticles for the co-delivery of ovalbumin and Toll-like receptor ligand monophosphoryl lipid A for the purpose of vaccine development.[30] They demonstrated that PLGA delivery of ovalbumin and ligand monophosphoryl lipid A to bone marrow-derived dendritic cells led to a marked increase in *in vitro* $CD8^+$ T cell proliferative responses (stimulation index > 3000) and more than a 13-fold increase in *in vivo* clonal expanded $CD4^+$ T cells.

The use of nanoparticles can also help to eliminate side effects of certain drugs, with one of the most important being bone marrow suppression. For example, high doses of ganciclovir, a drug used in the treatment of ocular viral infections, result in dose-related toxicity including bone marrow suppression and neutropenia. Akhter and colleagues created three different types of nanomaterials for loading with ganciclovir — mucoadhesive nanoemulsions, chitosan nanoparticles, and mucoadhesive niosomal dispersion, and demonstrated the non-irritant and non-toxic nature of these formulations.[31] The authors suggested that such ganciclovir-loaded particles could be useful in the treatment of ocular infections by topical instillation. Taken together, these results demonstrated the usefulness of nanoparticles in helping to induce suppression or stimulation of the bone marrow, as well as avoiding the immunomodulatory effects of certain drugs altogether.

6. Radioprotection

Bone marrow toxicity is the dose-limiting factor for both external beam radiation therapy (EBRT) and radioimmunotherapy (RIT) in cancer patients, such that doses over 200 cGy cause significant acute and long-term toxicity. The protection of the bone marrow against harmful effects of radiation in such patients would permit the administration of significantly higher doses of EBRT or RIT, which could translate into increased efficacy and safety of treatment. The most widely evaluated agents in radioprotection are known chemoprotectors such as amifostine, fullerenes, ascorbic acid, and melanin.

Amifostine (Ethyol^{TM}) is an approved cytoprotective agent prescribed to reduce certain side effects in the chemotherapy of ovarian or non-small cell lung cancer, or in the radiation treatment of head-and-neck cancer. The major drawback of this drug is its ineffectiveness when given orally. Pamujula *et al.* attempted to circumvent this problem by using amisfostine-loaded nanoparticles.[32] The nanoparticles were administered orally to mice 1 h before 8 or 9 Gy TBI and bone marrow hematopoietic progenitor cell survival was evaluated together with other parameters at 30 days. The authors reported that hematopoietic progenitor cell survival was significantly enhanced in the group of mice given oral amifostine nanoparticles prior to irradiation. Later on, the same group extended this approach to the active metabolite of amifostine, N-(2-mercaptoethyl)1,3-diaminopropane (WR-1065), and again demonstrated the protection of the bone marrow in mice after lethal and sublethal doses of radiation when WR-1065-loaded PLGA nanoparticles were given 1 h prior to radiation.[33]

Another class of chemoprotectors evaluated in radioprotection in the form of nanoparticles are fullerenes. Fullerene compounds are known to possess antioxidant properties, a common property of chemical radioprotectors.[34] Brown *et al.* hypothesized that dendrofullerene DF-1, which is a nanoparticle with antioxidant properties, would be protective in a murine model of lethal external TBI.[35] The researchers observed only a modest radioprotective effect of DF-1 *in vivo* — the LD50/30 for C57Bl6/Ncr mice treated with a single 300 mg/kg dose of DF-1 pre-irradiation was 10.09 Gy versus 8.29 Gy for unprotected mice (control). No increase in pimonidazole uptake was appreciated in the bone marrows of mice treated with DF-1 compared to vehicle controls. In a completely different approach from using fullerenes for radioprotection, Diener and co-workers utilized C_{60} fullerene for the stable encapsulation of radionuclides used in RIT of cancer, especially where conventional chelation chemistry is inadequate due to the physical and/or chemical properties of the radionuclide.[36] One of the radionuclides useful in RIT that requires improved containment *in vivo* is ^{212}Pb ($t_{½}$ = 10.6 h), the β-emitting parent to α-emitting ^{212}Bi ($t_{½}$ = 60.6 min) as myelotoxicity resulting from the accumulation of ^{212}Pb in the bone marrow has limited the use of this radionuclide for clinical trials despite its favorable decay characteristics. The researchers prepared ^{212}Pb@C_{60} and its malonic ester derivatives and performed biodistribution in mice. Unlike the case of conventional polyaminocarboxylate chelators for ^{212}Pb, fullerene-bound ^{212}Pb did not accumulate in the bone which should result in sparing the bone marrow from radiation. The researchers suggested that fullerenes may have an important role in the targeted delivery of ^{212}Pb.

Furthermore, Chandrasekharan and colleagues employed silver nanoparticles for the oral delivery of ascorbic acid derivative 6-palmitoyl ascorbic acid-2-glucoside.[37] When orally administered to mice 1 h prior to whole-body radiation, ascorbic acid-loaded silver nanoparticles protected cellular DNA from radiation-induced strand breaks in various tissues, including bone marrow cells and blood leukocytes. In addition, the particles enhanced the rate of repair of cellular DNA in blood leukocytes and bone marrow cells when administered immediately after radiation exposure, pointing to their potential use not only as a radioprotector, but also as a mitigator of radiation injury.

Finally, Schweitzer *et al.* made use of the radioprotective properties of melanin, a naturally occurring pigment, for bone marrow protection in mice.[38] They hypothesized that melanin, which is insoluble, could be delivered to the bone marrow before irradiation by the i.v. administration of melanin-covered nanoparticles because of the human body's "self-sieving" ability. The melanin particles were created via the enzymatic polymerization of 3,4-dihydroxyphenylalanine and/or 5-S-cysteinyl-3,4-dihydroxyphenylalanine on the surface of 20-nm plain silica nanoparticles. Intact CD-1 mice or human melanoma tumor-bearing nude mice were given melanin nanoparticles (i.v.), 50 mg/kg of body weight, 3 h before either whole-body exposure to 125 cGy or treatment with 1 mCi of ^{188}Re-labeled 6D2 melanin-binding antibody. The platelet and white blood cells counts in nanoparticle-treated mice attested to the protection of the bone marrow from both external and internal ionizing radiation.

7. Conclusions

Nanoparticles have found wide range of applications in relation to the bone marrow. They are used for imaging of the bone marrow to help in cancer treatment, to distinguish infection from inflammation, and for the delivery of numerous drugs to this organ. Depending on the specific need, the drug-loaded nanoparticles are also utilized for the immunosuppression or immunostimulation of the bone marrow, as well as for radioprotection from the side effects of radiation treatment of cancer. In terms of toxicity of nanoparticles to the bone marrow, much more work needs to be done as nanomaterials have only relatively recently entered the consumer market and healthcare. At this point in time, the biodegradable nanoparticles seem to be non-toxic to the bone marrow, while metal and metal oxide-based nanoparticles, especially TiO_2 and ZnO, do present some toxicity concerns.

References

1. Rosenthall L. Visualization of the bone marrow with technetium-99m sulfur colloid. *J Can Assoc Radiol* 1967;18(3):407–411.
2. Stern HS, McAfee JG, Subramanian G. Preparation, distribution and utilization of technetium-99m-sulfur colloid. *J Nucl Med* 1966;7(9):665–675.
3. Agool A, Glaudemans AW, Boersma HH, Dierckx RA, Vellenga E, Slart RH. Radionuclide imaging of bone marrow disorders. *Eur J Nucl Med Mol Imaging* 2011; 38(1): 166–178.
4. Sanchez VC, Weston P, Yan A, Hurt RH, Kane AB. A 3-dimensional *in vitro* model of epithelioid granulomas induced by high aspect ratio nanomaterials. *Part Fibre Toxicol* 2011;18(8):17.

5. Aitken R, Tran L. Toxicity and Safety. Institute of Occupational Medicine, http://www.iom-world.org/research/nonoparticles.php, 2012.
6. Banerjee T, Mitra S, Kumar Singh A, Kumar Sharma R, Maitra A. Preparation, characterization and biodistribution of ultrafine chitosan nanoparticles. *Int J Pharm* 2002; 243(1–2):93–105.
7. Valdivia Uríac JG, Ibarra García MR, Fernández Pacheco R, Viloria A, Higuera T, Laborda A, García A, García de Jalón JA, Gutiérrez M, Romero MS, Cornudella R, Arruebo M, Marquina C, Arbiol J, Santamaría J. Research study on renal targeted chemotherapy with magnetic harpoons and intravenous administration of ferro-carbon nanoparticles. *Arch Esp Urol* 2007;60(1):5–14.
8. Halder KK, Mandal B, Debnath MC, Bera H, Ghosh LK, Gupta BK. Chloramphenicol-incorporated poly lactide-co-glycolide (PLGA) nanoparticles: Formulation, characterization, technetium-99m labeling and biodistribution studies. *J Drug Target* 2008;16(4):311–320.
9. Wang J, Chen Y, Chen B, Ding J, Xia G, Gao C, Cheng J, Jin N, Zhou Y, Li X, Tang M, Wang XM. Pharmacokinetic parameters and tissue distribution of magnetic Fe(3)O(4) nanoparticles in mice. *Int J Nanomed* 2010;5:861–866.
10. Wang YX, Hussain SM, Krestin GP. Superparamagnetic iron oxide contrast agents: Physicochemical characteristics and applications in MR imaging. *Eur Radiol* 2001;11(11): 2319–2331.
11. Metz S, Lohr S, Settles M, Beer A, Woertler K, Rummeny EJ, Daldrup-Link HE. Ferumoxtran-10-enhanced MR imaging of the bone marrow before and after conditioning therapy in patients with non-Hodgkin lymphomas. *Eur Radiol* 2006;16(3):598–607.
12. Daldrup-Link HE, Rummeny EJ, Ihssen B, Kienast J, Link TM. Iron-oxide-enhanced MR imaging of bone marrow in patients with non-Hodgkin's lymphoma: Differentiation between tumor infiltration and hypercellular bone marrow. *Eur Radiol* 2002;12(6):1557–1566.
13. Daldrup-Link HE, Reinländer C, Link TM, Richter KJ, Könemann S, Rummeny EJ. Experimental studies of the value of SPIO for MRI of bone marrow before and after whole body irradiation. *Rofo* 2001;173(6):547–553.
14. Azoulay R, Olivier P, Baud O, Verney C, Santus R, Robert P, Gressens P, Sebag G. USPIO (ferumoxtran-10)-enhanced MRI to visualize reticuloendothelial system cells in neonatal rats: Feasibility and biodistribution study. *J Magn Reson Imaging* 2008 Oct; 28(4):1046–1052.
15. Fukuda Y, Ando K, Ishikura R, Kotoura N, Tsuda N, Kato N, Yoshiya S, Nakao N. Superparamagnetic iron oxide (SPIO) MRI contrast agent for bone marrow imaging: Differentiating bone metastasis and osteomyelitis. *Magn Reson Med Sci* 2006 Dec; 5(4):191–196.
16. Bierry G, Jehl F, Boehm N, Robert P, Dietemann JL, Kremer S. Macrophage imaging by USPIO-enhanced MR for the differentiation of infectious osteomyelitis and aseptic vertebral inflammation. *Eur Radiol* 2009;19(7):1604–1611.
17. Bodo J, Durkin L, Hsi ED. Quantitative *in situ* detection of phosphoproteins in fixed tissues using quantum dot technology. *J Histochem Cytochem* 2009;57(7):701–708.
18. Jaetao JE, Butler KS, Adolphi NL, Lovato DM, Bryant HC, Rabinowitz I, Winter SS, Tessier TE, Hathaway HJ, Bergemann C, Flynn ER, Larson RS. Enhanced leukemia cell detection using a novel magnetic needle and nanoparticles. *Cancer Res* 2009; 69(21):8310–8316.

19. Chatterjee DK, Rufaihah AJ, Zhang Y. Upconversion fluorescence imaging of cells and small animals using lanthanide doped nanocrystals. *Biomaterials* 2008;29(7):937–943.
20. Shinohara N, Matsumoto K, Endoh S, Maru J, Nakanishi J. *In vitro* and *in vivo* genotoxicity tests on fullerene C60 nanoparticles. *Toxicol Lett* 2009;191(2–3): 289–296.
21. Kim YS, Kim JS, Cho HS, Rha DS, Kim JM, Park JD, Choi BS, Lim R, Chang HK, Chung YH, Kwon IH, Jeong J, Han BS, Yu IJ. Twenty-eight-day oral toxicity, genotoxicity, and gender-related tissue distribution of silver nanoparticles in Sprague-Dawley rats. *Inhal Toxicol* 2008;20(6):575–583.
22. Cenni E, Granchi D, Avnet S, Fotia C, Salerno M, Micieli D, Sarpietro MG, Pignatello R, Castelli F, Baldini N. Biocompatibility of poly(D,L-lactide-co-glycolide) nanoparticles conjugated with alendronate. *Biomaterials* 2008;29(10):1400–1411.
23. Amaral AC, Bocca AL, Ribeiro AM, Nunes J, Peixoto DL, Simioni AR, Primo FL, Lacava ZG, Bentes R, Titze-de-Almeida R, Tedesco AC, Morais PC, Felipe MS. Amphotericin B in poly(lactic-co-glycolic acid) (PLGA) and dimercaptosuccinic acid (DMSA) nanoparticles against paracoccidioidomycosis. *J Antimicrob Chemother* 2009;63(3):526–533.
24. Bregoli L, Chiarini F, Gambarelli A, Sighinolfi G, Gatti AM, Santi P, Martelli AM, Cocco L. Toxicity of antimony trioxide nanoparticles on human hematopoietic progenitor cells and comparison to cell lines. *Toxicology* 2009;262(2):121–129.
25. Trouiller B, Reliene R, Westbrook A, Solaimani P, Schiestl RH. Titanium dioxide nanoparticles induce DNA damage and genetic instability *in vivo* in mice. *Cancer Res* 2009;69(22):8784–8789.
26. Moon EY, Yi GH, Kang JS, Lim JS, Kim HM, Pyo S. An increase in mouse tumor growth by an *in vivo* immunomodulating effect of titanium dioxide nanoparticles. *J Immunotoxicol* 2011;8(1):56–67.
27. Heng BC, Zhao X, Tan EC, Khamis N, Assodani A, Xiong S, Ruedl C, Ng KW, Loo JS. Evaluation of the cytotoxic and inflammatory potential of differentially shaped zinc oxide nanoparticles. *Arch Toxicol* 2011; 85(12):1517–1528.
28. Zhang XD, Wu HY, Wu D, Wang YY, Chang JH, Zhai ZB, Meng AM, Liu PX, Zhang LA, Fan FY. Toxicologic effects of gold nanoparticles *in vivo* by different administration routes. *Int J Nanomed* 2010;5:771–781.
29. Beauchesne PR, Chung NS, Wasan KM. Cyclosporine A: A review of current oral and intravenous delivery systems. *Drug Dev Ind Pharm* 2007;33(3):211–220.
30. Hamdy S, Elamanchili P, Alshamsan A, Molavi O, Satou T, Samuel J. Enhanced antigen-specific primary CD4+ and CD8+ responses by codelivery of ovalbumin and toll-like receptor ligand monophosphoryl lipid A in poly(D,L-lactic-co-glycolic acid) nanoparticles. *J Biomed Mater Res A* 2007;81(3):652–662.
31. Akhter S, Talegaonkar S, Khan ZI, Jain GK, Khar RK, Ahmad FJ. Assessment of ocular pharmacokinetics and safety of ganciclovir loaded nanoformulations. *J Biomed Nanotechnol* 2011;7(1):144–1445.
32. Pamujula S, Kishore V, Rider B, Fermin CD, Graves RA, Agrawal KC, Mandal TK. Radioprotection in mice following oral delivery of amifostine nanoparticles. *Int J Radiat Biol* 2005;81(3):251–257.
33. Pamujula S, Kishore V, Rider B, Agrawal KC, Mandal TK. Radioprotection in mice following oral administration of WR-1065/PLGA nanoparticles. *Int J Radiat Biol* 2008;84(11):900–908.

34. Jena G, Vikram A, Tripathi DN, Ramarao P. Use of chemoprotectants in chemotherapy and radiation therapy: The challenges of selecting an appropriate agent. *Integr Cancer Ther* 2010;9(3):253–258.
35. Brown AP, Chung EJ, Urick ME, Shield WP 3rd, Sowers AL, Thetford A, Shankavaram UT, Mitchell JB, Citrin DE. Evaluation of the fullerene compound DF-1 as a radiation protector. *Radiat Oncol* 2010;5:34.
36. Diener MD, Alford JM, Kennel SJ, Mirzadeh S. (212)Pb@C(60) and its water-soluble derivatives: Synthesis, stability, and suitability for radioimmunotherapy. *J Am Chem Soc* 2007;129(16):5131–5138.
37. Chandrasekharan DK, Khanna PK, Kagiya TV, Nair CK. Synthesis of nanosilver using a vitamin C derivative and studies on radiation protection. *Cancer Biother Radiopharm* 2011;26(2):249–257.
38. Schweitzer AD, Revskaya E, Chu P, Pazo V, Friedman M, Nosanchuk JD, Cahill S, Frases S, Casadevall A, Dadachova E. Melanin-covered nanoparticles for protection of bone marrow during radiation therapy of cancer. *Int J Radiat Oncol Biol Phys* 2010; 78(5):1494–1502.

Chapter 15

Nanoparticles, Immunomodulation and Vaccine Delivery

Sue D. Xiang,**, Martina Fuchsberger*, Tanya De L. Karlson*, Charles L. Hardy*, Cordelia Selomulya† and Magdalena Plebanski**

**Department of Immunology, Central Clinical School*
Faculty of Medicine, Nursing and Health Sciences
Monash University, 89 Commercial Rd
Melbourne, VIC 3004, Australia
†Department of Chemical Engineering
Monash University, Clayton
VIC 3800, Australia

Interest in nanoparticles and their use as vaccine carriers and adjuvants has greatly increased in recent times. However, despite current intense research in this field, the ways in which the immune system responds to nanoscale particulates are still being defined. This chapter will review the physical and chemical characteristics of nanoparticles 1–1000 nm in diameter, considering size, shape, surface charge and chemistry, and their effects on the immune system, including drainage to the lymph nodes (LNs), uptake by antigen-presenting cells (APCs) and the triggering of intracellular signalling pathways. We examine how particle size affects nanoparticle uptake by the key innate stimulators of the immune system, i.e., dendritic cells (DCs), and how nanoparticles modulate DCs and the induction of multiple arms of the immune response, including antibody production and CD4 and CD8 T cell responses via conventional and cross-priming pathways. We further discuss how inert nanoparticles, which by themselves may not necessarily promote the significant inflammation usually associated with adjuvants, can nevertheless induce

**Corresponding author. E-mail: Sue.Xiang@monash.edu

powerful immunity, suggesting nanotechnology has outstanding potential to deliver safe synthetic vaccines against today's major diseases such as cancer and malaria. Biodegradable or biocompatible nanoparticles, such as polymeric particles, chitosan, polystyrene, gold/silver particles and magnetic/metallic particles, are discussed in relation to the induction of immune responses and vaccine formulations. An in-depth understanding of how nanoparticles physicochemically modulate the immune system supports the rational development of nanoparticle-based vaccines, as well as safe nanoparticulate drug delivery systems.

1. Introduction

Since the first recorded vaccination by British physician and scientist Edward Jenner over two centuries ago (in 1796), vaccines have been used to prevent and treat a range of infectious diseases. For the most part, these vaccines have been based on live attenuated or inactivated pathogen-based formulations, which have the tendency to vary in quality or to induce adverse side effects. Over the years, many individual pathogen-derived target antigens have been discovered for potential inclusion into the new generation of non-live vaccines in purified or recombinant form (e.g., recombinant proteins, synthetic peptides, carbohydrates, lipids, DNA). Although these antigens present many safety advantages relative to immunisation with live attenuated vaccines, they are likely to have a short half-life *in vivo* and are not always processed and presented efficiently by the immune system. These antigens therefore require adjuvants and delivery systems to induce the desired immunological responses. The ideal adjuvants and delivery systems should enhance the uptake of the antigen, protect the antigen from degradation and activate immune cells without inducing adverse effects by the adjuvant itself. Currently, of the numerous adjuvants which have been developed, only very few have been approved for human use worldwide (Table 1).[1,2] Of these, the most widely used are: 1) aluminium hydroxide (alum), which has been shown to induce strong T helper 2 (Th2)-biased cellular and antibody responses (IgG1 and IgE), but is ineffective in stimulating T helper 1 (Th1) cellular responses;[3–6] 2) MF59, an oil-in-water emulsion, which has been used as an adjuvant for flu vaccines[7,8] and shown to promote both humoral (Th2)[9,10] and T cell proliferative[10,11] immune responses; and 3) AS04, a combination of MPL (lipopolysaccharide, LPS analogue monophosphoryl lipid A) and alum based adjuvant for viral vaccines, which has also been approved for human use in recent years.[2,12] Unlike alum, AS04 induces both Th1 and antibody responses.[13] However, both these oil emulsions and alum are poor inducers of the cytotoxic T lymphocyte (CTL) response. Yet, many of today's most common

Table 1. Currently licensed human vaccine adjuvants.

Common name and licensed company	Composition	Nanoparticles	Mechanisms of action and functions	Applications	References
Alum (various)	Aluminium hydroxide or aluminium phosphate or potassium aluminum sulfate	No	Relatively weak adjuvant, but induce Th2-biased cellular and antibody responses (e.g., IgG1 and IgE). Poor CTL response. Good safety record, but can induce local inflammation. Difficult to manufacture in a physiochemically reproducible way.	HAV/HBV, diphtheria, tetanus, pertussis, HPV, influenza, polio virus, etc.	4–6, 169, 170
MF59 (Novartis)	Squalene-based oil-in-water emulsions	Yes (200–300 nm in size)	Promote T cell proliferation and induce Th2 response. Strong antibody response and cytokine production. Induce CTL response. Good safety profile, but can induce local reactogenicity and systemic reactions in young children.	Influenza (Fluad®)	8, 171, 172

(Continued)

Table 1. (*Continued*)

Common name and licensed company	Composition	Nanoparticles	Mechanisms of action and functions	Applications	References
MPL® (Corixa/GSK)	Hydrolysed lipid A from LPS	No	Activates through TLR-4, induces Th1-biased responses. Strong antibody (IgG1 and IgG2a), cytokines (e.g., IL12, IFN-γ, TNF, etc.) responses. Poor CTL response. Difficulty to keep batch-to-batch consistency for formulations, limiting their usage. Reduced toxicity but can induce local inflammation at injection site.	HBV (Fendrix®), HPV (Cervarix®), EBV	173–175
AS03 (GSK)	Squalene oil-in-water emulsion + tocopherol	Yes (150–155 nm in size)	Good safety and acceptable reactogenicity profile. High titre and persistence of antibody production (over six months). Induce persistent vaccine specific $CD4^+$ T cell responses (over six months).	Pre-pandemic H5N1 vaccine (Pandemrix®)	176–179

(*Continued*)

Table 1. (*Continued*)

Common name and licensed company	Composition	Nanoparticles	Mechanisms of action and functions	Applications	References
AS04 (GSK)	MPL + Alum	No	Induce broad and long lasting responses (for both Th1 and antibody responses). Poor CTL activity. Local inflammation.	HBV (Fendrix®) and HPV (Cervarix®) vaccines	13, 180, 181
Liposome/Virosome (Crucell)	Oil-in-water emulsion	Yes (~150 nm in size)	Phospholipid vesicles that are taken up by APCs and mediate MHC class I and class II antigen presentation. Induce excellent humoral immune response. Stability, manufacturing and quality assurance problems are the major factors that limit their use	Influenza (Inflexal®V) and Hepatis A (Epaxal®)	182, 183

Alum: aluminium salt, CTL: cytotoxic T lymphocyte, EBV: Epstein–Barr virus, HAV: hepatitis A virus, HBV: hepatitis B virus, HPV: human papillomavirus, LPS: lipopolysaccharide, MPL: monophosphoryl lipid A, Th1: T helper 1, Th2: T helper 2.

diseases, such as cancer, HIV and malaria, require a strong CTL response for effective therapeutic treatment.[14] In addition, some of these adjuvants have the potential to cause severe local and potentially systemic side effects, although fortunately, serious cases are rare.[15] Hence, the challenge still remains to develop safer and more efficient adjuvants for effective vaccine delivery.

The use of nanoparticles in medical applications has increased markedly in recent decades. For example, synthetic polymer micro- and nanoparticles, such as poly(D,L-lactide-co-glycolide) (PLG) and poly(D,L-lactic-coglycolic acid) (PLGA) have long been used in drug delivery systems to encapsulate drugs and protect them from extracellular enzymatic degradation, provide vehicles for delivering drugs to the target area and are being investigated as powerful platforms for vaccine delivery.[16,17] The formulation of vaccines with particles of different materials in the viral (20–200 nm) or bacterial (200–5,000 nm) size ranges can often provide a stable half-life, prevent the enzymatic degradation of the antigen and help initiate antigen uptake and processing by specifically targeting specialist antigen-presenting cells (APCs), such as dendritic cells (DCs).[18–21] The unique properties of nanoparticles, which can be designed and modified according to their intended application, make them excellent tools for vaccine and drug delivery in the battle against a range of different diseases. The use of nanoparticles as vaccine adjuvants and carriers has opened up a whole new field of enquiry.

The idea that the immune system has evolved to differentially handle particles in a unique size-dependent fashion has been growing in acceptance. However, the spectrum of immunomodulatory activities induced by diverse nanoparticles remains largely unmapped. New understanding in this area may help drive successful and safe synthetic nanovaccine design.[22] In this chapter, we will discuss nanoparticle characteristics which affect their uptake by immune cells and the responses induced, as well as nanoparticle effects on intracellular signalling pathways and their applications in vaccine development.

2. Vaccine Design and Immune Responses

It is well known that an infectious milieu promotes DC activation and the development of immunity, while absence of infection in the homeostatic state fails to do so. DCs are the "sentinels" of the immune system and can both induce and modulate the immune response. DCs are key T cell activators and regulators, and for this reason, they are the target of most vaccines. Traditional vaccine design has been focused on the use of "danger signals" (signals from non-host cells which can activate host DCs) in combination with a specific

Table 2. The advantages of particulate adjuvants.[38–40]

- Improving the solubility of poorly soluble antigens
- Releasing antigens for a sustained and controlled period of time, therefore reducing the number of administrations
- Targeting antigens to specific areas in the body, resulting in diminished side effects
- Preventing the degradation of the antigen by using encapsulation methods
- Delivering multiple antigens concurrently
- Lower therapeutic toxicity

antigen to induce the desired types of immune responses.[23–26] For example, antibody-mediated responses are preferred for viral and bacterial infections; therefore, adjuvants such as alum, which can promote the induction of antibody responses, are used.[27–29] T cell-mediated responses (including both CD4 and CD8 T cells) are generally required for the control of "intracellular" pathogens, such as virus-infected or cancerous cells; hence, adjuvants such as Quil A or CpG, which induce potent T cell responses, are used.[30–32] Moreover, DCs are uniquely capable of cross-priming, a process by which exogenous antigens taken up via endocytic pathways can be processed into peptides for presentation on cell surface major histocompatiblity complex (MHC) class I molecules,[33,34] leading to the activation of $CD8^+$ T cells and differentiation into fully functional CTLs.[34] CTL responses are important for the elimination of infections by intracellular pathogens such as liver stage malaria[35,36] or tumour cells.[37] For more information on DCs, please refer to Chapter 13.

The use of nanoparticles as vaccine carriers has generated much interest of late. Particulate adjuvants can offer great flexibility, stability and advantages in the design of vaccines (summarised in Table 2 and also reviewed in Xiang *et al.*[38] and others[39,40]). Therefore, nanoparticles have become an attractive option for the design of antigen carrier elements in vaccines. The particles used in vaccine formulations are usually comparable in size to pathogens recognised by the immune system (<5,000 nm). It has become apparent that the size, surface charge and shape of the particles are crucial in determining the success of an immune response.[33,41,42]

3. Nanoparticle Modulation of Immune Cells

3.1. *Size dependence of nanoparticle uptake*

The size-dependent uptake of nanoparticles by immune cells has been reported in various studies.[4,43,44] Depending on their size, particles are internalised by APCs via different mechanisms.[41,45] Virus-sized particles (20–200 nm) are usually taken up by endocytosis via clathrin-coated vesicles (pits), caveolae or their independent receptors,[46] and are preferentially ingested by DCs. Larger-sized particles (>500 nm) however, are primarily taken up by macrophages via macropinocytosis or phagocytosis. Particle size therefore can determine the type of APC targeted.[33,41] Many studies have shown that immunisation with nanoparticles in the viral size range (20–200 nm) elicits stronger immune responses than larger-sized particles.[41,42,45,47,49]

In our studies using polystyrene particles, size was shown to be a critical parameter affecting immunogenicity. Smaller particles (<500 nm), in particular 40–50 nm, were significantly more potent at inducing $CD4^+$ and $CD8^+$ T cell responses than larger ones (>500 nm). However, the latter were better at inducing IFN-γ production and antibody responses.[41,42,47,48] Smaller nanoparticles (20–50 nm) were found to travel rapidly to the draining lymph nodes (LNs) as early as 2 h post-injection (intradermal),[49] likely due to direct lymphatic drainage and targeted the resident LN DCs.[25] These small particles had not only been taken up by LN resident DCs, but also by B cells and macrophages. In contrast, larger particles (>500 nm) were only found 48 h post-injection and were primarily associated with DCs, which were derived from peripheral phagocytic monocytes and also Langerhans cells. This suggested that large particles rely mostly on cell-associated migration to reach the lymphatic organs, whereas smaller particles utilise both cell-associated migration and lymphatic drainage.[49] In addition, a small proportion of the 20 nm particles was associated with plasmacytoid DCs (pDCs), while none of the larger (1,000 nm) particles were. Both nano- and microparticles could still be detected in their respective LN compartments eight days after injection. Eventually, smaller nanoparticles (<30 nm) were eliminated by renal excretion;[50] 30–150 nm nanoparticles have been found to accumulate in the bone marrow, heart, kidney and stomach,[51] while larger nanoparticles (>200 nm) were rapidly taken up by the cells of the reticuloendothelial system (RES) and accumulated mostly in the liver.[52]

Various strategies have been employed to enhance the potential uptake by DCs, such as modifications of nanoparticle size, shape, surface chemistry and charge. Our own studies indicated that inert particles of 40–100 nm were taken up more efficiently by $DEC205^+$ DCs, while 1 μm particles were predominantly taken up by $F4/80^+$ macrophages *in vivo*.[41] In agreement, others have demonstrated that the optimal nanoparticle diameter for uptake by DCs is ~50 nm[53–55] independent of materials used, including gold,[56] silver,[57] single-walled carbon nanotubes,[58] or inorganic metal hydroxide (LDH).[59]

The size of nanoparticles can also determine their immunostimulatory profile,[60] with <200 nm preferentially taken up by pDCs, and inducing IFN-α production, while 500–1,000 nm particles were phagocytosed by monocytic DCs, which induced TNF-α. Furthermore, nanoparticles can be used to induce a specific type of immunity,[42] with 40 nm particles promoting Th1 and $CD8^+$ T cell responses, and 100 nm particles promoting Th2 responses.[41,42,47] Microparticles do not target the cross-presenting $CD8^+$ DC subset, yet they are still able to induce strong $CD4^+$ and $CD8^+$ T cell responses.[61] This is most likely due to their capacity to recruit inflammatory monocytes, which then can differentiate into DCs capable of cross-priming. Therefore, not only could the use of nanoparticles within the correct size range result in enhanced antigen uptake, but also lead to the induction of the preferred immune response.

3.2. *Nanoparticle surface charge and chemistry*

Surface charge and chemistry also play a significant role in determining the type of immune response generated, since the chemical nature of the particles strongly influences phagocytosis.[38,62–64] Manipulating the surface charge can be used to modify the immune response, extend the half-life of the drug *in vivo,* guide nanoparticles to specific targets within the body or help nanoparticles to cross barriers like mucus,[65] cell membranes and even endosomal compartments. For example, positively charged cationic particles in general have greater affinity towards the cell surface than negatively charged or neutral particles,[62] presumably due to the strong ionic attraction between the positively charged particles and the negatively charged cell surface.[33] Therefore, positively charged particles can enhance particle uptake by DCs[62,63,66] and induce DC maturation.[25,67–70] The beneficial ionic attraction between the positively charged nanoparticles and the negatively charged surface of cells is likely to represent a strong stimulus that initiates binding and subsequent internalisation.[71] It has been reported that PLG particles of ~300 nm with a cationic surface improved the delivery of adsorbed DNA into APCs after intramuscular injection and significantly enhanced antibody and CTL responses at a low dose in comparison to naked DNA.[72] Cationic liposomes have been shown to enhance the delivery efficiency of DNA–gold nanoparticles (30–80 nm) into cancer cells.[73] In our study, DNA adsorbed to cationic polystyrene nanoparticles (40–50 nm) enhanced antibody and CD8 T cell responses after intradermal administration compared to naked DNA.[48] Furthermore, cationic nanoparticles (e.g., PEG–PLA, ~70–90 nm) have the ability to escape endolysosomal degradation compared to anionic nanoparticles of the same size.[74]

Nanoparticles with a neutral or hydrophilic surface charge can significantly reduce opsonisation and thus clearance rate by the RES.[75] Polyethylene glycol (PEG)ylation of the surface of nanoparticles appears to be the most widely used strategy with the least unwanted side effects to minimise opsonisation.[75] The effects of PEGylation are diverse and depend on a number of factors such as the type of nanoparticles used, animal species and the length and density of the PEG chain on the nanoparticle surface.[76,77] PEGylation can also extend the circulation time in the blood by minutes or even days.[75] In contrast to the initial belief that PEG was not immunogenic,[78] a recent study showed that repeated injections of PEGylated liposomes induced anti-PEG IgM antibodies that led to accelerated blood clearance of the nanoparticles.[79] It was furthermore demonstrated that the extent of this clearance depended not only on the physiochemical properties of the injected liposomes, but also on the time between injections, lipid dose and drug encapsulation.[79] Poly(ethylene carbonate) (PEC) nanoparticles (~200 nm) have also been shown to be excellent in repelling opsonins and binding significantly less serum proteins compared to polystyrene nanoparticles.[80] Furthermore, due to their unique hydrophilic properties, these particles showed a significantly diminished clearance by macrophages. As such, these particles present a promising candidate as long-lasting circulating nanocarriers.

3.3. *Nanoparticle shape*

The shape of the nanoparticle also affects its cellular uptake[64] and distribution.[81] Spherical particles have been shown to be taken up by cells (such as cancer cell lines) more efficiently than rod-shaped particles,[64,82] and oblate ellipsoids have a greater internalisation rate into cells than prolate ellipsoids or spheres.[83] Cylindrically shaped nanoparticles have been shown to evade non-specific uptake by macrophages of the RES. Such nanoparticles can be found circulating in the blood for up to one week after injection.[84] A similar effect was also observed for disc-shaped nanoparticles.[85] Non-spherical nanoparticles have been shown to have higher accumulation and retention rates in cancer tissues *in vivo* compared to spherical nanoparticles (as reviewed by Tao *et al.*).[86] Another elegant study by Decuzzi *et al.*[81] showed that particle shape influences particle distribution in various organs, including the lung, liver, spleen, heart and brain. Interestingly, disc-shaped particles tended to accumulate more than cylindrical or hemi-spherical particles in most organs with the exception of the liver. This suggested that discoidal particles can avoid clearance by the RES and subsequent sequestration within the liver. Other studies have shown that short-rod mesoporous silica nanoparticles (MSNPs) were found mostly in the liver, while long-rod MSNPs accumulated preferentially in the spleen within 2 h post-injection (i.v.). When coated with PEG, the MSNPs migrated preferentially to the lung, regardless of their shape.[86] Furthermore, clearance of particles is also shape-dependent, with short-rod MSNPs cleared more rapidly than long-rod MSNPs.[87] A study on nanoparticle shapes and their effect on flow and adhesion of drug carriers in blood vessels showed that discs and rods adhered with a higher efficiency compared to spheres of the same volume.[88]

4. Nanoparticles and Cell Signalling Pathways

In order to further understand the adjuvant/carrier effects of nanoparticles and optimise the vaccine formulation to obtain the desired outcomes, the effects of nanoparticles on the activation of cell signalling pathways have been recently investigated. However, such studies are still in their infancy, and much remains to be resolved. Preliminary data studying airborne nanoparticles suggested that they have an impact in the expression of reactive oxygen species, causing inflammatory responses.[89,90] The study of oxidative stress by nanoparticles has demonstrated that particles such as titanium dioxide, iron oxide, silver and carbon can induce mitogen-activated protein kinase (MAPK), extracellular signal-regulating kinase (ERK), p38, c-Jun N-terminal kinase (JNK) and nuclear factor-κB (NF-κB) pathways.[91–95] It has been reported that ovalbumin (OVA)-conjugated poly(γ-glutamic acid) (γ-PGA) nanoparticles can induce the activation of CTL activity measured by the production of OVA-specific IFN-γ. These responses were attributed to γ-PGA efficiency in activating signalling cascades through the engagement of myeloid differentiation factor 88 (MyD88)-mediated NF-κB and p38 MAPK.[96,97] A recent study also showed that murine DCs treated with NF-κB stimulators TLR3 and TLR9 agonists poly(I:C) and CpG ODN together with OVA and PLGA nanoparticles, respectively, resulted in prolonged and enhanced MHC class I presentation.[98] Layered double hydroxide nanoparticles (~57 nm) have also been reported to activate bone marrow-derived DCs by inducing the expression of TNF-α, IL-12, CD86 and CD40 through the engagement of the NF-κB signalling pathway.[99]

Comprehensive insights into signalling activation pathways have encouraged the development of vaccines which target specific components of signalling pathway cascades. This knowledge has been especially useful in the design of cancer vaccines. Signalling pathways involving NF-κB, MAPK, PI3K/AKT and the Src family of non-receptor tyrosine kinases (SFK) are especially important with regard to site-specific targeting by nanoparticles due to their activation in various types of cancer.[100–104] These signalling pathways are involved in cell proliferation, apoptosis and cell survival, making them key players in the regulation of immune responses and therefore, excellent targets for development of targeted drug/vaccine delivery. For example, PD98059, a MAPK inhibitor, was coated onto PLGA nanoparticles and shown to inhibit tumour growth *in vivo* and the proliferation of melanoma and lung carcinoma cells *in vitro*.[105] Polyhydroxyalkanoate (PHA) nanoparticles loaded with a PI3K inhibitor (TGX221 or LY294002) were shown to inhibit the proliferation of prostate cancer cells,[106] proliferation and apoptosis of mouse melanoma cells and block angiogenesis *in vitro* and *in vivo*.[107]

5. Nanoparticle-based Vaccines

Currently, there are number of biodegradable or biocompatible nanoparticles such as polymeric particles, chitosan, gold/silver particles and magnetic/metallic particles, which have shown great promise for improving antigen-specific immune responses.[38,108] These delivery systems can be used to target APCs directly or indirectly.[109] Features of these nanoparticles and their uses in vaccine formulations are discussed below.

5.1. *Polymeric particles*

The use of polymeric particles has recently increased in popularity, as they can be used to encapsulate the antigen, protecting it from degradation and controlling its extended release.[109] The most widely used polymeric particles are PLGA, PLG, poly(D,L-lactide) (PLA) and poly(ortho-esters) (POE).[33,109,110] These polymers generally form microparticles of size >200 nm,[33] are biodegradable or biocompatible and have been approved for use in humans.[110] As well as encapsulation, the antigen can be adsorbed onto the surface of the polymeric particles.[110] Other advantages of polymeric particles include their capacity to encapsulate large-sized agents, passive targeting to APCs, and the improvement of uptake and targeting via conjugation to suitable functional groups.[111] However, the main disadvantages of using polymeric particles are issues with antigen stability during encapsulation, storage and release.[112]

Comparative studies have shown that nanoparticles are more efficient as antigen carriers/adjuvants in increasing antigen uptake and inducing an immune response as compared to alum.[113,114] PLGA and PLA nanoparticles with size >200 nm are the most extensively investigated and used in human vaccines because of their biocompatibility and biodegradability. Their favoured use as antigens/drug carriers is due to their capacity to escape endosomal–lysosomal degradation, allowing efficient delivery of their antigenic cargo to the appropriate target cell.[115] PLGA and PLA nanoparticles can also be modified to protect antigens/drugs from enzymatic degradation. For example, PLGA particles were shown to enhance and prolong cross-presentation of encapsulated OVA by MHC-I on bone marrow-derived DCs.[116,117] Similarly, as mentioned above, γ-PGA nanoparticles were more efficient at promoting the activation of antigen-specific $CD8^+$ T cells by bone marrow-derived DCs compared to OVA–alum and OVA–Freund's adjuvant.[118] Furthermore, the conjugation of specific ligands to the nanoparticle surface can target uptake to specific APCs and tissues. For example, transferrin-coated PLGA nanoparticles have been shown to be highly absorbed by brain endothelial cells,[119] while microparticles coated with anti-CD205 were taken up by $CD8^+$ DCs much more efficiently then their naïve counterparts.[120] Therefore, nanoparticles can be engineered to be highly selective for the intended targets by choosing the appropriate ligand. The receptor–ligand binding can then facilitate internalisation through endocytic pathways, and the strength of these interactions can be controlled by the type of ligand used (e.g., affinity) and the density at which the ligand is attached to the nanoparticle surface (e.g., avidity). Using the right targeting agents, nanoparticles can be directed with high selectivity towards the diseased tissues *in vivo*.[121,122]

A number of studies have shown that PEGylated nanoparticles of <100 nm in diameter have a reduced propensity to absorb plasma proteins onto their surface and consequently, have lower hepatic filtration. This leads not only to a longer half-life within the bloodstream, but also to much higher extravasation into permeable tissues. This can be exploited in passively targeting[123] nanoparticle-based vaccine delivery systems towards tumour sites. For example, it was shown that biodegradable γ-PGA nanoparticles were more efficient in inducing an adaptive immune response through the induction of the innate immunity. DCs exposed to these nanoparticles produced pro-inflammatory cytokines such as IL-12p70, TNF and upregulated co-stimulatory molecules, such as CD40, CD80 and CD86.

5.2. *Chitosan*

Chitosan, a linear polysaccharide derived from chitin obtained from crustacean shells, has emerged as a useful drug delivery matrix because of its polycationic nature, biodegradability, biocompatibility, mucoadhesiveness and ease of physical and chemical modification.[124–127] Chitosan contains D-glucosamine and N-acetyl-D-glucosamine units, which are linked by glycosidic bonds.[128]

It can be easily linked to other polymers such as PEG,[129] tripolyphosphate (TPP)[130] or PLGA[131] to form chitosan nanoparticles and enhance protein/antigen loading. Chitosan has recently gained much attention as a potential DNA vaccine delivery vehicle[109,132,133] as it is naturally positively charged, making it easy to encapsulate the negatively charged DNA. In addition, its inherent mucoadhesiveness[131] facilitates the transport of peptides and proteins across mucosal barriers,[134,135] providing advantages for the delivery of antigens via the oral and nasal routes.

5.3. *Non-degradable but biocompatible nanoparticles*

Non-degradable particles of latex, gold, silica or polystyrene have successfully been used as antigen carriers and even adjuvants.[41,42,47,48,136–139] Antigens coupled to these particles induce long-term immune reactivity, possibly because the particles persist in the tissues, providing a reservoir of antigens for continued stimulation. Our own studies have shown that polystyrene nanoparticles, while gradually cleared from the lung, can still be detected up to one month after instillation.[140] Gold nanoparticles have been shown to enhance the efficiency of DNA vaccination by promoting both Th1 and Th2 responses, in addition to the predominant Th1 response observed after vaccination with DNA alone.[138,139] As indicated above, our studies have shown that 40–50 nm polystyrene particles specifically localise to DCs and induce high levels of IFN-γ, as well as high antibody titres when coupled with the OVA protein,[41,108,141] and such responses were superior or comparable to Freund's adjuvant, which has been used as a "standard" adjuvant, in a size-dependent manner. We have further extended the potential use of these nanoparticle carriers to the development of effective DNA cancer vaccines,[48] which have the rare and beneficial property of eliciting both high levels of CD8 T cell responses (IFN-γ) and high antibody titres.

Latex and silica nanoparticles have also been tested *in vivo* for diagnostic and biomedical applications. For example, exogenous antigens, when attached to latex particles, were presented 1,000–10,000-fold more efficiently by the MHC class I cross-priming pathway than soluble antigen[142] or antigen presented via MHC class II.[143] Silica nanoparticles have also been used as gene delivery vectors in both basic research and clinical trials.[144–146]

5.4. *Magnetic nanoparticles*

Magnetic nanoparticles have emerged as promising agents in many biomedical applications, including in magnetic resonance imaging,[147] targeted drug delivery,[148] and as vaccine carriers.[149–152] Different forms of these nanoparticles can be engineered to specific morphologies (spherical, rod-like, hollow, core-shell), formed as alloys or into hybrid forms with other oxides or polymers.[153] Among the metallic particles (Fe, Ni, Co, Cu) and their oxides, alloys (FePt, FePd, FeCo) and ferromagnets ($CoFe_2O_4$, $MgFe_2O_4$), only iron oxide nanoparticles such as magnetite (Fe_3O_4) and maghemite (γ-Fe_2O_3) are classified as non-toxic and biocompatible by the U.S. Food and Drug Administration.[154]

When Fe_3O_4 and γ-Fe_2O_3 nanoparticles are sufficiently small (<20 nm), they typically exhibit superparamagnetic behaviour, though this size limit also depends on the morphology, crystallinity and particle–particle or particle–matrix interactions.[155] Superparamagnetic iron oxide nanoparticles (SPIONs) only display magnetic behaviour in the presence of an external magnetic field, and have no remnant magnetisation when the field is removed.[156] This unique characteristic is useful when utilising these particles for *in vivo* applications, as the tendency to aggregate due to magnetic interactions can be minimised, while there is also the potential for direct targeting to specific organs using external magnets.

In the preparation of magnetic nanoparticles with a specific size and morphology, organic surfactants and metal precursors (metal acetylacetonates, metal oleates) are often needed, thus producing magnetic nanoparticles covered with long-chain hydrocarbons and hydrophobic surfaces.[153] Consequently, surface modifications of magnetic nanoparticles are required to improve their biocompatibility and dispersion in aqueous media, as well as to enable interaction and recognition with targeted biomolecules. Several strategies to modify the surface of SPIONs have been explored, including the direct adsorption of amphiphilic molecules such as PEG to improve aqueous solubility,[158] the addition of a secondary layer such as silanisation,[157] the formation of core-shell structures, ligand or surfactant exchange and ionic interactions.[148] Examples of non-polymeric coating materials include gold, silica and carbon, while polymeric materials such as chitosan, dextran, pullulan, PLGA, PVA, PEG and lipids have been used to provide long-term stability of these particles in biological solutions.[159,160]

In gene therapy, the concept of magnetic drug delivery was introduced in the late 1970s, inspiring the term "magnetofection" to enhance the delivery of viral and non-viral vectors coupled with superparamagnetic nanoparticles under the influence of an external magnetic field at low doses *in vitro* and site-specifically *in vivo*.[151] The transfection efficiency can be significantly enhanced with a lower dose of the vector, while the duration of the transfection process can be reduced to several minutes, rather than a few hours of incubation period as with standard protocols.[161] Chapman *et al.* demonstrated that the transfection efficiency in osteoblasts and macrophages could be enhanced at least four- to eight-fold under a static magnetic field for 5 min, as compared to conventional transfection methods.[162] The additional application of a pulsating magnetic field significantly increased the transfection efficiency in the primary cell lines, as the movements of the particles could assist in their translocation across the cell membrane.[162] Although the higher uptake of magnetic nanoparticles has been attributed to better sedimentation of particles attracted by the magnets to the cells, the gene expression efficiency would still depend on the design of the carriers to protect and release DNA upon internalisation.[163] Recently, we demonstrated the use of SPIONs/polyethylenimine polymer complexes to enhance the condensation and delivery of malaria DNA vaccine (MSP_{1-19}) into eukaryotic cells *in vitro*, as indicated by significantly higher MSP_{1-19} expressions under an external magnetic field, while the cytotoxicity of the complexes was comparable to a benchmark non-viral reagent of Lipofectamine 2000.[164]

6. Challenges of Nanoparticulate Vaccines

Although the use of nanoparticle-based adjuvants and delivery systems is undoubtedly beneficial, concerns over nanoparticles toxicity and safety still remain. The fact that nanoparticles have such large surface areas increases the potential for adverse interactions and are highly reactive.[110] Furthermore, nanoparticles are able to cross many biological membranes, and slow degradation rates could lead to the accumulation of nanoparticles in organs, although the use of biodegradable particles would eliminate this risk.[110] However, it is practical to assume that small doses and/or only occasional administration of nanoparticles may render them safe and non-toxic.[110] Indeed, in contrast to expectations based on the literature, our recent studies showed that inert polystyrene nanoparticles instilled into the lung can inhibit, rather than promote, experimental asthma.[140] Ideally, the adjuvant or delivery system should induce cellular, humoral and mucosal immune responses for an extended period of time.[110] The adjuvant components should also be easily broken down by the body.[110] Additionally, routes of administration which are less invasive and more effective should be identified, and the cost-effective production and distribution of the vaccines should be explored.[110]

"New-generation" nanoparticles are being designed with multi-functional properties that will enable them to selectively target specific cell types and be internalised more efficiently. For example, to improve the efficacy of drug delivery, novel nanoparticles have been developed that release the drug only in response to external stimuli at the target site.[165] Such stimuli can be physical signals such as temperature, electrical or magnetic fields, ultrasound or chemical signals such as pH, ionic strength or enzymatic activities. Examples of such nanoparticles used to develop these multifunctional platforms are liposomes, polymeric micelles and polyplexes.[166,167] A different approach to increase the uptake of nanoparticles by peripheral DCs and monocytes is to use nanoparticles that can gradually release chemoattractants at the site of injection over an extended period of time (>seven days) for the selective recruitment of the target cells.[168]

In summary, nanoparticles provide great advantages over non-particle–based adjuvants in vaccine development. The efficiency of such vaccines depends on the interactions between the nanoparticles and the target cells, which are generally influenced by particle characteristics such as size, shape and surface properties, including surface charges and functional groups. As such, nanoparticle formulation can be rationally designed in order to elicit the desired immune responses.

References

1. Lambrecht BN, Kool M, Willart MAM, Hammad H. Mechanism of action of clinically approved adjuvants. *Curr Opin Immunol* 2009;21(1):23–29.
2. Mbow ML, De Gregorio E, Valiante NM, Rappuoli R. New adjuvants for human vaccines. *Curr Opin Immunol* 2010;22(3):411–416.
3. Rajesh KG. Aluminum compounds as vaccine adjuvants. *Adv Drug Delivery Rev* 1998;32(3):155–172.
4. Brewer JM, Conacher M, Hunter CA, Mohrs M, Brombacher F, Alexander J. Aluminium hydroxide adjuvant initiates strong antigen-specific Th2 responses in the absence of IL-4- or IL-13-mediated signaling. *Immunology* 1999;163(12):6448–6454.
5. HogenEsch H. Mechanisms of stimulation of the immune response by aluminum adjuvants. *Vaccine* 2002;20(Suppl 3):S34–9.
6. McKee AS, Munks MW, Marrack P. How do adjuvants work? Important considerations for new generation adjuvants. *Immunity* 2007;27(5):687–690.
7. Tagliabue A, Rappuoli R. Vaccine adjuvants: The dream becomes real. *Hum Vaccines* 2008;4(5):347–349.
8. Vesikari T, Karvonen A, Tilman S, Borkowski A, Montomoli E, Banzhoff A, *et al.* Immunogenicity and safety of MF59-adjuvanted H5N1 influenza vaccine from infancy to adolescence. *Pediatrics* 2010;126(4):e762–e770.
9. Higgins DA, Carlson JR, Van Nest G. MF59 adjuvant enhances the immunogenicity of influenza vaccine in both young and old mice. *Vaccine* 1996;14(6):478–484.
10. O'Hagan DT, Rappuoli R, De Gregorio E, Tsai T, Del Giudice G. MF59 adjuvant: The best insurance against influenza strain diversity. *Expert Rev Vaccines* 2011;10(4):447–462.
11. Wack A, Baudner BC, Hilbert AK, Manini I, Nuti S, Tavarini S, *et al.* Combination adjuvants for the induction of potent, long-lasting antibody and T-cell responses to influenza vaccine in mice. *Vaccine* 2008;26(4):552–561.
12. Coffman RL, Sher A, Seder RA. Vaccine adjuvants: Putting innate immunity to work. *Immunity* 2010;33(4):492–503.

13. Didierlaurent AM, Morel S, Lockman L, Giannini SL, Bisteau M, Carlsen H, *et al.* AS04, an aluminum salt- and TLR4 agonist-based adjuvant system, induces a transient localized innate immune response leading to enhanced adaptive immunity. *J Immunol* 2009;183(10):6186–6197.
14. Rappuoli R. Bridging the knowledge gaps in vaccine design. *Nat Biotechnol* 2007;25(12):1361–1366.
15. Petrovsky N, Aguilar JC. Vaccine adjuvants: Current state and future trends. *Immunol Cell Biol* 2004;82(5):488–496.
16. Kumari A, Yadav SK, Yadav SC. Biodegradable polymeric nanoparticles based drug delivery systems. *Colloids Surf B* 2010;75(1):1–18.
17. Pinto Reis C, Neufeld RJ, Ribeiro AJ, Veiga F. Nanoencapsulation I. Methods for preparation of drug-loaded polymeric nanoparticles. *Nanomed Nanotechnol Biol Med* 2006;2(1):8–21.
18. Cruz LJ, Rueda F, Cordobilla B, Simón L, Hosta L, Albericio F, *et al.* Targeting nanosystems to human DCs via Fc receptor as an effective strategy to deliver antigen for immunotherapy. *Mol Pharm* 2009;8(1):104–116.
19. Yoshida M, Babensee JE. Poly(lactic-co-glycolic acid) enhances maturation of human monocyte-derived dendritic cells. *J Biomed Materials Res Part A* 2004;71A(1):45–54.
20. Cruz LJ, Tacken PJ, Fokkink R, Figdor CG. The influence of PEG chain length and targeting moiety on antibody-mediated delivery of nanoparticle vaccines to human dendritic cells. *Biomaterials* 2011;32(28):6791–6803.
21. Fischer S, Uetz-von Allmen E, Waeckerle-Men Y, Groettrup M, Merkle HP, Gander B. The preservation of phenotype and functionality of dendritic cells upon phagocytosis of polyelectrolyte-coated PLGA microparticles. *Biomaterials* 2007;28(6):994–1004.
22. Oberdorster G, Oberdorster E, Oberdorster J. Nanotoxicology: An emerging Discipline evolving from studies of ultrafine particles. *Environ Health Perspect* 2005;113(7):823–839.
23. Leroux-Roels G. Unmet needs in modern vaccinology: Adjuvants to improve the immune response. *Vaccine* 2010;28(Suppl 3):C25–C36.
24. Reddy ST, Swartz MA, Hubbell JA. Targeting dendritic cells with biomaterials: Developing the next generation of vaccines. *Trends Immunol* 2006;27(12):573–579.
25. Reddy ST, van der Vlies AJ, Simeoni E, Angeli V, Randolph GJ, O'Neil CP, *et al.* Exploiting lymphatic transport and complement activation in nanoparticle vaccines. *Nat Biotech* 2007;25(10):1159–1164.
26. Caskey M, Lefebvre F, Filali-Mouhim A, Cameron MJ, Goulet JP, Haddad EK, *et al.* Synthetic double-stranded RNA induces innate immune responses similar to a live viral vaccine in humans. *J Exp Med* 2011;208(12):2357 2366.
27. Pulendran B, Ahmed R. Immunological mechanisms of vaccination. *Nat Immunol* 2011;131(6):509–517.
28. Bungener L, Geeraedts F, Ter Veer W, Medema J, Wilschut J, Huckriede A. Alum boosts Th2-type antibody responses to whole-inactivated virus influenza vaccine in mice but does not confer superior protection. *Vaccine* 2008;26(19):2350–2359.
29. Kang SJ, Locksley RM. The inflammasome and alum-mediated adjuvanticity. *F1000 Biol Rep* 2009;1:15.
30. Kensil CR. Saponins as vaccine adjuvants. *Crit Rev Ther Drug Carrier Syst* 1996; 13(1–2):1–55.
31. Yu H, Babiuk LA, van Drunen Littel-van den Hurk S. Priming with CpG-enriched plasmid and boosting with protein formulated with CpG oligodeoxynucleotides and Quil A

induces strong cellular and humoral immune responses to hepatitis C virus NS3. *J Gen Virol* 2004;85(6):1533–1543.
32. Li WM, Dragowska WH, Bally MB, Schutze-Redelmeier M-P. Effective induction of CD8+ T-cell response using CpG oligodeoxynucleotides and HER-2/neu-derived peptide co-encapsulated in liposomes. *Vaccine* 2003;21(23):3319–3329.
33. Xiang SD, Scholzen A, Minigo G, David C, Apostolopoulos V, Mottram PL, *et al.* Pathogen recognition and development of particulate vaccines: Does size matter? *Methods* 2006;40(1):1–9.
34. Yewdell JW. Designing CD8+ T cell vaccines: It's not rocket science (yet). *Curr Opin Immunol* 2010;22(3):402–410.
35. Chakravarty S, Cockburn IA, Kuk S, Overstreet MG, Sacci JB, Zavala F. CD8+ T lymphocytes protective against malaria liver stages are primed in skin-draining lymph nodes. *Nat Med* 2007;13(9):1035–1041.
36. Reece JC, Loh L, Alcantara S, Fernandez CS, Stambas J, Sexton A, *et al.* Timing of immune escape linked to success or failure of vaccination. *PLoS One* 2010;5(9):el.2774.
37. Palucka K, Ueno H, Banchereau J. Recent developments in cancer vaccines. *J Immunol* 2011;186(3):1325–1331.
38. Xiang SD, Scalzo-Inguanti K, Minigo G, Park A, Hardy CL, Plebanski M. Promising particle-based vaccines in cancer therapy. *Expert Rev Vaccines* 2008;7(7):1103–1119.
39. Kilppstein R, Pozo D. Nanotechnology-based manipulation of dendritic cells for enhanced immunotherapy strategies. *Nanomedicine* 2010;6:523–529.
40. Zhang L, Gu FX, Chan JM, Wang AZ, Langer RS, Farokhzad OC. Nanoparticles in medicine: Therapeutic applications and developments. *Clin Pharmacol Ther* 2008;83 (5):761–769.
41. Fifis T, Gamvrellis A, Crimeen-Irwin B, Pietersz GA, Li J, Mottram PL, *et al.* Size-dependent immunogenicity: Therapeutic and protective properties of nano-vaccines against tumors. *J Immunol* 2004;173(5):3148–3154.
42. Mottram P, Leong D, Crimeen-Irwin B, Gloster S, Xiang SD, Meanger J, *et al.* Type 1 and Type 2 immunity following vaccination is influenced by nanoparticle size: Formulation of a model vaccine for respiratory syncytial virus. *Mol Pharm* 2007;4(1):73–84.
43. Belz GT, Bedoui S, Kupresanin F, Carbone FR, Heath WR. Minimal activation of memory CD8+ T cell by tissue-derived dendritic cells favors the stimulation of naive CD8+ T cells. *Nat Immunol* 2007;8(10):1060–1066.
44. Smith CM, Wilson NS, Waithman J, Villadangos JA, Carbone FR, Heath WR, *et al.* Cognate CD4(+) T cell licensing of dendritic cells in CD8(+) T cell immunity. *Nat Immunol* 2004;5(11):1143–1148.
45. O'Hagan DT, MacKichan ML, Singh M. Recent developments in adjuvants for vaccines against infectious diseases. *Biomol Eng* 2001;18(3):69–85.
46. Pelkmans L. Secrets of caveolae- and lipid raft-mediated endocytosis revealed by mammalian viruses. *Biochim Biophys Acta* 2005;1746(3):295–304.
47. Fifis T, Mottram P, Bogdanoska V, Hanley J, Plebanski M. Short peptide sequences containing MHC class I and/or class II epitopes linked to nano-beads induce strong immunity and inhibition of growth of antigen-specific tumour challenge in mice. *Vaccine* 2004;23(2):258–266.
48. Minigo G, Scholzen A, Tang CK, Hanley JC, Kalkanidis M, Pietersz GA, *et al.* Poly-L-lysine-coated nanoparticles: A potent delivery system to enhance DNA vaccine efficacy. *Vaccine* 2007;25(7):1316–1327.

49. Manolova V, Flace A, Bauer M, Schwarz K, Saudan P, Bachmann MF. Nanoparticles target distinct dendritic cell populations according to their size. *Eur J Immunol* 2008;38(5):1404–1413.
50. Moghimi SM, Hunter AC, Murray JC. Long-circulating and target-specific nanoparticles: Theory to practice. *Pharmacol Rev* 2001;53(2):283–318.
51. Banerjee T, Mitra S, Kumar Singh A, Kumar Sharma R, Maitra A. Preparation, characterization and biodistribution of ultrafine chitosan nanoparticles. *Int J Pharm* 2002; 243(1–2):93–105.
52. Owens III DE, Peppas NA. Opsonization, biodistribution, and pharmacokinetics of polymeric nanoparticles. *Int J Pharm* 2006;307(1):93–102.
53. Aoyama Y, Kanamori T, Nakai T, Sasaki T, Horiuchi S, Sando S, *et al.* Artificial viruses and their application to gene delivery. Size-controlled gene coating with glycocluster nanoparticles. *J Am Chem Soc* 2003;125(12):3455–3457.
54. Nakai T, Kanamori T, Sando S, Aoyama Y. Remarkably size-regulated cell invasion by artificial viruses. Saccharide-dependent self-aggregation of glycoviruses and its consequences in glycoviral gene delivery. *J Am Chem Soc* 2003;125(28):8465–8475.
55. Wang J, Fu L, Gu F, Ma Y. Notch1 is involved in migration and invasion of human breast cancer cells. *Oncol Rep* 2011;26(5):1295–1303.
56. Chithrani DB. Intracellular uptake, transport, and processing of gold nanostructures. *Mol Membr Biol* 2010;27(7):299–311.
57. Xu X-HN, Brownlow WJ, Kyriacou SV, Wan Q, Viola JJ. Real-time probing of membrane transport in living microbial cells using single nanoparticle optics and living cell imaging. *Biochemistry* 2004;43(32):10400–10413.
58. Jin H, Heller DA, Sharma R, Strano MS. Size-dependent cellular uptake and expulsion of single-walled carbon nanotubes: Single particle tracking and a generic uptake model for nanoparticles. *ACS Nano* 2009;3(1):149–158.
59. Oh J-M, Choi S-J, Lee G-E, Kim J-E, Choy J-H. Inorganic metal hydroxide nanoparticles for targeted cellular uptake through clathrin-mediated endocytosis. *Chem Asian J* 2009;4(1):67–73.
60. Rettig L, Haen SP, Bittermann AG, von Boehmer L, Curioni A, Krämer SD, *et al.* Particle size and activation threshold: A new dimension of danger signaling. *Blood* 2010;115(22):4533–4541.
61. Waeckerle-Men Y, Allmen EU, Gander B, Scandella E, Schlosser E, Schmidtke G, *et al.* Encapsulation of proteins and peptides into biodegradable poly(d,l-lactide-co-glycolide) microspheres prolongs and enhances antigen presentation by human dendritic cells. *Vaccine* 2006;24(11):1847–1857.
62. Foged C, Brodin B, Frokjaer S, Sundblad A. Particle size and surface charge affect particle uptake by human dendritic cells in an *in vitro* model. *Int J Pharm* 2005;298(2): 315–322.
63. Thiele L, Merkle HP, Walter E. Phagocytosis and phagosomal fate of surface-modified microparticles in dendritic cells and macrophages. *Pharm Res* 2003;20(2):221–228.
64. Verma A, Stellacci F. Effect of surface properties on nanoparticle-cell interactions. *Small* 2010;6(1):12–21.
65. Lieleg O, Vladescu I, Ribbeck K. Characterization of particle translocation through mucin hydrogels. *Biophys J* 2010;98(9):1782–1789.

66. Villanueva A, Cañete M, Roca AG, Calero M, Veintemillas-Verdaguer S, Serna CJ, *et al.* The influence of surface functionalization on the enhanced internalization of magnetic nanoparticles in cancer cells. *Nanotechnology* 2009;20(11):115103.
67. Little SR, Lynn DM, Ge Q, Anderson DG, Puram SV, Chen J, *et al.* Poly-beta amino ester-containing microparticles enhance the activity of nonviral genetic vaccines. *Proc Natl Acad Sci U S A* 2004;101(26):9534–9539.
68. Thiele L, Rothen-Rutishauser B, Jilek S, Wunderli-Allenspach H, Merkle HP, Walter E. Evaluation of particle uptake in human blood monocyte-derived cells *in vitro*. Does phagocytosis activity of dendritic cells measure up with macrophages? *J Controlled Release* 2001;76(1–2):59–71.
69. Jilek S, Merkle HP, Walter E. DNA-loaded biodegradable microparticles as vaccine delivery systems and their interaction with dendritic cells. *Adv Drug Delivery Rev* 2005;57(3): 377–390.
70. Jilek S, Ulrich M, Merkle HP, Walter E. Composition and surface charge of DNA-loaded microparticles determine maturation and cytokine secretion in human dendritic cells. *Pharm Res* 2004;21(7):1240–1247.
71. Josephson L, Tung CH, Moore A, Weissleder R. High-efficiency intracellular magnetic labeling with novel superparamagnetic-Tat peptide conjugates. *Bioconjug Chem* 1999;10(2):186–191.
72. Singh M, Briones M, Ott G, O'Hagan D. Cationic microparticles: A potent delivery system for DNA vaccines. *Proc Natl Acad Sci U S A* 2000;97(2):811–816.
73. Rhim W-K, Kim J-S, Nam J-M. Lipid-gold-nanoparticle hybrid-based gene delivery. *Small* 2008;4(10):1651–1655.
74. Harush-Frenkel O, Rozentur E, Benita S, Altschuler Y. Surface charge of nanoparticles determines their endocytic and transcytotic pathway in polarized MDCK cells. *Biomacromolecules* 2008;9(2):435–443.
75. Chrastina A, Massey KA, Schnitzer JE. Overcoming *in vivo* barriers to targeted nanodelivery. *Wiley Interdiscip Rev Nanomed Nanobiotechnol* 2011;3(4):421–437.
76. van den Berg JH, Oosterhuis K, Hennink WE, Storm G, van der Aa LJ, Engbersen JFJ, *et al.* Shielding the cationic charge of nanoparticle-formulated dermal DNA vaccines is essential for antigen expression and immunogenicity. *J Controlled Release* 2010;141(2):234–240.
77. Jokerst JV, Lobovkina T, Zare RN, Gambhir SS. Nanoparticle PEGylation for imaging and therapy. *Nanomedicine* 2011;6(4):715–728.
78. Ryan SaM, Mantovani G, Wang X, Haddleton DM, Brayden DJ. Advances in PEGylation of important biotech molecules: Delivery aspects. *Expert Opin Drug Delivery* 2008;5(4): 371–383.
79. Ishida T, Kiwada H. Accelerated blood clearance (ABC) phenomenon upon repeated injection of PEGylated liposomes. *Int J Pharm* 2008;354(1–2):56–62.
80. Bege N, Renette T, Jansch M, Reul R, Merkel O, Petersen H, *et al.* Biodegradable poly(ethylene carbonate) nanoparticles as a promising drug delivery system with "stealth" potential. *Macromol Biosci* 2011;11(7):897–904.
81. Decuzzi P, Godin B, Tanaka T, Lee SY, Chiappini C, Liu X, *et al.* Size and shape effects in the biodistribution of intravascularly injected particles. *J Controlled Release* 2010;141(3):320–327.

82. Chithrani BD, Chan WC. Elucidating the mechanism of cellular uptake and removal of protein-coated gold nanoparticles of different sizes and shapes. *Nano Lett* 2007;7(6):1542–1550.
83. Sharma G, Valenta DT, Altman Y, Harvey S, Xie H, Mitragotri S, *et al.* Polymer particle shape independently influences binding and internalization by macrophages. *J Controlled Release* 2010;147:408–412.
84. Champion JA, Mitragotri S. Shape induced inhibition of phagocytosis of polymer particles. *Pharm Res* 2009;26(1):244–249.
85. Champion JA, Mitragotri S. Role of target geometry in phagocytosis. *Proc Natl Acad Sci U S A* 2006;103(13):4930–4934.
86. Tao L, Hu W, Liu Y, Huang G, Sumer BD, Gao J. Shape-specific polymeric nanomedicine: Emerging opportunities and challenges. *Exp Biol Med* 2011;236(1):20–29.
87. Huang XL, Li LL, Liu TL, Hao NJ, Liu HY, Chen D, *et al.* The shape effect of mesoporous silica nanoparticles on biodistribution, clearance, and biocompatibility *in vivo*. *ACS Nano* 2011;5(7):5390–5399.
88. Doshi N, Prabhakarpandian B, Rea-Ramsey A, Pant K, Sundaram S, Mitragotri S. Flow and adhesion of drug carriers in blood vessels depend on their shape: A study using model synthetic microvascular networks. *J Controlled Release* 2010; 146(2):196–200.
89. Sydlik U, Bierhals K, Soufi M, Abel J, Schins RP, Unfried K. Ultrafine carbon particles induce apoptosis and proliferation in rat lung epithelial cells via specific signaling pathways both using EGF-R. *Am J Physiol Lung Cell Mol Physiol* 2006; 291(4):L725–L733.
90. Kennedy T, Ghio AJ, Reed W, Samet J, Zagorski J, Quay J, *et al.* Copper-dependent inflammation and nuclear factor-kappaB activation by particulate air pollution. *Am J Respir Cell Mol Biol* 1998;19(3):366–378.
91. Park EJ, Yi J, Chung KH, Ryu DY, Choi J, Park K. Oxidative stress and apoptosis induced by titanium dioxide nanoparticles in cultured BEAS-2B cells. *Toxicol Lett* 2008;180(3):222–229.
92. Liu X, Sun J. Endothelial cells dysfunction induced by silica nanoparticles through oxidative stress via JNK/P53 and NF-kappaB pathways. *Biomaterials* 2010;31 (32):8198–8209.
93. Park EJ, Kim H, Kim Y, Yi J, Choi K, Park K. Inflammatory responses may be induced by a single intratracheal instillation of iron nanoparticles in mice. *Toxicology* 2010;275(1–3): 65–71.
94. Eom HJ, Choi J. p38 MAPK activation, DNA damage, cell cycle arrest and apoptosis as mechanisms of toxicity of silver nanoparticles in Jurkat T cells. *Environ Sci Technol* 2010;44(21):8337–8342.
95. Unfried K, Sydlik U, Bierhals K, Weissenberg A, Abel J. Carbon nanoparticle-induced lung epithelial cell proliferation is mediated by receptor-dependent Akt activation. *Am J Physiol Lung Cell Mol Physiol* 2008;294(2):L358–L367.
96. Uto T, Akagi T, Yoshinaga K, Toyama M, Akashi M, Baba M. The induction of innate and adaptive immunity by biodegradable poly(gamma-glutamic acid) nanoparticles via a TLR4 and MyD88 signaling pathway. *Biomaterials* 2011;32(22):5206–5212.
97. Uto T, Wang X, Sato K, Haraguchi M, Akagi T, Akashi M, *et al.* Targeting of antigen to dendritic cells with poly(gamma-glutamic acid) nanoparticles induces antigen-specific humoral and cellular immunity. *J Immunol* 2007;178(5):2979–2986.

98. Lee YR, Lee YH, Im SA, Yang IH, Ahn GW, Kim K, *et al.* Biodegradable nanoparticles containing TLR3 or TLR9 agonists together with antigen enhance MHC-restricted presentation of the antigen. *Arch Pharm Res* 2010;33(11):1859–1866.
99. Li A, Qin L, Zhu D, Zhu R, Sun J, Wang S. Signalling pathways involved in the activation of dendritic cells by layered double hydroxide nanoparticles. *Biomaterials* 2010;31(4):748–756.
100. Okamoto T, Sanda T, Asamitsu K. NF-kappa B signaling and carcinogenesis. *Curr Pharm Des* 2007;13(5):447–462.
101. Lee CH, Jeon YT, Kim SH, Song YS. NF-kappaB as a potential molecular target for cancer therapy. *Biofactors* 2007;29(1):19–35.
102. Dhillon AS, Hagan S, Rath O, Kolch W. MAP kinase signalling pathways in cancer. *Oncogene* 2007;26(22):3279–3290.
103. Wagner EF, Nebreda AR. Signal integration by JNK and p38 MAPK pathways in cancer development. *Nat Rev Cancer* 2009;9(8):537–549.
104. Carnero A, Blanco-Aparicio C, Renner O, Link W, Leal JF. The PTEN/PI3K/AKT signalling pathway in cancer, therapeutic implications. *Curr Cancer Drug Targets* 2008;8(3):187–198.
105. Basu S, Harfouche R, Soni S, Chimote G, Mashelkar RA, Sengupta S. Nanoparticle-mediated targeting of MAPK signaling predisposes tumor to chemotherapy. *Proc Natl Acad Sci U S A* 2009;106(19):7957–7961.
106. Lu XY, Ciraolo E, Stefenia R, Chen GQ, Zhang Y, Hirsch E. Sustained release of PI3K inhibitor from PHA nanoparticles and *in vitro* growth inhibition of cancer cell lines. *Appl Microbiol Biotechnol* 2011;89(5):1423–1433.
107. Harfouche R, Basu S, Soni S, Hentschel DM, Mashelkar RA, Sengupta S. Nanoparticle-mediated targeting of phosphatidylinositol-3-kinase signaling inhibits angiogenesis. *Angiogenesis* 2009;12(4):325–338.
108. Xiang SD, Kalkanidis M, Pietersz GA, Mottram P, Crimeen-Irwin B, Ardipradja K, *et al.* Methods for nano-particle based vaccine formulation and evaluation of their immunogenicity. *Methods* 2006;40:20–29.
109. Gamvrellis A, Leong D, Hanley JC, Xiang SD, Mottram P, Plebanski M. Vaccines that facilitate antigen entry into dendritic cells. *Immunol Cell Biol* 2004;82:506–516.
110. Peek LJ, Middaugh CR, Berkland C. Nanotechnology in vaccine delivery. *Adv Drug Delivery Rev* 2008;60:915–928.
111. Xiang SD, Selomulya C, Ho J, Apostolopoulos V, Plebanski M. Delivery of DNA vaccines: An overview on the use of biodegradable polymeric and magnetic nanoparticles. *Nanomed Nanobiotechnol* 2010;2:205–218.
112. Uchida T, Yagi A, Oda Y, Nakada Y, Goto S. Instability of bovine insulin in poly(lactide-co-glycolide) (PLGA) microspheres. *Chem Pharma Bull* 1996;44(1):235–236.
113. Uto T, Akagi T, Toyama M, Nishi Y, Shima F, Akashi M, *et al.* Comparative activity of biodegradable nanoparticles with aluminum adjuvants: Antigen uptake by dendritic cells and induction of immune response in mice. *Immunol Lett* 2011;140(1–2):36–43.
114. Sun H, Pollock KG, Brewer JM. Analysis of the role of vaccine adjuvants in modulating dendritic cell activation and antigen presentation *in vitro*. *Vaccine* 2003;21(9–10):849–855.
115. Panyam J, Zhou WZ, Prabha S, Sahoo SK, Labhasetwar V. Rapid endo-lysosomal escape of poly(DL-lactide-co-glycolide) nanoparticles: Implications for drug and gene delivery. *FASEB J* 2002;16(10):1217–1226.

116. Han R, Zhu J, Yang X, Xu H. Surface modification of poly(D,L-lactic-co-glycolic acid) nanoparticles with protamine enhanced cross-presentation of encapsulated ovalbumin by bone marrow-derived dendritic cells. *J Biomed Mater Res A* 2011;96(1):142–149.
117. Shen H, Ackerman AL, Cody V, Giodini A, Hinson ER, Cresswell P, *et al.* Enhanced and prolonged cross-presentation following endosomal escape of exogenous antigens encapsulated in biodegradable nanoparticles. *Immunology* 2006;117(1):78–88.
118. Hamasaki T, Uto T, Akagi T, Akashi M, Baba M. Modulation of gene expression related to Toll-like receptor signaling in dendritic cells by poly(gamma-glutamic acid) nanoparticles. *Clin Vaccine Immunol* 2010;17(5):748–756.
119. Chang J, Jallouli Y, Kroubi M, Yuan X-B, Feng W, Kang C-S, *et al.* Characterization of endocytosis of transferrin-coated PLGA nanoparticles by the blood-brain barrier. *Int J Pharm* 2009;379(2):285–292.
120. Kwon YJ, James E, Shastri N, Fréchet JMJ. *In vivo* targeting of dendritic cells for activation of cellular immunity using vaccine carriers based on pH-responsive microparticles. *Proc Natl Acad Sci U S A* 2005;102(51):18264–18268.
121. Praetorius NP, Mandal TK. Engineered nanoparticles in cancer therapy. *Recent Pat Drug Delivery Formulation* 2007;1(1):37–51.
122. Minchin RF, Martin DJ. Minireview: Nanoparticles for molecular imaging — an overview. *Endocrinology* 2010;151(2):474–481.
123. Danhier F, Feron O, Préat V. To exploit the tumor microenvironment: Passive and active tumor targeting of nanocarriers for anti-cancer drug delivery. *J Controlled Release* 2010;148(2):135–146.
124. Hu Y, Jiang X, Ding Y, Ge H, Yuan Y, Yang C. Synthesis and characterization of chitosan-poly(acrylic acid) nanoparticles. *Biomaterials* 2002;23(15):3193–3201.
125. Janes KA, Fresneau MP, Marazuela A, Fabra A, Alonso MJ. Chitosan nanoparticles as delivery systems for doxorubicin. *J Controlled Release* 2001;73(2–3):255–267.
126. Shu XZ, Zhu KJ. A novel approach to prepare tripolyphosphate/chitosan complex beads for controlled release drug delivery. *Int J Pharm* 2000;201(1):51–58.
127. van der Lubben IM, Verhoef JC, Borchard G, Junginger HE. Chitosan and its derivatives in mucosal drug and vaccine delivery. *Eur J Pharma Sci* 2001;14(3):201–207.
128. Mao S, Sun W, Kissel T. Chitosan-based formulations for deliver of DNA and siRNA. *Adv Drug Delivery Rev* 2010;62:12–27.
129. Calvo P, Remunan-Lopez C, Vila-Jato JL, Alonso MJ. Novel hydrophilic chitosan-polyethylene oxide nanoparticles as protein carriers. *J Appl Polym Sci* 1997;63:125–132.
130. Xu Y, Du Y. Effect of molecular structure of chitosan on protein delivery properties of chitosan nanoparticles. *Int J Pharm* 2003;250:215–226.
131. Vila A, Sanchez A, Tobio M, Calvo P, Alonso MJ. Design of biodegradable particles for protein delivery. *J Controlled Release* 2002;78:15–24.
132. Yang X, Yuan X, Cai D, Wang S, Zong L. Low molecular weight chitosan in DNA vaccine delivery via mucosa. *Int J Pharm* 2009;375(1–2):123–132.
133. Xu J, Dai W, Wang Z, Chen B, Li Z, Fan X. Intranasal vaccination with chitosan-DNA nanoparticles expressing pneumococcal surface antigen A protects mice against nasopharyngeal colonization by *Streptococcus pneumoniae. Clin Vaccine Immunol* 2011;18(1):75–81.
134. Vila A, Sanchez A, Janes K, Behrens I, Kissel T, Vila-Jato JL, *et al.* Low molecular weight chitosan nanoparticles as new carriers for nasal vaccine delivery in mice. *Eur J Pharm Biopharm* 2004;57:123–131.

135. Amidi M, Romeijn SG, Borchard G, Junginger HE, Hennink WE, Jiskoot W. Preparation and characterization of protein-loaded N-trimethyl chitosan nanoparticles as nasal delivery system. *J Controlled Release* 2006;111:107–116.
136. Pavelic K, Hadzija M, Bedrica L, Pavelic J, Dikic I, Katic M, *et al.* Natural zeolite clinoptilolite: New adjuvant in anticancer therapy. *J Mol Med (Berl)* 2001;78(12):708–720.
137. Walsh MC, Banas JA, Mudzinski SP, Preissler MT, Graziano RF, Gosselin EJ. A two-component modular approach for enhancing T-cell activation utilizing a unique anti-FcgammaRI-streptavidin construct and microspheres coated with biotinylated-antigen. *Biomol Eng* 2003;20(1):21–33.
138. Tighe H, Corr M, Roman M, Raz E. Gene vaccination: Plasmid DNA is more than just a blueprint. *Immunol Today* 1998;19(2):89–97.
139. Weiss R, Scheiblhofer S, Freund J, Ferreira F, Livey I, Thalhamer J. Gene gun bombardment with gold particles displays a particular Th2-promoting signal that over-rules the Th1-inducing effect of immunostimulatory CpG motifs in DNA vaccines. *Vaccine* 2002;20(25–26):3148–3154.
140. Hardy CL, LeMasurier JS, Belz GT, Scalo-Inguanti K, Yao J, Xiang SD, *et al.* Inert 50 nm polystyrene nanoparticles that modify pulmonary dendritic cell function and inhibit allergic airway inflammation. *J Immunol* 2012;188(3):1431–1441.
141. Scheerlinck JP, Gloster S, Gamvrellis A, Mottram PL, Plebanski M. Systemic immune responses in sheep, induced by a novel nano-bead adjuvant. *Vaccine* 2006;24(8):1124–1131.
142. Kovacsovics-Bankowski M, Rock KL. A phagosome-to-cytosol pathway for exogenous antigens presented on MHC class I molecules. *Science* 1995;267(5195):243–246.
143. Vidard L, Kovacsovics-Bankowski M, Kraeft SK, Chen LB, Benacerraf B, Rock KL. Analysis of MHC class II presentation of particulate antigens of B lymphocytes. *J Immunol* 1996;156(8):2809–2818.
144. Luo D, Saltzman WM. Nonviral gene delivery: Thinking of silica. *Gene Ther* 2005;13(7):585–586.
145. Roy I, Ohulchanskyy TY, Bharali DJ, Pudavar HE, Mistretta RA, Kaur N, *et al.* Optical tracking of organically modified silica nanoparticles as DNA carriers: A nonviral, nanomedicine approach for gene delivery. *Proc Natl Acad Sci U S A* 2005;102(2):279–284.
146. Benezra M, Penate-Medina O, Zanzonico PB, Schaer D, Ow H, Burns A, *et al.* Multimodal silica nanoparticles are effective cancer-targeted probes in a model of human melanoma. *J Clin Invest* 2011;121(7):2768–2780.
147. Kim BH, Lee N, Kim H, An K, Park YI, Choi Y, *et al.* Large-scale synthesis of uniform and extremely small-sized iron oxide nanoparticles for high-resolution T1 magnetic resonance imaging contrast agents. *J Am Chem Soc* 2011;133(32):12624–12631.
148. Oh JK, Park JM. Iron oxide-based superparamagnetic polymeric nanomaterials: Dosing, preparation, and biomedical application. *Prog Polym Sci* 2011;36:168–189.
149. Dobson J. Gene therapy progress and prospects: Magnetic nanoparticle-based gene delivery. *Gene Ther* 2006;13(4):283–287.
150. Gersting SW, Schillinger U, Lausier J, Nicklaus P, Rudolph C, Plank C, *et al.* Gene delivery to respiratory epithelial cells by magnetofection. *J Gene Med* 2004;6(8): 913–922.
151. Scherer F, Anton M, Schillinger U, Henkel J, Bergemann C, Kruger A, *et al.* Magnetofection: Enhancing and targeting gene delivery by magnetic force *in vitro* and *in vivo*. *Gene Ther* 2002;9(2):102–109.

152. Zhou XF, Liu B, Yu XH, Zha X, Zhang XZ, Wang XY, *et al.* Using magnetic force to enhance immune response to DNA vaccine. *Small* 2007;3(10):1707–1713.
153. Hao R, Xing R, Xu Z, Hou Y, Gao S, Sun S. Synthesis, functionalization, and biomedical applications of multifunctional magnetic nanoparticles. *Adv Mater* 2010;22(25): 2729–2742.
154. Lattuada M, Hatton TA. Functionalization of monodisperse magnetic nanoparticles. *Langmuir* 2007;23(4):2158–2168.
155. Li D, Teoh WY, Woodward RC, Cashion JD, Selomulya C, Amal R. Evolution of morphology and magnetic properties in silica/maghemite nanocomposites. *J Phys Chem C* 2009;113(28):12040–12047.
156. Jun Y-W, Seo J-W, Cheon J. Nanoscaling laws of magnetic nanoparticles and their applicabilities in biomedical sciences. *Acc Chem Res* 2008;41(2):179–189.
157. Seo WS, Lee JH, Sun X, Suzuki Y, Mann D, Liu Z, *et al.* FeCo/graphitic-shell nanocrystals as advanced magnetic-resonance-imaging and near-infrared agents. *Nat Mater* 2006;5(12):971–976.
158. Li D, Teoh WY, Gooding JJ, Selomulya C, Amal R. Functionalisation strategies for protease immobilisation on magnetic nanoparticles. *Adv Funct Mater* 2010;20:1767–1777.
159. Massia SP, Stark J, Letbetter DS. Surface-immobilized dextran limits cell adhesion and spreading. *Biomaterials* 2000;21(22):2253–2261.
160. Suh WH, Suslick KS, Stucky GD, Suh YH. Nanotechnology, nanotoxicology, and neuroscience. *Prog Neurobiol* 2009;87(3):133–70.
161. Plank C, Scherer F, Schillinger U, Bergemann C, Anton M. Magnetofection: Enhancing and targeting gene delivery with superparamagnetic nanoparticles and magnetic fields. *J Liposome Res* 2003;13(1):29–32.
162. Kamau Chapman SW, Hassa PO, Koch-Schneidemann S, von Rechenberg B, Hofmann-Amtenbrink M, Steitz B, *et al.* Application of pulsed-magnetic field enhances non-viral gene delivery in primary cells from different origins. *J Magn Magn Mater* 2008;320(8):1517–1527.
163. Arsianti M, Lim M, Marquis CP, Amal R. Polyethylenimine based magnetic iron-oxide vector: The effect of vector component assembly on cellular entry mechanism, intracellular localization, and cellular viability. *Biomacromolecules* 2010;11(9):2521–2531.
164. Al-Deen FN, Ho J, Selomulya C, Ma C, Coppel R. Superparamagnetic nanoparticles for effective delivery of malaria DNA vaccine. *Langmuir* 2011;27(7):3703–3712.
165. Stuart MAC, Huck WTS, Genzer J, Muller M, Ober C, Stamm M, *et al.* Emerging applications of stimuli-responsive polymer materials. *Nat Mater* 2010;9(2):101–113.
166. Li Y, Dong H, Wang K, Shi D, Zhang X, Zhuo R. Stimulus-responsive polymeric nanoparticles for biomedical applications. *Sci China Chem* 2010;53(3):447–457.
167. Paleos CM, Tsiourvas D, Sideratou Z, Tziveleka L-A. Drug delivery using multifunctional dendrimers and hyperbranched polymers. *Expert Opin Drug Delivery* 2010;7(12):1387–1398.
168. Hori Y, Winans AM, Huang CC, Horrigan EM, Irvine DJ. Injectable dendritic cell-carrying alginate gels for immunization and immunotherapy. *Biomaterials* 2008;29(27):3671–3682.
169. Li H, Nookala S, Re F. Aluminum hydroxide adjuvants activate caspase-1 and induce IL-1β and IL-18 release. *J Immunol* 2007;178(8):5271–5276.
170. Baylor NW, Egan W, Richman P. Aluminum salts in vaccines — US perspective. *Vaccine* 2002;20(Suppl 3):S18–S23.

171. Tritto E, Mosca F, De Gregorio E. Mechanism of action of licensed vaccine adjuvants. *Vaccine* 2009;27(25–26):3331–3334.
172. Khurana S, Verma N, Yewdell JW, Hilbert AK, Castellino F, Lattanzi M, *et al.* MF59 adjuvant enhances diversity and affinity of antibody-mediated immune response to pandemic influenza vaccines. *Sci Transl Med* 2011;3(85):85ra48.
173. De Becker G, Moulin V, Pajak B, Bruck C, Francotte M, Thiriart C, *et al.* The adjuvant monophosphoryl lipid A increases the function of antigen-presenting cells. *Int Immunol* 2000;12(6):807–815.
174. MacLeod MK, McKee AS, David A, Wang J, Mason R, Kappler JW, *et al.* Vaccine adjuvants aluminum and monophosphoryl lipid A provide distinct signals to generate protective cytotoxic memory CD8 T cells. *Proc Natl Acad Sci U S A* 2011;108(19):7914–7419.
175. Martin M, Michalek SM, Katz J. Role of innate immune factors in the adjuvant activity of monophosphoryl lipid A. *Infect Immune* 2003;71(5):2498–2507.
176. Carmona A, Omeñaca F, Tejedor JC, Merino JM, Vaman T, Dieussaert I, *et al.* Immunogenicity and safety of AS03-adjuvanted 2009 influenza A H1N1 vaccine in children 6–35 months. *Vaccine* 2010;28(36):5837–5844.
177. Chu DW-S, Hwang S-J, Lim FS, Oh HML, Thongcharoen P, Yang P-C, *et al.* Immunogenicity and tolerability of an AS03A-adjuvanted prepandemic influenza vaccine: A phase III study in a large population of Asian adults. *Vaccine* 2009;27(52):7428–7435.
178. Manuel O, Pascual M, Hoschler K, Giulieri S, Alves D, Ellefsen K, *et al.* Humoral response to the influenza A H1N1/09 monovalent AS03-adjuvanted vaccine in immunocompromised patients. *Clin Infect Dis* 2011;52(2):248–256.
179. Leroux-Roels G. Prepandemic H5N1 influenza vaccine adjuvanted with AS03: A review of the pre-clinical and clinical data. *Expert Opin Biol Ther* 2009;9(8):1057–1071.
180. Harper DM. Currently approved prophylactic HPV vaccines. *Expert Rev Vaccines* 2009;8(12):1663–1679.
181. Einstein MH, Baron M, Levin MJ, Chatterjee A, Edwards RP, Zepp F, *et al.* Comparison of the immunogenicity and safety of Cervarix and Gardasil human papillomavirus (HPV) cervical cancer vaccines in healthy women aged 18–45 years. *Hum Vaccines* 2009;5(10):705–719.
182. Frosner G, Steffen R, Herzog C. Virosomal hepatitis a vaccine: Comparing intradermal and subcutaneous with intramuscular administration. *J Travel Med* 2009;16(6):413–419.
183. Herzog C, Hartmann K, Kunzi V, Kursteiner O, Mischler R, Lazar H, *et al.* Eleven years of Inflexal V-a virosomal adjuvanted influenza vaccine. *Vaccine* 2009;27(33): 4381–4387.

Chapter 16

Nanoparticles as Drug Delivery Vehicles for the Therapy of Inflammatory Disorders

Deepthy Menon, J. Gopikrishna, Dhanya Narayanan, and Shantikumar V. Nair*

Amrita Centre for Nanosciences and Molecular Medicine
Amrita Institute of Medical Sciences and Research Centre
Amrita Vishwa Vidyapeetham University
Kochi-682041, Kerala, India

The ever-exploding scientific advancements in the field of nanotechnology have enabled its integration into biosciences and medicine. Such interdisciplinary approaches allow researchers to engineer novel tools for diagnosis, therapy, and prognosis of different clinical complications. One of the most successful nano-tools that have emerged in recent years are nano-drug delivery devices. The use of such nano-sized carriers to load therapeutic agents (drugs, vaccines, biomolecules, and enzymes) aid in overcoming their pharmacological and toxicological hurdles through site-specific and controlled drug delivery. This chapter is a comprehensive account on the recent developments in the delivery of anti-inflammatory drugs using nanocarriers. Here, various kinds and types of nanocarriers and their countless inherent and engineered properties which make them ideal for delivering anti-inflammatory drugs, and the special characteristics of inflammation and inflammatory cells and tissues, which make them vulnerable to these nano-tools, will be discussed. The potential active and passive targeting approaches using drug-loaded nanocarriers for various inflammatory disorders are also elaborated.

*Corresponding author. E-mail: deepthymenon@aims.amrita.edu

1. Introduction

Nanopharmaceuticals are widely expected to transform the landscape of pharmaceutical and biotechnology industries in the foreseeable future.[1–3] The diminutive size (at least one dimension typically in the range 1–100 nm) of nanoparticles used in such applications allow for direct cell interaction, uptake by endocytosis, and interference with specific subcellular components, enabling less toxic and more efficient therapeutic action. Furthermore, the large surface area of nanocarriers allows enhanced bioactivity and efficient surface modification, facilitating their solubility, targeting, circulation, and stability.[2,4–10] A variety of nanocarriers, which include polymeric nanoparticles, micelles, dendrimers, liposomes, solid lipid nanoparticles (SLNs), proteins, carbohydrates, and inorganics have been developed for drug delivery applications.[8–10] The biocompatible nanocarriers serve the purpose of conferring protection to the body from the toxic action of the drug payloads on normal tissues, while facilitating their therapeutic action on diseased cells.[11–13] Conversely, the carrier protects the drug from efflux or degradation in the biological environment of the body, hence preserving their efficacy.[14] The biodegradability of the carrier precludes the potentially harmful long-term accumulation of the carrier in tissues and also provides for the controlled release of the drug through carrier degradation, improving pharmacokinetics and aiding reduction in the frequency and dose of drug administration.[8,9,15–18] Another very critical advantage of drug encapsulation in nanocarriers is the ability of "site-specific drug delivery" by targeting the carrier to specific sites or cells,[19] either through active[20–22] or passive[23–25] targeting strategies.

While the bulk of nanomedicines have focused on cancers and other infectious diseases such as tuberculosis and malaria, its application for inflammatory disorders is still emerging. Inflammatory disorders span the range from serious ailments such as atherosclerosis, rheumatoid arthritis (RA), psoriasis, multiple sclerosis to more common ailments such as hay fever, inflammatory bowel diseases (IBD), and periodontitis.[26] Furthermore, inflammatory disorders have been linked to cancers, ischemic heart diseases, and other immune system disorders. Both early stage detection of incipient inflammation and its treatment can therefore have benefits in the treatment of other such disorders as well. In this chapter, we will review the important developments in the treatment of inflammatory disorders using various nanomedicines.

2. The Inflammatory Process

Inflammatory disorders result from a cascade of highly regulated events propagated upon stimulation, and are known to activate the cellular immune response.[27–29] While acute inflammation results from an external chemical, mechanical, or pathogenic influence and is relatively short-lived,[30] chronic inflammation does not require any external stimulus and may have painful long-term effects.[31,32] Such uncontrolled inflammation can lead to more serious pathological implications such as opportunistic infections, autoimmune, neurological, or cardiovascular disorders, or even cancer.[33–35] At the early stages of inflammation, only histological changes with aberrant cell distribution in diseased tissues would be dominant.[36–38] A histological analysis would reveal that any site of chronic inflammation would have a localized population of immune-related cells such as macrophages, lymphocytes, neutrophils, natural killer cells, mast cells, regulatory T cells,[36] as well as compromised blood vessels, fibrosis, and tissue necrosis.[39] These immune cells and the cytokines produced are also the primary defense line in the innate immune mechanism.

3. Active *vs.* Passive Targeting in Inflammatory Disorders

A hydrophobic nanoparticle, when presented to the circulation, is usually phagocytosed and cleared by macrophages in the bloodstream, following foreign body recognition by the mononuclear phagocytic system (MPS) mainly in the liver and spleen.[40] This sequential antigen recognition and removal mechanism by the innate immune system are explored in passive targeting of nanocarriers.[41,42] Nanoparticles, owing to their small size, have the ability to extravasate through the endothelium in inflammatory sites, the epithelium in the intestinal tract and liver, macrophages, and tumors, or penetrate microcapillaries.[43–47] Increased permeability and endothelial activation accompany vessel dilation in the initial stages of inflammation. The permeabilized blood vessels allow for the accumulation of the intravenously administered nanocarriers at the site of inflammation that can be regulated to a great extent by the size and charge of nanoparticles.[48,49] Thus, by appropriately tuning the nanoparticle size and its surface characteristics, an efficient uptake into the immune-related cells in inflammatory diseases can be realized.[50,51] In the pathogenesis of a variety of inflammatory disorders, the disruption of the balance between pro- and anti-inflammatory cytokines has also been implicated.[52] RNA interference of cytokines using nanocarriers has evolved to be another powerful approach to treat inflammatory diseases.[53] Although passive targeting to the immune-related cells appears rather easy, selective targeting to organs different from the liver and spleen would ultimately be essential for the success of a potential anti-inflammatory therapy.

Nanoparticles can be targeted to the sites of inflammation by utilizing the unique phenotypic features of diseased cells, thereby enabling the delivery of drugs in specific doses solely at the desired location. While passive targeting attempts to minimize non-specific interactions between the drug carrier and non-target sites in the body through alterations in the physico-chemical properties (size, morphology, hydrophilicity, surface charge, etc.), active targeting involves biologically specific interactions such as antigen–antibody and ligand–receptor binding, in addition to nanoparticle characteristics.[54,55] A variety of molecules such as peptides, monoclonal antibodies, and carbohydrates could be used as targeting entities depending upon the type of the cell/tissue intended to be targeted.[56] In chronic inflammation, the body naturally enhances local angiogenesis by the proliferation of endothelial cells which overexpress a family of cell surface receptors called cell adhesion molecules (CAMs).[39] These receptors form the most attractive targets for active targeting of endothelial cells in inflammatory disorders.[57,58] Leukocytes, upon migration to the site of inflammation, roll on endothelial cells, and subsequently get arrested on the overexpressed CAMs of the inflammatory endothelium.[59] The major CAMs present on the endothelial surface include P-selectins and E-selectins, which will bind to their reciprocal ligands on leukocyte surface L-selectins. Other cell surface markers that are overexpressed on endothelial cells at sites of pathological inflammation include the intercellular adhesion molecules (ICAM-1, ICAM-2), vascular CAMs (VCAM-1), mucosal addressin CAMs (MAdCAM-1), and integrins.[60] It is well established that VCAM-1 is localized on the aortic endothelium overlying early foam cell lesions as well as in colitis;[61,62] P-selectin and E-selectin are upregulated in ischemia-reperfusion injury, arthritis, colitis, and in several other pathological conditions;[63–65] folate and mannose receptors are overexpressed on activated macrophages in RA.[66,67] These markers can be chosen as active targets for delivering drug-loaded nanoparticles to inflammatory regions by appropriately modifying nanoparticles with specific targeting moieties, either through covalent or non-covalent linkage.[68,69] Active targeting demands that the surface of the nanocarrier possess reactive functional groups, which enable the chemical coupling of the targeting ligand in the right conformation for the appropriate linkage to the defined cell surface markers. The concentration of the ligands at the nanoparticle surface should be also high enough to enable selective targeting of diseased cells and sparing the normal cells. This specificity is indeed one of the most prominent advantages of drug-loaded nanocarriers over conventional drug delivery systems. A schematic of the various modes of targeting immune cells as well as endothelial cells in inflammation-related diseases by nanocarriers is depicted in Figure 1.

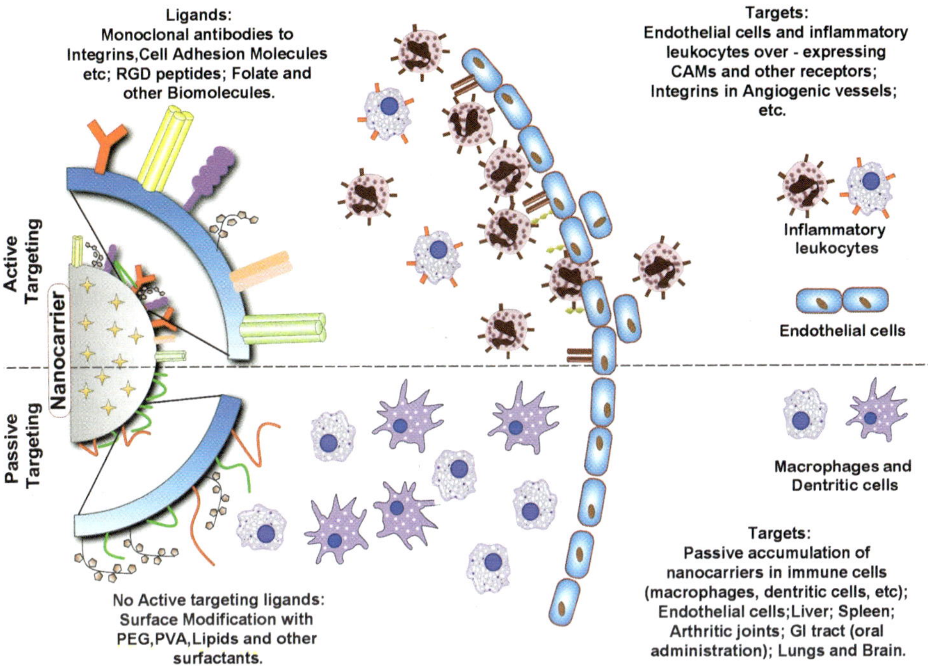

Figure 1. Various strategies for targeting the immune and blood cells by nanocarriers through active and passive modes. The top panel represents the drug-loaded nanocarrier (shown as a solid hemisphere) conjugated with active targeting ligands that can bind to receptors overexpressing immune and endothelial cells. The bottom panel represents the drug-loaded nanocarrier surface modified to facilitate the passive targeting approaches as described in the text.

4. Nanoconstructs for Inflammatory Diseases

Nanoscale drug delivery systems allow the integration of various entities including biologically active molecules, imaging contrast agents (fluorescence, magnetic, etc.), biocompatible surface coatings, and targeting sequences, all within a single carrier as depicted in Figure 2. Furthermore, alterations in the nanocarrier size and shape or choice of material can offer an extra dimension of physical control on the specificity of its function.[70–74] These nanoconstructs prepared using polymers, proteins, or lipids, should ideally be submicron-sized, with particle size ranging from 10 to 200 nm.[49,75] In addition to the choice of the nanocarrier material, the nanoconstruct also accommodates the active pharmaceutical ingredients (APIs) such as the conventional non-steroidal anti-inflammatory drugs (NSAIDs) and glucocorticoids, genes, vaccines, and immunotherapy. These APIs can be dissolved, entrapped, adsorbed, and/or encapsulated into or onto the selected nanomatrix.[76–78] The

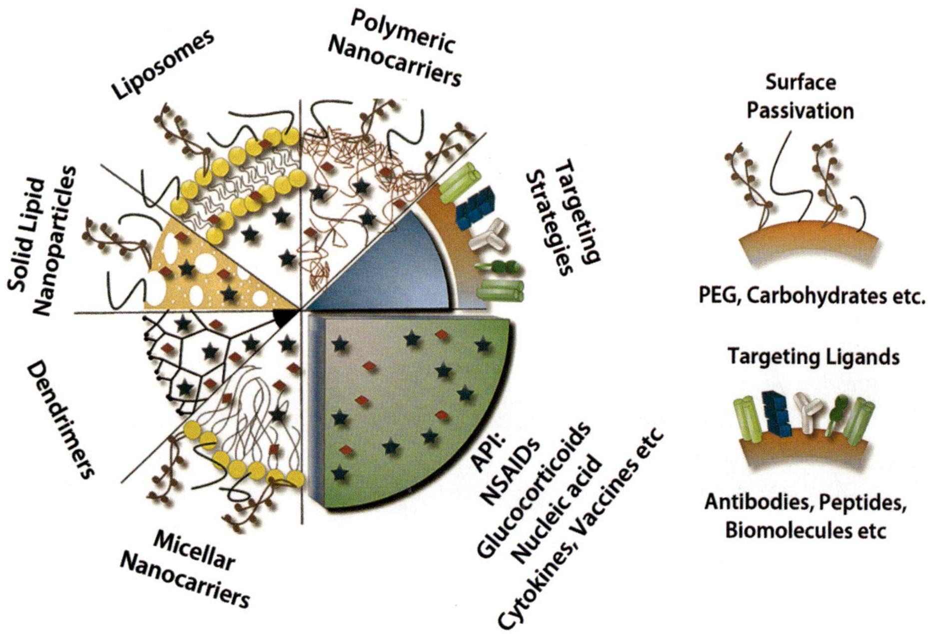

Figure 2. Schematic representation of potential nanoconstruct designs in a nutshell. Nanocarriers formed from polymers, micelles, liposomes, dendrimers, solid lipid nanoparticles (SLNs) are depicted, along with the choice of targeting ligands, surface modifiers, and APIs.

choice of the nanomatrix should be based on the route of administration and the properties of the drug (stability, hydrophilicity/hydrophobicity) to be administered. The targets selected for nanotherapy can be either immune or non-immune cells according to the stage of inflammation and the choice of target site.[79]

To avoid the recognition of the submicron-sized particles by the MPS, the surface of nanocarriers can be intentionally modified to prevent opsonization by plasma proteins and increase their systemic circulation time, thus developing "stealthy nanocarriers."[80–82] Poly(ethylene glycol) (PEG), a hydrophilic polymer, is the most commonly used and well-established polymer providing steric stabilization when grafted, conjugated, or adsorbed to the particle surface.[83] Protein adsorption studies on nanocarriers surface modified with PEG of varying molecular weights in the range of 2–10 kDa have established that long-chain PEG molecules (typically 10 kDa) showed minimal plasma protein adsorption.[84,85] However, the unexpected immunogenicity of PEG molecules by eliciting IgM-mediated complement activation, resulting in the accelerated clearance and enhanced hepatic uptake of the nanocarriers, was reported.[86,87] Hence, surface modification of nanocarriers is an area that demands careful optimization.

5. Drug Loading and Release from Nanoconstructs

The type of material selected for the nanocarrier, its structure and composition, and the methods employed for drug entrapment (preparatory routes) within the nanoparticles are the decisive factors that define the drug release kinetics from the nanomedicine. Several researchers have elaborated on the different methods of preparing and characterizing drug-loaded nanocarriers to have the desired release characteristics.[56,88] Ideally, the selected drug should be entrapped within the nanoconstructs or be conjugated to them via cleavable linkers, enabling the selective release of the therapeutic agents only at the diseased site.[89] Additionally, the nanoconstruct must be designed to resist any hydrophobic/hydrophilic or metabolic/biochemical barriers, degradation, or clearance mechanisms at the target site or in circulation such that the integrity of the construct is intact *in vivo*.[83] With the advent of bioabsorbable materials, the dream of having a material that would carry out its job and then be degraded after drug delivery, thereby eliminating the risk of toxicity related to carrier accumulation, was realized. It is important to note that the harsh chemical conditions utilized for nanoconstruct preparation should not hamper the bioactivity of the API.

Drug release from the nanoconstruct can occur through either an abrupt release (burst release) or a sustained release mechanism, depending on the type of carrier material and the nature of the drug–carrier interaction.[90] The rate limiting step for drug release kinetics could be twofold: 1) diffusion of the drug out of the carrier or 2) dissolution of the carrier itself. An appropriate selection of the carrier material and the methods adopted for drug encapsulation are the determinant factors for tuning drug release to achieve a therapeutic dosage.

6. Types of Nanocarriers

The nanocarriers selected for drug delivery can be classified based on the type of material selected for preparation. The pre-requisite features for these materials include minimal toxicity, good blood compatibility, and low immunogenicity, with tunable degradation and mechanical properties as demanded by the application. In the present section, we will elaborate on the different nanosystems that have been developed to deliver drugs to inflammatory sites through several routes of administration (oral, intravenous, nasal, dermal, etc.), either actively or passively.

6.1 *Polymeric nanoparticles*

The essential characteristics that make polymeric nanoparticles ideal for drug delivery are: 1) biocompatibility, 2) biodegradability, 3) ease of surface modification through chemical transformations, 4) high entrapment efficiency, 5) excellent pharmacokinetic control, and 6) controlled drug release. Polymeric nanoparticles can be based either on natural or synthetic polymers and prepared in two different forms: 1) nanospheres (solid matrix wherein the drug is physically and uniformly dispersed) and 2) nanocapsules (vesicular systems wherein drug is confined to a cavity surrounded by a polymer membrane).[91–93]

6.1.1 *Natural polymeric nanocarriers*

Naturally derived polymers, with special thrust on polysaccharides and proteins, have emerged as attractive options for controlled drug release systems owing to their resemblances with the extracellular matrix in the human body and various other favorable physicochemical properties. Among them, those of natural origin such as albumin, gelatin, heparin, starch, and chitosan, have been the choice for delivering proteins, drugs, DNA, and genes. As a major plasma protein, albumin has a distinct edge over other materials for nanoparticle preparation, owing to its biodegradability, non-toxicity, and non-immunogenicity.[94] Both bovine serum albumin (BSA) and human serum albumin (HSA) have been used for nanoparticle formulations.[95–97] Albumin-bound preparations of various anti-inflammatory drugs, such as aspirin, ganciclovir, hydrocortisone, pilocarpine, and Rose Bengal, have been reported to provide sustained release formulations, ideal for topical or intravenous delivery.[98–103] Ganciclovir-loaded BSA nanoparticles, when tested for their ocular toxicity induced by the prolonged residence of the drug-loaded albumin nanocarriers in the eye after their intravitreal injection, appeared to be safe and non-toxic.[99] It was inferred that vision was not affected by any alterations in the behavior of ophthalmic cells due to the intravitreal injection of ganciclovir–albumin nanocarriers, proving its utility in the treatment of cytomegalovirus infection of the eye.

Another protein of interest which is biodegradable, non-toxic, less antigenic, easy to cross-link and chemically modify, and has immense potential for the preparation of colloidal drug delivery systems such as microspheres and nanoparticles, is gelatin.[104,105] Nanocarriers of gelatin have been prepared mainly by the double desolvation method for the controlled delivery of various anti-inflammatory drugs including indomethacin.[106] Bhavsar and

Amiji have utilized a novel approach of developing a nanoparticles-in-microsphere system for oral interleukin-10 (IL-10) gene therapy for the treatment of IBD.[107] This oral system formulated with IL-10–expressing plasmid DNA in type-B gelatin nanoparticles, which were further encapsulated in a poly(ε-caprolactone) microsphere matrix, provides good evidence of oral gene delivery and transfection for IBD therapy. Hoffmann *et al.* have successfully established selective targeting of nuclear factor κB (NF-κB) in the liver resident Kupffer cells that play an important role in inflammatory liver signaling. This was demonstrated through the use of solid gelatin nanoparticles loaded with NF-κB decoy oligodeoxynucleotides injected intraportally.[108] Thus, gelatin nanoparticles proved to be an appropriate carrier in delivering the decoy oligonucleotides to resident liver macrophages and being selectively uptaken into Kupffer cells, while hepatocytes remained completely unaffected in various *in vivo* settings like ischemia and reperfusion liver injury.

Ocular drug delivery systems based on nanoparticles that would prolong drug retention, allowing the drug to remain in contact with the cornea for long durations and showing increased bioavailability, are also being rigorously pursued. Gelatin nanoparticles possessing good bioadhesion properties, encapsulating ophthalmic drugs such as pilocarpine HCl or hydrocortisone have demonstrated success for use as efficient drug delivery carriers to the ocular tissues.[109]

Starch, which is composed of amylose and amylopectin, through appropriate cross-linking techniques, can yield an ideal bioadhesive matrix for delivering various types of drugs, including ophthalmic drugs such as timolol and flurbiprofen, and vaccines for nasal delivery.[110,111] Microparticles of starch as well as its grafts with polymers have been used for the nasal delivery of drugs as well as for vaccine delivery when administered orally and intramuscularly.[112,113] Nanoparticles made from starch derivatives such as propyl-starch have displayed efficient transdermal delivery using flufenamic acid and testosterone.[95] To improve on the enzymatic degradation of starch by pancreatic enzymes after oral ingestion, combinations of starch with other polymers (cyclodextrin derivatives, chitosan, etc.) encapsulating various hydrophobic drugs have been demonstrated.[96,97]

Chitosan, a polycationic polysaccharide derived from naturally occurring chitin, is endowed with good absorption, controlled release, and bioadhesive properties, all of which can be tuned by varying the degree of deacetylation and chemical modifications for specific drug delivery applications. Intranasal administration of chitosan/interferon-γ (IFN-γ) pDNA nanoparticles successfully inhibited allergen-induced airway inflammation as well as acute respiratory syncytial viral infection.[114,115] Nanoparticle therapy carried out in allergen-challenged mice resulted in increased IFN-γ expression by epithelial cells, which facilitated a reduction in inflammation and an immediate restoration of lung morphology within 3–6 h.

Derivatives of chitosan such as N-succinyl chitosan, due to the presence of carboxyl groups, exhibit pH-dependent swelling behavior. Due to the negative charge, they adhere easily to inflamed tissues that have a large number of positively charged proteins, especially for colon-specific delivery. Anti-inflammatory drugs such as 5-aminosalicylic acid (5-ASA), which is used for the long-term therapy of IBD, have been encapsulated within N-succinyl chitosan matrices including microspheres, hydrogels, and nanoparticles to achieve colon-specific controlled drug release.[116–119] Lee *et al.* proved that the absorption of theophylline, a well-acclaimed drug that reduces the inflammatory effects of allergic asthma, was considerably enhanced *in vivo* when administered with thiolated chitosan nanoparticles because of their greater mucoadhesiveness and permeability properties. This in turn improved theophylline absorption by the bronchial epithelium and enhanced its anti-inflammatory effects.[120]

Intra-articular (i.a.) drug delivery systems based on chitosan for the controlled release of drugs following i.a. injection showed an enhanced uptake of the nanoparticles by synovial macrophages at large joints (knees and hips), thereby opening avenues for efficient treatment strategies in osteoarthritis.[121] A novel approach for inflammatory reduction in RA has been achieved through clinically relevant RNAi-based anti-inflammatory treatments. This was accomplished through the intraperitoneal administration of chitosan/small interfering RNA (siRNA) nanoparticles that helped in silencing tumor necrosis factor-α (TNF-α) in systemic macrophages, which in turn down-regulated systemic and local inflammation.[122] Gene therapy is proposed as a new treatment strategy for common joint disorders like osteoarthritis for which non-toxic, targeted non-viral gene carriers are essential. Lu *et al.* have expounded the feasibility of using hybrids of hyaluronic acid/chitosan nanoparticles as vectors capable of transferring exogenous genes into primary chrondocytes for the treatment of these joint diseases.[123]

The mucoadhesive and penetration-enhancing properties across the mucosal epithelia as well as its antibacterial activity make chitosan an ideal nanocarrier for ocular drug delivery. A wide range of hydrophobic drugs, intended for the treatment of different extra- and intraocular diseases (such as antibiotics and anti-inflammatory and anti-glaucoma drugs) have been encapsulated within different chitosan nanostructures including chitosan-based nanoemulsions, chitosan-coated nanocapsules, chitosan nanoparticles, chitosan–polysaccharide nanoparticles, and chitosan–lipid nanoparticles.[124–130]

6.1.2 *Synthetic polymeric nanocarriers*

Pertinent nanoparticle formulations based on synthetic polymers include those made from polyesters such as polyglycolic acid (PGA), polylactic acid (PLA), the copolymer poly(lactic-co-glycolic acid) (PLGA), poly(caprolactone) (PCL), poly(methylmethacrylate) (PMMA), poly(alkylcyanoacrylate) (PCA), and poly(butylcyanoacrylate).[131–135] To avoid the rapid recognition of these intravenously injected polymeric nanocarriers from the blood by Kupffer cells, studies have particularly focused on the stealth-type long-circulating particles prepared by modifying their surfaces with PEG.[9,136]

Among the synthetic polymers, PLGA and PLA have been approved by the U.S. Food and Drug Administration (FDA) for human use.[137,138] Particles formed from such biodegradable polymers can help to attain a long-term therapeutic effect by the slow release of drugs, concomitant with the degradation of the polymer.[139] Numerous investigators have demonstrated the use of such biodegradable polymers individually as well as in combination as copolymers to successfully entrap several poorly water-soluble anti-inflammatory drugs that are widely used in clinical practice. The drugs are generally incorporated within the solid polymeric matrix through hydrophobic interactions using a variety of methods including oil-in-water emulsion/diffusion, nanoprecipitation, electrospraying, dialysis, and salting-out, wherein the extent of drug entrapment and its retention are considerably influenced by the physicochemical properties of the drug as well as the hydrophobic core.[7,77,18,140,141]

The anti-inflammatory drugs (clodronate, ibuprofen, dexamethasone, indomethacin, rolipram, 5-ASA, glucocorticoids) that have been encapsulated within polymeric nanocarriers such as PLGA, PLA, PLA–PEG, and PLGA–PEG, vary to include those that target the joint synovium as in RA, ocular mucosa for conjuctival inflammation as in uveitis, inflammatory demyelinating diseases of the central nervous system (CNS) as in multiple sclerosis (MS), inflamed colonic mucosa in IBD, Crohn's disease (CD), and ulcerative colitis, ulcerated inflamed tissues in gastric ulcers, type I allergy, etc.[50,142–149,150] In all of the above cases of passive-targeted delivery of drug-encapsulated nanocarriers through oral or intravenous routes, an enhanced and selective drug penetration into the inflammation site was achieved, mainly owing to the protection of the drug from efflux systems and mucosal metabolism.

An active targeting approach for inflammatory disorders such as colitis using polymeric PLA–PEG nanocarriers targeted to the inflamed endothelium was validated by Sakhalkar *et al.* by exploiting the leukocyte–endothelial cell adhesion chemistry.[151] Likewise, PLGA–PEG nanoparticles conjugated with peptides displayed rapid binding to human umbilical vein endothelial cells (HUVECs) overexpressing ICAM-1.[152] Such strategies can be utilized to target therapeutic agents to the inflamed endothelium having specific overexpression of CAMs.

PCL, in comparison to PLGA, has high permeability to small drug molecules and a relatively slower degradation rate. This makes PCL ideally suited for preparing microspheres, nanospheres, or implants that aid long-term controlled or targeted drug delivery extending to more than a year.[153] In contrast to the acidic environment generated by PLGA during its degradation, which can lead to peptide/protein instability, PCL has the ability to evade acidic conditions, making it suitable for such delivery applications. Nanospheres can be prepared either by PCL alone, or by using copolymers with PCL or PCL blends.[154,155] Various categories of drugs have been encapsulated in PCL for targeted drug delivery and controlled drug release. Of these, the use of PCL devices in the prolonged and controlled intravitreous release of dexamethasone showed good promise as extended release implants for long-term sustained intraocular delivery of corticosteroids.[156] Other studies have also demonstrated good success with the use of this sustained delivery polymer for ocular applications.[130,157,158] Different surface coatings on PCL nanoparticles were also investigated as an approach to modulate its interaction with the oral mucosa, thereby enhancing the bioavailability of entrapped drugs such as dexamethasone, indomethacin, cartelol, and flurbiprofen.[159–161]

PLA nanoparticles encapsulating betamethasone phosphate were found to be efficiently localized in the retina after a single intravenous injection to rats with experimental autoimmune uveoretinitis.[162] This strategy provides an alternate means for ocular targeting from the endothelial side, with sustained release in ocular inflammation using polymeric nanocarriers.

6.2 *Liposomes*

Liposomes are self-assembling closed colloidal vesicular structures of globular shape, composed of lipid bilayers, with the outer lipid bilayer surrounding a central aqueous space. They are very widely used in drug delivery owing mainly to their colloidal size, biocompatibility, biodegradability, non-immunogenicity, and their dual characters of hydrophilicity and hydrophobicity. Drugs with widely varying lipophilicities can be encapsulated in liposomes, in the phospholipid bilayer, the entrapped aqueous volume, or at the bilayer interface.[163,164] The structure of the synthetic bilayers in liposomes is very similar to that of the biological membranes in the body. Liposomes vary in size (ranging from 100 nm to several micrometers), phospholipid composition, and surface characteristics. Classical/conventional liposomes enable selective drug delivery to mononuclear phagocytic tissues, primarily the liver and spleen, mainly because they are easily recognized by the macrophages and are cleared from the circulation in a dose-dependent manner.[165] Physicochemical properties such as size, charge, and lipid composition significantly influence the efficiency with which liposomes target the wide range of receptors (integrins, scavenger receptors, mannose, folate receptors) expressed on cells of the MPS (monocytes, macrophages, dendritic cells).[166] In addition to the alterations in the liposomal surface through the selection of lipids, the amalgamation and linkage of glycoproteins and synthetic polymers to the bilayer also alter their biophysical and surface characteristics.[165,167,168] "Stealth" liposomes with prolonged blood circulation times are a real advancement in drug delivery, with reduced opsonization conferred by the choice of the lipid, other surface-modifying molecules (dextran, poly-N-vinylpyrrolidones), or through the use of hydrophilic PEG.[169–171] The pivotal role of MPS cells in inflammatory disorders makes them important targets for intracellular drug delivery. The use of ligands including peptides, antibodies, and lectins, increases target specificity and avoids the need for cationic lipids to trigger intracellular delivery.[166] However, it must be noted that liposomal drug delivery also has certain shortcomings such as poor scale-up, high cost, short shelf-life, and in some cases, off-target effects, which can be maneuvered through appropriate modifications in synthesis strategies.

In the past decade, liposomal nanocarrier-based delivery of anti-inflammatory drugs such as glucocorticoids, NSAIDs, and antioxidant enzymes, have been explored for delivering payloads directly to the site of inflammation. Liposomal corticosteroids have shown to exert strong beneficial effects by inhibiting macrophage activity in a number of preclinical models of inflammatory diseases such as RA, MS, and cancer.[173–176] Liposomal formulations of two potent corticosteroids, viz., prednisolone phosphate and liposomal 6α-methylprednisolone phosphate have demonstrated success in attenuating the inflammatory response in experimental autoimmune encephalitis (EAE) induced in rats.[173,175] However, Crielaard *et al.* very recently reported potentially diverse therapeutic efficacies for liposomal dexamethasone phosphate (DXP) in mouse models of MS and CD.[176] While the treatment with liposomal DXP in mice with EAE revealed decreased disease activity, the same treatment did not induce any improvements in mice with chronic colitis, but showed aggravated symptoms including fecal blood loss. Such contrasting disease outcomes in models of CD and MS after treatment with liposomal DXP, which changed the macrophage phenotype, imply the different inflammatory context of both diseases, with CD not responding favorably to liposomal corticosteroid treatment.

Liposomes have also demonstrated a remarkable capacity to be used as delivering and targeting anti-rheumatic drugs at lower doses with reduced toxicity.[177] The efficacy of long-circulating liposomes loaded with prednisolone disodium phosphate for arthritis is being evaluated in a phase II clinical trial (http://clinicaltrials.gov/ct2/show/NCT00241982). Cationic liposomal formulations that target the vascular endothelium and their effects in alleviating inflammatory arthritis, especially RA, have also been established.[177,178] Such clinically active liposomal delivery systems would most likely display enhanced efficacy and selective delivery with the inclusion of site-directed surface ligands (receptors, CAMs, etc.) to facilitate the *in vivo* transport and delivery of macromolecules and drugs.[179] Metselaar *et al.* have observed that the encapsulation of prednisolone phosphate in PEG liposomes demonstrated a drastic enhancement in therapeutic efficacy in adjuvant-induced arthritic rats.[172,180,181] Methotrexate and its lipophilic analog, a commonly used disease-modifying anti-rheumatic drug (DMARD), when loaded within liposomes also showed enhanced efficacy and better retention time in the joints.[182] Liposome-encapsulated clodronate and glucocorticoid drugs have proven successful in targeting macrophages in adjuvant-induced arthritic rats. Liposome-based drug delivery proved to be more effective than the "free" control, with complete remission observed after six days.[172] RGD peptide-conjugated PEG liposomes loaded with DXP, targeting the angiogenic vascular endothelial cells at the inflammatory sites of arthritis, have also revealed good potential in inhibiting experimental arthritis in rat models.[183] Thus, peptide targeting of endothelial cells would be an efficient approach in delivering drugs specifically to the endothelium of inflammatory regions. Active targeting of sialyl Lewis X (SLX) glycoliposomes to the inflammatory region was achieved in collagen antibody-induced arthritic mice in an inflammatory-dependent manner.[184] SLX is an E-selectin ligand expressed on leukocyte surfaces that could be targeted using glycoliposomes to enable anti-inflammatory effects.

Likewise, ocular drug delivery has also benefited immensely with the use of liposomes for topical application to the eye, aiding in maintaining the drug at the site of application for a longer period of time, thereby improving its therapeutic efficacy.[185,186] SLX-conjugated liposomes loaded with dexamethasone and injected into experimental autoimmune uveoretinitis-induced mice showed nearly double the concentration of the drug in the eyes in comparison to the free drug.[187] Drug-loaded SLX liposomes targeted the selectins on the activated vascular endothelial cells of the inflamed eye, hence establishing the benefits of using liposomal formulations for intraocular delivery at reduced dosing frequencies. PEG-coated liposomes after intravitreal injections in healthy rats were found to accumulate in the vitreous humour, inner retina, etc. via internalization by the macrophages and neutrophils in the conjunctiva, as revealed through experiments using fluorescent liposomes.[188] Vasoactive intestinal peptide (VIP) loaded within sterically stabilized PEG-coated liposomes enabled controlled release of the drug with concomitant efficacy, when injected intravitreally to rats suffering from endotoxin-induced uveitis.[189]

A liposomal formulation of dexamethasone 21-palmitate as the active ingredient (Lipotalon®, Merckle, Germany), is available for i.a. injection. The increased anti-inflammatory activity of this injection is attributed to the prolonged residence in the joint following injection, where liposomes accumulate in the synovial membrane and slowly release the active compound.[190] To date, a range of potential therapies for inflammatory-related conditions including nerve injury-associated hyperalgesia, endometriosis, lung cancer cell metastasis, restenosis, arthritis, and hyperlipidemia have been assessed using liposome-mediated bisphosphonate delivery.[191–194]

Additionally, drug delivery to the brain, which is greatly hampered by the extremely selective permeability of the blood–brain barrier (BBB), can be achieved by targeting circulating mononuclear cells that are able to cross the BBB.[195] Liposomes are readily taken up by macrophages, microglia, and astrocytes in the CNS. The binding selectivity of charged liposomes, mainly cationic liposomes, to the spinal cord of rats with EAE after an intravenous injection was illustrated by Cavaletti *et al.*[196] This specificity could be attributed to the enhanced permeability of BBB in neuroinflammatory processes, wherein cationic particles bind to the proliferating vasculature in chronic inflammation and angiogenesis. PEGylated liposomes also accumulate more rapidly in brain when the BBB is compromised such as in EAE.[195,197]

Liposomal glucocorticoids also provide a new approach for relapse therapy in MS, with PEG minocycline liposomes ameliorating CNS autoimmune diseases. Treatment with intravenous PEG minocycline liposomes every five days showed similar effects on the clinical signs of EAE compared to treatment with daily intraperitoneal injections of minocycline, enabling reduced drug dosage and potential adverse risks.[175] Phosphatidylglycerol-composed liposomes were successfully applied for the delivery of the neurotransmitter serotonin, and *in vivo* studies showed that liposomal delivery improved the localization of the drug to the brain.[198]

6.3 *Dendrimers*

Dendrimers are highly branched, monodisperse, and uniformly distributed polymeric macromolecules with a unique spatial organization of molecular groups and host–guest entrapment properties.[199] Owing to the presence of a hydrophobic core and a hydrophilic surface, dendrimers can be loaded with drug molecules both in the interior as well as on the surface groups, based on their solubility.[200] In comparison to linear polymers, nanoparticles, and liposomes, the multivalency and nano-size of dendrimers confer substantial advantages in drug delivery such as ease of conjugation for targeting, rapid cell entry, and easier passage across biological barriers.[201,202] Among the various dendritic polymers investigated, polyamidoamine (PAMAM) dendrimers were the first to be commercialized and continue to be the most widely studied group. The pharmacokinetics and biodistribution of dendrimers are considerably influenced by their physical attributes such as molecular size, surface charge, and chemical modification.[203]

Various kinds of drugs have been encapsulated in or covalently conjugated with dendrimers, with the aim of enhancing the solubility of hydrophobic drugs in aqueous biological fluids,[204] augmenting the transdermal permeability of drugs, such as the NSAID indomethacin,[205] facilitating the intestinal absorption of poorly absorbable hydrophilic drugs and macromolecular compounds,[206] or improving the pulmonary absorption of peptide and protein drugs.[207] PEGylation is considered to be ideally suited for the amendment of dendrimers for numerous reasons, the primary being their stability (long circulation) in the body, in addition to reducing drug leakage, hemolytic toxicity, and improving drug-loading capacity, with sustained drug delivery. One such PEGylated polypropyleneimine (PPI) dendrimeric system for the sustained delivery of aceclofenac has demonstrated nearly 77% inhibition of paw edema in albino rats at the seventh hour of drug administration.[208] In an effort to understand the allergen-suppressive action of nanoparticles, Wegmann *et al.* showed that synthetic peptide dendrimers may block experimental allergic encephalomyelitis.[209] Chauhan *et al.* have patented the development of a composition comprising monodisperse dendrimers having a plurality of terminal groups complexed to drug moieties having anti-inflammatory activity.[210]

The biodistribution of dendrimers is also critically altered by the biological factors related to specific diseases. Lesniak *et al.* observed that dendritic polymers targeted using peptides specific for integrin receptors showed a significant accumulation at the sites of angiogenesis in a mouse model of hindlimb ischemia.[211] Likewise, folate-conjugated dendrimers were preferentially accumulated at sites of inflammation in animal models of arthritis.[212] Thus, dendrimers find use in the transport of various therapeutic agents to intracellular target sites through topical, intravenous, pulmonary, transdermal, ocular, or oral routes, for various diseased states to control pain and inflammation.

Hui *et al.* in a recent study demonstrated that 4-nm, neutral generation 4 PAMAM dendrimers without any targeting ligands were localized in activated microglial and astrocyte cells (responsible for neuroinflammation), in a rabbit model of cerebral palsy following subarachnoid administration.[213] This intrinsic ability of dendrimers to localize passively in the neuroinflammatory cells imply that they have the potential for prospective use as efficient drug or gene delivery vehicles for targeting therapy in several neuroinflammatory or neurodegenerative disorders.

Of the various biological and medical applications of dendrimers that have evolved in the nanomedicine landscape, the *in vivo* anti-inflammatory properties of surface-functionalized dendrimers *per se* is a relatively recent discovery.[214] The intrinsic anti-inflammatory properties of dendrimers such as PAMAM are mainly due to the inhibition of pro-inflammatory cytokine secretion by macrophages and dendritic cells.[215,216] Among the various classes of dendrimers, PAMAM-based, phosphorus-based, and PEO/PEG dendrimers are polyanionic nanodevices and display anti-inflammatory properties.[217,218] Very recently, Hayder *et al.* tested the potential usefulness of an anionic azabisphosphonate (ABP) dendrimer in two established inflammatory arthritis models resembling human RA. It was noted that the ABP dendrimer inhibited osteoclastogenesis and the secretion of pro-inflammatory cytokines, both of which are underlying monocyte-dependent processes in RA inflammation and bone damage.[215] This study thus illustrated the potential of phosphorous-containing dendrimers, particularly ABP, as a prospective nanotherapeutic that can modulate the innate immunity through its direct dual action on both inflammation and bone erosion, and thus offers itself as a new cost-effective drug candidate for RA therapy in lieu of current biological agents.[219]

6.4 *Micelles*

Polymeric micelles are nano-sized, spherical, colloidal particles, and are molecular assemblies of amphiphilic diblock or triblock copolymers. Micelles have a core-shell architecture and form spontaneously in aqueous solutions of amphiphilic block copolymers, with a core comprising hydrophobic polymer blocks [e.g., poly(propylene glycol), PCL, poly(D,L-lactide), etc.] and a shell of hydrophilic polymer blocks (PEG).[220,221] This core-shell structure of micelles with size ranging from ~1–100 nm accommodates hydrophobic drugs in its core, and the hydrophilic brush-like corona shell stabilizes micelles in dispersion.[222] This helps to mask the drug from interactions with serum proteins and untargeted cells, making the particle an appropriate candidate for intravenous administration.[223,224] Drugs can be loaded into a polymeric micelle either through physical encapsulation or via chemical covalent attachment.[225,226] Micellar size is typically ~50 nm, which provides obvious advantages over liposomes. Polymeric micelles with stabilizing PEG coronas (stealth micelles) could circulate in the bloodstream for a prolonged time period by evading host defenses. While micelles allow a great depth of tissue penetration for targeted drug delivery, they usually disintegrate rapidly in the body, thereby making sustained drug delivery from micellar nanocarriers a challenge. Hence, researchers have explored various strategies to provide sustained drug release from nanomicellar carriers through the use of prodrugs, drug–polymer conjugates, polymers with low critical micellar concentration, reverse micelles, multi-layer micelles with layer-by-layer assembly, micelle coats on a solid support, etc.[227] The drug-loading capacity of micelles is found to be influenced by the copolymer composition, the chemical structure of the drug, as well as the hydrogen bonding interaction between the drug and hydrophilic corona.[228] Among various polymeric micelles, lipid-core micelles formed by the conjugation of soluble copolymers with lipids as in PEG–phosphatidyl ethanolamine (PEG–PE) have emerged important. Such lipid-core mixed micelles formed by positively charged lipids have the ability to accumulate within endosomes and deliver drugs directly into the cell cytoplasm.[229] Polymeric multifunctional micelles based on thermo-responsive biotin-P(NIPAAm-co-NDAPM)-b-PCL that can hold therapeutic agents together with imaging agents as well as targeting ligands, aid the simultaneous sustained release of drugs and facilitate *in vivo* imaging.[225]

Amphiphilic polyphosphazenes (PNIPAAm/EAB-PPPs) with poly (N-isopropylacrylamide) (PNIPAAm) and ethyl 4-aminobenzoate (EAB) loaded with indomethacin showed sustained therapeutic efficacy when administered to arthritic rat models through the topical injection of an aqueous solution of indomethacin-loaded micelles.[232] Such a local delivery of indomethacin can help to avoid severe gastrointestinal stimulation, which is frequently associated with oral administration as evidenced by ulceration evaluation. Dendritic unimolecular micelles with a hydrophobic core surrounded by a hydrophilic shell prepared by coupling dendritic hypercores with PEG mesylate have also demonstrated success in the encapsulation of indomethacin, providing sustained release characteristics.[233] Targeting local macrophages in the liver and ankle joints in an induced arthritis mouse model with 100-nm clodronate-encapsulated small unilamellar vesicles (SUVs) resulted in the depletion of the local macrophage population, while an untargeted control macrophage population in the spleen remained unaffected.[193] This removal of macrophages from the site inhibited inflammatory pathogenesis. It has been also shown that the effect of an anti-inflammatory drug encapsulated in SUVs had a 90% enhanced effect on hepatic macrophages compared to systemically administered free drug.

Jelena *et al.* have established that amphiphilic star-like macromolecules (ASMs), as against conventional polymeric micelles (e.g., Pluronic P-85) have the ability to incorporate sparingly water-soluble NSAID drugs such as indomethacin, piroxicam, and ketoprofen into the micellar core and provide delayed dermal delivery, as well as prevent compound permeation through the skin from aqueous formulations.[234] Polymeric nanomicelles encapsulating 3-acetyl-11-keto-beta-boswellic acid (AKBA) developed for topical delivery showed considerably enhanced skin permeability as well as anti-inflammatory and anti-arthritic activity in comparison with AKBA gel when tested through excised abdominal skin.[235] Significant anti-inflammatory activity of diclofenac in lecithin bile-salt mixed micellar formulations with cyclic monoterpenes was observed on rat hind paw edema due to the higher permeation rate of diclofenac through skin.[236] Thus, a mixed micellar formulation with cyclic monoterpenes was found to be more effective and safer for the transdermal delivery of diclofenac. Similarly, Li *et al.* have recently evaluated the combined effect of a micellar solution (Tween 80) and inclusion compound (β-cyclodextrin) in improving the solubility and bioavailability of orally administered flurbiprofen.[237] pH-sensitive copolymer micelles capable of inhibiting premature burst drug release under intestinal conditions can aid in the oral drug delivery of poorly water-soluble drugs.[238] A micellar nanocarrier based on A2B-type miktoarm polymers of PEG and PCL, developed through "click" chemistry followed by ring-opening polymerization, have been efficiently loaded with nimodipine (NIM, a drug used for the treatment of delayed ischemic neurological disorders) to provide sustained drug release.[239] Micelle-encapsulated NIM reduced the release of the pro-inflammatory cytokine TNF-α from microglia, suggesting the potential of these miktoarm micelles for the treatment of neuroinflammation.

Polymeric micelles prepared from the copolymers viz., NIPAAM, vinyl pyrrolidone (VP), and acrylic acid (AA), were used to entrap ketorolac, an anti-inflammatory drug for ocular applications.[240] Corneal permeation studies through an excised rabbit cornea indicated high ocular availability with no corneal damage compared to an aqueous suspension of the drug. Likewise, a nanomicellar system composed of the polyoxyethylated nonionic surfactant

Pluronic F127 (F127) and cationic polyelectrolyte chitosan loaded with dexamethasone showed an enhanced drug release rate due to the addition of the permeability enhancer, viz., chitosan.[241] Topical administration of cross-linked polymeric nanomicelles of NIPAAM–VP–MAA loaded with dexamethasone also showed significantly reduced inflammation of the eye for a longer duration compared to the aqueous drug suspension.[242] All these studies suggested that nanomicellar formulations, due to their small particle size and mucoadhesiveness, help to achieve clearly reduced uveitis symptoms and hence, are efficient for ocular drug delivery. Polymeric micelles have also been indicated for pulmonary drug delivery in the treatment of asthma and chronic obstructive pulmonary disease.[243]

Recently, nano-sized micelles of amphiphilic PEG–bl–poly(propylene sulfide) (PEG–bl–PPS) block copolymers have been successfully utilized for the local delivery of immunomodulatory drugs directly to the lymph nodes, thereby targeting local immune cells. Immunosuppressive drug-loaded micellar particles of 50 nm were found to be efficiently transported through lymphatic vessels to draining lymph nodes, and were found to be effective in prolonging mismatched allograft survival due to the increased drug bioavailability to local immune cells.[244]

Amino acid-based anionic amphiphilic copolymers of poly(sodium N-acryloyl-l-valinate-co-alkylacrylamide) self-assembled to form micellar aggregates and proved to be potential carriers for entrapping several NSAIDs of differing hydrophobicities.[245] It was observed that the solubility of the drugs increased nearly 10-fold in the presence of the copolymers, thus confirming that amino acid-based copolymer micelles can be potential reservoirs for hydrophobic drugs. Likewise, Khalfin *et al.* have demonstrated the ability of the amphiphilic molecule β-casein, the main protein found in milk, to form stable micelles that efficiently entrap the anti-inflammatory drugs celecoxib and budesonide.[246] Thus, being a natural food product and a protein which is easily biodegradable and does not provoke immune response, casein micelles are ideal candidates for delivering drugs to inflammatory sites.

Lecithin organogels (LOs) are stable, viscoelastic, and biocompatible jelly-like phases, mainly constituted by hydrated phospholipids and an appropriate organic liquid. These have been used to load various therapeutic agents for topical (dermal or transdermal) drug delivery.[247] Topical administration of an LO formulation containing a therapeutic dose of digoxin was found effective for the treatment of muscle spasms as well in the management of neuropathic pain.[248] Pluronic LOs have been successfully used for the topical delivery of local anesthetics and NSAIDs in regional orofacial disorders, where the systemic administration of drugs does not cause a rapid onset of action.[249] The analgesic action of ketamine hydrochloride and amitryptiline hydrochloride formulated in LOs showed significant improvement owing to their enhanced skin penetration when delivered in micellar form.[250] The transdermal delivery of indomethacin and diclofenac incorporated in LOs was observed to be considerably high, including remarkable therapeutic effects in the treatment of pain associated with mild to moderate osteoarthritis.[251,252] A novel transdermal delivery system with a sustained release profile based on a thermosensitive hydrogel loaded with Pluronic F127 micelles incorporating honokiol, a hydrophobic drug, has been reported by Zheng *et al.*[253]

Polymer-based or polymer–lipid micellar assemblies can also be designed as multimodal receptor-targeted blockers or drug carriers whose activity can be tuned by varying polymer hydrophobicity, charge, and architecture. Such multifunctional, drug-loaded micellar nanostructures that modify the atheroinflammatory cascade within vascular intima by targeting atheroinflammatory lesions and atherosclerotic plaques would find immense applications in the management of atherosclerosis, thereby combating coronary artery disease.[254]

6.5 *Solid lipid nanoparticles*

SLNs are lipid-based submicron colloidal carriers and represent an alternative carrier system to the traditional colloidal carriers such as liposomes, emulsions, and polymeric micro- and nanoparticles.[255] SLNs bear the positive aspects of traditional systems, overcoming some of their major disadvantages. The relatively rigid core of SLNs consisting of hydrophobic lipids (triglycerides, glyceride mixtures, or waxes) that remain solid at room (or body) temperature and surrounded by a monolayer of phospholipids, make them more stable than liposomes in biological systems. SLNs with their solid lipid core reduce drug mobility, thereby enhancing the controlled release of loaded drugs.[256] SLN stability can be further improved through the appropriate use of a surfactant.[257,258] Their ease of biodegradation, lipid constitution, and organic solvent-free assembly make them less toxic than polymeric nanoparticles. SLNs can be engineered for drug loading to have either a homogeneous matrix or a drug-enriched shell or core.[259,260] Compared to other drug delivery systems, SLNs have emerged to be potentially attractive marketable choices for drug delivery owing to their easily scalable synthesis process, mainly using high pressure homogenization techniques. However, SLNs are plagued with undesirable particle growth by aggregation, which can lead to a rapid "burst release" of drugs. Additionally, the formation and enhancement of the crystal lipid structure during both production and storage of SLNs often result in the expulsion of the loaded molecular drug solution, which is a major drawback of this nanoformulation.[258] Various research groups have focused on improving the stability of SLNs in body fluids after administration, by coating them with hydrophilic molecules such as PEG. This helps to improve its plasma stability, biodistribution, and the subsequent bioavailability of entrapped drugs when administered through various routes.[261]

Similar to other nanoforms used for drug delivery, numerous SLN formulations have been described since the early 1990s. Various anti-inflammatory drugs ranging from NSAIDs to specific drugs that target the BBB have been encapsulated within SLNs and delivered parenterally, orally, transdermally, pulmonarily, etc. The role of SLNs in enhancing the anti-inflammatory activity of drugs such as dexamethasone and butyrate in human IBD revealed that inflammatory cytokine secretion is considerably regulated *in vitro* in comparison to bare drugs, with no toxicity observed even at the highest concentration tested.[262] In the anti-inflammatory therapy of ulcerative colitis where high doses of butyric acid have been shown to ameliorate inflammation, cholesteryl-butyrate SLNs (Chol-but SLNs) as a prodrug proved ideal for delivering butyric acid.[263] Chol-but SLNs exert rapid and potent anti-inflammatory effects on polymorphonuclear cells, that can be ascribed to the rapid internalization of SLNs in the cytoplasmic compartment of cells. Thus, Chol-but SLNs can be a viable alternative to sodium butyrate in the anti-inflammatory therapy of IBD, evading complications related to the administration of sodium butyrate.

SLNs encapsulating diclofenac sodium, triptolide, etc. with relatively high entrapment efficiencies exhibited improved localization of the drug when applied transdermally.[264,265] SLNs have also proved to be ideal carriers for the delivery of beclomethasone dipropionate (BDP) via the pulmonary route. SLNs were found to be efficiently nebulized, yielding aerosols of a suitable particle size for BDP deep lung delivery, providing avenues for lipophilic drug-targeting strategies by nebulization.[266] Moreover, SLNs have been suggested to enhance drug penetration into the skin for topically applied drugs. To validate this, Schlupp *et al.* tested three glucocorticoids loaded onto SLNs — prednisolone, the diester prednicarbate, and the monoester betamethasone 17-valerate, all having varied structures and lipophilicities.[267] SLNs were found to influence skin penetration of the glucocorticords through a specific interaction of the drug–carrier complex with the skin surface, due to their lipid nature and nanosize, with prednisolone showing the maximum skin penetration profile.

SLNs are suggested as promising alternatives to conventional and popular ocular carrier systems such as nanoemulsions, liposomes, and polymeric nanoparticles, with topical applications of both steroidal or NSAIDs inducing minimal corneal irritation and maximal anti-inflammatory effects.[268] SLNs have also been used to encapsulate curcumin, a molecule with pluripharmacological properties to improve its oral bioavailability.[269] The *in vivo* pharmacokinetics of this hydrophobic drug after oral administration in rats was found to be considerably improved with respect to solubilized curcumin. This enhanced bioavailability would enhance its therapeutic usefulness for neurodegenerative and cancerous disorders in humans. Corticosteroids, such as betamethasone 17-valerate, find immense use in the pharmacological treatment of skin diseases, e.g., in eczema or psoriasis. The SLN system was found to be an ideal nanocarrier for the prolonged and localized delivery of these active drugs into the skin, with minimal systemic drug absorption, which was dependent on the composition of the lipid.[270,271]

Passive targeting of the spleen by drug-loaded SLNs mediated by macrophage uptake would be an efficient way of improving the therapeutic efficacy of drugs with reduced side effects. In one such study, injectable actarit-loaded SLNs yielded promising results for the therapy of RA.[272] The pharmacokinetics, tissue distribution, and bioavailability of drug-loaded SLNs after intravenous and intraduodenal administration were considerably improved.[273] SLNs prepared using various triglycerides and charge modifiers encapsulating the drug showed high bioavailability, with charged SLNs better taken up by the liver and brain, respectively. By taking advantage of the ability of SLNs to cross the BBB, nanoparticles complexed with oligonucleotides have been tested in an *in vivo* mouse model and *in vitro* endothelial cell model aimed at developing novel treatments for neurodegenerative disorders.[274] A nanogel formed by cross-linked PEG and polyethyleneimine that spontaneously encapsulated negatively charged oligonucleotides after intravenous administration resulted in a 15-fold accumulation of oligonucleotides in the brain in comparison to its free form. SLNs encapsulating iron oxide nanoparticles have also been shown to cross the BBB and accumulate in the brain with long circulation properties.[275] These superparamagnetic contrast agents would provide a novel approach to imaging the CNS by magnetic resonance imaging. Stealth SLNs, prepared by PEGylating the surface of SLNs, showed lower clearance, a higher distribution volume, and a significantly higher area under the curve than non-stealth SLNs in many cases.[261] Thus, SLN formulations are being extensively developed and characterized for their *in vitro* and *in vivo* applications to deliver various anti-inflammatory drugs as well as macromolecules such as proteins and oligonucleotides.[276]

Table 1 depicts several representative combinations of targeted and non-targeted nanocarriers with anti-inflammatory drugs for different diseased states administered through oral, intravenous, cutaneous/subcutaneous, and

Table 1. Representative combinations of nanocarriers and anti-inflammatory drugs for different inflammatory disorders.

Nanocarrier system	Material–drug combination	Targeting strategy	Route of administration	Potential application	Reference
Polymeric Nanocarriers	Albumin: ganciclovir	Passive: epithelium	Ocular	Not tested	99
	Gelatin: indomethacin	Passive: epithelium	Oral	Experimental paw edema	106
	Type-B gelatin: IL-10 gene	Passive: mucosal epithelium	Oral	IBD	107
	Gelatin: piplocarpine HCL	Passive: epithelium	Ocular	Not tested	109
	Starch: vaccines	Passive: immune cells	Oral	Adjuvants	112
	Chitosan: anti-TNF–α siRNA	Passive: peritoneal macrophages	Intravenous	Experimental arthritis	122
	Chitosan: indomethacin	Passive: cornea	Ocular	Improve bioavailability	129
	DL-PLGA: betamethasone sodium phosphate	Passive	Intra-articular	RA	143
	PLA/PEG: betamethasone sodium phosphate	Passive: inflamed joints	Intravenous	Experimental arthritis	136
	PLA:betamethasone phosphate	Passive: endothelium	Intravenous	Uveitis	162
Liposomes	Liposome: alendronate	Passive	Stenting	Not tested	194
	PEG and non-PEG liposomes: prednisolone	Active: endothelium	Intravenous	RA	172
	RGD peptide–PEG liposomes: dexamethasone phosphate	Active: angiogenic endothelial cells	Intravenous	RA	183
	Liposomes: prednisolone	Passive: Brain	Intravenous	MS	175

(*Continued*)

Table 1. (*Continued*)

Nanocarrier system	Material–drug combination	Targeting strategy	Route of administration	Potential application	Reference
Dendrimers	G4–OH and G4–NH_2 dendrimers: indomethacin	Passive	Transdermal	Experimental paw edema	205
	PAMAM dendrimer: 5(6)-carboxyflourescin/ isothiocyanate-dextrans	Passive: mucosal epithelium	Oral	Improving small intestinal absorption	206
	Folate–PAMAM dendrimers: indomethacin	Active: folate receptors	Intravenous	RA	212
	Dendritic polyglycerol sulfate	Passive	Intravenous	Targeting leukocyte trafficking	217
Micellar nanocarriers	Small unilamellar vesicles: clodronate	Passive: endothelial cells	Intravenous	RA	193
	Polyphosphazene micelles: indomethacin	Passive: endothelial cells	Intravenous	RA	232
	Micelle: Fluribuprofen	Passive: small intestine	Oral	To improve bioavailability	237
	Pluronic F127–chitosan micelles: dexamethasone	Passive	Ocular	Not tested	241
	PEG–bi–PPS micelles: glucocorticoid mometasone	Passive: lymph nodes	Intravenous	Allograft survival	244
Solid lipid nanoparticles	SLN: cholesteryl butyrate	Passive epithelium	Intravenous	IBD	263
	SLN: glucocorticoid	Passive: skin	Topical	To improve skin penetration	267

pulmonary routes. From the table, it is evident that the same API, when encapsulated in diverse nanosystems, can deliver the drugs to different target sites, either passively or actively.

7. Conclusions

Therapeutics currently used for the treatment of inflammatory disorders provide efficacious disease maintenance, giving relief of symptoms, but are recurrently accompanied by intolerable side effects upon prolonged drug administration. The use of nano-sized carriers for such therapies would enhance the specificity of drugs to target sites, with concomitant decrease in the drug dosage and thereby, its side effects. Several studies on autoimmune diseases and allergic models demonstrated the feasibility of nanodrugs in alleviating the diseases. Varieties of nanomaterials such as natural and synthetic polymeric nanoparticles, protein nanocarriers, and micelles, in combination with currently administered drugs, drug targets, and/or novel gene-targeted approaches present a broad spectrum of opportunities in an attempt to unravel the best alternative(s) for the treatment of inflammatory disorders.

References

1. Xia Y. Nanomaterials at work in biomedical research. *Nat Mater* 2008;7:758–760.
2. Ferrari M. Cancer nanotechnology: Opportunities and challenges. *Nat Rev Cancer* 2005;5:161–171.
3. Rothaermel F, Thursby M. The nanotech versus the biotech revolution: Sources of productivity in incumbent firm research. *Res Policy* 2007;36:832–849.
4. Dobrovolskaia MA, McNeil SE. Immunological properties of engineered nanomaterials. *Nat Nanotechnol* 2007;2:469–478.
5. Cao G. *Nanostructures & Nanomaterials: Synthesis, Properties & Applications*, Imperial College Press, 2004.
6. Jain PK, Lee KS, El-Sayed IH, El-Sayed MA. Calculated absorption and scattering properties of gold nanoparticles of different size, shape, and composition: Applications in biological imaging and biomedicine. *J Phys Chem B* 2006;110:7238–7248.
7. Goodman CM, McCusker CD, Yilmaz T, Rotello VM. Toxicity of gold nanoparticles functionalized with cationic and anionic side chains. *Bioconjug Chem* 2004;15: 897–900.
8. Panyam J, Labhasetwar V. Biodegradable nanoparticles for drug and gene delivery to cells and tissue. *Adv Drug Delivery Rev* 2003;55:329–347.
9. Soppimath KS, Aminabhavi TM, Kulkarni AR, Rudzinski WE. Biodegradable polymeric nanoparticles as drug delivery devices. *J Controlled Release* 2001;70:1–20.
10. Vasir JK, Labhasetwar V. Biodegradable nanoparticles for cytosolic delivery of therapeutics. *Adv Drug Delivery Rev* 2007;59:718–728.

11. Kingsley JD, Dou H, Morehead J. Nanotechnology: A focus on nanoparticles as a drug delivery system. *J Neuroimmune Pharmacol* 2006;1:340–350.
12. Hu L, Tang X, Cui F. Solid lipid nanoparticles (SLNs) to improve oral bioavailability of poorly soluble drugs. *J Pharm Pharmacol* 2004;56:1527–1535.
13. Zhang L, Gu F, Chan J. Nanoparticles in medicine: Therapeutic applications and developments. *Clin Pharmacol Ther* 2008;83:761–769.
14. Jabr-Milane LS, van Vlerken LE, Yadav S, Amiji MM. Multi-functional nanocarriers to overcome tumor drug resistance. *Cancer Treat Rev* 2008;34:592–602.
15. Faraji AH, Wipf P. Nanoparticles in cellular drug delivery. *Bioorg Med Chem* 2009;17:2950–2962.
16. Cho K, Wang X, Nie S, Chen ZG, Shin DM. Therapeutic nanoparticles for drug delivery in cancer. *Clin Cancer Res* 2008;14:1310–1316.
17. Olivier J-C. Drug transport to brain with targeted nanoparticles. *NeuroRx* 2005;2:108–119.
18. Pinto Reis C, Neufeld RJ, Ribeiro AJ, Veiga F. Nanoencapsulation ii. Biomedical applications and current status of peptide and protein nanoparticulate delivery systems. *Nanomed Nanotechnol Biol Med* 2006;2:53–65.
19. Freeman AI, Mayhew E. Targeted drug delivery. *Cancer.* 1986;58:573–583.
20. Levy-Nissenbaum E, Radovic-Moreno AF, Wang AZ, Langer R, Farokhzad OC. Nanotechnology and aptamers: Applications in drug delivery. *Trends Biotechnol* 2008;26:442–449.
21. Sudimack J, Lee RJ. Targeted drug delivery via the folate receptor. *Adv Drug Delivery Rev* 2000;41:147–162.
22. Brannon-Peppas L, Blanchette JO. Nanoparticle and targeted systems for cancer therapy. *Adv Drug Delivery Rev* 2004;56:1649–1659.
23. Maeda H, Wu J, Sawa T, Matsumura Y, Hori K. Tumor vascular permeability and the EPR effect in macromolecular therapeutics: A review. *J Controlled Release* 2000;65: 271–284.
24. Duncan R. Polymer conjugates for tumour targeting and intracytoplasmic delivery. The EPR effect as a common gateway? *Pharm Sci Technol Today* 1999;2:441–449.
25. Maeda H, Bharate GY, Daruwalla J. Polymeric drugs for efficient tumor-targeted drug delivery based on EPR-effect. *Eur J Pharm Biopharm* 2009;71:409–419.
26. Calder PC. N-3 polyunsaturated fatty acids, inflammation, and inflammatory diseases. *Am J Clin Nutr* 2006;83:S1505–1519.
27. Steinhoff M, Buddenkotte, DM. Proteinase-activated receptors: Transducers of proteinase-mediated signaling in inflammation and immune response. *Endocr Rev* 2005; 26:1–43.
28. Dantzer R, O'Connor JC, Freund GG, Johnson RW, Kelley KW. From inflammation to sickness and depression: When the immune system subjugates the brain. *Nat Rev Neurosci* 2008;9:46–56.
29. Karin M, Greten FR. NF-kappab: Linking inflammation and immunity to cancer development and progression. *Nat Rev Immunol* 2005;5:749–759.
30. Luster AD, Alon R, von Andrian UH. Immune cell migration in inflammation: Present and future therapeutic targets. *Nat Immunol* 2005;6:1182–1190.
31. Baniyash M. Chronic inflammation, immunosuppression and cancer: New insights and outlook. *Semin Cancer Biol* 2006;16:80–88.

32. Feghali CA, Wright TM. Cytokines in acute and chronic inflammation. *Front Biosci* 1997;2:d12–d26.
33. Itzkowitz SH, Yio X. Inflammation and cancer iv. colorectal cancer in inflammatory bowel disease: The role of inflammation. *Am J Physiol Gastrointest Liver Physiol* 2004; 287:G7–G17.
34. Murtaugh M, Baarch M, Zhou Y, Scammura R, Lin G. Inflammatory cytokines in animal health and disease. *Vet Immunol Immunopathol* 1996;54:45–55.
35. Clark IA, Alleva LM, Budd AC, Cowden WB. Understanding the role of inflammatory cytokines in malaria and related diseases. *Travel Med Infect Dis* 6:67–81.
36. Fex E. Tissue-derived macromolecules and markers of inflammation in serum in early rheumatoid arthritis: Relationship to development of joint destruction in hands and feet. *Rheumatology* 1997;36:1161–1165.
37. Movat HZ, Fernando NV. Allergic inflammation. i. The earliest fine structural changes at the blood-tissue barrier during antigen-antibody interaction. *Am J Pathol* 1963;42:41–59.
38. Mazzucchelli L, Hauser C, Mueller C. Expression of interleukin-8 gene in inflammatory bowel disease is related to the histological grade of active inflammation. *Am J Pathol* 1994;144:997–1007.
39. Fujiwara N, Kobayashi K. Macrophages in inflammation. *Curr Drug Targets Inflamm Allergy* 2005;4:281–286.
40. Pinto-Alphandary H, Andremont A, Couvreur P. Targeted delivery of antibiotics using liposomes and nanoparticles: Research and applications. *Int J Antimicrob Agents* 2000; 13:155–168.
41. Tabata Y. Size effect on systemic and mucosal immune responses induced by oral administration of biodegradable microspheres. *Vaccine* 1996;14:1677–1685.
42. Pinto-Alphandary H, Balland O, Couvreur P. Intracellular visualization of ampicillin-loaded nanoparticles in peritoneal macrophages infected *in vitro* with *Salmonella typhimurium*. *Pharm Res* 1994;11:38–46.
43. Ulbrich W, Lamprecht A. Targeted drug-delivery approaches by nanoparticulate carriers in the therapy of inflammatory diseases. *J R Soc Interface* 2010;7:S55–66.
44. Lamprecht A. *Nanotherapeutics: Drug Delivery Concepts in Nanoscience*, Pan Stanford Publishing, 2009.
45. Jani P, Halbert AT. Nanoparticle uptake by the rat gastrointestinal mucosa: Quantitation and particle size dependency. *J Pharm Pharmacol* 1990;42:821–826.
46. Florence AT, Hillery AM, Hussain N, Jani PU. Nanoparticles as carriers for oral peptide absorption: Studies on particle uptake and fate. *J Controlled Release* 1995;36:39–46.
47. Reddy LH, Sharma RK, Murthy RSR. Enhanced tumour uptake of doxorubicin-loaded poly(butyl cyanoacrylate) nanoparticles in mice bearing dalton's lymphoma tumour. *J Drug Target* 2004;12:443–451.
48. Desai MP, Labhasetwar V, Walter E, Levy RJ, Amidon GL. The mechanism of uptake of biodegradable microparticles in Caco-2 cells is size-dependent. *Pharm Res* 1997;14: 1568–1573.
49. Desai MP, Labhasetwar V, Amidon GL, Levy RJ. Gastrointestinal uptake of biodegradable microparticles: Effect of particle size. *Pharm Res* 1996;13:1838–1845.
50. Lamprecht A, Ubrich N, Lehr C-M. Biodegradable nanoparticles for targeted drug delivery in treatment of inflammatory bowel disease. *J Pharmacol Exp Ther* 2001;299: 775–781.

51. Lamprecht A, Schäfer U, Lehr C-M. Size-dependent bioadhesion of micro- and nano-particulate carriers to the inflamed colonic mucosa. *Pharm Res* 2001;18:788–793.
52. McInnes IB, Schett G. Cytokines in the pathogenesis of rheumatoid arthritis. *Nat Rev Immunol* 2007;7:429–442.
53. Wilson DS, Dalmasso G, Murthy N. Orally delivered thioketal nanoparticles loaded with TNF-α-siRNA target inflammation and inhibit gene expression in the intestines. *Nat Mater* 2010;9:923–928.
54. Conner SD, Schmid SL. Regulated portals of entry into the cell. *Nature* 2003;422:37–44.
55. Yokoyama M, Okano T. Targetable drug carriers: Present status and a future perspective. *Adv Drug Delivery Rev* 1996;21:77–80.
56. Nobs L, Buchegger F, Gurny R, Allémann E. Current methods for attaching targeting ligands to liposomes and nanoparticles. *J Pharm Sci* 2004;93:1980–1992.
57. Ulbrich H, Eriksson EE, Lindbom L. Leukocyte and endothelial cell adhesion molecules as targets for therapeutic interventions in inflammatory disease. *Trends Pharmacol Sci* 2003;24:640–647.
58. Banquy X, Leclair G, Giasson S. Selectins ligand decorated drug carriers for activated endothelial cell targeting. *Bioconjug Chem* 2008;19:2030–2039.
59. Peltz G. Inflammatory disease: Where immunology and adhesion meet? *Curr Opin Biotechnol* 1993;4:745–750.
60. Adelsman AM, Shimizu Yoji. *Encyclopedia of Immunology*, Elsevier, 1998.
61. Cybulsky M, Gimbrone M. Endothelial expression of a mononuclear leukocyte adhesion molecule during atherogenesis. *Science* 1991;251:788–791.
62. Soriano A, Salas A, Salas A. VCAM-1, but not ICAM-1 or MadCAM-1, immunoblockade ameliorates DSS-induced colitis in mice. *Lab Invest* 2000;80:1541–1551.
63. Jones SP, Trocha SD, Lefer DJ. Leukocyte and endothelial cell adhesion molecules in a chronic murine model of myocardial reperfusion injury. *Am J Physiol Heart Circ Physiol* 2000;279:H2196–H2201.
64. Chapman PT, Haskard DO. Use of a radiolabeled monoclonal antibody against E-selectin for imaging of endothelial activation in rheumatoid arthritis. *Arthritis Rheum* 1996;39:1371–1375.
65. Schürmann GM, Bishop AE, Polak JM. Increased expression of cell adhesion molecule P-selectin in active inflammatory bowel disease. *Gut* 1995;36:411–418.
66. Paulos CM, Turk MJ, Breur GJ, Low PS. Folate receptor-mediated targeting of therapeutic and imaging agents to activated macrophages in rheumatoid arthritis. *Adv Drug Delivery Rev* 2004;56:1205–1217.
67. Chellat F, Merhi Y, Moreau A, Yahia L. Therapeutic potential of nanoparticulate systems for macrophage targeting. *Biomaterials* 2005;26:7260–7275.
68. Moghimi SM, Hunter AC, Murray JC. Long-circulating and target-specific nanoparticles: Theory to practice. *Pharmacol Rev* 2001;53:283–318.
69. Ruoslahti E, Bhatia SN, Sailor MJ. Targeting of drugs and nanoparticles to tumors. *J Cell Biol* 2010;188:759–768.
70. Emerich DF, Thanos CG. Nanotechnology and medicine. *Expert Opin Biol Tther* 2003;3:655–663.
71. Farokhzad OC, Langer R. Nanomedicine: Developing smarter therapeutic and diagnostic modalities. *Adv Drug Delivery Rev* 2006;58:1456–1459.

72. Caruthers SD, Wickline SA, Lanza GM. Nanotechnological applications in medicine. *Curr Opin Biotechnol* 2007;18:26–30.
73. Roco MC. Nanotechnology: Convergence with modern biology and medicine. *Curr Opin Biotechnol* 2003;14:337–346.
74. Sahoo SK, Labhasetwar V. Nanotech approaches to drug delivery and imaging. *Drug Discovery Today* 2003;8:1112–1120.
75. Yokoyama M. Drug targeting with nano-sized carrier systems. *J Artif Org* 2005;8: 77–84.
76. Pinto Reis C, Neufeld RJ, Ribeiro AJ, Veiga F. Nanoencapsulation i. Methods for preparation of drug-loaded polymeric nanoparticles. *Nanomed Nanotechnol Biol Med* 2006;2:8–21.
77. Peer D, Karp JM, Hong S. Nanocarriers as an emerging platform for cancer therapy. *Nat Nanotechnol* 2007;2:751–760.
78. Nishiyama N. Nanomedicine: Nanocarriers shape up for long life. *Nat Nanotechnol* 2007;2:203–204.
79. Davda J, Labhasetwar V. Characterization of nanoparticle uptake by endothelial cells. *Int J Pharm* 2002;233:51–59.
80. Moghimi S. Chemical camouflage of nanospheres with a poorly reactive surface: Towards development of stealth and target-specific nanocarriers. *BBA Mol Cell Res* 2002;1590:131–139.
81. Huynh NT, Roger E, Lautram N, Benoît J-P, Passirani C. The rise and rise of stealth nanocarriers for cancer therapy: Passive versus active targeting. *Nanomedicine* 2010;5:1415–1433.
82. Paillard A, Passirani C, Betbeder D. Positively-charged, porous, polysaccharide nanoparticles loaded with anionic molecules behave as "stealth" cationic nanocarriers. *Pharm Res* 2010;27:126–133.
83. Alexis F, Pridgen E, Molnar LK, Farokhzad OC. Factors affecting the clearance and biodistribution of polymeric nanoparticles. *Mol Pharm* 2008;5:505–515.
84. Gref R, Lück M, Quellec P, *et al.* "Stealth" corona-core nanoparticles surface modified by polyethylene glycol (PEG): Influences of the corona (PEG chain length and surface density) and of the core composition on phagocytic uptake and plasma protein adsorption. *Colloid Surf B* 2000;18:301–313.
85. Fang C, Shi B, Chen H-Z. *In vivo* tumor targeting of tumor necrosis factor-alpha-loaded stealth nanoparticles: Effect of mePEG molecular weight and particle size. *Eur J Pharm Sci* 2006;27:27–36.
86. Ishida T, Ichihara M, Wang X, *et al.* Injection of PEGylated liposomes in rats elicits PEG-specific IgM, which is responsible for rapid elimination of a second dose of pegylated liposomes. *J Controlled Release* 2006;112:15–25.
87. Harding JA, Engbers CM, Newman MS, Goldstein NI, Zalipsky S. Immunogenicity and pharmacokinetic attributes of poly(ethylene glycol)-grafted immunoliposomes.*Biochim Biophys Acta* 1997;1327:181–192.
88. Pinto Reis C, Neufeld RJ, Ribeiro AJ, Veiga F. Nanoencapsulation i. Methods for preparation of drug-loaded polymeric nanoparticles. *Nanomedicine* 2006;2:8–21.
89. West KR, Otto S. Reversible covalent chemistry in drug delivery. *Curr Drug Discovery Technol* 2005;2:38.
90. Uhrich KE, Cannizzaro SM, Langer RS, Shakesheff KM. Polymeric systems for controlled drug release. *Chem Rev* 1999;99:3181–3198.

91. Barratt G. Therapeutic applications of colloidal drug carriers. *Drug Discovery Today Technol* 2000;3:163–171.
92. Barratt G. Colloidal drug carriers: Achievements and perspectives. *Cell Mol Life Sci* 2003;60:21–37.
93. Couvreur P, Barratt G, Fattal E, Legrand P, Vauthier C. Nanocapsule technology: A review. *Crit Rev Ther Drug Carrier Syst* 2002;19:99–134.
94. A. Fraass B. Antitumour activity of acid labile transferrin and albumin doxorubicin conjugates in *in vitro* and *in vivo* human tumour xenograft models. *Eur J Cancer* 1997;33:S138.
95. Santander-Ortega MJ, Stauner T, Lehr CM. Nanoparticles made from novel starch derivatives for transdermal drug delivery. *J Controlled Release* 2010;141:85–92.
96. Thiele C, Auerbach D, Wenz G. Nanoparticles of anionic starch and cationic cyclodextrin derivatives for the targeted delivery of drugs. *Polym Chem* 2011;2:209.
97. Kumari K, Rani U. Controlled release of metformin hydrochloride through crosslinked blends of chitosan-starch. *Adv Appl Sci Res* 2011;2:48–54.
98. Lin W, Garnett MC, Illum L. Preparation and characterisation of rose bengal-loaded surface-modified albumin nanoparticles. *J Controlled Release* 2001;71:117–126.
99. Merodio M, Arnedo A, Renedo MJ, Irache JM. Ganciclovir-loaded albumin nanopartiles: Characterization and *in vitro* release properties. *Eur J Pharm Sci* 2001;12:251–259.
100. Merodio M, Irache JM, Valamanesh F, Mirshahi M. Ocular disposition and tolerance of ganciclovir-loaded albumin nanoparticles after intravitreal injection in rats. *Biomaterials* 2002;23:1587–1594.
101. Zimmer AK, Maincent P, Thouvenot P, Kreuter J. Hydrocortisone delivery to healthy and inflamed eyes using a micellar polysorbate 80 solution or albumin nanoparticles. *Int J Pharm* 1994;110:211–222.
102. Das S, Banerjee R, Bellare J. Aspirin loaded albumin nanoparticles by coacervation: Implications in drug delivery. *Drugs* 2005;18:203–212.
103. Zimmer AK, Chetoni P, Saettone MF, Zerbe H, Kreuter J. Evaluation of pilocarpine-loaded albumin particles as controlled drug delivery systems for the eye. ii. Co-administration with bioadhesive and viscous polymers. *J Controlled Release* 1995;33:31–46.
104. Jahanshahi M, Sanati MH, Hajizadeh S, Babaei Z. Gelatin nanoparticle fabrication and optimization of the particle size. *Phys Status Solidi A* 2008;205:2898–2902.
105. Jahanshahi M, Sanati MH, Babaei Z. Optimization of parameters for the fabrication of gelatin nanoparticles by the taguchi robust design method. *J Appl Stat* 2008;35: 1345–1353.
106. Kumar R, Nagarwal RC, Dhanawat M, Pandit JK. *In vitro* and *in vivo* study of indomethacin loaded gelatin nanoparticles. *J Biomed Nanotechnol* 2011;7:325–333.
107. Bhavsar MD, Amiji MM. Oral IL-10 gene delivery in a microsphere-based formulation for local transfection and therapeutic efficacy in inflammatory bowel disease. *Gene Ther* 2008;15:1200–1209.
108. Hoffmann F, Sass G, Zillies I, Zahler S, *et al.* A novel technique for selective NF-κB inhibition in Kupffer cells — contrary effects in fulminant hepatitis and ischemia/reperfusion. *Gait* 2009;58:1670–1678.
109. Vandervoort J, Ludwig A. Preparation and evaluation of drug-loaded gelatin nanoparticles for topical ophthalmic use. *Eur J Pharm Biopharm* 2004;57:251–261.

110. Coucke D, Schotsaert M, Libert C, *et al.* Spray-dried powders of starch and crosslinked poly(acrylic acid) as carriers for nasal delivery of inactivated influenza vaccine. *Vaccine.* 2009;27:1279–1286.
111. Rodrigues A, Emeje M. Recent applications of starch derivatives in nanodrug delivery. *Carbohydr Polym* 2011; 87: 987–994.
112. Rydell N, Stertman L, Sjöholm I. Starch microparticles as vaccine adjuvant. *Expert Opin Drug Delivery* 2005;2:807–828.
113. McDermott MR, Heritage PL, Bartzoka V, Brook MA. Polymer-grafted starch microparticles for oral and nasal immunization. *Immunol Cell Biol* 1998;76:256–262.
114. Kumar M, Behera AK, De La Cruz CP, Mohapatra SS. Intranasal gene transfer by chitosan-DNA nanospheres protects balb/c mice against acute respiratory syncytial virus infection. *Hum Gene Ther* 2002;13:1415–1425.
115. Kumar M, Kong X, Mohapatra SS. Chitosan IFN-γ-pDNA nanoparticle (cin) therapy for allergic asthma. *Genet Vaccines Ther* 2003;1:1–10.
116. Hou Z, Han J, Zhan C, Zhang Q. Synthesis and evaluation of N-succinyl-chitosan nanoparticles toward local hydroxycamptothecin delivery. *Carbohydr Polym* 2010; 81:765–768.
117. Jam SK, Jain A, Gupta Y, Khare P, Kannadasan M. Targeted delivery of 5-ASA to colon using chitosan hydrogel microspheres. *J Drug Delivery Sci Technol* 18:315–321.
118. Mura C, Manconi M, Fadda AM. *In vitro* study of N-succinyl chitosan for targeted delivery of 5-aminosalicylic acid to colon. *Carbohydr Polym* 2011;85:578–583.
119. Park JH, Saravanakumar G, Kim K, Kwon IC. Targeted delivery of low molecular drugs using chitosan and its derivatives. *Adv Drug Delivery Rev* 2010;62:28–41.
120. Lee D-W, Shirley SA, Lockey RF, Mohapatra SS. Thiolated chitosan nanoparticles enhance anti-inflammatory effects of intranasally delivered theophylline. *Resp Res* 2006; 7:112.
121. Gerwin N, Hops C, Lucke A. Intraarticular drug delivery in osteoarthritis. *Adv Drug Delivery Rev* 2006;58:226–742.
122. Howard KA, Paludan SR, Kjems J. Chitosan/siRNA nanoparticle — mediated TNF-α knockdown in peritoneal macrophages for anti-inflammatory treatment in a murine arthritis model. *Mol Ther* 2009;17:162–168.
123. Lu H-D, Zhao H-Q, Wang K, Lv L-L. Novel hyaluronic acid-chitosan nanoparticles as non-viral gene delivery vectors targeting osteoarthritis. *Int J Pharm* 2011; 420: 358–365.
124. de la Fuente M, Raviña M, Paolicelli P, *et al.* Chitosan-based nanostructures: A delivery platform for ocular therapeutics. *Adv Drug Delivery Rev* 2010;62:100–117.
125. Diebold Y, Calonge M. Applications of nanoparticles in ophthalmology. *Prog Retin Eye Res* 2010;29:596–609.
126. Diebold Y, Jarrín M, Alonso MJ. Ocular drug delivery by liposome-chitosan nanoparticle complexes (LCS-NP). *Biomaterials* 2007;28:1553–1564.
127. de Campos AM, Diebold Y, Carvalho ELS, Sánchez A, José Alonso M. Chitosan nanoparticles as new ocular drug delivery systems: *In vitro* stability, *in vitro* fate, and cellular toxicity. *Pharm Res* 2004;21:803–810.
128. Alonso MJ, Sánchez A. The potential of chitosan in ocular drug delivery. *J Pharm Pharmacol* 2003;55:1451–1463.
129. Badawi AA, El-Laithy HM, El-Qidra RK, El-Mofty H, El-Dally M. Chitosan-based nanocarriers for indomethacin ocular delivery. *Arch Pharm Res* 2008;31:1040–1049.

130. Araújo J, Gonzalez E, Egea MA, Garcia ML, Souto EB. Nanomedicines for ocular NSAIDS: Safety on drug delivery. *Nanomed Nanotechnol Biol Med* 2009;5:394–401.
131. Kumari A, Yadav SK, Yadav SC. Biodegradable polymeric nanoparticles-based drug delivery systems. *Colloid Surf B* 2010;75:1–18.
132. Duncan R. The dawning era of polymer therapeutics. *Nat Rev Drug Discovery* 2003;2: 347–360.
133. Peracchia MT, Gref R, Langer R. PEG-coated nanospheres from amphiphilic diblock and multiblock copolymers: Investigation of their drug encapsulation and release characteristics. *J Controlled Release* 1997;46:223–231.
134. Pitt G, Grazil M, Sohindler A. Aliphatic polyesters ii. The degradation of poly (dl-lactide), poly (ε-caprolactone), and their copolymers *in vitro*. *Biomaterials* 1981;2:215–220.
135. Redhead HM, Davis SS, Illum L. Drug delivery in poly(lactide-co-glycolide) nanoparticles surface modified with poloxamer 407 and poloxamine 908: *In vitro* characterisation and *in vitro* evaluation. *J Controlled Release* 2001;70:353–363.
136. Ishihara T, Kubota T, Choi T, Higaki M. Treatment of experimental arthritis with stealth-type polymeric nanoparticles encapsulating betamethasone phosphate. *Pharmacology* 2009;329:412–417.
137. Conn RE, Kolstad JJ, Borzelleca JF, *et al.* Safety assessment of polylactide (PLA) for use as a food-contact polymer. *Food Chem Toxicol* 1995;33:273–283.
138. Wang C, Pham P-T. Polymers for viral gene delivery. *Expert Opin Drug Delivery* 2008; 5:385–401.
139. Okada H, Toguchi H. Biodegradable microspheres in drug delivery. *Crit Rev Ther Drug Carrier Syst* 1995;12:1–99.
140. Midhun BT, Shalumon KT, Manzoor K, *et al.* Preparation of budesonide-loaded polycaprolactone nanobeads by electrospraying for controlled drug release. *J Biomater Sci Polym Ed* 2011;22:2431–2444.
141. Torchilin V. Multifunctional and stimuli-sensitive pharmaceutical nanocarriers. *Eur J Pharm Biopharm* 2009;71:431–444.
142. Bazile D, Prud'homme C, Bassoullet M-T, *et al.* Stealth Me.PEG-PLA nanoparticles avoid uptake by the mononuclear phagocytes system. *J Pharm Sci* 1995;84: 493–498.
143. Horisawa E, Hirota T, Kawazoe S, *et al.* Prolonged anti-inflammatory action of dl-lactide/glycolide copolymer nanospheres containing betamethasone sodium phosphate for an intra-articular delivery system in antigen-induced arthritic rabbit. *Pharm Res* 2002;19:403–410.
144. Horisawa E, Kubota K, Tuboi I, *et al.* Size-dependency of dl-lactide/glycolide copolymer particulates for intra-articular delivery system on phagocytosis in rat synovium. *Pharm Res* 2002;19:132–139.
145. Agnihotri SM, Vavia PR. Diclofenac-loaded biopolymeric nanosuspensions for ophthalmic application. *Nanomed Nanotechnol Biol Med* 2009;5:90–95.
146. Bourges J-L. Ocular drug delivery targeting the retina and retinal pigment epithelium using polylactide nanoparticles. *Invest Ophthalmol Vis Sci* 2003;44:3562–3569.
147. Schöll I, Kopp T, Bohle B, Jensen-Jarolim E. Biodegradable PLGA particles for improved systemic and mucosal treatment of type I allergy. *Immunol Allergy Clin North Am* 2006;26:349–64, ix.
148. Vega E, Egea MA, Valls O, Espina M, García ML. Flurbiprofen loaded biodegradable nanoparticles for ophtalmic administration. *J Pharm Sci* 2006;95:2393–2405.

149. Meissner Y, Pellequer Y, Lamprecht A. Nanoparticles in inflammatory bowel disease: Particle targeting versus PH-sensitive delivery. *Int J Pharm* 2006;316:138–143.
150. Lamprecht A, Yamamoto H, Takeuchi H, Kawashima Y. Nanoparticles enhance therapeutic efficiency by selectively increased local drug dose in experimental colitis in rats. *J Pharmacol Exp Ther* 2005;315:196–202.
151. Sakhalkar HS, Dalal MK, Salem AK, *et al.* Leukocyte-inspired biodegradable particles that selectively and avidly adhere to inflamed endothelium *in vitro* and *in vivo*. *Proc Natl Acad Sci* 2003;100:15895–15900.
152. Zhang N, Chittasupho C, Duangrat C, Siahaan TJ, Berkland C. PLGA nanoparticle — peptide conjugate effectively targets intercellular cell-adhesion molecule-1. *Bioconjugugate Chem* 2008;19:145–152.
153. Sinha VR, Bansal K, Kaushik R, Kumria R, Trehan A. Poly-epsilon-caprolactone microspheres and nanospheres: An overview. *Int J Pharm* 2004;278:1–23.
154. Dash TK, Konkimalla VB. Poly-ε-caprolactone based formulations for drug delivery and tissue engineering: A review. *J Controlled Release* 2012;158:15–33.
155. Kim SY, Shin IG, Lee YM, Cho CS, Sung YK. Methoxy poly(ethylene glycol) and epsilon-caprolactone amphiphilic block copolymeric micelle containing indomethacin. ii. Micelle formation and drug release behaviours. *J Controlled Release* 1998;51:13–22.
156. Fialho SL, Behar-Cohen F, Silva-Cunha A. Dexamethasone-loaded poly(epsilon-caprolactone) intravitreal implants: A pilot study. *Eur J Pharm Biopharm* 2008;68: 637–646.
157. Losa C, Marchal-Heussler L, Orallo F, Jato JLV, Alonso MJ. Design of new formulations for topical ocular administration: Polymeric nanocapsules containing metipranolol. *Pharm Res* 1993;10:80–87.
158. Lacoulonche F, Gamisans F, Chauvet A, *et al.* Stability and *in vitro* drug release of flurbiprofen-loaded poly-ε-caprolactone nanospheres. *Drug Delivery Ind Pharm* 1999;25: 983–993.
159. Calvo P, Vila-Jato JL, Alonso MJ. Comparative *in vitro* evaluation of several colloidal systems, nanoparticles, nanocapsules, and nanoemulsions, as ocular drug carriers. *J Pharm Sci* 1996;85:530–536.
160. Marchal-Heussler L, Sirbat D, Hoffman M, Maincent P. Poly(ε-caprolactone) nanocapsules in carteolol ophthalmic delivery. *Pharm Res* 1993;10:386–390.
161. Beck RCR, Guterres SS, Freddo RJ, Michalowski CB. Nanoparticles containing dexamethasone: Physicochemical properties and anti-inflammatory activity. *Acta Farm Bonaerense* 2003;22:11–16.
162. Sakai T, Kohno H, Kitahara K. Treatment of experimental autoimmune uveoretinitis with poly(lactic acid) nanoparticles encapsulating betamethasone phosphate. *Exp Eye Res* 2006;82:657–663.
163. Fenske DB, Cullis PR. Liposomal nanomedicines. *Expert Opin Drug Deliv* 2008;5:25–44.
164. Lian T, Ho RJ. Trends and developments in liposome drug delivery systems. *J Pharm Sci* 2001;90:667–680.
165. Gregoriadis G. Engineering liposomes for drug delivery: Progress and problems. *Trends Biotechnol* 1995;13:527–537.
166. Kelky C, Jefferies C, Cryan SA. Targeted liposomal drug delivery to monocytes and macrophages. *J Drug Delivery* 2011;727–741.

167. Lasic D, Papahadjopoulos D. Liposomes revisited. *Science* 1995;267:1275–1276.
168. Torchilin VP. Affinity liposomes *in vitro*: Factors influencing target accumulation. *J Mol Recognit* 9:335–346.
169. Lasic DD, Vallner JJ, Working PK. Sterically stabilized liposomes in cancer therapy and gene delivery. *Curr Opin Mol Ther* 1999;1:177–185.
170. Pain D, Das PK, Ghosh P, Bachhawat BK. Increased circulatory half-life of liposomes after conjunction with dextran. *J Biosci* 1984;6:811–816.
171. Torchilin V, Levchenko T, Whiteman K, *et al.* Amphiphilic poly-n-vinylpyrrolidones. *Biomaterials* 2001;22:3035–3044.
172. Metselaar JM, Wauben MHM, Wagenaar-Hilbers JPA, Boerman OC, Storm G. Complete remission of experimental arthritis by joint targeting of glucocorticoids with long-circulating liposomes. *Arthritis Rheum* 2003;48:2059–2066.
173. Schmidt J, Metselaar JM, Gold R. Drug targeting by long-circulating liposomal glucocorticosteroids increases therapeutic efficacy in a model of multiple sclerosis. *Brain* 2003;126:1895–1904.
174. Schiffelers RM, Metselaar JM, Storm G. Liposome-encapsulated prednisolone phosphate inhibits growth of established tumors in mice. *Neoplasia* 2005;7:118–127.
175. Linker RA, Metselaar JM, Gold R. Liposomal glucocorticosteroids in treatment of chronic autoimmune demyelination: Long-term protective effects and enhanced efficacy of methylprednisolone formulations. *Exp Neurol* 2008;211:397–406.
176. Crielaard BJ, Lammers T, Morgan ME, *et al.* Macrophages and liposomes in inflammatory disease: Friends or foes? *Int J Pharm* 2011;416:499–506.
177. Vanniasinghe AS, Bender V, Manolios N. The potential of liposomal drug delivery for the treatment of inflammatory arthritis. *Semin Arthritis Rheum* 2009;39:182–196.
178. Funk M, Schulze B, Schmitt-Sody M. Cationic liposomal preparations for the treatment of rheumatoid arthritis. European Patent Specification EP2040674B1, 2010.
179. Forssen E, Willis M. Ligand-targeted liposomes. *Adv Drug Delivery Rev* 1998;29: 249–271.
180. Metselaar JM. Liposomal targeting of glucocorticoids to synovial lining cells strongly increases therapeutic benefit in collagen type II arthritis. *Ann Rheum Dis* 2004;63: 348–353.
181. Banciu M, Metselaar JM, Schiffelers RM, Storm G. Antitumor activity of liposomal prednisolone phosphate depends on the presence of functional tumor-associated macrophages in tumor tissue. *Neoplasia* 2008;10:108–117.
182. Willialms AS, Camilleri JP, Williams BD. Suppression of adjuvant-induced arthritis by liposomally conjugated methotrexate in the rat. *Rheumatology* 1994;33:530–533.
183. Koning GA, Schiffelers RM, Wauben MHM, *et al.* Targeting of angiogenic endothelial cells at sites of inflammation by dexamethasone phosphate-containing RGD peptide liposomes inhibits experimental arthritis. *Arthritis Rheum* 2006;54:1198–1208.
184. Minaguchi J, Oohashi T, Inagawa K, Ohtsuka A, Ninomiya Y. Transvascular accumulation of Sialyl Lewis X conjugated liposome in inflamed joints of collagen antibody-induced arthritic (CAIA) mice. *Arch Histol Cytol* 2008;71:195–203.
185. Singh K, Mezei M. Liposomal ophthalmic drug delivery system i. Triamcinolone acetonide. *Int J Pharm* 1983;16:339–344.
186. Lee VHL, Urrea PT, Smith RoE, Schanzlin DJ. Ocular drug bioavailability from topically applied liposomes. *Surv Ophthalmol* 1985;29:335–348.

187. Hashida N, Ohguro N, Yamazaki N, *et al.* High-efficacy site-directed drug delivery system using Sialyl-Lewis X conjugated liposome. *Exp Eye Res* 2008;86:138–149.
188. Camelo S, Lajavardi L, Bochot A, *et al.* Drainage of fluorescent liposomes from the vitreous to cervical lymph nodes via conjunctival lymphatics. *Ophthalmic Res* 2008;40: 145–150.
189. Lajavardi L, Bochot A, Camelo S, *et al.* Downregulation of endotoxin-induced uveitis by intravitreal injection of vasoactive intestinal peptide encapsulated in liposomes. *Invest Ophthalmol Vis Sci* 2007;48:3230–3238.
190. Lopez-Garcia. Intra-articular therapy of experimental arthritis with a derivative of triamcinolone acetonide incorporated in liposomes. *J Pharm Pharmacol* 1993;45: 576–578.
191. Haber E, Danenberg HD, Koroukhov N, *et al.* Peritoneal macrophage depletion by liposomal bisphosphonate attenuates endometriosis in the rat model. *Hum Reprod* 2009;24:398–407.
192. Hiraoka K, Zenmyo M, Watari K, *et al.* Inhibition of bone and muscle metastases of lung cancer cells by a decrease in the number of monocytes/macrophages. *Cancer Sci* 2008;99:1595–1602.
193. Richards PJ. Suppression of chronic streptococcal cell wall-induced arthritis in Lewis rats by liposomal clodronate. *Rheumatology* 2001;40:978–987.
194. Danenberg HD, Golomb G, Groothuis A, *et al.* Liposomal alendronate inhibits systemic innate immunity and reduces in-stent neointimal hyperplasia in rabbits. *Circulation* 2003;108:2798–2804.
195. Chen Y, Liu L. Modern methods for delivery of drugs across the blood-brain barrier. *Adv Drug Delivery Rev* 2012;64:640–665.
196. Cavaletti G, Cassetti A, Canta A, *et al.* Cationic liposomes target sites of acute neuroinflammation in experimental autoimmune encephalomyelitis. *Mol Pharm* 2009;6: 1363–1370.
197. Yang H. Nanoparticle-mediated brain-specific drug delivery, imaging, and diagnosis. *Pharm Res* 2010;27:1759–1771.
198. Afergan E, Epstein H, Dahan R, *et al.* Delivery of serotonin to the brain by monocytes following phagocytosis of liposomes. *J Controlled Release* 2008;132:84–90.
199. Tomalia DA, Fréchet JMJ. Discovery of dendrimers and dendritic polymers: A brief historical perspective. *J Polym Sci A Polym Chem* 2002;40:2719–2728.
200. Gillies ER, Fréchet JMJ. Dendrimers and dendritic polymers in drug delivery. *Drug Discov Today* 2005;10:35–43.
201. Svenson S, Tomalia DA. Dendrimers in biomedical applications — reflections on the field. *Adv Drug Delivery Rev* 2005;57:2106–2129.
202. Menjoge AR, Kannan RM, Tomalia DA. Dendrimer-based drug and imaging conjugates: Design considerations for nanomedical applications. *Drug Discovery Today* 2010; 15:171–185.
203. Malik N, Wiwattanapatapee R, Klopsch R, *et al.* Dendrimers. *J Controlled Release* 2000;65:133–148.
204. Gupta U, Agashe HB, Asthana A, Jain NK. Dendrimers: Novel polymeric nanoarchitectures for solubility enhancement. *Biomacromolecules* 2006;7:649–658.
205. Chauhan AS, Sridevi S, Chalasani KB, *et al.* Dendrimer-mediated transdermal delivery: Enhanced bioavailability of indomethacin. *J Controlled Release* 2003;90:335–343.

206. Lin Y, Fujimori T, Yamamoto A. Polyamidoamine dendrimers as novel potential absorption enhancers for improving the small intestinal absorption of poorly absorbable drugs in rats. *J Controlled Release* 2011;149:21–28.
207. Dong Z, Hamid KA, Gao Y, *et al.* Polyamidoamine dendrimers can improve the pulmonary absorption of insulin and calcitonin in rats. *J Pharm Sci* 2011;100:1866–1878.
208. Gajbhiye V, Kumar PV, Sharma A, Jain NK. Novel PEGylated PPI dendritic nanostructures for sustained delivery of anti-inflammatory agent. *Curr Nanosci* 2008;4:11.
209. Wegmann KW, Wagner CR, Whitham RH, Hinrichs DJ. Synthetic peptide dendrimers block the development and expression of experimental allergic encephalomyelitis. *J Immunol* 2008;181:3301–3309.
210. Chauhan A, Diwan P, Jain N. Compositions and complexes containing a macromolecular compound as potential anti-inflammatory agents. US Patent App, No.: 2003/0180250/A1, 2003.
211. Lesniak WG, Kariapper MST, Nair BM, *et al.* Synthesis and characterization of PAMAM dendrimer-based multifunctional nanodevices for targeting alphavbeta3 integrins. *Bioconjug Chem* 2007;18:1148–1154.
212. Chandrasekar D, Sistla R, Ahmad FJ, Khar RK, Diwan PV. The development of folate-pamam dendrimer conjugates for targeted delivery of anti-arthritic drugs and their pharmacokinetics and biodistribution in arthritic rats. *Biomaterials* 2007;28:504–512.
213. Dai H, S Navath R, K Mishra M. Intrinsic targeting of inflammatory cells in the brain by polyamidoamine dendrimers upon subarachnoid administration. *Nanomedicine (Lond)* 2010;5:1317–1329.
214. Shaunak S, Thomas S, Gianasi E, *et al.* Polyvalent dendrimer glucosamine conjugates prevent scar tissue formation. *Nat Biotechnol* 2004;22:977–984.
215. Hayder M, Poupot M, Baron M, *et al.* A phosphorus-based dendrimer targets inflammation and osteoclastogenesis in experimental arthritis. *Sci Transl Med* 2011;3:81ra35.
216. Chauhan AS, Diwan PV, Jain NK, Tomalia DA. Unexpected *in vitro* anti-inflammatory activity observed for simple, surface functionalized poly(amidoamine) dendrimers. *Biomacromolecules* 2009;10:1195–1202.
217. Dernedde J, Rausch A, Weinhart M, *et al.* Dendritic polyglycerol sulfates as multivalent inhibitors of inflammation. *Proc Natl Acad Sci* 2010;107:19679–19684.
218. Hayder M, Fruchon S, Fournié J-J, Poupot M, Poupot R. Anti-inflammatory properties of dendrimers per se. *Sci World J* 2011;11:1367–1382.
219. Bosch X. Dendrimers to treat rheumatoid arthritis. *ACS Nano* 2011;:6779–6785.
220. Kataoka K, Kwon GS, Sakurai Y. Block copolymer micelles as vehicles for drug delivery. *J Controlled Release* 1993;24:119–132.
221. Torchilin VP. Micellar nanocarriers: Pharmaceutical perspectives. *Pharm Res* 2007;24:1–16.
222. Kataoka K, Harada A, Nagasaki Y. Block copolymer micelles for drug delivery: Design, characterization and biological significance. *Adv Drug Delivery Rev* 2001;47:113–131.
223. Adams ML, Lavasanifar A, Kwon GS. Amphiphilic block copolymers for drug delivery. *J Pharm Sci* 2003;92:1343–1355.
224. Kwon GS. Polymeric micelles for delivery of poorly water-soluble compounds. *Crit Rev Ther Drug Carrier Syst* 20:357–403.

225. Batrakova EV, Dorodnych TY, Klinskii EY, *et al.* Anthracycline antibiotics non-covalently incorporated into the block copolymer micelles: *In vitro* evaluation of anti-cancer activity. *Br J Cancer* 1996;74:1545–1552.
226. Nakanishi T, Fukushima S, Okamoto K, *et al.* Development of the polymer micelle carrier system for doxorubicin. *J Controlled Release* 2001;74:295–302.
227. Trivedi R, Kompella UB. Nanomicellar formulations for sustained drug delivery: Strategies and underlying principles. *Nanomedicine (Lond)* 2010;5:485–505.
228. Zhang JX, Li XJ, Zhu KJ. Indomethacin-loaded polymeric nanocarriers based on amphiphilic polyphosphazenes with poly (N-isopropylacrylamide) and ethyl tryptophan as side groups: Preparation, *in vitro* and *in vivo* evaluation. *J Controlled Release* 2006;116: 322–329.
229. Torchilin VP. Lipid-core micelles for targeted drug delivery. *Curr Drug Delivery* 2005;2: 319–327.
230. Torchilin VP. Targeted polymeric micelles for delivery of poorly soluble drugs. *Cell Mol Life Sci* 2004;61:2549–2559.
231. Li Y-Y, Zhang X-Z, Cheng H, *et al.* Fluorescent, thermo-responsive biotin-p(nipaam-co-ndapm)-b-pcl micelles for cell-tracking and drug delivery. *Nanotechnology* 2007; 18:505101.
232. Zhang JX, Yan MQ, Li XH, *et al.* Local delivery of indomethacin to arthritis-bearing rats through polymeric micelles based on amphiphilic polyphosphazenes. *Pharm Res* 2007;24:1944–1953.
233. Liu M, Kono K, Fréchet JM. Water-soluble dendritic unimolecular micelles. *J Controlled Release* 2000;65:121–131.
234. Djordjevic J, Michniak B, Uhrich KE. Amphiphilic star-like macromolecules as novel carriers for topical delivery of nonsteroidal anti-inflammatory drugs. *AAPS Pharm Sci* 2003;5:E26.
235. Goel A, Ahmad FJ, Singh RM, Singh GN. 3-acetyl-11-keto-beta-boswellic acid loaded-polymeric nanomicelles for topical anti-inflammatory and anti-arthritic activity. *J Pharm Pharmacol* 2010;62:273–278.
236. Endradi EH, Bata YO, Sowa KI, Agai TN, Akayama KT. Effect of mixed micelle formulations including terpenes on the transdermal delivery of diclofenac. *Society* 2003;26:1739–1743.
237. Li DX, Han MJ, Balakrishnan P, *et al.* Enhanced oral bioavailability of flurbiprofen by combined use of micelle solution and inclusion compound. *Arch Pharmacol Res* 2010;33:95–101.
238. Yang YQ, Zheng LS, Guo XD, Qian Y, Zhang LJ. pH-sensitive micelles self-assembled from amphiphilic copolymer brush for delivery of poorly water-soluble drugs. *Biomacromolecules* 2011;12:116–122.
239. Soliman GM, Sharma R, Choi AO, *et al.* Tailoring the efficacy of nimodipine drug delivery using nanocarriers based on A2B miktoarm star polymers. *Biomaterials* 2010; 31:8382–8392.
240. Gupta A. Ketorolac entrapped in polymeric micelles: Preparation, characterisation and ocular anti-inflammatory studies. *Int J Pharm* 2000;209:1–14.
241. Pepić I, Hafner A, Lovrić J, Pirkić B, Filipović-Grcić J. A nonionic surfactant/chitosan micelle system in an innovative eye drop formulation. *J Pharm Sci* 2010;99: 4317–4325.

242. Rafie F, Javadzadeh Y, Javadzadeh AR, *et al. In vitro* evaluation of novel nanoparticles containing dexamethasone for ocular drug delivery on rabbit eye. *Curr Eye Res* 2010; 35:1081–1089.
243. Gaber NN, Darwis Y, Peh K-K, Tan YT-F. Characterization of polymeric micelles for pulmonary delivery of beclomethasone dipropionate. *J Nanosci Nanotechnol* 2006;6: 3095–3101.
244. Dane KY, Nembrini C, Hubbell JA. Nano-sized drug-loaded micelles deliver payload to lymph node immune cells and prolong allograft survival. *J Controlled Release* 2011; 156:154–160.
245. Dutta P, Dey J, Perumal V, Mandal M. Amino acid based amphiphilic copolymer micelles as carriers of non-steroidal anti-inflammatory drugs: Solubilization, *in vitro* release and biological evaluation. *Int J Pharm* 2011;407:207–216.
246. Khalfin RL, Bachar M, Danino D. Beta-casein micelles and nanocapsules in drug delivery. *Nanotechnology* 2011;3:306–309.
247. Kumar R, Katare OP. Lecithin organogels as a potential phospholipid-structured system for topical drug delivery: A review. *AAPS PharmSciTech* 2005;6:298–310.
248. Laudadio C. Cardiac glycosides for treating muscle pain and spasm. US Patent App, No.: 10/454,065, 2003.
249. Padilla M, Clark GT, Merrill RL. Topical medications for orofacial neuropathic pain: A review. *J Am Dent Assoc* 2000;131:184–195.
250. Ford P. Topical pain relief composition and carrier. US Patent App, No.: 6,461,600, 2002.
251. Dreher F, Walde P, Walther P, Wehrli E. Interaction of a lecithin microemulsion gel with human stratum corneum and its effect on transdermal transport. *J Controlled Release* 1997;45:131–140.
252. Roth SH, Shainhouse JZ. Efficacy and safety of a topical diclofenac solution (pennsaid) in the treatment of primary osteoarthritis of the knee: A randomized, double-blind, vehicle-controlled clinical trial. *Arch Int Med* 2004;164:2017–2023.
253. Zheng X, Wang X, Gou M, *et al.* A novel transdermal honokiol formulation based on pluronic F127 copolymer. *Drug Delivery* 2010;17:138–144.
254. Lewis DR, Kamisoglu K, York AW, Moghe PV. Polymer-based therapeutics: Nanoassemblies and nanoparticles for management of atherosclerosis. *Wiley Interdiscip Rev Nanomed Nanobiotechnol* 3:400–420.
255. Müller R. Solid lipid nanoparticles (SLN) for controlled drug delivery — a review of the state of the art. *Eur J Pharm Biopharm* 2000;50:161–177.
256. Uner M. Preparation, characterization and physico-chemical properties of solid lipid nanoparticles (SLN) and nanostructured lipid carriers (NLC): Their benefits as colloidal drug carrier systems. *Die Pharmazie* 2006;61:375–386.
257. Murthy SK. Nanoparticles in modern medicine: State of the art and future challenges. *Int J Nanomed* 2007;2:129–141.
258. Martins S, Sarmento B, Ferreira DC, Souto EB. Lipid-based colloidal carriers for peptide and protein delivery — liposomes versus lipid nanoparticles. *Int J Nanomed* 2007;2: 595–607.
259. Wissing SA, Kayser O, Müller RH. Solid lipid nanoparticles for parenteral drug delivery. *Adv Drug Delivery Rev* 2004;56:1257–1272.
260. Puri A, Loomis K, Smith B, *et al.* Lipid-based nanoparticles as pharmaceutical drug carriers: From concepts to clinic. *Crit Rev Ther Drug Carrier Syst* 2009;26:523–580.

261. Uner M, Yener G. Importance of solid lipid nanoparticles (SLN) in various administration routes and future perspectives. *Int J Nanomed* 2007;2:289–300.
262. Serpe L, Canaparo R, Daperno M, *et al.* Solid lipid nanoparticles as anti-inflammatory drug delivery system in a human inflammatory bowel disease whole-blood model. *Eur J Pharm Sci* 2010;39:428–436.
263. Dianzani C, Cavalli R, Zara GP, *et al.* Cholesteryl butyrate solid lipid nanoparticles inhibit adhesion of human neutrophils to endothelial cells. *Br J Pharm* 2006;148: 648–656.
264. Liu D, Ge Y, Tang Y, *et al.* Solid lipid nanoparticles for transdermal delivery of diclofenac sodium: Preparation, characterization and *in vitro* studies. *J Microencapsul* 2010;27: 726–734.
265. Mei Z. Solid lipid nanoparticle and microemulsion for topical delivery of triptolide. *Eur J Pharm Biopharm* 2003;56:189–196.
266. Jaafar-Maalej C, Andrieu V, Elaissari A, Fessi H. Beclomethasone-loaded lipidic nanocarriers for pulmonary drug delivery: Preparation, characterization and *in vitro* drug release. *J Nanosci Nanotechnol* 2011;11:1841–1851.
267. Schlupp P, Blaschke T, Kramer KD, *et al.* Drug release and skin penetration from solid lipid nanoparticles and a base cream: A systematic approach from a comparison of three glucocorticoids. *Skin Pharmacol Physiol* 2011;24:199–209.
268. Souto EB, Doktorovova S, Gonzalez-Mira E, Egea MA, Garcia ML. Feasibility of lipid nanoparticles for ocular delivery of anti-inflammatory drugs. *Curr Eye Res* 2010;35: 537–552.
269. Kakkar V, Singh S, Singla D, Kaqul IP. Molecular nutrition & food research abstracts: Lipid technology 6/2011. *Lipid Technol* 2011;23:134–137.
270. Zhang J, Smith E. Percutaneous permeation of betamethasone 17-valerate incorporated in lipid nanoparticles. *J Pharm Sci* 2011;100:896–903.
271. Schäfer-Korting M, Mehnert W, Korting H-C. Lipid nanoparticles for improved topical application of drugs for skin diseases. *Adv Drug Delivery Rev* 2007;59:427–443.
272. Ye J, Wang Q, Zhou X, Zhang N. Injectable actarit-loaded solid lipid nanoparticles as passive targeting therapeutic agents for rheumatoid arthritis. *Int J Pharm* 2008; 352:273–279.
273. Manjunath K, Venkateswarlu V. Pharmacokinetics, tissue distribution and bioavailability of nitrendipine solid lipid nanoparticles after intravenous and intraduodenal administration. *J Drug Target* 2008;14:632–645.
274. Vinogradov SV, Batrakova EV, Kabanov AV. Nanogels for oligonucleotide delivery to the brain. *Bioconjug Chem* 2003;15:50–60.
275. Peira E, Marzola P, Podio V, *et al.* *In vitro* and *in vivo* study of solid lipid nanoparticles loaded with superparamagnetic iron oxide. *J Drug Target* 2008;11:19–24.
276. Sawant KK, Dodiya SS. Recent advances and patents on solid lipid nanoparticles. *Recent Pat Drug Delivery Formulation* 2008;2:120–135.

Chapter 17

Nanostructures and Allergy

Silvia Lorenzo-Abalde and África González-Fernández*

Immunology, Biomedical Research Centre (CINBIO)
University of Vigo, Campus Lagoas Marcosende
Vigo 36310, Spain

Nanostructures can interact with biological systems, e.g., the immune system, inducing a response to these foreign elements in different ways, including the induction of phagocytosis and activation or inhibition of the immune cells. However, it can also respond by triggering hypersensitivity or allergy reactions. The process of allergy to nanostructures is still relatively unknown, but the process of pseudoallergy (which involves complement activation and inflammatory responses) is the one most described. This chapter summarizes the types of hypersensitivity reactions, the most frequent symptoms, several examples of allergy induced by nanomaterials, and the most common methods of diagnosis.

1. Introduction

In recent years, the large variety of nanostructures (liposomes, polymeric nanospheres, dendrimers, micelles, carbon nanotubes, nanocrystals, nanoshells, scaffolds, etc.) designed and produced by various research groups and companies has expanded rapidly. They offer new elements with interesting properties for various purposes (energy, textiles, cosmetics, biomedicine, computing, etc.). In the field of nanomedicine, many nanoparticles (NPs) have been administered into the body or are intended for use in numerous applications, among others, as vehicles for drug delivery,[1–5] in hyperthermia,[6–9]

*Corresponding author. E-mail: silvi@uvigo.es

phototherapy,[10,11] vaccines,[12–20] nanodevices,[21–23] and as diagnostic agents.[24–27] Nanostructures can be biodegradable or non-biodegradable, and their characteristics (composition, size, shape, charge, and stability in physiological media) affect their possible use. In addition, the deliberately engineered NPs for medical purposes can also enter the organism inadvertently or accidentally (environment, workplace) by unsuspected routes. Therefore, their potential undesirable (i.e., toxic) effects must be comprehensively studied.

In this context, one important issue to be carefully addressed is the potential capacity of NPs to activate the immune system. This is a very complex system dedicated, among other functions, to the recognition and elimination of foreign elements, such as viruses, bacteria, parasites, and fungi ("antigens").[28] The immune system includes a large variety of cells and many types of soluble elements. Non-specific immune cells, such as macrophages, granulocytes, dendritic, or natural killer cells, together with specific cells — T helper (Th) 1, 2, and 17, T cytotoxic, T regulatory, and B lymphocytes, are the most important cellular components. With regard to humoral factors, there are many types of non-specific soluble elements (complement factors, enzymes, acute phase proteins, cytokines), but also antibodies or immunoglobulins (IgG, IgM, IgA, IgE, IgD), which are highly specific molecules.[29]

As a defense mechanism, the immune system is ready to detect and destroy foreign elements. Thus, during an infection or after vaccination, the immune system tries to eliminate the pathogen or its components in an efficient way. The cooperation between non-specific cells with specific ones (B and T lymphocytes) and the secretion of soluble factors are crucial for a good adaptive immune response. Consequently, B and T lymphocytes become activated, proliferate, and differentiate into memory cells, and are able to improve their responses after further encounters with the antigen. Some of the B lymphocytes differentiate into plasma cells (able to secrete soluble specific antibodies) after their activation, mediated by specific antigens. With help from other components of the immune system (phagocytic cells, natural killer cells, complement, etc.), these antibodies are able to neutralize or destroy the foreign elements. However, on some occasions, the immune system in vertebrate animals can generate deleterious immune responses, such as allergic (or hypersensitivity) reactions or autoimmune diseases. This chapter will provide a review of the types of allergy, the hypersensitivity reactions induced by nanomaterials, and the methods of diagnosis.

2. Allergy

Although the main purpose of the immune system is the recognition of pathogens, it is possible to develop immune responses against non pathogens or foreign elements. Some people develop quite strong immune responses to certain components. These responses are called "allergy" or "hypersensitivity" reactions, which include exaggerated humoral or cell-mediated responses to non-pathogen components. Several common allergens that induce hypersensitivity are pollen, poison plants, dust, insects, mold/mildew, animal dander, latex, bee sting, medication, or food (cow milk, nuts, chicken egg, fish and seafood, some fruits, gluten) (Table 1).

In recent years, the number of people suffering from allergy has increased exponentially, mainly in economically developed regions.[30,31] Several theories are attempting to explain the reasons for this rise. The pathogenesis of the hypersensitivity reaction is very complex, and it is associated with numerous factors, e.g., genetic background, altered gastroduodenal permeability (for example, in chronic urticaria induced by food),[32] infections, exercise and obesity, air pollution, absence or low exposure to pathogens, altered immune tolerance, and the use of antibiotics for minor illnesses in early life (the theory of "too much cleaning" claims that the absence of immune responses to pathogens at early ages can affect the balance of T cell helper subclasses with predominant Th2 over Th1 responses).[31,33]

The clinical manifestations of hypersensitivity reactions are quite diverse; cutaneous, systemic, respiratory, and/or cardiovascular symptoms appear with different severities. On many occasions, the reactions are weak, for example, eczema in contact allergy or drippy nose and watering eyes in allergic rhinitis. However, in some cases, they are more severe (asthma, anaphylactic shock) or may even have life-threatening effects.[34] Asthma is a very complex and heterogeneous illness, and allergic, non-allergic, or intrinsic pathways,[35] involving many immune cells and cytokines, have all been described in its pathogenesis.[36]

Despite many years of intensive research, it is still not known why a particular person develops an allergy and others do not. Yet, the impossibility of

Table 1. Common allergens.

Contact	Ingestion	Inhalation	Injection
Animal dander	Nuts	Pollen	Bee sting
Nickel	Milk	Dust	Medication
Poison plants	Eggs	Mould	
Latex	Fish and sea food	Animal dander	
	Cow milk		
	Gluten		
	Medication		

predicting the type of reaction that a person may develop still remains a more crucial issue, particularly for medication purposes. It is also the reason why any new drug to be used in human therapy must undergo intensive study and careful trials to demonstrate that it does not induce systemic, contact, or respiratory hypersensitivity reactions.

3. Types of Hypersensitivity

Gell and Coombs subdivided the types of hypersensitivity into four categories: I, II, III, and IV.[37,38] Although this classification was proposed more than 30 years ago, it is still widely used. Yet, many hypersensitivity reactions cannot be explained in this context, and some drug allergies do not fit into this classification. Hence, for preclinical safety evaluation of new therapeutic agents, some authors have defined three different situations namely, pseudoallergic reactions, primarily antibody-mediated reactions, and cell-mediated reactions.[39] In particular, special attention should be paid to pseudoallergy, a hypersensitivity reaction that shares many features with hypersensitivity type I.[40]

The main types of hypersensitivity reactions are:

- Type I: mediated by specific immunoglobulin E (IgE).
- Pseudoallergy: similar to type I, but in the absence of specific IgE. The complement cascade plays a crucial role.
- Type II: cytotoxic hypersensitivity, mediated by IgM and IgG antibodies.
- Type III: mediated by antigen–antibody immunocomplexes (IgM and IgG antibodies).
- Type IV: mediated by cells (mainly Th cells and macrophages).

The subtypes and mediators of the five types of hypersensitivity reactions are summarized in Table 2.

3.1. *Type I hypersensitivity*

This is the most prevalent type of and an increasingly common hypersensitivity, also known as *acute* or *immediate hypersensitivity*, but many people simply refer to it as *allergy*.[41] Once developed, the reaction occurs very rapidly (in the span of minutes) due to the pre-existing allergen-specific IgE antibodies.

The whole process begins with the entry and recognition of an allergen (e.g., pollen) by antigen-presenting cells (macrophages, dendritic cells, or B

Table 2. Main characteristics of the different types of hypersensitivity reactions.

Characteristics	Type I (anaphylactic or immediate hypersensitivity)	Pseudoallergy	Type II (cytotoxic hypersensitivity)	Type III (immune complex mediated hypersensitivity)	Type IV (delayed hypersensitivity)
Main mediators	IgE, mast cells	Complement	Mainly IgG, natural killer cells, complement	IgG, IgM, complement	T helper cells and macrophages
Antigen	Exogenous	Exogenous	Cell surfaces	Soluble antigens	Bacteria, tissues
Time	15–30 min	15–30 min	minutes–hours	3–8 h	48–72 h
Skin reaction	Skin prick positive wheal and flare	Skin prick-negative		Intradermal injection (swelling and redness)	Mantoux-positive (erythema and induration)
Clinical manifestations	Allergic asthma, hay fever, anaphylactic shock	Urticaria, angioedema	Pemphigus, nephritis, autoimmune haemolytic anaemia, Goodpasture's syndrome	Serum sickness, fever, glomerulitis, vasculitis	Tuberculin test, poison ivy, contact dermatitis, maculopapular rashes, granuloma

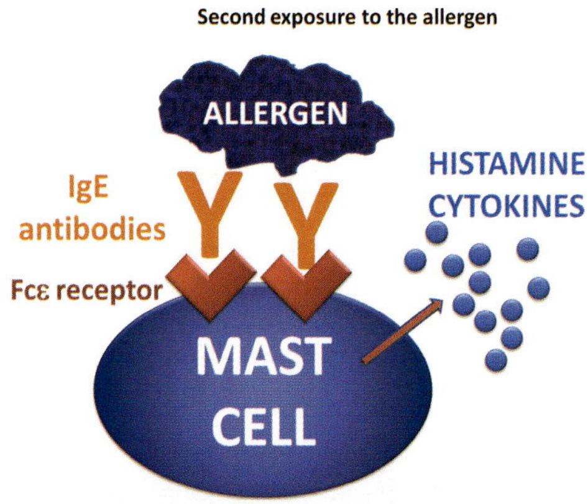

Figure 1. Schematic process of a second exposure to an allergen inducing type I hypersensitivity.

cells), which process the antigen and present allergen peptides to Th cells. As a consequence, a subtype of Th cells, Th2 cells, become activated and are able to induce B cells to differentiate into secretory plasma cells, producing allergen-specific IgE antibodies. As a result of this first contact, allergen-specific soluble IgE antibodies can bind to receptors (Fcε) on the surface of mast cells. Upon further exposure to the same allergen, this allergen will bind to the specific IgE antibodies attached to the mast cells, which will release their granules containing histamine, prostaglandins, and leukotrienes that will mediate the inflammatory response (cytokine production, increased vascular permeability, bronchoconstriction, edema, etc.) (Figure 1).

As described in Table 1, a variety of allergens (pollen, dust, food, drugs) can induce type I hypersensitivity. The severity of symptoms varies from person to person and in some cases, is so severe that it can cause the death of the allergic patient. When the allergen is chemically reactive and capable of covalent conjugation to carrier proteins, it has the capacity for inducing allergy. For example, this occurs with penicillins and related β-lactam antibiotics (which form amide linkages with the side chain of lysine residues) and with D-penicillamine (an anti-arthritic agent that forms conjugates by disulfide bonds to cysteine residues).[42] Moreover, allergy can also be induced by similar elements shared among different allergens because of immunological cross-reactivity.[43]

3.2. *Pseudoallergy*

This is a hypersensitivity reaction that is very similar to type I hypersensitivity, but it does not involve specific IgE antibodies. Complement cascade activation is the crucial factor in a pseudoallergy and for this reason, it is also called *complement activation-related pseudoallergy* (CARPA) or *anaphylactoid reaction.*

The complement system is a very important arm of both the innate and acquired immunity. The complement cascade can be activated by at least three different pathways: classical, alternative, and lectin. Although the pathways differ in the initial steps of activation, they converge at the level of the C3 factor and also share the final steps in the cascade, with the formation of the membrane attack complex (MAC) that includes C5b, C6, C7, C8, and C9 factors.

For the classical pathway, activation is mediated by immunocomplexes (antigen–IgM or antigen–IgG), which induces proenzyme C1, followed by the activation of proenzymes C4, C2, C3, and C5, until the formation of the MAC. The alternative pathway is induced directly by the surface of pathogens or by other pathogen-like elements. In this case, the factors involved in the initial activation are C3b, B, D, and properdin. Finally, the lectin pathway is induced by mannan and other carbohydrate residues, and the cascade of activation is quite similar to its classical counterpart, except that it does not involve antibodies. Chapter 11 provides further details on complement activation.

The activation of the complement cascade has several important consequences. First, anaphylotoxins (C3a, C4a, and C5a in both classical and lectin pathways, but only C3a and C5a in the alternative pathway) that trigger mast cells and basophils to degranulate and recruit inflammatory cells are generated. Second, the opsonization of the foreign elements is induced, increasing their recognition and clearance (mediated by neutrophils and macrophages). Finally, the MAC is formed, which opens pores on cell surfaces or on target elements (such as liposomes), releasing their content in the blood. Thus, the consequences of the activation of the complement cascade can be deleterious, inducing inflammation, edema, and even cell death.

The immune response underlying pseudoallergy is still not completely understood, but it can be distinguished from type I hypersensitivity by using the skin prick test. Patients with type I hypersensitivity (mediated by allergen-specific IgE) will show a positive prick test against the allergen, whereas those with the pseudoallergy will be negative (Table 2).

It has been shown that numerous components are able to induce pseudoallergy, for example, food ingredients,[44] radiocontrast media, medicines such as liposomal drugs (Doxil®, Ambisome®, DaunoXome®) or micellar solvents containing amphiphilic lipids (e.g., Cremophor® EL, the vehicle of Taxol®),[40] nonsteroidal anti-inflammatory drugs,[40,45,46] and several vaccines.[47] Non-allergic asthma, angioedema, and chronic urticaria are a few examples of non-IgE–mediated pseudoallergic reactions.[48]

3.3. *Type II hypersensitivity*

Type II hypersensitivity is also called *cytotoxic hypersensitivity* because the immune response destroys normal host cells. This type of reaction occurs in blood transfusions with incompatible erythrocytes, or in Rhesus-positive (Rh+) fetuses from sensitized Rh- mothers, where maternal antibodies destroy foreign fetal erythrocytes. There are reports of type II hypersensitivity to medicines and to some pathogenic elements (lipopolysaccharides, lectins, viruses).[49–51] It is mediated by IgM and IgG antibodies, and by other elements of the immune system such as complement, macrophages, and natural killer cells. The interaction of specific antibodies with target cells or drugs activates complement by the classical cascade (which mediates inflammatory responses and the lysis of the target cell). Antibodies also recruit natural killer cells (with lytic granules containing perforin and granzyme) and they collaborate in the killing of the target cell. In addition, macrophages contribute to the elimination of the opsonized cells (Figure 2).

In cases of hypersensitivity to medicines, blood cells can be destroyed by three mechanisms:

1) The drug is able to induce the production of specific antibodies that bind to the drug and form immunocomplexes. The complement cascade becomes activated by the classical pathway and kills erythrocytes because although they are not the real target, they are close to the complement factors.
2) Some drugs remain unspecifically adsorbed to the surface of erythrocytes, granulocytes, or platelets. These elements are then recognized as foreign

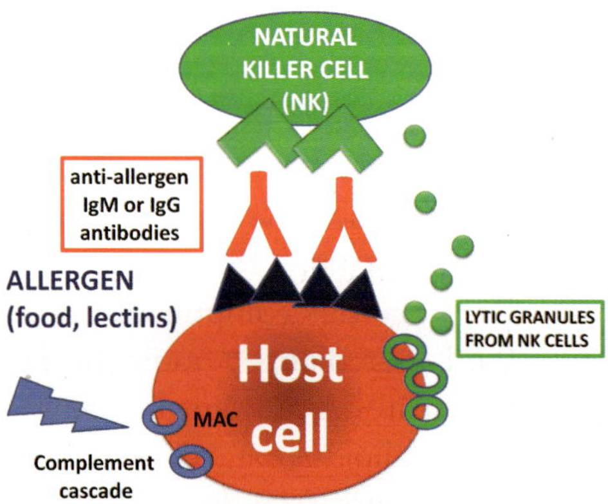

Figure 2. Schematic process of type II hypersensitivity mediated by antibodies, complement, and natural killer (NK) cells.

elements by the immune system, producing antibodies that will try to eliminate them. This is the case for hemolytic anaemia, which occurs in some patients receiving penicillin, L-dopa, aminosalicylic acid, or quinine.[51–53] Some examples of drugs that can decrease the number of granulocytes are phenylbutazone, phenothiazine, aminopyrine, and sulphonamides.[54,55] Moreover, phenylbutazone, quinine, chloramphenicol, and sulphamides can destroy platelets by this mechanism.[56]

3) Some drugs are capable of modifying the erythrocyte surface and the antibodies can be passively absorbed, hence red blood cells are eliminated by phagocytic cells.

Type II hypersensitivity has not been very well documented for nanostructures. Although some of them (for example, large dendrimers) induce hemolysis, antibodies on the surface of red blood cells have to be demonstrated. Much more research is needed to determine whether nanomaterials can mediate this type of hypersensitivity in animal models *in vivo*.

3.4. *Type III hypersensitivity*

Type III hypersensitivity is mediated by immunocomplexes (antigen–antibody, mainly IgG) (Figure 3). These immunocomplexes cause severe inflammation when deposited in large quantities in tissues or in circulation. The severity of the hypersensitivity reaction depends on the quantity and site of deposit of the immunocomplexes. If the antigen enters via the respiratory pathway, the immunocomplexes could induce pneumonitis. However, if they are formed in the systemic circulation, the kidneys, synovial capsules, and arteries are the main targets, with the development of glomerulonephritis, arthritis, and arteritis, respectively.

The whole process starts with the entry, either locally or systemically, of large amounts of antigen. Specific antibodies are generated and bind to that antigen, producing immunocomplexes, which induce several immune reactions:

1) Complement activation: production of anaphylotoxins (C3a, C4a, C5a) and recruitment of inflammatory cells (inflammation).
2) Accumulation of neutrophils, which release their granules full of proteases and oxidant elements (tissular damage).
3) Recruitment of macrophages: production of inflammatory cytokines, such as tumor necrosis factor-α (TNF-α) and interleukin-1 (IL-1), and production of nitric oxide (tissular destruction).

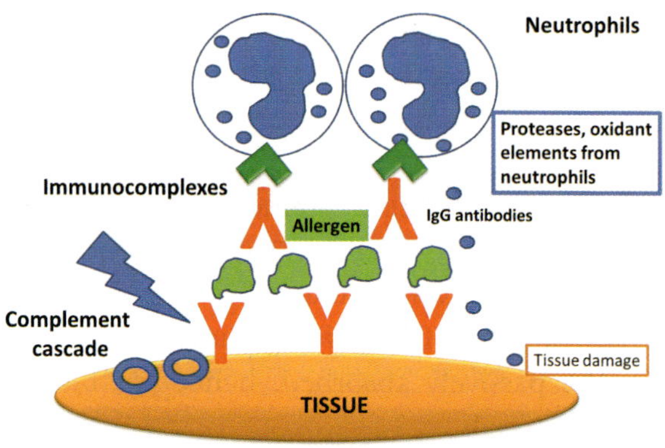

Figure 3. Schematic process of type III hypersensitivity, mediated by immunocomplexes (antigen–antibody).

Many years ago, when anti-sera was used in passive immunization, some patients receiving large doses of anti-tetanic or anti-diphteric horse serum developed a strong hypersensitivity reaction 10 days later, called "serum illness," with symptoms of generalized vasculitis, rash, neutropenia, arthritis, and glomerulonephritis. Pulmonary problems have been described in people chronically exposed to antigens, such as farmers (to moldy hay), librarians (to dust from books), bird fanciers (to bird feces), and mushroom farmers (to fungal spores).

In addition to these processes, type III hypersensitivity can be induced in infectious processes, autoimmune diseases (rheumatoid arthritis, lupus), or by food and medicines. In the case of food, it has been shown in very young calves that the addition of soybean as a food supplement can induce type III hypersensitivity, resulting in slow growth and weight loss.[57]

Some medicines induce the formation of specific antibodies and if the immunocomplexes attach to blood cells, the opsonized cells will be phagocytosed. Thus, some medicines can induce anemia, granulocytopenia, or thrombocytopenia. In some cases, the immunocomplexes containing drug antibodies are concentrated in the skin, inducing several cutaneous lesions with the deposition of immunocomplexes in the dermal vessels. However, although it is sometimes difficult to distinguish if the effect is mediated by the immunocomplexes or by a direct toxic effect of the drug, specific antibodies can at least be separated from the cells and analyzed. It is also difficult to distinguish this type of hypersensitivity from type II.

3.5. *Type IV hypersensitivity*

Type IV hypersensitivity is also called *delayed* or *cell-mediated immune reaction*, and is mediated by Th cells and antigen-presenting cells (Figure 4).

This is a rather common type of allergy; indeed, it is the most common occupational and environmental skin disease.[58] Although not a life-threatening condition, it is sometimes very unpleasant and requires modifications in lifestyle or a change in work locations. The most ubiquitous allergens are poison from plants, metallic salts (nickel sulfate, cobalt chloride), organic dyes, rubber additives, germicides, plastic resins (epoxy resin), chemicals (thymerosal, formaldehyde, fragances), and medicines (neomycin, bacitracin).[58–62]

The process typically starts with topical exposure to a contact allergen (metal, creams, textiles, shampoos). The allergen is then absorbed by the skin, and epidermal Langerhans cells internalize the antigen and migrate to the draining lymph nodes,[63] which will present antigen peptides in class II molecules of the major histocompatibility complex (MHC) to specific Th cells[58] (Figure 4). These T lymphocytes become activated, expand the population clonally, and establish memory. Hence, the sensitized person will respond more aggressively and rapidly in any future encounter with the same antigen.

Skin penetration is a key issue in type IV hypersensitivity because the first obstacle for a successful skin penetration is the stratum corneum barrier. Several characteristics such as molecular weight (<500 Da), charge, and physicochemical properties (lipophilic compounds penetrate the skin more easily) are crucial for a chemical compound to be an allergen in this type of hypersensitivity.

As Th cells predominantly recognize peptides via their specific T cell receptor (TCR), it is not very clear how a chemical substance, which in itself cannot generate an immune response (called hapten), could induce type IV

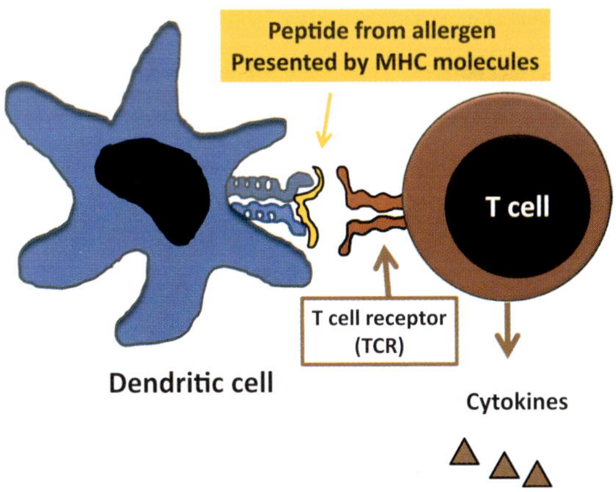

Figure 4. Schematic process of the initiation of type IV hypersensitivity. The immune response is mounted via the activation of antigen-presenting cells and Th lymphocytes. Type IV hypersensitivity is associated with occupational and environmental skin diseases and the MHC. MHC molecules: major histocompatibility complex molecules.

hypersensitivity when applied cutaneously. Several mechanisms have been proposed:[64]

1) The allergen modifies soluble self-proteins that could be phagocytosed by Langerhans cells, and some self-peptides are presented to Th cells.[65,66]
2) The allergen binds directly to peptides that are already being presented by MHC molecules in antigen-presenting cells.[66,67]
3) The allergen penetrates cell membranes and modifies intracytoplasmic proteins that can be processed and presented by Langerhans cells.
4) The allergen directly activates T cells through non-specific receptors, such as Toll-like receptors (TLR4, Marco).[67,68]

Skin sensitization is followed by the elicitation phase.[69] The second or further exposure to the allergen induce memory T cells that become activated and produce cytokines. Inflammatory cells, mostly macrophages, also produce inflammatory cytokines. As a consequence, there is a cutaneous allergic reaction with erythema, pruritus, vesiculation, papules, induration, and edema.[59] The process usually terminates after one to three weeks. However, if the exposure is chronic, other skin problems can materialize, such as scaling, hyperpigmentation, and fissures.[58,70]

Approximately 3,700 chemical substances have been described as potential allergens[71] and they can be used in humans for skin patch testing. In addition to allergy, metals such as nickel, copper, zinc, vanadium, chromium, cobalt, or iron can induce other direct or indirect toxic effects, but this is outside the scope of the present chapter.

4. Allergic Diseases Induced by Nanomaterials

4.1. *Pseudoallergy*

Since the introduction of nano-sized components in clinical practice, there are several reports of pseudoallergy in humans and animals induced by nanostructures.[40,72–79] The main evidence that nanostructures can activate the complement cascade comes from cholesterol-rich and polyethylene glycol (PEG)ylated liposomes.[79,80]. The following substances have been described as inducers of hypersensitivity reactions: liposomal drugs (Doxil®, Ambisome®, DaunoXome®), micellar solvents containing amphiphilic lipids (Cremophor® EL, the vehicle of Taxol®), radiocontrast media, carbon nanotubes, liposome- and polymer-based nanomedicines, PEG-coated vesicles, and other related lipid-based and phospholipid–mPEG conjugate stabilized entities.[81–85]

Although in some cases where other hypersensitivity reactions can occur (antibodies have been found in patients receiving PEGylated drugs), most seem to induce pseudoallergic reactions. In many cases, the activation of complement is mediated through both the classical (with the participation of specific antibodies) and the alternative pathways. Using large multilamellar liposomes and the pig as the animal model, Szebeni *et al.* showed that liposome-induced pulmonary hypertension is due to complement activation, because the administration of human C3a, C5a, and immunoglobulins reproduced the vasoactivity induced by liposomes.[86]

Furthermore, Moghimi *et al.* have shown that both naked and PEGylated liposomes were capable of activating the human complement system, proving that PEG was not responsible for the PEGylated liposome-mediated complement activation.[80] The authors claimed that it is due to the anionic charge on the phosphate moiety of the phospholipid–mPEG conjugate, which is able to activate both the classical and the alternative pathways of the complement system.

Small micelles (8–25 nm) (Cremophor® EL, CrEL) and poloxamer 188 (a monomeric block copolymer) also activate complement.[87] The mechanism of activation is not clear because lipoproteins can not only interact with the micelles, inducing the formation of large lipoprotein-like structures, but also mediate the inhibition of poloxamer 188-induced complement activation.[87]

Cremophor® EL, a non-ionic emulsifier vehicle of paclitaxel (an antitumoral drug), forms spherical micelles with diameters in the range of 8–22 nm. However, after incubation with human serum, the micelles show structural transformation in blood with the presence of crystalline paclitaxel, and the formation of a needle-like structure extended to the multimicron range.[88]

4.2. *Asthma*

Although there are not yet many studies investigating the toxicity of NPs in humans, it has been largely demonstrated that ultrafine particles are associated with adverse health effects, such as cardiovascular disease and asthma, as they present a high surface area per unit mass ratio, resulting in high lung deposition efficiency and an increased potential for biological interactions.[89–92] Some NPs have been linked to allergic sensitization in animals, triggering specific IgE production. They can act as adjuvants or haptens by inducing a specific pattern of cytokines, antibodies, and cells that favor allergic sensitization to environmental allergens. Some NPs have demonstrated pro-inflammatory effects in the lung in experimental animals with the increased expression of the following: IL-1β, macrophage inflammatory protein 1-alpha (MIP-1α), monocyte chemoattractant protein 1 (MCP-1), macrophage inflammatory protein 2 (MIP-2), keratinocyte chemoattractant, thymus and activation-regulated chemokine (TARC), and granulocyte-macrophage colony-stimulating factor (GM-CSF). In addition, the activation of stress-activated protein kinases, SAPK/Jun N-terminal kinases (JNK), and mitogen-activated protein kinases (MAPK) p38[93] has been reported.

Exposure to certain titanium dioxide (TiO_2) NPs induces emphysema-like lung injury in mice,[94] and upregulates the expression of lung neurotrophins (key regulatory elements of neuronal development and responsiveness of airway sensory neurons) in an age-dependent manner. This effect is associated with airway hyper-responsiveness and inflammation.[91]

Morimoto *et al.* measured the concentration of 21 cytokines and the presence of inflammation, fibrosis, and allergy-related responses in the lungs of rats, following the intratracheal instillation of nickel oxide NPs.[95] They observed persistent infiltration of alveolar macrophages in the nickel oxide-exposed group and the expression of MIP-1α in lung tissue and broncho-alveolar lavage fluid (BALF). The results suggest that nano-agglomerates of nickel oxide NPs could have a persistent pulmonary inflammatory effect.[95]

It has also been shown that photocatalytic TiO_2 NPs can enhance allergy to aerosolized ovalbumin in mice,[96] and similar findings have been reported with single-walled (SW) and multi-walled (MW) carbon nanotubes.[97]

4.3. *Acute dermatitis*

There are several references in current literature with regard to NPs causing or aggravating atopic dermatitis (AD). For example, Yanagisawa *et al.* have shown that TiO_2 NPs of different sizes aggravate AD-like skin lesions related to a mite allergen in a model of mice.[98] The enhancing effects are paralleled by the overproduction of IL-4 in the skin, as well as the levels of total IgE and histamine in serum in relation to the overall trend. In contrast, in the presence of an allergen, TiO_2 NPs decreased the local expression of IFN-γ. These authors concluded that exposure to TiO_2 NPs under skin barrier dysfunction/defect can exacerbate AD symptoms through Th2-biased immune responses.[98]

4.4. *Contact dermatitis*

Several articles have shown that some NPs can cause or enhance contact dermatitis. Boscolo *et al.* have reported that palladium (Pd) NPs (5–10 nm) enhance the release of IFN-γ, a typical Th1 cytokine in delayed allergic reactions. This result is interesting because of the increase of allergic contact dermatitis to Pd in people exposed to Pd NPs in urban environments.[99] Di Gioacchino *et al.* recently confirmed these results, and extended the study to other types of NPs, such as platinum, nickel, and cobalt, which can modulate cytokine production toward Th1 responses, while Ti nanostructures and MW and SW carbon nanotubes induce a pattern more similar to Th2 production.[93]

The mechanisms underlying inflammatory responses are being comprehensively studied, and it has been shown that nickel, which is the most frequent cause of contact hypersensitivity in industrialized countries,[100] triggers an inflammatory response by directly activating human TLR4 in leukocyte cells.[68]

5. Beneficial Anti-Inflammatory Effects

NPs not only induce inflammatory effects, but also anti-inflammatory effects. To avoid inflammation by preventing leukocyte trafficking, Dernedde *et al.* investigated synthetic dendritic polyglycerol sulphates (dPGS) as macromolecular inhibitors that operate via a multivalent binding mechanism, mimicking naturally occurring ligands.

The dPGS inhibited both leukocytic L-selectin and endothelial P-selectin with high efficacy. The administration of dPGS in a contact dermatitis mouse model dampened leukocyte extravasation as effectively as glucocorticoids, and edema formation was significantly reduced. In addition, dPGS interacted with the complement factors C3 and C5 and reduced C5a levels in a mouse model of complement activation.[101]

Other authors have found anti-inflammatory effects with topical nanocrystalline silver cream for allergic contact dermatitis, in several animal models (guinea pig,[102] murine,[103] and rat[104]). In these studies, nanocrystalline silver was shown to significantly suppress the expression of inflammatory elements, such as interleukins (IL-1β, IL-12), matrix metalloproteinase 9, and TNF-α.[105]

The best example of demonstrating the beneficial effects of nanostructures in clinical practice is through the use of the new drug Abraxane®, which includes paclitaxel (an antitumoral drug) in human albumin.[106,107] The coating with human albumin avoids many of the problems that Taxol® (paclitaxel in Cremophor®) causes, such as hypersensitivity reactions (which require premedication and a low flow rate of administration) due to complement activation,[106] and also improves its antitumoral efficacy.

In this area, other improvements are ongoing. A novel versatile nanocarrier system based on micelles formed by a well-defined linear PEGylated two-arm oligomer of cholic acids, has been developed for the targeted delivery of hydrophobic anticancer drugs.[108] Showing similar antitumor activity as Abraxane®, these micellar formulations are much more stable and their drug-loading capacity is shown to be superior.[108]

Another promising application of NPs is their use in vaccines, either as adjuvants or as the vehicle of the product.[109] Colloidal carriers, e.g., polymeric NPs, improve the oral delivery of some vaccines, since they are stable in the gastrointestinal tract, protective for encapsulated substances, and able to modulate physicochemical characteristics, drug release, and biological behavior.[110] Different nanocarriers are also suitable for the nasal delivery of antigenic molecules. Besides improved protection and facilitated transport of the antigen, NP delivery systems could also provide more effective antigen recognition by immune cells. The design of optimized vaccine nanocarriers offers a promising way for nasal mucosal vaccination.[111]

6. Hypersensitivity Tests and Models

The diagnosis of immediate and delayed hypersensitivities is based on clinical history (symptomatology and others) and if possible, on skin and biological tests.[112] Currently, there are no accepted *in vitro* or *in vivo* models available for predicting or measuring all types of hypersensitivity responses in humans. There are some animal tests for topical and inhalation medicines, but the existing preclinical models have limited capability for the prediction of allergy in humans. Therefore, it is necessary to find new *in vitro* and *in vivo* methods.[113]

6.1. *Diagnosis of type I hypersensitivity*

One of the most commonly used methods to identify type I hypersensitivity in humans is the skin prick test.[114] Usually, a combination of several allergens and positive (histamine solution) and negative controls are used in the skin prick test. Approximately 15–30 min after skin administration of the allergen, it is possible to identify the presence of rash, edema, and inflammation in allergic individuals.[115,116] A positive reaction is indicated by erythema, itching, and a palpable weal (the diameter is measured in millimeters, and compared with positive and negative controls). This test could help to differentiate between type I allergy and pseudoallergy, because people with an IgE-mediated allergy will develop a positive prick test, whereas for cases of pseudoallergy the test will be negative.[44]

Regarding animal models, currently, no validated animal model is available for determining the sensitizing threshold of food allergens that can be extrapolated to or be predictive of the human allergic response.[117] Mice are not appropriate for skin tests, and although guinea pigs are more suitable, larger animals that closely mimic allergic diseases in humans (dog[118] and swine[119]) can be more useful models.[117,120]

The passive cutaneous anaphylaxis (PCA) test performed in mice or rats has been widely used to measure allergen-specific IgE antibody levels in animal models for half a century.[121–125] This test involves the inoculation of several serum dilutions (presumably with allergen-specific IgE) in different parts of the skin of a normal animal and 24–48 h later, the allergen is injected intravenously. To increase the visibility of the inflammation, a colourant (Evans blue) is also injected systemically. The colourant binds to serum albumin and usually remains in the circulation, except in those areas with increased vascular permeability, such as the inflamed skin. The size of the blue skin region can be used to measure the intensity of the inflammatory response. Although the PCA test is highly sensitive, it suffers practical disadvantages — high cost, animal use that is labor-intensive, and concerns about animal welfare,[121] thus alternative, inexpensive *in vitro* methods that are easy to perform and comparable in sensitivity to the PCA test are needed. In serum samples, it is possible to identify antibodies against the allergen using very sensitive methods to measure total and specific IgE antibodies. The radioactive radioallergosorbent test (RAST), or the enzymatic ELISA and Western blot assays are the most common methods used for type I hypersensitivity.[121]

The RAST is routinely used in clinical practice for the measurement of specific IgE antibodies in the human serum. In fact, this technique is filed under the regulation ISO 13485:2003, but although it is applicable to a mouse system, its utility is limited by the use of radioisotopes and its low level of precision.[121]

Regarding ELISA-based methods, there are several authors who have described the measurement of IgE specific to individual purified proteins such as ovalbumin, milk proteins, peanut major allergens, or haptens,[122,126–131] and Birminghan *et al.* dictated an ELISA-based method for food extract-specific IgE antibody detection.[121] Although these techniques are quite sensitive, there is not always a good correlation among the serology, the cutaneous test, and the severity of the immune response. In fact, serological studies give frequent false-positives (low specificity), but a negative ELISA is sufficient for confirming negative allergy. Other tests, such as basophil histamine release assays,[132] are not routinely used and are generally poorly reproducible,[133] although in some cases they can be useful.[134]

Moreover, one problem is that allergens may contain isoforms or contaminants, and the identification of all the potential isoforms of each allergen could be very difficult, laborious, and expensive.

Another test to measure type I allergy and pseudoallergy reactions is the analysis of serum tryptase levels, a marker of mast cell degranulation, after the administration of the allergen.[135,136] It is also possible to measure the activation of the complement cascade, by analyzing serum C3a desArg, iC3b, and sC5b-9, or by analyzing the degradation of C3 using ELISA or Western blot.[85] However, C3 and C4 plasma levels are not sufficient to discriminate between IgE-mediated allergy and pseudoallergy.[137]

In some cases, an *in vivo* provocation test is needed to assure identification of the type of allergy. The allergen can be administered to the nasal or respiratory mucosa, and nasal airflow (rhinomanometry) or pulmonary functions [forced expiratory volume (FEV), peak expiratory flow (PEF), etc.] are then measured.[138]

With regard to animal models for acute reactions, pigs are a good model because in minutes, it reveals cardiovascular symptoms and analytical changes that mimic some of the human symptoms. An *in vivo* bronchial provocation test is used with different nanostructures in mice. Some reports focusing on exposure to carbon nanotubes have shown that they can induce pulmonary toxicity, inflammation, and bronchoconstriction.[139] Szebeni *et al.* have studied several models of complement-mediated hypersensitivity reactions to liposomal drugs, diagnostic agents, micelles, and other lipid-based NPs, and concluded that different animals display different responses.[78] For instance, pigs, dogs, and rats differ in sensitivity and symptoms. The pig model is the most sensitive, and a few milligrams of liposomes can cause anaphylactoid shock, characterized by pulmonary hypertension, systemic hypotension, decreased cardiac output, and major cardiac arrhythmias. In contrast, the rat model was found to be relatively insensitive.[78] The same authors suggested that porcine and dog models are applicable for measuring and predicting the (pseudo)allergic activity of particulate "nanodrugs."

6.2. *Diagnosis of type II hypersensitivity*

Hemolysis and agglutination are the most commonly used tests for type II hypersensititivy diagnosis.[140] Hemolysis involves the use of erythrocytes and rabbit complement, and the red color indicates the hemolytic level (several dilutions of serum are tested, and the dilution that is able to kill 50% of the erythrocytes is analyzed). In the case of hemolytic anemia, the

Coombs test shows the presence of antibodies on the surface of erythrocytes, but the addition of anti-immunoglobulins (IgG) is required to be able to observe the agglutination of red blood cells with IgG attached on their surfaces.[141]

For hypersensitivity to drugs, it has to be demonstrated that the eluted antibodies from blood cells are directed specifically against the drug, but not to the blood cells.

6.3. *Diagnosis of type III hypersensitivity*

A similar process to "serum illness" in humans can be experimentally induced in animals such as rabbits. Rabbits are immunized intravenously with the antigen, using a single high dose or small quantities in several doses, to study acute or chronic exposure, respectively. Approximately 10 days later, they may develop glomerulonephritis, arthritis, and spleen lesions.[142]

After excluding the other types of hypersensitivity (via the negative skin prick test, low levels of IgE, negative Mantoux test), type III hypersensitivity can be diagnosed by the detection of immunocomplexes. Several methods, most of them applied to autoimmune diseases, can be used to detect immunocomplexes, such as immunofluorescence on tissues from affected organs,[143] PEG-mediated turbidimetric method,[144] Raji cell assays,[145,146] zone electrophoresis on agarose,[147] or ELISA.[148]

6.4. *Diagnosis of type IV hypersensitivity*

Type IV hypersensitivity can be measured by the induration on skin testing after the inoculation of the allergen or patch (usually on the back).[149] The skin patch test including several agents (nickel sulfate, medicines, fragances) are conducted by healthcare professionals, and should never be conducted with unknown substances.[150]

The local lymph node assay (LLNA) is a test for contact-sensitizing potential, which can be used in animal models such as mice.[151] It is the approved method for the assessment of skin sensitization potential.[151] In 2002, the LLNA was incorporated into a new test guideline (Guideline No. 429; Skin Sensitization: Local Lymph Node Assay) by the OECD (Organization for Economic Cooperation and Development). The test starts with the administration of the allergen in one mouse ear, followed by the administration of radioactive thymidine. Proliferating lymphocytes on the draining lymph node are measured (counts per minute), thereby indicating the activation of the immune response.[152,153]

A summary of the currently available methods for the different types of hypersensitivity is shown in Table 3.

7. Conclusion

Allergy has been and is still increasing worldwide, mainly in economically developed regions. Hypersensitivity to many components such as drugs, contrast agents, cosmetics, metals, and food has been very well described. However, it is impossible to predict in whom a hypersensitivity reaction will develop and when this reaction will occur, or the extent of severity of the immune response. The recent arrival of various nanostructures has opened numerous possibilities in energy, computing, biomedicine, fabrics, and

Table 3. Diagnostic methods for the different types of hypersensitivity reactions.

Hypersensitivity type	Diagnostic methods	References
Type I hypersensitivity	Skin tests (skin prick and/or intradermal test)	114–116
	Passive cutaneous anaphylaxis (PCA) test	122–125
	Radioallergosorbent testing (RAST) (allergen-specific IgE detection)	154
	Western blot/ELISA (allergen-specific IgE detection)	121, 122, 126–131
	Serum tryptase levels and or histamine release	135, 136
	In vivo provocation test	138
Pseudoallergy	Complement cascade activation (with negative skin test)	44
Type II (cytotoxic)	Agglutination	141
	Hemolysis	140
Type III (immune complex)	Immunofluorescence on tissues	143
	PEG-mediated turbidimetry	144
	ELISA	148
	Zone electrophoresis on agarose	147
	Raji cell assays (RIA, flow cytometry)	145, 146
Type IV (delayed type)	Patch test	149
	Local lymph node assay (LLNA)	151

cosmetics, all with considerable potential. Nevertheless, we should be aware that the exposure to these nanomaterials (either intentional or accidental) can have harmful effects such as cell toxicity or the triggering of allergic responses. Yet, there is evidence that some nanostructures could also be used to reduce hypersensitivity reactions. More in-depth research is required, focusing on the analysis of the potentially hazardous effects mediated by nanostructures on allergy, and on the development of new *in vitro* and *in vivo* methods of prediction and measurement.

Acknowledgements

We greatly appreciate the support from INBIOMED (2009/063, the *Xunta de Galicia*), IMMUNONET (SOE1/P1/E014, Immunotherapy network, SUDOE-Feder), and HINAMOX (7th EU programme) projects.

Conflict of Interests

The authors declare no conflict of interest.

References

1. Mudshinge SR, Deore AB, Patil S, Bhalgat CM. Nanoparticles: Emerging carriers for drug delivery. *Saudi Pharm J* 2011;19:129–141.
2. Nanjwade BK, Bechra HM, Derkar GK, Manvi FV, Nanjwade VK. Dendrimers: Emerging polymers for drug-delivery systems. *Eur J Pharm Sci* 2009;38:185–196.
3. Scholes PD, Coombes AGA, Illum L, Daviz SS, Vert M, Davies MC. The preparation of sub-200 nm poly(lactide-co-glycolide) microspheres for site-specific drug delivery. *J Controlled Release* 1993;25:145–153.
4. Zeng W, Qian X-F, Zhang Y-B, Yin J, Zhu Z-K. Organic modified mesoporous MCM-41 through solvothermal process as drug delivery system. *Mater Res Bull* 2005;40:766–772.
5. Zhang L, Gu FX, Chan JM, Wang AZ, Langer RS, Farokhzad OC. Nanoparticles in medicine: Therapeutic applications and developments. *Clin Pharmacol Ther* 2008;83:761–769.
6. Cherukuri P, Glazer ES, Curley SA. Targeted hyperthermia using metal nanoparticles. *Adv Drug Delivery Rev* 2010;62:339–345.
7. Hirsch L, Gobin A, Lowery A, Tam F, Drezek R, Halas N, West J. Metal nanoshells. *Ann Biomed Eng* 2006;34:15–22.
8. Hirsch LR, Stafford RJ, Bankson JA, Sershen SR, Rivera B, Price RE, Hazle JD, Halas NJ, West JL. Nanoshell-mediated near-infrared thermal therapy of tumors under magnetic resonance guidance. *Proc Natl Acad Sci U S A* 2003;100:13549–13554.
9. Loo C, Lowery A, Halas N, West J, Drezek R. Immunotargeted nanoshells for integrated cancer imaging and therapy. *Nano Lett* 2005;5:709–711.

10. Lal S, Clare SE, Halas NJ. Nanoshell-enabled photothermal cancer therapy: Impending clinical impact. *Acc Chem Res* 2008;41:1842–1851.
11. Wieder ME, Hone DC, Cook MJ, Handsley MM, Gavrilovic J, Russell DA. Intracellular photodynamic therapy with photosensitizer-nanoparticle conjugates: Cancer therapy using a 'Trojan horse'. *Photochem Photobiol Sci* 2006;5:727–734.
12. Amidi M, Romeijn SG, Verhoef JC, Junginger HE, Bungener L, Huckriede A, Crommelin DJA, Jiskoot W. N-trimethyl chitosan (TMC) nanoparticles loaded with influenza subunit antigen for intranasal vaccination: Biological properties and immunogenicity in a mouse model. *Vaccine* 2007;25:144–153.
13. Csaba N, Garcia-Fuentes M, Alonso MJ. The performance of nanocarriers for transmucosal drug delivery. *Expert Opin Drug Delivery* 2006;3:463–478.
14. Friede M, Aguado MT. Need for new vaccine formulations and potential of particulate antigen and DNA delivery systems. *Adv Drug Delivery Rev* 2005;57:325–331.
15. Jung T, Kamm W, Breitenbach A, Hungerer K-D, Hundt E, Kissel T. Tetanus toxoid loaded nanoparticles from sulfobutylated poly(vinyl alcohol)-graft-poly(lactide-co-glycolide): Evaluation of antibody response after oral and nasal application in mice. *Pharm Res* 2001;18:352–360.
16. Lincopan N, Santana M, Faquim-Mauro E, da Costa M, Carmona-Ribeiro A. Silica-based cationic bilayers as immunoadjuvants. *BMC Biotechnol* 2009;9:5.
17. Peek LJ, Middaugh CR, Berkland C. Nanotechnology in vaccine delivery. *Adv Drug Delivery Rev* 2008;60:915–928.
18. Prego C, Paolicelli P, Díaz B, Vicente S, Sánchez A, González-Fernández Á, Alonso MJ. Chitosan-based nanoparticles for improving immunization against hepatitis B infection. *Vaccine* 2010;28:2607–2614.
19. Prego C, Torres D, Alonso MJ. The potential of chitosan for the oral administration of peptides. *Expert Opin Drug Delivery* 2005;2:843–854.
20. Tobío M, Gref R, Sánchez A, Langer R, Alonso MJ. Stealth PLA-PEG nanoparticles as protein carriers for nasal administration. *Pharm Res* 1998;15:270–275.
21. Kim H, Lee HBR, Maeng WJ. Applications of atomic layer deposition to nanofabrication and emerging nanodevices. *Thin Solid Films* 2009;517:2563–2580.
22. Orcutt KM, Wells ML. A liposome-based nanodevice for sequestering siderophore-bound Fe. *J Membr Sci* 2007;288:247–254.
23. Park WI, Kim D-W, Jung, SW, Yi, G-C. Catalyst-free growth of ZnO nanorods and their nanodevice applications. *Int J Nanotechnol* 2006;3:372–395.
24. Agasti SS, Rana S, Park M-H, Kim CK, You C-C, Rotello VM. Nanoparticles for detection and diagnosis. *Adv Drug Delivery Rev* 2010;62:316–328.
25. Azzazy HME, Mansour MMH, Kazmierczak SC. Nanodiagnostics: A new frontier for clinical laboratory medicine. *Clin Chem* 2006;52:1238–1246.
26. Chirathaworn C, Chantaramalai T, Sereemaspun A, Kongthong N, Suwancharoen D. Detection of Leptospira in urine using anti-Leptospira-coated gold nanoparticles. *Comp Immuno Microbiol* 2011;34:31–34.
27. Huang X, El-Sayed MA. Gold nanoparticles: Optical properties and implementations in cancer diagnosis and photothermal therapy. *J Adv Res* 2010;1:13–28.
28. Jiang H, Chess L. How the immune system achieves self-nonself discrimination during adaptive immunity. *Adv Immunol* 2009;102:95–133.
29. Parkin J, Cohen B. An overview of the immune system. *Lancet* 2001;357:1777–1789.

30. CDC Fast Facts A-Z. Vital health statistics, from http://www.cdc.gov/nchs/fastats/allergies.htm, 2003.
31. Kay AB. Allergy and allergic diseases. *N Engl J Med* 2001a;344:30–37.
32. Buhner S, Reese I, Kuehl F, Lochs H, Zuberbier T. Pseudoallergic reactions in chronic urticaria are associated with altered gastroduodenal permeability. *Allergy* 2004;59:1118–1123.
33. Wills-Karp M, Santeliz J, Karp CL. The germless theory of allergic disease: Revisiting the hygiene hypothesis. *Nat Rev Immunol* 2001;1:69–75.
34. Kay AB. Allergy and allergic diseases. *N Engl J Med* 2001b;344:109–113.
35. Humbert M, Menz G, Ying S, Corrigan CJ, Robinson DS, Durham SR, Kay AB. The immunopathology of extrinsic (atopic) and intrinsic (non-atopic) asthma: More similarities than differences. *Immunol Today* 1999;20:528–533.
36. Kim HY, DeKruyff RH, Umetsu DT. The many paths to asthma: Phenotype shaped by innate and daptive immunity. *Nat Immunol* 2010;11:577–584.
37. Coombs RRA, Gell P.G.H. Classification of allergic reactions responsible for drug hypersensitivity reactions. In: Coombs RRA, Gell PGH (eds.), *Clinical Aspects of Immunology*, Davis, 1968, pp. 575–596.
38. Gell PGH, Coombs RRA. The classification of allergic reactions underlying disease. In: Gell PGH, Coombs RRA (eds.), *Clinical Aspects of Immunology*, Blackwell, 1963.
39. Descotes J, Choquet-Kastylevsky G. Gell and Coombs's classification: Is it still valid? *Toxicology* 2001;158:43–49.
40. Szebeni J. Complement activation-related pseudoallergy: A new class of drug-induced acute immune toxicity. *Toxicology* 2005a;216:106–121.
41. HayGlass KT. Allergy: Who, why and what to do about it? *Immunol Today* 1995;16:505–507.
42. Coleman JW, Blanca M. Mechanisms of drug allergy. *Immunol Today* 1998;19:196–198.
43. Valenta R, Steinberger P, Duchene M, Kraft D. Immunological and structural similarities among allergens: Prerequisite for a specific and component-based therapy of allergy. *Immunol Cell Biol* 1996;74:187–194.
44 Reese I, Zuberbier T, Bunselmeyer B, Erdmann S, Henzgen M, Fuchs T, Jäger L, Kleine-Tebbe J, Lepp U, Niggemann B, *et al.* Diagnostic approach for suspected pseudoallergic reaction to food ingredients. *J Dtsch Dermatol Ges* 2009;7:70–77.
45. Czech WS, Schöpf E, Kapp A. Release of sulfidoleukotrienes *in vitro*: Its relevance in the diagnosis of pseudoallergy to acetylsalicylic acid. *Inflammation Res*1995;44:5.
46. Groneberg DA, Heppt W, Welker P, Peiser C, Thai Dinh Q, Cryer A, Zweng M, Witt C, Fischer A. Aspirin-sensitive rhinitis-associated changes in upper airway innervation. *Eur Respir J* 2003;22:986–991.
47. Catelain A, Cousin F, Freymond N, Nancey S, Bérard F, Nicolas JF. Vaccine-induced urticaria: How to differentiate allergy (IgE) from pseudoallergy? *Ann Dermatol Venereol* 2004;131:239–243.
48. Bircher AJ. Drug-induced urticaria and angioedema caused by non-IgE mediated pathomechanisms. *Eur J Dermatol* 1999;9:657–663.
49. Chenoweth CE, Judd JW, Steiner AE, Kauffman CA. Cefotetan-induced immune hemolytic anemia. *Clin Infect Dis* 1992;15:863–865.
50. Garratty G, Petz LD. Drug-induced immune hemolytic anemia. *Am J Med* 1975;58:398–407.

51. Ries CA, Rosenbaum TJ, Garratty G, Petz LD, Fudenberg HH. Penicillin-induced immune hemolytic anemia. *Am Med Assoc* 1975;233:432–435.
52. Freedman AL, Barr PS, Brody EA. Hemolytic anemia due to quinidine: Observations on its mechanism. *J Med* 1956;20:806–816.
53. Petz LD, Mueller-Eckhardt C. Drug-induced immune hemolytic anemia. *Transfusion* 1992;32:202–204.
54. Means RT, Sandidge DR, Rankin KM, Robichaux MR. Treatment of phenothiazine-induced agranulocytosis with recombinant granulocyte colony-stimulating factor. *Am J Hematol* 1992;41:296–296.
55. Pisciotta A. Agranulocytosis induced by certain phenothiazine derivatives. *J Am Med Assoc* 1969;208:1862–1868.
56. Comaish S. The blood platelets in immune hypersensitivity. *Acta Derm Venereol* 1968;48:592–599.
57. Barratt MEJ, Strachan PJ, Porter P. Immunologically mediated nutritional disturbances associated with soya-protein antigens. *Proc Nutr Soc* 1979;38:143–150.
58. Imbesi S, Minciullo PL, Isola S, Gangemi S. Allergic contact dermatitis: Immune system involvement and distinctive clinical cases. *Allergol Immunopathol* 2011;39:374–377.
59. Saint-Mezard P, Rosieres A, Krasteva M, Berard F, Dubois B, Kaiserlian D, Nicolas JF. Allergic contact dermatitis. *Eur J Dermatol* 2004;14:284–295.
60. Fowler JF, Jr. Current problems in dermatology. Occupational dermatology. *Curr Probl Dermatol* 1998;10:211–244.
61. Marks JG, Jr., Elsner P, Deleo VA. Standard allergens. In: *Contact & Occupational Dermatology*, 3rd edn. Mosby, 2002, pp. 65–139.
62. Scheman A, Jacob S, Zirwas M, Warshaw E, Nedorost S, Katta R, Cook J, Castanedo-Tardan MP. Contact allergy: Alternatives for the 2007 North American contact dermatitis group (NACDG) standard screening tray. *Dis-a-Month* 2008;54:7–156.
63. Kimber I, Dearman RJ, Cumberbatch M, Huby RJD. Langerhans cells and chemical allergy. *Curr Opin Immunol* 1998;10:614–619.
64. Schnyder B, Pichler WJ. Mechanisms of drug-induced allergy. *Mayo Clin Proc* 2009;84:268–272.
65. Naisbitt DJ, Gordon S, Pirmohamed M, Park B. Immunological principles of adverse drug reactions: The initiation and propagation of immune responses elicited by drug treatment. *Drug Saf* 2000;23:483–507.
66. Naisbitt DJ. Drug hypersensitivity reactions in skin: Understanding mechanisms and the development of diagnostic and predictive tests. *Toxicology* 2004;194:179–196.
67. Pichler WJ. Delayed drug hypersensitivity reactions. *Ann Intern Med* 2003;139:683–693.
68. Schmidt M, Raghavan B, Muller V, Vogl T, Fejer G, Tchaptchet S, Keck S, Kalis C, Nielsen PJ, Galanos C, *et al.* Crucial role for human Toll-like receptor 4 in the development of contact allergy to nickel. *Nat Immunol* 2010;11:814–819.
69. Kimber I, Basketter DA, Gerberick GF, Dearman RJ. Allergic contact dermatitis. *Int Immunopharmacol* 2002;2:201–211.
70. Usatine RP, Riojas M. Diagnosis and management of contact dermatitis. *Am Fam Physician* 2010;82:249–255.
71. Luebke R, House R, Kimber I. *Immunotoxicology and Immunopharmacology,* 3rd edn. Boca Raton, CRC Press/Taylor & Francis, 2007.

72. Grosen E, Siitari E, Larrison E, Tiggelaar C, Roecker E. Paclitaxel hypersensitivity reactions related to bee-sting allergy. *Lancet* 2000;355:288–289.
73. Hamad I, Hunter AC, Szebeni J, Moghimi SM. Poly(ethylene glycol)s generate complement activation products in human serum through increased alternative pathway turnover and a MASP-2-dependent process. *Mol Immunol* 2008a;46:225–232.
74. Hamad I, Christy Hunter A, Rutt KJ, Liu Z, Dai H, Moein Moghimi S. Complement activation by PEGylated single-walled carbon nanotubes is independent of C1q and alternative pathway turnover. *Mol Immunol* 2008b;45:3797–3803.
75. Laing RBS, Milne LJR, Leen CLS, Malcolm GP, Steers AJW. Anaphylactic reactions to liposomal amphotericin. *Lancet* 1994;344:682.
76. Moghimi SMH, I. Liposome-mediated triggering of complement cascade. *J Liposome Res* 2008;18:195–209.
77. Sculier JP, Coune A, Brassinne C, Laduron C, Atassi G, Ruysschaert JM, Fruhling J. Intravenous infusion of high doses of liposomes containing NSC 251635, a water-insoluble cytostatic agent. A pilot study with pharmacokinetic data. *J Clin Oncol* 1986;4:789–797.
78. Szebeni J, Alving CR, Rosivall L, Bünger R, Baranyi L, Bedöcs P, Tóth M, Barenholz Y. Animal models of complement-mediated hypersensitivity reactions to liposomes and other lipid-based nanoparticles. *J Liposome Res* 2007;17:107–117.
79. Szebeni J, Bedöcs P, Rosivall L, Bünger R, Horkay F, Tóth M. Animal models of complement-mediated hypersensitivity reactions to lipid nanoparticle-based drugs and drug solvents. *Mol Immunol* 2007;44:247–247.
80. Moghimi SM, Hamad I, Bünger R, Andresen TL, Jørgensen K, Christy Hunter A, Baranji L, Rosivall L, Szebeni J. Activation of the human complement system by cholesterol-rich and PEGylated liposomes-modulation of cholesterol-rich liposome-mediated complement activation by elevated serum LDL and HDL levels. *J Liposome Res* 2006;16:167–174.
81. Andersen AJ, Hashemi SH, Andresen TL, Hunter Ac, Moghimi SM. Complement: Alive and kicking nanomedicines. *J Biomed Nanotechnol* 2009;5:364–372.
82. Moghimi SM, Andersen AJ, Hashemi SH, Lettiero B, Ahmadvand D, Hunter AC, Andresen TL, Hamad I, Szebeni J. Complement activation cascade triggered by PEG-PL engineered nanomedicines and carbon nanotubes: The challenges ahead. *J Controlled Release* 2010;146:175–181.
83. Szebeni J, Baranyi L, Savay S, Chanan-Khan AA, Klein P, Koenig K, Muggia FM, Barenholz Y, Danino D, Talmon Y, *et al.* Hypersensitivity to Taxol and Doxil: Experimental and clinical evidence for a causal role of complement activation. *Immunopharmacology* 2000a;49:71–71.
84. Szebeni J, Baranyi L, Savay S, Milosevits J, Bunger R, Laverman P, Metselaar JM, Storm G, Chanan-Khan A, Liebes L, *et al.* Role of complement activation in hypersensitivity reactions to doxil and hynic PEG liposomes: Experimental and clinical studies. *J Liposome Res* 2002;12:165–172.
85. Zolnik BS, González-Fernández Á, Sadrieh N, Dobrovolskaia MA. Minireview: Nanoparticles and the immune system. *Endocrinology* 2010;151:458–465.
86. Szebeni J, Baranyi L, Savay S, Bodo M, Morse DS, Basta M, Stahl GL, Bünger R, Alving CR. Liposome-induced pulmonary hypertension: Properties and mechanism of a complement-mediated pseudoallergic reaction. *Am J Physiol Heart Circ Physiol* 2000b;279:H1319–H1328.

87. Szebeni J. Complement activation-related pseudoallergy caused by amphiphilic drug carriers: The role of lipoproteins. *Curr Drug Delivery* 2005b;2:443–449.
88. Szebeni J, Alving CR, Savay S, Barenholz Y, Priev A, Danino D, Talmon Y. Formation of complement-activating particles in aqueous solutions of Taxol: Possible role in hypersensitivity reactions. *Int Immunopharmacol* 2001;1:721–735.
89. Kreyling WG, Semmler-Behnke M, Möller W. Ultrafine particle-lung interactions: Does size matter? *J Aerosol Med* 2006;19:74–83.
90. Schmid O, Möller W, Semmler-Behnke M, A. Ferron G, Karg E, Lipka J, Schulz H, Kreyling WG, Stoeger T. Dosimetry and toxicology of inhaled ultrafine particles. *Biomarkers* 2009;14:67–73.
91. Scuri M, Chen BT, Catranova V, Reynolds JS, Johnson VJ, Samsell L, Walton C, Piedimonte G. Effects of titanium dioxide nanoparticle exposure on neuroimmune responses in rat airways. *J Toxicol Environ Health A* 2010;73:1353–1369.
92. Simeonova PP, Erdely A. Engineered nanoparticle respiratory exposure and potential risks for cardiovascular toxicity: Predictive tests and biomarkers. *Inhalation Toxicol* 2009;21:68–73.
93. Di Gioacchino M, Petrarca C, Lazzarin F, Di Giampaolo L, Sabbioni E, Boscolo P, Mariani-Costantini R, Bernardini G. Immunotoxicity of nanoparticles. *Int J Immunopathol Pharmacol* 2011;24:65S-71S.
94. Chen H-W, Su S-F, Chien C-T, Lin W-H, Yu S-L, Chou C-C, Chen JJW, Yang P-C. Titanium dioxide nanoparticles induce emphysema-like lung injury in mice. *FASEB J* 2006;20:2393–2395.
95. Morimoto Y, Ogami A, Todoroki M, Yamamoto M, Murakami M, Hirohashi M, Oyabu T, Myojo T, Nishi K-I, Kadoya C, *et al.* Expression of inflammation-related cytokines following intratracheal instillation of nickel oxide nanoparticles. *Nanotoxicology* 2010;4:161–176.
96. Larsen ST, Roursgaard M, Jensen KA, Nielsen GD. Nano titanium dioxide particles promote allergic sensitization and lung inflammation in mice. *Basic Clin Pharmacol Toxicol* 2010;106:114–117.
97. Nygaard UC, Hansen JS, Samuelsen M, Alberg T, Marioara CD, Løvik M. Single-walled and multi-walled carbon nanotubes promote allergic immune responses in mice. *Toxicol Sci* 2009;109:113–123.
98. Yanagisawa R, Takano H, Inoue K, Koike E, Kamachi T, Sadakane K, Ichinose T. Titanium dioxide nanoparticles aggravate atopic dermatitis-like skin lesions in NC/Nga mice. *Exp Biol Med* 2009;234:314–322.
99. Boscolo P, Bellante V, Leopold K, Maier M, Di Giampaolo L, Antonucci A, Iavicoli I, Tobia L, Paoletti A, Montalti M, *et al.* Effects of palladium nanoparticles on the cytokine release from peripheral blood mononuclear cells of non-atopic women. *J Biol Regul Homeost Agents* 2010;24:207–214.
100. Lu LK, Warshaw EM, Dunnick CA. Prevention of nickel allergy: The case for regulation? *Dermatol Clin* 2009;27:155–161.
101. Dernedde J, Rausch A, Weinhart M, Enders S, Tauber R, Licha K, Schirner M, Zügel U, von Bonin A, Haag R. Dendritic polyglycerol sulfates as multivalent inhibitors of inflammation. *Proc Natl Acad Sci U S A* 2010;107:19679–19684.
102. Bhol KC, Alroy J, Schechter PJ. Anti-inflammatory effect of topical nanocrystalline silver cream on allergic contact dermatitis in a guinea pig model. *Clin Exp Dermatol* 2004;29:282–287.

103. Bhol KC, Schechter PJ. Topical nanocrystalline silver cream suppresses inflammatory cytokines and induces apoptosis of inflammatory cells in a murine model of allergic contact dermatitis. *Br J Dermatol* 2005;152:1235–1242.
104. Bhol KC, Schechter PJ. Effects of nanocrystalline silver (NPI 32101) in a rat model of ulcerative colitis. *Dig Dis Sci* 2007;52:2732–2742.
105. Park HS, Kim KH, Jang S, Park JW, Cha HR, Lee JE, Kim JO, Kim SY, Lee CS, Kim JP, Jung SS. Attenuation of allergic airway inflammation and hyperresponsiveness in a murine model of asthma by silver nanoparticles. *Int J Nanomed* 2010;5:505–515.
106. Green MR, Manikhas GM, Orlov S, Afanasyev B, Makhson AM, Bhar P, Hawkins MJ. Abraxane®, a novel Cremophor®-free, albumin-bound particle form of paclitaxel for the treatment of advanced non-small-cell lung cancer. *Ann Oncol* 2006;17:1263–1268.
107. Nie S, Xing Y, Kim GJ, Simons JW. Nanotechnology applications in cancer. *Ann Rev Biomed Eng* 2007;9:257–288.
108. Li Y, Xiao K, Luo J, Lee J, Pan S, Lam KS. A novel size-tunable nanocarrier system for targeted anticancer drug delivery. *J Controlled Release* 2010;144:314–323.
109. Li N, Peng L-H, Chen X, Nakagawa S, Gao J-Q. Transcutaneous vaccines: Novel advances in technology and delivery for overcoming the barriers. *Vaccine* 2011;29:6179–6190.
110. des Rieux A, Fievez V, Garinot M, Schneider Y-J, Préat V. Nanoparticles as potential oral delivery systems of proteins and vaccines: A mechanistic approach. *J Controlled Release* 2006;116:1–27.
111. Csaba N, Garcia-Fuentes M, Alonso MJ. Nanoparticles for nasal vaccination. *Adv Drug Delivery Rev* 2009;61:140–157.
112. Demoly P, Bousquet J. Drug allergy diagnosis work up. *Allergy* 2002;57:37–40.
113. Bala S, Weaver J, Hastings KL. Clinical relevance of preclinical testing for allergic side effects. *Toxicology* 2005;209:195–200.
114. Basomba A, Sastre A, Peláez A, Romar A, Campos A, García-Villalmanzo A. Standardization of the prick test. *Allergy* 1985;40:395–399.
115. Ruëff F, Bergmann KC, Brockow K, Fuchs T, Grübl A, Jung K, Klimek L, Müsken H, Pfaar O, Przybilla B, *et al.* Hauttests zur Diagnostik von allergischen Soforttyp-Reaktionen (Skin tests for diagnostics of allergic immediate-type reactions. Guideline of the German Society for Allergology and Clinical Immunology). *Pneumologie* 2011;65:484–495.
116. Seibert SM, King TS, Kline D, Mende C, Craig T. Reliability of skin test results when read at different time points. *Allergy Asthma Proc* 2011;32:203–205.
117. Helm RM, Ermel RW, Frick OL. Nonmurine animal models of food allergy. *Environ Health Perspect* 2003;111:239–244.
118. Buchanan BB, Frick OL. The dog as a model for food allergy. *Ann N Y Acad Sci* 2002;964:173–183.
119. Helm RM, Furuta GT, Stanley JS, Ye J, Cockrell G, Connaughton C, Simpson P, Bannon GA, Burks AW. A neonatal swine model for peanut allergy. *J Allergy Clin Immunol* 2002;109:136–142.
120. McClain S, Bannon G. Animal models of food allergy: Opportunities and barriers. *Curr Allergy Asthma Rep* 2006;6:141–144.
121. Birmingham N, Payankaulam S, Thanesvorakul S, Stefura B, HayGlass K, Gangur V. An ELISA-based method for measurement of food-specific IgE antibody in mouse

serum: An alternative to the passive cutaneous anaphylaxis assay. *J Immunol Methods* 2003;275:89–98.
122. Gavériaux C, Renard P, Cammisuli S, Loor F. A comparison of five different methods for the detection of TNP specific mouse IgE: ELISA, ELISA on cells, rosetting, granule enzyme release assay and passive cutaneous anaphylaxis. *J Immunol Methods* 1986;93:107–114.
123. Gieni RS, Yang X, HayGlass KT. Allergen-specific modulation of cytokine synthesis patterns and IgE responses *in vivo* with chemically modified allergen. *J Immunol* 1993;150:302–310.
124. Gaboriau-Routhiau VR, Moreau M-C. Oral tolerance to ovalbumin in mice: Induction and long-term persistence unaffected by *Staphylococcus aureus* enterotoxin B and *Clostridium perfringens* Type A enterotoxin. *Pediatr Res* 1997;42:503–508.
125. Chen YL, Simons FE, Peng Z. A mouse model of mosquito allergy for study of antigen-specific IgE and IgG subclass responses, lymphocyte proliferation, and IL-4 and IFN-gamma production. *Int Arch Allergy Immunol* 1998;116:269–277.
126. Hill PN, Liu FT. A sensitive enzyme-linked immunosorbent assay for the quantitation of antigen-specific murine immunoglobulin E. *J Immunol Methods* 1981;45:51–63.
127. Hall TJ, Rittenberg MB. A novel *in vitro* assay for mouse IgE. *J Immunol Methods* 1985;80:145–154.
128. Hanashiro K, Nakamura M, Tamaki N, Kosugi T. Production of a monoclonal dinitrophenyl-specific rat IgE and establishment of an IgE capture ELISA for estimating the concentration of rat IgE antibodies to dinitrophenyl-Ascaris suum. *Int Arch Allergy Immunol* 1996;110:371–379.
129. Li XM, Schofield BH, Huang CK, Kleiner GI, Sampson HA. A murine model of IgE-mediated cow's milk hypersensitivity. *J Allergy Clin Immunol* 1999;103:206–214.
130. Li XM, Serebrisky D, Lee SY, Huang CK, Bardina L, Schofield BH, Stanley JS, Burks AW, Bannon GA, Sampson HA. A murine model of peanut anaphylaxis: T- and B-cell responses to a major peanut allergen mimic human responses. *J Allergy Clin Immunol* 2000;106:150–158.
131. Adel-Patient K, Créminon C, Bernard H, Clément G, Négroni L, Frobert Y, Grassi J, Wal JM, Chatel JM. Evaluation of a high IgE-responder mouse model of allergy to bovine β-lactoglobulin (BLG): Development of sandwich immunoassays for total and allergen-specific IgE, IgG1 and IgG2a in BLG-sensitized mice. *J Immunol Methods* 2000;235:21–32.
132. Hook WA, Powers K, Siraganian RP. Skin tests and blood leukocyte histamine release of patients with allergies to laboratory animals. *J Allergy Clin Immunol* 1984;73:457–465.
133. American College of Physicians. Allergy testing. *Ann Intern Med* 1989;110:317–320.
134. Sato S, Tachimoto H, Shukuya A, Ogata M, Komata T, Imai T, Tomikawa M. Ebisawa M. Utility of the peripheral blood basophil histamine release test in the diagnosis of hen's egg, cow's milk, and wheat allergy in children. *Int Arch Allergy Immunol* 2011;155(Suppl 1):96–103.
135. Kucharewicz I, Bodzenta-Lukaszyk A, Symanski W, Mroczko B, Szmitkowski M. Basal serum tryptase level correlates with severity of hymenoptera sting and age. *J Invest Allergol Clin Immunol* 2007;17:65–69.
136. Peterson CG, Hansson T, Skott A, Bengtsson U, Ahlstedt S, Magnussons J. Detection of local mast-cell activity in patients with food hypersensitivity. *J Invest Allergol Clin Immunol* 2007;17:314–320.

137. Anania A, Massobrio Am, Cascio B, Bosso E, Cascio G. Components of complement in patients with immediate hypersensitivity. *Minerva Med* 1998;89:77–81.
138. Gooi HC, Chapel H. *Clinical Immunology: A Practical Approach*. Oxford University Press, 1990.
139. Inoue K, Yanagisawa R, Koike E, Nishikawa M, Takano H. Repeated pulmonary exposure to single-walled carbon nanotubes exacerbates allergic inflammation of the airway: Possible role of oxidative stress. *Free Radical Biol Med* 2010;48:924–934.
140. Jesús M, Yuste JRPV. *Balcells. La Clinicay el laboratorio*, Elsevier España, 1989.
141. Coombs RRA, Mourant AE, Race RR. Detection of weak and "incomplete" Rh agglutinins: A new test. *Lancet* 1945;246:15–16.
142. Germuth Jr. FG. A comparative histologic and immunologic study in rabbits of induced hypersensitivity of the serum sickness type. *J Exp Med* 1953;97:257–282.
143. Vargas L, Sewell R, Marshall A, Galatioto J, Tsong Y-Y, Catterall JF, Hunnicutt GR. Immune complex deposition in adult male Sprague–Dawley rats chronically immunized with GnRH. *Am J Reprod Immunol* 2005;54:292–310.
144. Chhajlani V, Zingole S, Adwani S, Gothoskar, B. Circulating immune complexes in chronic myeloid leukemia: Turbidimetric measurement and two-dimensional electrophoretic analysis. *Leukocyte Res* 1983;7:565–573.
145. Theofilopoulos AN, Dixon FJ. Complement receptors on Raji cells as *in-vitro* detectors of immune complexes in human sera. In: Rose NR, Friedman H (eds.), *Manual of Clinical Immunology*, American Society for Microbiology, 1976, pp. 932.
146. Lightfoote M, Folks TM, Redfield R, Gold J, Marti GE, Kelly J, Sell KW. Flow-cytometric detection of circulating immune complexes. *J Immunol Methods* 1986;95:107–112.
147. Kelly RH, Scholl MA, Harvey VS, Devenyi AG. Qualitative testing for circulating immune complexes by use of zone electrophoresis on agarose. *Clin Chem* 1980;26:396–402.
148. Kurata Y, Hayashi S, Aochi H, Nagamine K, Oshida M, Mizutani H, Tomiyama Y, Tsubakio T, Yonezawa T, Tarui S. Analysis of antigen involved in circulating immune complexes in patients with idiopathic thrombocytopenic purpura. *Clin Exp Immunol* 1987;67:293–299.
149. Barbaud A. Skin testing in delayed reactions to drugs. *Immunol Allergy Clin North Am* 2009;29:517–535.
150. Barbaud A, Gonçalo M, Bruynzeel D, Bircher A. Guidelines for performing skin tests with drugs in the investigation of cutaneous adverse drug reactions. *Contact Dermatitis* 2001;45:321–328.
151. Gerberick GF, Ryan CA, Dearman RJ, Kimber I. Local lymph node assay (LLNA) for detection of sensitization capacity of chemicals. *Methods* 2007;41:54–60.
152. Cockshott A, Evans P, Ryan CA, Gerberick GF, Betts CJ, Dearman RJ, Kimber I, Basketter DA. The local lymph node assay in practice: A current regulatory perspective. *Hum Exp Toxicol* 2006;25:387–394.
153. Kimber I, Dearman RJ, Basketter DA, Ryan CA, Gerberick GF. The local lymph node assay: Past, present and future. *Contact Dermatitis* 2002;47:315–328.
154. U.S. Food and Drug Administration. Radioallergo sorbent Test (RAST) methods for Allergen-Specific Immunoglobin E (IgE) 510(k)s; Final Guidance for industry and FOA, 2001.

Chapter 18

Nanoparticles and Antigenicity

Marina A. Dobrovolskaia

Nanotechnology Characterization Laboratory
Advanced Technology Program, SAIC-Frederick, Inc.
NCI-Frederick, Frederick, MD 21702, USA
marina@mail.nih.gov

The immune system is highly versatile at recognizing foreign substances and mounting a multi-stage response against them. Antigenicity is a subtype of the immunogenic response characterized by the formation of an antibody specific to the given type of foreign substance, called an antigen. Due to their small size, nanoparticles are not antigenic themselves. However, some of them act as haptens and become antigenic when conjugated to a protein carrier. Antibodies against several types of engineered nanomaterials have been produced by the conjugation of the nanomaterial to a protein carrier and immunization in the presence of strong adjuvants. Conversely, engineered nanomaterials not conjugated to a protein, and those specifically designed to carry therapeutics proteins, were shown to be non-antigenic and moreover, helped to reduce the antigenicity of therapeutic proteins attached to them. However, accidental non-engineered nanomaterials which contaminate clinical formulations of therapeutic proteins are believed to contribute to their antigenicity. This chapter will review the scientific literature regarding the antigenic response to engineered nanomaterials, describe nanoparticle physicochemical properties contributing to antigenicity, with focus on the role of nanoparticle surface coating in antibody generation, and discuss the role of accidental nanoparticles in the antigenicity of therapeutic proteins.

1. Introduction

In health, the immune system functions to defend the host from pathogenic microorganisms (viruses, bacteria, fungi, etc.) and other invaders. The immune system effectively recognizes invading substances and builds a multi-stage response to inactivate and eliminate the foreign substances. Antibody generation is one form of the immune response against invading pathogens. An antibody is a protein produced by terminally differentiated B cells called plasma cells. Antibodies, also known as immunoglobulins, circulate in blood and lymph where they bind antigens. Antigens can have a diverse variety of natures; they can be proteins, lipids, polysaccharides, or other small and macromolecules which are foreign to the host (or in the case of autoimmune diseases, part of the host's own body). Each antibody is specific to one antigen; however, in instances where several antigens contain similar epitopes (the portion of the antigen recognized by the antibody), an antibody may cross-react with multiple antigens. When an antibody binds an antigen, the resulting immune complex (antibody–antigen complex) is eliminated from the circulation primarily through phagocytosis by the cells of the mononuclear phagocytic system (MPS), such as macrophages. The formation of an immune complex contributes to immunity in three main ways: 1) clearance of the foreign substance from circulation, 2) inactivation of the foreign substance, and 3) activation of the complement system to destroy the foreign substance. In traditional immunology, common foreign substances are pathogens (e.g., bacteria, viruses, fungi, worms). For the purpose of this chapter, I will consider nanoparticles and therapeutic proteins as foreign substances. There are different classes of antibodies, a review of all of which is beyond the scope of this chapter. Here, I will focus on immunoglobulin classes G (IgG) and M (IgM), because these two types of antibodies are the most well-studied with the most extensive literature, with respect to nanoparticles and therapeutic proteins.

In the field of therapeutic proteins, the term "immunogenicity" is often used to refer to the antigenic response to protein products. As the same term is often used with a broader meaning, with reference to the inflammatory properties of engineered nanomaterials, in this chapter, I will use word "antigenicity" to refer to antibody generation against both nanoparticles and therapeutic proteins.

The antigenicity of nanoparticles is drawing interest for several reasons: 1) some nanoparticles can serve as adjuvants due to their ability to increase the antigenicity of weak antigens, 2) many engineered nanoparticles are used as drug delivery vehicles and thus, are intentionally introduced into the human body, 3) many of these so-called nanomedicines (nanoparticle drug delivery systems) carry proteins, peptides, lipids, and nucleic acids as targeting moieties or as active therapeutic ingredients (APIs), 4) many nanoparticles with "bare" surfaces — surfaces which are not protected by polyethylene glycol (PEG) or other polymers to avoid immune recognition — interact with and bind to plasma proteins, 5) there is data showing that some nanoparticles behave as haptens and become antigenic after binding to a protein carrier, and lastly, 6) some nanoparticles have been shown to enhance the antigenicity of therapeutic proteins.

Growing evidence supports that nanoparticle adjuvanticity is determined by the particle's physicochemical properties such as size and surface charge.[1–3] This area is being actively explored for the development of vaccines, as discussed in Chapter 15. Here, the focus is on nanoparticles, which are neither intended to be antigenic nor to be adjuvants. By "intended," I annotate specifically designed or engineered to stimulate antibody generation to either themselves or to the therapeutic payload they carry. Currently, there is sparse data in the literature describing engineered nanomaterials approved for clinical use, which can cause an antigenic response. Likewise, there is no data demonstrating that engineered nanomaterials carrying therapeutic protein payloads result in the formation of anti-nanoparticle or anti-therapeutic protein antibodies. Several studies have reported that the injection of some nanomaterials, e.g., poly(amido amine) (PAMAM) dendrimers, gold nanoparticles, fullerene derivatives, did not result in antibody formation.[4–8] However, the conjugation of these engineered nanoparticles to a protein carrier (serum albumin or cytokine) and the injection of this conjugate into animals in the presence of a strong adjuvant resulted in the formation of an anti-nanoparticle antibody.[9–13] There are a few studies demonstrating that engineered nanomaterials which carry therapeutic proteins not only do not cause formation of anti-therapeutic protein antibodies, but actually assist in problems with anti-therapeutic protein-neutralizing antibodies formed by the injection of the non-nano formulation in the same protein.[14–17] There is a growing body of data suggesting that some accidental nanoparticles (e.g., glass and cellulose fibers, tungsten, silicon oil) contaminating therapeutic protein formulations contribute to patients' antigenic response to the therapeutic proteins.[18–23] A summary of this data is presented in Figure 1 and will be reviewed in more detail below.

2. Antigenic Response to Engineered Nanomaterials

An antigenic response to engineered nanomaterials is a concern because in theory, the generation of antibodies to nanoparticles may 1) affect the efficacy of the particle-based nanomedicines, 2) lead to immunotoxicities associated with the formation of antibody–antigen complexes, 3) lead to allergic reactions in the case of IgE-type responses, and 4) in the case of nanoparticles which carry therapeutic biomolecules, cause neutralization of a patient's own biomolecules. The last effect is of great concern because it may result in the

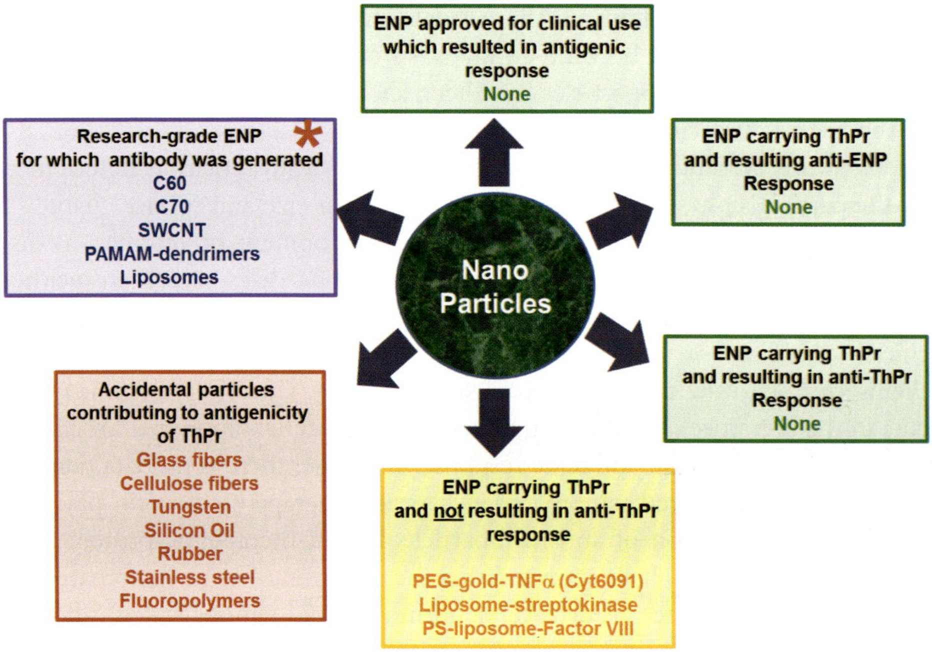

Figure 1. Antigenicity of nanoparticles. Each nanoparticle is unique. Most engineered nanoparticles are not antigenic. In order to generate particle-specific antibodies, these nanomaterials have to be conjugated to a protein carrier and injected in the presence of strong adjuvants. Nanoparticles can be engineered to carry therapeutic proteins and prevent anti-therapeutic protein antibody formation, while some accidental nanoparticles contaminating non-nano therapeutic protein formulations contribute to the antigenicity of these therapeutic proteins.

ENP: engineered nanoparticle, (*these particles were either conjugated to a protein carrier such as bovine serum albumin (BSA) and administered with a strong adjuvant or administered with an adjuvant without a protein carrier), ThPr: therapeutic protein, PS: phosphatidylserine, none: currently there is no data in the public domain. This diagram is prepared based on Refs. 9–19, 21, 22, 24–40.

elimination of both particle-based and endogenous ligands, and lead to life-threatening conditions akin to effects observed with biotechnology-derived pharmaceutics.

The current literature suggests that some engineered nanoparticles behave as haptens, i.e., they are not antigenic unless they are conjugated to a protein carrier (Table 1), though the successful generation of antibodies using a protein-free formulation of particulate antigens has been described for liposomes (Table 1). However, strong adjuvants were required in both protein-containing and protein-free formulations, which caused the generation of particle-specific antibodies (Table 1). These studies are reviewed in greater

Table 1. Summary of studies describing generation of nanoparticle specific antibodies. Conjugation to a protein carrier and use of adjuvants was necessary to induce an antibody response to all engineered nanomaterials except for some liposomes.

Nanoparticle	Protein carrier	Adjuvant	Animal	Antibody	Specificity/Cross-reactivity	Reference
Chol-rich liposome	None	Lipid A	Rabbits	IgG and IgM	Cholesterol/UKN	35
PAMAM dendrimers, G0, amine-term.*	BSA	CFA/IFA	Mice	IgG	NH_2-terminal groups/G0-G10 unmodified PAMAM dendrimers	9
PAMAM dendrimers G5, sulfhydryl-term.*	IL-3	CFA/IFA	Mice	IgG	Sulfhydryl terminal groups/UKN	9
PAMAM dendrimers, G5, oxiamine-term.*	IL-3	CFA/IFA	Mice	IgG	Oxiamine terminal groups/UKN	9
(1,2-methanofullerene C_{60})-61-carboxylic acid	BSA, PTG, or STI	None	Mice	IgG	Methanofullerene C_{60}/C_{60} fullerene aminocaproic acid; fullerenol; C_{70} fullerene; MWCNT	24
Various fullerene C_{60}-derivatives	BTG, BSA, or RSA	CFA	Mice	IgG1 kappa	C_{60}/C_{70}, SWCNT	10, 11, 13
Gold NP*,#	NA	NA	Humans	Fv	Gold	41
DMPC:Chol: DCP-liposomes	None	Lipid A	Mice	IgM	PC	36–38
PI:DMPC: Chol	None	None	Rabbits infected with *T. rhodesiense*	IgM	PI	40

Abbreviations

BSA	bovine serum albimun	DMPC	dimyristoyl phosphatidylcholine	PI	phosphatidylinositol
BTG	bovine thyroglobulin	IFA	incomplete Freund's adjuvant	PTG	porcine thyroglobulin
CFA	complete Freund's adjuvant	MWCNT	multiwalled carbon nanotubes	RSA	rabbit serum albumin
Chol	cholesterol	NA	not applicable; these antibodies were selected from phage display library	STI	soy trypsin inhibitor
DCP	dicetylphosphate	NP	nanoparticles	Term	terminated
				UKN	unknown

*The same particles not conjugated to a protein carrier and administered with adjuvant were not antigenic.

The antibody is not the result of immunization, but discovered through phage display.

detail later in the chapter, and I will thoroughly discuss two hypothetical models on the possibility of generating of nanoparticle-specific antibodies. The reader should note that both these models are logic-driven and based on reviewing experimental details and published data demonstrating successful generation of nanoparticle-specific antibodies. These models have not been experimentally verified.

Chen *et al.* immunized BALB/c mice with fullerene (C_{60}) conjugated with thyroglobulin, bovine serum albumin (BSA), rabbit serum albumin, or 3L peptide (Lys–Lys–Lys*3HCl).[11] This immunization resulted in a polyclonal response consisting of antibodies specific to C_{60} fullerene and a subpopulation that cross-reacted with a C_{70} fullerene derivative. The same group demonstrated later that an anti-C_{60} specific monoclonal antibody may be produced using a standard immunization procedure, but conjugation to the protein carrier was required.[10] Monoclonal 1–10F-A8 anti-C_{60} antibodies generated in this study were determined to be IgG1 kappa with affinity to fullerene in the nanomolar range and a fullerene binding site consisting of four hydrophobic amino acids. In a parallel study conducted by another group in rabbits, four water-soluble fullerene derivatives (C_{60}–serine, C_{60}–alanine, C_{60}–aminocapronic acid, and C_{60}–alanine dipeptide) were tested for immunogenic, allergenic, and adjuvant properties.[5] The C_{60} derivatives used in this study did not induce any antibody response even when emulsified in complete Freud's adjuvant. The C_{60} derivatives also did not induce any immune response when the experiments were repeated in mice.[5] The discordance of results from studies by Erlanger's lab[10,11,13] with those published by Andreev *et al.*[5] may be partly explained by the different choice of C_{60} derivatives. Variation in the genetic backgrounds of animal models — Braden *et al.*[10] and Chen[11] used BALB/c mice whereas Andreev *et al.*[5] used C57BL/6 — may also contribute to the different outcomes. However, the key difference is that Erlanger's[10,11,13] studies used a protein carrier, while Andreev's[5] did not. This is further confirmed by a recent study published by Hendrickson *et al.* who also were able to induce the generation of the anti-C_{60} antibody by using C_{60}–thyroglobulin conjugates and complete Freud's adjuvant.[24] In both Erlanger's lab studies[10,11,13] and the most recent study by Hendrickson *et al.*,[24] antibodies generated against C_{60} also cross-reacted with other carbon nanomaterials, such as C_{70} fullerene and single-walled and multi-walled carbon nanotubes, but not with polymeric nanoparticles containing similar surface groups. This data suggested the existence of a common epitope within the carbon core, which is not related to the surface functional groups.

Dendrimers are another class of nanomaterials facing controversy over antigenicity. An early study on PAMAM dendrimers did not reveal any overt immunogenicity of generation 3 (G3), G5, and G7 amino-terminated dendrimers.[8] However, later, G0 amine-terminated, and G5 sulfhydryl-terminated and oxiamine-terminated PAMAM dendrimers conjugated to BSA or to cytokines were used for immunization in the presence of Freund's adjuvant and generated a strong antibody response.[9] Consistent with the previous study, Lee *et al.* observed no antigenic response to plain, unconjugated dendrimers, irrespective of their size or surface functionalization.[9] In contrast to anti-fullerene antibodies, the epitopes recognized by the antibodies to PAMAM dendrimers were related to surface functional groups rather than to the core. For example, the antibody generated using G0 amine-terminated PAMAM dendrimers cross-reacted with amine-terminated PAMAM dendrimers of greater generations, but did not bind to PAMAM dendrimers containing sulfhydryl or oxiamine terminal groups.[9] The antibody generated to G5 sulfhydryl-terminated PAMAM dendrimers bound only to G5 sulfhydryl-terminated PAMAM dendrimers and did not cross-react with G5 oxiamine-terminated PAMAM dendrimers. Likewise, the antibody generated to G5 oxiamine terminated dendrimers did not cross-react with PAMAM dendrimers with other terminal groups.[9] Of interest is the data demonstrating that terminal groups *per se* are not enough for the anti-dendrimer antibody to bind and that dendrimer architecture contributes to epitope formation. Lee *et al.* have shown that anti-G0 amine-terminated PAMAM dendrimer antibody did not bind to polytriethyleneamine (POPAM) dendrimers, which also has primary amines on its surface, but a dendrimer architecture distinct from that of PAMAM dendrimers.[9] Similar to PAMAM dendrimers, polypropyleneimine (PPI) dendrimers were also not antigenic.[4] This study did not attempt to generate an anti-PPI dendrimer antibody using dendrimer–protein conjugates.

In contrast to fullerenes and dendrimers, antibodies to gold nanoparticles were obtained by the screening of combinatorial human antibody libraries using a phage display method.[41] This approach allowed for the selection of A14P-b2 Fv fragments, which showed strong binding to the gold surface and did not cross-react with other noble metals (e.g., silver, platinum, palladium). Other antibody fragments isolated from the same library (anti-lysozyme fragment HyHEL10, anti-CD16 fragment 3G8, and anti-polyhydroxybutirate fragment PH7–3d3) did not bind to the gold surface, further confirming the specificity of the A14P-b2 fragment to gold.[41] Another important message from the study by Watanabe *et al.*[41] is that our immune system is highly versatile and may contain antibodies to virtually all nanomaterials; however, the isolation of such antibodies is only possible using precise and highly specialized techniques such as phage display.

The presence of these antibodies is undetectable when nanoparticles are administered into the body because the magnitude of this response is not strong enough to initiate the activation of T cells and B cells, class-switch recombination (CSR), and expansion. However, a completely different situation exists when the same nanoparticles (e.g., dendrimers and fullerenes as described above) are conjugated to protein carriers. Antibodies located on the surface of B cells recognize nanoparticles and take them up along with the protein carrier, and the latter is then digested, processed, and presented to T helper cells along with MHC-II (major histocompatibility complex class II). Both T helper and B cells become activated, and this results in the production of cytokines by T lymphocytes and antibodies against nanoparticles by differentiated B cells (plasma cells). This hypothetical sequence of events is shown in Figure 2 and was also proposed and discussed previously.[42] This response is strong enough to be detected; the protein carrier is essential for anti-nanoparticle antibody generation at detectable levels.

The antigenicity of liposomes is completely different from that of other nanomaterials studied to date. This difference may be partially explained by their composition, in that liposomes are made of common, naturally occurring lipids, which are also present in the membrane of most cells. For example, antibodies are not only generated in response to dimyristoyl phosphatidylcholine (DMPC)/cholesterol/dicetylphosphate (DCP)/lipid A liposomes bound to phosphatidylcholine and cholesterol, but also reacted with membrane phospholipids of cultured macrophages.[39] This area was extensively studied by Alving *et al.*[35–40,43] Unlike fullerenes and dendrimers, which required the conjugation of the nanoparticle to a protein carrier, the generation of an anti-liposome antibody was possible using protein-free formulations of liposomes as antigens[35] and even occurred spontaneously under certain inflammatory conditions.[40] Similar to fullerenes and dendrimers, the induction of the anti-liposome antibody required the presence of adjuvants during immunization. In one study, antibodies were generated against the cholesterol component of the liposome.[35] These antibodies belong to both IgG and IgM classes, suggesting the activation of the classical immune response, in that a first, low affinity IgM antibody was formed, followed by the maturation of the immune response and production of IgG. Both antibodies resulted in complement-dependent lysis of cholesterol-based liposomes, were able to bind non-oxidized crystalline cholesterol *in vitro,* and neutralize blood cholesterol in diet-induced hyperlipidemia *in vivo.*[35] The approach was so successful in animal models of hypercholesterolemia that these liposomes were proposed for use as a cholesterol-lowering vaccine. In another study, antibodies against the following liposomal constituents were detected in rabbits infected with

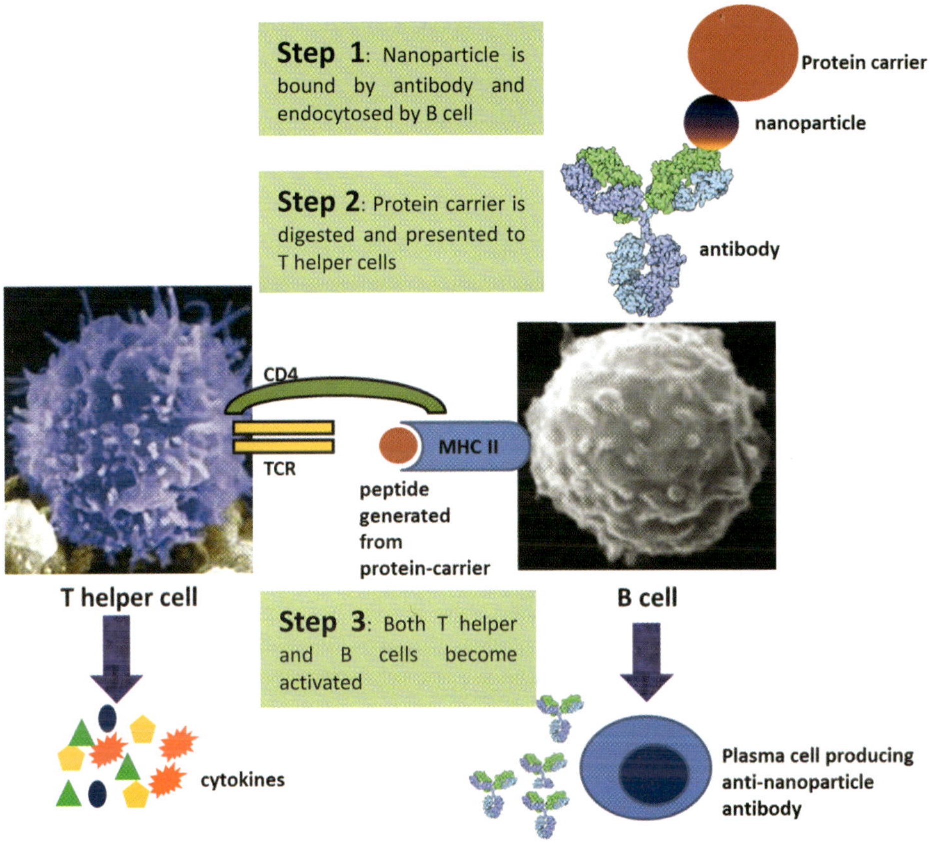

Figure 2. Hypothetical model for the role of a protein carrier in the generation of nanoparticle-specific antibodies. Nanoparticles are not antigenic, although antibodies recognizing nanoparticles exist. The number of B cells carrying nanoparticle-specific antibodies is low, and not sufficient for a potent response when a nanoparticle is introduced into the circulation. However, when nanoparticles are bound to a protein carrier, the endocytosis of such particles through particle-reactive antibodies on B cell surfaces results in the uptake of the protein carrier and generation of peptides. The latter stimulates T helper cells, which produce cytokines. As a result, both T helper cells and B cells become activated and this leads to the expansion of B cells, producing nanoparticle-specific antibodies. (This model is a logic-driven interpretation of published studies and has not been validated experimentally.)

Trypanosoma rhodesiense: DMPC, phosphatidylinositol (PI), phosphatidylinositol phosphate (PIP), and cardiolipin.[40] The liposomes in this study did not contain adjuvants or proteins, but an acute infection in animals caused by trypanosome primed the immune system and thus, can be considered like an adjuvant. Interestingly, in another study, anti-liposome antibodies were induced in both immunologically normal and immune-compromised mice

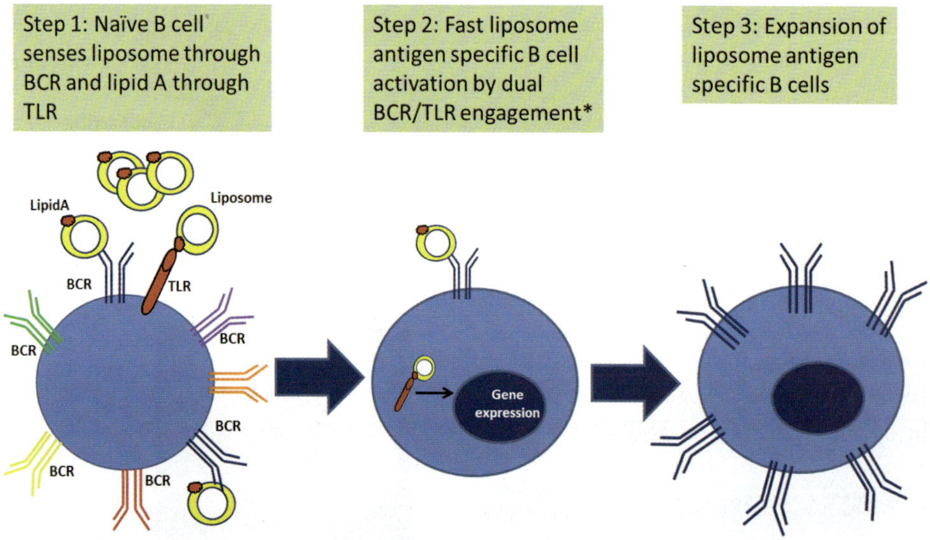

Figure 3. Hypothetical model explaining generation of anti-liposome antibodies through the T cell-independent route. Naïve B cells recognize liposomal components through the BCR and the lipid A component through TLR. This dual signal leads to an antigen-specific activation and expansion of the liposome–antigen-specific B cells.

BCR: B cell receptor (IgM), TLR: Toll-like receptor. (This model is logic-driven and has not been proven experimentally.)

injected with the lipid A portion of bacterial lipopolysaccharide as an adjuvant.[38] The induction of antibodies in athymic mice suggests that this antibody response is independent of T cells. The fact that antibody generation was not possible without lipid A and the type of antibody was IgM, may suggest that anti-liposome antibodies result from a type 1 T cell-independent activation of B cells. The mechanism of this response may be similar to that described by Zinkernagel *et al.*[44] for artificial virus-like particulate vaccines — the liposome–antigen is bound by the IgM B cell receptor and Toll-like receptor (TLR) signaling provides a second stimulus, leading to the activation and expansion of B cells specific to the given liposome–antigen, except that CSR does not occur (Figure 3). Lipopolysaccharide and its lipid A portion are known B cell mitogens,[45] which can further explain the necessity of lipid A for the positive antibody outcome.

3. Antigenic Response to Nanoparticle Surface Coatings

All published data describing antigenic responses to nanoparticle surface coatings come from studies of PEGylated liposomes. In repeated administration studies with PEG-coated liposomes, several research groups observed the rapid clearance of the second dose from the circulation.[30,46,47] The observed phenomenon was called accelerated blood clearance (ABC). However, there is a large discrepancy among research groups regarding the mechanism of this phenomenon. Multiple studies by Kiwada's group[25–33,48–53] and a study by Kaminskas *et al.*[54] attributed ABC to the generation of PEG-specific antibodies of the IgM class, while a study by Dams *et al.*[46] concluded that the protein responsible for the ABC of PEG–liposomes does not belong to any immunoglobulin class. Independent studies by Laverman in rats, rabbits, and humans,[55] and by Ishida *et al.*[29] in rats, suggested that splenic and hepatic macrophages play a key role in the accelerated clearance of PEGylated liposomes. The summary of these studies is presented in Table 2 and will be further discussed below.

In all studies, ABC was accompanied by the accumulation of liposomes in the organs of the MPS (the liver and the spleen) and was reported for both PEGylated and conventional liposomes.[28,46,51–53] Moreover, the depletion of phagocytic cells attenuated accelerated clearance, suggesting that the MPS may be involved.[29,55] In animal models (mice, rats, monkeys, rabbits) *in vivo*, it has been shown that ABC 1) is inhibited by the presence of cytotoxic agents,[27,53] 2) is enhanced by the presence of an immunogenic payload (e.g., siRNA),[48–50] 3) is independent of T cells,[27,53] 4) requires functional hepatic and splenic macrophages,[27,47,55] 5) depends on liposomal composition,[26] 6) depends on particle size and surface charge,[52,54,57] 7) depends on the time interval between the first and second injections,[54,58] 8) depends on PEG surface density,[26] 9) depends on the lipid dose[26,55] and formation of anti-PEG IgM, and 10) depends on the species.[25–33,54,55,58] Some of these factors (doses, interval between first and second doses, particle size, charge, liposomal composition, PEG density, anti-PEG IgM formation, and species specificity) are reviewed in more detail below, mainly because there is some controversy among studies from different research groups regarding these factors.

3.1. *Dose*

Several studies by Kiwada *et al.* suggested that anti-PEG antibody generation and ABC are achievable at concentrations below 1 μM/kg, and that higher concentrations inhibit ABC.[26,27,52] In contrast, reports from two other independent groups demonstrated ABC and antibody generation at higher liposomal concentrations (5–7 μM/kg).[47,54] Interestingly, one study from Kiwada *et al.* also observed ABC at high concentrations (10 and 25 μM/kg).[28]

Table 2. Summary of studies describing accelerated blood clearance (ABC) of PEG-modified liposomes.

Particle	Composition (molar ratio)	Polymer	Loaded API	Synthesis method	Size, nm/ charge, mV	ABC (1st dose/ 2nd dose)	Anti-polymer IgM	Reference
Liposome	PEG2000–DSPE: L-a-PC:Chol (0.15:1.5:1.0)	mPEG	none	Extrusion through polycarbonate filters	96/NT	Yes (liposome/ liposome) and (liposome/ micelle)	Yes	54
Micelle	PEG2000–DSPE: soyPC (75:25)	mPEG	none	Extrusion through polycarbonate filters	18/NT	Yes (micelle/ liposome) No (micelle/ micelle)	No*	54
PLA–NP	PEG–PLA	mPEG	none	Ring-opening polymerization	133/–1.3	Yes (PEG–PLA/ PEG–PLA) No (PEG–PLA/ PVP–PLA)	Yes	32
PLA–NP	PVP–PLA	(t)PVP	none	Radical polymerization	122/–3.0	No (PVP–PLA/ PVP–PLA) No (PVP–PLA/ PEG–PLA)	No	32
PLA–NP	PAcM–PLA	(t)PAcM	none	Radical polymerization	125/–2.7	No (PAcM–PLA/ PAcM–PLA)	NT	32
PLA–NP	PDMAA–PLA	(t)PDMAA	none	Radical polymerization	131/–2.6	No (PDMAA–PLA/ PDMAA–PLA)	NT	32

(*Continued*)

Table 2. (*Continued*)

Particle	Composition (molar ratio)	Polymer	Loaded API	Synthesis method	Size, nm/ charge, mV	ABC (1st dose/ 2nd dose)	Anti-polymer IgM	Reference
Lipid-wrapped lipoplex (WS)	DC-6-14:POPC (50:50):mPEG–DSPE:siRNA	mPEG	siRNA	Extrusion through polycarbonate filters followed by incubation with si-RNA in 9% sucrose	284.1/ +11.5	Yes (PEG–WS/ PEG–WS)	Yes	50
Liposome	PEG–DSPE:PHEPC: chol:HYNIC–DSPE (0.15:1.85:1:0.07)	PEG	None	Extrusion through polycarbonate filters	80–85/NT	Yes (liposome/ liposome)	No	46
Liposome	PHEPC:Chol (1.85:1)	none	None	Extrusion through polycarbonate filters	100/NT	Yes (liposome/ liposome)	No	46
Liposome	PEG–DSPE:PHEPC: Chol (0.15:1.85:1)	PEG	None	Extrusion through polycarbonate filters	400/NT	Yes (liposome/ liposome)	No	46
Liposome	PEG:DSPE:PHEPC: Chol (0.15:1.85:1)	PEG	None	Extrusion through polycarbonate filters	85/NT	Yes (liposome/ liposome)	No	46

(*Continued*)

Table 2. (*Continued*)

Particle	Composition (molar ratio)	Polymer	Loaded API	Synthesis method	Size, nm/ charge, mV	ABC (1st dose/ 2nd dose)	Anti-polymer IgM	Reference
Liposome	DPPC:Chol:PEG–DSPE (10:5:1)	mPEG	None	Extrusion through polycarbonate filters	100/NT	Yes (liposome/ liposome)	Yes	53
Liposome	HSPC:Chol:DC-6-14:PEG–DSPE (2:1:0.2:0.2)	mPEG	1-OHP	Extrusion through polycarbonate filters	201/11.3	No	NA	56
Liposome	HSPC:Chol:PEG–DSPE (2:1:0.2)	mPEG	1-OHP	Extrusion through polycarbonate filters	202/–6.3	No	NA	56
Lipid NP	L-PLA(20,000)	PEG	PGE1	Oil-in-water solvent diffusion	110–154/ NT	Yes (lipid NP/ lipid NP)	Yes	33
Liposome	HEPC:Chol:mPEG 2000–DSPE (1.85:1:0.15)	mPEG	None	Extrusion through polycarbonate filters	NT/NT	Yes (liposome/ liposome)	Yes	30
Liposome	DC-6-14:POPC: Chol:DOPE (10:30:30:30)	mPEG	CpG-DNA	Extrusion through polycarbonate filters	315.3/13.1	Yes (liposome/ liposome)	Yes	49

(*Continued*)

Table 2. (*Continued*)

Particle	Composition (molar ratio)	Polymer	Loaded API	Synthesis method	Size, nm/ charge, mV	ABC (1st dose/ 2nd dose)	Anti-polymer IgM	Reference
Liposome	HEPC:Chol:mPEG 2000–DSPE (1.85:1.0: 0.15)	mPEG	None	Extrusion through polycarbonate filters	112/NT	Yes (liposome/ liposome)	Yes	31
Liposome	HEPC:Chol:mPEG 2000–DSPE (1.85:1.0: 0.15)	mPEG	None	Extrusion through polycarbonate filters	115/NT	Yes (liposome/ liposome)	Yes	51
Liposome	HSPC:Chol:mPEG 2000–DSPE (1.49:1:0.134)	mPEG	None	Extrusion through polycarbonate filters	110/NT	Yes (liposome/ liposome)	Yes	25
Liposome	HEPC:Chol:mPEG 2000–DSPE (1.85:1.0: 0.15)	mPEG	None	Extrusion through polycarbonate filters	107/–1.51	Yes (liposome/ liposome)	Yes	27
Liposome	HEPC:Chol:mPEG 2000–DSPE (1.85:1.0: 0.15)	mPEG	None	Extrusion through polycarbonate filters	113/NT	Yes (liposome/ liposome)	NT	26
Liposome	HEPC:Chol:mPEG 5000–DSPE (1.85:1.0: 0.15)	mPEG	None	Extrusion through polycarbonate filters	109/NT	Yes (liposome/ liposome)	NT	26

(*Continued*)

Table 2. (*Continued*)

Particle	Composition (molar ratio)	Polymer	Loaded API	Synthesis method	Size, nm/ charge, mV	ABC (1st dose/ 2nd dose)	Anti-polymer IgM	Reference
Liposome	HEPC:Chol:mPEG 2000–DSPE (1.85:1.0: 0.15)	mPEG	None	Extrusion through polycarbonate filters	60/−5 & 110/−1.8	Yes (PEG–liposome/PEG–liposome)	NT	52
	HEPC:Chol (2:1)				60/−2.4 & 110/−2.7	Yes (un-PEGylated liposome/PEG–liposome		
	HEPC:Chol:DCP (5:4:1)				60/−18.9 & 110/−21.5	Yes (negatively charged liposome/PEG–liposome)		
	HEPC:Chol: SA (5:4:1)				60/3.48 & 110/4.93	No (PEG-liposome/un PEGylated liposome		
						Yes (positively charged lispome/ PEG–liposome)		
						Yes (PEG–liposome/ positively charged liposome)		

(*Continued*)

Table 2. (*Continued*)

Particle	Composition (molar ratio)	Polymer	Loaded API	Synthesis method	Size, nm/ charge, mV	ABC (1st dose/ 2nd dose)	Anti-polymer IgM	Reference
						Yes (small PEGylated liposome/large PEGylated liposome) No (small unPEGylated liposome/large PEGylated liposome)		
Liposome	HEPC:Chol:mPEG 2000–DSPE (1.85:1.0: 0.15) HEPC:Chol:mPEG 5000-DSPE (1.85:1:0.15)	mPEG	None	Extrusion through polycarbonate filters	100/−1.5	Yes (dose-, interval-, lipid content- and PEG length dependent; size- and charge-independent)	NT	28

Abbreviations

API	active pharmaceutical ingredient
Chol	cholesterol
DPPC	dipalmitoylphosphatidylcholine
DSPE	1,2-diastearoyl-sn-glyceto-3-phosphoethanolamine
HYNIC	N-hydroxysuccinimidyl hydrazine nicotinate hydrochloride
IgM	immunoglobulin M
mPEG	methoxy(polyethylene glycol)
No*	micelles stimulate limited IgM production but are not substrate for IgM binding
NT	not tested
OHP	oxaliplatin
PAcM	poly(4-acryloylmorpholine)
PC	phosphatidylcholine
PDMAA	poly(N,N-dimethylacrylamide)
PEG	polyethylene glycol
PGE1	prostaglandin E1
PHEPC	partially hydrogenated egg phosphatidylcholine
PLA	poly(lactic acid)
PVA	poly(vinyl alcohol)
PVP	poly(N-vinyl-2-pyrrolidone)
(t)	thiol was the functional group at the end of PLA to attach coating
WS	wrapsome referring to lipid wrapped lipoplex complex

3.2. *Interval between first and second doses*

With a one-week interval between injections, more than an 85-fold decrease in plasma levels of PEG-liposomes 4 h after injection of the second dose was observed; however, such an enhancement in blood clearance was less pronounced after the third dose and was not observed at all after the fourth dose.[46] Another study reported an approximately 40-fold increase in the clearance of the second dose administered one week after the first dose.[54] In contrast, Ishida *et al.* reported only a slight decrease in blood liposomal concentration when the second injection was performed on Day 7.[28] In this study, ABC was observed when the second injection was performed 10 days after the first dose, and no change in clearance was seen on Day 14.[28] The prolonged interval (six weeks) between the doses may explain the lack of ABC observed in a rabbit study.[59]

3.3. *Size*

Koide *et al.* reported the accelerated clearance of PEG–liposomes (large particles) after the initial administration of PEG–micelles (small particles),[57] while the most recent study by Kaminskas *et al.* demonstrated that pre-treatment with PEG–liposomes induced antibody formation and the ABC of the second dose of PEG–liposomes and PEG–micelles, while prior exposure to PEG-micelles did not lead to the ABC of either PEG–micelles or PEG–liposomes, though a low titer of anti-PEG antibody was observed.[54] This data led the authors of the study to suggest that the anti-PEG antibody generated in response to micelles do not bind PEG associated with the micelles, but rather, recognize the more defined epitope present in PEG–liposomes.[54] Another potential explanation of the differences between the studies may be in the study design — the former study relied on liposomal distribution to the liver as a measure of clearance, while the latter study assessed plasma PK and measured anti-PEG IgM by ELISA. This second approach may be more sensitive for detecting clearance than the assessment of liposomal distribution to liver.

3.4. *Charge*

Wang *et al.* reported that the surface charge of the first dose of liposomes is not critical for the induction of ABC, as there was no difference in clearance rates when cationic, anionic, and neutral liposomes were tested. However, the surface charge of the second dose was important, in that only neutral PEGylated liposomes were affected.[52]

3.5. *Liposomal composition*

ABC and IgM formation were reported in response to only PEGylated liposomes, while the clearance of liposomes coated with another hydrophilic polymer, polyvinylpyrrolidone (PVP), was not affected.[32] Of interest, different reactive groups were used to incorporate different polymers into liposomes, in that methoxy-PEG–hydroxyl was used to make PEG–liposomes and PLA-thiol was used to make PVP-coated particles. Different immunogenicities of thiol- and methoxy-terminated polymers were demonstrated in other studies[60,61] and will be further discussed later.

3.6. *PEG length and density*

No differences in accelerated clearance were reported for liposomes coated with PEG of various lengths.[26] An increase in the concentration of PEG–lipid from 5 mol% to 10 or 15 mol%, which presumably resulted in an increased density of the PEG coating, attenuated accelerated clearance.[26]

3.7. *Anti-PEG IgM*

The original study by Dams *et al.* verified the need for a serum factor in the ABC of PEGylated liposomes by transfusing serum from animals injected with PEGylated liposomes one week after injection into naïve animals receiving a first dose of PEG liposomes.[46] Serum from primed animals was capable of inducing ABC similar to that induced by prior injection of PEGylated liposomes into the same animals. They further separated proteins from sera of treated and untreated animals using traditional polyacrylamide gel electrophoresis (PAGE), and identified a band with molecular mass of 150 kD in the sera of animals with accelerated clearance. However, the transfusion of sera to treated animals after the depletion of IgG and IgM from naïve animals still induced accelerated clearance, which led the authors to conclude that anti-PEG IgG and IgM are not involved in the observed ABC phenomenon.[46] In contrast to this study, all studies conducted by Kiwada *et al.*[25–33,58] and that by Kaminskas *et al.*[54] reported the formation of anti-PEG IgM. However, the approach used to verify the involvement of immunogolubulins in ABC was different in these later studies. To detect IgM and IgG, these researchers used traditional ELISA. For plate coating, PEGylated lipid (e.g., PEG–DSPE) was dissolved in an organic solvent (ethanol) and the coating included the incubation of the test-plate at room temperature until the organic solvent evaporated. The immobilization of PEGylated lipid on the ELISA plate was not verified. The evaluation of specificity and sensitivity, as well as verification with PEG-standard, were not performed in these assays either. An increase in OD values was observed in the well containing sera from treated animals, and this was interpreted as a PEG-specific antibody response with the assumption that PEG was immobilized on the ELISA plate and was not lifted during the blocking and washing steps. The OD values of positive sera were relatively low (0.4 *vs.* 0.1 in control). Therefore, it is very likely that the observed positive response may be the result of matrix interference commonly experienced with ELISAs.[62–68] Studies reporting the induction of accelerated clearance of PEG–liposomes by conventional (unPEGylated liposomes) further suggested that the reported antibody response is not likely to be specific to PEG.[28]

3.8. *Species specificity*

Most studies by Kiwada's group,[25,26,29,31,32,51] the study by Kaminskas *et al.*,[54] and that by Leverman *et al.*[47] were conducted in Wistar or Sprague–Dawley rats. Accelerated clearance was also reported in mice.[28,48–50,53,56] Interestingly, the initial study by Dams *et al.* reported that accelerated clearance is observed in rats and rhesus monkeys, but not in mice.[46] The reason for this discrepancy is unclear at the moment. Accelerated clearance of submicromolar doses of no-drug-carrying ^{99m}Tc–PEGylated liposomes was reported in four human volunteers.[55] No accelerated clearance was observed upon repeated administration of PEG–liposomes in rabbits.[59]

In contrast to the animal experiments discussed above, clinical trial studies utilizing PEGylated liposomes did not observe the ABC phenomenon. For example, PEG–liposome-encapsulated doxorubicin (Doxil®) did not cause ABC after repeated administration,[69,70] but the ABC phenomenon was observed when Doxil® was combined with another cytotoxic agent, cisplatin.[70] The reason for discrepancies between animal and clinical studies is unclear; however, it could be related to endotoxin contamination of research-grade liposomes. Endotoxin is strictly regulated in clinical formulations and often overlooked in research-grade formulations, yet the latter are highly prone to containing high levels of endotoxin[71–73] (see also Chapters 4 and 5). None of the studies describing the anti-PEG antibody-mediated ABC phenomenon for PEG–liposomes assessed the level of contaminating endotoxin in the liposomal formulations. Endotoxin is highly immunogenic and thus, may assume the role of an adjuvant and contribute to the antigenicity of PEGylated liposomes, similarly described for other immunogenic substances, such as siRNA.[48–50]

Interestingly, several studies have reported the presence of naturally occurring antibodies specific to PEG in the blood of healthy donor volunteers.[74–78] The clinical significance of these antibodies is largely unclear, though two studies suggested that naturally occurring PEG antibodies contribute to the neutralization of PEGylated therapeutic proteins.[76,77] There is also a large discrepancy among studies with regard to the frequency of the anti-PEG response in a healthy population. For example, Richter and Akerblom reported an anti-PEG response in 0.2% of healthy donors,[74] while Leger *et al.* and Armstrong *et al.* estimated this level to be as high as 22–25%.[75,78] The reason for this disagreement in experimental results is most likely due to matrix and drug interference issues commonly experienced with *in vitro* antigenicity assays.[64,66,79,80] The evaluation of assay specificity and drug interference controls are essential steps in assay validation, which is required for assays intended for clinical studies, but often ignored by basic research studies. The use of an *in vitro* assay that did not undergo rigorous validation may produce erroneous data, and may complicate the understanding of important topics such as the antigenicity of nanomaterials and their surface coatings.

Two other factors which need to be considered when dealing with the antigenic response to PEG are the type of PEG and the synthetic procedure used to attach PEG to a protein or a nanoparticle surface.[61,81] For example, in contrast to methoxy-PEG, PEG-thiol was shown to be non-immunogenic.[60] In an antigenicity study focused on the generation of PEG-specific antibody response, hydroxyl-PEG was found to be non- or only weakly antigenic, while methoxy-PEG and t-butoxy-PEG were antigenic.[61] The role of the terminal group is critical as an increase in the number of methyl groups at the end of the polymer contributed significantly to its antigenicity, in that t-butoxy-PEG was more antigenic than methoxy-PEG (Figure 4).

4. Antigenic Response to Therapeutic Proteins

As proven through many years of preclinical and clinical research, biological therapeutics are effectively recognized by the immune system, which builds up a multi-stage response against them. Several factors, including protein structure, folding architecture, formulation, storage conditions, protein (in)stability, as well as the presence of degradation byproducts, are thought to play crutial roles in the generation of antibodies in response to the administration of a protein-based therapeutic.[72,82–85] Other factors, such as the route of administration and the dosage, can also contribute to the antigenicity of therapeutic proteins by modulating the staging and amplitude of the immune response. The consequences of antigenic response to a therapeutic protein-containing formulation depend on the nature of antibodies formed against

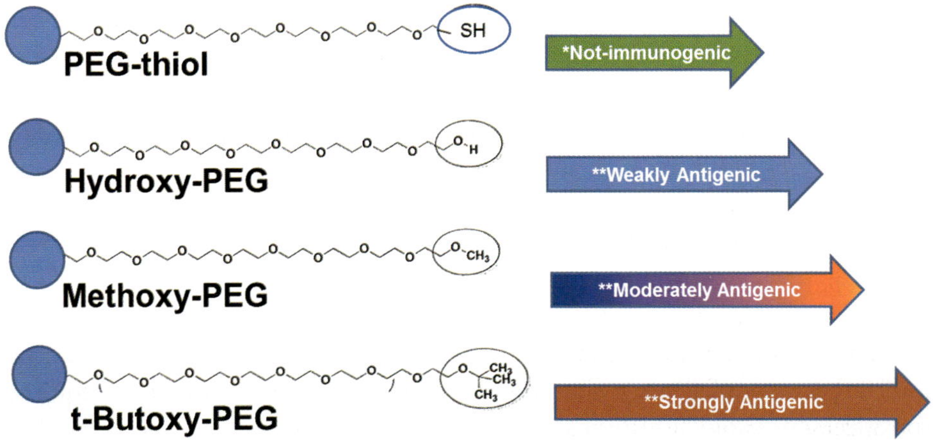

Figure 4. Antigenicity of PEG coating depends on its terminal groups. Various terminal groups of PEG polymers were shown to have different immunogenicity profiles. Immunogenicity here refers to complement activation and antigenic response. The studies were designed to have terminal groups shown in circles at the distal end of the polymers, while the methoxy group was used at the proximal end attached to the protein.

*From Ref. 60; **From Ref. 61.

the particular therapeutic product. If the antibody is not cross-reacting and does not neutralize the patient's endogenous protein, such immune response is generally considered as "benign" because it affects the product's efficacy and pharmacokinetics (PK), and may also result in general immune and hypersensitivity reactions.[86,87] However, the antigenic response characterized by the formation of antibodies neutralizing both the pharmacological product and the patient's own endogenous protein, is a completely different story. In this case, serious clinical consequences, including fatalities, may be experienced.[82–85,88–90] The best way to explain this point is to consider the experience with the clinical use of recombinant hematopoietic growth factors such as thrombopoietin and erythropoietin. For example, the generation of antibodies neutralizing both the thrombopoietin drug product and endogenous thrombopoietin resulted in the elimination of the drug product and an inhibition of the endogenous protein, which led to severe thrombocytopenia in patients.[89] Even more serious consequences were experienced when the recombinant erythropoietin product Eprex® was used in cancer patients to treat chemotherapy-induced anemia, but led to the formation of antibodies neutralizing both the therapeutic product and the patient's own erythropoietin, resulting in pure red cell aplasia.[91–98] Antibodies generated in response to Eprex® also cross-reacted with and neutralized other formulations of erythropoetin (Epogen®, NeoRecormon®, and Aranesp®), which resulted in even greater safety concerns as many patients remained transfusion-dependent.[82] Recombinant interferons are another class of biotechnology-derived pharmaceuticals which underwent extensive characterization in terms of immunogenicity. Clinical use of IFN-α and IFN-β products resulted in neutralizing antibody formation in ≥50% of cases, with a higher incidence in infectious disease indications.[99] Fortunately, due to the considerable functional redundancy among interferons, the generation of the anti-interferon immune response in patients was less deleterious than that experienced with the clinical use of thrombopietin and erythropoietin products.[99]

Due to the clinical consequences of anti-therapeutic protein antibodies, the antigenicity of these pharmacological products is an extremely extensive and a rapidly progressing area of research. A PubMed search using the key words "therapeutic proteins" and "immunogenicity" results in more than 7,500 hits. Those who desire to learn more about this topic are advised to read seminal publications by Drs. Paul Chamberlain of BioLOGICA Consulting, Anthony Mire-Sluis and Steven Swanson of Amgen, and Amy Rosenberg of the U.S. FDA. For the purpose of this chapter, I have only provided a brief overview of the immunogenicity of therapeutic proteins as background information and to explain the significance of this issue. However, as the main focus of this chapter is on nanoparticles, below I will focus more on the rather limited available data on the relationship between nanoparticles and the immunogenicity of therapeutic proteins.

4.1. *Antigenicity of therapeutic proteins in the presence of accidental particulate contaminants*

There is a growing concern that particles of various sizes (0.1–10 μm) are contaminating formulations of recombinant protein therapeutics.[18] These particles are believed to contribute to the antigenicity of therapeutic proteins by inducing protein aggregation[23,100] and adsorbing proteins onto the particle surface, thus creating highly immunogenic repeated protein structures.[23,100] However, studies elucidating the exact mechanism where this occurs are limited. Liu *et al.* recently reported the presence of tungsten microparticles originated from tungsten pins used during the syringe barrel manufacturing process.[22] These tungsten microparticles resulted in protein aggregation and increased the immunogenicity of a recombinant protein product. Another study has demonstrated that recombinant therapeutic proteins adsorb onto surfaces of hydrophobic metal, glass, and polystyrene particle contaminants, which contributed to protein aggregation and immunogenicity.[100] Among the particles contaminating therapeutic proteins are cellulose and glass fibers, silicon oil, rubber, stainless steel, fluoropolymers, and plastics.[18,22,100] Gowns and other materials used in clean rooms during the manufacturing of therapeutic proteins, containers, filling pumps, closures, vial stoppers, etc. serve as sources of such particulate contaminants.[18] The detection and quantification of particulates in therapeutic protein formulations are now requirements in the drug industry; however, there is no general agreement as to what methods should be used.[18] Another challenge in understanding the mechanism of how accidental particulates contribute to the antigenicity of therapeutic proteins are the limited quantities of these particulate materials. In a recent meeting report, Carpenter *et al.* mentioned a case study from a drug company which was able to detect and isolate contaminating particles from their therapeutic protein drug product; however, the amount of the isolated particulate material was insufficient to conduct follow-up mechanistic analyses.[18] It is clear that more studies are needed to address this critical issue.

4.2. *Antigenicity of therapeutic proteins attached to nanoparticle surfaces*

Because engineered nanoparticles share size similarity with viruses, the attachment of therapeutic proteins to nanoparticle surfaces raised concerns over possible increases in the antigenicity of these proteins due to the potential formation of repeated protein structures on nanoparticle surfaces, which could be processed by the immune system as virus-like particles (VLPs).[23] This concern was fueled by studies demonstrating increased antigenicity of protein aggregates and experiences with attempting to use microsphere particulate delivery systems for the human growth hormone, which resulted in increased unwanted antigenicity of this product.[18,23] It is a common perception that protein aggregates are more antigenic due to the appearance of new epitopes and/or formation of highly repetitive structures, disrupting B cell tolerance.[86,87] Although this has been proven for monomeric therapeutic proteins and their aggregated forms, e.g., rhIL-2, rhGH, HSA, IFN-α2b, and other proteins,[23] there is no concrete data showing that engineered nanomaterials indeed enhance the antigenicity of therapeutic proteins.

Here, I would like to emphasize the difference between engineered nanomaterials, i.e., materials with nanometer sizes which are specifically designed to possess certain surface properties and perform certain functions, and accidental, submicron particulate and protein aggregates. One of the main advantages of nanotechnology is that nanoparticles may be engineered such that their physicochemical properties such as size, charge, surface chemistry, and architecture can be specifically selected to modulate a biological response. For instance, these properties can be tuned to either enhance immune recognition (e.g., nanoparticle adjuvants) or to mask the particle from the immune system (e.g., nanoparticle drug delivery). Specifically, colloidal gold nanoparticles are quickly eliminated from the circulation by cells of the MPS, while the same particles with PEG surface coatings avoid immune recognition and MPS uptake.[101] When colloidal gold particles were used to deliver rhTNF-α, the therapeutic protein was also masked by PEG, and this allowed the intravenous injection of PEG–gold–TNF-α formulation Cyt6091® at TNF-α dose levels far beyond those possible for free rhTNF-α.[14] This study clearly demonstrated that reformulating a cytokine onto a nanotechnology platform helped to decrease the immunotoxicity of this cytokine. Of special interest for this chapter is the study by Libutti *et al.* describing the lack of antigenic response to rhTNF-α in patients receiving repeated intravenous administration of the PEG–gold–TNF-α nanomedicine.[14] Another example where the encapsulation of a recombinant protein helped to overcome hurdles associated with patients developing neutralizing antibodies in response to a protein is the liposomal delivery of streptokinase.[15] In this study, loss of thrombolytic efficacy of the recombinant streptokinase was observed in animals receiving repeated administration of this protein. The loss of efficacy was attributed to the formation of neutralizing antibodies similar to that observed in patients receiving this protein therapy. However, the encapsulation of the same therapeutic protein into liposomes helped to overcome the neutralization of streptokinase activity by the antibody.[15] Most recently, Ramani *et al.* used phosphatidylserine (PS)-containing liposomes for the delivery of recombinant factor VIII.[16,17] Similar to the streptokinase study, the encapsulation of this therapeutic protein into liposomes helped to reduce its antigenicity. The mechanism is still not completely understood, but the authors believed that PS plays an important role in antigenicity reduction as it negatively regulates $CD4^+$ T cells and contributes to maintaining immunotolerance to this essential endogenous protein.[16,17]

Though limited, these studies emphasize the advantages of nanotechnology for the delivery of therapeutic proteins. To further support this concept, studies using engineered nanoparticles for the delivery of immunogenic therapeutic proteins are needed.

5. Summary and the Road Ahead

The antigenicity of therapeutic proteins is a well-known issue because it may change PK, clearance, efficacy of therapeutic proteins, as well as contribute to their toxicity.[82,83,86,87,93] Since proteins may be used as targeting moieties on nanoparticle surfaces, the antigenicities of these proteins and of nanoparticles *per se*, are of interest to the scientific community and drug industry. Studies conducted to date with engineered nanomaterials demonstrated the hapten-like behavior of some nanoparticles, and the importance of the protein carrier and strong adjuvants in generating particle-specific immune responses.[10–13,24] However, no study has been reported which demonstrates the antigenicity of engineered therapeutic nanoparticles. Moreover, a few studies have shown no antigenicity of nanoparticles and therapeutic proteins incorporated onto the nanoparticle platform.[14,15] The only class of nanomaterials for which antigenicity in the absence of a protein carrier and adjuvant was suspected are PEGylated liposomes.[31] However, most of these studies are descriptive in nature and there is no solid evidence for anti-liposome antibodies. The antigenicity of PEG, commonly used as a coating for engineered nanomaterials to provide stealth properties, i.e., protecting them from immune recognition and extending their blood circulation, is another interesting issue. However, the literature is divided, and suggests that PEG itself is not antigenic, but that the PEGylation method and linkers used to attach PEG to a protein or a particle surface may be antigenic.[60,61,81] Clearly, more studies are necessary in this area to further understand this important issue.

There is an increasing number of studies demonstrating the attenuation of immunotoxicity of legacy drugs after their reformulation onto a nanotechnology platform (e.g., Abraxane® *vs.* Taxol®, doxorubicin *vs.* Doxil®, rhTNF *vs.* Cyt6091®)[14,102] (see Chapter 19). It would be compelling to examine if the same holds true for the antigenicity of therapeutic proteins. For example, EPO and PEG–EPO were associated with the induction of pure red blood cell aplasia due to the induction of anti-EPO antibody response. It would be interesting to see if the incorporation of EPO onto a nanotechnology platform similar to Cyt6091® (recombinant TNF α bound to the surface of a PEGylated colloidal gold nanoparticle) would decrease the antigenicity of this therapeutic protein. Careful design of the nanotechnology delivery platform and exhaustive physicochemical characterization would be needed to support such a study.

Furthermore, there is a growing concern that accidental (not engineered) nanoparticles (e.g., leachables from plastic tubing, components of rubber stoppers, environmental particulate contaminants from manufacturing clean rooms) may contribute to the immunogenicity of therapeutic proteins.[18,23] Yet, there are no systematic structure–activity relationship studies addressing this issue. Understanding the differences between these accidental nanoparticles and engineered nanoparticles is critical for the further exploration in this area.

In conclusion, future studies directed at the antigenicity of engineered nanomaterials should focus on the following areas: 1) endotoxin contamination as a potential adjuvant, 2) opsonization with serum proteins as a potential source of protein carrier, 3) the role of PEG linkers and synthesis methods, 4) the role of nanoparticle composition (e.g., lipids, metals, polymers), 5) reformulation of antigenic therapeutic proteins onto nanotechnology platforms, and 6) structure–activity relationship studies aimed at identifying nanoparticle physicochemical properties (e.g., size, charge, hydrophobicity) critical for an antigenic response to the nanoparticles themselves, to biological ligands attached to the particle surface for targeted delivery (e.g., nucleic acids, proteins, antibodies), and to recombinant protein therapeutics.

References

1. Fifis T, Gamvrellis A, Crimeen-Irwin B, Pietersz GA, Li J, Mottram PL, McKenzie IF, Plebanski M. Size-dependent immunogenicity: Therapeutic and protective properties of nano-vaccines against tumors. *J Immunol* 2004;173:3148–3154.
2. Manolova V, Flace A, Bauer M, Schwarz K, Saudan P, Bachmann MF. Nanoparticles target distinct dendritic cell populations according to their size. *Eur J Immunol* 2008;38:1404–1413.
3. Reddy ST, van der Vlies AJ, Simeoni E, Angeli V, Randolph GJ, O'Neil CP, Lee LK, Swartz MA, Hubbell JA. Exploiting lymphatic transport and complement activation in nanoparticle vaccines. *Nat Biotechnol* 2007;25:1159–1164.
4. Agashe HB, Dutta T, Garg M, Jain NK. Investigations on the toxicological profile of functionalized fifth-generation poly(propylene imine) dendrimer. *J Pharm Pharmacol* 2006;58:1491–1498.
5. Andreev SM, Babakhin AA, Petrukhina AO, Romanova VS, Parnes ZN, Petrov RV. Immunogenic and allergenic properties of fulleren conjugates with aminoacids and proteins. *Dokl Biochem* 2000;370:4–7.
6. Dykman LA, Sumaroka MV, Staroverov SA, Zaitseva IS, Bogatyrev VA. [Immunogenic properties of the colloidal gold]. *Izv Akad Nauk Ser Biol* 2004:86–91.
7. Masalova OV, Shepelev AV, Atanadze SN, Parnes ZN, Romanova VS, Vol'pina OM, Semiletov Iu A, Kushch AA. [Immunostimulating effect of water-soluble fullerene derivatives — perspective adjuvants for a new generation of vaccine]. *Dokl Akad Nauk* 1999;369:411–413.
8. Roberts JC, Bhalgat MK, Zera RT. Preliminary biological evaluation of polyamidoamine (PAMAM) starburst dendrimers. *J Biomed Mater Res* 1996;30:53–65.
9. Lee SC, Parthasarathy R, Duffin TD, Botwin K, Zobel J, Beck T, Lange G, Kunneman D, Janssen R, Rowold E, *et al.* Recognition properties of antibodies to PAMAM dendrimers and their use in immune detection of dendrimers. *Biomed Microdevices* 2001;3:53–59.
10. Braden BC, Goldbaum FA, Chen BX, Kirschner AN, Wilson SR, Erlanger BF. X-ray crystal structure of an anti-Buckminsterfullerene antibody fab fragment: Biomolecular recognition of C(60). *Proc Natl Acad Sci U S A* 2000;97:12193–12197.

11. Chen BX, Wilson SR, Das M, Coughlin DJ, Erlanger BF. Antigenicity of fullerenes: Antibodies specific for fullerenes and their characteristics. *Proc Natl Acad Sci U S A* 1998;95:10809–10813.
12. Lee SC, Parthasarathy R, Botwin K, Kunneman D, Rowold E, Lange G, Klover J, Abegg A, Zobel J, Beck T, *et al.* Biochemical and immunological properties of cytokines conjugated to dendritic polymers. *Biomed Microdevices* 2004;6:191–202.
13. Erlanger BF, Chen BX, Zhu M, Brus L. Binding of an anti-fullerene IgG monoclonal antibody to single wall carbon nanotubes. *Nano Lett* 2001;1:465–467.
14. Libutti SK, Paciotti GF, Byrnes AA, Alexander HR, Jr, Gannon WE, Walker M, Seidel GD, Yuldasheva N, Tamarkin L. Phase I and pharmacokinetic studies of CYT-6091, a novel PEGylated colloidal gold-rhTNF nanomedicine. *Clin Cancer Res* 2010;16: 6139–6149.
15. Perkins WR, Vaughan DE, Plavin SR, Daley WL, Rauch J, Lee L, Janoff AS. Streptokinase entrapment in interdigitation-fusion liposomes improves thrombolysis in an experimental rabbit model. *Thromb Haemost* 1997;77:1174–1178.
16. Ramani K, Miclea RD, Purohit VS, Mager DE, Straubinger RM, Balu-Iyer SV. Phosphatidylserine containing liposomes reduce immunogenicity of recombinant human factor VIII (rFVIII) in a murine model of hemophilia A. *J Pharm Sci* 2008;97: 1386–1398.
17. Ramani K, Purohit V, Miclea R, Gaitonde P, Straubinger RM, Balu-Iyer SV. Passive transfer of polyethylene glycol to liposomal-recombinant human FVIII enhances its efficacy in a murine model for hemophilia A. *J Pharm Sci* 2008;97:3753–3764.
18. Carpenter J, Cherney B, Lubinecki A, Ma S, Marszal E, Mire-Sluis A, Nikolai T, Novak J, Ragheb J, Simak J. Meeting report on protein particles and immunogenicity of therapeutic proteins: Filling in the gaps in risk evaluation and mitigation. *Biologicals* 2010;38:602–611.
19. Fradkin AH, Carpenter JF, Randolph TW. Glass particles as an adjuvant: A model for adverse immunogenicity of therapeutic proteins. *J Pharm Sci* 2011;100:4953–4964.
20. Jiskoot W, Randolph TW, Volkin DB, Middaugh CR, Schoneich C, Winter G, Friess W, Crommelin DJ, Carpenter JF. Protein instability and immunogenicity: Roadblocks to clinical application of injectable protein delivery systems for sustained release. *J Pharm Sci* 2012;101:946–954.
21. Jiang Y, Nashed-Samuel Y, Li C, Liu W, Pollastrini J, Mallard D, Wen ZQ, Fujimori K, Pallitto M, Donahue L, *et al.* Tungsten-induced protein aggregation: Solution behavior. *J Pharm Sci* 2009;98:4695–4710.
22. Liu W, Swift R, Torraca G, Nashed-Samuel Y, Wen ZQ, Jiang Y, Vance A, Mire-Sluis A, Freund E, Davis J, *et al.* Root cause analysis of tungsten-induced protein aggregation in pre-filled syringes. *PDA J Pharm Sci Technol* 2010;64:11–19.
23. Mire-Sluis A, Cherney B, Madsen R, Polozova A, Rosenberg A, Smith H, Arora T, Narhi L. Analysis and immunogenic potential of aggregates and particles: A practical approach, part 2. *BioProc Int* 2011;9:38–43.
24. Hendrickson O, Fedyunina N, Zherdev A, Solopova O, Sveshnikov P, Dzantiev B. Production of monoclonal antibodies against fullerene C60 and development of a fullerene enzyme immunoassay. *Analyst* 2012;137:98–105.
25. Ishida T, Atobe K, Wang X, Kiwada H. Accelerated blood clearance of PEGylated liposomes upon repeated injections: Effect of doxorubicin-encapsulation and high-dose first injection. *J Controlled Release* 2006;115:251–258.

26. Ishida T, Harada M, Wang XY, Ichihara M, Irimura K, Kiwada H. Accelerated blood clearance of PEGylated liposomes following preceding liposome injection: Effects of lipid dose and PEG surface-density and chain length of the first-dose liposomes. *J Controlled Release* 2005;105:305–317.
27. Ishida T, Ichihara M, Wang X, Yamamoto K, Kimura J, Majima E, Kiwada H. Injection of PEGylated liposomes in rats elicits PEG-specific IgM, which is responsible for rapid elimination of a second dose of PEGylated liposomes. *J Controlled Release* 2006;112:15–25.
28. Ishida T, Ichikawa T, Ichihara M, Sadzuka Y, Kiwada H. Effect of the physicochemical properties of initially injected liposomes on the clearance of subsequently injected PEGylated liposomes in mice. *J Controlled Release* 2004;95:403–412.
29. Ishida T, Kashima S, Kiwada H. The contribution of phagocytic activity of liver macrophages to the accelerated blood clearance (ABC) phenomenon of PEGylated liposomes in rats. *J Controlled Release* 2008;126:162–165.
30. Ishida T, Kiwada H. Accelerated blood clearance (ABC) phenomenon upon repeated injection of PEGylated liposomes. *Int J Pharm* 2008;354:56–62.
31. Ishida T, Wang X, Shimizu T, Nawata K, Kiwada H. PEGylated liposomes elicit an anti-PEG IgM response in a T cell-independent manner. *J Controlled Release* 2007;122:349–355.
32. Ishihara T, Maeda T, Sakamoto H, Takasaki N, Shigyo M, Ishida T, Kiwada H, Mizushima Y, Mizushima T. Evasion of the accelerated blood clearance phenomenon by coating of nanoparticles with various hydrophilic polymers. *Biomacromolecules* 2010;11:2700–2706.
33. Ishihara T, Takeda M, Sakamoto H, Kimoto A, Kobayashi C, Takasaki N, Yuki K, Tanaka K, Takenaga M, Igarashi R, *et al.* Accelerated blood clearance phenomenon upon repeated injection of PEG-modified PLA-nanoparticles. *Pharm Res* 2009;26:2270–2279.
34. Alving CR. Antibodies to liposomes, phospholipids and phosphate esters. *Chem Phys Lipids* 1986;40:303–314.
35. Alving CR, Swartz GM, Jr., Wassef NM, Ribas JL, Herderick EE, Virmani R, Kolodgie FD, Matyas GR, Cornhill JF. Immunization with cholesterol-rich liposomes induces anti-cholesterol antibodies and reduces diet-induced hypercholesterolemia and plaque formation. *J Lab Clin Med* 1996;127:40–49.
36. Banerji B, Alving CR. Anti-liposome antibodies induced by lipid A. I. Influence of ceramide, glycosphingolipids, and phosphocholine on complement damage. *J Immunol* 1981;126:1080–1084.
37. Banerji B, Alving CR. Antibodies to liposomal phosphatidylserine and phosphatidic acid. *Biochem Cell Biol* 1990;68:96–101.
38. Banerji B, Kenny JJ, Scher I, Alving CR. Antibodies against liposomes in normal and immune-defective mice. *J Immunol* 1982;128:1603–1607.
39. Fogler WE, Swartz GM, Jr., Alving CR. Antibodies to phospholipids and liposomes: Binding of antibodies to cells. *Biochim Biophys Acta* 1987;903:265–272.
40. Richards RL, Aronson J, Schoenbechler M, Diggs CL, Alving CR. Antibodies reactive with liposomal phospholipids are produced during experimental Trypanosoma rhodesiense infections in rabbits. *J Immunol* 1983;130:1390–1394.
41. Watanabe H, Nakanishi T, Umetsu M, Kumagai I. Human anti-gold antibodies: Biofunctionalization of gold nanoparticles and surfaces with anti-gold antibodies. *J Biol Chem* 2008;283:36031–36038.

42. Zolnik BS, Gonzalez-Fernandez A, Sadrieh N, Dobrovolskaia MA. Nanoparticles and the immune system. *Endocrinology* 2010;151:458–465.
43. Alving CR, Swartz GM, Jr. Antibodies to cholesterol, cholesterol conjugates and liposomes: Implications for atherosclerosis and autoimmunity. *Crit Rev Immunol* 1991;10:441–453.
44. Fehr T, Skrastina D, Pumpens P, Zinkernagel RM. T cell-independent type I antibody response against B cell epitopes expressed repetitively on recombinant virus particles. *Proc Natl Acad Sci U S A* 1998;95:9477–9481.
45. Zagon IS, Donahue RN, Bonneau RH, McLaughlin PJ. B lymphocyte proliferation is suppressed by the opioid growth factor-opioid growth factor receptor axis: Implication for the treatment of autoimmune diseases. *Immunobiology* 2011;216:173–183.
46. Dams ET, Laverman P, Oyen WJ, Storm G, Scherphof GL, van Der Meer JW, Corstens FH, Boerman OC. Accelerated blood clearance and altered biodistribution of repeated injections of sterically stabilized liposomes. *J Pharmacol Exp Ther* 2000;292: 1071–1079.
47. Laverman P, Carstens MG, Boerman OC, Dams ET, Oyen WJ, van Rooijen N, Corstens FH, Storm G. Factors affecting the accelerated blood clearance of polyethylene glycol-liposomes upon repeated injection. *J Pharmacol Exp Ther* 2001;298:607–612.
48. Tagami T, Nakamura K, Shimizu T, Ishida T, Kiwada H. Effect of siRNA in PEG-coated siRNA-lipoplex on anti-PEG IgM production. *J Controlled Release* 2009;137: 234–240.
49. Tagami T, Nakamura K, Shimizu T, Yamazaki N, Ishida T, Kiwada H. CpG motifs in pDNA-sequences increase anti-PEG IgM production induced by PEG-coated pDNA-lipoplexes. *J Controlled Release* 2010;142:160–166.
50. Tagami T, Uehara Y, Moriyoshi N, Ishida T, Kiwada H. Anti-PEG IgM production by siRNA encapsulated in a PEGylated lipid nanocarrier is dependent on the sequence of the siRNA. *J Controlled Release* 2011;151:149–154.
51. Wang X, Ishida T, Kiwada H. Anti-PEG IgM elicited by injection of liposomes is involved in the enhanced blood clearance of a subsequent dose of PEGylated liposomes. *J Controlled Release* 2007;119:236–244.
52. Wang XY, Ishida T, Ichihara M, Kiwada H. Influence of the physicochemical properties of liposomes on the accelerated blood clearance phenomenon in rats. *J Controlled Release* 2005;104:91–102.
53. Koide H, Asai T, Hatanaka K, Akai S, Ishii T, Kenjo E, Ishida T, Kiwada H, Tsukada H, Oku N. T cell-independent B cell response is responsible for ABC phenomenon induced by repeated injection of PEGylated liposomes. *Int J Pharm* 2010;392:218–223.
54. Kaminskas LM, McLeod VM, Porter CJ, Boyd BJ. Differences in colloidal structure of PEGylated nanomaterials dictate the likelihood of accelerated blood clearance. *J Pharm Sci* 2011;100:5069–5077.
55. Laverman P, Brouwers AH, Dams ET, Oyen WJ, Storm G, van Rooijen N, Corstens FH, Boerman OC. Preclinical and clinical evidence for disappearance of long-circulating characteristics of polyethylene glycol liposomes at low lipid dose. *J Pharmacol Exp Ther* 2000;293:996–1001.
56. Abu Lila AS, Doi Y, Nakamura K, Ishida T, Kiwada H. Sequential administration with oxaliplatin-containing PEG-coated cationic liposomes promotes a significant delivery of subsequent dose into murine solid tumor. *J Controlled Release* 2010;142:167–173.

57. Koide H, Asai T, Hatanaka K, Urakami T, Ishii T, Kenjo E, Nishihara M, Yokoyama M, Ishida T, Kiwada H, *et al.* Particle size-dependent triggering of accelerated blood clearance phenomenon. *Int J Pharm* 2008;362:197–200.
58. Ishida T, Maeda R, Ichihara M, Irimura K, Kiwada H. Accelerated clearance of PEGylated liposomes in rats after repeated injections. *J Controlled Release* 2003;88: 35–42.
59. Coins B, Philips WT, Klipper R. Repeat injection studies of technetium-99m-labeled PEG-liposomes in the same animal. *J Liposome Research* 1998;8:265–281.
60. Arima Y, Toda M, Iwata H. Complement activation on surfaces modified with ethylene glycol units. *Biomaterials* 2008;29:551–560.
61. Sherman MR, Williams LD, Sobczyk MA, Michaels SJ, Saifer MG. Role of the methoxy group in immune responses to mPEG-protein conjugates. *Bioconjug Chem* 2012;23: 485–499.
62. DeSilva B, Smith W, Weiner R, Kelley M, Smolec J, Lee B, Khan M, Tacey R, Hill H, Celniker A. Recommendations for the bioanalytical method validation of ligand-binding assays to support pharmacokinetic assessments of macromolecules. *Pharm Res* 2003; 20:1885–1900.
63. Smolec J, DeSilva B, Smith W, Weiner R, Kelly M, Lee B, Khan M, Tacey R, Hill H, Celniker A, *et al.* Bioanalytical method validation for macromolecules in support of pharmacokinetic studies. *Pharm Res* 2005;22:1425–1431.
64. Araujo J, Zocher M, Wallace K, Peng K, Fischer SK. Increased rheumatoid factor interference observed during immunogenicity assessment of an Fc-engineered therapeutic antibody. *J Pharm Biomed Anal* 2011;55:1041–1049.
65. Bautista AC, Wullner D, Moxness M, Swanson SJ, Chirmule N, Jawa V. Impact of matrix-associated soluble factors on the specificity of the immunogenicity assessment. *Bioanalysis* 2010;2:721–731.
66. Hart MH, de Vrieze H, Wouters D, Wolbink GJ, Killestein J, de Groot ER, Aarden LA, Rispens T. Differential effect of drug interference in immunogenicity assays. *J Immunol Methods* 2011;372:196–203.
67. Neubert H, Grace C, Rumpel K, James I. Assessing immunogenicity in the presence of excess protein therapeutic using immunoprecipitation and quantitative mass spectrometry. *Anal Chem* 2008;80:6907–6914.
68. Zhong ZD, Dinnogen S, Hokom M, Ray C, Weinreich D, Swanson SJ, Chirmule N. Identification and inhibition of drug target interference in immunogenicity assays. *J Immunol Methods* 2010;355:21–28.
69. Gabizon A, Isacson R, Rosengarten O, Tzemach D, Shmeeda H, Sapir R. An open-label study to evaluate dose and cycle dependence of the pharmacokinetics of PEGylated liposomal doxorubicin. *Cancer Chemother Pharmacol* 2008;61:695–702.
70. Lyass O, Hubert A, Gabizon AA. Phase I study of Doxil-cisplatin combination chemotherapy in patients with advanced malignancies. *Clin Cancer Res* 2001;7: 3040–3046.
71. Dobrovolskaia MA, Neun BW, Clogston JD, Ding H, Ljubimova J, McNeil SE. Ambiguities in applying traditional Limulus Amebocyte Lysate tests to quantify endotoxin in nanoparticle formulations. *Nanomedicine (Lond)* 2010;5:555–562.
72. Jones CF, Grainger DW. *In vitro* assessments of nanomaterial toxicity. *Adv Drug Delivery Rev* 2009;61:438–456.

73. Neun BW, Dobrovolskaia MA. Detection and quantitative evaluation of endotoxin contamination in nanoparticle formulations by LAL-based assays. *Methods Mol Biol* 2011;697:121–130.
74. Richter AW, Akerblom E. Polyethylene glycol reactive antibodies in man: Titer distribution in allergic patients treated with monomethoxy polyethylene glycol modified allergens or placebo, and in healthy blood donors. *Int Arch Allergy Appl Immunol* 1984;74:36–39.
75. Leger RM, Arndt P, Garrathy, G, *et al.* Normal donor sera contain antibodies to polyethyleneglycol (PEG). *Transfusion* 2001;41:29S.
76. Armstrong JK, Hempel G, Koling S, Chan LS, Fisher T, Meiselman HJ, Garratty G. Antibody against poly(ethylene glycol) adversely affects PEG-asparaginase therapy in acute lymphoblastic leukemia patients. *Cancer* 2007;110:103–111.
77. Ganson NJ, Kelly SJ, Scarlett E, Sundy JS, Hershfield MS. Control of hyperuricemia in subjects with refractory gout, and induction of antibody against poly(ethylene glycol) (PEG), in a phase I trial of subcutaneous PEGylated urate oxidase. *Arthritis Res Ther* 2006;8:R12.
78. Armstrong JK, Leger RM, Wanby RB, Meiselman HJ, Garratty G, Fisher TC. Occurence of an antibody to poly(ethylene glycol) in normal donors. *Blood* 2003;102.
79. Mikulskis A, Yeung D, Subramanyam M, Amaravadi L. Solution ELISA as a platform of choice for development of robust, drug tolerant immunogenicity assays in support of drug development. *J Immunol Methods* 2011;365:38–49.
80. Oh S, Tsai AK, Ohlfest JR, Panoskaltsis-Mortari A, Vallera DA. Evaluation of a bispecific biological drug designed to simultaneously target glioblastoma and its neovasculature in the brain. *J Neurosurg* 2011;114:1662–1671.
81. Francis GE, Fisher D, Delgado C, Malik F, Gardiner A, Neale D. PEGylation of cytokines and other therapeutic proteins and peptides: The importance of biological optimisation of coupling techniques. *Int J Hematol* 1998;68:1–18.
82. Chamberlain P, Mire-Sluis AR. An overview of scientific and regulatory issues for the immunogenicity of biological products. *Dev Biol (Basel)* 2003;112:3–11.
83. Rosenberg AS. Immunogenicity of biological therapeutics: A hierarchy of concerns. *Dev Biol (Basel)* 2003;112:15–21.
84. Shankar G, Shores E, Wagner C, Mire-Sluis A. Scientific and regulatory considerations on the immunogenicity of biologics. *Trends Biotechnol* 2006;24:274–280.
85. Swanson SJ. Immunogenicity issues in drug development. *J Immunotoxicol* 2006;3: 165–172.
86. Buttel IC, Chamberlain P, Chowers Y, Ehmann F, Greinacher A, Jefferis R, Kramer D, Kropshofer H, Lloyd P, Lubiniecki A, *et al.* Taking immunogenicity assessment of therapeutic proteins to the next level. *Biologicals* 2011;39:100–109.
87. Buttel IC, Voller K, Schneider CK. Immunogenicity and its impact on benefit/risk considerations in the authorisation of biopharmaceuticals. *Curr Drug Saf* 2010;5: 287–292.
88. Koren E, Smith HW, Shores E, Shankar G, Finco-Kent D, Rup B, Barrett YC, Devanarayan V, Gorovits B, Gupta S, *et al.* Recommendations on risk-based strategies for detection and characterization of antibodies against biotechnology products. *J Immunol Methods* 2008;333:1–9.

89. Koren E, Zuckerman LA, Mire-Sluis AR. Immune responses to therapeutic proteins in humans — clinical significance, assessment and prediction. *Curr Pharm Biotechnol* 2002;3:349–360.
90. Schellekens H, Casadevall N. Immunogenicity of recombinant human proteins: Causes and consequences. *J Neurol* 2004;251(Suppl 2):II4–9.
91. Casadevall N. Antibodies against rHuEPO: Native and recombinant. *Nephrol Dial Transplant* 2002;17(Suppl 5):42–47.
92. Casadevall N, Eckardt KU, Rossert J. Epoetin-induced autoimmune pure red cell aplasia. *J Am Soc Nephrol* 2005;16(Suppl 1):S67–69.
93. Casadevall N, Rossert J. Importance of biologic follow-ons: Experience with EPO. *Best Pract Res Clin Haematol* 2005;18:381–387.
94. Jacob A, Sandhu K, Nicholas J, Jones H, Odum J, Rylance P, Carmichael P, Jackson M, Handa S, Macwhannell A, *et al.* Antibody-mediated pure red cell aplasia in a dialysis patient receiving darbepoetin alfa as the sole erythropoietic agent. *Nephrol Dial Transplant* 2006;21:2963–2965.
95. McKoy JM, Stonecash RE, Cournoyer D, Rossert J, Nissenson AR, Raisch DW, Casadevall N, Bennett CL. Epoetin-associated pure red cell aplasia: Past, present, and future considerations. *Transfusion* 2008;48:1754–1762.
96. Rossert J, Casadevall N, Eckardt KU. Anti-erythropoietin antibodies and pure red cell aplasia. *J Am Soc Nephrol* 2004;15:398–406.
97. Smalling R, Foote M, Molineux G, Swanson SJ, Elliott S. Drug-induced and antibody-mediated pure red cell aplasia: A review of literature and current knowledge. *Biotechnol Annu Rev* 2004;10:237–250.
98. Swanson SJ, Ferbas J, Mayeux P, Casadevall N. Evaluation of methods to detect and characterize antibodies against recombinant human erythropoietin. *Nephron Clin Pract* 2004;96:c88–95.
99. Bendtzen K. Critical review: Assessment of interferon-beta immunogenicity in multiple sclerosis. *J Interferon Cytokine Res* 2010;30:759–766.
100. Van Beers MM, Gilli F, Schellekens H, Randolph TW, Jiskoot W. Immunogenicity of recombinant human interferon beta interacting with particles of glass, metal, and polystyrene. *J Pharm Sci* 2012;101:187–199.
101. Paciotti GF, Myer L, Weinreich D, Goia D, Pavel N, McLaughlin RE, Tamarkin L. Colloidal gold: A novel nanoparticle vector for tumor directed drug delivery. *Drug Delivery* 2004;11:169–183.
102. Hawkins MJ, Soon-Shiong P, Desai N. Protein nanoparticles as drug carriers in clinical medicine. *Adv Drug Delivery Rev* 2008;60:876–885.

Chapter 19

In Vitro Assays for Monitoring Nanoparticle Interaction with Components of the Immune System

Marina A. Dobrovolskaia and Scott E. McNeil*

Nanotechnology Characterization Laboratory
Advanced Technology Program, SAIC-Frederick, Inc.
NCI-Frederick, Frederick, MD 21702, USA

Preclinical immunotoxicity studies are usually conducted to identify potential causes for concern before a new drug or a medical device is used in clinical trials. Traditionally, these studies consist of *in vivo* standard toxicity studies and specialized *in vivo* immune function tests. In immunotoxicology, *in vitro* assays are used to support and further explore the findings of *in vivo* studies in cases where adverse effects were triggered by the test material. This strategy, established over multiple years of use in the pharmaceutical industry, is also applicable to engineered nanomaterials. However, because nanoparticles are complex and may be expensive and time-consuming to synthesize, they are often only available in extremely limited quantities during early preclinical development. There is a need to rapidly screen small amounts of nanomaterials to identify toxic candidates and understand the physicochemical properties that contribute to toxicity. This allows for the fine-tuning of these properties before the nanomaterial is produced at large scales. A battery of reliable and predictive *in vitro* tests would meet this need, but would require a firmly established correlation between *in vitro* and *in vivo* results. In this chapter, we summarize the current literature on this subject and also share our own experiences with *in vitro* immunological screening of engineered nanomaterials for toxicity to the immune system.

*Corresponding author. E-mail: marina@mail.nih.gov

1. Introduction

Nanoparticle interactions with the immune system can result in two principal undesirable effects: 1) immunosuppression, where the function of the immune system is lowered and 2) immunostimulation, where the immune system is over-activated.[1] Depending on the mechanism of induction, immunostimulation can be further subdivided into: 1) overwhelming immunological response to a foreign material (e.g., hypersensitivity reactions to nanomaterials as foreign substances) and 2) immune dysregulation (e.g., autoimmunity).[2] Since *in vivo* manifestations of immunotoxicity may include changes in the cellularity of the bone marrow (BM), peripheral blood, and/or lymphoid organs, traditional toxicological *in vivo* studies include the analysis of the weights of lymphoid organs, histology, clinical chemistry, and hematology.[3] While strong immunosuppression and immunostimulation can be, and often are, identified through these studies, moderate immunotoxicity resulting from immune dysregulation will only manifest at the functional level and may be missed. For this reason, it has become an accepted practice in drug development to supplement standard toxicity studies with immune function tests (IFTs).[1,2,4,5]

Due to significant structural and organizational differences between the human immune system and those of animal species used for testing, traditional toxicology studies conducted in rodents (commonly rats) and non-rodents (commonly dogs) are often not predictive, especially of moderate and functional effects. One highly specialized set of *in vivo* IFTs in rodents has been shown to be predictive of immunoxicity in humans,[4–6] but these *in vivo* IFTs require significant investments of time, cost, and test materials.[7] Furthermore, there are ethical concerns regarding animal use and a worldwide 3R effort, where 3R stands for the *replacement* of animals in research with non-animal alternatives, the *reduction* in the number of animals used in experiments, and the *refinement* of scientific procedures and animal husbandry with the aim of minimizing animal suffering.[8] Some examples of 3R efforts include the establishment of the European Center for Validation of Alternative Test Methods (ECVAM) (http://ecvam.jrc.it/) and the Interagency Coordinating Committee on the Validation of Alternative Methods (ICCVAM) started in the United States in 1994, and the National Centre for Replacement, Refinement, and Reduction of Animals in Research (NC3Rs) in May 2004 in the United Kingdom (www.nc3rs.org.uk). The setting up of such organizations support the increasing need for reliable and predictive *in vitro* assays.

Although the likelihood of identifying immunotoxicity increases with the progression through the phases of drug development (from *in vitro* preclinical, to *in vivo* preclinical, to the clinical phase), the high cost of *in vivo* preclinical and clinical studies often preclude their use despite their advantages in predictability. The high throughput nature and lower costs of preclinical *in vitro* tests makes them a very attractive alternative (Figure 1).

There are four main challenges in the *in vitro* testing of nanoparticle immunotoxicity: 1) the selection of an appropriate model, 2) the selection of an appropriate endpoint, 3) the selection of appropriate positive and negative controls, and 4) nanoparticle interference with *in vitro* assays.

The first two challenges involve decisions on the choice of appropriate cell models and toxicity endpoints. Many laboratories have tried using *in vitro* cytotoxicity tests to estimate nanoparticle immunotoxicity. For instance, most studies in environmental immunotoxicology use *in vitro* cytotoxicity assays to study nanoparticle effects on the immune system.[9–12] Unfortunately, this is based on a common misconception that *in vitro* cytotoxicity assays on an immune cell line are a good indicator for nanoparticle immunotoxicity. There are many immune cell lines representing various types of immune cells, such as monocytes, macrophages, T cells, B cells, and natural killer (NK) cells. By definition, these cells are immortalized and their individual sensitivities to cytotoxic agents vary.[13,14] Therefore, it is impossible to select one cell line to represent the entire immune system, and even the representation of one type of immune cell (e.g., monocytes, NK cells) by a cell line may not be accurate because cell viability *in vitro* may not be a true reflection of their *in vivo* function. Due to this disparity, creating a battery of *in vitro* immunotoxicity assays

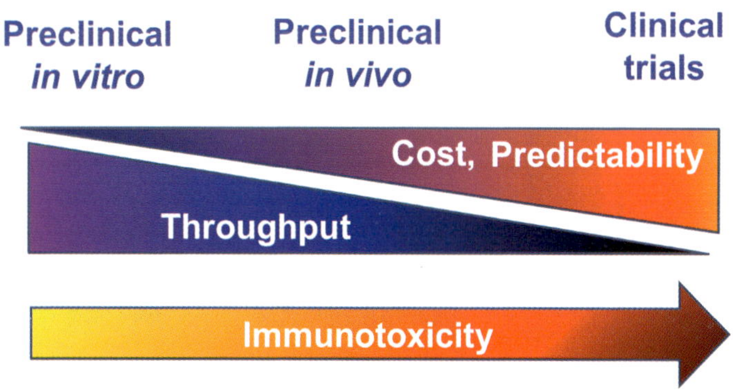

Figure 1. The increasing need for reliable *in vitro* preclinical immunotoxicity tests. Immunotoxicity can be uncovered during all phases of the drug development process; however, the likelihood of discovery dramatically increases with progress toward the clinical phase. This is due to the higher predictability of *in vivo* preclinical and clinical studies compared to *in vitro* assays. However, high costs and the low throughput nature of *in vivo* studies cause many developers to conduct *in vitro* assays first, due to their high throughput nature and lower costs.

for the assessment of multiple primary cell types and estimation of both the integrity of the cells and their function is a better approach to understanding the immunotoxicological potential of engineered nanomaterials.

The selection of appropriate positive and negative controls is another challenge as it requires the use of different materials specific to the endpoint of interest. For example, cytotoxic compounds are acceptable controls for cell viability assays, but they are not optimal for assays assessing immune cell function. Similarly, immunostimulatory compounds are applicable for functional tests, but are unsuitable as controls for cell viability tests. Even within the same category, such as for immunostimulation assays, different controls are needed depending on the assay endpoint. For example, bacterial lipopolysaccharide (LPS) should be used for T helper 1 (Th1) cytokine induction, phytohemaglutinin-M (PHA-M) for leukocyte proliferation, zymosan for phagocytosis, etc. In addition to these controls traditionally used in immunology, it is always encouraged to include a control which is specifically relevant to nanoparticles (i.e., a nanoparticle known to induce a particular response). For example, polyethylene glycol (PEG)ylated nano-liposome Doxil® is known to cause complement activation in patients, thus Doxil® can be used as a nano-relevant positive control in an *in vitro* complement activation assay.

Another, and perhaps the biggest challenge encountered in the use of *in vitro* immunotoxicity tests for nanomaterials, is related to their unique physicochemical properties. We have found that these properties may change biodistribution and interfere with traditional *in vitro* tests (Figure 2 and Table 1).

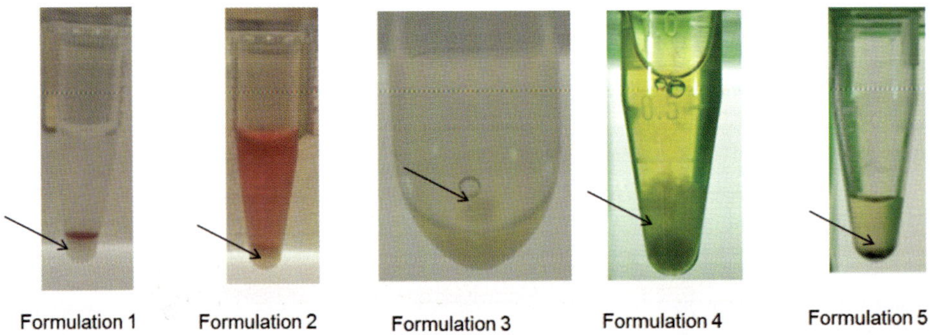

Figure 2. Nanoparticle interference with traditional *in vitro* tests. Due to their large surface-to-volume ratios and novel physicochemical properties, engineered nanomaterials often interfere with traditional *in vitro* tests. The spectrum and mechanism of interference depend largely on the particular nanomaterial and *in vitro* assay (i.e., composition- and assay-dependent). Some examples from the NCL's *in vitro* assay cascade are shown, which has been used to evaluate more than 250 engineered nanomaterials.

Table 1. Nanoparticle interference with traditional *in vitro* tests. Due to their large surface-to-volume ratios and novel physicochemical properties, engineered nanomaterials often interfere with traditional *in vitro* tests. The spectrum and mechanism of interference largely depend on the nanomaterial composition and the particular *in vitro* assay. Some examples from the NCL's *in vitro* assay cascade are shown, which has been used to evaluate more than 250 engineered nanomaterials.

False-Positive Results		False-Negative Results	
Mechanism	**Nanoparticles/assay**	**Mechanism**	**Nanoparticles/assay**
Presence of surfactants	Polystyrene/hemolysis Gold nanorods/ cytotoxicity assays	Adsorbing or binding assay specific test substance	Polystyrene/ hemolysis Gold nanoparticles/ LAL test
Absorbance at assay wavelength	Gold, fullerenes, some liposomes, some nanoemulsions/multiple assays Iron oxide/hemolysis	Presence of surfactants	Polystyrene/LAL test
Oxidation of assay reagents	Iron oxide/hemolysis	Quenching luminescence, quenching fluorescence, or interfere with absorbance	Gold nanoshells, liposomes/multiple assays
Intrinsic fluorescence	Quantum dots, liposomes, doxorubicin-loaded carriers/fluorescence-based assays	Blocking protein binding and interaction	Liposomes/cytokine test Carboxy-terminated dendrimers/ ELISA-based assays
Catalysis of enzymatic reactions	Carboxy-terminated dendrimers/LAL test	Protein binding	Colloidal gold/LAL test

Abbreviations

ELISA enzyme-linked immunosorbent assay LAL Limulus Amebocyte Lysate

In vitro immunotoxicity testing of engineered nanomaterials is conducted with the assumption that all nanoparticles administered to a body remain in circulation or accumulate in the target organ represented by a given cell line. Initially, it was naïvely thought that commercial kits designed for various *in vitro* assays may be applied to engineered nanomaterials the same way these kits and reagents are used in traditional cell biology, immunology, and toxicology. However, this myth was quickly debunked by *in vitro* studies using widely available commercial reagents for nanoparticle testing.[9–12,15] Nanomaterial composition, size, surface chemistry, surface area, color, charge, and catalytic properties may all interfere with common *in vitro* assays.[12,16–19] For example, Pfaller *et al.* recently reported on a large interlaboratory study conducted by several European laboratories to understand the suitability of various *in vitro* cytotoxicity methods, as well as on more specific monocyte and T cell immunostimulatory assays.[11] Their study demonstrated the need for method harmonization and the standardization of reagents and procedures. Another European collaborative study reported problems with validating cytotoxicity and cytokine *in vitro* assays due to the contamination of commercial metal oxide nanomaterials with bacterial endotoxin and chemical synthesis by-products, as well as particle agglomeration in cell culture media and optical interference, limiting the utility of these assays for immunotoxicity testing.[10] Some nanomaterials, such as carbon nanotubes, were shown to interact with MTT-formazan crystals and alter the amount of soluble formazan, thus leading to false-positive results.[20] Monteiro-Riviere and Inman reported that carbon black nanoparticles adsorb cytokine molecules and block them from recognition by cytokine-specific antibodies, causing an underestimation of released cytokines.[21] Our laboratory has reported both false-positive and false-negative test results with an *in vitro* hemolysis assay due to nanoparticle absorbance at the assay wavelength and adsorption of plasma-free hemoglobin, respectively.[16] The presence of surfactants in some nanoformulations also led to both false-positive and false-negative assay results, depending on the assay read-out (Figure 2 and Table 1). In addition, intrinsic luminescence and fluorescent properties, as well as catalytic properties, can result in overestimated toxicity for some engineered nanomaterials (Table 1).

In addition to nanoparticle interference, another problem commonly experienced with the *in vitro* testing of engineered nanomaterials is their instability in the biological matrix (Figure 2). Such instability may lead to a change in nanoparticle properties such as size, aggregation, release of payload, and undesirable interference with *in vitro* assays. For example, nanoparticles may aggregate in the plasma and precipitate out of solution (Figure 2; formulations 1, 2, 3, and 5) or induce the coagulation of plasma proteins (Figure 2; formulation 4). If nanoparticles are unstable in the assay media, the results of *in vitro* testing will be inconclusive and cannot be relied on to make decisions regarding the immunotoxicity of engineered nanoparticles.

Another important lesson learned from more than five years of experience with the NCL *in vitro* immunology assay cascade, is the importance of including inhibition/enhancement controls (IECs) and cell-free nanoparticle-only controls, in addition to traditional controls. Nevertheless, even the inclusion of these controls may not be sufficient to detect interference.[17] This is why the verification of test results using multiple assays estimating the same toxicity endpoint but with different detection systems is another important step in *in vitro* nanoparticle characterization. For example, nanoparticle uptake by immune cells is often used to estimate the potential for capture by the mononuclear phagocytic system and elimination from circulation. Nanoparticles containing electron-dense elements such as gold nanoshells and metal oxide particles can be studied using electron microscopy. Nanoparticles with radioactive labels or containing unique chemical elements can be studied using dosimetry or elemental analysis, respectively. Nanoparticles with intrinsic fluorescence can be analyzed using confocal microscopy. However, "soft" water-soluble nanomaterials (e.g., liposomes, dendrimers, polymeric formulations) that do not contain specific unique chemical elements or radioactive labels may not be easily detected in cells using these methods.

One of the common *in vitro* methods used to study nanoparticle uptake by immune cells is based on luminol uptake during phagocytosis in a luminescence assay. In some cases, the results of the luminol phagocytosis assay are in agreement with the results of orthogonal methods of evaluation of the same nanoparticle formulation, but in other cases, the results may differ. In the example shown in Figure 3, no uptake of either of the two formulations of gold-coated nanomaterials was detected by the luminol assay, but transmission electron microscopic (TEM) analysis showed the uptake of one formulation. This example demonstrates the reliability of the luminol assay for formulation 2, but not for formulation 1, and reinforces that multiple different techniques should be included in the validation of an *in vitro* method.

2. *In Vitro–In Vivo* Correlation as a Measure of Predictability of *In Vitro* Assays

The *in vitro* immunology assay cascade by the NCL was established in 2005. In determining the types of tests to include in this cascade, we relied on several FDA guidance documents describing various considerations for the immunotoxicity assessment of small molecules, macromolecules, and medical devices.[22–28] Our assay cascade is designed for the evaluation of nanomaterials intended for systemic administration (i.e., which will come into contact with blood). It is not optimal for testing nanoparticles that are administered through routes other than intravenously (i.v.) unless they eventually undergo systemic distribution.

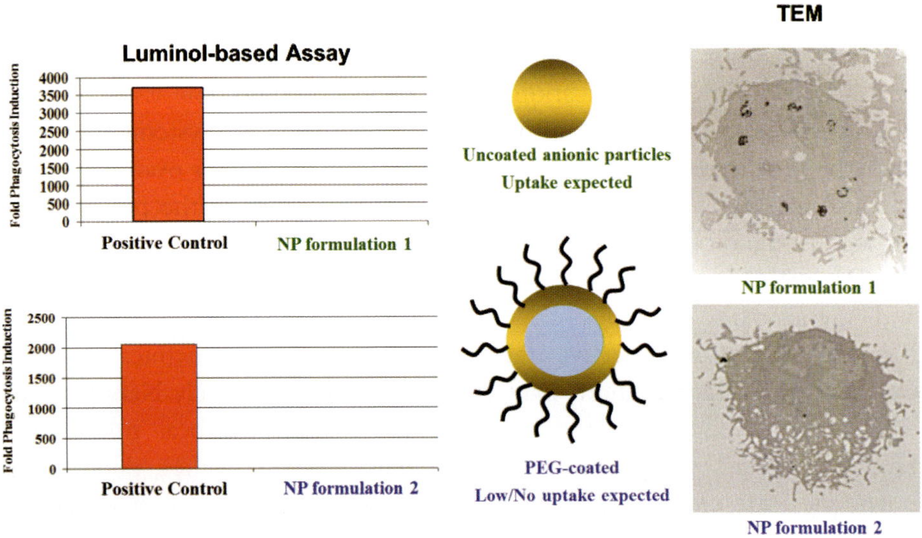

Figure 3. Assessing one immunological read-out using *in vitro* assays using different detection systems is important to verify *in vitro* data. Two nanoparticle formulations containing gold were studied *in vitro* with a phagocytosis assay using the murine macrophage cell line RAW264.7. Formulation 1 was composed of anionic, citrate-stabilized gold nanoparticles, and formulation 2 consisted of gold nanoshells with a PEG-grafted surface. The luminol-based traditional phagocytosis assay showed the uptake of the positive control, yeast particle zymosan A, but not for either of the two tested nanoformulations. TEM analysis confirmed the results of the *in vitro* luminol assay for formulation 2 (i.e., also showed no uptake of this formulation), but clearly showed the uptake of formulation 1 as was theoretically expected based on the surface properties of this formulation. This data shows that the luminol-based phagocytosis assay may be insensitive to certain types of nanoparticles and their uptake by macrophages has to be verified by additional methods wherever possible.

As a general rule, the NCL does not use the *in vitro* assay cascade to make go/no-go decisions about nanomaterials; instead, we use these assays to inform our *in vivo* toxicity studies. A long-term goal is the validation of our *in vitro* assays and a thorough understanding of their predictability with respect to *in vivo* immunological toxicity. Toward this, we follow two approaches: 1) comparison of *in vitro* results with that of standard toxicity studies and 2) retrospective analysis of FDA-aproved formulations that have demonstrated immunotoxicity in patients. This allows us to understand the sensitivity and specificity of our *in vitro* assays with respect to certain types of immunotoxicity. After several years of using those approaches, we have established a good correlation between *in vivo* and *in vitro* data for some assays, which we will share below, and are continuing to add more data points. For example, if the percentage of hemolysis observed *in vitro* using human whole blood treated with certain nanomaterials is low (4–5%), a decrease in hematocrit and hemoglobin levels in the blood of animals injected with the same formulations was also generally observed *in vivo*. Similarly, when the *in vitro* percentage hemolysis is greater than 50%, we observed animal death and necropsy findings consistent with hemolysis (Table 2).

Another example involves the use of bioassays to measure the pyrogenicity of test materials — the *in vitro* macrophage activation test (MAT) and the *in vivo* rabbit pyrogen test (RPT). When a nanomaterial was positive in the MAT *in vitro*, it also was pyrogenic in the RPT, and resulted in fever when injected into animals; similarly, when the result of the *in vitro* pyrogen test was negative, no fever was observed *in vivo* (Table 3).

Table 2. Correlation between *in vitro* hemolysis assays and *in vivo* toxicity studies. Several formulations which demonstrated various degrees of hemolysis *in vitro* were also tested *in vivo*. The analysis of blood samples from animals or tissues at necropsy showed a correlation between the *in vitro* and *in vivo* data.

Nanoparticles	*In vitro*	*In vivo*
Formulation A	5% hemolysis	Low RBC count: low hematocrit in both males and females rats
Formulation B	4.4% hemolysis	Low RBC count: low hematocrit in athymic females
Formulation C	60% hemolysis	Animals died after injection; histology consistent with hemolysis
Formulation D	80% hemolysis	Animals died after injection; histology consistent with hemolysis

Abbreviations

RBC red blood cell

Table 3. Correlation between *in vitro* and *in vivo* pyrogenicity tests. Nanoparticles with different composition were tested for pyrogenic response using an *in vitro* MAT and an *in vivo* RPT. A correlation between *in vitro* and *in vivo* tests was observed for all tested formulations.

Nanoparticles	*In vitro*	*In vivo*
Formulation 1	BLOD	Negative
Formulation 2	BLOD	Negative
Formulation 3	Enhancement or contamination	Positive

Abbreviations

BLOD below limit of detection

Measuring cytokine release in cultured immune cells or whole blood is a widely accepted way to estimate the inflammatory properties of a test material, because all immune responses are accompanied by cytokine release. Cytokines are often used as biomarkers of inflammation, and careful evaluation of cytokine levels and types of cytokines may provide meaningful insight into the mechanism of immunostimulation. While peripheral blood mononuclear cell (PBMC) primary cultures and various monocytic/macrophage cell lines are the most common models used in basic research, whole blood cultures are more popular in the drug development industry. Cytokine analysis in PBMCs, monocytic cell line MonoMac6 (MM-6), and whole blood was validated by the ECVAM and ICCVAM for studies of potential endotoxin contamination of various small molecule-type test materials.[29] This validation effort was limited to interleukin 1 (IL-1) and IL-6 because these inflammatory cytokines are considered to be the primary cause of fever, and the detection of these cytokines is considered as an indication of pyrogenic material properties. In our laboratory, we have used all three models (PBMCs, whole blood cells, and MM-6 cells), and observed consistent results among these models for the same type of engineered nanomaterial. While the cell models are more or less standardized, there is no platform generally agreed upon which can be used for the analysis of culture supernatants for the presence of cytokines. The approaches vary from single analyte traditional ELISA, single cytokine or multiplex flow cytometry methods, and various 96-well plate multiplex platforms. Depending on their physicochemical properties, nanoparticles may interfere with one or more of these platforms. Moreover, even within the same multiplex platform, nanoparticles may enhance the detection of one cytokine, inhibit the detection of another, and have no effect on the detection of others. This is why validating a single analyte immunoassay for the detection of cytokines in culture supernatants containing nanoparticles is easier than for a multiplex platform. After the interference challenges are overcome, the next important question is the relevance of the *in vitro* cytokine assay to nanoparticle toxicity *in vivo*. In our laboratory, we use the following approach: 1) screen for cytokine secretion using a multiplex platform, 2) verify multiplex data using a single analyte traditional ELISA, and 3) analyze plasma samples from a single-dose acute toxicity study for the presence of cytokines exhibiting high expression levels *in vitro*. Below is an example demonstrating a correlation between *in vitro* cytokine secretion by human PBMCs and *in vivo* cytokine secretion in the plasma of animals injected with the same formulation (Figure 4).

Hypersensitivity reactions are not uncommon and can be induced by a wide variety of substances including medications. Types of hypersensitivity reactions and tests for their identification are described in Chapter 17. One

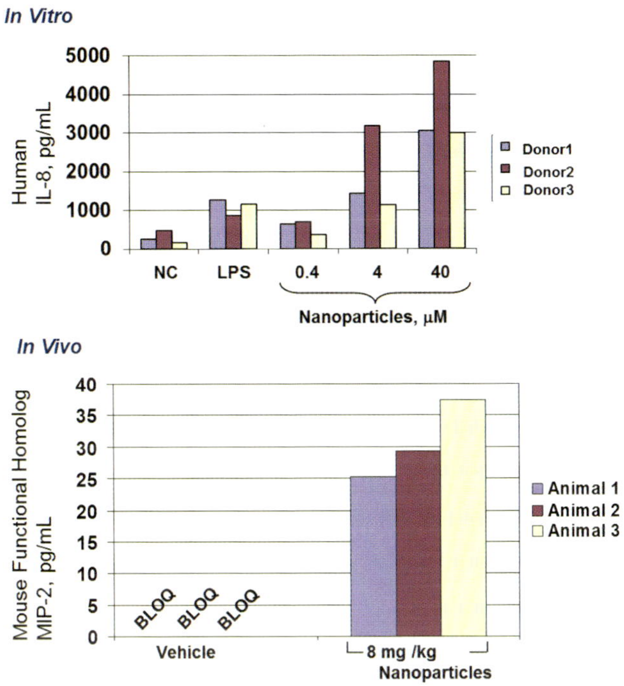

Figure 4. Correlation between *in vitro* and *in vivo* cytokine response tests. Human PBMCs were treated *in vitro* with various concentrations of nanoparticles. Culture supernatants collected 24 h after treatment and were subsequently analyzed by ELISA for the presence of IL-8. Since rodents do not have IL-8, the plasma from animals injected with the same nanoparticle formulation was tested for the presence of the IL-8 functional homolog MIP-2. BLOQ: below limit of quantification.

such reaction is mediated by the activation of the complement system, known as complement activation-related pseudoallergy or CARPA (see Chapter 11). Complement activation can be monitored *in vitro* using several methods including Western blot[30] and ELISA. Some formulations (e.g., Taxol® and Doxil®) have been shown to activate the complement system in patients,[31–34] while others (e.g., Abraxane®) did not demonstrate this activity and is associated with toxicity.[35] This is why we use Taxol® and Doxil® as additional positive controls and Abraxane® as an additional negative control in our *in vitro* complement activation assay to challenge the assay sensitivity for the detection of clinically relevant complement activation. Cobra venom factor, which is also used as a positive control in this assay, is a very potent inducer of complement activation and is not relevant to clinical settings. With this approach, we have established a correlation between the results of our *in vitro* complement activation assay and *in vivo* clinical studies of Taxol® and Abraxane® (Figure 5).

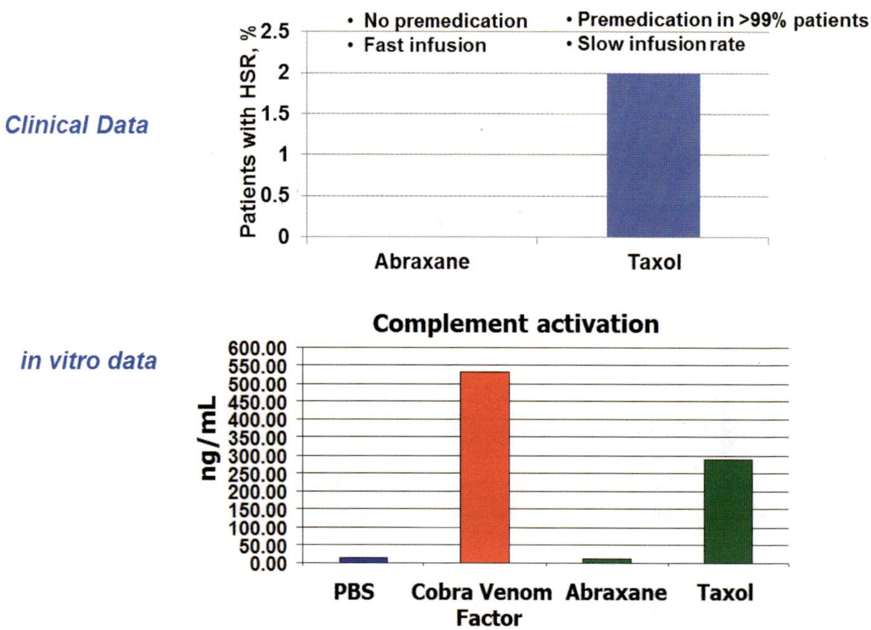

Figure 5. ***In vitro–in vivo* correlation for complement activation assay.** Complement activation-related pseudoallergy (CARPA) is one of the several types of hypersensitivity reactions which can be observed in patients receiving the Cremophor®-EL formulation of the anti-cancer drug paclitaxel (Taxol®). The frequency and severity of the hypersensitivity reaction to the nano-albumin formulation of the same anticancer drug (Abraxane®) is much lower. Upper panel: Rate of hypersensitivity reactions including CARPA is higher in Taxol than in Abraxane® patients. Clinical data are plotted from Gradishar *et al.*[35] Lower panel: Human plasma from healthy donors was treated *in vitro* with Taxol® or Abraxane® at equivalent paclitaxel doses. Complement activation was monitored by ELISA specific to the human iC3b component of complement.

Leukocyte procoagulant activity (PCA) is another endpoint which can be monitored *in vitro* in combination with platelet aggregation and plasma coagulation assays to understand the potential for a test nanomaterial to induce disseminated intravascular coagulation (DIC). DIC is a coagulation disorder characterized by blood clotting and abnormal bleeding, and is a common dose-limiting toxicity in sepsis[36–39] and cancer.[40–44] In cancer patients, DIC often starts after the initiation of chemotherapy. Cytotoxic oncology drugs that act by altering DNA replication (e.g., doxorubicin, daunorubicin, vinblastin, and vincristin) are common culprits in DIC complications in clinic.[42,45] Interestingly, cytotoxic oncology drugs acting via other mechanisms (e.g., methotrexate and paclitaxel) do not induce DIC.[46,47] The link between PCA in leukocytes *in vitro* and DIC *in vivo* has been established for oncology drugs (doxorubicin, vincristin, daunorubicin) and endotoxin, and is related to the expression of PCA complex composed of tissue factor and phosphatidylserine on the cell surface.[42,48–55] Platelet activation/aggregation for some engineered nanomaterials *in vitro*[56] has been linked to DIC-like toxicity *in vivo*.[57] DIC is repeatedly reported for engineered nanomaterials and thus, there is a need for an *in vitro* assay which can provide information regarding nanomaterial properties related to procoagulant toxicity. To verify the sensitivity of this *in vitro* test, one can use the same approach described above for other assays (i.e., test nanoformulations and/or other drugs and compounds known to cause DIC in the clinic). For example, doxorubicin was shown to induce DIC in acute promyelocytic leukemia (APL) patients, but this type of toxicity was not reported for the liposomal formulation of the same drug Doxil®.[58] When doxorubicin and Doxil® were tested *in vitro* in a leukocyte procoagulant assay using the APL cell line as a model, PCA was observed with doxorubicin, but not with Doxil® (Figure 6).

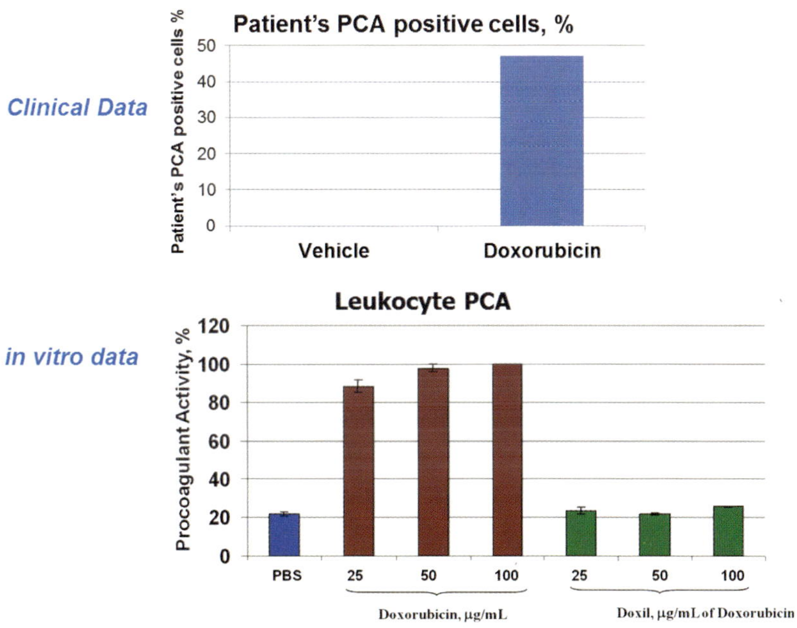

Figure 6. ***In vitro–in vivo* correlation for the leukocyte PCA test.** DIC is a common complication in cancer especially after the onset of chemotherapy. The traditional formulation of the anti-cancer agent doxorubicin is known to cause PCA and DIC complications in cancer patients. Upper panel: Data demonstrating the high percentage of PCA-positive cells in patients treated with doxorubicin from Schneider *et al.*[58] Lower panel: Results of the *in vitro* PCA assay shows the induction of PCA in acute promyelocytic leukemia (APL) cell line HL-60 after treatment with doxorubicin salt (Adriamycin®). The nano-liposomal formulation of the same anticancer agent (Doxil®) when used at equivalent doxorubicin concentrations did not cause this toxicity. The protocol for the PCA assay is presented in detail in this chapter.

3. *In Vitro* Assays to Monitor Immunomodulatory Properties of Nanoparticles

The *in vitro* assays summarized in Table 3 were developed to monitor the immunomodulatory properties of nanoparticles injected into blood or known to distribute to the systemic circulation. Some of these tests have been published before, while some are described in more detail in this chapter. Two of these tests, for hemolysis[59] and granulocyte macrophage colony-forming units (CFU-GM),[64] have also been standardized by ASTM International for the analysis of nanoparticle immunotoxicity. The methods presented in the table and described below can be used either alone or in combination with other *in vitro* methods to estimate the immunostimulatory, and in some cases, immunosuppressive properties of nanomaterials. Regardless of whether these methods are used alone or in combination, it is very important that a test nanomaterial meet two conditions prior to analysis in these assays — 1) it has passed physicochemical characterization (see Chapter 2) and 2) it has been screened for endotoxin contamination and proven that endotoxin is at acceptable levels (see Chapter 4).

3.1. *Effects of nanoparticles on the formation of CFU-GM in vitro*

Myelosuppression is a common dose-limiting toxicity of cytotoxic oncology drugs. Nanoparticles may be distributed to the BM and/or release a drug that is distributed to the BM. Therefore, understanding the potential toxicity of nanoparticles or drugs which they carry is an important step in preclinical safety evaluation. Below, we present a protocol which describes the quantitative analysis of CFU-GM by employing the murine BM. Hematopoietic stem cells of the BM proliferate and differentiate to form discrete cell clusters or colonies. This protocol can be used for both *in vitro* and *ex vivo* analyses. The *in vitro* protocol involves the isolation of BM cells from healthy animals, followed by treatment *in vitro* with nanoparticle formulations. In the *ex vivo* version, the BM is isolated from animals injected with the nanoparticle formulation. The *in vitro* protocol does not account for nanoparticle biodistribution; however, in cases when dose information is not available and nanoparticle formulation is in the early phase of development, the *in vitro* protocol allows for the rapid screening of potentially toxic nanoparticle formulations. The *in vitro* protocol can also give a quick estimation of the myelosuppressive potential of a cytotoxic oncology drug bound to a nanoparticle surface, in comparison to a traditional formulation of the same cytotoxic oncology drug (for an

Table 4. *In vitro* assays to study immunotoxicity of engineered nanomaterials administered to or distributed into systemic circulation.

Assay	Purpose	Reference
Hemolysis	To test nanoparticle potential for damaging erythrocytes	59
Platelet aggregation	To test nanoparticle potential for activating platelets and inducing platelet aggregation; *in vitro* surrogate for thrombogenic properties	60
Plasma coagulation	To test nanoparticle potential for affecting plasma coagulation cascade; *in vitro* surrogate for thrombogenic properties	60
Complement activation	To test nanoparticle potential for triggering activation of complement system; potential to mediate CARPA *in vivo*; immunostimulation test	30
Interaction with plasma proteins	To test binding of opsonins and other proteins to nanoparticle surface which may affect their biodistribution and clearance from systemic circulation	http://ncl.cancer.gov/NCL_Method_ITA-4.pdf
Leukocyte proliferation	To test nanoparticle ability to activate leukocyte and induce their proliferation; *in vitro* test for immunostimulation	This chapter
Chemotaxis	To test nanoparticle ability to induce migration of immune cells to the sites of distribution; *in vitro* test for immunostimulation	61
Phagocytosis	To test nanoparticle uptake by the immune cells; *in vitro* test for immunostimulation	62, 63
CFU-GM	To test nanoparticle effects on proliferation and differentiation of BM cells into granulocytes and macrophages; important for nanoparticles loaded with cytotoxic oncology drugs and nanoparticles with known distribution to the BM; an immunosuppression test	64 and this chapter
Nitric oxide secretion	To test nanoparticle ability to induce oxidative burst of macrophages; *in vitro* test for immunostimulation	This chapter
Effects on maturation of dendritic cells	To test nanoparticle ability to induce maturation of monocyte-derived dendritic cells *in vitro* or to inhibit endotoxin-induced maturation	This chapter
Cytotoxicity of NK cells	To test nanoparticle ability to inhibit cytotoxic properties of natural killer cells; *in vitro* test for immunosuppression	This chapter
Leukocyte procoagulant activity	To test nanoparticle ability to induce expression of PCA complex on leukocyte surface; in combination with platelet activation and plasma coagulation time assays can be used to estimate nanoparticle potential in causing DIC	This chapter

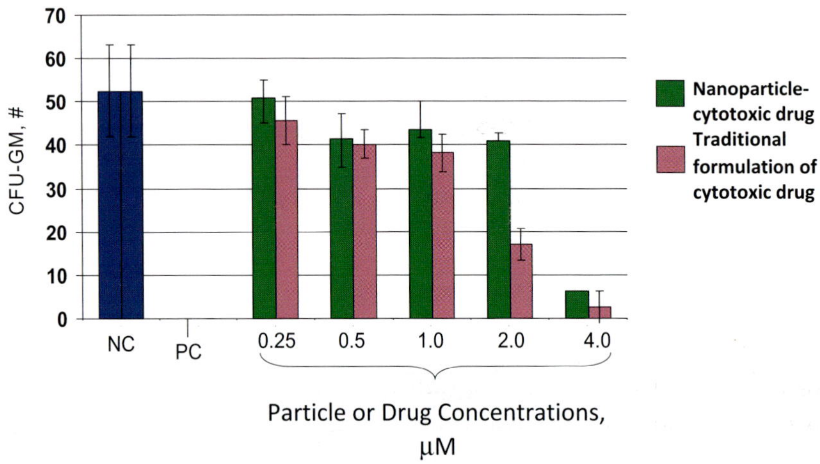

Figure 7. ***In vitro* comparison of traditional and nanoformulations of cytotoxic oncology drug in an CFU-GM assay.** BM cells were treated *in vitro* with a nanoparticle-bound cytotoxic oncology drug or with a traditional formulation of the same cytotoxic oncology drug. The effects of the drug on CFU-GM formation were studied. Cisplatin was used as the positive control. PBS (phosphate buffered saline) was used as the negative control. The API (active pharmaceutical ingredient) was used to compare the cytotoxic oncology drug in equimolar concentrations.

example, see Figure 7). The protocol principle involves the isolation of BM cells from eight- to 12-week-old mice and the culture of these cells in methylcellulose-based medium supplemented with cytokines (mSCF, mIL-3, and hIL-6) either untreated (baseline) or treated with nanoparticles (test). The cytokines used in the MethoCult media promote the formation of CFU-GM. After a 12-day incubation of the BM cells, the number of colonies is quantified in both the baseline and test samples. A percent inhibition of CFU is then calculated for each test sample. The basic protocol for BM isolation and culture was adopted from technical manual #28405 developed by StemCell Technologies, Inc. (Mouse Colony-Forming Cell Assays Using MethoCult. Technical manual. StemCell Technologies, Inc., cat #28405; Pifat, Dominique. Assay Validation. Center for Biologics Evaluation and Research, FDA, 2003.).

If the *ex vivo* study is required, the *in vitro* addition of nanoparticles and controls into the cell culture medium is not necessary (outlined in Section 3.1.6).

3.1.1. *Reagents, materials, and equipment*

- MethoCult medium (StemCell Technologies, Inc., 03534)
- Fetal bovine serum (FBS) prescreened for hematopoietic stem cells (StemCell Technologies, Inc., 06200)

- Iscove's MDM with 2% FBS (StemCell Technologies, Inc., 07700)
- Sterile distilled water
- Cisplatin (positive control) (Sigma, P4394)
- Sterile Ca^{2+}-/Mg^{2+}-free DPBS (negative control) (Sigma, D8537)
- 3% acetic acid with methylene blue (StemCell Technologies, Inc., 07060)
- Pipettes covering the range of 0.05 to 10 mL
- Prescreened 35-mm culture dishes (StemCell Technologies, Inc., 27100)
- 16-gauge blunt-end needles (StemCell Technologies, Inc., 03534)
- 100-mm Petri dishes
- Plastic beakers
- Polypropylene tubes, 5 and 15 mL
- Scissors for tissue dissection
- Forceps
- Centrifuge
- Refrigerator, 2–8°C
- Freezer, –20°C
- Cell culture incubator with 5% CO_2 and 95% humidity
- CO_2 euthanasia box or appropriate equipment approved by institution
- Biohazard safety cabinet approved for level II handling of biological material
- Inverted microscope
- Vortex
- Hemocytometer

Note: The NCL does not endorse any of the suppliers listed above; their inclusion is for informational purposes only. Equivalent supplies from alternative vendors can be used in place.

3.1.2. *Animals*

This protocol utilizes six- to 12-week-old C56BL6 males or females. The use of pooled cells derived from at least two animals is highly recommended. Prior to carrying out this protocol, one has to receive approval from their institutional Animal Care and Use Committee.

3.1.3. *Preparation of reagents and controls*

- MethoCult Medium

The MethoCult medium is supplied in 100-mL size batches. It is recommended by the manufacturer that the medium be thawed at room temperature or in a refrigerator overnight, vortexed to mix well, then left at a room temperature for approximately 5 min to allow air bubbles to dissipate. A 16-gauge blunt-end needle is used to aliquot MethoCult medium, 3 mL into sterile 15-mL tubes. Store the aliquoted medium at a nominal temperature of –20°C. Before the test, thaw the required number of aliquots at room temperature for approximately 20 min and keep on ice prior to use. Repeated freeze/thaw of the medium should be avoided.

- 50 mM cisplatin (positive control)

Cisplatin is supplied in a lyophilized form. Reconstitute the lyophilized powder by adding an appropriate amount of DMSO to make a stock solution with a nominal concentration of 50 mM. Prepare small aliquots and store at a nominal temperature of –80°C. Prior to use in the assay, thaw an aliquot of the stock solution at room temperature and dilute in IMDM supplemented with 2% FBS to bring the concentration to 2 mM. 150 μL of this intermediate solution is then added to 3 mL of MethoCult medium. The final concentration of cisplatin in the positive control sample is 50 μM.

Note: this control is not required if the ex vivo study is conducted because the ex vivo study has its own negative and positive controls.

3.1.4. *Preparation of study samples*

This assay requires 0.6 mL of nanoparticle solution at 22× the desired final test concentration (i.e., three 0.15-mL replicates per test sample). The nanoparticle should be dissolved/resuspended in IMDM medium.

The concentration is selected based on the intended therapeutic concentration of the nanoparticle. If the intended therapeutic concentration is 0.1 mg/mL, then the highest final concentration of particle tested in the assay is 10 mg/mL (i.e., 100× above the intended therapeutic concentration). If the particle solubility does not allow for the testing of the 100× concentration, then the highest final concentration is 30× the intended therapeutic concentration.

For example, to achieve the highest final test concentration of 10 mg/mL, the stock nanoparticles should be prepared at 220 mg/mL to account for a 1:22 dilution in the test well. Ideally, the nanoparticle concentration should be given in the amount of active component API. When the concentration of the active component is unknown, the weight of the total formulation can be used. When the intended therapeutic concentration is unknown, the highest final concentration tested is 1 mg/mL, or the highest feasible concentration.

Nanoparticles are tested in this assay at four concentrations: the highest concentration and three 1:5 serial dilutions. For example, if the highest final concentration to be tested is 1 mg/mL, then a stock of 22 mg/mL is used. Three serial dilutions are made from this sample to yield samples with concentrations of 4.4, 0.88, and 0.18 mg/mL. When 0.15 mL of each of these samples is added to 3.0 mL of cell culture media (from Section 3.1.3) and mixed with 0.15 mL of cell suspension (from Section 3.1.5), the final nanoparticle concentrations tested in the assay are 1, 0.2, 0.04, and 0.008 mg/mL.

3.1.5. *Isolation and counting of BM Cells*

- Position euthanized mouse on its back and rinse fur thoroughly with 70% alcohol. (Euthanize animals according to the protocol approved by your institution.)
- Cut a slit in the fur just below the rib cage without damaging the peritoneal membrane.
- Firmly grasp the skin and peel back to expose the hind limbs.
- Using sterile sharp dissecting scissors, cut the knee joint in the center. Cut through ligaments and excess tissue.
- Grasp the femur with forceps and cut the femur near the hip joint.
- Free the tibia by cutting near the ankle joint.
- Trim the ends of the long bones to expose the interior marrow shaft. Put the bones in sterile Petri dish or in sterile culture medium, and place on ice. Bones can be collected from multiple animals.
- Using a 3cc syringe with 21- or 22-gauge needle, draw up to 1–3 mL of cold Iscove's MDM supplemented with 2% FBS.
- Insert the bevel of the needle into the marrow shaft and flush the marrow into a 15-mL tube. Repeat this procedure for all bones. The same medium can be used to isolate the marrow from one to three animals. The bone shallow should appear white once all the marrow has been expelled.
- Keeping the needle below the medium surface, gently draw the medium with cells up and down with a 3cc syringe and 21-gauge needle three to four times to make a single cell suspension.
- Keep the cells in medium, on ice until use.
- Perform a nucleated cell count. To do so, first dilute the cells 1:100 with 3% acetic acid with methylene blue (e.g., 10 μL cells + 990 μL 3% acetic acid/methylene blue). Then, use either a hemocytometer or an automatic cell counter to obtain counts. An average cell count is expected to be 1–2 $\times 10^7$ from the femur and 0.6–1 $\times 10^7$ from the tibia.
- If cell viability (at least 90%) and count are acceptable, proceed to the next step.

3.1.6. *Experimental procedure*

- Label the lids of 35-mm culture dishes at the edge using a permanent, fine-tip felt marker.
- Thaw the MethoCult medium at room temperature or in a refrigerator overnight.
- Vortex the tubes to ensure all components are thoroughly mixed.
- Dilute the cells isolated according to the procedure described in Section 3.1.5 with Iscove's medium supplemented with 2% FBS to 4×10^5 cells/mL.
- Add 150 μL of cell suspension and 150 μL of either Iscove's medium with 2% FBS (baseline), PBS (negative control), cisplatin (positive control), or nanoparticles (test sample) to 3 mL of MethoCult medium.

  (Note: If the *ex vivo* study is conducted, add 150 μL of cell suspension from individual animals and 150 μL of Iscove's medium with 2% FBS to 3 mL of Methocult medium. No addition of PBS, nanoparticles, or cisplatin is required because animals were previously injected with various formulations representing vehicle control, nanoparticle treatment, and/or traditional cytotoxic oncology drug formulation.)
- Vortex the tubes to ensure all cells and medium components are mixed thoroughly.
- Let the tubes stand for 5 min to allow bubbles to dissipate.
- Attach 16-gauge blunt-ended needle to a 3cc syringe; place the needle below the surface of the solution containing MethoCult media with cells in it from steps described in Section 3.1.5 and draw up approximately 1 mL. Gently depress the plunger and expel the medium completely. Repeat until no air space is visible.
- Draw up MethoCult medium with cells into the syringe and dispense 1.1 mL per 35-mm dish. All samples are tested in duplicate ($n = 2$) (i.e., prepare two 35-mm dishes for each sample tested).
- Distribute the medium evenly by gently tilting and rotating each dish.
- Place the two covered dishes with cells and one uncovered dish filled with 3 mL of sterile water into a 150-mm Petri dish.
- Place the cultures in an incubator maintained at 37°C, 5% CO_2, and 95% humidity.
- Incubate for 12 days. On the 12th day, remove dishes from the incubator, and identify and count colonies as described below. The representative values of CFU-GM for C57BL6 mice at eight to 12 weeks of age is 64 ± 16.

3.1.7. *Description of CFU-GM*

This classification includes CFU-granulocyte (CFU-G), CFU-macrophage (CFU-M), and CFU-GM. The colonies contain 30 to thousands of CFU-G,

CFU-M, or both cell types (CFU-GM). CFU-GM colonies often contain multiple clusters and appear as a dense core surrounded by cells. The monocytic lineage cells are large cells with an oval to round shape and appear to have a drainy or grey center. The granulocytic lineage cells are round, bright, and are much smaller and more uniform in size than macrophages. It is easy to distinguish among individual cells of a CFU-GM colony, especially in the periphery of the colony. See Figure 8 for a depiction of CFU-GM, CFU-M, and CFU-G colonies.

3.1.8. *Calculations*

The following parameters should be calculated for each control and test sample:

$$\text{Percent Coefficient of Variation (\%CV)} = \frac{\text{SD}}{\text{Mean}} \times 100\%$$

and

$$\text{Percent CFU Inhibition} = \frac{(\text{Baseline CFU-GM} - \text{Test CFU-GM})}{\text{Baseline CFU-GM}} \times 100\%.$$

3.1.9. *Acceptance criteria*

- %CV for each control and test sample should be less than 30%.
- If the positive control or negative control fails to meet the acceptance criterion for %CV described above, the assay should be repeated.
- Within the acceptable assay, if two of three replicates of the unknown sample fail to meet the acceptance criterion for %CV, this unknown sample should be re-analyzed.
- If two duplicates of the same study sample give results showing a difference of more than 25%, this sample should be re-analyzed.

3.2. *Effects of nanoparticles on the cytotoxicity of NK cells in vitro*

NK cells are a type of lymphocyte which play a major role in the host-rejection of both cancer cells and cells infected by viruses. NK cells carry small granules in their cytoplasm which contain special proteins such as perforin and granzymes. When NK cells release perforin in close proximity to target cells (i.e., tumorous or virus-infected cells), it forms pores in the cell membrane of the target cell through which the granzymes and

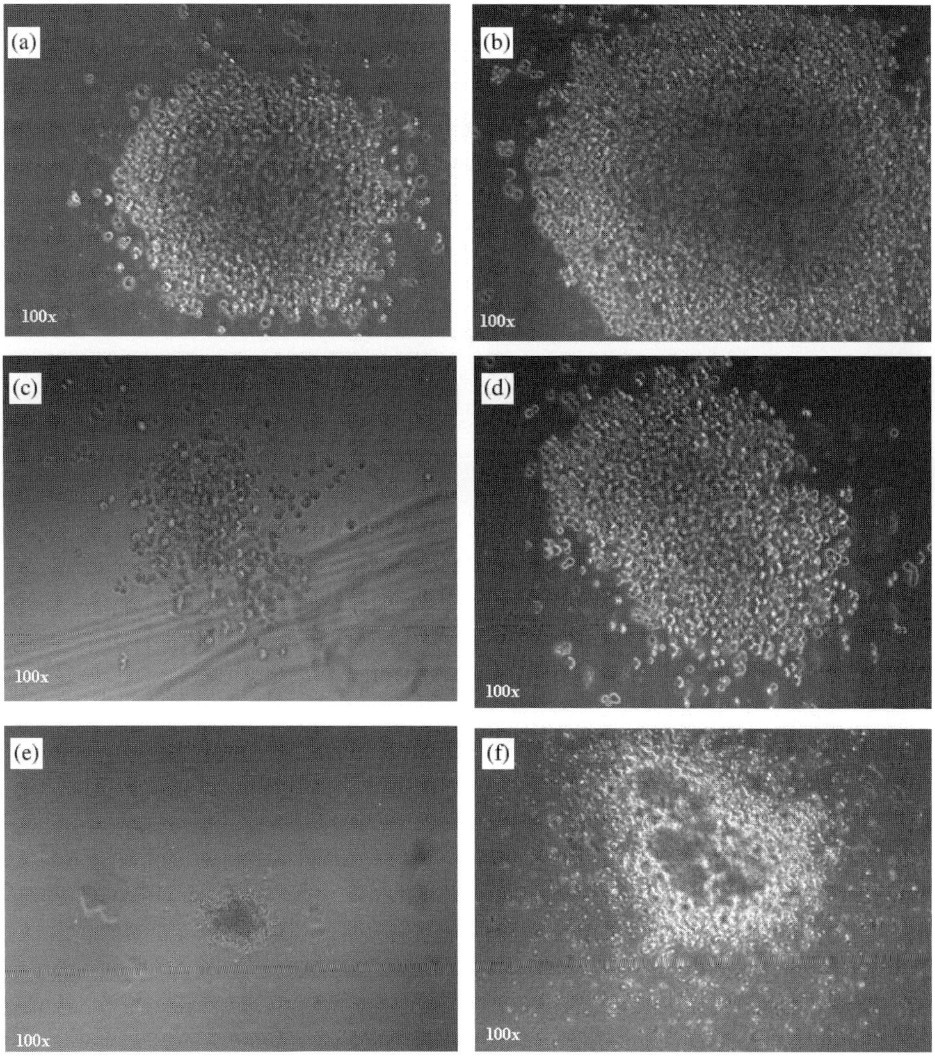

Figure 8. Depiction of CFU-GM, CFU-M, and CFU-G colonies. (a) and (b) are CFU-GM colonies. (c) and (d) are CFU-M colonies. (e) depicts a single CFU-G colony. (f) shows a few CFU-G colonies growing together.

associated molecules can enter, inducing apoptosis. The cytotoxic activity of NK cells is an important component of the innate immunity, which provides a quick body response to cancerous or virus-infected cells before the more specialized adaptive immunity is generated. Understanding a drug's effect on the cytotoxicity of NK cells is thus an important part of immunotoxicity studies aimed at identifying potential immunosuppression. Below, we describe a protocol for assessing the effect of nanoparticles on the capacity of human NK cells to lyse tumorous target cells under *in vitro* conditions. In this method, the NK92 cell line is used as the model for NK cells, and the hepatocellular carcinoma HepG2 cell line is used as the model for target cells. The viability of target cells is monitored through a label-free real-time cell-electronic system (RT-CES), which was reported by Abashi and Zhu.[65]

3.2.1. *Reagents, materials, and equipment*

- PBS (HyClone, AQB 22934)
- FBS (Hyclone, SH30070.03)
- Horse serum (Hyclone, SH30074.03)
- MEM, alpha modification (Hyclone, SH30568.01)
- RPMI-1640 (Hyclone, SH30096.01)
- L-glutamine (Hyclone, SH30034.01)
- Myo-inositol (Sigma, I7508)
- Folic acid (Sigma, 235873)
- β-mercaptoethanol (Invitrogen, 21985–023)
- Recombinant human IL-2 (Peprotech, 200–02)
- Trypan blue solution (Invitrogen, 15250–061)
- Pipettes ranging from 0.05 mL to 10 mL
- Flat-bottomed 96-well plates
- Polypropylene tubes, 5 and 15 mL
- RT-CES plates (ACEA Biosciences)
- Centrifuge
- Refrigerator, 2–8°C
- Freezer, –20°C
- Cell culture incubator with 5% CO_2 and 95% humidity
- Biohazard safety cabinet approved for level II handling of biological material
- Inverted microscope
- Vortex
- Hemacytometer
- RT-CES instrument (ACEA Biosciences)

Note: The NCL does not endorse any of the suppliers listed above; their inclusion is for informational purposes only. Equivalent supplies from alternative vendors can be used in place.

3.2.2. *Preparation of reagents*

- Complete RPMI medium (to maintain HepG2 cells)
  The complete RPMI medium should contain the following reagents:

  10% FBS (heat-inactivated)
  2 mM L-glutamine

Store at 2–8°C, protected from light, for no longer than one month. Before using, warm in a water bath.

- Complete alpha-MEM (to maintain NK92 cells)

  2 mM L-glutamine
  1.5 g/L sodium bicarbonate
  0.2 mM inositol
  0.1 mM β-mercaptoethanol
  0.02 mM folic acid
  100–200 U/mL recombinant IL-2
  10% horse serum
  10% FBS

- Heat-inactivated FBS

  Thaw a bottle with FBS at room temperature or overnight at 2–8°C, and allow it to equilibrate to room temperature. Incubate for 30 min at 56°C in a water bath, mixing every 5 min. Single-use aliquots may be stored at 2–8°C for up to one month or at a nominal temperature of −20°C.

3.2.3. *Preparation of study samples*

This assay requires 500 μL of the nanoparticle at a concentration 10× above the highest final concentration to be tested; the final dilution in the cell culture media is 1:10. Nanoparticles should be dissolved/resuspended in alpha-MEM medium. The following questions have to be considered when selecting the concentration: 1) solubility of nanoparticles in a biocompatible buffer, 2) pH within physiological range, 3) availability of nanomaterial, 4) stability, and 5) cytotoxicity of nanoparticles to NK cells. For example, many nanoparticles containing cytotoxic oncology drugs may result in 100% death of NK92 cells; in such cases, it is recommended to use nanoparticles at their IC_{50} concentration as the highest concentration. A minimum of three dilutions of the highest possible/reasonable concentration is required. For example, if the highest test concentration is 1.0 mg/mL, then three serial 1:5 dilutions of this stock are tested (0.2, 0.04, and 0.008 mg/mL).

3.2.4. *Preparation of effector and target cells*

- Preparation of NK92 effector cells

Grow cells in complete medium. Start new cultures at 1×10^6 viable cells/mL. Subculture at 5×10^6 cells/mL. Replace or add medium every two to three days.

Cultures can be maintained by the addition or replacement of medium. When replacing media, centrifuge cells and resuspend cell pellet in fresh medium at $2–3 \times 10^5$ viable cells/mL. Pipette the cells up and down on the back of the flask every two to three days to produce a single-cell suspension. NK92 cells are extremely sensitive to overgrowth and media exhaustion. Replace with fresh medium every two to three days (depending on cell density).

- Preparation of HepG2 target cells

Grow cells in complete medium. Renew growth media twice a week. A subcultivation ratio of 1:4 or 1:6 is recommended. To split the cells, remove and discard the culture medium; briefly rinse the cell layer with 0.25% (w/v) trypsin- 0.53 mM EDTA solution to remove all traces of serum (contains a trypsin inhibitor). Then, add 2.0 to 3.0 mL of trypsin-EDTA solution to the flask and observe cells under an inverted microscope until the cell layer is dispersed (usually within 2 to 5 min).

To avoid clumping, do not agitate the cells by hitting or shaking the flask while waiting for the cells to detach. Cells that are difficult to detach may be placed at 37°C to facilitate dispersal. Add 6.0 to 8.0 mL of complete growth medium and aspirate cells by gentle pipetting. Add appropriate aliquots of the cell suspension to new culture flasks and incubate the cultures at 37°C in a 5% CO_2 environment.

3.2.5. *Experimental procedure (this will require three days)*

Day 1

- Adjust effector cell (NK92) number to 1×10^6 cells/mL using alpha-MEM. Prepare 15 mL of cell suspension for each sample and negative control.
- Treat NK92 cells with test nanoparticles or negative control for 24 ± 0.5 h.
- Adjust target cell (HepG2) number to 0.5×10^5 cells/mL using RPMI.
- Plate 50 μL of media to all wells, attach the plate to RT-CES and begin program. Following background measurement, plate 50 μL of HepG2 cells from Section 3.2.4 per well in RT-CES plates, attach to RT-CES, and start data acquisition. HepG2 cells are in culture for ~ 16–20 h prior to the addition of NK92 effector cells. The acquisition program is described in Table 5.

Day 2

- Using Trypan blue, determine the viability of NK92 cells prepared on Day 1.

Table 5. RT-CES acquisition protocol.

Step #	Step name	Interval (min)	Sweeps
1	Step 1	1.00	
2	Step 2	30.0	1
3	Step 3	2.00	500
	Step 3	10.0	30
4	Step 4	IDLE	1000

- Concentrate NK92 cells from Day 1 treatment by centrifugation (5 M, 400 *g*); remove and discard supernatants, and reconstitute cells in each sample with fresh alpha-MEM media to a final concentration of 2.5×10^6 viable cells/mL. This concentration will allow for an effector-to-target (E:T) ratio of 5:1. An E:T ratio of 2.5:1 or 1.25:1 is also acceptable and in this case, the NK92 concentration should be adjusted to 1.25×10^6 or 0.625×10^6 viable cells/mL. If the viability of NK cells treated with the negative control is ≥97%, proceed to the next step.

  Note: If nanoparticles were cytotoxic to NK92 cells and resulted in more than 50% cell death, it may not be possible to evaluate the cytotoxicity of NK92 cells treated with these nanoparticles due to the lack of the required number of effector cells.
- Pause the RT-CES data acquisition program, remove RT-CES plates from the instrument, and add 100 μL of NK cells from Day 1 to the designated wells on RT-CES plate. An example of a template is provided in Figure 9.
- Return RT-CES plates containing NK92 effector cells treated with either the nanoparticle or negative control and target (HepG2) cells to the instrument and resume data acquisition for another 24 h.

Day 3

- Stop the acquisition program on the RT-CES instrument and analyze the data.
- If a decrease in NK cytotoxicity is observed in the nanoparticle-treated sample, remove an aliquot of NK cells and count cells using Trypan blue. This control will clarify whether the decrease in NK cell cytotoxicity is attributed to NK cell death induced by particles residing inside these cells and therefore, not detectable by cell count performed on Day 2 after NK cell incubation with nanoparticles and before adding these cells to RT-CES plates containing growing HepG2 cultures (Day 2).

For each nanoparticle, the test should be repeated at least once.

HepG2 only (no NK92)	HepG2 only (no NK92)
HepG2 Untreated NK92	HepG2 Untreated NK92
HepG2 NK92 + NC	HepG2 NK92 + NC
HepG2 NK92 + NP (1.0 mg/mL)	HepG2 NK92 + NP (1.0 mg/mL)
HepG2 NK92 + NP (0.2 mg/mL)	HepG2 NK92 + NP (0.2 mg/mL)
HepG2 NK92 + NP (0.04 mg/mL)	HepG2 NK92 + NP (0.04 mg/mL)
HepG2 NK92 + NP (0.008 mg/mL)	HepG2 NK92 + NP (0.008 mg/mL)
NP (1.0 mg/mL)	NP (1.0 mg/mL)

Figure 9. Example of an RT-CES plate template. In this example, 1.0 mg/mL is the highest tested concentration and the three serial 1:5 dilutions are 0.2, 0.04, and 0.008 mg/mL. The testing of the "nanoparticle-only" sample is recommended to identify potential nanoparticle interference with the RT-CES instrument acquisition system. During the validation of gold nanoshells, citrated gold colloids, colloidal silver, iron oxide, polyamidoamine (PAMAM) dendrimers, water-soluble fullerene derivatives, TiO_2 particles, and conductive materials such as gadolinium and ruthenium, no interference was observed.

3.2.6. *Calculations*

An example of cell index plot is shown in Figure 10.

A %CV is used to control precision and is calculated for each control or test sample according to the following formula:

$$\%\text{CV} = \frac{\text{SD}}{\text{Mean}} \times 100\%$$

and

$$\text{Percent Cytotoxicity} = \frac{\text{AUC}_{\text{particles}}}{\text{AUC}_{\text{negative control}}} \times 100\%,$$

where AUC is the area under the curve determined for HepG2 growth from the time of the addition of NK2 effector cells to the time when data acquisition was stopped (i.e., 24 ± 0.5 h later), and normalized to the number of cells plated in each individual well.

3.2.7. *Acceptance criteria*

- %CV of test samples and control should be within 25%.
- The assay is acceptable if percent cytotoxicity in the negative control is ≥50%.

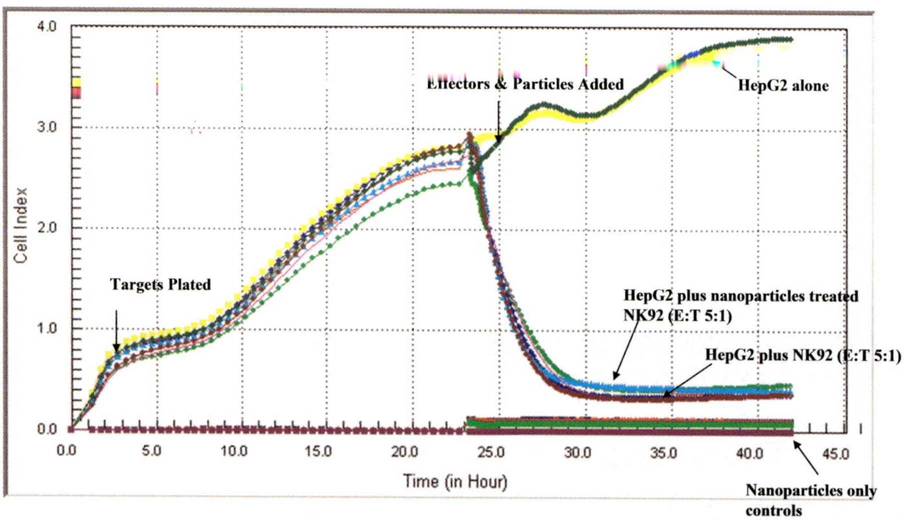

Figure 10. Example of a cell index plot. The raw data demonstrates changes in the HepG2 cell index after co-incubation with NK92 cells.

3.3. *Effects of nanoparticles on the maturation of monocyte-derived dendritic cells in vitro*

Dendritic cells (DCs) are antigen-presenting cells. They are involved in the initiation and regulation of the cellular immune response. The maturation of dendritic cells is achieved by inflammatory cytokines (e.g., TNF-α) or pathogen-associated molecular patterns (PAMPs) (e.g., bacterial LPS). This protocol can be used to study the *in vitro* efficacy of nanoparticle-based vaccine formulations. In addition, it can also be used to assess the potential cytotoxicity of nanoparticle formulations to DCs and the nanoparticle effects on DC maturation in response to traditional inflammatory stimuli such as endotoxin. Below, we describe a protocol for assessing the effects of nanoparticles on the maturation of DCs derived from PBMCs. This protocol was adopted from that described by Nelson *et al.*[66] and optimized to suit nanoparticle analysis. Monocytes are isolated from heparinized peripheral blood (PBMCs) and cultured for six days in the presence of IL-4 and GM-CSF to allow differentiation into immature DCs. On Day 5 of culture, LPS and/or the nanoparticle are added to the cell cultures and incubation is continued for an additional 24–48 h to stimulate the maturation of DCs. To test for the particles' potential ability to interfere with inflammation-induced DC maturation, LPS can be combined with the nanoparticles prior to the addition to the cells.

3.3.1. *Reagents, materials, and equipment*

- Human blood from normal donors
- Ficoll-Paque PLUS (GE-HealthCare, 17-1440–03)
- PBS (Gibco, 14190)
- LPS O55.B5 or equivalent (Sigma, L6529)
- FBS (heat-inactivated, low endotoxin concentration) (Lonza, 14–507F)
- RPMI-1640 (Gibco, 21870)
- Pen-Strep-L-Glutamine solution (Gibco, 10378-016)
- β-mercaptoethanol (Sigma, M7522)
- Trypan blue solution (Gibco, 15250-061)
- HEPES (Gibco, 15630)
- Sodium pyruvate (Gibco, 11360)
- MEM NEAA (Gibco, 11140)
- Human AB serum (GemCell, 100-512)
- DNAse I, 10 units/μL (Roche, G8101782)
- Human IL-4 (Peprotech, 200–04)
- Human GM-CSF (Leukine Sargramostim) (Immunex-Berlex, 58406-33)
- Simultest Isotype control (BD, 340041)

- HLA-ABC-FITC (BD Pharmingen, 555552)
- HLA-DR-PE (BD, 347367)
- CD86-FITC (BD Pharmingen, 555657)
- CD80-PE (BD Pharmingen, 557227)
- CD14-FITC (BD Pharmingen, 55397)
- CD83-PE (Beckman Coulter, PN IM2218U)
- Paraformaldehyde (Sigma)
- Pipettes covering the range of 5 to 25 mL (Corning)
- Pipetters covering the range of 2 to 1,000 μL (Ranin)
- Pipette tips covering the range of 2 to 1,000 μL (Ranin)
- Six-, 24-, and 96-well TC plates (Corning)
- 96-well polypropylene plate (Corning)
- T100 TC flasks, low profile, CELLBIND surface (Corning, 3073)
- 15-, 50-, and 250-mL polypropylene tubes (Falcon)
- Vacuum disposable filtration unit (Millipore)
- Centrifuge
- Refrigerator
- Freezer, –20°C and –80°C
- Cell culture incubator with 5% CO_2 and 95% humidity
- Biohazard safety cabinet approved for level II handling of biological material
- Inverted microscope
- Vortex
- Hemacytometer
- Coulter particle counter
- Flow cytometer, FACScan
- CellQuest software for data analysis

Note: The NCL does not endorse any of the suppliers listed above; their inclusion is for informational purposes only. Equivalent supplies from alternative vendors can be used in place.

3.3.2. *Media, buffers, and cytokines*

- 20% FBS freeze/thaw medium (500 mL):

  FBS (20%)
  Pen-Strep-L-Glutamine (1%)
  HEPES (2.5%)
  Adjust volume to 500 mL with RPMI-1640 and filter

- DC medium (500 mL)

  FBS (10%) (each new lot screened for good DC generation)
  Pen-Strep-L-Glutamine (1%) (100 μg/mL Pen/100 μg/mL Strep final)
  HEPES (2.5%) (25 mM final)
  MEM NEAA (1%) (0.1 mM final)
  Sodium pyruvate (1%) (1 mM final)
  β-mercaptoethanol (50 μM final)
  Adjust volume to 500 mL with RPMI-1640 and filter

- FACS blocking buffer

  PBS + 5% human AB serum

- FACS staining buffer

  PBS + 0.5% BSA + 0.1% NaN_3

- 1% paraformaldehyde in PBS

- GM-CSF stock solution (1.4×10^6 IU/mL)

Prepare the stock solution by reconstituting commercial lyophilized cytokine Leukine Sargramostim in water to a concentration of 1.4×10^6 IU/mL (250 μg/mL). Store in glass vials at 4°C. Do not freeze. Add to DC medium prior to use. The working concentration is 100 IU/mL.

- Human IL-4 stock solution (100 μg/mL)

Prepare the stock solution by reconstituting commercial lyophilized cytokine to a concentration of 100 μg/mL in PBS. Prepare small aliquots and store them at –80°C. Add to DC medium prior to use at the final concentration of 50 ng/mL.

3.3.3. *Preparation of study materials*

This assay requires 1.1 mL of nanoparticle solution, at a concentration 2× the highest concentration tested, dissolved/resuspended in complete culture medium, i.e., two 0.5-mL replicates per test sample. The concentration is selected based on the intended therapeutic concentration of the nanoparticle. If the intended therapeutic concentration is 0.1 mg/mL, then the highest final concentration of particle tested in the assay is 0.1 mg/mL (i.e., 100× above the intended therapeutic concentration). If the particle solubility does not allow for testing of 100× concentration, then the highest final concentration tested is 30× of the intended therapeutic concentration. To achieve the highest final concentration of 10 mg/mL, the stock nanoparticles should be prepared at 20 mg/mL to account for a 1:2 dilution in the test well.

Ideally, the nanoparticle concentration should be given in the amount of the active component/API. When the concentration of the active component is unknown, the weight of total formulation can be used. When the intended therapeutic concentration is unknown, the highest final concentration is 1 mg/mL or the highest feasible concentration.

Nanoparticles are tested in this assay at four concentrations: the highest concentration and three 1:5 serial dilutions. For example, if the highest final concentration to be tested is 1 mg/mL, then a stock of 2 mg/mL and three serial dilutions with concentration 0.4, 0.08, and 0.016 mg/mL are prepared. When 0.5 mL of each of these samples is added to cell culture plates and mixed with 0.5 mL of cell suspension, the final nanoparticle concentrations tested in the assay are 1, 0.2, 0.04, and 0.008 mg/mL.

3.3.4. *Isolation of human PBMCs*

- Aliquot 15 mL of Ficoll-Hypaque into each 50-mL centrifuge tube.
- Dilute the leucopack 1:2 with PBS and whole blood 1:3 with PBS.
- Slowly layer 35 mL of diluted blood over Ficoll-Hypaque.
- Centrifuge for 15 min, 2,000 rpm at room temperature, without brakes.
- Remove the mononuclear cell layer at the interface and add it to a sterile 250-mL tube containing 125 mL of PBS. Depending on the total cell number, this layer varies from thin and translucent to thick and milky in appearance.
- Centrifuge for 10 min at 1,500 rpm.
- Discard the supernatants and resuspend the pellets in PBS. Wash the pellet with PBS.
- Combine all washed cells into one tube and adjust the volume with PBS.
- Perform cell count using the Coulter particle counter and confirm cell viability using a hemocytometer and Trypan blue exclusion.
- Use isolated lymphocytes, resuspended in DC medium, for further DC generation or freeze them in 20% freeze/thaw medium with 7.5% of DMSO for further storage.

3.3.5. *Rapid thaw method of cryopreserved PBMCs*

- Retrieve cryovial containing PBMCs from storage area (–70°C or liquid nitrogen).

- Rapidly thaw cells in 37°C water bath while shaking gently.
- Wipe off cryovials with 70% ethanol and carefully open cap.
- Transfer cells to a labeled 15-mL polypropylene round bottom tube.
- Immediately add 10 mL of freeze/thaw medium in a drop-wise manner, continuously swirling the tube contents.
- Centrifuge the cell suspension at 1,000 rpm for 5 min.
- Discard the supernatant and resuspend the cell pellet in warm DC medium.
- Perform cell count on the Coulter particle counter and determine viability on a hemocytometer, using a microscope and Trypan blue exclusion.
- Adjust the volume of cell suspension to the desired concentration (3×10^6 cells/mL).

3.3.6. *Generation of immature DCs*

- Isolate PBMCs or thaw frozen PBMCs using the rapid thaw method according to procedures described in Sections 3.3.4 or 3.3.5, respectively.
- Resuspend the cell pellet in 5 mL of DC medium.
- Add 15 μL of DNase (150 units). Incubate the cell suspension in a polypropylene tube with a loosened cap at 37°C for 1–2 h.
- After incubation, spin the cells and resuspend the pellet in 5–10 mL of DC medium.
- Perform cell count on the Coulter counter and viability on a hemocytometer, and adjust the cell concentration to 3×10^6 cells/mL in DC medium.
- Add 3 mL of cell suspension to wells in a six-well TC plate. A T75 or T100 TC flask can be used. In this case, do not add more than 20 mL of cell suspension per flask.
- Incubate plate/flask for 1–2 h at 37°C to allow the monocytes to adhere.
- For each well of the six-well plate utilized, prepare 5 mL of DC medium containing IL-4 (50 ng/mL) and GM-CSF (100 IU/mL). Prepare 25 mL of complete DC medium for each T75 or T100 flask. (Do not use previously thawed IL-4 aliquots for the initial set up.)
- After incubation, gently pipette up and down, aspirate, and collect the media containing non-adherent cells. Wash each well (flask) three times with sterile PBS; combine all washes with the initial supernatant.
- Add 4 mL of complete DC medium (prepared in Section 3.3.2) to each well with adhered cells (20 mL for flask).
- Perform cell count on the total non-adherent cells collected. Approximate the number of adherent monocytes by subtracting the number of non-adherent cells from the total number of plated PBMCs.

- Incubate plate (flask) for five to seven days at 37°C.
- Feed DC culture on Days 3 to 4 (can be late Day 2). Mix cells gently and aspirate half of the volume of each well/flask. Centrifuge at 1,000 rpm for 5 min. Resuspend the cell pellet in the same volume of fresh DC medium with 2× IL-4 and GM-CSF. Add this cell suspension to initial wells or flasks.
- At the end of Day 6, monocytes (large, round, $CD14^+$) will differentiate to immature DCs (large, irregular-shaped, $CD14^-$).

3.3.7. *Maturation of DCs with or without nanomaterials*

- On Day 6, harvest the immature DCs into a 50-mL polypropylene tube. Wash each well or flask with PBS two times and collect all washes in one tube. Some DCs may be attached to the plate; in this case, add 2 mL of 5 mM sterile EDTA to each well and incubate the plate for 10–15 min at room temperature. Collect the floating cells and add them to the previous washes.
- Centrifuge the tube at 1,500 rpm for 5 min. Discard the supernatant and resuspend the cell pellet in 1 mL of fresh DC medium (without IL-4 and GM-CSF).
- Perform cell count on the hemocytometer using Trypan blue exclusion.
- Adjust the cell concentration to 0.4×10^6 cells/mL.
- Add 0.5 mL of cell suspension to each well of a 24-well TC plate.
- Prepare the nanomaterial in DC medium at 2× the desired concentration.
- Add 0.5 mL of the nanomaterial to appropriate wells. Add 0.5 mL of DC medium to control wells.
- Add LPS (1 ng/mL final concentration) to the appropriate wells. Use LPS at a standard concentration of 20 ng/mL as a positive control.
- Generally, you will have wells with immature DCs (in DC medium only) as a negative control; wells with LPS only, as a standard for maturation; wells with the nanomaterial only, as an internal control; and experimental wells with the nanomaterial and LPS.
- Incubate DCs for 24–48 h at 37°C.
- Harvest DCs and perform flow analysis of DC phenotype, staining for HLA-ABC HLA-DR, CD86, CD80, CD83, and CD14.

3.3.8. *Analysis of DC maturation by flow cytometry*

- Centrifuge harvested DCs and resuspend the cell pellet in FACS blocking buffer.

- Transfer an equal amount of cell suspension (not more than 200 μL) into each well of a 96-well polypropylene plate. Ideally, you will have 100,000–200,000 cells/well. However, as few as 10,000 cells/well can be used as well.
- Keep the plate at 4°C for 15–30 min.
- Centrifuge the plate at 1,500 rpm for 5 min. Decant the supernatant from each well by flicking the plate. Wash the cells once with FACS staining buffer. Decant the supernatant again.
- Add 20 μL of an appropriate antibody to each well. Add 150 μL of FACS staining buffer to each well.
- You will have four wells in total for each DC sample: Isotype control, HLA-ABC/HLA-DR, CD86/CD80, and CD14/CD83.
- Keep the plate at 4°C for 15–30 min.
- Wash the plate with FACS buffer three times, flicking the supernatant every time.
- After the last wash, add 100 μL of 1% paraformaldehyde in PBS to each well. Wrap the plate in foil and store at 4°C. Analyze the samples within one week. Before flow cytometry, transfer the cell suspension from each well to a FACS tube and add 200 μL of PBS.
- The samples can be analyzed "fresh" (without fixation) if time permits.
- For analysis, create forward/side scatter dot plot for each sample and make R1 gate (live cells) to exclude dead cells and debris. Create histograms for each staining in R1 gate. Set up M1 marker on Isotype control histograms and transfer this marker to all other histograms. Perform histogram statistics, looking for percent positive cells in M1, as well as mean channel fluorescence (MCF).
- Matured DCs should have HLA-DR MCF increased, percent positive for CD86 and CD80 cells increased (although MCF may stay the same), and percent positive for CD83 cells increased significantly. Some immature DCs may still be $CD14^+$, but this percentage should decrease significantly after maturation with LPS.

3.4. *Effects of nanoparticles on the production of nitric oxide by macrophages in vitro*

Macrophage activation by inflammatory stimuli results in an oxidative burst. One of the markers of such an activation is nitric oxide. Below, we will describe a protocol for the quantitative determination of nitrite (NO_2^-), a stable oxidative end-product of the antimicrobial effector molecule nitric oxide in cell culture medium.[67,68] The protocol is used to evaluate the ability of nanomaterials to induce nitric oxide production by macrophages. Nitric oxide secreted by macrophages has a half-life of seconds; it interacts with a number of different molecular targets, resulting in cytotoxicity. In the presence of oxygen and water, nitric oxide interacts with itself to generate other reactive nitrogen oxide intermediates and ultimately decomposes to form nitrite (NO_2^-) and nitrate (NO_3^-). In this assay, nitrite is measured in tissue culture medium using the Greiss reagent. This measurement provides a surrogate marker and quantitative indicator of nitric oxide production. The murine macrophage cell line RAW264.7 is used as the model in this assay. The upper limit of quantification is 250 μM and the lower limit of quantification is 1.95 μM.

3.4.1. *Reagents, materials, cell lines, and equipment*

- PBS (HyClone, AQB 22934)
- LPS O55:B5 or equivalent (Sigma, L6529)
- FBS (Hyclone, SH30070.03)
- RPMI-1640 without phenol red (Invitrogen, 11835–055)
- Hanks' balanced salt solution (HBSS) (Invitrogen, 24020–117)
- Pen/Strep solution (Invitrogen, 15140–148)
- β-mercaptoethanol (Sigma, M7522)
- Trypan blue solution (Invitrogen, 15250–061)
- Naphthylethylenediamine dihydrochloride (Sigma, N9125)
- Sulfanilamide (Sigma, S9251)
- Phosphoric acid (Aldrich, 22,248–8)
- Sodium nitrite ($NaNO_2$) standard, 0.1 M stock solution (Sigma, 35271)
- Pipettes covering the range of 0.05 to 10 mL
- Flat-bottomed 96-well plates
- 24-well plates
- Polypropylene tubes, 5 and 15 mL
- RAW264.7 murine macrophages
- Centrifuge
- Refrigerator, 2–8°C
- Freezer, –20°C
- Cell culture incubator with 5% CO_2 and 95% humidity
- Biohazard safety cabinet approved for level II handling of biological material
- Inverted microscope
- Vortex
- Hemocytometer
- Plate reader

Note: The NCL does not endorse any of the suppliers listed above; their inclusion is for informational purposes only. Equivalent supplies from alternate vendors can be used in place.

3.4.2. *Preparation of reagents and controls*

- Complete RPMI-1640 medium

The complete RPMI medium should contain the following reagents:

10% FBS (heat-inactivated)
2 mM L-glutamine
50 μM β-mercaptoethanol
100 U/mL penicillin
100 μg/mL streptomycin sulfate

Store at 2–8°C protected from light for no longer than one month. Before use, warm in a water bath.

- LPS (stock) (1mg/mL)

Add 1 mL of sterile PBS or cell culture medium per 1 mg of LPS to the vial and vortex to mix. Store daily-use aliquots at a nominal temperature of –20°C. Avoid repeated freezing/thawing.

- Positive control

Dilute the stock LPS solution in cell culture medium to a final concentration of 100 ng/mL. Store at room temperature. Discard the unused portion after the experiment.

- Negative control

Use PBS as a negative control.

- Heat-inactivated FBS

Thaw a bottle of FBS at room temperature or overnight at 2–8°C and allow it to equilibrate to room temperature. Incubate 30 min at 56°C in a water bath, mixing every 5 min. Single-use aliquots may be stored at 2–8°C for up to one month or at a nominal temperature of –20°C indefinitely.

- Greiss reagent

Dissolve sulfanilamide in 2.5% phosphoric acid (H_3PO_4) to a final concentration of 1% (w/v), e.g.,

A. dissolve 1 g of sulfanilamide in 100 mL of 2.5% H_3PO_4.
B. Dissolve naphthylethylenediamine dihydrochloride in 2.5% H_3PO_4 to a final concentration of 0.1% (w/v), e.g., dissolve 100 mg of naphthylethylenediamine dihydrochloride in 100 mL of 2.5% H_3PO_4.

Store both solutions in glass bottles at 4°C; discard if discoloration occurs or if solutions are not clear. Equal volumes of reagents A and B will be combined just prior to use to form the Griess reagent. This solution should be used immediately after preparation and the remainder should be discarded.

- $NaNO_2$ stock solution (used for the preparation of calibration curves and quality controls)

This stock is supplied from the commercial source as a 0.1 M solution. Store at room temperature.

- Preparation of $NaNO_2$ calibration standards

An example is shown in Table 6. Volumes can be adjusted based on need. The stock is the $NaNO_2$ reagent prepared as described above.

Table 6. Preparation of $NaNO_2$ calibration standards.

Level	Nominal concentration (μM)	Preparation procedure
Int. A	10,000	100 μL Stock + 900 μL complete medium
Int. B	1,000	100 μL Int. A + 900 μL complete medium
Cal 1	250	200 μL Int. B + 600 μL complete medium
Cal 2	125	400 μL Cal 1 + 400 μL complete medium
Cal 3	62.5	400 μL Cal 2 + 400 μL complete medium
Cal 4	31.3	400 μL Cal 3 + 400 μL complete medium
Cal 5	15.6	400 μL Cal 4 + 400 μL complete medium
Cal 6	7.81	400 μL Cal 5 + 400 μL complete medium
Cal 7	3.91	400 μL Cal 6 + 400 μL complete medium
Cal 8	1.95	400 μL Cal 7 + 400 μL complete medium

Table 7. Preparation of $NaNO_2$ quality controls.

Level	Nominal concentration (μM)	Preparation procedure
Int. A	10,000	100 μL Stock + 900 μL complete medium
Int. B	1,000	100 μL Int. A + 900 μL complete medium
QC 1	100	100 μL Int. B + 900 μL complete medium
QC 2	50	400 μL QC 1 + 400 μL complete medium
QC 3	5	100 μL QC 2 + 900 μL complete medium

- Preparation of $NaNO_2$ quality controls

An example is shown in Table 7. Volumes can be adjusted based on need. The stock is $NaNO_2$ reagent prepared as described above.

3.4.3. *Preparation of study samples*

This assay requires 2,000 μL of nanoparticles (at 1× the highest final test concentration) dissolved/resuspended in complete culture medium. The following questions have to be considered when selecting the concentration: 1) solubility of nanoparticles in a biocompatible buffer, 2) pH within the physiological range, 3) availability of nanomaterial, and 4) stability.

The nanoparticles are tested at four concentrations. The highest tested concentration is generally 1.0 mg/mL, or if maximum plasma concentra tion is known from an *in vivo* study, use 10×, 30×, or 100× of that concentration (depending on material solubility) for the highest tested concentration. If a nanoparticle carries a drug, the concentrations should be based on the drug (API) and not based on the total concentration of the formulation.

The highest tested concentration is also used to prepare three serial 1:5 dilutions. For example, if the highest tested concentration is 1.0 mg/mL, the stock nanoparticle concentration should also be 1.0 mg/mL. The other test concentrations are 0.2, 0.04, and 0.008 mg/mL. Three 0.5 mL samples will be used for each test concentration (two for the assay test samples and one for a cell-free control).

3.4.4. *Cell preparation*

Grow cells in complete medium. Dislodge the cells using trypsin-EDTA solution and resuspend in complete medium. A subcultivation ratio of 1:3 to 1:6 is recommended. Replace or add medium every two to three days.

3.4.5. *Experimental procedure*

- Adjust the cell concentration to 1×10^5 cells/mL using complete RPMI medium.
- Plate 1,000 μL of cell suspension per well in a 24-well plate. Prepare triplicate wells for each sample and duplicate wells for each control. Always leave one cell-free well per nanoparticle per plate. These wells will be used to assay potential nanoparticle interference with the assay. This is the "Culture Plate" (Figure 11).
- Incubate the "Culture Plate" for 24 h in a humidified 37°C, 5% CO_2 incubator.
- Remove the culture medium and add 500 μL of study samples, controls, or medium blank to appropriate wells. Position samples on the plate such that study samples are bracketed by controls and blank medium.

	1	2	3	4	5	6
A	Blank	NC	PC		TS 1a 1.0 mg/mL	TS 1a 0.2 mg/mL
B	TS 1a 0.04 mg/mL	TS 1a 0.008 mg/mL	TS 1b 1.0 mg/mL	TS 1b 0.2 mg/mL	TS 1b 0.04 mg/mL	TS 1b 0.008 mg/mL
C	Blank	NC	PC			
D	TS 1 1.0 mg/mL (No cells)	TS 1 0.2 mg/mL (No cells)	TS 1 0.04 mg/mL (No cells)	TS 1 0.008 mg/mL (No cells)		

Figure 11. Example of a "Culture Plate" map. A 24-well plate is used to prepare triplicate wells for each sample and duplicate cells for each control. Potential nanoparticle interference will be assayed with the "Culture Plate."

NC: negative control, PC: positive control, TS: test sample.

- Incubate the "Culture Plate" 48 ± 1 h in a humidified 37°C, 5% CO_2 incubator.
- To a fresh 96-well plate, add 50 μL per well of reagent blank (culture medium used to prepare calibration standards and quality controls), calibration standards, quality controls, and medium from each well of the "Culture Plate." Load duplicate wells for each sample and control. This is the "NO^- Test Plate" (Figure 12).

  Note: The removal of nanoparticles from culture medium may be required prior to this step if there is the possibility of nanoparticle interference with the assay, e.g., if particles react with either or both components of the Greiss reagent, or have absorbance at or close to 550 nm. If particle removal is not feasible, the results obtained for the "minus particles" control may be subtracted from that obtained for particle test sample to correct for particle background interference.

- In a separate tube, combine equal volumes of reagent A and reagent B; this is the Greiss reagent.
- Add 100 μL of the Greiss reagent to each well of the "NO^- Test Plate."
- Place the plate on a shaker for 2–3 min, allowing all ingredients to mix.
- Measure the absorbance at 550 nm.

3.4.6. *Calculations*

- A percent coefficient of variation is used to control precision and is calculated for each control or test sample according to the following formula:

$$\%\mathrm{CV} = \frac{\mathrm{SD}}{\mathrm{Mean}} \times 100\%.$$

- Percent difference from theoretical (PDFT) is used to control the accuracy of the assay calibration standards and quality controls, and is calculated according to the following formula:

$$\mathrm{PDFT} = \frac{\text{Calculated NaNO}_2\text{ concentration} - \text{Theoretical NaNO}_2\text{ concentration}}{\text{Theoretical NaNO}_2\text{ concentration}} \times 100\%.$$

3.4.7. *Acceptance criteria*

- %CV for each control and test sample should be within 30%.
- If the positive control fails to meet the acceptance criterion for %CV described above, the assay should be repeated.

	1	2	3	4	5	6	7	8	9	10	11	12
A	Blank	QC 1	QC 2	QC 3	Cal 1	Cal 2	Cal 3	Cal 4	Cal 5	Cal 6	Cal 7	Cal 8
B	Blank	QC 1	QC 2	QC 3	Cal 1	Cal 2	Cal 3	Cal 4	Cal 5	Cal 6	Cal 7	Cal 8
C	NC	PC	TS 1a 1.0 mg/mL	TS 1a 0.2 mg/mL	TS 1a 0.04 mg/mL	TS 1a 0.008 mg/mL	TS 1b 1.0 mg/mL	TS 1b 0.2 mg/mL	TS 1b 0.04 mg/mL	TS 1b 0.008 mg/mL	NC	PC
D	NC	PC	TS 1a 1.0 mg/ mg/mL	TS 1a 0.2 mg/mL	TS 1a 0.04 mg/mL	TS 1a 0.008 mg/mL	TS 1b 1.0 mg/mL	TS 1b 0.2 mg/mL	TS 1b 0.04 mg/mL	TS 1b 0.008 mg/mL	NC	PC
E												
F												
G	TS 1 1.0 mg/mL (No cells)	TS 1 0.2 mg/mL (No cells)	TS 1 0.04 mg/mL (No cells)	TS 1 0.008 mg/mL (No cells)					Blank	QC 1	QC 2	QC 3
H	TS 1 1.0 mg/mL (No cells)	TS 1 0.2 mg/mL (No cells)	TS 1 0.04 mg/mL (No cells)	TS 1 0.008 mg/mL (No cells)					Blank	QC 1	QC 2	QC 3

Figure 12. Example of a "NO^- Test Plate." A 96-well plate is used to prepare calibration standards and quality controls.

Cal: calibration standard, NC: negative control, PC: positive control, QC: quality control, TS: test sample.

- Within the acceptable assay, if two of three replicates of the unknown sample fail to meet the acceptance criterion for %CV described above, this unknown sample should be re-analyzed.
- If two duplicates of the same study sample demonstrate results >30% different, this sample should be re-analyzed.
- %CV and PDFT of calibration standards and quality controls should be within 20%. At least five calibrators should be available. Four of six quality controls and at least one of each level should be acceptable. If not, a new set of calibration standards and quality controls should be prepared and test samples re-loaded onto a new plate.

3.5. *Effects of nanoparticles on leukocyte PCA in vitro*

Leukocyte PCA is accepted as an important component in the onset of DIC.[37,40–44,47,52,69] This following section describes a protocol for assessing the ability of a nanoparticle formulation to induce leukocyte PCA. This protocol includes two cell models: normal leukocytes, represented by PBMCs isolated from the blood of healthy donor volunteers, and acute promyelocytic leukemia cells, represented by the HL-60 cell line. Briefly, cells are treated with nanoparticles, and then undergo a wash step to remove excess particles. Isolated cells are then used to initiate plasma coagulation, which is measured using an aggregometer (following NCL protocol ITA-12 from http://ncl.cancer.gov/NCL_Method_ITA-4.pdf) for the analysis of prothrombin time.

3.5.1 *Reagents, materials, cell lines, and equipment*

- Human blood from at least three donors, anti-coagulated with Li-heparin for PBMC isolation and anti-coagulated with Na-citrate for plasma coagulation test
- Ficoll-Paque PLUS (Amersham Biosciences, 17-1440-02)
- PBS (HyClone, AQB 22934)
- Ultrapure LPS (InvivoGen, tlrl-peklps)
- Doxorubicin hydrochloride (Bedford Labs, NIH DVR pharmacy)
- FBS (Hyclone, SH30070.03)
- RPMI-1640 (Invitrogen, 11835-055)
- Hanks' balanced salt solution (HBSS) (Invitrogen, 24020-117)
- Pen/Strep solution (Invitrogen, 15140–148)
- β-mercaptoethanol (Sigma, M7522)
- L-glutamine (Hyclone, SH30034.01)
- Trypan blue solution (Invitrogen, 15250–061)

- Neoplastine Cl (Diagnostica Stago, 00666)
- CoagControl N+ABN (Diagnostica Stago, 00676)
- SystemControl N+P (Diagnostica Stago, 00678)
- Pipettes covering the range of 0.05 to 10 mL
- 96-well plates
- Polypropylene tubes, 5 and 15 mL
- HL-60 promyelocytic cells (ATCC, CCL-240)
- Centrifuge
- Refrigerator, 2–8°C
- Freezer, –20°C
- Cell culture incubator with 5% CO_2 and 95% humidity
- Biohazard safety cabinet approved for level II handling of biological material
- Inverted microscope
- Vortex
- Hemocytometer
- Coagulometer STart4 (Diagnostica Stago) or comparable

Note: The NCL does not endorse any of the suppliers listed above; their inclusion is for informational purposes only. Equivalent supplies from alternate vendors can be used in place.

3.5.2. *Preparation of reagents and controls*

- Complete RPMI-1640 medium

The complete RPMI medium should contain the following reagents:

10% FBS (heat-inactivated)
2 mM L-glutamine
50 μM β-mercaptoethanol
100 U/mL penicillin
100 μg/mL streptomycin sulfate

Store at 2–8°C protected from light for no longer than one month. Before use, warm in a water bath.

- LPS stock (1 mg/mL)

Add 1 mL of sterile PBS or cell culture medium per 1 mg of LPS to the vial and vortex to mix. Store daily-use aliquots at a nominal temperature of –20°C.

- Doxorubicin (stock)

Doxorubicin (Adriamycin®) is provided as a solution at 2 mg/mL and stored at 2–8°C.

- Positive controls for cell culture

Positive control for PBMCs

Dilute the stock LPS solution in cell culture medium to a final concentration of 1 μg/mL. Store at room temperature. Discard unused portion after experiment.

Positive control for HL-60 cells

Dilute the doxorubicin stock solution in cell culture medium to a final concentration of 50 μg/mL. For example, if using doxorubicin at 2 mg/mL stock concentration, add 300 μL of media to 100 μL of doxorubicin. Store at room temperature. Discard unused portion after experiment.

- Negative control

Use PBS as a negative control.

- Heat-inactivated FBS

Thaw a bottle of FBS at room temperature or overnight at 2–8°C and allow to equilibrate to room temperature. Incubate 30 min at 56°C in a water bath, mixing every 5 min. Single-use aliquots may be stored at 2–8°C for up to one month or at a nominal temperature of –20°C indefinitely.

- Buffer A

Prepare Buffer A by dissolving NaCl to a final concentration of 150 mM and $CaCl_2$ to a final concentration of 6.6 mM in 20 mM HEPES, pH 7.4.

- Preparation of coagulation controls (System N+P and Coag N+ABN)

Reconstitute lyophilized control plasmas with 2 mL of distilled water. Let the solutions stand at room temperature 30 min prior to use. Mix thoroughly before use. Keep unused portion refrigerated and use within 48 h after reconstitution. These plasma samples are used as instrument controls.

- Preparation of plasma for coagulation test

Use freshly collected whole blood anti-coagulated with Na-citrate (blue top vacutainers) within 1 h after collection. Spin the blood 10 min at 2500*g* at 20–22°C; collect plasma and pool. Pooled plasma is stable for 8 h at room temperation. Do not refrigerate or freeze.

Analyze two duplicates of test plasma in each coagulation assay. Run one duplicate before nanoparticle-treated plasma samples and the second duplicate at the end of each run.

3.5.3. *Preparation of study samples*

This assay requires 1.1 mL of nanoparticles, at 10× the highest tested concentration, dissolved/resuspended in complete culture medium. For the original screen, we recommend using the highest concentration of nanoparticles possible. The following have to be considered when selecting the concentration: 1) solubility of nanoparticles in a biocompatible buffer, 2) pH within physiological range, 3) availability of nanomaterial, 4) stability, and 5) potential cytotoxicity of nanomaterial. The material is tested at the highest concentration and at three serial 1:5 dilutions. For example, if the highest tested concentration is 1.0 mg/mL, the other tested concentrations are 0.2, 0.04, and 0.008 mg/mL; the stock concentration should be 10 mg/mL. Two 0.4 mL replicates ($n = 2$) will be used for each test concentration.

3.5.4. *Cell preparation*

3.5.4.1. Isolation of human lymphocytes

- Place freshly drawn blood anticoagulated with Li-heparin (green top vacutainers) into 15- or 50-mL conical centrifuge tubes; add an equal volume of room-temperature PBS and mix well.
- Slowly layer the Ficoll-Paque solution underneath the blood/PBS mixture by placing the tip of the pipette containing the solution at the bottom of the blood sample tube. Alternatively, the blood/PBS mixture may be slowly layered over the Ficoll-Paque solution. Use 3 mL of Ficoll-Paque solution per 4 mL of blood/PBS mixture.

Note: To maintain the Ficoll–blood interface, it is helpful to hold tube at a 45° angle.

- Centrifuge 30 min at 900*g*, 18–20°C, without brakes.
- Using a sterile pipette, remove the upper layer containing plasma and platelets and discard.

- Using a fresh sterile pipette, transfer the mononuclear cell layer into another centrifuge tube.
- Wash cells by adding an excess of HBSS; centrifuge for 10 min at 400*g*, 18–20°C. The HBSS volume should be ~three times the volume of the mononuclear layer.

Note: Typically, 4 mL of blood/PBS mixture results in ~2 mL of the mononuclear layer and requires 6 mL of HBSS for the wash step.

- Discard the supernatant and repeat the wash step once more.
- Resuspend the cells in complete RPMI-1640 medium. Count an aliquot of cells and determine viability using Trypan blue exclusion.
- If cell viability is ≥99%, dilute cells in complete culture medium to the concentration of 3×10^6 cells/mL and proceed to Section 3.5.5.1.

3.5.4.2. Preparation of HL-60 cells

- Grow cells in complete culture medium; avoid a cell density greater than 1×10^6 cells/mL. On the day of experiment, count cells; if cell viability is ≥99%, concentrate cells in complete culture medium to the concentration of 3×10^6 cells/mL and proceed to Section 3.5.5.1.

3.5.5. *Experimental procedure*

3.5.5.1. Cell treatment with nanoparticles and controls

- Aliquot 3.6 mL of cell suspension into each well of a six-well plate.
- Add 400 μL of test nanoparticle, positive control, and negative control to the respective wells. The negative control for both cell models is PBS. The positive control for PBMCs is LPS at a final concentration of 1 μg/mL. The positive control for HL-60 cells is doxorubicin at a final concentration of 50 μg/mL. Prepare one well each for the positive and negative controls. Prepare two wells each, at a minimum, for each concentration of the test sample.
- Incubate cells with nanoparticles and controls. The times of incubation are 4 h for the HL-60 model and 24 h for PBMCs.
- At the end of incubation, remove the cells from the incubator and wash the cells twice with 1 mL of PBS.
- After the last wash, reconstitute the cell pellet in Buffer A and adjust the cell volume to a concentration of 5×10^6 cells/mL.
- Keep the cell suspensions at room temperature; place them in the incubator (37°C) for 5–10 min prior to testing for warming up. The sample will

then be transferred to the 37°C chamber on the coagulometer when the experiment is ready to be started.

3.5.5.2. Determining plasma coagulation time

- Set up the instrument test parameters as shown below:
  Max. time: 360 s
  Incubation time: 120 s
  Single/duplicate: Duplicate
  Precision: 5%

  Allow the instrument to warm up for 5–10 min prior to use.
- Prepare all reagents and cells, and warm them up to 37°C before use. Note that lyophilized reagents should be reconstituted at least 30 min prior to use. It is not advisable to keep more than 10 cell samples at 37°C at one time.
- Place cuvettes into A, B, C, and D test rows on the coagulometer. (Note that this protocol is based on semi-automatic STart4 coagulometer from Diagnostica Stago; if a different instrument is used, please follow the operation guidelines recommended by the instrument manufacturer.)
- Add one metal ball into each cuvette and allow to warm for at least 3 min before use.
- Add 100 μL of control plasma or test plasma to a cuvette. Prepare duplicate cuvettes for each plasma sample.
- Start timer for each of the test rows by pressing A, B, C, or D buttons. The timer will beep 10 s before the time is up. When this happens, immediately transfer the cuvettes to the PIP row and press the PIP button to activate the pipettor.
- Add 100 μL of Neoplastin reagent to control plasma samples or 100 μL of cell suspension in lieu of coagulation activation reagent to corresponding cuvettes, and record the coagulation time.

3.5.6. *Calculations*

- A percent coefficient of variation should be calculated for each control or test sample according to the following formula:

$$\%CV = \frac{SD}{Mean} \times 100\%.$$

- A percent PCA of a nanoparticle test sample is calculated according to the following formula:

$$\%\,\mathrm{PCA} = \frac{\text{Mean Time}_{\text{positive control}}}{\text{Time}_{\text{test sample}}} \times 100\%.$$

3.5.7. *Acceptance criteria*

- %CV between replicates of each plasma control sample should be within 5%.
- %CV between replicates of test plasma samples in which nanoparticle-treated cells were used as coagulation initiation reagents should be within 25%.
- If two duplicates of the same study sample demonstrate results with >25% difference, this test sample should be re-analyzed.
- The positive control is only considered positive if the coagulation time in this sample is less than 360 s. The coagulation time induced by leukemia cells treated with doxorubicin and those induced by PBMCs treated with LPS may vary from donor to donor. Typical coagulation times observed in our laboratory with HL-60 cells treated with 50 μg/mL doxorubicin are 60–90 s. Typical coagulation times observed in our laboratory with PBMCs treated with 1 μg/mL LPS are 100–200 s.

3.6. *Effects of nanoparticles on leukocyte proliferation in vitro*

In this section, we describe a protocol for assessing the effects of a nanoparticle formulation on the basic immunological function of human lymphocytes, i.e., the measurement of lymphocyte proliferative responses. This assay will allow for the measurement of the ability of a nanoparticle to trigger a proliferative response of human lymphocytes or to suppress those induced by PHA-M.

Lymphocytes are isolated from pooled human blood and anti-coagulated with Li-heparin using Ficoll-Paque PLUS solution. The isolated cells are incubated with or without PHA-M in the presence or absence of nanoparticles and analyzed spectrophotometrically via the MTT [3-(4,5-dimethyl-2-thiazolyl)-2,5-diphenyl-2H-tetrazolium bromide] assay.

3.6.1. *Reagents, materials, and equipment*

- Human blood from at least three donors, anti-coagulated with Li-heparin
- Ficoll-Paque PLUS (Amersham Biosciences, 17-1440-02)
- PBS (HyClone, AQB 22934)
- Phytohemaglutinin-M (PHA-M) (Sigma, L8902)

- FBS (Hyclone, SH30070.03)
- RPMI-1640 (Invitrogen, 11875-119)
- Hanks' balanced salt solution (HBSS) (Invitrogen, 24020-117)
- Pen/Strep solution (Invitrogen, 15140-148)
- β-mercaptoethanol (Sigma, M7522)
- Trypan blue solution (Invitrogen, 15250-061)
- MTT [3-(4, 5-dimethyl-2-thiazolyl)-2,5-diphenyl-2H-tetrazolium bromide] (Sigma, M5655)
- L-glutamine (Hyclone, SH30034.01)
- Glycine (Sigma, 67403)
- Sodium chloride (Sigma, 57653)
- Pipettes covering a range of 0.05 to 10 mL
- 96-well round-bottomed plates
- Polypropylene tubes, 5 and 15 mL
- Centrifuge
- Refrigerator, 2–8°C
- Freezer, –20°C
- Cell culture incubator with 5% CO_2 and 95% humidity
- Biohazard safety cabinet approved for level II handling of biological material
- Inverted microscope
- Vortex
- Hemocytometer

Note: The NCL does not endorse any of the suppliers listed above; their inclusion is for informational purposes only. Equivalent supplies from alternate vendors can be used in place.

3.6.2. *Preparation of reagents and controls*

- Complete RPMI-1640 medium

The complete RPMI medium should contain the following reagents:

10% FBS (heat-inactivated)
2 mM L-glutamine
100 U/mL penicillin
100 μg/mL streptomycin sulfate

Store at 2–8°C, protected from light for no longer than one month. Before use, warm in a water bath.

- PHA-M stock (1 mg/mL)

Add 1 mL of sterile PBS or cell culture medium per 1 mg of PHA-M to the vial and gently rotate to mix. Store daily-use aliquots at a nominal temperature of –20°C. Avoid repeated freezing/thawing.

- Positive control

Dilute the stock PHA-M solution in cell culture medium to a final concentration of 100 μg/mL and use it to prepare three serial (1:2) working dilutions (1:5, 1:10 and 1:20 denoted as PC1, PC2, and PC3, respectively) of the positive control. The working dilution is prepared in complete RPMI-1640 medium.

- Negative control

Use PBS as a negative control.

- Heat-inactivated FBS

Thaw a bottle of FBS at room temperature or overnight at 2–8°C and allow to equilibrate to room temperature. Incubate 30 min at 56°C in a water bath mixing every 5 min. Single-use aliquots may be stored at 2–8°C for up to one month or at a nominal temperature of –20°C indefinitely.

- MTT solution

Prepare MTT solution in PBS at a final concentration of 5 mg/mL. Store for up to one month at 4°C in the dark.

- Glycine buffer

Prepare the buffer by dissolving glycine and NaCl in water to a final concentration of 0.1 M for glycine (MW 75.07 g/mol), and 0.1 M for NaCl (MW 58.44 g/mol). Adjust the pH to 10.5. Store at room temperature.

3.6.3. *Preparation of study samples*

This assay requires 3.0 mL of nanoparticles dissolved/resuspended in complete culture medium to a concentration of 4× the highest tested concentration (i.e., if the highest test concentration is 1.0 mg/mL, the stock concentration of nanoparticles should be 4.0 mg/mL). Half (1.5 mL) of the stock sample will be used at the 4× concentration, for preparation of the nanoparticle + positive control treatments. The remaining half of the stock sample will be diluted to 2×(i.e., 2.0 mg/mL) and will be used as the stock for the nanoparticle-only treatments.The stock solutions will be used to test the particles at the highest final concentration of 1.0 mg/mL, and to prepare three serial 1:5 dilutions in cell culture medium. The final test concentrations of nanoparticles analyzed in this assay are 1.0, 0.2, 0.04, and 0.008 mg/mL. Each dilution is analyzed three times in duplicate ($n = 6$). Sample volumes are 100 µL of 2× stock for nanoparticle-only treatment or 50 µL of 4× stock for nanoparticle + positive control treatments.

The following questions have to be considered when selecting the concentration: 1) solubility of the nanoparticle in a biocompatible buffer, 2) pH within physiological range, 3) availability of nanomaterials, and 4) stability. If the highest final test concentration of 1.0 mg/mL is not feasible, the sample should be analyzed at concentrations which are more suitable for the given formulation.

3.6.4. *Isolation of human lymphocytes*

- Place freshly drawn blood into 15- or 50-mL conical centrifuge tubes, add an equal volume of room-temperature PBS, and mix well.
- Slowly layer the Ficoll-Paque solution underneath the blood/PBS mixture by placing the tip of the pipette containing the Ficoll-Paque at the bottom of the blood sample tube. Alternatively, the blood/PBS mixture may be slowly layered over the Ficoll-Paque solution. Use 3 mL of Ficoll-Paque solution per 4 mL of blood/PBS mixture.

Note: To maintain the Ficoll–blood interface, it is helpful to hold the tube at a 45° angle.

- Centrifuge 30 min at 900*g*, 18–20°C, without brakes.
- Using a sterile pipette, remove the upper layer containing the plasma and platelets and discard it.
- Using a fresh sterile pipette, transfer the mononuclear cell layer into another centrifuge tube.
- Wash cells by adding an excess of HBSS and centrifuging for 10 min at 400*g*,18–20°C. The HBSS volume should be ~three times the volume of the mononuclear layer.

Note: Usually 4 mL of blood/PBS mixture results in ~2 mL of the mononuclear layer and requires at least 6 mL of HBSS for the wash step. We use 10 mL of HBSS per 2 mL of cells.

- Discard the supernatant and repeat the wash step once more.
- Resuspend cells in complete RPMI-1640 medium. Dilute cells 1:5 or 1:10 with Trypan blue, count cells, and determine the viability using Trypan blue exclusion. If the viability is at least 90%, proceed to the next step.

3.6.5. *Experimental procedure*

- Adjust cell concentration to 1×10^6 cells/mL using complete RPMI medium.
- Dispense a) 100 µL of controls, b) 100 µL of test samples (at 2× the final test concentration, for nanoparticle treatment only), and c) 50 µL of test samples (at 4× the final test concentration, for nanoparticle + positive control treatment) and 50 µL of PC2 per well on a 96-well round bottom plate. Each dilution is analyzed three times in duplicate. Always include cell-free controls as well, i.e., 100 µL of nanoparticles at 2× final test concentration and 100 µL of cell culture medium.
- Dispense 100 µL of cell suspension (or cell culture medium for cell-free controls) per well. Gently shake the plate to allow all components to mix.
- Incubate for three days in a humidified 37°C, 5% CO_2 incubator.
- Centrifuge the plate for 5 min at 700*g*. Aspirate the medium, leaving cells and approximately 50 µL of medium behind. Add 150 µL of fresh medium to each well.
- Add 50 µL of MTT to all wells.
- Cover in aluminum foil and incubate in a humidified 37°C, 5% CO_2 incubator for 4 h.
- Remove the plate from the incubator and spin at 700*g* for 5 min.
- Aspirate media and MTT.
- Add 200 µL of DMSO to all wells.
- Add 25 µL of glycine buffer to all wells.
- Read at 570 nm on a plate reader.

3.6.6. *Calculations*

- A percent coefficient of variation should be calculated for each control or test according to the following formula:

$$\%CV = \frac{SD}{Mean} \times 100\%.$$

- A percent viability is calculated as follows:

$$\%\text{Cell Proliferation} = \left(\text{Mean OD}_{\text{sample}} - \text{Mean OD}_{\text{negative control}}\right) \times 100\%$$

and

$$\%\text{Proliferation Inhibition} = \frac{\text{Mean OD}_{\text{positive control}} - \text{Mean OD}_{\text{positive control+nanoparticles}}}{\text{Mean OD}_{\text{positive control}}} \times 100\%.$$

3.6.7. *Acceptance Criteria*

- %CV for each control and test sample should be less than 30%.
- If the positive control or negative control fails to meet the acceptance criterion described for %CV above, the assay should be repeated.
- Within the acceptable assay, if two of three replicates of the unknown sample fail to meet the acceptance criterion described for %CV above, this unknown sample should be re-analyzed.
- If two duplicates of the same study sample demonstrate results >30% different, this sample should be re-analyzed.

References

1. Luebke R. Immunotoxiant screening and prioritization in the twenty-first century. *Toxicol Pathol* 2012;40:294–299.
2. Dean JH, Cornacoff JB, Haley PJ, Hincks JR. The integration of immunotoxicology in drug discovery and development: Investigative and *in vitro* possibilities. *Toxicol In Vitro* 1994;8:939–944.
3. Germolec DR, Kashon M, Nyska A, Kuper CF, Portier C, Kommineni C, Johnson KA, Luster MI. The accuracy of extended histopathology to detect immunotoxic chemicals. *Toxicol Sci* 2004;82:504–514.
4. Dean JH. Issues with introducing new immunotoxicology methods into the safety assessment of pharmaceuticals. *Toxicology* 1997;119:95–101.
5. Dean JH, Hincks JR, Remandet B. Immunotoxicology assessment in the pharmaceutical industry. *Toxicol Lett* 1998;102–103:247–255.
6. Spanhaak S. The ICH S8 immunotoxicity guidance. Immune function assessment and toxicological pathology: Autonomous or synergistic methods to predict immunotoxicity? *Exp Toxicol Pathol* 2006;57:373–376.
7. Collinge M, Cole SH, Schneider PA, Donovan CB, Kamperschroer C, Kawabata TT. Human lymphocyte activation assay: An *in vitro* method for predictive immunotoxicity testing. *J Immunotoxicol* 2010;7:357–366.
8. Daneshian M, Akbarsha MA, Blaauboer B, Caloni F, Cosson P, Curren R, Goldberg A, Gruber F, Ohl F, Pfaller W, *et al.* A framework program for the teaching of alternative methods (replacement, reduction, refinement) to animal experimentation. *ALTEX* 2011;28:341–352.

9. Dhawan A, Pandey A, Sharma V. Toxicity assessment of engineered nanomaterials: Resolving the challenges. *J Biomed Nanotechnol* 2011;7:6–7.
10. Oostingh GJ, Casals E, Italiani P, Colognato R, Stritzinger R, Ponti J, Pfaller T, Kohl Y, Ooms D, Favilli F, *et al.* Problems and challenges in the development and validation of human cell-based assays to determine nanoparticle-induced immunomodulatory effects. *Part Fibre Toxicol* 2011;8:8.
11. Pfaller T, Colognato R, Nelissen I, Favilli F, Casals E, Ooms D, Leppens H, Ponti J, Stritzinger R, Puntes V, *et al.* The suitability of different cellular *in vitro* immunotoxicity and genotoxicity methods for the analysis of nanoparticle-induced events. *Nanotoxicology* 2010;4:52–72.
12. Wilhelmi V, Fischer U, van Berlo D, Schulze-Osthoff K, Schins RP, Albrecht C. Evaluation of apoptosis induced by nanoparticles and fine particles in RAW 264.7 macrophages: Facts and artefacts. *Toxicol In Vitro* 2012;26:323–334.
13. Galluzzi L, Chiarantini L, Pantucci E, Curci R, Merikhi J, Hummel H, Bachmann PK, Manuali E, Pezzotti G, Magnani M. Development of a multilevel approach for the evaluation of nanomaterials toxicity. *Nanomedicine (Lond)* 2012;7:393–409.
14. Laurent S, Burtea C, Thirifays C, Hafeli UO, Mahmoudi M. Crucial ignored parameters on nanotoxicology: The importance of toxicity assay modifications and "cell vision" *PLoS One* 2012;7:e29997.
15. McNeil SE. Challenges for nanoparticle characterization. *Methods Mol Biol* 2011;697:9–15.
16. Dobrovolskaia MA, Clogston JD, Neun BW, Hall JB, Patri AK, McNeil SE. Method for analysis of nanoparticle hemolytic properties *in vitro*. *Nano Lett* 2008;8:2180–2187.
17. Dobrovolskaia MA, Neun BW, Clogston JD, Ding H, Ljubimova J, McNeil SE. Ambiguities in applying traditional Limulus amebocyte lysate tests to quantify endotoxin in nanoparticle formulations. *Nanomedicine (Lond)* 2010;5:555–562.
18. Keene AM, Allaway RJ, Sadrieh N, Tyner KM. Gold nanoparticle trafficking of typically excluded compounds across the cell membrane in JB6 Cl 41–5a cells causes assay interference. *Nanotoxicology* 2011;5:469–478.
19. Monteiro-Riviere NA, Inman AO, Zhang LW. Limitations and relative utility of screening assays to assess engineered nanoparticle toxicity in a human cell line. *Toxicol Appl Pharmacol* 2009;234:222–235.
20. Worle-Knirsch JM, Pulskamp K, Krug HF. Oops they did it again! Carbon nanotubes hoax scientists in viability assays. *Nano Lett* 2006;6:1261–1268.
21. Monteiro-Riviere NA, Inman AO. Challenges for assessing carbon nanomaterial toxicity to the skin. *Carbon* 2006;44:1070–1078.
22. ANSI/AAM/ISO 10993-4:2002&A1/2006. Biological evaluation of medical devices — part 4: Selection of tests for interactions with blood. American National Standards Institute, 2002.
23. ASTM International. Standard Terminology Relating to Nanotechnology. ASTM Int E 2456–06 2006:1–4.
24. U.S. FDA. Guidance for Industry. ICH S6 Preclinical Safety Evaluation of Biotechnology — Derived Pharmaceuticals. Center for Drug Evaluation and Research, 1997.
25. U.S. FDA. Guidance for Industry. "Developing Medical Imaging Drug and Biological Products. Part 1: Conducting Safety Assessments. Center for Drug Evaluation and Research, 2004.

26. U.S. FDA. Guidance for Industry. ICH S8 Immunotoxicity Studies for Human Pharmaceuticals. Center for Drug Evaluation and Research, 2006.
27. U.S. FDA/CDER. Guidance for Industry. "Immunotoxicology Evaluation of Investigational New Drugs". Center for Drug Evaluation and Research, 2002.
28. U.S. FDA. Guidance for Industry. and FDA Reviewers. "Immunotoxicity Testing Guidance". Center for Devices and Radiological Health, 1999.
29. Daneshian M, von Aulock S, Hartung T. Assessment of pyrogenic contaminations with validated human whole-blood assay. *Nat Protoc* 2009;4:1709–1721.
30. Neun BW, Dobrovolskaia MA. Qualitative analysis of total complement activation by nanoparticles. *Methods Mol Biol* 2011;697:237–245.
31. Chanan-Khan A, Szebeni J, Savay S, Liebes L, Rafique NM, Alving CR, Muggia FM. Complement activation following first exposure to pegylated liposomal doxorubicin (Doxil): Possible role in hypersensitivity reactions. *Ann Oncol* 2003;14:1430–1437.
32. Price KS, Castells MC. Taxol reactions. *Allergy Asthma Proc* 2002;23:205–208.
33. Szebeni J, Baranyi L, Savay S, Milosevits J, Bunger R, Laverman P, Metselaar JM, Storm G, Chanan-Khan A, Liebes L, *et al.* Role of complement activation in hypersensitivity reactions to Doxil and hynic PEG liposomes: Experimental and clinical studies. *J Liposome Res* 2002;12:165–172.
34. Weiszhar Z, Czucz J, Revesz C, Rosivall L, Szebeni J, Rozsnyay Z. Complement activation by polyethoxylated pharmaceutical surfactants: Cremophor-EL, Tween-80 and Tween-20. *Eur J Pharm Sci* 2012;45:492–498.
35. Gradishar WJ, Tjulandin S, Davidson N, Shaw H, Desai N, Bhar P, Hawkins M, O'Shaughnessy J. Phase III trial of nanoparticle albumin-bound paclitaxel compared with polyethylated castor oil-based paclitaxel in women with breast cancer. *J Clin Oncol* 2005;23:7794–7803.
36. Khemani RG, Bart RD, Alonzo TA, Hatzakis G, Hallam D, Newth CJ. Disseminated intravascular coagulation score is associated with mortality for children with shock. *Intensive Care Med* 2009;35:327–333.
37. Lando PA, Edgington TS. Lymphoid procoagulant response to bacterial endotoxin in the rat. *Infect Immun* 1985;50:660–666.
38. Oh D, Jang MJ, Lee SJ, Chong SY, Kang MS, Wada H. Evaluation of modified non-overt DIC criteria on the prediction of poor outcome in patients with sepsis. *Thromb Res* 2010;126:18–23.
39. Slofstra SH, ten Cate H, Spek CA. Low dose endotoxin priming is accountable for coagulation abnormalities and organ damage observed in the Shwartzman reaction. A comparison between a single-dose endotoxemia model and a double-hit endotoxin-induced Shwartzman reaction. *Thromb J* 2006;4:13.
40. Barbui T, Falanga A. Disseminated intravascular coagulation in acute leukemia. *Semin Thromb Hemost* 2001;27:593–604.
41. Franchini M, Dario Di Minno MN, Coppola A. Disseminated intravascular coagulation in hematologic malignancies. *Semin Thromb Hemost* 2010;36:388–403.
42. Higuchi T, Toyama D, Hirota Y, Isoyama K, Mori H, Niikura H, Yamada K, Omine M. Disseminated intravascular coagulation complicating acute lymphoblastic leukemia: A study of childhood and adult cases. *Leuk Lymphoma* 2005;46:1169–1176.
43. Levi M. Cancer and DIC. *Haemostasis* 2001;31(Suppl 1):47–48.
44. Levi M. Disseminated intravascular coagulation in cancer patients. *Best Pract Res Clin Haematol* 2009;22:129–136.

45. Uchiumi H, Matsushima T, Yamane A, Doki N, Irisawa H, Saitoh T, Sakura T, Jimbo T, Handa H, Tsukamoto N, *et al.* Prevalence and clinical characteristics of acute myeloid leukemia associated with disseminated intravascular coagulation. *Int J Hematol* 2007;86:137–142.
46. Napoleone E, Zurlo F, Latella MC, Amore C, Di Santo A, Iacoviello L, Donati MB, Lorenzet R. Paclitaxel downregulates tissue factor in cancer and host tumour-associated cells. *Eur J Cancer* 2009;45:470–477.
47. Swystun LL, Shin LY, Beaudin S, Liaw PC. Chemotherapeutic agents doxorubicin and epirubicin induce a procoagulant phenotype on endothelial cells and blood monocytes. *J Thromb Haemost* 2009;7:619–626.
48. Fibach E, Treves A, Korenberg A, Rachmilewitz EA. *In vitro* generation of procoagulant activity by leukemic promyelocytes in response to cytotoxic drugs. *Am J Hematol* 1985;20:257–265.
49. Gralnick HR, Abrell E. Studies of the procoagulant and fibrinolytic activity of promyelocytes in acute promyelocytic leukaemia. *Br J Haematol* 1973;24:89–99.
50. Hiller E, Saal JG, Ostendorf P, Griffiths GW. The procoagulant activity of human granulocytes, lymphocytes and monocytes stimulated by endotoxin. Coagulation and electron microscopic studies. *Klin Wochenschr* 1977;55:751–757.
51. Kwaan HC, Wang J, Boggio LN. Abnormalities in hemostasis in acute promyelocytic leukemia. *Hematol Oncol* 2002;20:33–41.
52. ten Cate H, Falanga A. Overview of the postulated mechanisms linking cancer and thrombosis. *Pathophysiol Haemost Thromb* 2008;36:122–130.
53. ten Cate H, Falanga A. The pathophysiology of cancer and thrombosis. Summary and conclusions. *Pathophysiol Haemost Thromb* 2008;36:212–214.
54. Walsh J, Wheeler HR, Geczy CL. Modulation of tissue factor on human monocytes by cisplatin and adriamycin. *Br J Haematol* 1992;81:480–488.
55. Wheeler HR, Geczy CL. Induction of macrophage procoagulant expression by cisplatin, daunorubicin and doxorubicin. *Int J Cancer* 1990;46:626–632.
56. Dobrovolskaia MA, Patri AK, Potter TM, Rodriguez JC, Hall JB, McNeil SE. Dendrimer-induced leukocyte procoagulant activity depends on particle size and surface charge. *Nanomedicine (Lond)* 2011;7:245–256.
57. Greish K, Thiagarajan G, Herd H, Price R, Bauer H, Hubbard D, Burckle A, Sadekar S, Yu T, Anwar A, *et al.* Size and surface charge significantly influence the toxicity of silica and dendritic nanoparticles. *Nanotoxicology* 2011 [Epub ahead of print].
58. Schneider P, van Dreden P, Rousseau A, Kassim Y, Legrand E, Vannier JP, Vasse M. Increased levels of tissue factor activity and procoagulant phospholipids during treatment of children with acute lymphoblastic leukaemia. *Br J Haematol* 2010; 148:582–592.
59. ASTM E2524-08. Standard test method for analysis of hemolytic properties of nanoparticles. American Society for Testing and Materials International, 2008.
60. Neun BW, Dobrovolskaia MA. Method for *in vitro* analysis of nanoparticle thrombogenic properties. *Methods Mol Biol* 2011;697:225–235.
61. Skoczen SL, Potter TM, Dobrovolskaia MA. Method for analysis of nanoparticle effects on cellular chemotaxis. *Methods Mol Biol* 2011;697:247–253.
62. Aggarwal P, Dobrovolskaia MA. Gold nanoparticle quantitation via fluorescence in solution and cell culture. *Methods Mol Biol* 2011;697:137–143.

63. Skoczen SL, Potter TM, Dobrovolskaia MA. *In vitro* analysis of nanoparticle uptake by macrophages using chemiluminescence. *Methods Mol Biol* 2011;697:255–261.
64. ASTM E2525-08. Standard test method for evaluation of the effect of nanoparticulate materials on the formulation of mouse granulocyte — macrophage colonies. American Society for Testing and Standards International, 2008.
65. Zhu J, Wang X, Xu X, Abassi YA. Dynamic and label-free monitoring of natural killer cell cytotoxic activity using electronic cell sensor arrays. *J Immunol Methods* 2006;309:25–33.
66. Nelson EL, Strobl S, Subleski J, Prieto D, Kopp WC, Nelson PJ. Cycling of human dendritic cell effector phenotypes in response to TNF-alpha: Modification of the current 'maturation' paradigm and implications for *in vivo* immunoregulation. *FASEB J* 1999;13:2021–2030.
67. Aldridge C, Razzak A, Babcock TA, Helton WS, Espat NJ. Lipopolysaccharide-stimulated RAW 264.7 macrophage inducible nitric oxide synthase and nitric oxide production is decreased by an omega-3 fatty acid lipid emulsion. *J Surg Res* 2008;149:296–302.
68. Chi DS, Qui M, Krishnaswamy G, Li C, Stone W. Regulation of nitric oxide production from macrophages by lipopolysaccharide and catecholamines. *Nitric Oxide* 2003;8:127–132.
69. Stein E, McMahon B, Kwaan H, Altman JK, Frankfurt O, Tallman MS. The coagulopathy of acute promyelocytic leukaemia revisited. *Best Pract Res Clin Haematol* 2009;22:153–163.

Chapter 20

Evaluating the Adverse Effects of Nanomaterials on the Immune System with Animal Models

Matthew J. Smith[*,‡], *Colleen E. McLoughlin*[*], *Kimber L. White, Jr.*[*], *and Dori R. Germolec*[†]

[*]*Immunotoxicology Group*
Virginia Commonwealth University
PO Box 980613, Richmond, VA 23298, USA
[†]*Division of the National Toxicology Program*
The National Institute for Environmental Health Sciences
530 Davis Drive, Morrisville, NC 27560, USA

The last decade has seen an explosion in the use of nanomaterials (NMs). From their increasing use in improving and developing new technologies for industrial purposes, to their potential in medical applications, these materials offer exciting promise to a variety of research fields due in large part to their novel properties, including small size, increased specific surface area, physicochemical properties (such as morphology, surface charge, and chemical makeup), and surface modifications. There is much concern, however, that NM interactions with biological systems can lead to harmful or toxic effects as a result of these novel properties. In particular, the small size of NMs makes them a target for uptake by phagocytic cells of the immune system and subsequent biodistribution into lymphoid tissues such as the spleen, lymph nodes, and bone marrow. Current *in vitro* screening techniques typically do not correlate well with observed *in vivo* toxicity. Therefore, evaluating

[‡]Corresponding author. E-mail: msmith@immunotox.com

the immunotoxic effects of NMs *in vivo* is increasingly important, as the use of these materials for industrial, research, and medical applications continues to increase. This chapter aims to discuss how the adverse effects of NMs on the immune system can be evaluated in light of their unique characteristics, to consider various *in vivo* models by which NM-mediated immune effects can be detected, and to review the immune effects of three different types of NMs with differing primary routes of human exposure.

1. Introduction

In the European Union (EU), nanomaterials (NMs) have been defined as materials with a length of between 1 and 100 nm in at least one dimension, although this definition remains the subject of debate in the United States.[1–3] There is a general consensus that the testing of NMs for their impact on human health is critical;[4] however, several differences of opinion exist regarding how such testing should be conducted. Some experts believe that current, "traditional" toxicological methods may not adequately assess NM toxicity and may therefore require novel methods for safety assessment.[4,5] However, the development, validation, and implementation of new methods are not trivial.[6] Furthermore, due to the vast number of different types of NMs (a number that increases exponentially when differences in size, surface modification, etc. are considered), it is not feasible to evaluate every possible NM for potential hazardous effects. This necessitates the development of alternative approaches, such as the identification of potential relationships between the bioactivity of NMs and their size, physicochemical properties, and surface modifications, to better identify which NMs are likely to result in toxic effects following exposure.[7]

The difficulty grows as a comprehensive characterization of the physicochemical properties of each NM is necessary in order to proceed with this type of assessment strategy.[7,8] As novel methods and approaches for NM toxicological assessments may require lengthy development and validation processes, experts have recommended that, in the meantime, "traditional" assessments be conducted to evaluate NM toxicity.[4] These assessments typically include both *in vitro* and *in vivo* evaluations. However, it has been suggested that traditional *in vitro* assays may not accurately assess the immunotoxic effects of NMs, and no *in vitro* test has been validated as a substitute for the *in vivo* evaluation of NM immunotoxicity.[9]

In vivo studies are believed by many to be a necessary component of nanotoxicology research, in part because the complexity of *in vivo* systems lends the possibility of unique biodistribution, clearance, metabolism, and immune responses.[10] Furthermore, NM distribution can affect the immune response (phagocytosis/uptake) and the subsequent clearance of NMs.[11] If unique biodistribution leads to altered immune responses, it may not be possible to adequately evaluate NM immunotoxicity *in vitro* without first identifying the target organs and affected cell types following *in vivo* NM exposure.[10] Furthermore, Sayes *et al.*[12] found little correlation between *in vitro* screening techniques and *in vivo* pulmonary toxicity following exposure to multiple types of fine ($\leq$250 nm in diameter[13]) nanoparticles. These authors also indicated that *in vitro* methods are in need of further development, standardization, and validation before they can be proven useful in assessing the toxicity of inhaled particles. Therefore, *in vitro* studies should be implemented to complement, rather than replace, *in vivo* immunotoxicity studies, particularly in light of the current absence of a correlation between *in vivo* results and *in vitro* screenings.[9,10]

The aims of this chapter are threefold: 1) to examine how the adverse effects of nanomaterials on the immune system can be evaluated by considering the potential unique interactions between NMs and the immune system, 2) to consider various models and methods by which NM-mediated immune effects can be detected, and 3) to review the immune effects of three different types of NMs, each with a different primary route of human exposure.

2. Immunotoxicological Assessments

2.1 *Nanomaterials and the immune system*

The immune system is a complex array of proteins, cells, and organs, which is responsible for protecting the host from invasion by pathogens and other foreign substances. The innate immune system, consisting of cells such as macrophages, neutrophils, and natural killer (NK) cells, and soluble factors such as cytolytic enzymes, signaling molecules, and the proteins of the complement system, responds rapidly upon activation, even in the absence of prior exposure to the specific invader. The acquired immune system, on the other hand, requires an initial exposure and the subsequent formation of effector and memory cells before it can provide effective protection. This branch of immunity is further subdivided into cell-mediated immunity (CMI), which includes effector, helper, and regulatory T cells, and humoral immunity, which includes B cells and plasma cells. The branches of the acquired immune system do not function in isolation from one another or from the innate immune system. Rather, their efforts are coordinated and directed by the production of signaling proteins (cytokines), including interleukins (IL), interferons (IFN), and tumor necrosis factors (TNF).

Cytokines perform a wide variety of tasks, including mediating/maintaining inflammation and directing acquired immune responses toward either cell-mediated or humoral responses. Cytokines that promote cell-mediated responses, including but not limited to IL-2, IL-12, and IFN-γ, are produced by T helper 1 (Th1) cells, while those that enhance humoral immune responses, including but not limited to IL-4, IL-5, and IL-10, are produced by T helper 2 (Th2) cells.

Exposure to NMs may result in either immunostimulation or immunosuppression. An inflammatory response and a subsequent influx of phagocytes may occur, unless the materials are not recognizable by cells of the immune system. In addition, a number of NMs have proven to be potent adjuvants, capable of augmenting the immune response directed against certain antigens.[14–16] Furthermore, NMs administered dermally are conceivably capable of eliciting a contact hypersensitivity response. Alternatively, the suppression of normal immune functions or even the induction of immune tolerance may occur following NM exposure.[14,15]

A thorough, working knowledge of the immune system can facilitate the design and engineering of NMs to avoid recognition by the immune system or even to allow the targeting of specific cell types to stimulate or inhibit specific immune responses.[15] The interaction of NMs with the components of the immune system may result in a variety of immune responses. However, due to their size and physical characteristics, the initial contact between NMs and the immune system is likely to be mediated by phagocytes such as polymorphonuclear cells (PMNs), or neutrophils, and macrophages. These interactions, which are dictated by size and physicochemical characteristics such as surface charge, hydrophobicity/hydrophilicity, and the presence or absence of surface functional groups, can influence NM biodistribution and may result in undesirable or unforeseen consequences.[11,15,17,18] For example, the activation of both the complement system and granulocytes in the blood has been reported to increase with decreasing NM size.[19,20] Also, surface modification by the addition of various functional groups to the NM surface can direct immune system–NM interactions.[10] For example, single-walled carbon nanotubes were readily internalized by phagocytic cells of the immune system when modified with phosphatidylserine on their surfaces,[21] but they evaded uptake when coated with the chelator diethylenetriaminepentaacetic acid.[22]

The phagocytic cells of the mononuclear phagocytic system (MPS) play an important role in the uptake of nanoparticles.[6] These cells have several mechanisms by which nanoparticles and other molecules can be internalized, including phagocytosis, macropinocytosis, and three different types of endocytosis.[14] The term "phagocytosis" is generally reserved to refer to the uptake of large

(>0.5 μm) particles,[23] and the internalization of nanoparticles has been described instead as "receptor-mediated endocytosis" by some.[14] However, there are reports of various NMs that are internalized by phagocytic mechanisms, such as via the Fcγ receptor in the uptake of C_{60} fullerenes. Furthermore, the mechanism of NM internalization by phagocytes may correlate with the production (or lack thereof) of pro-inflammatory cytokines.[14]

Clearly, in light of both the complexity of the immune system and the inevitable interaction of NMs with its various components, evaluating potential immune effects of NMs is critical. Accomplishing this task, however, will require a well-defined and systematic approach, including the selection of suitable methods that can adequately and appropriately answer the scientific questions being considered.

2.2 *Model and study designs*

The choice of animal models in toxicological testing requires the consideration of a number of factors, including the scientific questions being explored, the route of exposure to be utilized and its relevance to human exposure, and local *vs.* systemic effects. These considerations can impact the endpoints to be evaluated and the battery of assays selected to conduct such evaluations. With regard to NMs, these choices can become increasingly important, as an incomplete study design or a poorly chosen route of exposure or dose selection can significantly and negatively impact on the ability to interpret results and may ultimately make the determination of a material's toxic potential more difficult.

Every effort should be made to ensure that the route of exposure is relevant to human exposure, in order to allow for a clearer interpretation of the results with regard to potential risk. In addition, the consideration of both local and systemic toxicities may be important, particularly when evaluating NMs. Specifically, some NMs may have associated local lung immunotoxic effects with no systemic immunosuppression. Alternatively, a material may have systemic immunosuppressive effects with little to no associated pulmonary immune effects. In short, a thorough evaluation of a material's potential effects on both systemic and local immunity should be conducted. Although not the subject of this chapter, it is also worth noting that considerable efforts are underway regarding how best to regulate the use of NMs and how to meet the inherent challenges of such regulations.[24–27]

Screening strategies for assessing NM toxicity have been suggested.[28,29] NM exposure can occur via the pulmonary, oral, dermal, and parenteral routes, and the *in vivo* testing of each of these routes is therefore important.[28]

Another issue to be considered is whether NMs should be evaluated in the aggregated form or alternatively, as well-dispersed suspensions, as many NMs can form aggregates much larger in size when placed in suspension. Although there are many varying opinions regarding what properties (size, dispersion/aggregation state, specific surface area, surface modification, chemical composition, surface charge, etc.) are the most relevant, each of these properties should be carefully considered and well characterized in order to promote a more thorough and appropriate evaluation of results.[4,28–30]

Also, there is continuing debate and indecision regarding the most appropriate dose metric to use when assessing NM toxicity. Three options have been proposed for quantifying doses: by mass, by specific surface area, and by particle number.[28] It has been suggested that specific surface area (surface area per unit mass, e.g., m^2/g) may be a more appropriate metric than mass.[31] Indeed, Oberdörster *et al.*[32, 33] have reported that the inflammatory responses following TiO_2 inhalation correlated better with particle specific surface area than with mass. However, the general consensus is that there is not enough data available to definitively select one of these metrics over another.[29] Therefore, sufficient information should be collected and maintained in order that the effects can be evaluated in the context of each of these dose metrics.[28]

2.3 *Methods: Inflammation, immunomodulation, and hypersensitivity*

Evaluations of the effects of a xenobiotic on the immune system can take on many forms. Frequently, inflammatory responses including cell infiltration and cytokine production are evaluated. The examination of a material's effects on other aspects of the immune system is also essential in order to adequately assess immunomodulatory potential or contact hypersensitivity reactions.

Local inflammatory responses can be evaluated following pulmonary exposure, for example, by collecting bronchoalveolar lavage fluid (BALF) and subsequently analyzing changes in cell phenotypes. In addition, cytokine profiles in the BALF can be examined for both dose–response and time-course effects. This, in particular, represents a very powerful technique for examining mechanisms of NM immunotoxicity, as tests can be tailored to focus on the specific cytokines of interest with the ability to evaluate a relatively large number of cytokines in a single sample if conducted using commercially available cytometric bead arrays.

Oxidative stress, which occurs when the host is unable to rapidly detoxify reactive metabolites (such as reactive oxygen and nitrogen species), is well established as a mechanism of NM toxicity.[6,34] Markers of oxidative stress such as glutathione and nitrate/nitrite levels in the BALF can provide information regarding the general toxicity of NMs following pulmonary exposure.[28] Although it has been suggested that oxidative stress mechanisms contribute to the progression of pro-inflammatory diseases of the lung,[35] correlations between NM-induced oxidative stress and NM-induced immunomodulatory effects remain to be defined and represent an important area for continued research.

Systemic immune responses can also be affected following NM exposure. Therefore, both innate and acquired (cell-mediated and humoral) immunity must be evaluated. Innate immune responses can be evaluated by assessing NK cell activity and also the phagocytic capability of the fixed tissue macrophages of the MPS. While assessments of NK activity are more routine than those of the MPS, the critical role of macrophages in the immune response to NMs[6] suggests that the evaluation of changes in MPS activity should be routinely employed. Indeed, Oberdörster *et al.*[33] have demonstrated a decreased alveolar macrophage clearance of TiO_2 nanoparticles as compared to microparticles following inhalation exposure.

The phagocytic activity of the fixed tissue macrophages of the MPS can be assessed by considering the vascular clearance and organ uptake of radio-labeled sheep erythrocytes (sRBCs).[36,37] In addition to standard dose–response evaluations, time-course effects may also need to be assessed, as the phagocytic activity of the MPS can demonstrate a biphasic response. Specifically, the suppression of MPS activity was observed one day after exposure to maleic vinyl ether (MVE), yet activity was enhanced six days post-exposure.[38] These authors suggested that the observed impairment of vascular clearance and tissue uptake of the radio-labeled sRBCs at one day post-exposure could have been due to the decreased synthesis of proteins necessary for phagocytosis, or alternatively, that the cells were physically saturated following the ingestion of the MVE polymer.[38] The potential for physical interference of macrophage function by nanoparticles has been proposed as a potential explanation for NM-mediated decreases in macrophage uptake.[39]

Humoral immunity is often evaluated by considering effects on the T-dependent antibody response (TDAR). Typical assays used to evaluate the TDAR include the antibody-forming cell (AFC) assay (also known as the plaque assay) and the enzyme-linked immunosorbent assay (ELISA). The AFC assay measures the ability of the host to produce antibody-forming cells (i.e., plasma cells) following immunization with sRBCs, while the ELISA is often used to measure primary and secondary serum antibody levels to T-dependent antigens such as keyhole limpet hemocyanin (KLH) or sRBCs.

CMI can be assessed using multiple *ex vivo* endpoints, including proliferation assays (mediated by the anti-CD3 antibody or mitogens such as concanavalin A), the mixed leukocyte response (MLR), and cytotoxic T lymphocyte (CTL) activity. Recently, evaluations of the delayed-type hypersensitivity (DTH) response have been suggested as an important addition to assessments of CMI,[40] particularly in light of evidence that this response is more sensitive than *ex vivo* CMI endpoints.[41] The AFC and DTH assays are considered "holistic" functional assays because they require the integrated response of both innate and acquired immunity. Specifically, multiple cell types (from both branches of the immune system) must function properly in an orchestrated manner under the direction and action of multiple cytokines, in order to mount an appropriate response. As a result, holistic functional assays are frequently extremely sensitive in detecting effects on the immune system.

The potential to induce hypersensitivity reactions (both immediate- and delayed-type) is another consideration when evaluating NMs. Recently, the use of nanoparticles as drug delivery devices has been shown to decrease allergic responses to the drug paclitaxel.[42] In contrast, some NMs, including carbon black and carbon nanotubes, promote allergic responses and IgE antibody production against co-administered allergens such as ovalbumin (OVA).[16,43]

Little research regarding the possibility of nanoparticles to induce contact hypersensitivity has been reported. As detailed in Section 3.1 of this chapter, TiO_2 nanoparticles can cause irritation when applied dermally to the ears of mice. Furthermore, they trigger increases in local lymph node proliferation following subcutaneous (s.c.) administration. This suggests that it is important to evaluate NMs for their contact hypersensitivity potential, particularly when the NMs of interest are intended for use in products that are applied dermally. However, there is significant debate regarding the ability of NMs to penetrate the stratum corneum layer of the skin.[44–46]

Many of the methods and assays discussed above have been well established and validated across many laboratories. However, the use of other assessment techniques may be necessary or recommended. Traditional contact hypersensitivity testing is often conducted using the local lymph node assay (LLNA). In this assay, the material of interest is applied dermally to the ears and some time later, the draining lymph nodes are excised and assessed for cell proliferation. However, Dobrovolskaia *et al.*[9] have indicated that the traditional LLNA is not recommended for evaluating the hypersensitivity potential of NMs. Instead, the authors suggested that the lymph node proliferation assay (LNPA) should be used as an alternative.

The LNPA utilizes s.c. administration of NMs in the area between the ears on the top of the head along the midline.[47] This allows NMs of sufficiently small size to travel through the lymphatic system and interact with immune cells of the draining lymph nodes. Indeed, Reddy *et al.*[48] have characterized the role of particle size in NM distribution into lymph nodes, wherein the authors delineated differences in lymphatic transport and dendritic cell targeting between 25- and 100-nm nanoparticles. Furthermore, Manolova *et al.*[49] have demonstrated that intradermally administered fluorescent beads with diameters ranging from 500 to 2000 nm remain at the injection site, where they are internalized, while those with diameters between 20 and 200 nm can drain into the lymphatic system and subsequently be taken up by lymph node resident dendritic cells. Although the penetration of NMs through the stratum corneum of healthy skin is a subject of debate, they may be able to penetrate through the skin when flexed, through hair follicles or damaged skin.[44–46] Thus, the testing of NM-induced contact hypersensitivity following s.c. administration is recommended.

The methods and procedures identified above are certainly not all-inclusive. However, with the exception of the LNPA, they represent a collection of "traditional," well established, and validated tools for identifying effects on the immune system. Table 1 summarizes the major events of several of these assays. Whether or not these methods are sufficiently sensitive to detect NM-mediated immunotoxicity remains to be determined. In the meantime, until a resolution is reached regarding the ability of these "traditional" methods to adequately evaluate nanotoxicity, these tools remain an important means of assessing immune responses following NM exposure. In spite of the availability of these methods, however, the majority of immunological testing in animals following NM exposure has centered primarily around the lung inflammatory response following pulmonary exposure, while other components of the immune system have largely been ignored.

3. *In Vivo* Immunotoxicological Evaluations of Nanomaterials: Three Case Studies

Here, we consider three different nanomaterials with differing primary routes of human exposure, as examples to demonstrate the current approaches being used and the current "state of the science" of evaluating the effects of NMs on the immune system.

Table 1. Major events of assays used in *in vivo* immunotoxicology assessments.

	Assay	Major events	Reference
Innate immunity	NK	*Ex vivo* cytotoxicity of tumor cells	50
	MPS	Phagocytosis of radio-labeled sRBCs by fixed tissue macrophages of the liver, spleen, lungs, and thymus	37
Humoral immunity	AFC	Antigen processing/presentation T cell co-stimulation of B cells Effector/Memory B cells formed against sRBC antigen Antigen-specific antibody secretion by splenic plasma cells	50
	ELISA	Antibody produced in multiple lymphoid organs Antigen-specific antibody levels measured in serum	51
Cell-mediated immunity	Proliferation	Mitogen or antibody-mediated T cell proliferation Concanavalin A or anti-CD3 antibody	50
	MLR	Recognition of allogeneic lymphocytes Proliferation response	50
	CTL	Recognition of target cells and formation of memory cells Proliferation response Formation of effector cells with cytotoxic capability	50
	DTH	Sensitization phase Antigen recognition, T cell proliferation Effector/memory T cells formed against *C. albicans* Challenge phase Effector/memory T cell migration to site of challenge Inflammation due to the production/release of cytokines	40, 41, 50
Hyper-sensitivity	LLNA and LNPA	Formation of memory against test chemical antigens Lymphocyte proliferation in draining lymph nodes	47, 52, 53
	MEST	Sensitization phase Antigen recognition, T cell proliferation Effector/memory T cells formed against test chemical antigens Challenge phase Effector/memory T cell migration to site of challenge Inflammation due to production/release of cytokines	54

Abbreviations

AFC	antibody-forming cell
CTL	cytotoxic T lymphocyte
DTH	delayed-type hypersensitivity
ELISA	enzyme-linked immunosorbent assay
LLNA	local lymph node assay
LNPA	lymph node proliferation assay
MEST	mouse ear swelling test
MLR	mixed leukocyte response
MPS	mononuclear phagocytic system

3.1 *TiO_2 nanoparticles*

TiO_2 is a naturally occurring mineral oxide of titanium(IV) and exists in four distinct crystal structures: rutile, anatase, brookite, and TiO_2(B).[55] Nanoparticulate TiO_2 is frequently used as a white pigment in a variety of commercial products, including paint, cosmetics, sunscreen, and food, and like many nanoparticles, it can form aggregates when suspended in biological media.[55] Although TiO_2 has long been considered to have little biological activity in animals and humans[56] and has been widely used as a negative control in particulate toxicity studies,[57] the International Agency for Research on Cancer (IARC) has classified TiO_2 as possibly carcinogenic to humans (Category 2B).[55] Because the primary routes of human exposure to nanoparticles are ingestion, respiration, and skin exposure,[10] exposure to nanoparticulate TiO_2 is most likely through the ingestion of foods and dermal application of cosmetic products that contain nanoparticulate TiO_2.

Oral exposure to a large dose (5 g/kg) of nanoparticulate TiO_2 (25 and 80 nm in diameter; crystal structure not specified) has been reported to induce liver injury in mice.[58] In addition, adverse effects have been demonstrated on multiple immune cell populations, platelets, and various cellular populations of the blood, following oral exposure to anatase nanoparticulate TiO_2.[56] Anatase TiO_2 nanoparticles (11 ± 5 nm) have also been shown to affect the kinetics of granule secretion by murine peritoneal mast cells in *in vitro* culture systems.[59] Furthermore, Sayes *et al.*[60] have reported that the *in vitro* culture of either human dermal fibroblasts or human lung epithelial cells with anatase TiO_2 (primary particle diameter of ~20 nm) at high concentrations (>1500 μg/mL) resulted in increased lactate dehydrogenase (LDH) release, increased IL-8 production, and decreased mitochondrial activity. In contrast, equal mass exposures of rutile TiO_2 (3.2-nm primary particle diameter) had no effects on IL-8 production or mitochondrial activity, with only mild increases in LDH being observed.[60] These authors also reported similar specific surface areas of both the anatase and rutile particle aggregates, indicating that the effects were not due to differences in this aspect. Rather, this suggests that different TiO_2 crystal structures can have different toxic effects.

Recent studies evaluating the immunotoxicity of nanoparticle anatase TiO_2 (primary particle diameter <25 nm) following both oral and dermal exposure in mice have been reported by Auttachoat *et al.*[61] No effects were observed using holistic and functional assays to evaluate humoral immunity (plaque assay, serum anti-sRBC IgM titers), CMI (DTH response to *C. albicans*), and innate immunity (MPS) following oral exposure to nanoparticulate TiO_2 at doses up to 250 mg/kg for 28 days.

Dermal exposure to anatase TiO_2 nanoparticles produced no effects in the LLNA or in the mouse ear swelling test (MEST), suggesting that it may not be a contact sensitizer.[61] However, dose-responsive irritant effects were observed following dermal treatment with 5–10% (w/v) nanoparticulate TiO_2. Furthermore, the use of the LNPA instead of the LLNA (i.e., s.c. administration instead of dermal) revealed significant increases in auricular lymph node cell proliferation following s.c. injection of 125 and 250 mg/kg nanoparticulate TiO_2. These results are consistent with reports that nanoparticulate TiO_2 does not penetrate the stratum corneum when applied dermally to healthy skin,[44] and they lend further support to the use of the LNPA in lieu of the LLNA, as recommended by Dobrovolskaia *et al.*[9] for evaluations of NM sensitizing potential.

In manufacturing and laboratory settings, inhalation exposure to nanoparticulate TiO_2 may be possible. Oberdörster *et al.*[33] have compared the pulmonary toxicity of ultrafine particles — which are particles with diameters ≤100 nm — and fine anatase TiO_2 particles having primary particle diameters of 20 and 250 nm, respectively. These authors found that when based on a mass dose metric, the inhalation of the ultrafine particles resulted in a greater inflammatory response (greater numbers of PMNs, alveolar macrophages, and total cells in the BALF) as compared to the fine particles. Furthermore, the authors reported that the ultrafine particles are more likely to promote the formation of lung interstitial fibrosis. However, when correlated with the specific surface area of the inhaled particles, there were no differences in PMN infiltration between the two particle sizes.

De Haar *et al.*[16] have reported that intranasal co-administration of OVA and ultrafine (29-nm diameter) TiO_2 nanoparticles resulted in increased pro-inflammatory cytokine production (TNF-α) in addition to increased levels of IL-4, IL-5, IL-10, and IL-13. Furthermore, allergy enhancement was demonstrated through observed increases in OVA-specific IgE and IgG1 antibody levels in serum. Liu *et al.*[62] have demonstrated that the inhalation of TiO_2 nanoparticles with primary diameters of <10 nm resulted in the impairment of the phagocytic and chemotactic functions of rat alveolar macrophages. In contrast,

Renwick *et al.*[63] have demonstrated increased chemotaxis of rat alveolar macrophages in response to complement fragment C5a, following a single intratracheal (i.t.) exposure to ultrafine TiO_2 (29-nm primary particle diameter). These authors also reported impaired alveolar macrophage phagocytic capability. A single i.t. exposure to 25-nm diameter nanoparticulate TiO_2 increased pro-inflammatory cytokine production (IL-1, IL-6, TNF-α), as well as increasing other T cell helper cytokines, including IL-4, IL-5, IL-10, IL-12, and IFN-γ.[64]

Recent studies conducted in our laboratory have demonstrated that systemic immunomodulation (as evaluated using the primary AFC response to sRBCs) can occur four days following a single pharyngeal aspiration (p.a.) of anatase TiO_2 nanoparticles. Specifically, anatase (primary particle diameter <25 nm) TiO_2 exposure produced a significant increase in the number of AFCs per million spleen cells [Figure 1(a)], although no significant effects

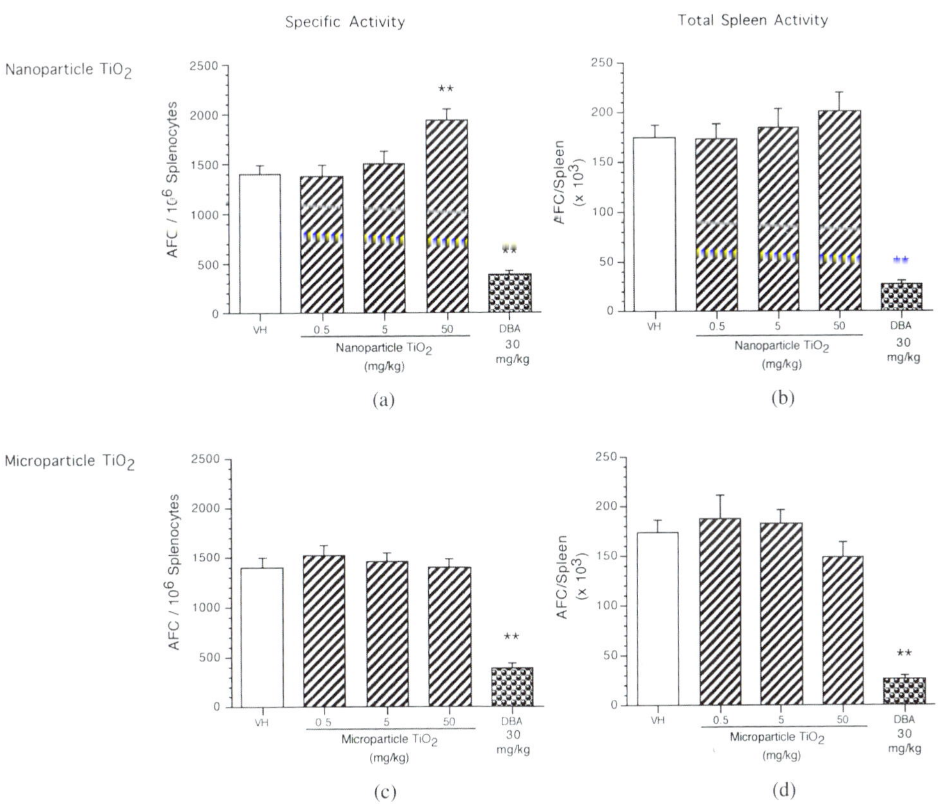

Figure 1. Spleen IgM antibody-forming cell (AFC) response to sRBCs four days following a single pharyngeal aspiration (p.a.) of nanoparticulate or microparticulate anatase TiO_2. Mice were immunized intravenously with 7.5×10^7 sRBCs on the same day as exposure to TiO_2. Results are presented as "Specific Activity" [AFC/10^6 splenocytes; (a) and (c)] and "Total Spleen Activity" [AFC/spleen; (b) and (d)]. DBA: 1,2:5,6-dibenzanthracene, the positive control, administered in a single p.a. on the same day as TiO_2 exposure.
$^{**}p < 0.01$ compared to VH.

were observed when data were expressed as total spleen activity [Figure 1(b)]. Furthermore, no effects were observed on spleen weight or on terminal body weights (data not shown).

In this same study, a single p.a. of microparticulate (< 45-μm primary particle diameter) anatase TiO_2 was also assessed for immunotoxic effects on the primary AFC response. In contrast to the effects observed following p.a. of nanoparticulate TiO_2, no effects were observed following exposure to TiO_2 microparticles [Figures 1(c) and 1(d)]. Whether the effects of nanoparticle anatase TiO_2 on the immune system are negated when the doses are expressed in terms of specific surface area is unclear. However, these results are consistent with those of Oberdörster *et al.*,[33] suggesting that under mass-based dose metrics, nanoparticles may have different effects on the immune system than their larger counterparts. Additionally, in light of the systemic and local immune effects observed following inhalation and p.a. exposure to TiO_2 particles, the widespread use of TiO_2 as a negative control in particulate toxicity studies[57] may not be advisable.

3.2 *Biodegradable electrospun materials for tissue engineering and drug delivery*

Large numbers of researchers are investigating NMs for applications in medicine. The fields exploring the use of NMs in medicine (commonly referred to as nanomedicine) include tissue engineering, regenerative medicine, drug delivery, and diagnostics. Researchers in these fields have increasingly looked to NMs with the intent of clinical or therapeutic applications, including drug delivery (including cytokines, growth factors, and drugs), tissue or organ regeneration or replacement (can be by means of a "scaffold," a temporary, functional replacement structure), and diagnostic tools (such as imaging with quantum dots).[65] The advantages of nanomedicines include the ability to effectively target sites such as tumors, increased surface area leading to increased solubility and bioavailability, and the potential for greater therapeutic efficacy with lower side effects than current formulations.[24]

Nanomedicines that are intended for clinical applications unrelated to the immune system must be examined for the possibility that they might produce unintentional or unanticipated immunomodulatory effects. However, depending on the desired application, the effects on the immune system may be desirable, and many researchers are investigating NMs for use in immunotherapy. Given the potential clinical uses and the typical parenteral route of exposure, these nanomedicines may require unique assessment. However, no unique methodology or approach to assess the immunotoxicity of these materials has been formally adopted.

Some NMs have been approved for use by the U.S. FDA, while others are currently in clinical trials. FDA-approved nanomedicines include iron oxide particles for use in magnetic resonance imaging (MRI), such as Feridex I.V.® and Resovist®. Several organic materials have been FDA-approved for use in cancer therapy, including the protein marketed under the trade name Abraxane®. Examples of NMs currently in clinical trials include gold particles (Aurimmune®) for cancer therapy and the dendrimer VivaGel® for use as a microbicide to prevent sexually transmitted diseases.[66]

Several agencies worldwide (e.g., the U.S. FDA and U.K. Medicines and Healthcare Products Regulatory Agency) have found insufficient evidence to suggest that nanomedicines pose a greater risk than conventional therapies.[24] Currently, the U.S. FDA assigns a nanomedicine to a center based on its primary mode of action, as no unique center exists for nanomedicines. However, several challenges exist in dealing with nanomedicines that may require a special regulatory framework in the future. Such challenges include the increasing complexity of nanomedical formulations, a lack of sufficient data on the risks of nanoparticles, and non-standard nomenclature to define "nano-sized" medicinal components. Many regulatory agencies in different countries have initiated task forces and review groups to determine if nanomedicines will require special regulation and assessment.[24]

Although nanomedicines can be made of metals, semiconductors, or various organic materials, biodegradable polymers are increasingly being explored. Nanomedicines utilizing such biodegradable materials can be manufactured using a wide variety of processes. One of these, the polymer-processing technique called electrospinning, is emerging as a relatively easy and inexpensive means by which nanofibrous materials can be manufactured in a variety of designs.[67, 68] However, a concern associated with the use of biodegradable materials is the potential for the formation of toxic degradation products and the possible effects of the material and its degradation products on the immune system. Recently, *in vitro* and *in vivo* immunotoxicology studies have been conducted on electrospun polymers.

Smith *et al.*[69, 70] have demonstrated that electrospun blends of polydioxanone (PDO), a synthetic biodegradable polymer with the natural polymers elastin (PDO-E) or collagen (PDO-C), can produce immunosuppressive effects following *in vitro* culture with mouse spleen cells. Figure 2 compares the immunosuppressive effects of these electrospun blends on humoral immunity, as evaluated by the B cell proliferative response mediated by

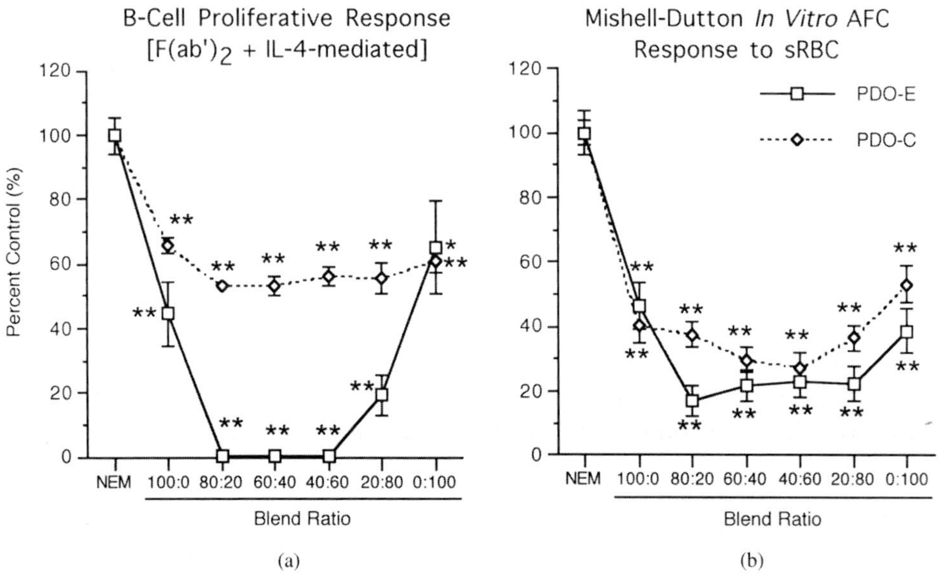

Figure 2. Effects of *in vitro* exposure to PDO-E and PDO-C electrospun materials on humoral immunity. (a) F(ab')$_2$ and IL-4-mediated B cell proliferative response combining two studies. Splenocytes (2×10^5 cells/culture) from B6C3F1 mice were cultured with 3-mm diameter circular punches of material with F(ab')$_2$ antibody fragments (5 μg/mL) and IL-4 (10 ng/mL) for three days. No electrospun material (NEM) means (counts per minute/2×10^5 splenocytes): 56,439 ± 6,410 and 61,971 ± 3,675 (PDO-E); 79,631 ± 896 and 75,476 ± 1,312 (PDO-C). (b) *In vitro* AFC response to sRBCs. Splenocytes (1×10^7 cells/culture) from B6C3F1 mice were cultured with 10-mm diameter circular punches of material in the presence of 1×10^7 sRBCs for four days in a steel box pressurized to 6 psi with a gas mixture of 83% N_2, 10% CO_2, and 7% O_2, and the number of AFCs were subsequently enumerated using the plaque assay. NEM means (AFCs/culture): 3,598 ± 136, 1,113 ± 118, and 2,425 ± 120 (PDO-E); 1,797 ± 209, 602 ± 185, and 1,158 ± 124 (PDO-C).
*$p < 0.05$ and **$p < 0.01$ as compared to NEM control.

F(ab)'$_2$ fragments and IL-4 [Figure 2(a)] and the Mishell–Dutton *in vitro* AFC response to sRBCs [Figure 2(b)].

Results of the Mishell–Dutton assay demonstrated decreased numbers of AFCs formed against sRBCs following exposure to all materials, when compared to no electrospun material (NEM) controls. Furthermore, PDO-E blends produced greater suppression than PDO-C blends. These results suggested that these materials may be targeting B cells, in light of the supporting data from the F(ab')$_2$ proliferative assay. The reasons for the differing immunosuppressive effects of elastin and collagen remain undetermined. These results have also demonstrated that PDO-E and PDO-C blends were each more immunosuppressive than either of the polymers alone. Given the absence of a documented correlation between *in vitro* immunosuppression and subsequent *in vivo* effects, and because no *in vitro* exposure regimen has been accepted as a substitute for *in vivo* exposure,[9] these results must be confirmed following the *in vivo* implantation of these biomaterials.

Another biodegradable synthetic polymer, polycaprolactone (PCL), has been evaluated in the nanofibrous electrospun form in our laboratory following 28 days of exposure *in vivo* in the B6C3F1 mouse. One, two, or four electrospun PCL (EPCL) disks, 6 mm in diameter each, were implanted s.c. in the ventral quadrant. Studies were conducted with either microfibrous or nanofibrous EPCL, and results indicated that neither material suppressed innate immunity, humoral immunity, or CMI. Results from two holistic assays, the AFC response to sRBCs [Figures 3(a) and 3(b)] and the DTH

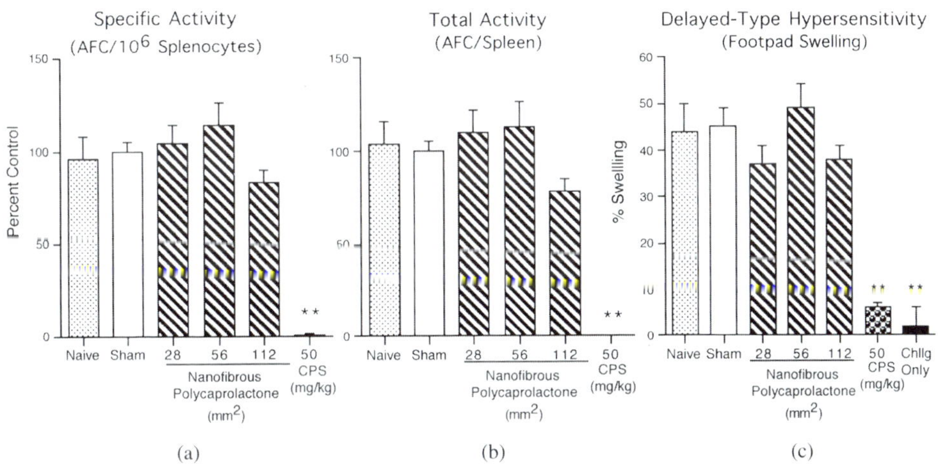

Figure 3. Effects of nanofibrous electrospun PCL implants on humoral immunity and CMI after 28 days of exposure. Naïve animals remained untreated and did not undergo surgical protocols. Sham control animals were anesthetized and received the same surgical treatment as the high dose (four implants or 112 mm^2) without the material being implanted. (a) and (b) Spleen IgM AFC response to sRBCs combined from two studies. Mice were immunized with 7.5×10^7 sRBCs by tail vein injection on Day 25 of exposure. Results are presented in terms of percent control for both (a) Specific Activity (AFC/10^6 splenocytes) and (b) Total Spleen Activity (AFC/spleen). Means for sham control: 3,077 ± 204 and 1,322 ± 232 (specific activity); 753 ± 53 and 261 ± 60 (total activity). (c) Delayed-type hypersensitivity response to *C. albicans*. Mice were immunized with 1×10^7 formalin-fixed *C. albicans* organisms on Day 21 of exposure. The right footpad was pre-measured on Day 29 and challenged with chitosan antigen. Footpad swelling was determined 24 h after challenge. The challenge-only (Chllg Only) group received the footpad challenge but not sensitization. Results are presented in terms of percent footpad swelling ([Post-measurement – Pre-measurement]/Pre-measurement). The positive control, cyclosphosphamide (CPS), was administered i.p. on Days 25–28.
** $p < 0.01$ as compared to sham control.

response to *C. albicans* [Figure 3(c)] are presented for nanofibrous EPCL, where one, two, and four implants correspond to areas of 28, 56, and 112 mm^2, respectively. Cyclophosphamide (CPS; 50 mg/kg on Days 25–28 of the study), injected intraperitoneally (i.p.), was used as the positive control.

No significant differences were observed in the AFC response between the sham control group and groups that received EPCL, indicating that nanofibrous EPCL did not suppress the humoral immune response [Figures 3(a) and 3(b)]. As expected, the response was decreased with the positive control, CPS. Similarly, the DTH results also indicated that nanofibrous EPCL did not affect CMI [Figure 3(c)]. Again, the positive control, CPS, produced the anticipated decrease in footpad swelling. The challenge-only group, in which animals received challenge but not sensitization, was included to control for the background swelling associated with the challenge injection.

The lack of effect of electrospun materials *in vivo* starkly contrasts the *in vitro* immunosuppressive effects reported by Smith *et al.*[69,70] Reasons for these differences, however, are unclear. The *in vitro* studies evaluated a different synthetic electrospun polymer (PDO) than the *in vivo* studies (PCL). Furthermore, the *in vitro* work of Smith *et al.* focused on evaluating different blend ratios of PDO with the natural polymers collagen and elastin, while the *in vivo* studies evaluated the dose-responsive effects of PCL. Another possible explanation for observed differences may be due to differences in cell–material exposure and interactions. Specifically, *in vitro*, there was prolonged direct contact between the material and the immune cells (including T cells, B cells, and macrophages) for up to five days. On the other hand, interactions between the material and immunocompetent cells, including the duration and kinetics of cell–material contact, may be different *in vivo*, where it is likely that only a small percentage of the cells of the immune system may have direct contact with the implanted material.

It should be noted that the PDO-E, PDO-C, and EPCL electrospun materials evaluated in the above studies do not meet the EU definition of NMs at the polymer concentrations at which they were electrospun. Instead, they are best considered to be in the submicron and/or micron size range. For the PDO-E blends, fiber diameters ranged from 627 ± 33 nm to 3.7 ± 0.4 μm and increased in proportion with increasing elastin content, while fiber diameters for PDO-C blends ranged from 546 ± 29 nm to 1,061 ± 85 nm and did not correlate with collagen content.[71] For the *in vivo* EPCL studies, the mean fiber diameter was 225 ± 25 nm. Although these materials are not NMs *per se*, they may be prepared with smaller fiber diameters by decreasing the concentrations of the polymers in solution prior to electrospinning.[72]

The use of such materials in biomedical research represents an emerging discipline and because these materials are being widely investigated, their effects on the various components of the immune system have been discussed here.

3.3 *Carbon nanomaterials*

The array of carbon NMs is vast, including carbon black nanoparticles, C_{60} fullerenes (also known as "buckyballs"), and carbon nanotubes (of both single- and multi-walled varieties). These materials, particularly C_{60} and carbon nanotubes, are being widely explored for many potential uses, including electrical conductivity, biosensors and medical imaging, and in biomaterials.[73,74] With increasing interest and use, the potential for human exposure to carbon NMs is of growing concern, thus necessitating the careful evaluation of their potential immune effects.

3.3.1 *Carbon black*

Carbon black (CB) is a particulate formed by the incomplete combustion of certain petroleum products and therefore, primary human exposure typically occurs via inhalation. IARC has classified CB under Category 2B, or "possibly carcinogenic to humans," due to well-documented carcinogenic effects in laboratory animals, but insufficient supporting evidence in humans.[55] There are a large number of studies in the literature evaluating the toxic effects of CB exposure; however, few studies have evaluated its effects on the immune system.

CB particles have diameters ranging from 10 nm to a few hundred nanometers, although they readily form aggregates with diameters ranging from 50 to 500 nm.[55] With an estimated worldwide production of approximately 10 million metric tons in 2005, the primary uses of CB are as a reinforcing component of rubber automobile parts including tires, belts, and hoses, and also as a pigment.[55]

Shwe *et al.*[75] have reported that an i.t. instillation of CB (14 or 95 nm in diameter) once weekly for four weeks resulted in significantly increased numbers of macrophages, lymphocytes, and neutrophils recovered from the BALF. In addition, exposure to the 14-nm diameter CB particles promoted significant increases in macrophage inflammatory protein 1-alpha (MIP-1α) as well as the pro-inflammmatory cytokines IL-1β, IL-6, and TNF-α.[75]

Immune effects other than inflammatory responses have also been evaluated. In a separate study from the same group, CB particles (14-nm and

56-nm diameters) were evaluated in an *in vitro* culture system.[76] The authors demonstrated that CB nanoparticles were able to promote the maturation and activation of mouse bone marrow-derived dendritic cells (BMDCs), with increased expression of CD80, CD86, and MHC class-II, although not in a size-dependent manner. Further, they evaluated the function of the BMDCs using an allogeneic MLR assay, where they demonstrated a greater enhancement of the response for 14-nm CB particles than for 56-nm CB particles. A single i.t. exposure to 500 μg of 14-nm CB nanoparticles increased pulmonary PMN infiltration and alveolar macrophage chemotaxis, while decreasing alveolar macrophage phagocytic function.[63] In the same study, decreased phagocytosis by alveolar macrophages (but not effects on chemotaxis or PMN infiltration) was observed for larger 260-nm CB particles when evaluated on an equal mass basis.

Fourteen-nanometer diameter CB particles have also been shown to act as an adjuvant following intranasal exposure in a mouse OVA respiratory allergy model.[16] However, these authors did not observe antigen-specific IgE or IgG1 antibody levels in the serum following ultrafine CB exposure, in contrast to the results obtained for ultrafine TiO_2 particles evaluated in the same study. This study also demonstrated that CB particles with a larger diameter (~250 nm) did not have adjuvant activity.

3.3.2 C_{60} *carbon fullerenes*

C_{60} fullerenes (or C_{60}, in short) are composed of 60 carbon atoms in a truncated icosahedron ("soccer ball") arrangement, containing 12 pentagonal faces and 20 hexagonal faces with a carbon atom at each vertex.[77,78] Also known as buckminsterfullerene, or "buckyballs," these particles have an estimated diameter of approximately 0.7 to 1 nm,[77,79] and due to their extremely small size, they have been considered for potential uses in a variety of applications. Rarely, however, are humans exposed to C_{60} particles of this size because C_{60}, like many NMs, can form aggregates of much larger sizes.

Due to the increased interest in C_{60} and also because the production of C_{60} can occur naturally via combustion processes,[79] concerns about their potential to exert toxic effects following exposure have been the subject of much study in the last decade. Exposure to C_{60} may occur orally, dermally, or by inhalation or injection.[79] While oral exposure to C_{60} appears to have no associated toxicity, i.p. exposure to C_{60} has been demonstrated to be toxic, with target organs being the kidney, liver, and spleen.[79]

C_{60} can be readily modified by the addition of a variety of functional groups. One of the more common surface modifications involves the addition of hydroxyl groups, producing polyhydroxylated fullerenes, or fullerenols. These nanoparticles have been reported to have fewer associated toxic effects than C_{60} without surface modification[80] and to offer protection against damage associated with exposure to ionizing radiation.[81,82]

There are conflicting reports regarding the potential of the inhalation of C_{60} to initiate an inflammatory response.[79,80,83–85] Although the number of studies evaluating the local lung inflammatory response following C_{60} exposure is relatively small, the majority of these studies suggest that C_{60} inhalation is neither toxic nor inflammatory, but rather results in an anti-inflammatory response.[79] Furthermore, 13-week nanoparticulate C_{60} inhalation studies in mice and rats have also confirmed the lack of an inflammatory response.[86,87] Specifically, no effects on IL-1, IL-6, TNF-α, MCP-1, or MIP-1α were observed following exposure to 50-nm diameter C_{60} aggregates at doses up to 2 mg/m^3.

Few studies evaluating the systemic immune effects of C_{60} exposure were located in the literature. Ryan *et al.*[88] have demonstrated that *in vitro* C_{60} exposure decreases the IgE-mediated allergic responses of human mast cells and peripheral blood basophils. This result has not been confirmed *in vivo*. In addition, Yamashita *et al.*[89] have demonstrated that colloidal C_{60} aggregates (165-nm mean diameter) suppressed the DTH response to methyl bovine serum albumin in mice after two i.v. administrations. The production of C_{60}-specific antibodies has also been reported.[90, 91] Finally, 13-week inhalation of C_{60} aggregates (50-nm diameter aggregates at doses up to 2 mg/m^3) did not affect the TDAR response (IgM AFCs to sRBCs) in either mice or rats.[86,87]

The effects of polyhydroxylated C_{60} particles on the immune system also have not been extensively reported in the literature. Cai *et al.*[81] have demonstrated that $C_{60}(OH)_{24}$ treatment (i.p.) was able to protect against decreases in splenocyte proliferation following exposure to gamma radiation. Zhu *et al.*[92] have reported that fullerenol exposure (i.p.) decreased nitric oxide production and increased TNF-α production, macrophage phagocytic activity, and anti-tumor activity. Similar enhancement of both TNF-α production and anti-tumor responses was also reported by Liu *et al.*[93] Decreased neutrophil functions (oxidative burst, degranulation) have also been reported in fish following i.p. treatment with fullerenols.[94]

Reports evaluating both the *in vitro* and *in vivo* immune effects of C_{60} have identified inconsistencies. Sayes *et al.*[85] have demonstrated that *in vitro* toxic effects of C_{60} exposure to lung epithelial cells did not extend to observations following *in vivo* inhalation exposure, where only transient effects were observed. Furthermore, Zogovic *et al.*[95] have demonstrated that nanocrystalline water-soluble C_{60} aggregates have contrasting effects *in vivo* and *in vitro*. Specifically, C_{60} aggregate exposure *in vitro* to mouse B16 melanoma cells induced oxidative stress and eventually cell death (both apoptotic and necrotic). *In vivo* exposure (i.p.), however, resulted in decreased nitric oxide production, decreased lymphocyte (T cell and B cell) proliferation, and significant tumor-promoting effects, in spite of the accumulation of C_{60} in the tumor tissue. This is in contrast with the well-documented anti-tumor properties of polyhydroxylated (water-soluble) C_{60} nanoparticles as discussed above, and suggests either particle size (i.e., in aggregated or non-aggregated states) or solubility modification processes as possible explanations for these differences.[95]

3.3.3 *Carbon nanotubes*

Carbon nanotubes (CNTs) are long, hollow tubes made entirely of carbon atoms. First discovered in 1953[96] and characterized later by Oberlin and Endo,[97] these tubes can be found in both single-walled (SWCNTs) and multi-walled (MWCNTs) varieties and are formed when carbon materials such as benzenes and hydrocarbons are burned.[97,98]

The handling and use of CNTs can often result in the aerosolizing of these materials.[99] A significant volume of research has reported on the pulmonary toxic effects of CNTs and several review articles (e.g., Helland *et al.*,[73] Johnston *et al.*,[100] van der Zande *et al.*[74]) summarizing these many studies are available in the literature.

The general consensus regarding the pulmonary toxic effects of SWCNTs is that they induce oxidative stress, granuloma formation, and fibrosis of the lung tissue.[73,74,101,102] MWCNT studies, on the other hand, have demonstrated contradictory results. MWCNT administered i.t. induced pulmonary toxicity and a local lung inflammatory response, but showed no signs of fibrosis or granuloma.[103] In contrast, Park *et al.*[104] reported not only an inflammatory response, but also the formation of granulomas in the lung tissue. Moreover, Muller *et al.*[105] found that MWCNT exposure (i.t.) resulted in lung inflammatory responses and fibrosis at 60 days post-instillation, while Carrero-Sanchez *et al.*[106] reported goblet cell hyperplasia following i.t. exposure. Similar results were also demonstrated following subchronic 13-week MWCNT inhalation exposure in rats, where elevated PMN numbers in the BALF were reported, in addition to exposure-related goblet cell hyperplasia/metaplasia and granuloma formation.[107,108] Contradicting each of these studies, Mitchell *et al.*[109] reported that MWCNT inhalation produced neither lung inflammation nor tissue damage. The discrepancies among these studies may be the result of physicochemical differences in the MWCNT preparations used, as structural defects of MWCNTs have been shown to affect their acute lung toxicity.[110] Alternatively, the presence or absence of transition metal contaminants in the CNT preparations may contribute to these differences, as a positive correlation has been demonstrated between metal contaminants of SWCNT preparations and increasing oxidative stress *in vitro*.[111]

In addition to the contradictory pulmonary effects following respiratory exposure to MWCNTs, there are varying reports regarding their toxicity when administered via other routes. Carrero-Sanchez *et al.*[106] reported that intranasal, oral, or i.p. exposure to MWCNTs resulted in neither inflammation nor tissue damage, while van der Zande *et al.*[74] and Shvedova and Kagan[112] have reviewed several studies with results indicating otherwise.

Effects of CNTs on the immune system have been reported, although they have not been thoroughly studied. Dermal SWCNT exposure in mice resulted in an irritancy response (i.e., increase in skin thickness) due to an infiltration of PMN and mast cells.[113] In addition, MWCNT exposure (s.c.) has been shown to activate the complement system and to increase the production of several cytokines, including IFN-γ, IL-1, IL-10, IL-12, MIP-1α, MCP-1, and GM-CSF.[114] Furthermore, these authors also reported the stimulation of macrophage phagocytosis and an increase in anti-tumor activity following MWCNT exposure, similar to reports of the effects of C_{60} exposure.

Both SWCNTs and MWCNTs were shown to promote allergy in a mouse model with the known allergen OVA.[43] These authors reported increased OVA-specific IgE antibody levels in the serum following either intranasal or s.c. exposure to CNTs. Furthermore, they reported increased Th2-associated cytokine secretion by mediastinal lymph node cells. Also, exposure to MWCNTs, but not SWCNTs, resulted in increased numbers of PMN cells and concentrations of IgG2a, MCP-1, and TNF-α in the BALF. Others have reported similar allergy-promoting effects. Specifically, i.t. MWCNT exposure produced increases in serum IgE levels and both serum and BALF levels of Th2 cytokines, including IL-4, IL-5, and IL-10.[104] These authors also evaluated ultrafine CB particles and found that the CNTs were able to induce a greater OVA-specific allergic response, higher Th2 cytokine levels, and a more pronounced eosinophilic infiltration into the lung than following exposure to ultrafine CB particles.

CNT exposure has also been demonstrated to result in both local and systemic immunomodulatory effects. Shvedova *et al.*[115] have shown that in spite of an increased pro-inflammatory cytokine response, exposure (p.a.) to SWCNTs three days prior to pulmonary infection with *Listeria monocytogenes* resulted in an increased bacterial presence in the lung six days after SWCNT exposure and a decreased rate of clearance of the infection. Both SWCNTs and double-walled CNTs (DWCNTs) have been shown to activate the complement system via the classical pathway, while the alternative pathway was activated by DWCNTs to a greater extent than by SWCNTs.[116]

MWCNT exposure can also result in systemic immunosuppression.[109] Specifically, the inhalation of MWCNTs for 14 days at concentrations greater than or equal to 0.3 mg/m^3 produced decreases in the T-dependent AFC response and the T cell proliferative response to concanavalin A stimulation, while the B cell proliferative response to lipopolysaccharide was unaffected. NK cytotoxic activity was moderately suppressed although the results were not dose-responsive, as no effect was observed at the high dose (5 mg/m^3). No systemic immunosuppressive effects were observed after only seven days of exposure. In subsequent studies, Mitchell *et al.*[117] demonstrated that these systemic immunosuppressive effects were due in part to the induction of cyclooxygenase 2 (COX-2) production following MWCNT inhalation.

4. Conclusion and Future Outlook

As the exploration of NMs and their uses in industry and medicine continue to increase, so does the level of risk of human exposure to these unique materials. The potential effects on the immune system following such exposure remain largely unknown, and conducting such evaluations are confounded by the ever-increasing number of differing NM designs (composition, size, surface modification, etc.). Furthermore, the uncertainty regarding the applicability of traditional methods to evaluations of NM immunotoxicity remains a concern, while the development and validation of alternative testing options in the short term are unlikely. While it may be desirable to relate physicochemical properties of NMs with their toxicity profiles, conflicting effects (both between *in vitro* and *in vivo* studies and discrepancies among interlaboratory *in vivo* evaluations) put this goal out of reach at the present time. In the meantime, however, the effects of NMs on the immune system must still be evaluated, utilizing currently acceptable techniques, so that local inflammatory responses, hypersensitivity potential, and both local and systemic immunomodulatory effects are characterized. Following this approach, a better understanding of the effects of NMs on human health can be achieved.

Acknowledgements

We would like to thank Drs. Cynthia Rider and Nigel Walker for their reviews and helpful suggestions. Also, we thank Dr. Gary Bowlin for his collaboration on electrospun polymers, and Dr. Wimolnut Auttachoat, Mrs. Ronnetta Brown, and Ms. Deborah Musgrove for their assistance with the immunotoxicology studies. This research was supported in part by the Intramural Research Program of the National Institute of Environmental Health Sciences (NIEHS), National Institutes of Health (NIH), and NIEHS contract N01-ES-55538. This article may be the work product of an employee or group of employees of the NIEHS and NIH; however, the statements, opinions, or conclusions contained therein do not necessarily represent the statements, opinions, or conclusions of NIEHS, NIH, or the United States government.

References

1. ASTM. Standard Terminology Relating to Nanotechnology. 2006;E2456–06.
2. Jang J, Lim DH, Choi IH. The impact of nanomaterials in immune system. *Immune Netw* 2010;(10):85–91.
3. Lidén G. The European commission tries to define nanomaterials. *Ann Occup Hyg* 2011;(55):1–5.
4. Balbus JM, Maynard AD, Colvin VL, Castranova V, Daston GP, Denison RA, Dreher KL, Goering PL, Goldberg AM, Kulinowski KM, Monterio-Riviere NA, Oberdorster G, Omenn GS, Pinkerton KE, Ramos KS, Rest KM, Sass JB, Silbergeld EK, Wong BA. Meeting report: Hazard assessment for nanoparticles — report from an interdisciplinary workshop. *Environ Health Perspect* 2007;115(11):1654–1659.
5. Hoet P, Legiest B, Geys J, Nemery B. Do nanomedicines require novel safety assessments to ensure their safety for long-term human use? *Drug Saf* 2009;32(8):625–636.
6. Patri A, Dobrovolskaia M, Stern S, McNeil S. Preclinical characterization of engineered nanoparticles intended for cancer therapeutics. In: Amiji MM (ed.), *Nanotechnology for Cancer Therapy*, CRC Press, 2007, pp. 105–137.
7. Geraci CL, Castranova V. Challenges in assessing nanomaterial toxicology: A personal perspective. *Wiley Interdiscip Rev Nanomed Nanobiotechnol* 2010;2(6):569–577.
8. Stratmeyer ME, Goering PL, Hitchins VM, Umbreit TH. What we know and don't know about the bioeffects of nanoparticles: Developing experimental approaches for safety assessment. *Biomed Microdevices* 2010;12(4):569–573.
9. Dobrovolskaia MA, Germolec DR, Weaver JL. Evaluation of nanoparticle immunotoxicity. *Nat Nanotechnol* 2009;4(7):411–414.
10. Fischer HC, Chan WC. Nanotoxicity: The growing need for *in vivo* study. *Curr Opin Biotechnol* 2007;18(7):565–571.
11. Kunzmann A, Andersson B, Thurnherr T, Krug H, Scheynius A, Fadeel B. Toxicology of engineered nanomaterials: Focus on biocompatibility, biodistribution and biodegradation. *Biochim Biophys Acta* 2011;1810(3):361–373.
12. Sayes CM, Reed KL, Warheit DB. Assessing toxicity of fine and nanoparticles: Comparing *in vitro* measurements to *in vivo* pulmonary toxicity profiles. *Toxicol Sci* 2007;97(1):163–180.
13. United States Environmental Proteetin Agency. Particulate Matter. *PM Standards Revision* 2006.

14. Dobrovolskaia MA, McNeil SE. Immunological properties of engineered nanomaterials. *Nat Nanotechnol* 2007;2(8):469–478.
15. Zolnik BS, González-Fernández Á, Sadrieh N, Dobrovolskaia MA. Nanoparticles and the immune system. *Endocrinology* 2010;151(2):458–465.
16. de Haar C, Hassing I, Bol M, Bleumink R, Pieters R. Ultrafine but not fine particulate matter causes airway inflammation and allergic airway sensitization to co-administered antigen in mice. *Clin Exp Allergy* 2006;36(11):1469–1479.
17. Shvedova AA, Kagan VE, Fadeel B. Close encounters of the small kind: Adverse effects of man-made materials interfacing with the nano-cosmos of biological systems. *Annu Rev Pharmacol Toxicol* 2010;50:63–88.
18. Sperling RA, Casals E, Comenge J, Bastus NG, Puntes VF. Inorganic engineered nanoparticles and their impact on the immune response. *Curr Drug Metab* 2009;10(8): 895–904.
19. Mayer A, Vadon M, Rinner B, Novak A, Wintersteiger R, Frohlich E. The role of nanoparticle size in hemocompatibility. *Toxicology* 2009;258(2–3):139–147.
20. Pedersen MB, Zhou X, Larsen EK, Sorensen US, Kjems J, Nygaard JV, Nyengaard JR, Meyer RL, Boesen T, Vorup-Jensen T. Curvature of synthetic and natural surfaces is an important target feature in classical pathway complement activation. *J Immunol* 2010;184(4):1931–1945.
21. Antonelli A, Serafini S, Menotta M, Sfara C, Pierige F, Giorgi L, Ambrosi G, Rossi L, Magnani M. Improved cellular uptake of functionalized single-walled carbon nanotubes. *Nanotechnology* 2010;21(42):425101.
22. Singh R, Pantarotto D, Lacerda L, Pastorin G, Klumpp C, Prato M, Bianco A, Kostarelos K. Tissue biodistribution and blood clearance rates of intravenously administered carbon nanotube radiotracers. *Proc Natl Acad Sci U S A* 2006;103(9):3357–3362.
23. Aderem A, Underhill DM. Mechanisms of phagocytosis in macrophages. *Annu Rev Immunol* 1999;17:593–623.
24. Hock SC, Ying YM, Wah CL. A review of the current scientific and regulatory status of nanomedicines and the challenges ahead. *PDA J Pharm Sci Technol* 2011; 65(2):177–195.
25. Powell MC, Griffin MP, Tai S. Bottom-up risk regulation? How nanotechnology risk knowledge gaps challenge federal and state environmental agencies. *Environ Manage* 2008;42(3):426–443.
26. Bowman DM, Fitzharris M. Too small for concern? Public health and nanotechnology. *Aust N Z J Public Health* 2007;31(4):382–384.
27. Wilson RF. Nanotechnology: The challenge of regulating known unknowns. *J Law Med Ethics* 2006;34(4):704–713.
28. Oberdorster G, Maynard A, Donaldson K, Castranova V, Fitzpatrick J, Ausman K, Carter J, Karn B, Kreyling W, Lai D Olin S, Monteiro-Riviere N, Warheit D, Yang H. Principles for characterizing the potential human health effects from exposure to nanomaterials: Elements of a screening strategy. *Part Fibre Toxicol* 2005;2:8.
29. Warheit DB, Borm PJ, Hennes C, Lademann J. Testing strategies to establish the safety of nanomaterials: Conclusions of an ECETOC workshop. *Inhal Toxicol* 2007;19(8):631–643.
30. Nel A, Xia T, Madler L, Li N. Toxic potential of materials at the nanolevel. *Science* 2006;311(5761):622–627.
31. Hoet PH, Bruske-Hohlfeld I, Salata OV. Nanoparticles — known and unknown health risks. *J Nanobiotechnol* 2004;2(1):12.

32. Oberdorster G, Ferin J, Gelein R, Soderholm SC, Finkelstein J. Role of the alveolar macrophage in lung injury: Studies with ultrafine particles. *Environ Health Perspect* 1992;97:193–199.
33. Oberdorster G, Ferin J, Lehnert BE. Correlation between particle size, *in vivo* particle persistence, and lung injury. *Environ Health Perspect* 1994;102 Suppl 5:173–179.
34. Li JJ, Muralikrishnan S, Ng CT, Yung LY, Bay BH. Nanoparticle-induced pulmonary toxicity. *Exp Biol Med (Maywood)* 2010;235(9):1025–1033.
35. Di Gioacchino M, Petrarca C, Lazzarin F, Di Giampaolo L, Sabbioni E, Boscolo P, Mariani-Costantini R, Bernardini G. Immunotoxicity of nanoparticles. *Int J Immunopathol Pharmacol* 2011;24(Suppl 1):65S-71S.
36. Munson AE, Regelson W, Wooles WR. Tissue localization studies in evaluating the functional role of the RES. *J Reticuloendothel Soc* 1970;7(3):366–374.
37. White KL, Jr. Sanders VM, Barnes DW, Shopp GM, Jr., Munson AE. Immunotoxicological investigations in the mouse: General approach and methods. *Drug Chem Toxicol* 1985;8(5):299–331.
38. Munson AE, Regelson W, Lawrence W, Jr., Wooles WR. Biphasic response of the reticuloendothelial system (RES) induced by pyran copolymer. *J Reticuloendothel Soc* 1970;7(3):375–385.
39. Moss OR, Wong VA. When nanoparticles get in the way: Impact of projected area on in vivo and *in vitro* macrophage function. *Inhal Toxicol* 2006;18(10):711–716.
40. Smith MJ, White KL, Jr. Establishment and comparison of delayed-type hypersensitivity models in the B6C3F1 mouse. *J Immunotoxicol* 2010;7(4):308–317.
41. White KL, Jr., McLoughlin CE, Auttachoat W, Smith MJ. Validation of the *Candida albicans* delayed-type hypersensitivity (DTH) model in the female B6C3F1 mouse for use in immunotoxicological investigations. *J Immunotoxicol* 2012;9(2):141–147.
42. Brahmachari B, Hazra A, Majumdar A. Adverse drug reaction profile of nanoparticle versus conventional formulation of paclitaxel: An observational study. *Indian J Pharmacol* 2011;43(2):126–130.
43. Nygaard UC, Hansen JS, Samuelsen M, Alberg T, Marioara CD, Lovik M. Single-walled and multi-walled carbon nanotubes promote allergic immune responses in mice. *Toxicol Sci* 2009;109(1):113–123.
44. Adachi K, Yamada N, Yamamoto K, Yoshida Y, Yamamoto O. *In vivo* effect of industrial titanium dioxide nanoparticles experimentally exposed to hairless rat skin. *Nanotoxicology* 2010;4(3):296–306.
45. Choksi AN, Poonawalla T, Wilkerson MG. Nanoparticles: A closer look at their dermal effects. *J Drugs Dermatol* 2010;9(5):475–481.
46. Robertson TA, Sanchez WY, Roberts MS. Are commercially available nanoparticles safe when applied to the skin? *J Biomed Nanotechnol* 2010;6(5): 452–468.
47. Weaver JL, Chapdelaine JM, Descotes J, Germolec D, Holsapple M, House R, Lebrec H, Meade J, Pieters R, Hastings KL, Dean JH. Evaluation of a lymph node proliferation assay for its ability to detect pharmaceuticals with potential to cause immune-mediated drug reactions. *J Immunotoxicol* 2005;2(1):11–20.
48. Reddy ST, van der Vlies AJ, Simeoni E, Angeli V, Randolph GJ, O'Neill CP, Lee LK, Swartz MA, Hubbell JA. Exploiting lymphatic transport and complement activation in nanoparticle vaccines. *Nat Biotechnol* 2007;25(10):1159–1164.

49. Manolova V, Flace A, Bauer M, Schwarz K, Saudan P, Bachmann MF. Nanoparticles target distinct dendritic cell populations according to their size. *Eur J Immunol* 2008;38(5):1404–1413.
50. Luster MI, Munson AE, Thomas PT, Holsapple MP, Fenters JD, White KL, Jr. Lauer LD, Germolec DR, Rosenthal GJ, Dean JH. Development of a testing battery to assess chemical-induced immunotoxicity: National Toxicology Program's guidelines for immunotoxicity evaluation in mice. *Fundam Appl Toxicol* 1988;10(1):2–19.
51. Temple L, Kawabata TT, Munson AE, White KL, Jr. Comparison of ELISA and plaque-forming cell assays for measuring the humoral immune response to SRBC in rats and mice treated with benzo[a]pyrene or cyclophosphamide. *Fundam Appl Toxicol* 1993;21(4):412–419.
52. Kimber I, Dearman RJ. The local lymph node assay and skin sensitization testing. *Methods Mol Biol* 2010;598:221–231.
53. Dean JH, Twerdok LE, Tice RR, Sailstad DM, Hattan DG, Stokes WS. ICCVAM evaluation of the murine local lymph node assay. Conclusions and recommendations of an independent scientific peer review panel. *Regul Toxicol Pharmacol* 2001;34(3):258–273.
54. Gad SC, Dunn BJ, Dobbs DW, Reilly C, Walsh RD. Development and validation of an alternative dermal sensitization test: The mouse ear swelling test (MEST). *Toxicol Appl Pharmacol* 1986;84(1):93–114.
55. IARC. IARC monographs on the evaluation of carcinogenic risks to humans: Carbon black, titanium dioxide, and talc. Vol. 93, 2010.
56. Duan Y, Liu J, Ma L, Li N, Liu H, Wang J, Zheng L, Liu C, Wang X, Zhao X, Yan J, Wang S, Wang H, Zhang X, Hong F. Toxicological characteristics of nanoparticulate anatase titanium dioxide in mice. *Biomaterials* 2010;31(5):894–899.
57. Wang H, Leigh J, Peters M, Bonin A. Apoptosis of pulmonary neutrophils induced by low dose titanium dioxide and silica. *Ann Occup Hyg* 2002;46 Suppl 1:210–214.
58. Wang J, Zhou G, Chen C, Yu H, Wang T, Ma Y, Jia G, Gao Y, Li B, Sun J, Li Y, Jiao F, Zhao Y, Chai Z. Acute toxicity and biodistribution of different sized titanium dioxide particles in mice after oral administration. *Toxicol Lett* 2007;168(6):176–185.
59. Maurer-Jones MA, Lin YS, Haynes CL. Functional assessment of metal oxide nanoparticle toxicity in immune cells. *ACS Nano* 2010;4(6):3363–3373.
60. Sayes CM, Wahi R, Kurian PA, Liu Y, West JL, Ausman KD, Warheit DB, Colvin VL. Correlating nanoscale titania structure with toxicity: A cytotoxicity and inflammatory response study with human dermal fibroblasts and human lung epithelial cells. *Toxicol Sci* 2006;92(1):174–185.
61. Auttachoat W, McLoughlin CE, Smith MJ, White KL, Jr. Titanium dioxide nanoparticles: Lack of immunosuppressive and contact hypersensitivity effects. In: *50th Annual Meeting of the Society of Toxicology,* 2011.
62. Liu R, Zhang X, Pu Y, Yin L, Li Y, Liang G, Li X, Zhang J. Small-sized titanium dioxide nanoparticles mediate immune toxicity in rat pulmonary alveolar macrophages *in vivo.* *J Nanosci Nanotechnol* 2010;10(8):5161–5169.
63. Renwick LC, Brown D, Clouter A, Donaldson K. Increased inflammation and altered macrophage chemotactic responses caused by two ultrafine particle types. *Occup Environ Med* 2004;61(5):442–447.
64. Park EJ, Yoon J, Choi K, Yi J, Park K. Induction of chronic inflammation in mice treated with titanium dioxide nanoparticles by intratracheal instillation. *Toxicology* 2009;260(1):37–46.

65. Verma S, Domb AJ, Kumar N. Nanomaterials for regenerative medicine. *Nanomedicine (Lond)* 2011;6(1):157–181.
66. Kim BY, Rutka JT, Chan WC. Nanomedicine. *N Engl J Med* 2010;363:2434–2443.
67. Ayres CE, Jha BS, Sell SA, Bowlin GL, Simpson DG. Nanotechnology in the design of soft tissue scaffolds: Innovations in structure and function. *Wiley Interdiscip Rev Nanomed Nanobiotechnol* 2010;2(1):20–34.
68. Barnes CP, Sell SA, Boland ED, Simpson DG, Bowlin GL. Nanofiber technology: Designing the next generation of tissue engineering scaffolds. *Adv Drug Delivery Rev* 2007;59(14):1413–1433.
69. Smith MJ, Smith DC, Bowlin GL, White KL. Modulation of murine innate and acquired immune responses following *in vitro* exposure to electrospun blends of collagen and polydioxanone. *J Biomed Mater Res A* 2010;93(2):793–806.
70. Smith MJ, White KL, Jr. Smith DC, Bowlin GL. *In vitro* evaluations of innate and acquired immune responses to electrospun polydioxanone — elastin blends. *Biomaterials* 2009;30(2):149–159.
71. Smith MJ. Acquired and innate immunomodulatory effects *in vitro* of electrospun polydioxanone, collagen, and soluble elastin (Ph.D. Dissertation), Richmond, Virginia Commonwealth University, 2008.
72. Boland ED, Coleman BD, Barnes CP, Simpson DG, Wnek GE, Bowlin GL. Electrospinning polydioxanone for biomedical applications. *Acta Biomater* 2005;1(1):115–123.
73. Helland A, Wick P, Koehler A, Schmid K, Som C. Reviewing the environmental and human health knowledge base of carbon nanotubes. *Environ Health Perspect* 2007;115(8):1125–1131.
74. van der Zande M, Junker R, Walboomers XF, Jansen JA. Carbon nanotubes in animal models: A systematic review on toxic potential. *Tissue Eng Part B Rev* 2011;17(1):57–69.
75. Shwe TT, Yamamoto S, Kakeyama M, Kobayashi T, Fujimaki H. Effect of intratracheal instillation of ultrafine carbon black on proinflammatory cytokine and chemokine release and mRNA expression in lung and lymph nodes of mice. *Toxicol Appl Pharmacol* 2005;209(1):51–61.
76. Koike E, Takano H, Inoue K, Yanagisawa R, Kobayashi T. Carbon black nanoparticles promote the maturation and function of mouse bone marrow-derived dendritic cells. *Chemosphere* 2008;73(3):371–376.
77. Kroto HW, Heath JR, O'Brien SC, Curl RF, Smalley RE. C60: Buckminsterfullerene. *Nature* 1985;318(6042):162–163.
78. Terrones H, Mackay AL. The geometry of hypothetical curved graphite structures. *Carbon* 1992;30:1251–1260.
79. Johnston HJ, Hutchison GR, Christensen FM, Aschberger K, Stone V. The biological mechanisms and physicochemical characteristics responsible for driving fullerene toxicity. *Toxicol Sci* 2010;114(2):162–182.
80. Usenko CY, Harper SL, Tanguay RL. *In vivo* evaluation of carbon fullerene toxicity using embryonic zebrafish. *Carbon N Y* 2007;45(9):1891–1898.
81. Cai X, Hao J, Zhang X, Yu B, Ren J, Luo C, Li Q, Huang Q, Shi X, Li W, Liu J. The polyhydroxylated fullerene derivative C60(OH)24 protects mice from ionizing-radiation-induced immune and mitochondrial dysfunction. *Toxicol Appl Pharmacol* 2010;243(1): 27–34.

82. Trajkovic S, Dobric S, Jacevic V, Dragojevic-Simic V, Milovanovic Z, Dordevic A. Tissue-protective effects of fullerenol C60(OH)24 and amifostine in irradiated rats. *Colloids Surf B Biointerfaces* 2007;58(1):39–43.
83. Kamat JP, Devasagayam TP, Priyadarsini KI, Mohan H. Reactive oxygen species mediated membrane damage induced by fullerene derivatives and its possible biological implications. *Toxicology* 2000;155(1–3):55–61.
84. Park EJ, Kim H, Kim Y, Yi J, Choi K, Park K. Carbon fullerenes (C60s) can induce inflammatory responses in the lung of mice. *Toxicol Appl Pharmacol* 2010;244(2):226–233.
85. Sayes CM, Marchione AA, Reed KL, Warheit DB. Comparative pulmonary toxicity assessments of C60 water suspensions in rats: Few differences in fullerene toxicity *in vivo* in contrast to *in vitro* profiles. *Nano Lett* 2007;7(8):2399–2406.
86. NTP: Range-Finding Report. *Immunotoxicity of C60 Fullerene in Female Wistar Han Rats.* In: Research Triangle Park, National Toxicology Program, 2012.
87. NTP: Range-Finding Report: *Immunotoxicity of C60 Fullerene in Female B6C3F1 Mice.* In: Research Triangle Park, National Toxicology Program, 2012.
88. Ryan JJ, Bateman HR, Stover A, Gomez G, Norton SK, Zhao W, Schwartz LB, Lenk R, Kepley CL. Fullerene nanomaterials inhibit the allergic response. *J Immunol* 2007;179(1):665–672.
89. Yamashita K, Sakai M, Takemoto N, Tsukimoto M, Uchida K, Yajima H, Oshio S, Takeda K, Kojima S. Attenuation of delayed-type hypersensitivity by fullerene treatment. *Toxicology* 2009;261(1–2):19–24.
90. Braden BC, Goldbaum FA, Chen BX, Kirschner AN, Wilson SR, Erlanger BF. X-ray crystal structure of an anti-Buckminsterfullerene antibody Fab fragment: Biomolecular recognition of C(60). *Proc Natl Acad Sci U S A* 2000;97(22):12193–12197.
91. Chen BX, Wilson SR, Das M, Coughlin DJ, Erlanger BF. Antigenicity of fullerenes: Antibodies specific for fullerenes and their characteristics. *Proc Natl Acad Sci U S A* 1998;95(18):10809–10813.
92. Zhu J, Ji Z, Wang J, Sun R, Zhang X, Gao Y, Sun H, Liu Y, Wang Z, Li A, Ma J, Wang T, Jia G, Gu Y. Tumor-inhibitory effect and immunomodulatory activity of fullerol C60(OH)x. *Small* 2008;4(8):1168–1175.
93. Liu Y, Jiao F, Qiu Y, Li W, Qu Y, Tian C, Li Y, Bai R, Lao F, Zhao Y, Chai Z, Chen C. Immunostimulatory properties and enhanced TNF-alpha mediated cellular immunity for tumor therapy by C60(OH)20 nanoparticles. *Nanotechnology* 2009;20(41):415102.
94. Jovanovic B, Anastasova L, Rowe EW, Palic D. Hydroxylated fullerenes inhibit neutrophil function in fathead minnow (Pimephales promelas Rafinesque, 1820). *Aquat Toxicol* 2011;101(2):474–482.
95. Zogovic NS, Nikolic NS, Vranjes-Djuric SD, Harhaji LM, Vucicevic LM, Janjetovic KD, Misirkic MS, Todorovic-Markovic BM, Markovic ZM, Milonjic SK, Trajkovic VS. Opposite effects of nanocrystalline fullerene (C(60)) on tumour cell growth *in vitro* and *in vivo* and a possible role of immunosupression in the cancer-promoting activity of C(60). *Biomaterials* 2009;30(36):6940–6946.
96. Davis WR, Slawson RJ, Rigby GR. An unusual form of carbon. *Nature* 1953;171(4356):756.
97. Oberlin A, Endo M. Filamentous growth of carbon through benzene decomposition. *J Crystal Growth* 1976;32(3):335–349.

98. Howard JB, Das Chowdhury K, Vander Sande JB. Carbon shells in flames. *Nature* 1994;370(6491):603.
99. Maynard AD, Baron PA, Foley M, Shvedova AA, Kisin ER, Castranova V. Exposure to carbon nanotube material: Aerosol release during the handling of unrefined single-walled carbon nanotube material. *J Toxicol Environ Health A* 2004;67(1):87–107.
100. Johnston HJ, Hutchison GR, Christensen FM, Peters S, Hankin S, Aschberger K, Stone V. A critical review of the biological mechanisms underlying the *in vivo* and *in vitro* toxicity of carbon nanotubes: The contribution of physico-chemical characteristics. *Nanotoxicology* 2010;4(2):207–246.
101. Shvedova AA, Kisin E, Murray AR, Johnson VJ, Gorelik O, Arepalli S, Hubbs AF, Mercer RR, Keohavong P, Sussman N, Jin J, Yin J, Stone S, Chen BT, Deye G, Maynard, A, Castranova V, Baron PA, Kagan VE. Inhalation *vs.* aspiration of single-walled carbon nanotubes in C57BL/6 mice: Inflammation, fibrosis, oxidative stress, and mutagenesis. *Am J Physiol Lung Cell Mol Physiol* 2008;295(4):L552–565.
102. Shvedova AA, Kisin ER, Porter D, Schulte P, Kagan VE, Fadeel B, Castranova V. Mechanisms of pulmonary toxicity and medical applications of carbon nanotubes: Two faces of Janus? *Pharmacol Ther* 2009;121(2):192–204.
103. Roda E, Coccini T, Acerbi D, Barni S, Vaccarone R, Manzo L. Comparative pulmonary toxicity assessment of pristine and functionalized multi-walled carbon nanotubes intratracheally instilled in rats: Morphohistochemical evaluations. *Histol Histopathol* 2011;26(3):357–367.
104. Park EJ, Cho WS, Jeong J, Yi J, Choi K, Park K. Pro-inflammatory and potential allergic responses resulting from B cell activation in mice treated with multi-walled carbon nanotubes by intratracheal instillation. *Toxicology* 2009;259(3):113–121.
105. Muller J, Huaux F, Moreau N, Misson P, Heilier JF, Delos M, Arras M, Fonseca A, Nagy JB, Lison D. Respiratory toxicity of multi-wall carbon nanotubes. *Toxicol Appl Pharmacol* 2005;207(3):221–231.
106. Carrero-Sanchez JC, Elias AL, Mancilla R, Arrellin G, Terrones H, Laclette JP, Terrones M. Biocompatibility and toxicological studies of carbon nanotubes doped with nitrogen. *Nano Lett* 2006;6(8):1609–1616.
107. Pauluhn J. Subchronic 13-week inhalation exposure of rats to multiwalled carbon nanotubes: Toxic effects are determined by density of agglomerate structures, not fibrillar structures. *Toxicol Sci* 2010;113(1):226–242.
108. Ma-Hock L, Treumann S, Strauss V, Brill S, Luizi F, Mertler M, Wiench K, Gamer AO, van Ravenzwaay B, Landsiedel R. Inhalation toxicity of multiwall carbon nanotubes in rats exposed for 3 months. *Toxicol Sci* 2009;112(2):468–481.
109. Mitchell LA, Gao J, Wal RV, Gigliotti A, Burchiel SW, McDonald JD. Pulmonary and systemic immune response to inhaled multiwalled carbon nanotubes. *Toxicol Sci* 2007;100(1):203–214.
110. Muller J, Huaux F, Fonseca A, Nagy JB, Moreau N, Delos M, Raymundo-Pinero E, Beguin F, Kirsch-Volders M, Fenoglio I, Fubini B, Lison D. Structural defects play a major role in the acute lung toxicity of multiwall carbon nanotubes: Toxicological aspects. *Chem Res Toxicol* 2008;21(9):1698–1705.
111. Kagan VE, Tyurina YY, Tyurin VA, Konduru NV, Potapovich AI, Osipov AN, Kisin ER, Schwegler-Berry D, Mercer R, Castranova V, Shevedova AA. Direct and indirect effects of single walled carbon nanotubes on RAW 264.7 macrophages: Role of iron. *Toxicol Lett* 2006;165(1):88–100.

112. Shvedova AA, Kagan VE. The role of nanotoxicology in realizing the 'helping without harm' paradigm of nanomedicine: Lessons from studies of pulmonary effects of single-walled carbon nanotubes. *J Intern Med* 2010;267(1):106–118.
113. Murray AR, Kisin E, Leonard SS, Young SH, Kommineni C, Kagan VE, Castranova V, Shvedova AA. Oxidative stress and inflammatory response in dermal toxicity of single-walled carbon nanotubes. *Toxicology* 2009;257(3):161–171.
114. Meng J, Yang M, Jia F, Kong H, Zhang W, Wang C, Xing J, Xie S, Xu H. Subcutaneous injection of water-soluble multi-walled carbon nanotubes in tumor-bearing mice boosts the host immune activity. *Nanotechnology* 2010;21:145104.
115. Shvedova AA, Fabisiak JP, Kisin ER, Murray AR, Roberts JR, Tyurina YY, Antonini JM, Feng WH, Kommineni C, Reynolds J, Barchowsky A, Castranova V, Kagan VE. Sequential exposure to carbon nanotubes and bacteria enhances pulmonary inflammation and infectivity. *Am J Respir Cell Mol Biol* 2008;38(5):579–590.
116. Salvador-Morales C, Flahaut E, Sim E, Sloan J, Green ML, Sim RB. Complement activation and protein adsorption by carbon nanotubes. *Mol Immunol* 2006;43(3):193–201.
117. Mitchell LA, Lauer FT, Burchiel SW, McDonald JD. Mechanisms for how inhaled multiwalled carbon nanotubes suppress systemic immune function in mice. *Nat Nanotechnol* 2009;4(7):451–456.

Chapter 21

Immunotoxicity Testing for Drug–Nanoparticle Conjugates: Regulatory Considerations

*Simona Bancos, Katherine M. Tyner, and James L. Weaver**

Division of Drug Safety Research
Center of Drug Evaluation and Research
Food and Drug Administration
Silver Spring, MD 20993, USA

Nanoscale materials are being developed for use in a wide variety of products including pharmaceuticals that are regulated by the U.S. Food and Drug Administration (FDA). Some of these represent reformulations of existing drugs or slow release degradable systems. A different category are drugs conjugated to durable nanoparticles (NPs), which do not degrade in the body. In all cases, these novel materials are subject to the same rules and regulations concerning drug development that apply to small molecules and therapeutic proteins. All drugs, including NP–drug conjugates, should be evaluated for immunotoxicity during development. This evaluation is governed by the ICH S8 guidance for small molecule drugs or by the ICH S6 guidance for protein drugs. These guidances continue to apply for drugs conjugated to NPs. Under S8, the primary level of evaluation is a weight of evidence review of the data collected in standard toxicology studies. If data suggestive of immunotoxicity are found, further focused studies may be needed. For proteins, the primary issue is drug hypersensitivity and samples should be collected from non-clinical studies to fully evaluate the role of any anti-drug antibodies. Several studies from CDER/FDA laboratories are used to illustrate how the principles described in the guidance documents could be applied in a real-world situation.

*Corresponding author. E-mail: james.weaver@fda.hhs.gov

Disclaimer:

This report is not an official U.S. FDA guidance or policy statement. No official support or endorsement by the U.S. FDA is intended or should be inferred.

1. Introduction

Manufactured nanomaterials are being incorporated into nearly all classes of products regulated by the U.S. Food and Drug Administration (FDA). There are multiple nanomaterial-containing drugs that are either FDA-approved or currently under review. These drugs include nanocrystals of small molecules (e.g., Emend®), lipid drug carriers (e.g., Doxil®), solid lipid nanoparticles (e.g., Leunesse®), inorganic drug delivery devices and imaging agents, nanoshells, and nanotubes. The majority of these particles are intended for dissolution or degradation within the body (e.g., nanocrystals and liposomes). A growing subset of nanoparticle (NP) therapeutics, however, consists of durable NPs (silica, gold, silver, quantum dots, etc.), which are not expected to dissolve or degrade rapidly inside the body. Data show that durable nanoplatforms accumulate within the macrophages of the reticuloendothelial system (RES) organs[1,2,3] and can remain in these tissues over an extended period of time.[4,5] Several NP platforms were also shown to have immunomodulatory effects.[6,7]

Products containing NPs are subject to the same regulatory processes as all other therapeutics. Due to their unique properties, there have been several concerns that gaps may exist in the conventional review process for these NPs[8,9,10] and therefore, nanomaterial drugs should be regulated differently. The FDA does not categorically judge all products containing nanomaterials or otherwise involving application of nanotechnology as intrinsically benign or harmful (http://www.fda.gov/RegulatoryInformation/Guidances/ucm-257698.htm). In the current chapter, we will provide a brief summary of the drug review process at the FDA, review the FDA guidances and studies that focus on drug immunotoxicity and how they apply to nanodrugs, and conclude the chapter with a case study.

2. Regulation of Drug Products

The U.S. FDA is the governmental agency responsible for protecting and promoting public health by monitoring and regulating biologic, food, drug, medical devices, tobacco, and cosmetic products. Drug products fall under the jurisdiction of the Center for Drug Evaluation and Research (CDER), which fulfills its role in the FDA by ensuring that both over-the-counter and prescription drugs are safe and effective.

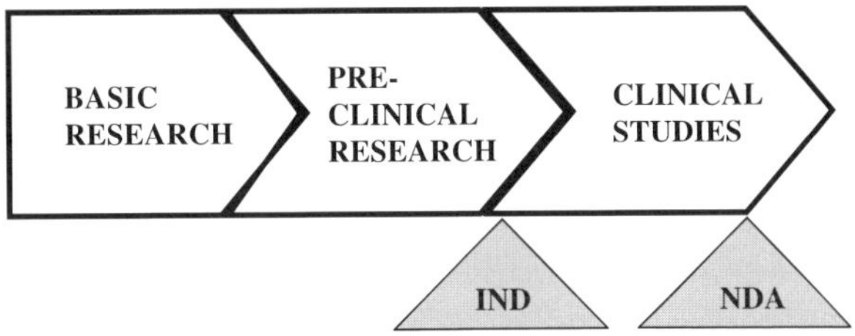

Figure 1. Schematic representation of the drug development and approval process. Data from basic, preclinical research, and clinical studies of a candidate drug conducted by the company are submitted as INDs or NDAs to the FDA.

IND: investigational new drug, NDA: new drug application.

Figure 1 illustrates the basic steps in drug development and submission to the FDA. Once a drug has been developed and its preclinical safety and effectiveness has been tested, the drug company submits to the FDA an investigational new drug (IND) application that contains all the basic and preclinical test data. IND approval allows the company to begin clinical trials. The new drug application (NDA) is a complex submission that contains all the data of the IND as well as the clinical trial outcomes and manufacturing and labeling information. If the NDA is approved, permission is granted to company to market the drug product in the United States.

All drug products, independent of their composition or intended use, are regulated by the CDER and follow the same regulatory path. The CDER has published multiple guidances that help companies in identifying the appropriate tests and data required for the characterization of their drug (available at http://www.fda.gov/RegulatoryInformation/Guidances/).

3. Regulation of Nanomaterial-Containing Drug Products

Nanomaterial-containing drugs must fulfill the same requirements as all other drugs. However, when submitting a nanomaterial-containing drug, there may be particular tests that need to be included in the submission package. The review division responsible for reviewing the IND and/or NDA will decide whether or not additional tests should be included for submission (Manual of Policies and Procedures, available at http://www.fda.gov/downloads/AboutFDA/CentersOffices/OfficeofMedicalProductsandTobacco/CDER/ManualofPoliciesProcedures/UCM214304.pdf).[11]

In 2011, FDA issued a "Draft Guidance for Industry" regarding the regulation of nanotechnology products (http://www.fda.gov/RegulatoryInformation/Guidances/ucm257698.htm).[12] This document focuses on how a sponsor may determine if their FDA-regulated product incorporates

nanotechnology. In this draft guidance, the FDA does not establish an official definition for nanotechnology, NPs, or other "nano"-related terms, but rather, include several "points to consider."
These points are:

1) whether an engineered material or end-product has at least one dimension in the nanoscale range (approximately 1 to 100 nm), or
2) whether an engineered material or its end-product(s) exhibits properties or phenomena, including physical or chemical properties or biological effects, that are attributable to its dimension(s), even if these dimensions fall outside the nanoscale range, up to 1 mm.

Despite the lack of an official FDA definition, the FDA has reviewed and approved multiple therapeutics that would traditionally be classified as "nano." Table 1 offers a list of examples of such therapeutics.

4. Immunotoxicological Regulation of Nanomaterial-Containing Drug Products

It is well established that drug products can have immunomodulatory effects by either suppressing or stimulating the immune system. A list of relevant FDA guidances focusing on immunotoxicity is provided in Table 2.

Examples of drugs that have demonstrated immunotoxicity effects are antitumor drugs,[17] some COX-2–selective inhibitors,[18] cytokine therapy,[19] and macrolide antibiotics.[20] As defined in "Guidance for Industry: S8 Immunotoxicity Studies for Human Pharmaceuticals" (ICH S8), immunotoxicity in most cases refers to "an unintended immunosuppression or enhancement," which means that the drug was intended for other purposes and the activation or repression of the immune system occured as an inadvertent side effect. Taking into consideration the potential adverse effects of the drugs on the immune system, it is proposed that immunotoxicity testing should be applied to all pharmaceutical products (including those that contain nanomaterials)

Immunotoxicity testing encompasses the non-clinical methods used to predict the ability of a drug to cause immunotoxicity in humans. Determining which guidance to follow depends on the type of drug. If the nanomaterial drug contains a drug with low molecular weight, then ICH S8 should be consulted. If the product contains a biotechnology-derived material (e.g., antibody or cytokine) than ICH S6 and its addendum should be the primary source of information.

5. Recommended Studies by ICH S8

The primary level of review for immunotoxicity is of relevant data from all sources and studies that are available. The data are subjected to a weight of

Table 1. List of FDA-approved drugs that contain NPs.

Platforms	Trade name	Indication
Nanocrystal	Rapamune	Immunosuppressant Prophylaxis of organ rejection in patients receiving renal transplants
	Emend	Antiemetic
	Tricor	Hypercholesterolemia Hypertriglyceridemia
	Triglide	Hypercholesterolemia Hypertriglyceridemia
	Megace ES	Anorexia Cachexia Unexplained significant weight loss in AIDS patients
	Invega Sustenna	Treatment of schizophrenia
Nanoparticle	Abraxane	Metastatic breast cancer
	Anthelios 20	Sunscreen
	Helioblock SX sunscreen cream	Sunscreen
Nanotube	Somatuline depot	Acromegaly
Superparamagnetic iron oxide	Feridex	MRI contrast agent
	GastroMARK	Imaging of abdominal structures
	Feraheme injection	Treatment of iron deficiency anemia in patients with chronic kidney disease (CKD)
Liposome	Doxil	Antineoplastic
	Abelcet	Fungal infections
	DaunoXome	Antineoplastic
	Amphotec	Fungal infections
	AmBisome	Fungal infections
	DepoCyt	Lymphomatous meningitis
	Visudyne (Verteporfin for injection)	Photodynamic therapy for age-related macular degeneration
	Definity	Echopharmaceutical Contrast enhancement
	DepoDur	Opioid local analgesic
Micelle	Taxotere	Antineoplastic
	Amphotec	Antifungal
	Estrasorb	Vasomotor symptoms associated with menopause
	Oraqix dental gel	Local dental anesthetic
	Luxacor	Hyperlipidemia Hyperglyceridemia

Table 2. List of U.S. FDA immunotoxocology guidance documents.

Year	Guidance	Reference
1997	Guidance for Industry: S6 Preclinical Safety Evaluation of Biotechnology-Derived Pharmaceuticals	13
2002	Guidance for Industry: Immunotoxicology Evaluation of Investigational New Drugs	14
2006	Guidance for Industry: S8 Immunotoxicology Studies for Human Pharmaceuticals	15
2012	Addendum to ICH S6: S6 Preclinical Safety Evaluation of Biotechnology-Derived Pharmaceuticals S6(R1)	16

evidence (WoE) review, which is used to determine if there is sufficient evidence of immunotoxicity to warrant either a firm conclusion or require additional testing (Figure 2).

The primary source of data for the WoE review is from the standard toxicity tests (STS) that are generally required for all drugs. The evaluations of specific data from the STS are used to indicate if the nanomaterial drug possesses immunotoxocity potential. The most commonly used data source is the 28-day consecutive daily dose study in rodents. However, data from all available studies should be included within the WoE review. The following types of data which should be specifically evaluated are discussed below.

5.1. *Hematological and clinical chemistry tests*

Hematological tests detect alterations (decreases or increases) in blood cell counts, including lymphopenia/lymphocytosis, granulocytopenia/granulocytosis, or leukocytopenia/leukocytosis. In cases of conjugates with known cytotoxic drugs, this may be an expected effect. In other cases, the changes in leukocyte populations may be due to the particle itself.[21] Note that it should be demonstrated that the peak *in vivo* particle concentration does not interfere with the hematology analyzer. Globulin levels as well as changes in serum albumin/globulin ratio can indicate large modifications in antibody synthesis.

When blood is harvested for hematological tests, it is important to keep in mind that nanomaterials can be quickly cleared by cells of the RES and may have a short circulation time (minutes to hours).[22–25] However, changes in white blood cell number and/or ratio can be caused when nanomaterial drugs interact with blood cell progenitors (e.g., stem cells) in which case, blood samples should be harvested at later points (days to weeks) to allow the effects on white cells progenitor populations to be observed in circulation.

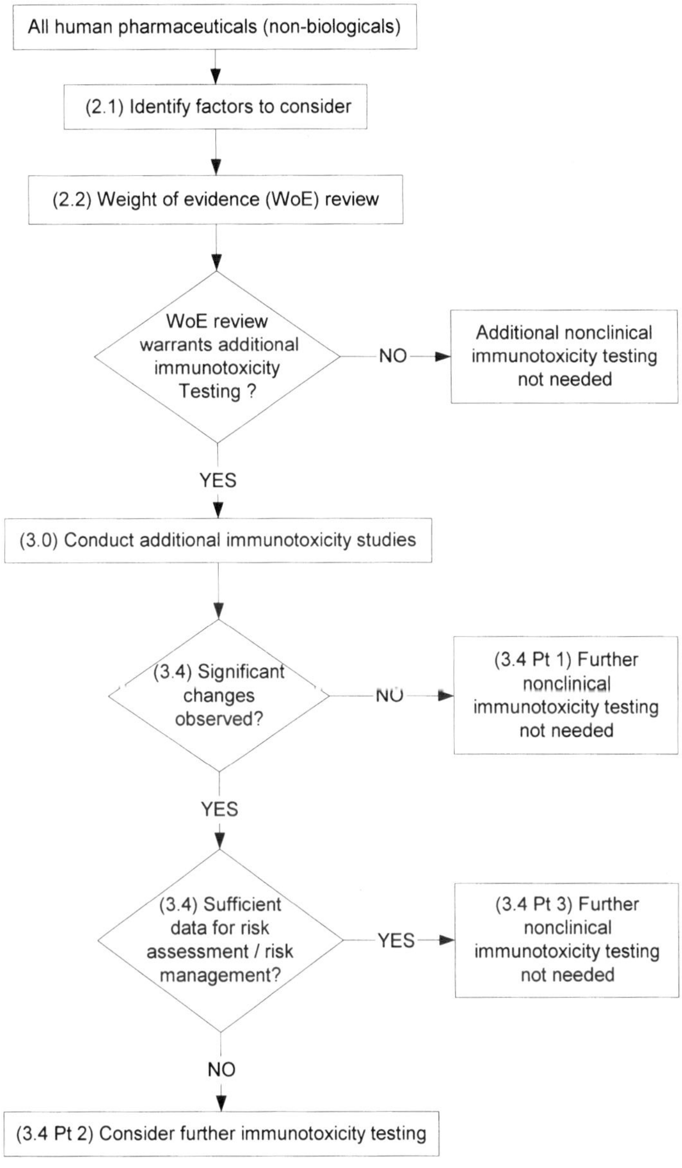

Figure 2. Flow diagram recommended for immunotoxicity testing (from ICH S8). Relevant data from all sources and studies are subjected to a weight of evidence (WoE) review.

5.2. *Histological tests and organ weights*

ICH S8 guidance recommends that the results of the histological examination of the organs of the immune system (spleen, lymph nodes, bone marrow, thymus) be evaluated in detail for signs of potential immunotoxicity. It is well established that nanomaterial drugs accumulate in organs of the RES; therefore, these studies should also include the detailed examination of the liver, bone marrow, and lung. The route of nanomaterial administration (inhalation, oral, intraperitoneal, or intravascular) helps to determine which organs and tissues should be examined. In case of inhalation studies, nanomaterials were reported to be primarily localized in the lungs,[26,27] whereas in the case of intraperitoneal or intravenous injections, nanomaterials were distributed throughout the RES organs.[28–30] Histology tests should, where possible, identify the organs in which nanomaterials are present, the tissue distribution of the nanomaterials (e.g., marginal zone of the spleen), as well as cellular co-localization (inside or in the vicinity of macrophages, B cells, T cells, etc.). For the types of particles that are not visible under microscopic examination, other methods may be needed to measure their accumulation in the RES organs. These data should be available as part of the standard ADME studies required for all drugs.

According to ICH S8, "If the compound and/or its metabolites are retained at high concentration in cells of the immune system, additional immunotoxicity testing should be considered." Therefore, histopathological analysis can be crucial in determining not only the distribution of the drug, but also the identification of the targeted cell population (through changes in morphology).

5.3. *Other signs of immunosuppression*

It is recommended that during preclinical and clinical studies, besides temperature and body weight, other health-related information should be monitored. For example, increased incidence in fever reaction or bacterial or viral infection, can point toward immunosuppression. In the case of animal models, increased occurrence of tumor formation can also indicate immunosuppression. In such instances, other plausible causes (stress, genotoxicity and sterility of the nanomaterial, hormonal effects, etc.) should be evaluated before a conclusion is drawn.

5.4. *Factors to consider in the WoE review*

If the results of one or more STS are positive for potential immunotoxicity, then additional criteria should be evaluated, including but not limited to:

1) dose that the immunotoxicity of the nanomaterial occurred,
2) severity of the effect caused by the drug,

3) stability and/or reversibility of the effect determined by the drug,
4) duration of the drug treatment, and
5) STS design (number of animals selected/groups, endpoints, appropriate controls, validating assays, etc.).

Conversely, important information regarding the immunotoxicity of the nanomaterial drug can be obtained from reviewing existing information on physicochemical properties, ADME, and clinical information from similar nanomaterial drugs. A WoE review of all of the relevant data from STS is used to decide whether or not additional studies are required. This review can result in the discovery of potential immunotoxicity from a single strong finding or from the aggregate of smaller effects in several areas.

6. Additional Recommended Studies by ICH S8

If the immunotoxicity assessment of the nanomaterial drug identifies causes of concern, additional immunotoxicity tests should be performed. Additional studies would determine the targeted cell population, functional changes of affected cells, the mechanism of action of the nanomaterial drug, etc. Unfortunately, there is no standard set of additional tests recommended by ICH S8. Additional tests will vary on a case-by-case basis, depending on nanomaterial characteristics, the observed immuntoxocity effects, and the intended use of the drug.

When the targeted population is not known or multiple populations appear to be affected, a standard assay such as T cell-dependent antibody response (TDAR) may be appropriate. As mentioned above, most NPs are localized in the macrophages of the RES. Therefore, tests assessing macrophage functions should be considered. Cytokine production was elevated in the serum of mice receiving silver NPs[31] and magnetic NPs.[32] When changes in cytokine synthesis are detected, it is preferable if the serum is harvested during a time-course study to evaluate cytokine levels during a wider time interval. In some cases, changes in the complement system was detected in response to polyethylene glycol (PEG)ylated liposomal doxorubicin, Doxil®, and phospholipid NP surfaces.[33,34]

Immunophenotyping by flow cytometry can be used to identify treatment-associated changes in the cell population number in the blood or organs (B cells, T cells, macrophages, etc.), cell ratio (e.g., $CD4^+$ *vs.* $CD8^+$ T cells), surface markers (activation markers), surface immunoglobulin expression (IgM, IgD), intracellular cytokine production (IL-6, TNF-α), etc. These results can complement or confirm data collected in tests such as STS and

ELISA. Changes in the number or ratio of white blood cells can result from the direct effect of nanomaterial drugs on circulating cells or on the maturation of stem cell progenitors from the bone marrow. If that is the case, a histological examination of the bone marrow may identify alterations in morphology associated with cell lineage changes resulting from exposure to the nanomaterial drug.

The development of antibodies against small molecule-conjugated NPs has the potential to lead to hypersensitivity. However, there are only a few reports that have identified an antibody response in mice to PEG-modified poly(lactic acid) (PLA) NPs.[35] The clinical relevance of these observations is not known.

Additionally, bacterial endotoxin can produce febrile and pro-inflammatory reactions and in extreme cases, cause sepsis. Hence, NP drug sterility and endotoxin load have to be confirmed prior to NP drug administration. Available tests to determine NP sterility are presented in Chapter 4.

The results of additional studies should be evaluated and conclusions about the immunotoxicological properties of the nanomaterial drug should be drawn. According to the ICH S8 guidance, three scenarios are possible:

1) additional studies show that there is no risk of immunotoxicity and no further studies are required,
2) additional studies show that there is a risk of immunotoxicity, but the risk is considered acceptable and no further studies are required, or
3) additional studies show that there is a risk of immunotoxicity, but a reasonable risk/benefit decision cannot be agreed upon. In this last case, additional studies are recommended. These studies can be conducted in phase 1 and phase 2 of a clinical trial and immune system parameters should be monitored.

7. Recommended Studies by ICH S6

This guidance applies to NP products containing biologic materials such as, but not limited to, "cytokines, plasminogen activators, recombinant plasma factors, growth factors, fusion proteins, enzymes, receptors, hormones, and monoclonal antibodies." Safety considerations take into account not only the direct effect of these compounds on the immune system, but also effects resulting from contamination during biologic compound purification. Therefore, in situations where nanomaterials are shown to contain biologic constituents, tests involving purity, stability, potency, pharmacokinetics, and pharmacodynamics may be needed.

The choice of the animal model used (rodent or non-rodent, wild-type or mutant) can be critical. The model selected must have a biological response to the therapeutic protein. ICH S6 recommends that if the nano-material biologic product in clinical trials is intended for short-term use (one week), then two weeks of repeated dose studies are considered acceptable. If the product is intended for chronic diseases, then repeated six-month studies are more appropriate.

The immunogenicity of nanomaterial-containing biologic products can be detected by antibody, cytokine, and/or complement measurements. Antibody detection can measure total as well as anti-drug antibody (ADA) formation. Immunophenotyping (surface staining of immune cells) can complement the serum measurements. If modifications in the antibody and/or cytokine synthesis are detected, the correlation between these results and pharmacokinetics/pharmacodynamics studies of the biologic constituent should be considered. On the other hand, immunologic responses in rodents do not guarantee a similar response in humans.

As mentioned above, a single finding does not indicate that the product has immunotoxic effects. WoE decision-making should be employed regardless of the guidance followed.

8. Summary

In conclusion, immunotoxicity studies for NP-conjugated drugs are regulated under the existing ICH S6 or S8 guidances by the FDA. The choice of guidance depends on whether the active drug is a small molecule (S8) or a protein therapeutic (S6). The basic studies conducted do not differ for both NP- and non-NP–containing drugs of each group. Additional studies may be needed depending on the results of the basic studies.

9. Nanoparticle Research Conducted at the FDA/CDER

Other than reviewing drug applications, the CDER also engages in regulatory research intended to inform and improve the drug review process. A case study of a regulatory research project involving immunotoxicity work is presented below.

In 2006, a Citizen's Petition was presented to the FDA requesting that the administration take regulatory action on products composed of engineered nanomaterials, specifically for sunscreen products containing ZnO and TiO_2 NPs. In response to this petition, the FDA initiated multiple studies to evaluate ZnO and TiO_2 NPs versus their larger counterparts. These studies included:

1) the identification and characterization of TiO_2 NPs in sunscreen formulations,
2) dermal penetration studies to evaluate the ability of TiO_2 NPs to cross the skin, and
3) *in vivo* studies to evaluate the toxicity effects of TiO_2 NPs in mice.

It was first confirmed that several commercial sunscreen products did indeed contain nanoscale TiO_2 NPs in their formulation.[36] To evaluate if TiO_2 NPs are able to penetrate the skin, a study was conducted to evaluate the dermal penetration of three types of TiO_2 NPs (sizes between 20–500 nm; coated or uncoated). These particles were constituted in sunscreen formulations that were topically applied to minipigs (four applications per day, five days per week for four weeks). The following tissues/organs were harvested at the time of necropsy: the skin (from different sites), lymph nodes (left and right inguinal, prescapular, and submandibular), liver, kidneys, gall bladder, heart, lungs, brain, and feces. Elevated levels of TiO_2 NPs were detected in the epidermis. There was also an increase in TiO_2 levels in the left inguinal lymph nodes following treatment with uncoated nano-TiO_2 and an increase in the right inguinal lymph nodes of minipigs treated with submicron TiO_2. However, when combining all the lymph nodes, TiO_2 levels in lymph nodes were not increased compared to controls. No increases in TiO_2 were seen in the other tissues and organs harvested (with the exception of the skin). It was concluded that TiO_2 NPs in sunscreen formulations lack significant dermal penetration.[37] In this study, the levels of material in the cells of the immune system was minimal and therefore did not raise immunotoxicology concerns.

In parallel, studies were also performed to observe the effects of direct systemic administration of TiO_2 NPs. For *in vivo* mouse models, large amounts of TiO_2 NPs were injected (560 mg/kg for intravenous administration or 5,600 mg/kg for subcutaneous injections for two consecutive days) and TiO_2 NPs were observed in the lung, liver, lymph node, spleen, and kidney of intravenous-treated mice, and in the liver, lymph node, and spleen of subcutaneous-treated animals, confirming RES localization.[38] Animals were monitored for 26 weeks after the injections, at which point TiO_2 NPs were still detected in the RES. An increased number of macrophages was detected in the lung and liver; however, there was no evidence of tissue damage.[5] It was concluded that TiO_2 NPs *in vivo* accumulate in the RES, but based on histopathology observations, there is no evidence of immunotoxicity.

The WoE review showed that the effects of TiO_2 NPs on the immune system was almost undetectable following dermal exposure. This, in combination with the lack of any immunotoxocity effects of large-scale systemic exposure to TiO_2 NPs, led to the conclusion that there was no cause for concern for TiO_2 NP immunotoxicity. According to the paradigm in ICH S8, no further studies are needed.

Additional regulatory research in the CDER is ongoing to determine appropriate immunotoxicity testing strategies for nanomaterial drugs. These efforts are primarily focused on the evaluation of methods to detect the effects of NP accumulation in the RES.

10. Conclusions

The current chapter focused on the U.S. FDA regulatory requirements for immunotoxicity tests and their applicability to nanomaterial drug research. Drugs conjugated to NPs are required to comply with the testing requirements applied to all drug products. Included in these requirements is the need for drugs to be tested for potential immunomodulatory effects as described in ICH S6 or S8 as applicable. Additional testing may be appropriate for nanomaterials with a proven tendency to accumulate in the RES.

All drugs, including NP drugs, are regulated by the CDER. It is therefore highly recommended that the company manufacturing the NP drug convene with the appropriate FDA review division and seek advice on how to prepare their drug application and the most suitable way to address potential immunotoxicity issues for their specific product. Although the existing FDA guidances (as listed in Table 2) are generally applicable to nanomaterial drug products, the application of these guidances may need to be clarified in relation to the particular drug product. Thus, early engagements with the specific FDA review division for their product class can prove beneficial to the sponsor.

References

1. Keene AM, Peters D, Rouse R, Stewart S, Rosen ET, Tyner KM. Tissue and cellular distribution of gold nanoparticles varies based on aggregation/agglomeration status. *Nanomedicine* 2011;7:199–209.
2. Gibaud S, Demoy M, Andreaux JP, Weingatren C, Gouritin B, Couvreur X. Cells involved in the capture of nanoparticles in hematopoietic organs. *J Pharm Sci* 1996;85:944–950.
3. Sadekar S, Ray A, Janat-Amsbury M, Peterson CM, Ghandehari H. Comparative biodistribution of PAMAM dendrimers and HPMA copolymers in ovarian-turmor-bearing mice. *Biomacromolecules* 2011;12:88–96.

4. Sadauskas E, Danscher G, Stoltenberg M, Vogel, U, Larsen A, Wallin H. Protracted elimination of gold nanoparticles from mouse liver. *Nanomedicine* 2009;5:162–169.
5. Umbreit TH, Francke-Carroll S, Weaver JL, Miller TJ, Goering PL, Sadrieh N, M. Stratmeyer ME. Tissue distribution and histopathological effects of titanium dioxide nanoparticles after intravenous or subcutaneous injection in mice. *J Appl Toxicol* 2012;32:350–357.
6. Di Gioacchino M, Petrarca C, Lazzarin F, Di Giampaolo L, Sabbioni E, Boscolo P, Mariani-Costantini R, Bernardini G. Immunotoxicity of nanoparticles. *Int J Immunopathol Pharmacol* 2011;24:65S-71S.
7. Moon E-Y, Yi G-H, Kang J-S, Lim J-S, Kim H-M, Py S. An increase in mouse tumor growth by an *in vivo* immunomodulating effect of titanium dioxide nanoparticles. *J Immunotoxicol* 2011;8:56–67.
8. Service RF. Nanotechnology. Calls rise for more research on toxicology of nanomaterials. *Science* 2005;310:1609.
9. Hoet P, Legiest B, Geys J, Nemery B. Do nanomedicines require novel safety assessments to ensure their safety for long-term human use? *Drug Saf* 2009;32:625–636.
10. Kunzmann A, Andersson B, Thurnherr T, Krug H, Scheynius A, Fadeel B. Toxicology of engineered nanomaterials: Focus on biocompatibility, biodistribution and biodegradation. *Biochim Biophys Acta* 2011;1810:361–373.
11. U.S. FDA. Manual of Polices and Procedures. Center for Drug Evaluation and Research, 2010.
12. U.S. FDA. Guidance for Industry. Considering Whether an FDA-Regulated Product Involves the Application of Nanotechnology, 2011.
13. U.S. FDA. Guidance for Industry. S6 Preclinical Safety Evaluation of Biotechnology-Derived Pharmaceuticals. Center for Drug Evaluation and Research, 1997.
14. U.S. FDA. Guidance for Industry. Immunotoxicology Evaluation of Investigational New Drugs. Center for Drug Evaluation and Research, 2002.
15. U.S. FDA. Guidance for Industry. S8 Immunotoxicity Studies for Human Pharamaceuticals. Center for Drug Evaluation and Research, 2006.
16. U.S. FDA. Guidance for Industry. S6 Adderium to Preclinical Safety Evaluation of Biotechnology-Derived Pharmaceuticals, Center for Drug Evaluation and Research, 2012.
17. Mihich E. On the immunomodulating effects of anti-cancer drugs and their therapeutic exploitation. *Jpn J Clin Oncol* 2000;30:469–471.
18. Ryan EP, Malboeuf CM, Bernard M, Rose RC, Phipps RP. Cyclooxygenase-2 inhibition attenuates antibody. *J Immunol* 2006;177:7811–7819.
19. Vial T, Descotes J. Immune-mediated side-effects of cytokines in humans. *Toxicology* 1995;105:31–57.
20. Labro, M-T. Macrolide antibiotics: Current and future uses. *Expert Opin Pharmacother* 2004;5:541–550.
21. Tiwari DK, Jin T, Behari J. Dose-dependent *in-vivo* toxicity assessment of silver nanoparticle in Wistar rats. *Toxicol Mech Methods* 2011;21:13–24.
22. Cheng L, Yang K, Shao M, Lu X, Liu Z. *In vivo* pharmacokinetics, long-term biodistribution and toxicology study of functionalized upconversion nanoparticles in mice. *Nanomedicine* 2010;6:1327–1340.
23. Almeida JP, Chen AL, Foster A, Drezek R. *In vivo* biodistribution of nanoparticles. *Nanomedicine* 2011;6:815–835.

24. He Q, Zhang Z, Gao F, Li Y, Shi J. *In vivo* biodistribution and urinary excretion of mesoporous silica nanoparticles: Effects of particle size and PEGylation. *Small*, 2011;7:271–280.
25. Morais T, Soares ME, Duarte JA, Soares L, Maia S, Gomes P, Pereira E, Fraga S, Carmoa H, de Lourdes Bastos M. Effect of surface coating on the biodistribution profile of gold nanoparticles in the rat. *Biopharmaceuticals* 2012;80:185–193.
26. Inoue KI, Takano H. Aggravating impact of nanoparticles on immune-mediated pulmonary inflammation. *Nanopart Inflamm* 2011;11:382–390.
27. Mansour HM, Rhee YS, Wu X. Nanomedicine and pulmonary delivery. *Int J Nanomed* 2009;4:299–319.
28. Lasagna-Reeves C, Gonzalez-Romero D, Barria MA, Olmedo I, Clos A, Sadagopa Ramanujam VM, Urayama A, Vergara L, Kogan MJ, Soto C. Bioaccumulation and toxicity of gold nanoparticles after repeated administration in mice. *Biochem Biophys Res Commun* 2010;393:649–655.
29. Sonavane G, Tomoda K, Makino K. Biodistribution of colloidal gold nanoparticles after intravenous administration: Effect of particle size. *Colloids Surf B Biointerfaces* 2008;66: 274–288.
30. Maincent P, Thouvenot P, Amicabile C, Hoffman M, Kreuter J, Couvreur P, Devissaguet JP. Lymphatic targeting of polymeric nanoparticles after intraperitoneal administration in rats. *Pharm Res* 1992;9:1534–1539.
31. Park EJ, Bae E, Yi J, Kim Y, Choi K, Lee SH, Yoon J, Lee BC, Park K.. Repeated-dose toxicity and inflammatory responses in mice by oral administration of silver nanoparticles. *Environ Toxicol Pharmacol* 2010;30:162–168.
32. Chen BA, Jin N, Wang J, Ding J, Gao C, Cheng J, Xia G, Gao F, Zhou Y, Chen Y, Zhou G, Li X, Zhang Y, Tang M, Wang X. The effect of magnetic nanoparticles of Fe(3)O(4) on immune function in normal ICR mice. *Int J Nanomed* 2010;5:593–599.
33. Chanan-Khan A, Szebeni J, Savay S, Liebes L, N. M. Rafique NM, Alving CR, Muggia FM. Complement activation following first exposure to PEGylated liposomal doxorubicin (Doxil®): Possible role in hypersensitivity reactions. *Ann Oncol* 2003;14:1430–1437.
34. Pham CT, Mitchell LM, Huang JL, Lubniewski CM, Schall OF, Killgore JK, Pan D, Wickline SA, Lanza GM, Hourcade DE. Variable antibody-dependent activation of complement by functionalized phospholipid nanoparticle surfaces. *J Biol Chem* 2011; 286:123–130.
35. Ishihara T, Takeda M, Sakamoto H, Kimoto A, Kobayashi C, Takasaki N, Yuki K, Tanaka K, Takenaga M, Rie Igarashi R, Maeda T, Yamakawa N, Okamoto Y, Otsuka M, Ishida T, Kiwada H, Mizushima Y, Mizushima T. Accelerated blood clearance phenomenon upon repeated injection of PEG-modified PLA-nanoparticles. *Pharm Res* 2009;26: 2270–2279.
36. Tyner KM, Wokovich AM, Doub WH, Buhse LF, Sung LP, Watson SS, Sadrieh N. Comparing methods for detecting and characterizing metal oxide nanoparticles in unmodified commercial sunscreens. *Nanomedicine* 2009;4:149–154.
37. Sadrieh N, Wokovich AM, Gopee NV, Zheng J, Haines D, Parmiter D, Siitonen PH, Cozart CR, Patri AK, McNeil SE, Howard PC, Doub WH, Buhse LF. Lack of significant dermal penetration of titanium dioxide from sunscreen formulations containing nano- and submicron-size TiO_2 particles. *Toxicol Sci* 2010;115:156–166.
38. Patri A, Umbreit T, Zheng J, Nagashima K, Goering P, Francke-Carroll S, Gordon E, Weaver J, Miller T, Sadrieh N, McNeil S, Stratmeyer M. Energy dispersive X-ray analysis of titanium dioxide nanoparticle distribution after intravenous and subcutaneous injection in mice. *J Appl Toxicol* 2009;29:662–72.

Index

accelerated blood clearance (ABC) 557–566
activated partial thromboplastin time (APTT) 263, 265, 271, 273, 274, 281–284
active targeting 480, 481, 487, 490
adjuvant 127–130, 135, 449–451, 455, 459, 460, 464, 462
AFC (antibody-forming cell) 645, 646, 648, 651, 652, 654–656, 659, 662
agglutination 175, 176, 180, 182–184
aggregation 175, 176, 182, 183, 188, 189, 192
allergen 519, 523, 527, 528, 532, 534, 536
allergy 517, 518, 522, 527, 530, 532, 536
analytical characterization 53
anaphylaxis 378
animal model 306, 315, 347
antibody 518, 520, 524, 525, 529, 533, 535, 547, 549–554, 556, 557, 564, 566–569, 571, 572
antigenicity 12
antigen-presenting cells (APCs) 417, 418, 422–425, 427, 428, 449, 454, 456, 460
anti-inflammatory drug 481, 484, 487, 489, 495, 496, 498–500
apoptosis 207, 215, 217, 224, 225, 227, 229, 233, 235, 237, 238, 240, 242, 250, 254
asthma 519, 523, 529
asymmetric flow field-flow fractionation (AFFFF) 34, 35
autoclaving 55
autophagy 207, 215, 217, 224, 225, 239, 251–253

batch-to-batch consistency 28, 30, 40
bioavailability 485, 488, 495, 496, 498, 500, 501
biocompatibility 208, 214, 215, 222, 228, 233, 244, 253
biodegradable 433, 440, 442, 445, 450, 460
biodistribution 151, 161, 162, 433–436, 444
bleeding 263–265
blood cell count 186
bone marrow 434–445

C1q 358, 360–363, 366, 368, 370, 371, 379

C3 358–361, 363, 364, 368, 371, 372, 375–377
C_{60} 657–661
carbon nanotubes (CNTs) 660–662
characteristics 436
chemistry 449, 457
chemokine 420, 427, 429
chitosan 461
clearance 387–389, 394–400, 402, 406–410
clinical effects 405
CMI (cell-mediated immunity) 646, 650, 656
collagen 295, 296, 308, 309, 313, 320, 324, 327, 328, 336, 340, 341, 346
complement activation 517, 523, 525, 529, 531
complement system 4, 7, 8, 13, 357
  alternative pathway 358, 360, 361, 364, 368, 370–376
  classical pathway 358, 360–364, 366, 370–374, 378
  lectin pathway 360, 361, 368, 371, 373, 376
composition 394, 396, 401, 410
configuration 364, 367–369
confocal immunomicroscopy 345
contamination 78, 79, 86, 94, 100, 102, 103, 105, 107, 110, 117, 120–123, 128–132, 138–140
coronas 120, 134
cross-priming 449, 455, 457, 462
cytokine 3, 4, 10–13, 420–422, 427, 429, 518, 519, 525, 528–530
cytotoxicity 207–209, 216, 223, 224, 229–233, 235–239, 241–244, 246, 247, 249, 518, 520, 521, 524, 536

dendrimer 478, 482, 492, 493, 501
dendritic cells (DCs) 449, 454, 457
dense tubular system (DTS) 319, 344
depyrogenation 98, 107–109
dermatitis 521, 530, 531
drug delivery systems 450, 454
drug development 671, 673
DTH (delayed-type hypersensitivity) 646, 648, 650, 655, 656, 659
dynamic light scattering (DLS) 28, 32, 34–37, 39–46, 48, 50

eczema 519
efficacy 478, 490, 491, 494, 499
ELISA kits 373, 375, 377
endothelial cell (EC) 207
  EC activation 211–214, 223, 227
  EC activation markers 207, 227, 242, 250
endotoxin *see* Chapter 4, 137–140
erythrocyte 173–177, 181, 182, 185, 189, 190, 192, 194
ethylene oxide 68
euglobulin lysis time (ELT) 271, 277
extracellular matrix 296, 340

fibrinogen (FBG) 266, 267, 271, 274, 277–279, 281
fibrinolysis 263–266, 270, 272, 274–277, 284
flow cytometry 302, 303, 305, 306, 307, 309–311, 317, 324, 326, 331–333, 337, 342–345
fluorescence correlation spectroscopy (FCS) 159, 160, 162
Food and Drug Administration (FDA) 671
  Center for Drug Evaluation and Research (CDER) 671–673, 681, 683
formaldehyde 69
free radical reactions 60

gamma irradiation 59
gel electrophoresis 155

gold 442
guidance 672–674, 676, 677, 680, 681, 683
 ICH S6 671, 674, 676, 680, 681, 683
 ICH S8 674, 677–680, 683

hematology 582
hemolysis 5, 6, 175–185, 188–192
hemolytic assay 370, 372–374, 376, 377
hemostasis 263, 264, 266–270, 278
high hydrostatic pressure sterilization 69
host response 119, 125, 126, 128, 137, 139, 142
humoral immunity 641, 645, 648, 650, 653, 654
hydrodynamic 32, 36, 37, 43, 45, 46
hypersensitivity 517–520, 642, 644, 646, 647, 648, 655, 662, 671, 680
 type I hypersensitivity 520
 type II hypersensitivity 524
 type III hypersensitivity 525
 type IV hypersensitivity 526

imaging 433–436, 438, 439, 445
immune cell function 584
immune function tests 15
immune responses 450, 454, 460, 465
immune system 449, 450, 454, 517, 518, 524, 525
immunity 450, 454, 457
immunogenicity 10, 128, 134, 548, 553, 568, 569, 572
immunomodulation 130, 433, 434, 442, 443, 449, 644, 651
 immunostimulation 1, 2, 3, 11
 immunosuppression 1, 2, 13, 642, 643, 655, 662, 674, 678
immunotoxicity 2, 3, 4, 16, 581–588, 594, 595, 602, 676, 682
inflammation 127, 131, 135–139, 142, 449, 451–453, 478–480, 482, 485–489, 491, 493, 495, 496, 498
inflammatory response 209–211, 214, 215, 236, 244, 522, 524, 531, 533
infusion reactions 378
inhibitor 266–270, 272, 274–278
instrumentation 26, 27, 30, 31, 35, 36
intracellular free Ca^{2+} 301, 307, 309, 320, 345
intracellular transport 162, 164
intravenous 483, 484, 487, 488, 491, 493, 499, 500–502
investigational new drug (IND) 673
in vitro 175–181, 186, 188–195
in vitro assay 581
in vitro–in vivo correlation 587, 592, 593
in vivo 178, 179, 182, 186–190, 192, 194, 195

kallikrein (KAL) 265, 266, 270–272, 276, 280
kallikrein–kinin system (KKS) 263–266, 270, 272, 276, 279, 280, 284

laser Doppler velocimetry (LDV) 32, 47
leukocyte adhesion molecules 214, 220
leukocytes 5, 6, 13
light microscopy 293, 301, 307
light transmission aggregometry (LTA) 307–309, 314, 320, 322, 331, 332

Limulus Amebocyte Lysate (LAL) 78, 81, 84, 86–104, 107, 109
lipopolysaccharide (LPS) 80–83, 89, 91, 99, 100–103, 105, 107, 109, 123, 138–140
lipoproteins 155, 156, 160
liposomes 54, 478, 482, 488–492, 494, 497, 498, 500
local 643, 644, 646, 652, 659–662
local lymph node assay (LLNA) 535, 536, 646, 648, 650
lymph node proliferation assay (LNPA) 646–648, 650
lymphocytes 518, 527

macrophage 388, 390–394, 396, 399–404, 641, 642, 645, 648, 650, 651, 656–659, 661
magnetic nanoparticles 462
magnetic resonance imaging (MRI) 435–438
major histocompatibility complex (MHC) 418, 420, 422, 423, 426, 427, 429
mechanical properties 175, 185, 186, 191
membrane microparticles (MPs) 294, 295, 298, 305, 310, 311, 319, 331, 342
micelles 478, 482, 493–497, 501, 502
microfiltration 57
microspheres 54
microvascular endothelium 247
moist heat sterilization 55
monocytes 388, 390, 391, 402–407
mononuclear phagocytic system (MPS) 9, 10, 425, 642, 645, 648, 650
morphogenesis 207, 219, 227, 238
morphology 175, 180, 183, 185, 186, 189, 194
mucoadhesiveness 462

nanomaterials (NMs) 173–178, 179–186, 188, 189, 191–195, 423, 428, 429
nanomedicine 152, 164, 517, 528, 652, 653
nanocarrier 477
nanoparticle-based vaccines 450, 460
nanoparticles (NPs) 449, 460, 517, 531
nanoparticle sterilization 53
nanostructures 517, 518, 525, 528, 531, 534, 536
nanotherapeutics (NTs) 53
necrosis 207, 215, 224, 233
new drug application (NDA) 673
nitric oxide (NO) 207, 210–212, 225, 226, 229, 230, 234, 240, 246
NK (natural killer) 641, 645, 648, 662
non-degradable but biocompatible nanoparticles 462

ocular 484–488, 490, 493, 495, 496, 498, 500, 501
opsonization 390, 391, 394, 395, 398, 399, 403
oral 483, 485, 487, 488, 493–495, 499–502
organ dysfunction 215
oxidative stress 175, 176, 181, 191, 192, 194

passive cutaneous anaphylaxis (PCA) test 533, 536
passive targeting 477, 479–481, 499
pathogen-associated molecular patterns (PAMPs) 127, 419
pattern recognition receptors (PRRs) 419
permeability 207, 209, 220, 225, 227–231, 233, 236, 243

phagocytic cells 518, 525
pharmaceuticals 671, 674, 676
pharmacokinetics (PK) 387, 389, 394–396, 399, 403, 405–407, 409, 410
physicochemical characterization (PCC) 27, 31, 35, 36, 50
plasma 154–157, 159
plasma coagulation system (PCS) 263
platelet (PLT) 293
  PLT activation 295, 296, 298, 310, 311, 315, 319, 320–329, 331–333, 340, 342, 345, 347
  PLT adhesion 293, 296, 301, 302, 304, 306, 307, 312, 313, 319, 321–323, 326, 329–331, 339, 340
  PLT aggregation 7, 293, 296, 297, 301–304, 306–309, 311, 314, 317–319, 320, 322–325, 327–329, 331–333, 335, 337, 340, 341
  PLT granules 295, 297, 298, 324
PMNs (polymorphonuclear cells) 642, 650, 658, 660, 661
polyaromatic hydrocarbons (PAHs) 123, 124, 130–133, 135
polydimethylsiloxanes (PDMS) 123, 124, 129, 130
polyethylene glycol (PEG) 362–372, 378, 379
polymer coatings 133–135
polymeric nanoparticles 450, 460, 478, 484, 497, 498, 502
polymyxin B (PMB) 102–106
prostanoids 211
protein adsorption 134–136, 151, 152, 154, 155, 159, 164
protein corona 151, 152, 154, 155, 159–164
protein C (PC) 266–278
protein monolayer 159
prothrombin time (PT) 263, 271, 273, 274, 281–284
pseudoallergy 517
pulmonary 492, 493, 496, 498, 502

rabbit pyrogen test (RPT) 83–85, 87, 88, 91, 98
radioprotection 433, 434, 443–445
reactive oxygen and nitrogen species (RONS) 210, 211, 214
receptor 295, 296, 298, 307, 308, 310, 317, 318, 320, 327, 328, 334, 336, 339, 341, 346
regulatory 671–673, 681, 683
reptilase time (RT) 271, 273, 274

scanning electron microscopy (SEM) 302, 305, 307, 302, 321, 325, 331, 338, 341, 343, 344
serum albumin 156, 157, 158, 159, 160
shape 449, 458
signalling pathways 449, 459
silver 178, 180, 187, 189, 440, 444
size 449, 455
size exclusion chromatography (SEC) 33–35
skin prick test 523, 532, 535
solid lipid nanoparticles (SLNs) 54, 497–499, 501
stability 27–33
sterile filtration 57
superparamagnetic iron oxides (SPIOs) 435–439
surface analysis 123, 125, 140, 141
surface charge 449, 455, 457
surface modification 134, 135, 478, 483, 484
sustained release 483, 484, 488, 494, 497
synthetic materials 358, 360, 363
synthetic substrate 276

synthetic vaccines 450
systemic 643, 645, 652, 659, 661, 662

therapeutic protein 547–550, 567, 569–573, 671, 681
thrombin generation assay 276
thromboelastography 271, 275–277, 280–282
thrombin 263
thrombosis 264, 266, 267, 268, 269, 280
TiO_2 441, 442, 445, 649, 650, 651, 652
tissue factor (TF) 267, 269, 274, 276
Toll-like receptors (TLRs) 127, 138, 140
toxicity 387, 400–402, 404, 405, 410, 433, 434, 439, 440–443, 445
transdermal 485, 492, 493, 495–498, 501
transferrin 157, 160, 162
transmission electron microscopy (TEM) 307
thrombin time (TT) 281

uptake 175, 176, 183, 184, 186, 191, 194
  cellular uptake 151, 153, 157, 161–164
  nanoparticle (NP) uptake 385, 388, 389, 390, 393, 394, 396, 399, 409, 449, 455

vaccine carriers 449, 455, 463
vaccine delivery 449
vaccine formulations 450, 455, 460
vascular endothelium 238
vasoactivity 207, 219, 227
von Willebrand factor (vWf) 210, 213, 218, 221, 227, 234, 241

Weibel–Palade (WP) bodies 213, 227, 241
whole blood aggregometry (WBA) 308

X-ray photoelectron spectroscopy (XPS) 129, 140, 141

zeta potential 27–29, 32, 35, 46–50
ZnO 442, 445